Renault 25
Service and Repair Manual

Steve Rendle and Mark Coombs

(1228-344)

Models covered
Renault 25 petrol and diesel models, including 1995 cc 12-valve, V6 Turbo and special/limited editions
Petrol engines - 2.0 litre (1995 cc), 2.2 litre (2165 cc), 2.5 V6 (2458 cc), 2.7 V6 (2664 cc), 2.9 V6 (2849 cc)
Diesel engine - 2.1 litre (2068 cc)

© Haynes Publishing 1997

A book in the **Haynes Service and Repair Manual Series**

All rights reserved. No part of this book may be reproduced or transmitted in any form or by any means, electronic or mechanical, including photocopying, recording or by any information storage or retrieval system, without permission in writing from the copyright holder.

ISBN 1 85960 289 4

British Library Cataloguing in Publication Data
A catalogue record for this book is available from the British Library.

ABCDE
FGHIJ
KLMNO
PQ
2

Printed in the USA

Haynes Publishing
Sparkford, Nr Yeovil, Somerset BA22 7JJ, England

Haynes North America, Inc
861 Lawrence Drive, Newbury Park, California 91320, USA

Editions Haynes S.A.
Tour Aurore - IBC, 18 Place des Reflets,
92975 Paris La Défense 2, Cedex, France

Haynes Publishing Nordiska AB
Box 1504, 751 45 Uppsala, Sverige

Contents

LIVING WITH YOUR RENAULT 25

Introduction to the Renault 25 — Page 0•4
Safety first! — Page 0•5

Roadside repairs

If your car won't start — Page 0•6
Jump starting — Page 0•7
Wheel changing — Page 0•8
Identifying leaks — Page 0•9
Towing — Page 0•9

Weekly checks

Introduction — Page 0•10
Underbonnet check points — Page 0•10
Engine oil level — Page 0•11
Coolant level — Page 0•12
Brake and clutch fluid level — Page 0•12
Power steering fluid level — Page 0•13
Screen washer fluid level — Page 0•13
Tyre condition and pressure — Page 0•14
Wiper blades — Page 0•15
Battery — Page 0•15
Bulbs and fuses — Page 0•16

Lubricants, fluids and tyre pressures — Page 0•17

MAINTENANCE

Routine maintenance and servicing

Renault 25 petrol models — Page 1A•1
    Maintenance schedule — Page 1A•4
    Maintenance procedures — Page 1A•7
Renault 25 diesel models — Page 1B•1
    Maintenance schedule — Page 1B•3
    Maintenance procedures — Page 1B•4

Contents

REPAIRS AND OVERHAUL

Engine and associated systems

Four-cylinder petrol engine	Page	2A•1
Six-cylinder petrol engine	Page	2B•1
Diesel engine	Page	2C•1
Cooling, heating and air conditioning systems	Page	3•1
Fuel and exhaust systems - carburettor petrol models	Page	4A•1
Fuel and exhaust systems - fuel-injected petrol models	Page	4B•1
Fuel and exhaust systems - diesel models	Page	4C•1
Starting and charging systems	Page	5A•1
Ignition system - petrol models	Page	5B•1
Preheating system - diesel models	Page	5C•1

Transmission

Clutch	Page	6•1
Manual transmission	Page	7A•1
Automatic transmission	Page	7B•1
Driveshafts	Page	8•1

Brakes and suspension

Braking system	Page	9•1
Suspension and steering	Page	10•1

Body equipment

Bodywork and fittings	Page	11•1
Body electrical systems	Page	12•1

Wiring diagrams

Page 12•16

REFERENCE

Dimensions and weights	Page	REF•1
Conversion factors	Page	REF•2
Buying spare parts	Page	REF•3
Vehicle identification	Page	REF•3
General repair procedures	Page	REF•4
Jacking and vehicle support	Page	REF•5
Radio/cassette unit anti-theft system	Page	REF•5
Tools and working facilities	Page	REF•6
MOT test checks	Page	REF•8
Fault finding	Page	REF•12
Glossary of technical terms	Page	REF•20

Index

Page REF•26

Introduction

The Renault 25 model range was introduced into the UK in 1984. All models are of five-door Hatchback design, although a Limousine version is also available.

Four-cylinder petrol and diesel engines, and six-cylinder petrol engines are available. The four-cylinder engines are of overhead camshaft design. The six-cylinder engines are of V6 overhead camshaft design. All engines are mounted longitudinally at the front of the vehicle.

Models may be fitted with four or five-speed manual, or three or four-speed automatic transmissions. The transmission is mounted at the rear of the engine.

All models have front-wheel-drive with fully-independent front and rear suspension.

Since its introduction, the Renault 25 range has continually been developed. All models have a high trim level, which is very comprehensive in the upper model range. Central locking, electric windows, an electric sunroof, a trip computer, air conditioning and cruise control are all available.

For the home mechanic, the Renault 25 is a straightforward vehicle to maintain and repair since design features have been incorporated to reduce the actual cost of ownership to a minimum, and most of the items requiring frequent attention are easily accessible.

Renault 25 V6i

Renault 25 Monaco

The Renault 25 Team

Haynes manuals are produced by dedicated and enthusiastic people working in close co-operation. The team responsible for the creation of this book included:

Authors	Mark Coombs Spencer Drayton
Sub-editors	Sophie Yar Carole Turk
Editor & Page Make-up	Bob Jex John Martin
Workshop manager	Paul Buckland
Photo Scans	John Martin Paul Tanswell Steve Tanswell
Cover illustration & Line Art	Roger Healing
Wiring diagrams	Matthew Marke

We hope the book will help you to get the maximum enjoyment from your car. By carrying out routine maintenance as described you will ensure your car's reliability and preserve its resale value.

Your Renault 25 Manual

The aim of this manual is to help you get the best value from your vehicle. It can do so in several ways. It can help you decide what work must be done (even should you choose to get it done by a garage), provide information on routine maintenance and servicing, and give a logical course of action and diagnosis when random faults occur. However, it is hoped that you will use the manual by tackling the work yourself. On simpler jobs, it may even be quicker than booking the car into a garage and going there twice, to leave and collect it. Perhaps most important, a lot of money can be saved by avoiding the costs a garage must charge to cover its labour and overheads.

The manual has drawings and descriptions to show the function of the various components, so that their layout can be understood. Then the tasks are described and photographed in a clear step-by-step sequence.

Acknowledgements

Thanks are due to Champion Spark Plug who supplied the illustrations showing spark plug conditions. Certain other illustrations are the copyright of the Renault (UK) Ltd, and are used with their permission. Thanks are also due to Sykes-Pickavant Limited, who provided some of the workshop tools, and to all those people at Sparkford who helped in the production of this manual.

We take great pride in the accuracy of information given in this manual, but vehicle manufacturers make alterations and design changes during the production run of a particular vehicle of which they do not inform us. No liability can be accepted by the authors or publishers for loss, damage or injury caused by any errors in, or omissions from, the information given.

Project vehicles

The vehicles used in the preparation of this manual, and which appear in many of the photographic sequences, were a Renault 25 2.0 GTS and a Renault 25 V6i.

Safety first! 0•5

Working on your car can be dangerous. This page shows just some of the potential risks and hazards, with the aim of creating a safety-conscious attitude.

General hazards

Scalding
• Don't remove the radiator or expansion tank cap while the engine is hot.
• Engine oil, automatic transmission fluid or power steering fluid may also be dangerously hot if the engine has recently been running.

Burning
• Beware of burns from the exhaust system and from any part of the engine. Brake discs and drums can also be extremely hot immediately after use.

Crushing
• When working under or near a raised vehicle, always supplement the jack with axle stands, or use drive-on ramps. *Never venture under a car which is only supported by a jack.*

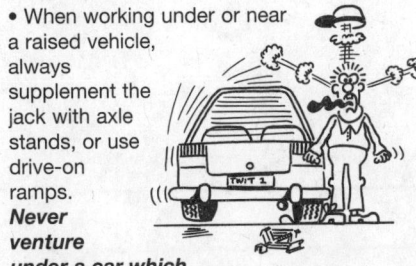

• Take care if loosening or tightening high-torque nuts when the vehicle is on stands. Initial loosening and final tightening should be done with the wheels on the ground.

Fire
• Fuel is highly flammable; fuel vapour is explosive.
• Don't let fuel spill onto a hot engine.
• Do not smoke or allow naked lights (including pilot lights) anywhere near a vehicle being worked on. Also beware of creating sparks (electrically or by use of tools).
• Fuel vapour is heavier than air, so don't work on the fuel system with the vehicle over an inspection pit.
• Another cause of fire is an electrical overload or short-circuit. Take care when repairing or modifying the vehicle wiring.
• Keep a fire extinguisher handy, of a type suitable for use on fuel and electrical fires.

Electric shock
• Ignition HT voltage can be dangerous, especially to people with heart problems or a pacemaker. Don't work on or near the ignition system with the engine running or the ignition switched on.

• Mains voltage is also dangerous. Make sure that any mains-operated equipment is correctly earthed. Mains power points should be protected by a residual current device (RCD) circuit breaker.

Fume or gas intoxication
• Exhaust fumes are poisonous; they often contain carbon monoxide, which is rapidly fatal if inhaled. Never run the engine in a confined space such as a garage with the doors shut.

• Fuel vapour is also poisonous, as are the vapours from some cleaning solvents and paint thinners.

Poisonous or irritant substances
• Avoid skin contact with battery acid and with any fuel, fluid or lubricant, especially antifreeze, brake hydraulic fluid and Diesel fuel. Don't syphon them by mouth. If such a substance is swallowed or gets into the eyes, seek medical advice.
• Prolonged contact with used engine oil can cause skin cancer. Wear gloves or use a barrier cream if necessary. Change out of oil-soaked clothes and do not keep oily rags in your pocket.
• Air conditioning refrigerant forms a poisonous gas if exposed to a naked flame (including a cigarette). It can also cause skin burns on contact.

Asbestos
• Asbestos dust can cause cancer if inhaled or swallowed. Asbestos may be found in gaskets and in brake and clutch linings. When dealing with such components it is safest to assume that they contain asbestos.

Special hazards

Hydrofluoric acid
• This extremely corrosive acid is formed when certain types of synthetic rubber, found in some O-rings, oil seals, fuel hoses etc, are exposed to temperatures above 400°C. The rubber changes into a charred or sticky substance containing the acid. *Once formed, the acid remains dangerous for years. If it gets onto the skin, it may be necessary to amputate the limb concerned.*
• When dealing with a vehicle which has suffered a fire, or with components salvaged from such a vehicle, wear protective gloves and discard them after use.

The battery
• Batteries contain sulphuric acid, which attacks clothing, eyes and skin. Take care when topping-up or carrying the battery.
• The hydrogen gas given off by the battery is highly explosive. Never cause a spark or allow a naked light nearby. Be careful when connecting and disconnecting battery chargers or jump leads.

Air bags
• Air bags can cause injury if they go off accidentally. Take care when removing the steering wheel and/or facia. Special storage instructions may apply.

Diesel injection equipment
• Diesel injection pumps supply fuel at very high pressure. Take care when working on the fuel injectors and fuel pipes.

⚠ *Warning: Never expose the hands, face or any other part of the body to injector spray; the fuel can penetrate the skin with potentially fatal results.*

Remember...

DO
• Do use eye protection when using power tools, and when working under the vehicle.
• Do wear gloves or use barrier cream to protect your hands when necessary.
• Do get someone to check periodically that all is well when working alone on the vehicle.
• Do keep loose clothing and long hair well out of the way of moving mechanical parts.
• Do remove rings, wristwatch etc, before working on the vehicle – especially the electrical system.
• Do ensure that any lifting or jacking equipment has a safe working load rating adequate for the job.

DON'T
• Don't attempt to lift a heavy component which may be beyond your capability – get assistance.
• Don't rush to finish a job, or take unverified short cuts.
• Don't use ill-fitting tools which may slip and cause injury.
• Don't leave tools or parts lying around where someone can trip over them. Mop up oil and fuel spills at once.
• Don't allow children or pets to play in or near a vehicle being worked on.

0•6 Roadside repairs

The following pages are intended to help in dealing with common roadside emergencies and breakdowns. You will find more detailed fault finding information at the back of the manual, and repair information in the main chapters.

If your car won't start and the starter motor doesn't turn

- ☐ If it's a model with automatic transmission, make sure the selector is in 'P' or 'N'.
- ☐ Open the bonnet and make sure that the battery terminals are clean and tight.
- ☐ Switch on the headlights and try to start the engine. If the headlights go very dim when you're trying to start, the battery is probably flat. Get out of trouble by jump starting (see next page) using a friend's car.

If your car won't start even though the starter motor turns as normal

- ☐ Is there fuel in the tank?
- ☐ Is there moisture on electrical components under the bonnet? Switch off the ignition, then wipe off any obvious dampness with a dry cloth. Spray a water-repellent aerosol product (WD-40 or equivalent) on ignition and fuel system electrical connectors like those shown in the photos. Pay special attention to the ignition coil wiring connector and HT leads. (Note that Diesel engines don't normally suffer from damp.)

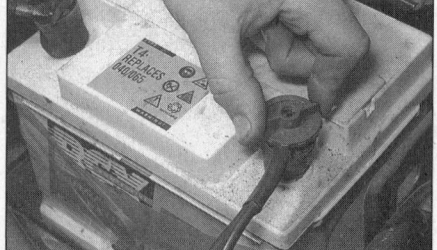

A Check the condition and security of the battery connections

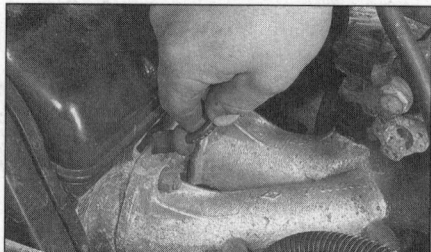

B Check that the spark plug HT leads are securely connected by pushing them onto the plugs

C Check that the HT lead is securely connected to the ignition coil

Check that electrical connections are secure (with the ignition switched off) and spray them with a water dispersant spray like WD40 if you suspect a problem due to damp

D Check that the wiring connectors are securely connected to the ignition coil

E Check the condition of the battery earth connection to the body. The connection shown has corroded badly, and requires cleaning

Roadside repairs 0•7

 HAYNES HiNT *Jump starting will get you out of trouble, but you must correct whatever made the battery go flat in the first place. There are three possibilities:*

1 The battery has been drained by repeated attempts to start, or by leaving the lights on.

2 The charging system is not working properly (alternator drivebelt slack or broken, alternator wiring fault or alternator itself faulty).

3 The battery itself is at fault (electrolyte low, or battery worn out).

When jump-starting a car using a booster battery, observe the following precautions:

✔ Before connecting the booster battery, make sure that the ignition is switched off.

✔ Ensure that all electrical equipment (lights, heater, wipers, etc) is switched off.

Jump starting

✔ Make sure that the booster battery is the same voltage as the discharged one in the vehicle.

✔ If the battery is being jump-started from the battery in another vehicle, the two vehcles MUST NOT TOUCH each other.

✔ Make sure that the transmission is in neutral (or PARK, in the case of automatic transmission).

1 Connect one end of the red jump lead to the positive (+) terminal of the flat battery

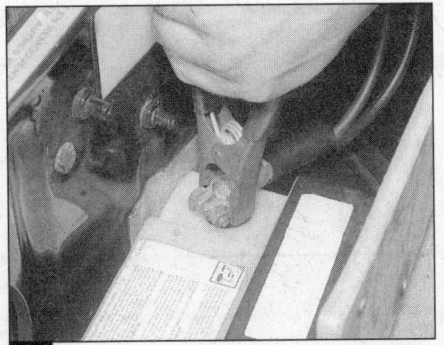
2 Connect the other end of the red lead to the positive (+) terminal of the booster battery.

3 Connect one end of the black jump lead to the negative (-) terminal of the booster battery

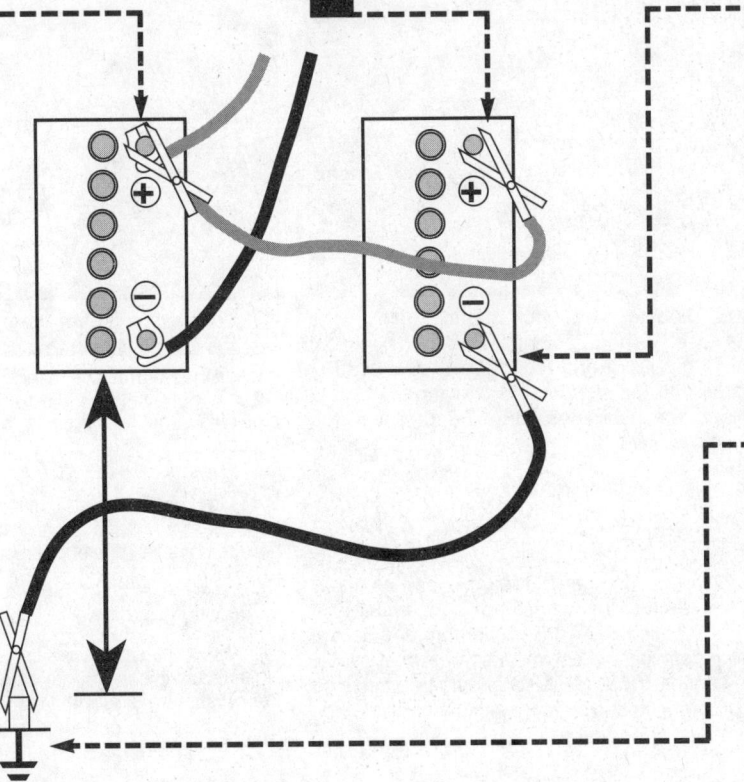

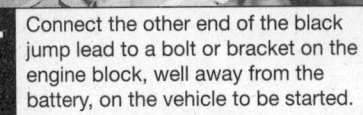
4 Connect the other end of the black jump lead to a bolt or bracket on the engine block, well away from the battery, on the vehicle to be started.

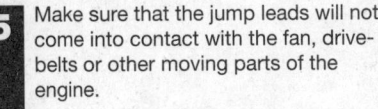
5 Make sure that the jump leads will not come into contact with the fan, drive-belts or other moving parts of the engine.

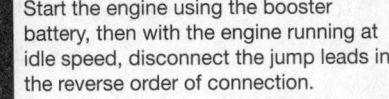
6 Start the engine using the booster battery, then with the engine running at idle speed, disconnect the jump leads in the reverse order of connection.

0•8 Roadside repairs

Wheel changing

Some of the details shown here will vary according to model. For instance, the location of the spare wheel and jack is not the same on all cars. However, the basic principles apply to all vehicles.

 Warning: *Do not change a wheel in a situation where you risk being hit by other traffic. On busy roads, try to stop in a lay-by or a gateway. Be wary of passing traffic while changing the wheel – it is easy to become distracted by the job in hand.*

Preparation

- When a puncture occurs, stop as soon as it is safe to do so.
- Park on firm level ground, if possible, and well out of the way of other traffic.
- Use hazard warning lights if necessary.
- If you have one, use a warning triangle to alert other drivers of your presence.
- Apply the handbrake and engage first or reverse gear (or Park on models with automatic transmission.
- Chock the wheel diagonally opposite the one being removed – a couple of large stones will do for this.
- If the ground is soft, use a flat piece of wood to spread the load under the jack.

Changing the wheel

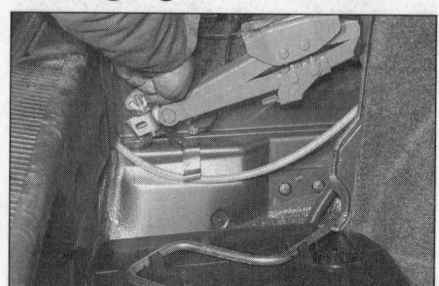

1 Working in the luggage compartment, unclip the trim panel for access to the jack and wheel brace. The jack is secured by a wing-nut

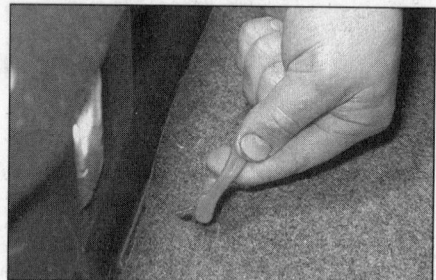

2 The spare wheel is stored in cradle under the rear of the vehicle. Working in the luggage compartment, pull up the cradle locking ring then, working under the vehicle, pull the safety catch back, and lower the spare wheel cradle

3 Lift the spare wheel from the cradle

4 Where applicable, remove the wheel trim, then loosen each wheel bolt by half a turn

5 Locate the jack head below the reinforced jacking point and on firm ground (don't jack the car at any other point on the sill). Ensure that the lug on the jack head engages with the cut-out in the jacking point

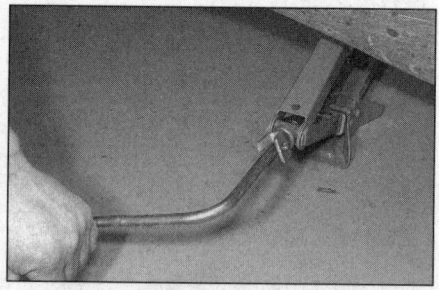

6 Engage the end of the wheel brace with the jack, ensuring that one of the pins on the wheel brace engages with the closed slot in the jack. Turn the wheel brace clockwise until the wheel is raised clear of the ground

7 Lower the car to the ground, then finally tighten the wheel bolts in a diagonal sequence. Fit the wheel trim and secure the damaged wheel in the spare wheel cradle. Note that the wheel bolts should be slackened and retightened to the specified torque at the earliest opportunity

Finally...

- Remove the wheel chocks.
- Stow the jack and tools in the correct locations in the car.
- Check the tyre pressure on the wheel just fitted. If it is low, or if you don't have a pressure gauge with you, drive slowly to the nearest garage and inflate the tyre to the right pressure.
- Have the damaged tyre or wheel repaired as soon as possible.

Roadside repairs 0•9

Identifying leaks

Puddles on the garage floor or drive, or obvious wetness under the bonnet or underneath the car, suggest a leak that needs investigating. It can sometimes be difficult to decide where the leak is coming from, especially if the engine bay is very dirty already. Leaking oil or fluid can also be blown rearwards by the passage of air under the car, giving a false impression of where the problem lies.

 Warning: Most automotive oils and fluids are poisonous. Wash them off skin, and change out of contaminated clothing, without delay.

 HAYNES HiNT *The smell of a fluid leaking from the car may provide a clue to what's leaking. Some fluids are distinctively coloured. It may help to clean the car carefully and to park it over some clean paper overnight as an aid to locating the source of the leak. Remember that some leaks may only occur while the engine is running.*

Sump oil

Engine oil may leak from the drain plug...

Oil from filter
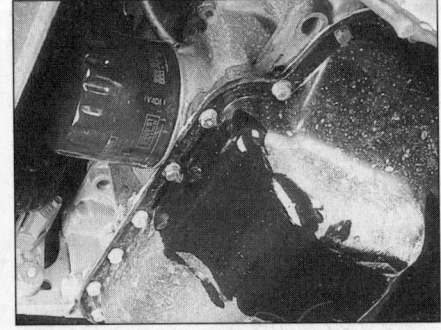
...or from the base of the oil filter.

Gearbox oil
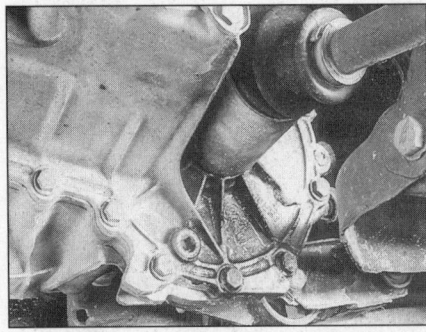
Gearbox oil can leak from the seals at the inboard ends of the driveshafts.

Antifreeze

Leaking antifreeze often leaves a crystalline deposit like this.

Brake fluid

A leak occurring at a wheel is almost certainly brake fluid.

Power steering fluid

Power steering fluid may leak from the pipe connectors on the steering rack.

Towing

When all else fails, you may find yourself having to get a tow home – or of course you may be helping somebody else. Long-distance recovery should only be done by a garage or breakdown service. For shorter distances, DIY towing using another car is easy enough, but observe the following points:

☐ Use a proper tow-rope – they are not expensive. The vehicle being towed must display an 'ON TOW' sign in its rear window.
☐ Always turn the ignition key to the 'on' position when the vehicle is being towed, so that the steering lock is released, and that the direction indicator and brake lights will work.
☐ Only attach the tow-rope to the towing eyes provided.
☐ Before being towed, release the handbrake and select neutral on the transmission.
☐ Note that greater-than-usual pedal pressure will be required to operate the brakes, since the vacuum servo unit is only operational with the engine running.
☐ On models with power steering, greater-than-usual steering effort will also be required.

☐ The driver of the car being towed must keep the tow-rope taut at all times to avoid snatching.
☐ Make sure that both drivers know the route before setting off.
☐ Only drive at moderate speeds and keep the distance towed to a minimum. Drive smoothly and allow plenty of time for slowing down at junctions.
☐ On models with automatic transmission, special precautions apply. If in doubt, do not tow, or transmission damage may result.

0•10 Weekly checks

Introduction

There are some very simple checks which need only take a few minutes to carry out, but which could save you a lot of inconvenience and expense.

These "Weekly checks" require no great skill or special tools, and the small amount of time they take to perform could prove to be very well spent, for example;

☐ Keeping an eye on tyre condition and pressures, will not only help to stop them wearing out prematurely, but could also save your life.

☐ Many breakdowns are caused by electrical problems. Battery-related faults are particularly common, and a quick check on a regular basis will often prevent the majority of these.

☐ If your car develops a brake fluid leak, the first time you might know about it is when your brakes don't work properly. Checking the level regularly will give advance warning of this kind of problem.

☐ If the oil or coolant levels run low, the cost of repairing any engine damage will be far greater than fixing the leak, for example.

Underbonnet check points

◀ **1995 cc 4-cyl petrol model**

- **A** Engine oil level dipstick
- **B** Engine oil filler cap
- **C** Coolant expansion tank
- **D** Brake fluid reservoir
- **E** Washer fluid reservoir
- **F** Power steering fluid reservoir
- **G** Battery

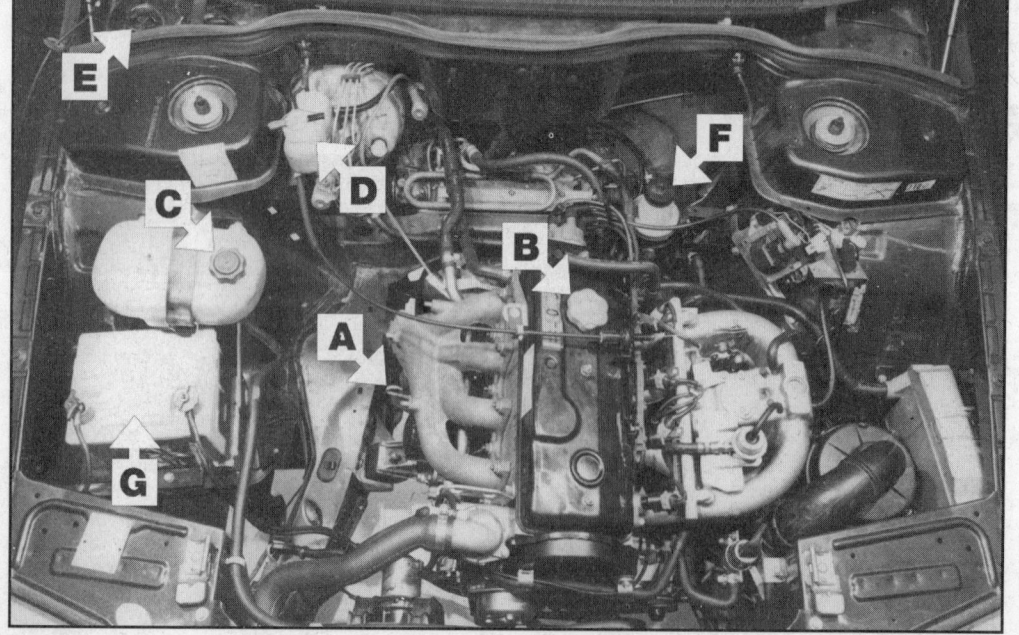

◀ **2165 cc 4-cyl petrol model**

- **A** Engine oil level dipstick
- **B** Engine oil filler cap
- **C** Coolant expansion tank
- **D** Brake fluid reservoir
- **E** Washer fluid reservoir
- **F** Power steering fluid reservoir
- **G** Battery

Weekly checks 0•11

◄ 2664 cc 6-cyl petrol model

A *Engine oil level dipstick*
B *Engine oil filler cap*
C *Coolant expansion tank*
D *Brake fluid reservoir*
E *Washer fluid reservoir*
F *Power steering fluid reservoir*
G *Battery*

Engine oil level

Before you start

✔ Make sure that your car is on level ground.
✔ Check the oil level before the car is driven, or at least 5 minutes after the engine has been switched off.

 **If the oil level is checked immediately after driving the vehicle, some of the oil will remain in the upper engine components, resulting in an inaccurate reading on the dipstick!**

The correct oil

Modern engines place great demands on their oil. It is very important that the correct oil for your car is used (See "Lubricants, fluids and tyre pressures").

Car Care

● If you have to add oil frequently, you should check whether you have any oil leaks. Place some clean paper under the car overnight, and check for stains in the morning. If there are no leaks, the engine may be burning oil (see "Fault Finding").

● Always maintain the level between the upper and lower dipstick marks (see photo 3). If the level is too low severe engine damage may occur. Oil seal failure may result if the engine is overfilled by adding too much oil.

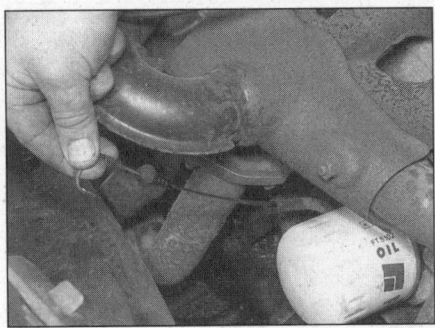

1 The dipstick is located in a tube on the right-hand side of the engine (see "Underbonnet Check Points" on pages 0-10 and 0-11 for exact location). Withdraw the dipstick

2 Using a clean rag or paper towel remove all oil from the dipstick. Insert the clean dipstick into the tube as far as it will go, then withdraw it again.

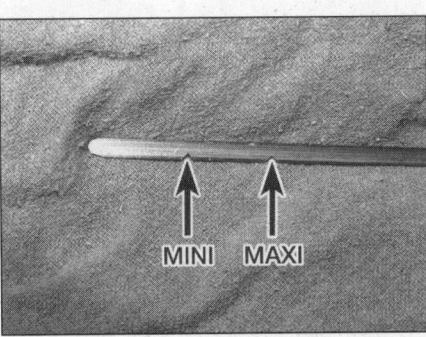

3 Note the oil level on the end of the dipstick, which should be between the upper ("MAX") mark and lower ("MIN") mark.

4 Oil is added through the filler cap. Where applicable, release the clips, then pull out the cap. Alternatively, twist the cap to release it. Top-up the level. A funnel may help to reduce spillage. Add the oil slowly, checking the level on the dipstick often. Don't overfill

0•12 Weekly checks

Coolant level

⚠️ **Warning:** *DO NOT attempt to remove the expansion tank pressure cap when the engine is hot, as there is a very great risk of scalding. Do not leave open containers of coolant about, as it is poisonous.*

Car Care

● Adding coolant should not be necessary on a regular basis. If frequent topping-up is required, it is likely there is a leak. Check the radiator, all hoses and joint faces for signs of staining or wetness, and rectify as necessary.

● It is important that antifreeze is used in the cooling system all year round, not just during the winter months. Don't top-up with water alone, as the antifreeze will become too diluted.

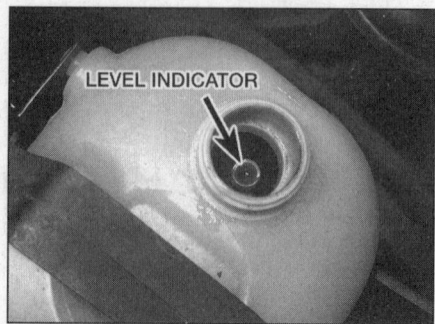

1 The coolant level varies with the temperature of the engine. The level should be checked in the expansion tank, which is at the rear right-hand corner of the engine compartment. The expansion tank may have "MAX" and "MIN" level markings, or there may be a single "filling" mark

2 When the engine is cold, the level should be between the marks, or up to the top of the level indicator. On some models, a level sensor unit is fitted in the bottom of the expansion tank.

3 Wait until the engine is cold, then unscrew the expansion tank cap. Add a mixture of water and antifreeze to the expansion tank, until the coolant is up to the appropriate level mark ("MAX" mark or "filling" mark). Refit the cap, turning it clockwise as far as it will go until it is secure

Brake and clutch fluid level

 Warning:
● Brake fluid can harm your eyes and will damage painted surfaces, so use extreme caution when handling and pouring it.
● Do not use fluid that has been standing open for some time, as it absorbs moisture from the air, which can cause a dangerous loss of braking effectiveness.

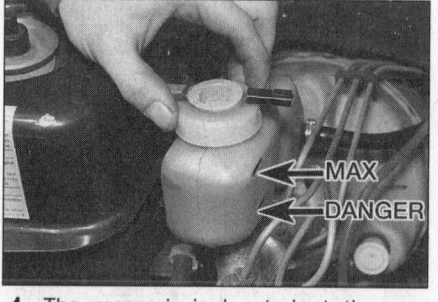

1 The reservoir is located at the rear driver's side of the engine compartment. The fluid level must be kept between the "MAX" and "DANGER" marks

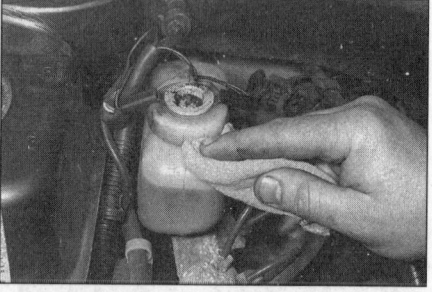

2 If topping-up is necessary, first wipe the area around the filler cap with a clean rag, then disconnect the fluid level sensor wires before removing the cap. The system should be drained and refilled if dirt is seen in the fluid (see Chapter 9 for details)

 **Haynes Hint**
● Make sure that your car is on level ground.
● The fluid level in the reservoir will drop slightly as the brake pads wear down, but the fluid level must never be allowed to drop below the "MIN" mark.

Safety First!

● If the reservoir requires repeated topping-up this is an indication of a fluid leak somewhere in the system, which should be investigated immediately.

● If a leak is suspected, the car should not be driven until the braking system has been checked. Never take any risks where brakes are concerned.

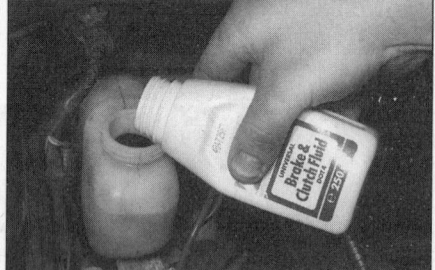

3 Carefully add fluid, avoiding spilling it on paintwork. Use only the specified hydraulic fluid. The level will rise slightly when the cap/float assembly is refitted. After filling to the correct level, refit the cap securely and reconnect the sensor wires. Wipe off any spilt fluid

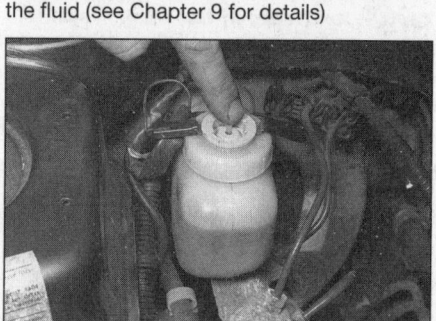

4 Check the operation of the low fluid level warning light. Switch on the ignition and ask an assistant to press the test button on top of the brake fluid reservoir cap. The brake warning light should come on - if not, the level switch, wiring or bulb may be faulty. Switch off the ignition after testing

Weekly checks 0•13

Power steering fluid level

Before you start:
✔ Park the vehicle on level ground.
✔ Set the steering wheel straight-ahead.
✔ The engine should be turned off.

HAYNES HiNT *For the check to be accurate, the steering must not be turned once the engine has been stopped.*

Safety First!
● The need for frequent topping-up indicates a leak, which should be investigated immediately.

1 The power steering fluid reservoir is located on the rear left-hand side of the engine. The fluid level should be checked with the engine stopped. Two types of fluid reservoir may be fitted; a translucent reservoir with "MAX" and "MIN" markings on the side of the reservoir, or a dark-coloured reservoir with a dipstick attached to the filler cap

2 On models with a translucent reservoir, the fluid level should be between the "MAX" and "MIN" marks. If topping-up is necessary, and before removing the cap, wipe the surrounding area so that dirt does not enter the reservoir. Unscrew the cap, allowing the fluid to drain from the bottom of the cap as it is removed. Top up the fluid level to the "MAX" mark, using the specified type of fluid (do not overfill the reservoir), then refit and tighten the filler cap

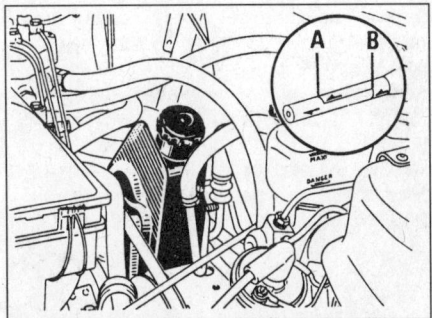

3 On models with a dipstick fitted to the filler cap, wipe around the filler cap, then unscrew the cap and wipe the dipstick using a clean rag. Refit the filler cap, then slowly unscrew and withdraw it, and read off the fluid level. The level should be between the "A" and "B" marks on the dipstick. If necessary, top up the fluid level, using the specified type of fluid (do not overfill the reservoir), then re-check the fluid level. Refit and tighten the filler cap on completion

Screen washer fluid level

Screenwash additives not only keep the windscreen clean during foul weather, they also prevent the washer system freezing in cold weather - which is when you are likely to need it most. Don't top up using plain water as the screenwash will become too diluted, and will freeze during cold weather. **On no account use coolant antifreeze in the washer system - this could discolour or damage paintwork.**

1 The windscreen/tailgate/headlight washer fluid reservoir is located in the scuttle at the rear right-hand corner of the engine compartment. If topping-up is necessary, open the cap

2 When topping-up the reservoir(s) a screenwash additive should be added in the quantities recommended on the bottle

Weekly checks

Tyre condition and pressure

It is very important that tyres are in good condition, and at the correct pressure - having a tyre failure at any speed is highly dangerous. Tyre wear is influenced by driving style - harsh braking and acceleration, or fast cornering, will all produce more rapid tyre wear. As a general rule, the front tyres wear out faster than the rears. Interchanging the tyres from front to rear ("rotating" the tyres) may result in more even wear. However, if this is completely effective, you may have the expense of replacing all four tyres at once!

Remove any nails or stones embedded in the tread before they penetrate the tyre to cause deflation. If removal of a nail does reveal that the tyre has been punctured, refit the nail so that its point of penetration is marked. Then immediately change the wheel, and have the tyre repaired by a tyre dealer.

Regularly check the tyres for damage in the form of cuts or bulges, especially in the sidewalls. Periodically remove the wheels, and clean any dirt or mud from the inside and outside surfaces. Examine the wheel rims for signs of rusting, corrosion or other damage. Light alloy wheels are easily damaged by "kerbing" whilst parking; steel wheels may also become dented or buckled. A new wheel is very often the only way to overcome severe damage.

New tyres should be balanced when they are fitted, but it may become necessary to re-balance them as they wear, or if the balance weights fitted to the wheel rim should fall off. Unbalanced tyres will wear more quickly, as will the steering and suspension components. Wheel imbalance is normally signified by vibration, particularly at a certain speed (typically around 50 mph). If this vibration is felt only through the steering, then it is likely that just the front wheels need balancing. If, however, the vibration is felt through the whole car, the rear wheels could be out of balance. Wheel balancing should be carried out by a tyre dealer or garage.

1 Tread Depth - visual check
The original tyres have tread wear safety bands (B), which will appear when the tread depth reaches approximately 1.6 mm. The band positions are indicated by a triangular mark on the tyre sidewall (A).

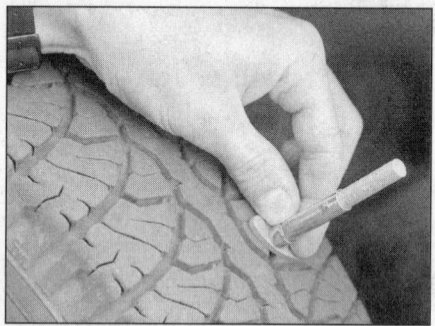

2 Tread Depth - manual check
Alternatively, tread wear can be monitored with a simple, inexpensive device known as a tread depth indicator gauge.

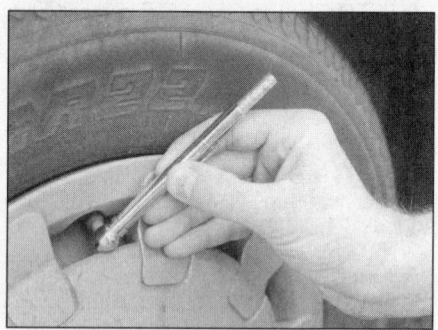

3 Tyre Pressure Check
Check the tyre pressures regularly with the tyres cold. Do not adjust the tyre pressures immediately after the vehicle has been used, or an inaccurate setting will result.

Tyre tread wear patterns

Shoulder Wear

Underinflation (wear on both sides)
Under-inflation will cause overheating of the tyre, because the tyre will flex too much, and the tread will not sit correctly on the road surface. This will cause a loss of grip and excessive wear, not to mention the danger of sudden tyre failure due to heat build-up.
Check and adjust pressures
Incorrect wheel camber (wear on one side)
Repair or renew suspension parts
Hard cornering
Reduce speed!

Centre Wear

Overinflation
Over-inflation will cause rapid wear of the centre part of the tyre tread, coupled with reduced grip, harsher ride, and the danger of shock damage occurring in the tyre casing.
Check and adjust pressures

If you sometimes have to inflate your car's tyres to the higher pressures specified for maximum load or sustained high speed, don't forget to reduce the pressures to normal afterwards.

Uneven Wear

Front tyres may wear unevenly as a result of wheel misalignment. Most tyre dealers and garages can check and adjust the wheel alignment (or "tracking") for a modest charge.
Incorrect camber or castor
Repair or renew suspension parts
Malfunctioning suspension
Repair or renew suspension parts
Unbalanced wheel
Balance tyres
Incorrect toe setting
Adjust front wheel alignment
Note: *The feathered edge of the tread which typifies toe wear is best checked by feel.*

Weekly checks 0•15

Wiper blades

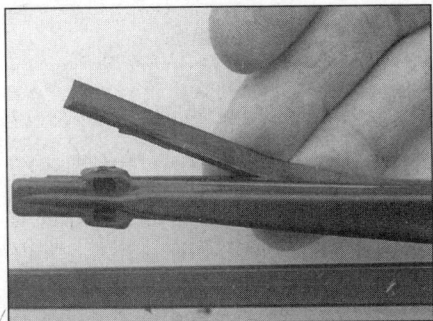

1 Check the condition of the wiper blades; if they are cracked or show any signs of deterioration, or if the glass swept area is smeared, renew them. Wiper blades should be renewed annually, as a matter of course.

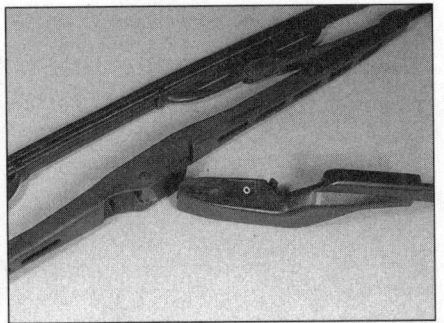

2 To remove a windscreen wiper blade, pull the arm fully away from the screen until it locks. Swivel the blade through 90°, then pull the blade sharply to release it from th end of the arm

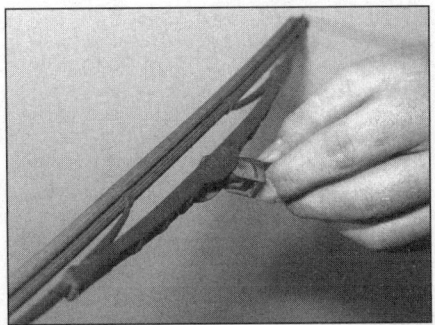

3 Don't forget to check the tailgate wiper blade as well. The blade can be removed in the same way as the windscreen wiper blade

Battery

Caution: *Before carrying out any work on the vehicle battery, read the precautions given in "Safety first" at the start of this manual.*

✔ Make sure that the battery tray is in good condition, and that the clamp is tight. Corrosion on the tray, retaining clamp and the battery itself can be removed with a solution of water and baking soda. Thoroughly rinse all cleaned areas with water. Any metal parts damaged by corrosion should be covered with a zinc-based primer, then painted.

✔ Periodically (approximately every three months), check the charge condition of the battery as described in Chapter 5A.

✔ If the battery is flat, and you need to jump start your vehicle, see **Roadside Repairs**.

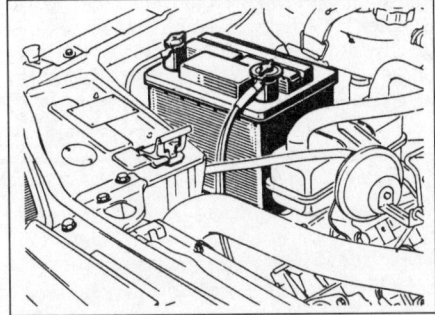

1 The battery is located on the left- or right-hand side of the engine compartment, depending on model. The exterior of the battery should be inspected periodically for damage such as a cracked case or cover

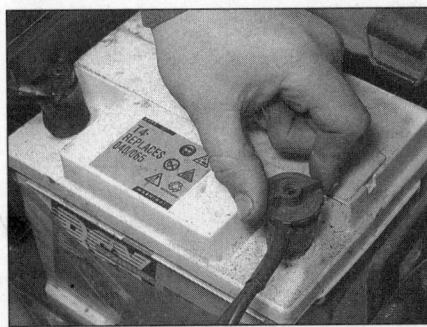

2 Check the tightness of the battery cable clamps to ensure good electrical connections. You should not be able to move them. Also check each cable for cracks and frayed conductors.

Battery corrosion can be kept to a minimum by applying a layer of petroleum jelly to the clamps and terminals after they are reconnected.

3 If corrosion (white fluffy deposits) is evident, remove the cables from the battery terminals, clean them with a small wire brush, then refit them. Automotive stores sell a useful tool for cleaning the battery post and terminals

4 Note that the battery terminal studs can be removed for cleaning or renewal. Unscrew the lead clamp, then pull off the plastic insulator, and lever off the stud and cover

0•16 Weekly checks

Bulbs and fuses

✔ Check all external lights and the horn. Refer to the appropriate Sections of Chapter 12 for details if any of the circuits are found to be inoperative.

✔ Visually check all accessible wiring connectors, harnesses and retaining clips for security, and for signs of chafing or damage.

 **If you need to check your brake lights and indicators unaided, back up to a wall or garage door and operate the lights. The reflected light should show if they are working properly.**

1 If a single indicator light, brake light or headlight has failed, it is likely that a bulb has blown and will need to be replaced. Refer to Chapter 12 for details. If both brake lights have failed, it is possible that the brake light switch operated by the brake pedal has failed. Refer to Chapter 9 for details

2 If more than one indicator light or headlight has failed, it is likely that either a fuse has blown or that there is a fault in the circuit (see *"Electrical fault finding"* in Chapter 12). The fuses are mounted in a panel located at the lower driver's side of the facia under a cover. Press the two catches to release the cover

3 To replace a blown fuse, remove it, where applicable, using the plastic tool provided. Fit a new fuse of the same rating, available from car accessory shops. It is important that you find the reason that the fuse blew (see *"Electrical fault finding"* in Chapter 12)

Lubricants, fluids and tyre pressures   0•17

Lubricants and fluids

Engine	Multigrade engine oil, viscosity SAE 10W/40 to 20W/50, to API SG/CD (Duckhams QXR, QS, Hypergrade Plus, or Hypergrade)
Cooling system	Ethylene glycol-based antifreeze (Duckhams Antifreeze and Summer Coolant)
Manual transmission	SAE 80W gear oil, to API GL5 (Duckhams Hypoid 80W/90S)
Automatic transmission	Refer to Chapter 7B
Braking system	Hydraulic fluid to SAE J1703F or DOT 4 (Duckhams Universal Brake and Clutch Fluid)
Power steering	Dexron-type ATF (Duckhams Uni-Matic)

Choosing your engine oil

Oils perform vital tasks in all engines. The higher the engine's performance, the greater the demand on lubricants to minimise wear as well as optimise power and economy. Duckhams tailors lubricants to the highest technical standards, meeting and exceeding the demands of all modern engines.

HOW ENGINE OIL WORKS

• **Beating friction**

Without oil, the surfaces inside your engine which rub together will heat, fuse and quickly cause engine seizure. Oil, and its special additives, forms a molecular barrier between moving parts, to stop wear and minimise heat build-up.

• **Cooling hot spots**

Oil cools parts that the engine's water-based coolant cannot reach, bathing the combustion chamber and pistons, where temperatures may exceed 1000°C. The oil assists in transferring the heat to the engine cooling system. Heat in the oil is also lost by air flow over the sump, and via any auxiliary oil cooler.

• **Cleaning the inner engine**

Oil washes away combustion by-products (mainly carbon) on pistons and cylinders, transporting them to the oil filter, and holding the smallest particles in suspension until they are flushed out by an oil change. Duckhams oils undergo extensive tests in the laboratory, and on the road.

Note: It is antisocial and illegal to dump oil down the drain. To find the location of your local oil recycling bank, call this number free.

Engine oil types

Mineral oils are the "traditional" oils, generally suited to older engines and cars not used in harsh conditions. *Duckhams Hypergrade Plus* and *Hypergrade* are well suited for use in most popular family cars.
Diesel oils such as *Duckhams Diesel* are specially formulated for Diesel engines, including turbocharged models and 4x4s.
Synthetic oils are the state-of-the-art in lubricants, offering ultimate protection, but at a fairly high price. One such is *Duckhams QS*, for use in ultra-high performance engines.
Semi-synthetic oils offer high performance engine protection, but at less cost than full synthetic oils. *Duckhams QXR* is an ideal choice for hot hatches and hard-driven cars.

For help with technical queries on lubricants, call Duckhams Oils on 0181 290 8207

Tyre pressures

Note: *Recommended tyre pressures are marked on a label attached to the driver's door edge or frame. Pressures apply to original-equipment tyres, and may vary if any other make or type of tyre is fitted; check with the tyre manufacturer or supplier for correct pressures if necessary.*

	Front*	Rear
Four-cylinder engine models	2.0 bar (29 psi)	2.2 bar (32 psi)
Six-cylinder non-turbo models	2.3 bar (33 psi)	2.5 bar (36 psi)
Six-cylinder turbo models	2.5 bar (36 psi)	2.3 bar (33 psi)

Notes

Chapter 1 Part A:
Routine maintenance and servicing - petrol models

Contents

Air conditioning refrigerant check 17	Idle speed and mixture check 10
Air filter element renewal 18	Intensive maintenance 2
Automatic transmission fluid and strainer renewal 26	Introduction 1
Automatic transmission fluid level check 9	Manual transmission oil level check 20
Auxiliary drivebelt check and renewal 8	Manual transmission oil renewal 25
Brake fluid renewal 27	Rear brake pad check - models with rear disc brakes 5
Clutch check 7	Rear brake shoe thickness check - models with rear drum brakes . 24
Coolant renewal 31	Road test 29
Electrical systems check 12	Roadwheel bolt check 15
Engine oil and filter renewal 3	Spark plug renewal 19
Exhaust system check 13	Spare fuse check 22
Fuel filter renewal 28	Suspension and steering check 14
Front brake pad check 4	Seat belt check 11
Front wheel alignment check 23	Timing belt renewal - 1995 cc and 2165 cc engines 30
Handbrake check 6	Turbocharger boost pressure check - 2458 cc turbo models 16
Hose and fluid leak check 21	

Degrees of difficulty

Easy, suitable for novice with little experience	**Fairly easy,** suitable for beginner with some experience	**Fairly difficult,** suitable for competent DIY mechanic	**Difficult,** suitable for experienced DIY mechanic 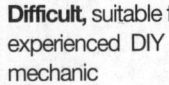	**Very difficult,** suitable for expert DIY or professional

1A•2 Servicing Specifications - petrol models

Lubricants and fluids
Refer to "Weekly checks"

Capacities

Engine oil
Excluding filter:
- 1995 cc 8-valve engines .... 6.0 litres
- 1995 cc 12-valve engines .... 5.0 litres
- 2165 cc engines .... 5.0 litres
- 2458 cc turbo engines .... 7.5 litres
- 2664 cc engines .... 5.0 litres
- 2849 cc engines .... 6.5 litres

Including filter:
- 1995 cc 8-valve engines .... 6.5 litres
- 1995 cc 12-valve engines .... 5.5 litres
- 2165 cc engines .... 5.5 litres
- 2458 cc turbo engines .... 8.0 litres
- 2664 cc engines .... 5.5 litres
- 2849 cc engines .... 7.0 litres

Cooling system (approximate)
- 1995 cc carburettor engines .... 7.5 litres
- 1995 cc fuel injection engines .... 7.8 litres
- 2165 cc engines .... 8.0 litres
- 2458 cc turbo engines .... 9.7 litres
- 2664 cc engines .... 9.5 litres
- 2849 cc engines .... 9.5 litres

Transmission*
Manual:
- NG1-type transmission .... 2.2 litres
- NG3-type transmission .... 2.2 litres
- UN1-type transmission .... 3.0 litres

Automatic:
From dry:
- MJ3 transmission .... 6.0 litres
- AR4 transmission:
  - Main transmission .... 6.6 litres
  - Final drive .... 0.85 litre

Drain and refill:
- MJ3 transmission .... 2.5 litres
- AR4 transmission:
  - Main transmission .... 4.0 litres
  - Final drive .... 0.80 litre

See Chapter 7A or 7B for transmission identification details

Fuel tank
Capacity .... 72.0 litres

Engine
Oil filter:
- Models up to 1994 models year .... Champion C104
- Models from 1995 models year .... Champion F104

Cooling system
Antifreeze mixture:
- 35% antifreeze .... Protection down to -23°C
- 50% antifreeze .... Protection down to -40°C

Fuel system
Idle speed:
- Carburettor models (see Chapter 4A):
  - Models with DARA 32 - 49 and 49 C carburettors .... 900 ± 50 rpm
  - All other models .... 800 ± 50 rpm
- Fuel-injected models (see Chapter 4B):
  - 1995 cc and 2165 cc models .... 800 ±50 rpm
  - 2458 cc turbo models .... 750 ± 50 rpm
  - 2664 cc models .... 900 ± 50 rpm
  - 2849 cc models .... 700 ± 50 rpm

Servicing Specifications - petrol models

Fuel system (continued)

Idle mixture (exhaust gas CO level) content:
- Models not equipped with a catalytic converter .... 1.5 ± 0.5 %
- Models with a catalytic converter .... Less than 1.5 %

Air filter element:
- Four-cylinder engine models:*
  - 2165 cc J7T-708 engines .... Champion W230
  - All other engines .... Champion W115
- Six-cylinder engine models:
  - 2664 cc engines up to 1985 model year .... Champion U603
  - All other engines .... Champion W230

*Refer to "Buying Spare Parts and Vehicle Identification" for details of engine code location

Fuel filter:
- 1995 cc carburettor engines .... Champion L101
- 1995 cc fuel-injected engines .... Champion L206
- 2165 cc engines .... Champion L206
- 2458 cc turbo engines .... Champion L206
- 2664 cc engines .... Champion L203
- 2849 cc engines .... Champion L206

Ignition system

Spark plugs:
- 1995 cc carburettor engines .... Champion S7YCC
- 1995 cc fuel-injected engines:
  - 8-valve engines .... Champion S6YCC
  - 12-valve engines .... Champion RC7BMC
- 2165 cc engines:
  - Models not fitted with a catalytic converter .... Champion S7YCC
  - Models with a catalytic converter .... Champion S6YCC
- 2458 cc turbo engines .... Champion S6YCC
- 2664 cc engines .... Champion RS9YCC
- 2849 cc engines .... Champion S7YCC

Spark plug electrode gap* .... 0.8 mm

*The spark plug gap quoted is that recommended by Champion for the specified plugs listed above. If spark plugs of any other type are fitted, refer to their manufacturer's recommendation.

Brakes

Front disc brakes:
- Pad thickness (including backing):
  - New .... 18.0 mm
  - Minimum thickness .... 6.0 mm

Rear disc brakes:
- Pad thickness (including backing):
  - New .... 14.0 mm
  - Minimum thickness .... 6.0 mm

Rear drum brakes:
- Shoe thickness (including backing):
  - New .... 7.0 mm
  - Minimum thickness .... 2.5 mm

Torque wrench settings

	Nm	lbf ft
Roadwheel bolts:		
4-bolt fixing	90	66
5-bolt fixing	100	74
Spark plugs	20	15

1A•4 Maintenance schedule - petrol models

1 The maintenance intervals in this manual are provided with the assumption that you, not the dealer, will be carrying out the work. These are the minimum maintenance intervals recommended by the manufacturer for vehicles driven daily. If you wish to keep your vehicle in peak condition at all times, you may wish to perform some of these procedures more often. We encourage frequent maintenance, because it enhances the efficiency, performance and resale value of your vehicle.

2 If the vehicle is driven in dusty areas, used to tow a trailer, or driven frequently at slow speeds (idling in traffic) or on short journeys, more frequent maintenance intervals are recommended.

3 When the vehicle is new, it should be serviced by a factory-authorised dealer service department, in order to preserve the factory warranty.

Every 250 miles (400 km) or weekly
Refer to *"Weekly Checks"*

Every 12 000 miles (20 000 km)
Carry out all the operations listed under the 6000 mile (10 000 km) service, along with the following:
- ☐ Renew the air filter element (Section 18)
- ☐ Renew the spark plugs (Section 19)
- ☐ Check the manual transmission oil level (Section 20)
- ☐ Check all underbonnet components and hoses for fluid leaks (Section 21)

Every 6000 miles (10 000 km)
- ☐ Renew the engine oil and filter (Section 3)

Note: *Renault recommend that the oil filter on non-turbo models is renewed every 12 000 miles, but it is advisable to renew the filter whenever the engine oil is renewed.*

- ☐ Check the front brake pad thickness (Section 4)
- ☐ Check the rear brake pad thickness - models with rear disc brakes (Section 5)
- ☐ Check the operation of the handbrake (Section 6)
- ☐ Check the operation of the clutch (Section 7)
- ☐ Check the condition of the auxiliary drivebelts (Section 8)
- ☐ Check the automatic transmission fluid level (Section 9)
- ☐ Check the idle speed and mixture setting (exhaust CO emissions) (Section 10)
- ☐ Check the condition of the seat belts (Section 11)
- ☐ Check the operation of all electrical systems (Section 12)
- ☐ Check the condition of the exhaust system and mountings (Section 13)
- ☐ Check the suspension and steering components (Section 14)
- ☐ Check the tightness of the roadwheel bolts (Section 15)
- ☐ Check the turbocharger boost pressure - turbo models (Section 16)
- ☐ Check the condition of the air conditioning refrigerant (Section 17)

Every 36 000 miles (60 000 km)
Carry out all the operations listed under the 6000 mile (10 000 km), and 12 000 mile (20 000 km) services, along with the following:
- ☐ Check the spare fuses are in place (Section 22)
- ☐ Check the front wheel alignment (Section 23)
- ☐ Check the rear brake shoe thickness - models with rear drum brakes (Section 24)
- ☐ Renew the manual transmission oil (Section 25)
- ☐ Renew the automatic transmission fluid and strainer (Section 26)
- ☐ Renew the brake fluid (Section 27)
- ☐ Renew the fuel filter (Section 28)
- ☐ Carry out a road test (Section 29)

Every 72 000 miles (120 000 km)
In addition to all the items listed previously, carry out the following:
- ☐ Renew the timing belt (Section 30)

Every 2 years
In addition to all the items listed previously, carry out the following:
- ☐ Renew the coolant (Section 31)

Maintenance - component location 1A•5

Underbonnet view of a 1995 cc four-cylinder engine model

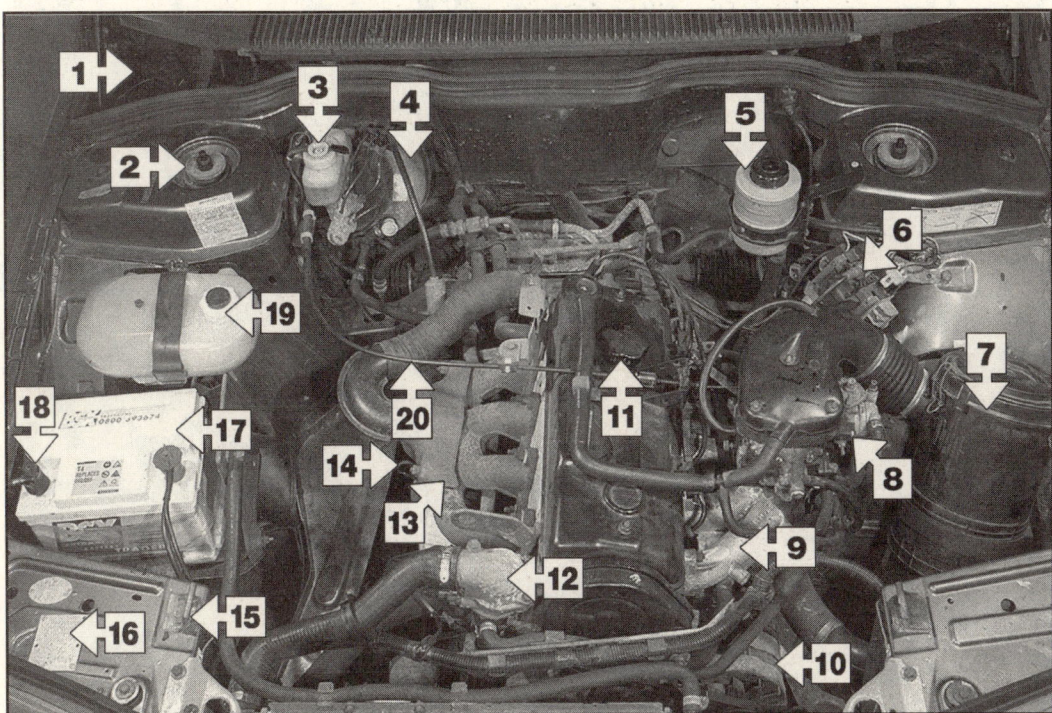

1. Washer fluid reservoir
2. Shock absorber top mounting
3. Brake fluid reservoir
4. Brake vacuum servo
5. Power steering fluid reservoir
6. Ignition coil
7. Air cleaner
8. Carburettor
9. Inlet manifold
10. Alternator
11. Engine oil filler cap
12. Thermostat housing
13. Exhaust manifold
14. Engine oil level dipstick
15. Bonnet lock striker
16. VIN plate
17. Battery
18. Battery negative lead
19. Coolant expansion tank
20. Throttle cable

Underbonnet view of a 2165 cc four-cylinder engine model

1. Heater control vacuum reservoir
2. Shock absorber top mounting
3. Brake fluid reservoir
4. Brake vacuum servo
5. Clutch master cylinder
6. Coolant expansion tank
7. Battery
8. Bonnet lock striker
9. Exhaust manifold
10. Engine oil filler cap
11. Power steering fluid reservoir
12. Ignition coil
13. Inlet manifold
14. Fuel rail and injectors
15. Idle speed control valve
16. Air cleaner
17. Engine ECU
18. Radiator
19. Electric cooling fan
20. Engine oil level dipstick

1A•6 Maintenance - component location

Underbonnet view of a typical 2664 cc six-cylinder engine

1 Heater control vacuum reservoir
2 Shock absorber top mounting
3 Brake fluid reservoir
4 Brake vacuum servo
5 Brake master cylinder
6 Fuel filter
7 Ignition coil
8 Diagnostic socket
9 Coolant expansion tank
10 Battery
11 Auxiliary air valve
12 Engine oil filler cap
13 Crankcase ventilation oil separator
14 Cruise control throttle actuator
15 Bonnet lock striker
16 Radiator
17 Throttle linkage
18 U-shaped inlet manifold section
19 Air cleaner
20 Engine ECU
21 Alternator
22 Clutch master cylinder
23 Power steering fluid reservoir

Front underbody view of a typical 2664 cc six-cylinder engine model

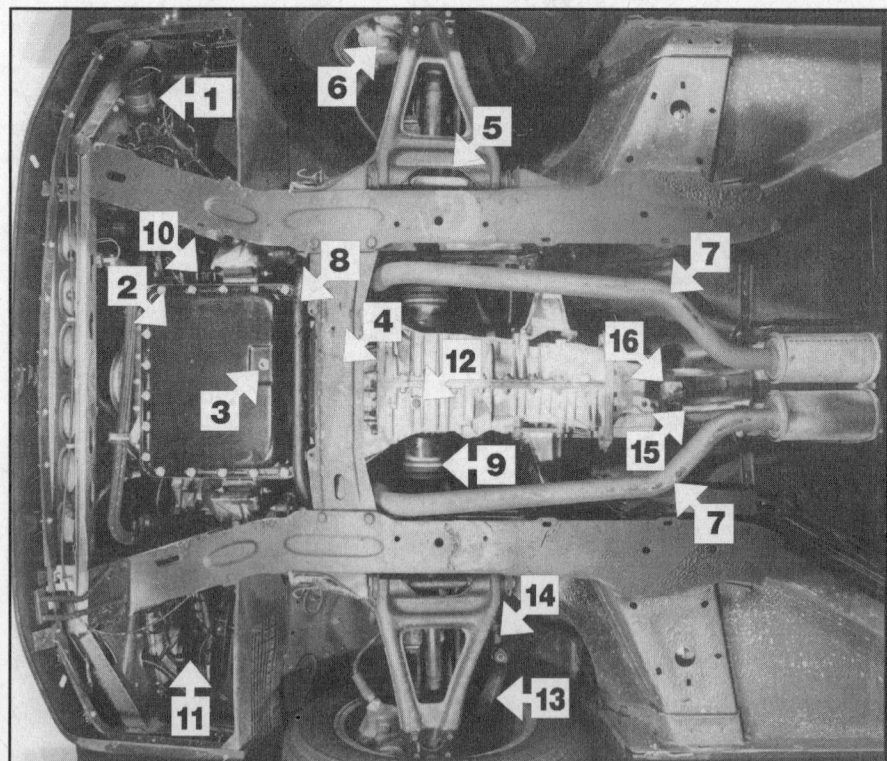

1 Horn compressor
2 Engine sump
3 Engine oil drain plug
4 Crossmember
5 Front suspension lower arm
6 Brake caliper
7 Exhaust pipes
8 Anti-roll bar
9 Driveshaft inboard joint
10 Oil filter
11 Headlight
12 Transmission oil drain plug
13 Steering arm/hub carrier
14 Track-rod
15 Gearchange linkage
16 Gearchange linkage

Maintenance - component location 1A•7

Rear underbody view of a typical 2664 cc six-cylinder engine model

1 Radius rod
2 Lower arm
3 Fuel tank support strap
4 Exhaust pipe
5 Handbrake cable equaliser
6 Fuel tank
7 Shock absorber
8 Exhaust silencer
9 Spare wheel carrier
10 Brake pipe

Maintenance procedures

1 Introduction

1 This Chapter is designed to help the home mechanic maintain his/her vehicle for safety, economy, long life and peak performance.
2 The Chapter contains a master maintenance schedule, followed by Sections dealing specifically with each task in the schedule. Visual checks, adjustments, component renewal and other helpful items are included. Refer to the accompanying illustrations of the engine compartment and the underside of the vehicle for the locations of the various components.
3 Servicing your vehicle in accordance with the mileage/time maintenance schedule and the following Sections will provide a planned maintenance programme, which should result in a long and reliable service life. This is a comprehensive plan, so maintaining some items but not others at the specified service intervals, will not produce the same results.
4 As you service your vehicle, you will discover that many of the procedures can - and should - be grouped together, because of the particular procedure being performed, or because of the close proximity of two otherwise-unrelated components to one another. For example, if the vehicle is raised for any reason, the exhaust can be inspected at the same time as the suspension and steering components.

5 The first step in this maintenance programme is to prepare yourself before the actual work begins. Read through all the Sections relevant to the work to be carried out, then make a list and gather together all the parts and tools required. If a problem is encountered, seek advice from a parts specialist, or a dealer service department.

2 Intensive maintenance

1 If, from the time the vehicle is new, the routine maintenance schedule is followed closely, and frequent checks are made of fluid levels and high-wear items, as suggested throughout this manual, the engine will be kept in relatively good running condition, and the need for additional work will be minimised.
2 It is possible that there will be times when the engine is running poorly due to the lack of regular maintenance. This is even more likely if a used vehicle, which has not received regular and frequent maintenance checks, is purchased. In such cases, additional work may need to be carried out, outside of the regular maintenance intervals.
3 If engine wear is suspected, a compression test (refer to Chapter 2A or 2B) will provide valuable information regarding the overall performance of the main internal components. Such a test can be used as a basis to decide on the extent of the work to be carried out. If,

for example, a compression test indicates serious internal engine wear, conventional maintenance as described in this Chapter will not greatly improve the performance of the engine, and may prove a waste of time and money, unless extensive overhaul work (Chapter 2A or 2B) is carried out first.
4 The following series of operations are those most often required to improve the performance of a generally poor-running engine:

Primary operations

a) Clean, inspect and test the battery (See "Weekly checks").
b) Check all the engine-related fluids (See "Weekly checks").
c) Check the condition and tension of the auxiliary drivebelt (Section 8).
d) Check the condition of the air filter element, and renew if necessary (Section 18).
e) Check the fuel filter (Section 28).
f) Check the condition of all hoses, and check for fluid leaks (Section 21).
g) Check the idle speed and mixture settings (Section 10).

5 If the above operations do not prove fully effective, carry out the following secondary operations:

Secondary operations

All items listed under "Primary operations", plus the following:

a) Check the charging system (Chapter 5A).
b) Check the ignition system (Chapter 5B).
c) Check the fuel system (Chapter 4A or 4B).

Every 6000 miles (10 000km)

3 Engine oil and filter renewal

Note: *On some early (1984) models, the oil filter thread has a 19 mm thread instead of the usual 20 mm thread. On these models, it is possible to fit a 20 mm filter onto the engine in error; if this is done the oil filter will vibrate off resulting in loss of oil pressure and serious engine damage. Ensure the oil filter thread is correct when fitting a new filter. Refer to your Renault dealer for further information.*

1 Frequent oil and filter changes are the most important preventative maintenance procedures which can be undertaken by the DIY owner. As engine oil ages, it becomes diluted and contaminated, which leads to premature engine wear.

2 Before starting this procedure, gather together all the necessary tools and materials. Also make sure that you have plenty of clean rags and newspapers handy, to mop up any spills. Ideally, the engine oil should be warm, as it will drain more easily, and more built-up sludge will be removed with it. Take care not to touch the exhaust or any other hot parts of the engine when working under the vehicle. To avoid any possibility of scalding, and to protect yourself from possible skin irritants and other harmful contaminants in used engine oils, it is advisable to wear gloves when carrying out this work. Access to the underside of the vehicle will be greatly improved if it can be raised on a lift, driven onto ramps, or jacked up and supported on axle stands. Whichever method is chosen, make sure that the vehicle remains level, or if it is at an angle, that the drain plug is at the lowest point. The drain plug is located at the rear of the sump.

3 Remove the oil filler cap.

4 Using a spanner, or preferably a suitable socket and bar, slacken the drain plug about half a turn. Position the draining container under the drain plug, then remove the plug completely **(see Haynes Hint)**.

5 Allow some time for the oil to drain, noting that it may be necessary to reposition the container as the oil flow slows to a trickle.

6 After all the oil has drained, wipe the drain plug and the sealing washer with a clean rag. Examine the condition of the sealing washer, and renew it if it shows signs of scoring or other damage which may prevent an oil-tight seal. Clean the area around the drain plug opening, and refit the plug complete with the washer and tighten it securely.

7 The oil filter is located at the right-hand side of the cylinder block on four cylinder engines and on the left-hand side of the cylinder block on six cylinder engines. Note that on models with an oil cooler, the filter is screwed onto the base of the oil cooler which is mounted onto the right-hand side of the engine compartment on four cylinder models, and the left-hand side of six cylinder models.

8 Move the container into position under the oil filter.

9 Use an oil filter removal tool to slacken the filter initially, then unscrew it by hand the rest of the way **(see illustration)**. Empty the oil from the old filter into the container.

10 Use a clean rag to remove all oil, dirt and sludge from the filter sealing area on the engine. Check the old filter to make sure that the rubber sealing ring has not stuck to the engine. If it has, carefully remove it.

11 Apply a light coating of clean engine oil to the sealing ring on the new filter, then screw the filter into position on the engine. Tighten the filter firmly by hand only - **do not** use any tools.

12 Remove the old oil and all tools from under the vehicle then, if applicable, lower the vehicle to the ground.

13 Fill the engine through the filler hole, using the correct grade and type of oil (refer to *"Weekly Checks"* for details of topping-up). Pour in half the specified quantity of oil first, then wait a few minutes for the oil to drain into the sump. Continue to add oil, a small quantity at a time, until the level is up to the lower mark on the dipstick. Adding approximately a further 1.0 litre will bring the level up to the upper mark on the dipstick.

14 Start the engine and run it for a few minutes, while checking for leaks around the oil filter seal and the sump drain plug. Note that there may be a delay of a few seconds before the low oil pressure warning light goes out when the engine is first started, as the oil circulates through the new oil filter and the engine oil galleries before the pressure builds up.

15 Stop the engine, and wait a few minutes for the oil to settle in the sump once more. With the new oil circulated and the filter now completely full, recheck the level on the dipstick, and add more oil as necessary.

16 Dispose of the used engine oil safely, with reference to *"General repair procedures"*.

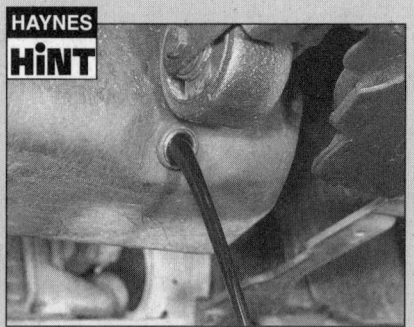

As the drain plug threads release, move it sharply away so the stream of oil issuing from the sump runs into the container, not up your sleeve!

3.9 Using an oil filter removal tool to slacken the oil filter

4 Front brake pad check

1 Apply the handbrake, then jack up the front of the car and support it securely on axle stands (see *"Jacking and vehicle support"*). Remove the front roadwheels.

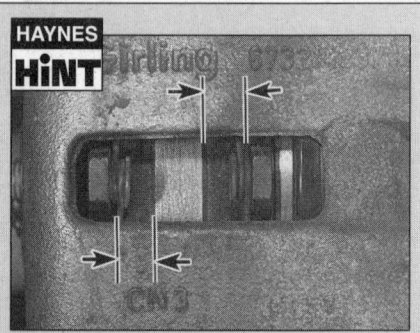

For a quick check, the thickness of friction material remaining on each brake pad can be measured through the aperture in the caliper body.

2 For a comprehensive check, the brake pads should be removed and cleaned. The operation of the caliper can then also be checked, and the condition of the brake disc itself can be fully examined on both sides. Refer to Chapter 9 for further information.

3 On completion refit the roadwheels and lower the car to the ground.

5 Rear brake pad check - models with rear disc brakes

1 Jack up the rear of the car and support it securely on axle stands (see *"Jacking and vehicle support"*). Remove the rear roadwheels.

Every 6000 miles - petrol models 1A•9

2 For a comprehensive check, the brake pads should be removed and cleaned. The operation of the caliper can then also be checked, and the condition of the brake disc itself can be fully examined on both sides. Refer to Chapter 9 for further information.
3 On completion refit the roadwheels and lower the car to the ground.

6 Handbrake check

General

1 The handbrake will normally be kept in adjustment by the action of the drum brakes automatic adjuster, or by the self-adjusting action of the rear disc calipers. Occasionally, the handbrake mechanism may require adjustment to compensate for cable stretch.
2 Chock the front wheels, then jack up the rear of the vehicle, and support securely on axle stands (see *"Jacking and vehicle support"*).
3 Fully release the handbrake and check that the wheels can be rotated easily by hand. The wheels may drag slightly, but there should be no binding.
4 Operate the handbrake, and check that the wheels are fully locked when the handbrake lever is pulled up by 9 notches on models with rear drum brakes, or 12 notches on models with rear disc brakes.

5 If the wheels bind, or if the handbrake lever travel is not as specified, it is likely that the handbrake mechanism is partially seized, or the mechanism is incorrectly adjusted. If the operation of the mechanism is not satisfactory, proceed as follows, according to type.

Models with rear drum brakes

6 Working under the vehicle, remove the cover from the vehicle floor (under the handbrake lever location) to reveal the handbrake adjuster sleeve.
7 Slacken the adjuster locknuts until there is no tension in the handbrake cables **(see illustration)**.
8 Remove the brake drums as described in Chapter 9, and check the operation of the handbrake lever on the trailing shoe **(see illustration)**. The lever should move freely.
9 If necessary, dismantle the components and clean them, then reassemble as described in Chapter 9.
10 On models with Bendix brakes, before refitting the brake drum, place the toothed adjuster segment on the edge of the adjuster lever **(see illustration)**.
11 On models with Girling brakes, check that the adjuster star wheel rotates freely in both directions, then back off the star wheel by five or six teeth.
12 Check that the handbrake cables slide freely in their sheaths.

13 Check that the handbrake operating levers are resting correctly on the shoes.
14 Working under the vehicle floor, turn the cable adjuster sleeve until the handbrake operating levers on the brake shoes begin to move when the handbrake lever is pulled beyond the second notch. The operating levers should be free with the handbrake lever on the first and second notches.
15 Tighten the adjuster locknuts.
16 Refit the brake drums as described in Chapter 9, then refit the roadwheels and the adjuster sleeve cover, and lower the vehicle to the ground.
17 With the vehicle resting on its wheels, depress the brake pedal repeatedly to operate the self-adjusting mechanism. It should be possible to hear the adjusting mechanism click as the brake pedal is depressed.

Models with rear disc brakes

18 Proceed as described in paragraphs 6 and 7.
19 Check that the handbrake cables slide freely in their sheaths, and that the operating levers on the calipers move freely.
20 Push the handbrake operating levers on the calipers as far as they will go towards the rear of the vehicle.
21 Working under the vehicle floor, turn the cable adjuster sleeve until the handbrake cable end fittings just contact the operating levers on the calipers, without moving the levers **(see illustration)**.
22 Continue to turn the adjuster sleeve until the handbrake operating levers on the calipers begin to move when the handbrake lever is pulled beyond the second notch. The operating levers should be free with the handbrake lever on the first and second notches.
23 Tighten the adjuster locknuts.
24 Refit the roadwheels and the adjuster sleeve cover, and lower the vehicle to the ground.

7 Clutch check

Models with a cable-operated clutch

On models with a cable-operated clutch, check and adjust the clutch cable as described in Chapter 6. Check that the cable moves freely and easily and lubricate its exposed section with multi-purpose grease.

Models with a hydraulically-operated clutch

Referring to Chapter 6, examine the clutch master cylinder and slave cylinder for signs of fluid leakage. Also check along the length of the clutch hydraulic hose/pipe looking for signs of damage or deterioration. Renew worn/damaged components as described in Chapter 6.

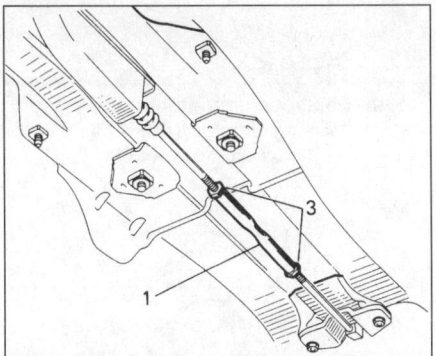
6.7 Handbrake adjuster sleeve (1) and locknuts (3)

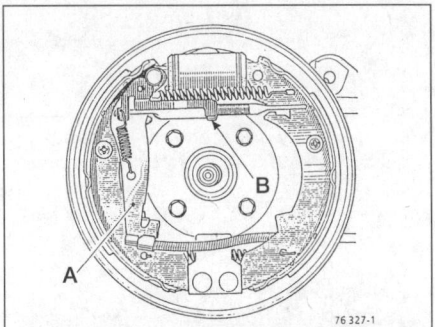
6.8 Handbrake operating lever (A) on trailing shoe, and adjuster star wheel (B) - Girling brakes

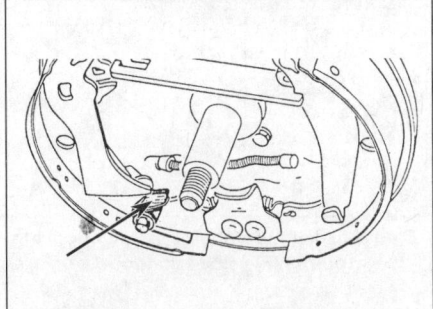
6.10 Place the toothed adjuster segment (arrowed) on the edge of the adjuster lever - Bendix brakes

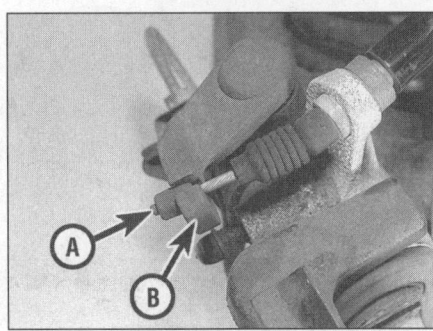
6.21 The handbrake cable end fitting (A) should just contact the operating lever (B) on the caliper

1A•10 Every 6000 miles - petrol models

8 Auxiliary drivebelt check and renewal

Checking

1 Vee, multi-toothed, or multi-grooved belts are used to drive the coolant pump, alternator, and air conditioning compressor (where fitted) from the crankshaft pulley. The power steering pump on four-cylinder and most six-cylinder models is driven from the crankshaft pulley. On six-cylinder non-turbo models, the power steering pump is driven from the rear of the left-hand camshaft.

2 Using a socket and extension bar fitted over the crankshaft pulley bolt, rotate the crankshaft so that the entire length of the relevant drivebelt can be examined. Examine the drivebelt for cracks, splitting, fraying or other damage. Check also for signs of glazing (shiny patches) and for separation of the belt plies. Renew the belt if worn or damaged.

3 If the condition of the belt is satisfactory, check the drivebelt tension as described in the following paragraphs.

4 Check the deflection of the belts at the points indicated (see illustrations). The belts should be tensioned so that under firm thumb pressure, the deflection of the belt is as specified in the following table. The belts should be tensioned when the engine is cold - if the engine is hot, add 1.0 mm to the figures in the table.

Drivebelt type	4-cyl engines	6-cyl engines
Vee-belt	5.5 to 6.5 mm	3.5 to 4.5 mm
Multi-toothed/ multi-grooved	4.5 to 5.5 mm	

Tensioning and removal

5 To tension the crankshaft pulley-driven drivebelts (with the exception of six-cylinder turbo engine models, and the air conditioning compressor drivebelt on six-cylinder non-turbo engines), release the relevant driven accessory adjuster link and mounting bolts, and move the accessory as necessary. Re-tighten the bolts (see illustrations).

6 To remove a crankshaft pulley-driven drivebelt (with the exception of six-cylinder turbo engine models, and the air conditioning compressor drivebelt on six-cylinder non-turbo engines), slacken the accessory adjuster link and mounting bolts, and push the relevant accessory fully towards the engine. With the drivebelt slack, slip it off the pulleys. If difficulty is experienced removing the belt, turn the crankshaft pulley and at the same time press the belt against the pulley rim, when it will ride up over the rim.

7 To tension the air conditioning compressor drivebelt, or the power steering pump drivebelt on six-cylinder non-turbo models, and the main drivebelt on six-cylinder turbo models, adjust the position of the tensioner pulley. If the belt is to be removed, fully slacken the pulley.

8 It will be obvious that before an inner belt can be removed, the outer belt will have to be removed.

Note: *On certain models with air conditioning, the air conditioning compressor drivebelt is located behind the timing belt. To remove the air conditioning compressor drivebelt, it will therefore be necessary to remove the timing belt as described in Chapter 2.*

Refitting

9 Fit the belt, again turning the crankshaft pulley if the belt is difficult to pass over the pulley rim.

10 Tension the belt in accordance with the figures given in the table at the end of paragraph 4, and tighten the accessory adjuster link and mounting bolts, or the tensioner pulley, as applicable.

> **HAYNES HiNT** *Correct tensioning of the drivebelt will ensure that it has a long life. A belt which is too slack will slip and squeal. Beware, however, of overtightening, as this can cause wear in the accessory bearings.*

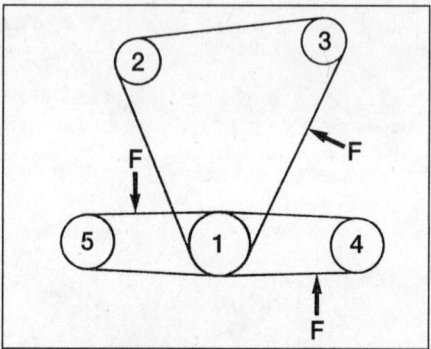

8.4a Auxiliary drivebelt tension checking points (F) - 1995 cc and 2165 cc engines

1 Crankshaft
2 Coolant pump
3 Alternator
4 Power steering pump
5 Air conditioning compressor

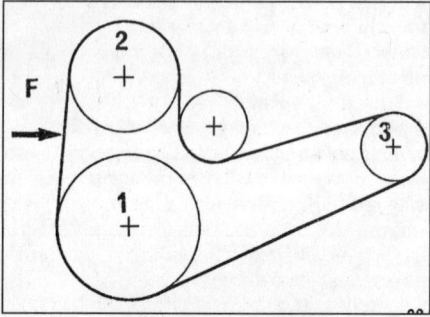

8.4b Coolant pump/alternator drivebelt tension checking point (F) - 2458 cc turbo models

1 Crankshaft
2 Coolant pump
3 Alternator

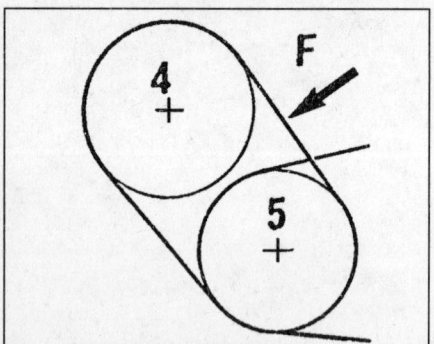

8.4c Air conditioning compressor drivebelt tension checking point (F) - 2458 cc turbo models

4 Power steering pump
5 Air conditioning compressor

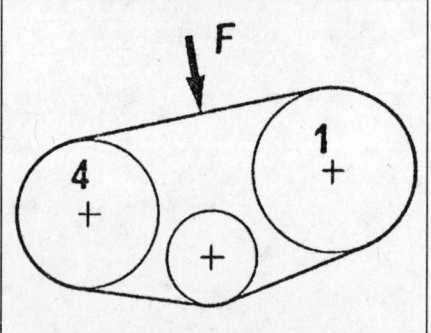

8.4d Power steering pump drivebelt tension checking point (F) - 2458 cc turbo models

1 Crankshaft
4 Power steering pump

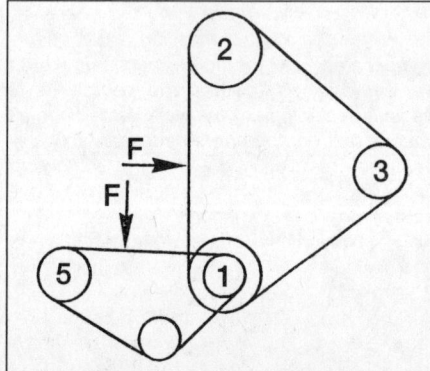

8.4e Auxiliary drivebelt tension checking points (F) - 2664 cc models

1 Crankshaft
2 Coolant pump
3 Alternator
5 Air conditioning compressor

Every 6000 miles - petrol models

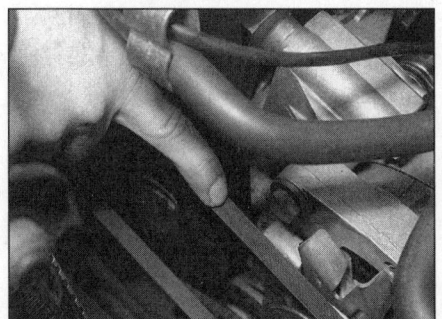

8.5a Checking the coolant pump/alternator drivebelt tension - 1995 cc engine shown

8.5b Adjusting the coolant pump/alternator drivebelt tension (adjuster arrowed) - 2664 cc model

8.5c Power steering pump adjuster bolt (arrowed) - 1995 cc engine

9 Automatic transmission fluid level check

4141 and MJ3 automatic transmissions

1 The automatic transmission fluid must be checked when the engine and transmission are cold.
2 Start the engine and allow it to run for a few minutes in order to fill the torque converter and cooler with fluid.
3 With the selector lever in P and the engine idling, withdraw the dipstick which is situated on the right-hand side of the engine compartment, wipe it clean, re-insert it and withdraw it for the second time. The fluid level should be between the "Min" and "Max" marks (see illustrations).
4 If necessary, top-up with the specified fluid poured in through the dipstick guide tube. Take care not to allow any dust or dirt to enter the tube.
5 Recheck the oil level. Once the level is correct, refit the dipstick and switch off the engine.

AR4 transmission

Note: *The fluid level checking procedure on the AR4 automatic transmission is particularly complicated, and the home mechanic would be well-advised to take the vehicle to a Renault dealer to have the work carried out, as special test equipment is necessary to carry out the check. However, the following procedure is given for those who may have access to this equipment.*

6 The transmission fluid level should be checked every 6000 miles, or immediately if a transmission malfunction develops.
7 Remove the plug from the top of the filler tube `D' on the right-hand side of the transmission (see illustration). Add 0.5 litre (1 pint) of the specified fluid to the transmission via the filler tube, using a clean funnel with a fine-mesh filter, then refit the plug.
8 Position the vehicle over an inspection pit, on a ramp, or jack it up and support it on axle stands, ensuring that the vehicle remains level.

9 Connect the Renault XR25 test meter to the diagnostic socket, and enter `DO4' then number `04'. With the selector lever in `Park', run the engine at idle speed until the fluid temperature reaches 60°C.
10 With the engine still running, unscrew the level plug from the transmission (see illustration). Allow the excess fluid to run out into a calibrated container for 20 seconds, then refit the plug. The amount of fluid should be more than 0.1 litre; if it is not, the fluid level in the transmission is incorrect.

9.3a Automatic transmission fluid dipstick location (2165 cc engine shown with MJ3 transmission)

9.7 AR4 transmission filler tube location (2165 cc engine shown)

11 If the level is incorrect, add an extra 1 litre (nearly 2 pints) of the specified fluid to the transmission, through the filler tube after removing the vent. Allow the transmission to cool down, then repeat the checking procedure again as described in the previous paragraphs. Repeat the procedure as required until more than the specified amount of fluid is drained as described in the previous paragraph, indicating that the transmission fluid level is correct, then securely tighten the level plug.

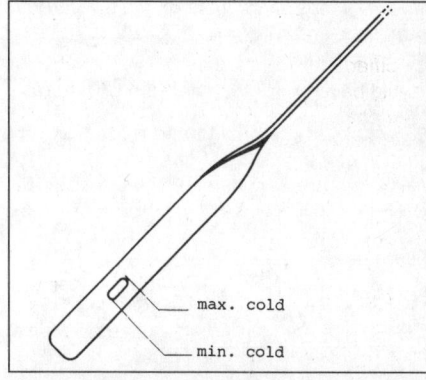

9.3b Automatic transmission fluid dipstick markings - 4141 and MJ3 transmission

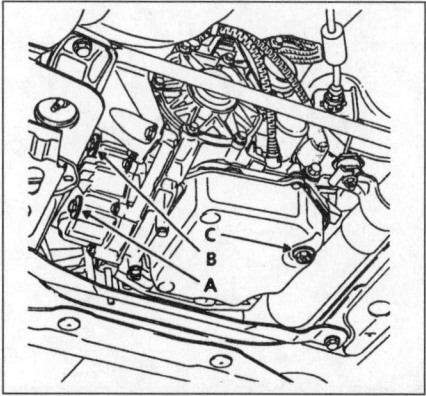

9.10 AR4 transmission drain and level plug locations

A Final drive drain plug
B Final drive filler/level plug
C Transmission level plug (see Text)

10 Idle speed and mixture check

1995 cc engines

Carburettor engines

1 A tachometer (rev counter) and an exhaust gas analyser (CO meter) are required for accurate adjustment.
2 Adjustment should be made with the air cleaner fitted, the accelerator cable correctly adjusted, and the engine at normal operating temperature. Ensure the adjustment screw is not touching the fast idle cam; on models equipped with automatic transmission, position the selector lever in the "N" position.
3 Connect the tachometer and exhaust gas analyser to the engine in accordance with the manufacturers' instructions. Start the engine and allow it to idle.
4 Read the idle speed on the tachometer and compare it with the value given in the *Specifications*. If adjustment is necessary, turn the idle speed adjustment screw as required **(see illustration)**.
5 Check that the CO content is as given in the *Specifications*. If not prise off the tamperproof cap and turn the idle mixture adjustment screw as required.
6 If necessary, repeat the procedure given in paragraph 4.
7 Switch off the engine and disconnect the tachometer and exhaust gas analyser.

Fuel injection engines

8 The idle speed is automatically controlled by the idle speed control valve and ECU and cannot be adjusted. If idle speed is incorrect, there must be a fault present. On models with a catalytic converter, the mixture adjustment is also controlled by the ECU with no manual adjustment of the mixture (exhaust gas CO level) being possible. If either is incorrect then a fault is present in the injection system and the vehicle should be taken to a Renault dealer for testing (see Chapter 4B). On models not equipped with a catalytic converter the mixture (exhaust gas CO level) can be adjusted as follows.
9 Warm the engine up to normal operating temperature then connect an exhaust gas analyser to the engine in accordance with the manufacturers instructions and check the CO level is within the specified limits. If necessary, adjustment is made using the screw on the mixture potentiometer which is mounted on the left-hand side of the engine compartment. Remove the tamperproof cap from the adjustment screw and rotate the screw in or out (as applicable) until the correct CO level is obtained **(see illustration)**. Once adjustment is correct switch off the engine and disconnect the exhaust gas analyser.

2165 cc engines

Early models

10 A tachometer (rev counter) and an exhaust gas analyser (CO meter) are required for accurate adjustment.
11 Have the engine at full operating temperature with an exhaust gas analyser connected in accordance with the manufacturer's instructions.
12 Remove the tamperproof cap from the mixture potentiometer screw **(see illustration 10.9)**.
13 With the engine idling, turn the air bypass screw until the specified idling speed is obtained **(see illustration)**.
14 Now turn the potentiometer adjustment screw in or out until the correct exhaust gas CO level is obtained. Once the CO level is correctly set, if necessary, re-adjust the idle speed.
15 Switch off the engine and disconnect the analyser.

Later models

16 Refer to the information given in paragraphs 8 and 9.

2458 cc turbo engines

17 Refer to the information given in paragraphs 8 and 9.

2664 cc engines

Early (pre 1986) models

18 A tachometer (rev counter) and an exhaust gas analyser (CO meter) are required for accurate adjustment.
19 Unscrew the take-off plug from each exhaust manifold and connect an exhaust gas analyser to the exhaust manifolds as shown **(see illustrations)**. Connect pipes and hoses using a three branch connector so that the probe which is normally inserted in the exhaust tailpipe can be located in the open branch of the three-way connector.
20 Warm the engine up to normal operating temperature and switch it off. Turn the three screws (A, B and C) adjacent to the airflow sensor fully in. Now unscrew screws (A) and (B) through two complete turns. If the screws are fitted with tamperproof caps they will have to be broken off **(see illustration)**.
21 Start the engine and allow it to idle. Turn screw (C) until the idle speed is at the specified level.

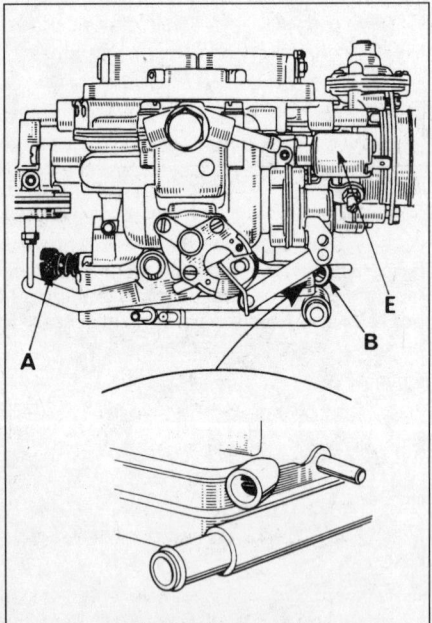

10.4 Carburettor adjustment screw locations

A Idle speed screw
B Idle mixture screw
E Fuel cut-off solenoid

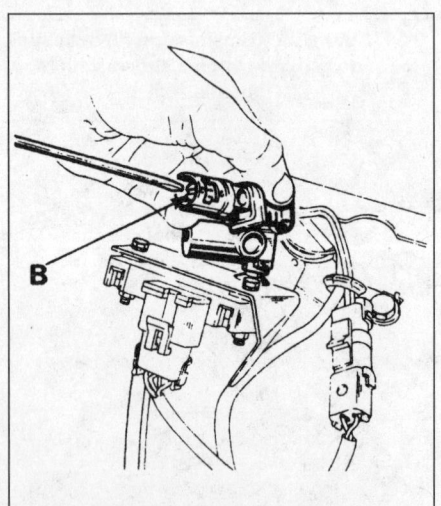

10.9 Mixture adjustment potentiometer screw (B)

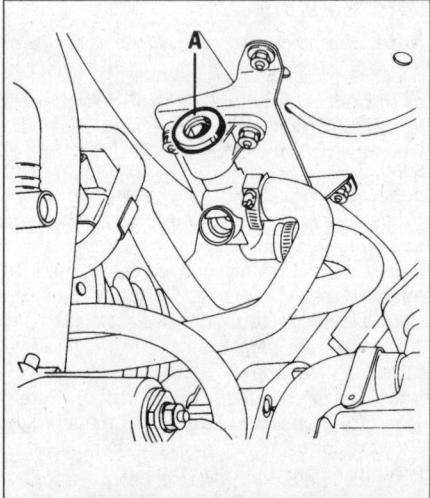

10.13 Air bypass screw (A) (idle speed adjustment screw) location - early 2165 cc engines

Every 6000 miles - petrol models 1A•13

22 Check the exhaust emission from the two exhaust manifolds simultaneously.
23 If the CO percentage is outside the specified tolerance, remove the tamperproof cap (I) extract the plug (M) and turn the mixture screw (D) using an Allen key, but without applying downward pressure. Screw in for richer, out for weaker mixture.
24 Readjust the idle speed and then place a finger over screw (D) and recheck the CO level.
25 The cylinder banks can be checked individually if small stop-cocks are used in the take off pipes and side adjustment carried out by turning screw (A) for the right-hand side and (B) for the left-hand side.

Later (1986 on) models
26 On later models, idle speed adjustment is carried out automatically by means of an electronic idling system. If the idle speed is incorrect, then a fault must be present and the vehicle should be taken to a Renault dealer for testing (see Chapter 4B). The mixture (exhaust gas CO level) can be adjusted as described above in paragraphs 18 to 25 ignoring all references to idle speed adjustment. Note also that the idle control valve hose should be clamped shut to ensure adjustment is accurate.

2458 cc turbo and 2849 cc engines
27 Refer to paragraphs 8 and 9.

11 Seat belt check

1 Carefully examine the seat belt webbing for cuts, or any signs of serious fraying or deterioration. If the belt is of the retractable type, pull the belt all the way out of the inertia reel, and examine the full extent of the webbing.
2 Fasten and unfasten the belt, ensuring that the locking mechanism holds securely, and releases properly when intended. If the belt is of the retractable type, check also that the retracting mechanism operates correctly when the belt is released.
3 Check the security of all seat belt mountings and attachments which are accessible without removing any trim or other components.

12 Electrical systems check

1 Check the operation of all electrical equipment, ie, lights, direction indicators, horn, etc. Refer to the appropriate Sections of Chapter 12 for details if any of the circuits are found to be inoperative.
2 Note that stop-light adjustment is described in Chapter 9.

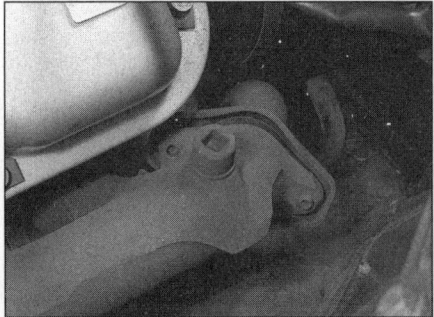

10.19a On 2664 cc engines, unscrew the take-off plug from each exhaust manifold to allow the exhaust gas analyser to be connected

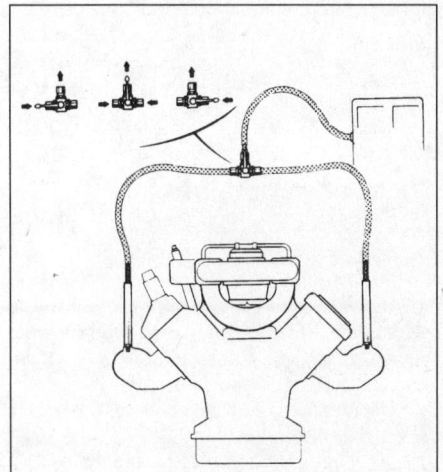

10.19b Typical exhaust gas analyser set-up on 2664 cc engine

3 Visually check all accessible wiring connectors, harnesses and retaining clips for security, and for signs of chafing or damage. Rectify any faults found.

13 Exhaust system check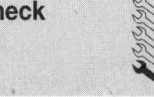

1 With the engine cold (at least an hour after the vehicle has been driven), check the complete exhaust system from the engine to the end of the tailpipe. The exhaust system is most easily checked with the vehicle raised on a hoist, or suitably supported on axle stands, so that the exhaust components are readily visible and accessible.
2 Check the exhaust pipes and connections for evidence of leaks, severe corrosion and damage. Make sure that all brackets and mountings are in good condition, and that all relevant nuts and bolts are tight. Leakage at any of the joints or in other parts of the system will usually show up as a black sooty stain in the vicinity of the leak.

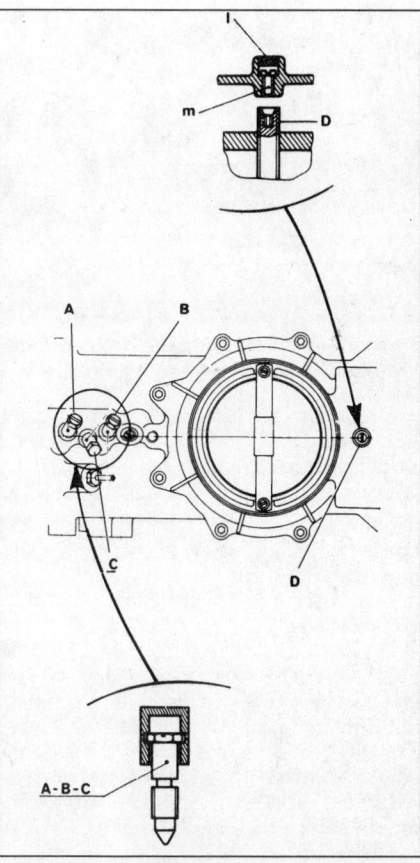

10.20 Idle speed and mixture adjustment screws - 2664 cc engine

A Right-hand idle speed (volume) screw
B Left-hand idle speed (volume) screw
C Air bypass screw
D Idle mixture screw
I Tamperproof cap
M Threaded plug

3 Rattles and other noises can often be traced to the exhaust system, especially the brackets and mountings. Try to move the pipes and silencers. If the components are able to come into contact with the body or suspension parts, secure the system with new mountings. Otherwise separate the joints (if possible) and twist the pipes as necessary to provide additional clearance.

14 Suspension and steering check

Front suspension and steering check

1 Raise the front of the vehicle, and securely support it on axle stands (see "Jacking and vehicle support").
2 Visually inspect the balljoint dust covers and the steering rack-and-pinion gaiters for splits, chafing or deterioration. Any wear of these components will cause loss of lubricant, together with dirt and water entry, resulting in

1A•14 Every 6000 miles - petrol models

14.4 Check for wear in the hub bearings by grasping the wheel and trying to rock it

rapid deterioration of the balljoints or steering gear.
3 On vehicles with power steering, check the fluid hoses for chafing or deterioration, and the pipe and hose unions for fluid leaks. Also check for signs of fluid leakage under pressure from the steering gear rubber gaiters, which would indicate failed fluid seals within the steering gear.
4 Grasp the roadwheel at the 12 o'clock and 6 o'clock positions, and try to rock it **(see illustration)**. Very slight free play may be felt, but if the movement is appreciable, further investigation is necessary to determine the source. Continue rocking the wheel while an assistant depresses the footbrake. If the movement is now eliminated or significantly reduced, it is likely that the hub bearings are at fault. If the free play is still evident with the footbrake depressed, then there is wear in the suspension joints or mountings.
5 Now grasp the wheel at the 9 o'clock and 3 o'clock positions, and try to rock it as before. Any movement felt now may again be caused by wear in the hub bearings or the steering track-rod balljoints. If the outer balljoint is worn, the visual movement will be obvious. If the inner joint is suspect, it can be felt by placing a hand over the rack-and-pinion rubber gaiter and gripping the track-rod. If the wheel is now rocked, movement will be felt at the inner joint if wear has taken place.
6 Using a large screwdriver or flat bar, check for wear in the suspension mounting bushes by levering between the relevant suspension component and its attachment point. Some movement is to be expected, as the mountings are made of rubber, but excessive wear should be obvious. Also check the condition of any visible rubber bushes, looking for splits, cracks or contamination of the rubber.
7 With the car standing on its wheels, have an assistant turn the steering wheel back and forth, about an eighth of a turn each way. There should be very little, if any, lost movement between the steering wheel and roadwheels. If this is not the case, closely observe the joints and mountings previously described. In addition, check the steering column universal joints for wear, and also check the rack-and-pinion steering gear itself.

Rear suspension check

8 Chock the front wheels, then jack up the rear of the vehicle and support securely on axle stands (see *"Jacking and vehicle support"*).
9 Working as described previously for the front suspension, check the rear hub bearings, the suspension bushes and the shock absorber mountings for wear.

Shock absorber check

10 Check for any signs of fluid leakage around the shock absorber body, or from the rubber gaiter around the piston rod. Should any fluid be noticed, the shock absorber is defective internally, and should be renewed.
Note: *Shock absorbers should always be renewed in pairs on the same axle.*
11 The efficiency of the shock absorber may be checked by bouncing the vehicle at each corner. Generally speaking, the body will return to its normal position and stop after being depressed. If it rises and returns on a rebound, the shock absorber is probably suspect. Also examine the shock absorber upper and lower mountings for any signs of wear.

15 Roadwheel bolt check

Where applicable, remove the wheel trims, and slacken the roadwheel bolts slightly.
Tighten the bolts to the specified torque, using a torque wrench.

17.2 Air conditioning drier bottle sight glass (arrowed)

16 Turbocharger boost pressure check - 2458 cc turbo models

Renault specify that the turbocharger boost pressure should be regularly checked at the specified intervals. Checking is a complex procedure, requiring the use of several special tools and adapters and should therefore be entrusted to a Renault dealer.

17 Air conditioning refrigerant check

1 Run the engine, and switch on the air conditioning.
2 After a few minutes, inspect the sight glass (located on top of the drier bottle at the front corner of the engine compartment), and check the fluid flow **(see illustration)**. Clear fluid should be visible - if not, the following will help to diagnose the problem:
a) *Clear fluid flow - the system is functioning correctly.*
b) *No fluid flow - have the system checked for leaks by a Renault dealer or air conditioning specialist.*
c) *Continuous stream of clear air bubbles in fluid - refrigerant level low - have the system recharged by a Renault dealer or air conditioning specialist.*
d) *Milky air bubbles visible - high humidity (have the system checked by a Renault dealer).*

Every 12 000 miles (20 000 km)

18 Air filter element renewal

1 On all models except early (pre 1986) 2664 cc engine models, the air filter element is located in the cylindrical housing mounted on the left-hand side of the engine compartment. On early 2664 cc models the filter is located on the housing mounted directly onto the airflow meter.
2 On early (pre 1986) 2664 cc models, undo the housing retaining bolts then release the retaining clips and separate the housing and lid. Release the filter element and remove it from the lid **(see illustrations)**. Wipe clean the air cleaner housing then fit the new filter, making sure it is correct seated in the housing lid. Reassemble the housing and lid and secure it in position with the retaining clips.

Every 12 000 miles - petrol models

18.2a On early 2664 cc models, undo the housing retaining screws . . .

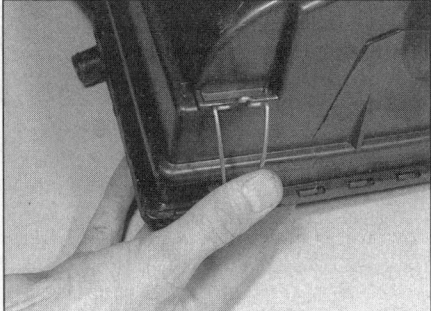

18.2b . . . then release the retaining clips and separate the housing and lid . . .

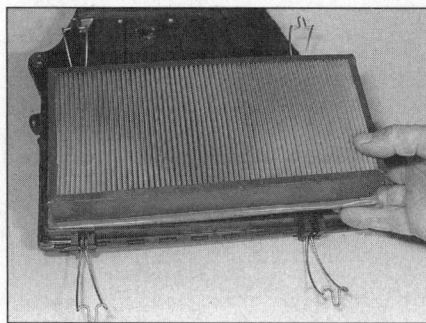

18.2c . . . and remove the air filter element

Make sure the sealing ring is in position on the airflow meter then locate the housing assembly in position and securely tighten its retaining bolts.

3 On all other models release the retaining clips or unscrew the wing nut (as applicable) securing the housing end cover in position **(see illustrations)**. Remove the cover then slide out the filter from the housing, if necessary release the retaining strap(s) (where fitted) and free the housing from its mounting bracket to gain the necessary clearance required to remove the filter. Wipe clean the filter housing and slide in the new filter. Refit the filter cover and secure it in position with the retaining clips/wing nut (as applicable). Where necessary, clip the filter housing back onto its mounting bracket and secure it in position with the retaining strap(s).

19 Spark plug renewal

1 The correct functioning of the spark plugs is vital for the correct running and efficiency of the engine. It is essential that the plugs fitted are appropriate for the engine (a suitable type is specified at the beginning of this Chapter). If this type is used and the engine is in good condition, the spark plugs should not need attention between scheduled replacement intervals. Spark plug cleaning is rarely necessary, and should not be attempted unless specialised equipment is available, as damage can easily be caused to the firing ends.

2 On 2458 cc turbo models, to gain access to the right-hand cylinder bank spark plugs, undo the screws securing the inlet duct to the throttle housing and reposition the duct slightly. Recover the sealing ring fitted to the throttle housing.

3 If the marks on the original-equipment spark plug (HT) leads cannot be seen, mark the leads to correspond to the cylinder the lead serves **(see illustrations)**. Pull the leads from the plugs by gripping the end fitting, not the lead, otherwise the lead connection may be fractured.

4 It is advisable to remove the dirt from the spark plug recesses using a clean brush,

18.3a On 2165 cc models, release the retaining clips . . .

vacuum cleaner or compressed air before removing the plugs, to prevent dirt dropping into the cylinders.

5 Unscrew the plugs using a spark plug spanner, suitable box spanner or a deep socket and extension bar **(see illustration)**. Keep the socket aligned with the spark plug - if it is forcibly moved to one side, the ceramic insulator may be broken off. As each plug is removed, examine it as follows.

6 Examination of the spark plugs will give a good indication of the condition of the engine. If the insulator nose of the spark plug is clean and white, with no deposits, this is indicative of a weak mixture or too hot a plug (a hot plug transfers heat away from the electrode slowly,

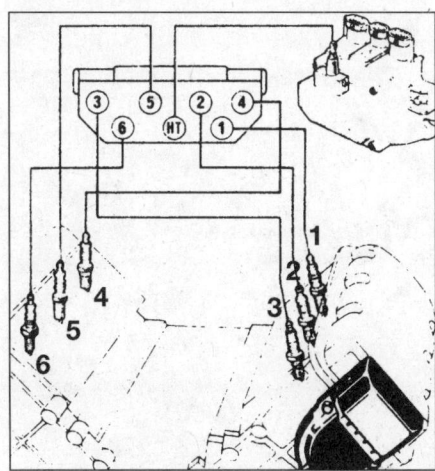

19.3a HT lead connections on 2458 cc turbo and 2849 cc engines

18.3b . . . then lift off the cover and withdraw the filter element

a cold plug transfers heat away quickly).

7 If the tip and insulator nose are covered with hard black-looking deposits, then this is indicative that the mixture is too rich. Should the plug be black and oily, then it is likely that the engine is fairly worn, as well as the mixture being too rich.

8 If the insulator nose is covered with light tan to greyish-brown deposits, then the mixture is correct and it is likely that the engine is in good condition.

9 The spark plug electrode gap is of considerable importance as, if it is too large or too small, the size of the spark and its efficiency will be seriously impaired. The gap should be set to the value given in the

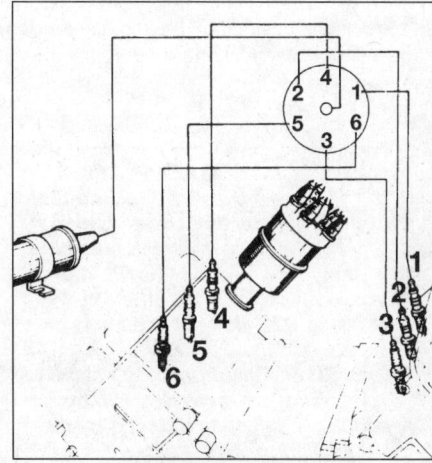

19.3b HT lead connections on 2664 cc engines

1A•16 Every 12 000 miles - petrol models

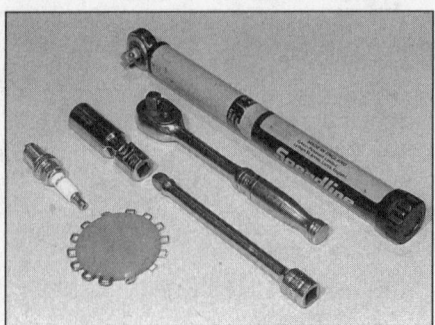

19.5 Tools required for spark plug removal, gap adjustment and refitting

19.10a Measuring the spark plug gap with a wire gauge

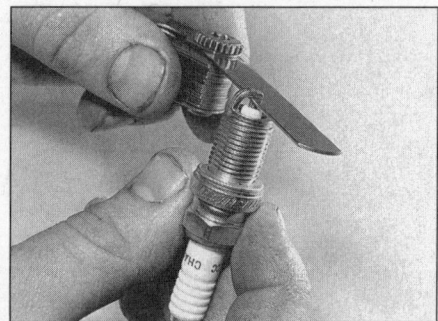

19.10b Measuring the spark plug gap with a feeler blade

Specifications at the beginning of this Chapter.

10 To set the gap, measure it with a feeler blade and then bend open, or closed, the outer plug electrode until the correct gap is achieved. The centre electrode should never be bent, as this may crack the insulator and cause plug failure, if nothing worse. If using feeler blades, the gap is correct when the appropriate-size blade is a firm sliding fit **(see illustrations)**.

11 Special spark plug electrode gap adjusting tools are available from most motor accessory shops, or from some spark plug manufacturers.

12 Before fitting the spark plugs, check that the threaded connector sleeves are tight, and that the plug exterior surfaces and threads are clean **(see Haynes Hint)**.

13 Remove the rubber hose (if used), and tighten the plug to the specified torque using the spark plug socket and a torque wrench. Refit the remaining spark plugs in the same manner.

14 Connect the HT leads in their correct order, and refit any components removed for access.

15 On 2458 cc turbo models ensure that the sealing ring is position on the throttle housing then refit the inlet duct and securely tighten its retaining screws.

> **HAYNES HINT** *It is very often difficult to insert spark plugs into their holes without cross-threading them. To avoid this possibility, fit a short length of 5/16 inch internal diameter rubber hose over the end of the spark plug. The flexible hose acts as a universal joint to help align the plug with the plug hole. Should the plug begin to cross-thread, the hose will slip on the spark plug, preventing thread damage to the aluminium cylinder head*

20 Manual transmission oil level check

1 Either position the vehicle over an inspection pit, or jack up the front and rear of the vehicle and support it on axle stands. The vehicle must be level for the check to be accurate.

2 Clean the area around the filler/level plug on the left-hand side of the transmission, then slacken and remove the plug from the transmission **(see illustration)**.

3 The transmission oil level should be up to the lower edge of the filler/level plug aperture.

4 If necessary, top-up using the specified type of lubricant until the transmission oil level is correct. Fill the transmission until oil starts to flow out and allow excess oil to drain out.

5 Once the transmission oil level is correct, refit the filler/level plug and tighten it securely.

6 The frequent need for topping-up indicates a leakage, possibly through an oil seal. The cause should be investigated and rectified.

21 Hose and fluid leak check

1 Visually inspect the engine joint faces, gaskets and seals for any signs of water or oil leaks. Pay particular attention to the areas around the cylinder head cover, cylinder head, oil filter and sump joint faces. Bear in mind that, over a period of time, some very slight seepage from these areas is to be expected - what you are really looking for is any indication of a serious leak. Should a leak be found, renew the offending gasket or oil seal by referring to the appropriate Chapters in this manual.

2 Also check the security and condition of all the engine-related pipes and hoses, and all braking system pipes and hoses **(see Haynes Hint)**. Ensure that all cable-ties or securing clips are in place, and in good condition. Clips which are broken or missing can lead to chafing of the hoses, pipes or wiring, which could cause more serious problems in the future.

3 Carefully check the radiator hoses and heater hoses along their entire length. Renew any hose which is cracked, swollen or deteriorated. Cracks will show up better if the hose is squeezed. Pay close attention to the hose clips that secure the hoses to the cooling system components. Hose clips can pinch and puncture hoses, resulting in cooling system leaks. If the crimped-type hose clips are used, it may be a good idea to replace them with standard worm-drive clips.

4 Inspect all the cooling system components (hoses, joint faces, etc) for leaks.

> **HAYNES HINT** *A leak in the cooling system will usually show up as white-or rust-coloured deposits on the area adjoining the leak*

5 Where any problems are found on system components, renew the component or gasket with reference to Chapter 3.

6 With the vehicle raised, inspect the fuel tank and filler neck for punctures, cracks and other damage. The connection between the filler neck and tank is especially critical. Sometimes a rubber filler neck or connecting

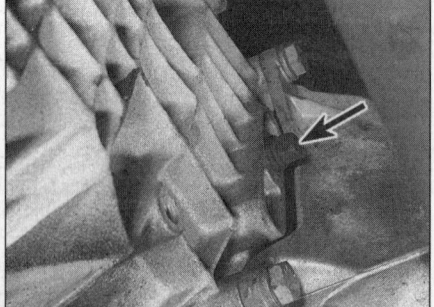

20.2 Manual transmission filler/level plug (arrowed)

Every 12 000 miles - petrol models 1A•17

hose will leak due to loose retaining clamps or deteriorated rubber.
7 Carefully check all rubber hoses and metal fuel lines leading away from the fuel tank. Check for loose connections, deteriorated hoses, crimped lines, and other damage. Pay particular attention to the vent pipes and hoses, which often loop up around the filler neck and can become blocked or crimped. Follow the lines to the front of the vehicle, carefully inspecting them all the way. Renew damaged sections as necessary. Similarly, whilst the vehicle is raised, take the opportunity to inspect all underbody brake fluid pipes and hoses.
8 From within the engine compartment, check the security of all fuel, vacuum and brake hose attachments and pipe unions, and inspect all hoses for kinks, chafing and deterioration.
9 Where applicable, check the condition of the power steering fluid pipes and hoses.

Every 36 000 miles (60 000 km)

22 Spare fuse check

Check that spare fuses are in place in the locations provided in the fusebox under the facia (see Chapter 12). It is advisable to carry at least one spare of each rating of fuse fitted. Spare fuses can be obtained from most car accessory shops, or from a Renault dealer.

23 Front wheel alignment check

Refer to the information given in Chapter 10.

24 Rear brake shoe thickness check - models with rear drum brakes

Remove the rear brake drums, and check the brake shoes for signs of wear or contamination. At the same time, also inspect the wheel cylinders for signs of leakage, and the brake drum for signs of wear. Refer to the relevant Sections of Chapter 9 for further information.

25 Manual transmission oil renewal

1 This operation is much quicker and more efficient if the car is first taken on a journey of sufficient length to warm the engine/transmission up to normal operating temperature.
2 Park the car on level ground, switch off the ignition and apply the handbrake firmly. For improved access, jack up the front of the car and support it securely on axle stands. Note that the car must be lowered to the ground and level, to ensure accuracy, when refilling and checking the oil level.
3 Wipe clean the area around the filler/level plug, which is situated on the left-hand side of the transmission. Unscrew the filler/level plug from the transmission.

4 Position a suitable container under the drain plug which is situated on the base of the transmission **(see illustration)**. Note that on some models it will be necessary to remove the protective plastic cover from the base of the transmission to gain access to the drain plug.
5 Unscrew the plug and allow the oil to drain completely into the container. If the oil is hot, take precautions against scalding. Clean both the filler/level and the drain plugs, being especially careful to wipe any metallic particles off the magnetic inserts.
6 When the oil has finished draining, clean the drain plug threads and those of the transmission casing and refit the drain plug, tightening it securely. Where necessary, refit the plastic cover to the transmission. It the car was raised for the draining operation, now lower it to the ground.
7 Refilling the transmission is an extremely awkward operation. Above all, allow plenty of time for the oil level to settle properly before checking it. Note that the car must be parked on flat level ground when checking the oil level.
8 Refill the transmission with the exact amount of the specified type of oil then check the oil level as described in Section 20; if the correct amount was poured into the transmission and a large amount flows out on checking the level, refit the filler/level plug and take the car on a short journey so that the new oil is distributed fully around the transmission components, then check the level again on your return.

25.4 Transmission drain plug

26 Automatic transmission fluid and strainer renewal

Refer to Chapter 7B, Sections 2 and 3.

27 Brake fluid renewal

⚠️ **Warning: Brake hydraulic fluid can harm your eyes and damage painted surfaces, so use extreme caution when handling and pouring it. Do not use fluid that has been standing open for some time, as it absorbs moisture from the air. Excess moisture can cause a dangerous loss of braking effectiveness.**

1 The procedure is similar to that for the bleeding of the hydraulic system as described in Chapter 9, except that the brake fluid reservoir should be emptied by siphoning, using a clean poultry baster or similar before starting, and allowance should be made for the old fluid to be expelled when bleeding a section of the circuit.
2 Working as described in Chapter 9, open the first bleed screw in the sequence, and pump the brake pedal gently until nearly all the old fluid has been emptied from the master cylinder reservoir. Top-up to the "MAX" level with new fluid, and continue pumping until only the new fluid remains in the reservoir, and new fluid can be seen emerging from the bleed screw. Tighten the screw, and top the reservoir level up to the "MAX" level line.

 *Old hydraulic fluid is invariably much darker in colour than the new, making it easy to distinguish the two.*

3 Work through all the remaining bleed screws in the sequence until new fluid can be seen at all of them. Be careful to keep the master cylinder reservoir topped-up to above the "MIN" level at all times, or air may enter

1A•18 Every 36 000 miles – petrol models

the system and greatly increase the length of the task.
4 When the operation is complete, check that all bleed screws are securely tightened, and that their dust caps are refitted. Wash off all traces of spilt fluid, and recheck the master cylinder reservoir fluid level.
5 Check the operation of the brakes before taking the car on the road.

28 Fuel filter renewal

Warning: *Before carrying out the following operation refer to the precautions given in Safety first! at the beginning of this Manual and follow them implicitly. Petrol is a highly dangerous and volatile liquid and the precautions necessary when handling it cannot be overstressed*

Carburettor models

1 The fuel filter is clipped onto the side of the inlet manifold and filters the fuel flowing from the fuel pump to the carburettor.
2 Slacken the retaining clips and disconnect both hoses. To minimise fuel loss clamp the hoses either side of the filter or be prepared to plug the hose ends as they are disconnected.
3 Withdraw the filter from its clamp. Note the orientation of the fuel flow direction indicator on the filter. This will be in the form of an arrow which points in the direction of the fuel flow (towards the carburettor).
4 Fit the new filter making sure the arrow is pointing towards the carburettor then reconnect the hoses and securely tighten their retaining clips.
5 Start the engine and check the disturbed hose connections for signs of leakage.

Fuel injection models

2664 cc engine

6 The fuel filter is located in the left-hand rear corner of the engine compartment **(see illustration)**.
7 Bearing in mind the information given on depressurising the fuel system (Chapter 4B, Section 7), carefully slacken the union nuts and disconnect the fuel hoses from the filter. To minimise fuel loss clamp the hoses either side of the filter or be prepared to plug the hose ends as they are disconnected.
8 Slacken the clamp bolt then slide the filter out of position, noting its correct fitted orientation. The arrow on the filter should point in the direction of fuel flow (towards the fuel distribution unit).
9 Slide the new filter into position making sure its arrow is pointing in the direction of fuel flow. Make sure the rubber mounting is correctly positioned then securely tighten the clamp bolt.
10 Reconnect the fuel pipes and securely tighten the union nuts.
11 Start the engine and check the filter for signs of fuel leakage.

All other models

12 The fuel filter is located underneath the rear of the vehicle, next to the fuel pump.
13 To improve access to the filter, chock the front wheels then jack up the rear of the vehicle and support on axle stands.
14 Bearing in mind the information given on depressurising the fuel system (Chapter 4B, Section 7), slacken the retaining clips and disconnect the fuel hoses from the filter. To minimise fuel loss clamp the hoses either side of the filter or be prepared to plug the hose ends as they are disconnected.
15 Slacken the clamp bolt then slide the filter out of position, noting its correct fitted orientation. The arrow on the filter should point in the direction of fuel flow (towards the fuel rail).
16 Slide the new filter into position, making sure its arrow is pointing in the direction of fuel flow, and securely tighten the clamp bolt.
17 Reconnect the fuel pipes and securely tighten the retaining clips.
18 Lower the vehicle to the ground then start the engine and check the filter for signs of fuel leakage.

29 Road test

Instruments and electrical equipment

1 Check the operation of all instruments and electrical equipment.
2 Make sure that all instruments read correctly, and switch on all electrical equipment in turn, to check that it functions properly.

Steering and suspension

3 Check for any abnormalities in the steering, suspension, handling or road "feel".
4 Drive the vehicle, and check that there are no unusual vibrations or noises.
5 Check that the steering feels positive, with no excessive "sloppiness", or roughness, and check for any suspension noises when cornering and driving over bumps.

Drivetrain

6 Check the performance of the engine, clutch, transmission and driveshafts.
7 Listen for any unusual noises from the engine, clutch and transmission.
8 Make sure that the engine runs smoothly when idling, and that there is no hesitation when accelerating.
9 Check that, where applicable, the clutch action is smooth and progressive, that the drive is taken up smoothly, and that the pedal travel is not excessive. Also listen for any noises when the clutch pedal is depressed.
10 Check that all gears can be engaged smoothly without noise, and that the gear lever action is not abnormally vague or "notchy".
11 On automatic transmission models, make sure that all gearchanges occur smoothly, without snatching, and without an increase in engine speed between changes. Check that all of the gear positions can be selected with the vehicle at rest. If any problems are found, they should be referred to a Renault dealer.
12 Listen for a metallic clicking sound from the front of the vehicle, as the vehicle is driven slowly in a circle with the steering on full-lock. Carry out this check in both directions. If a clicking noise is heard, this indicates wear in a driveshaft joint (see Chapter 8).

Check the operation and performance of the braking system

13 Make sure that the vehicle does not pull to one side when braking, and that the wheels do not lock prematurely when braking hard.
14 Check that there is no vibration through the steering when braking.
15 Check that the handbrake operates correctly, without excessive movement of the lever, and that it holds the vehicle stationary on a slope.
16 Test the operation of the brake servo unit as follows. Depress the footbrake four or five times to exhaust the vacuum, then start the engine. As the engine starts, there should be a noticeable "give" in the brake pedal as vacuum builds up. Allow the engine to run for at least two minutes, and then switch it off. If the brake pedal is now depressed again, it should be possible to detect a hiss from the servo as the pedal is depressed. After about four or five applications, no further hissing should be heard, and the pedal should feel considerably harder.

28.6 Fuel filter location - 2664 cc engine

Every 72 000 miles (120 000 km)

30 Timing belt renewal -
1995 cc and 2165 cc engines

Refer to Chapter 2A.

Every 2 years

31 Coolant renewal

Cooling system draining

Warning: Wait until the engine is cold before starting this procedure. Do not allow antifreeze to come in contact with your skin, or with the painted surfaces of the vehicle. Rinse off spills immediately with plenty of water. Never leave antifreeze lying around in an open container, or in a puddle in the driveway or on the garage floor. Children and pets are attracted by its sweet smell, but antifreeze can be fatal if ingested.

1 With the engine completely cold, remove the expansion tank filler cap. Turn the cap anti-clockwise, wait until any pressure remaining in the system is released, then unscrew it and lift it off.
2 Position a suitable container beneath the radiator bottom hose connection.
3 Slacken the hose clip, pull off the hose and allow the coolant to drain into the container.
4 To assist draining, open the cooling system bleed screws. These are located in various places according to engine and model type. Most models have a bleed screw in the thermostat housing, or in the radiator top hose. Carburettor engines may have a bleed screw located on the top of the automatic choke housing on the side of the carburettor. Note that some later engines have no bleed screws.
5 When the flow of coolant stops, reposition the container below the cylinder block drain plug. This is located at the rear of the cylinder block on the right-hand side **(see illustration)**.
6 Remove the drain plug, and allow the coolant to drain into the container.
7 If the coolant has been drained for a reason other than renewal, then provided it is clean and less than two years old, it can be re-used, though this is not recommended.
8 Refit the radiator bottom hose and cylinder block drain plug on completion of draining.

Cooling system flushing

9 If coolant renewal has been neglected, or if the antifreeze mixture has become diluted, then in time, the cooling system may gradually lose efficiency, as the coolant passages become restricted due to rust, scale deposits, and other sediment. The cooling system efficiency can be restored by flushing the system clean.
10 The radiator should be flushed independently of the engine, to avoid unnecessary contamination.

Radiator flushing

11 Disconnect the top and bottom hoses and any other relevant hoses from the radiator, with reference to Chapter 3.
12 Insert a garden hose into the radiator top inlet. Direct a flow of clean water through the radiator, and continue flushing until clean water emerges from the radiator bottom outlet.
13 If after a reasonable period, the water still does not run clear, the radiator can be flushed with a good proprietary cleaning agent. It is important that their manufacturer's instructions are followed carefully. If the contamination is particularly bad, insert the hose in the radiator bottom outlet, and reverse-flush the radiator.

Engine flushing

14 To flush the engine, first refit the cylinder block drain plug, and tighten the cooling system bleed screw(s).

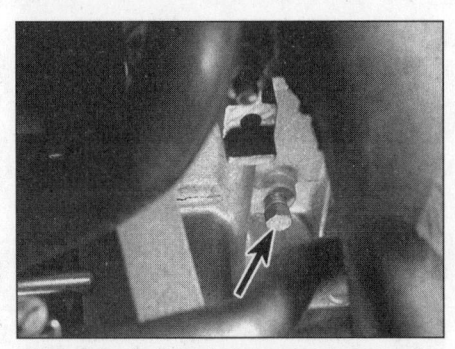

31.5 Cylinder block drain plug (arrowed) - four-cylinder engine

15 Remove the thermostat as described in Chapter 3, then temporarily refit the top hose at its engine connection.
16 With the top and bottom hoses disconnected from the radiator, insert a garden hose into the radiator top hose. Direct a clean flow of water through the engine, and continue flushing until clean water emerges from the radiator bottom hose.
17 On completion of flushing, refit the thermostat and reconnect the hoses with reference to Chapter 3.

Cooling system filling

18 Before attempting to fill the cooling system, make sure that all hoses and clips are in good condition, and that the clips are tight. Note that an antifreeze mixture must be used all year round, to prevent corrosion of the engine components (see Chapter 3). Also check that the cylinder block drain plug is in place and tight.
19 Remove the expansion tank filler cap.
20 Open the cooling system bleed screw(s) (see paragraph 4), where applicable.
21 Place a container under the vehicle, below the expansion tank, to catch any coolant which may be spilt during the topping up procedure. Also place a wad of rags around the expansion tank.
22 Slowly fill the system until the coolant level reaches the top of the expansion tank filler neck.
23 Where applicable, close the bleed screw(s) when coolant free from air bubbles emerges. If more than one bleed screw is fitted, close the screws in sequence, starting with the lowest screw in the system.
24 Start the engine, and run it at a fast idle speed (do not exceed 1500 rpm) for approximately 4 minutes. Keep the level topped up to the top of the expansion tank filler neck.
25 Refit and tighten the expansion tank filler cap.
26 Allow the engine to run for 10 minutes or so (until the cooling fan cuts in and out on models with an electric fan).
27 Stop the engine and check the coolant level, which should be up to the "MAX" mark on the side of the tank.

28 Allow the engine to cool, then re-check the coolant level with reference to "Weekly checks". Top-up the level if necessary and refit the expansion tank filler cap.

Antifreeze mixture

29 The antifreeze should always be renewed at the specified intervals. This is necessary not only to maintain the antifreeze properties, but also to prevent corrosion which would otherwise occur as the corrosion inhibitors become progressively less effective.

30 Always use an ethylene-glycol based antifreeze which is suitable for use in mixed-metal cooling systems. The quantity of antifreeze and levels of protection are given in the *Specifications*.

31 Before adding antifreeze, the cooling system should be completely drained, preferably flushed, and all hoses checked for condition and security.

32 After filling with antifreeze, a label should be attached to the expansion tank, stating the type and concentration of antifreeze used, and the date installed. Any subsequent topping-up should be made with the same type and concentration of antifreeze.

33 Do not use engine antifreeze in the washer system, as it will cause damage to the vehicle paintwork. A screenwash additive should be added to the washer system in the quantities stated on the bottle.

Chapter 1 Part B:
Routine maintenance and servicing - diesel models

Contents

Air conditioning refrigerant check	17
Air filter renewal	21
Auxiliary drivebelt check and renewal	9
Brake fluid renewal	26
Coolant renewal	29
Clutch check	8
Electrical systems check	12
Engine oil and filter renewal	3
Exhaust system check	13
Front brake pad check	5
Front wheel alignment check	23
Fuel filter renewal	18
Fuel filter water draining	4
Handbrake check	7
Hose and fluid leak check	20
Idle speed and anti-stall speed check and adjustment	10
Intensive maintenance	2
Introduction	1
Manual transmission oil level check	19
Manual transmission oil renewal	25
Rear brake pad check - models with rear disc brakes	6
Rear brake shoe thickness check - models with rear drum brakes	24
Road test	27
Roadwheel bolt check	15
Seat belt check	11
Spare fuse check	22
Suspension and steering check	14
Timing belt renewal	28
Turbocharger boost pressure check - turbo models	16

Degrees of difficulty

Easy, suitable for novice with little experience	Fairly easy, suitable for beginner with some experience	Fairly difficult, suitable for competent DIY mechanic	Difficult, suitable for experienced DIY mechanic	Very difficult, suitable for expert DIY or professional

1B•2 Servicing specifications - diesel models

Lubricants and fluids
Refer to "Weekly checks"

Capacities
Engine oil*
Excluding filter:
   Engines without crankcase stiffener (see Chapter 2C) . . . . 4.8 litres
   Engines with crankcase stiffener (see Chapter 2C) . . . . 5.5 litres
Including filter:
   Engines without crankcase stiffener (see Chapter 2C) . . . . 5.3 litres
   Engines with crankcase stiffener (see Chapter 2C) . . . . 6.0 litres

Cooling system (approximate)
All engines . . . . 7.5 litres

Transmission
NG1-type transmission . . . . 2.2 litres
NG3-type transmission . . . . 2.2 litres
UN1-type transmission . . . . 3.0 litres
Fuel tank . . . . 67.0 litres

Engine
Oil filter . . . . Champion F105

Cooling system
Antifreeze mixture:
  35% antifreeze . . . . Protection down to -23°C
  50% antifreeze . . . . Protection down to -40°C

Fuel system
Idle speed:
  Turbo models . . . . 800 ± 50 rpm
  Normally-aspirated models . . . . 750 ± 50 rpm
Fast idle speed:
  Later type Bosch pump . . . . 1000 ± 50 rpm
  All other pumps . . . . Not adjustable (factory set)
Anti-stall speed:
  Early Bosch pump . . . . Not adjustable (factory set)
  Later Bosch pump . . . . See Text
  Lucas pump (4 mm shim inserted) . . . . 1300 ± 50 rpm
Air filter element:*
  All except J8S-706 engines . . . . Champion W132
  J8S-706 engines . . . . Champion U503
*Refer to "Buying Spare Parts and Vehicle Identification" for details of engine code location
Fuel filter:
  Models with Lucas fuel injection pump . . . . Champion L132
  Models with Bosch fuel injection pump . . . . Champion L111

Preheating system
Glow plugs . . . . Champion CH137

Brakes
Front disc brakes:
  Pad thickness (including backing):
    New . . . . 18.0 mm
    Minimum thickness . . . . 6.0 mm
Rear disc brakes:
  Pad thickness (including backing):
    New . . . . 14.0 mm
    Minimum thickness . . . . 6.0 mm
Rear drum brakes:
  Shoe thickness (including backing):
    New . . . . 7.0 mm
    Minimum thickness . . . . 2.5 mm

Torque wrench settings
	Nm	lbf ft
Roadwheel bolts:		
4-bolt fixing	90	66
5-bolt fixing	100	74

Maintenance schedule - diesel models

1 The maintenance intervals in this manual are provided with the assumption that you, not the dealer, will be carrying out the work. These are the minimum maintenance intervals recommended by the manufacturer for vehicles driven daily. If you wish to keep your vehicle in peak condition at all times, you may wish to perform some of these procedures more often. We encourage frequent maintenance, because it enhances the efficiency, performance and resale value of your vehicle.

2 If the vehicle is driven in dusty areas, used to tow a trailer, or driven frequently at slow speeds (idling in traffic) or on short journeys, more frequent maintenance intervals are recommended.

3 When the vehicle is new, it should be serviced by a factory-authorised dealer service department, in order to preserve the factory warranty.

Every 250 miles (400 km) or weekly
Refer to *"Weekly Checks"*

Every 5000 miles (8000 km)
- [ ] Renew the engine oil and filter (Section 3)
- [ ] Drain any water from the fuel filter (Section 4)
- [ ] Check the front brake pad thickness (Section 5)
- [ ] Check the rear brake pad thickness - models with rear disc brakes (Section 6)
- [ ] Check the operation of the handbrake (Section 7)
- [ ] Check the operation of the clutch (Section 8)
- [ ] Check the condition of the auxiliary drivebelts (Section 9)
- [ ] Check the idle speed and anti-stall speed (Section 10)
- [ ] Check the condition of the seat belts (Section 11)
- [ ] Check the operation of all electrical systems (Section 12)
- [ ] Check the condition of the exhaust system and mountings (Section 13)
- [ ] Check the suspension and steering components (Section 14)
- [ ] Check the tightness of the roadwheel bolts (Section 15)
- [ ] Check the turbocharger boost pressure - turbo models (Section 16)
- [ ] Check the condition of the air conditioning refrigerant (Section 17)

Every 10 000 miles (16 000 km)
Carry out all the operations listed under the 5000 mile (8000 km) service, along with the following:
- [ ] Renew the fuel filter (Section 18)
- [ ] Check the manual transmission oil level (Section 19)
- [ ] Check all underbonnet components and hoses for fluid leaks (Section 20)

Every 15 000 miles (24 000 km)
Carry out all the operations listed under the 5000 mile (8000 km), and 10 000 mile (16 000 km) services, along with the following:
- [ ] Renew the air filter (Section 21)

Every 40 000 miles (64 000 km)
Carry out all the operations listed under the 5000 mile (8000 km), 10 000 mile (16 000 km), and 15 000 mile (24 000 km) services along with the following:
- [ ] Check the spare fuses are in place (Section 22)
- [ ] Check the front wheel alignment (Section 23)
- [ ] Check the rear brake shoe thickness - models with rear drum brakes (Section 24)
- [ ] Renew the manual transmission oil (Section 25)
- [ ] Renew the brake fluid (Section 26)
- [ ] Carry out a road test (Section 27)

Every 70 000 miles (112 000 km)
In addition to all the items listed previously, carry out the following:
- [ ] Renew the timing belt (Section 28)

Every 2 years
In addition to all the items listed previously, carry out the following:
- [ ] Renew the coolant (Section 29)

1B•4 Maintenance procedures - diesel models

1 Introduction

1 This Chapter is designed to help the home mechanic maintain his/her vehicle for safety, economy, long life and peak performance.
2 The Chapter contains a master maintenance schedule, followed by Sections dealing specifically with each task in the schedule. Visual checks, adjustments, component renewal and other helpful items are included. Refer to the accompanying illustrations of the engine compartment and the underside of the vehicle for the locations of the various components.
3 Servicing your vehicle in accordance with the mileage/time maintenance schedule and the following Sections will provide a planned maintenance programme, which should result in a long and reliable service life. This is a comprehensive plan, so maintaining some items but not others at the specified service intervals, will not produce the same results.
4 As you service your vehicle, you will discover that many of the procedures can - and should - be grouped together, because of the particular procedure being performed, or because of the close proximity of two otherwise-unrelated components to one another. For example, if the vehicle is raised for any reason, the exhaust can be inspected at the same time as the suspension and steering components.
5 The first step in this maintenance programme is to prepare yourself before the actual work begins. Read through all the Sections relevant to the work to be carried out, then make a list and gather together all the parts and tools required. If a problem is encountered, seek advice from a parts specialist, or a dealer service department.

2 Intensive maintenance

1 If, from the time the vehicle is new, the routine maintenance schedule is followed closely, and frequent checks are made of fluid levels and high-wear items, as suggested throughout this manual, the engine will be kept in relatively good running condition, and the need for additional work will be minimised.
2 It is possible that there will be times when the engine is running poorly due to the lack of regular maintenance. This is even more likely if a used vehicle, which has not received regular and frequent maintenance checks, is purchased. In such cases, additional work may need to be carried out, outside of the regular maintenance intervals.
3 If engine wear is suspected, a compression test or leakdown test (refer to Chapter 2C) will provide valuable information regarding the overall performance of the main internal components. Such a test can be used as a basis to decide on the extent of the work to be carried out. If, for example, a compression test indicates serious internal engine wear, conventional maintenance as described in this Chapter will not greatly improve the performance of the engine, and may prove a waste of time and money, unless extensive overhaul work (Chapter 2C) is carried out first.
4 The following series of operations are those most often required to improve the performance of a generally poor-running engine:

Primary operations

a) Clean, inspect and test the battery (See "Weekly checks").
b) Check all the engine-related fluids (See "Weekly checks").
c) Check the condition and tension of the auxiliary drivebelt (Section 9).
d) Check the condition of the air filter element, and renew if necessary (Section 21).
e) Check the fuel filter (Section 18).
f) Check the condition of all hoses, and check for fluid leaks (Section 20).
g) Check the idle speed and anti-stall speed (Section 10)

5 If the above operations do not prove fully effective, carry out the following secondary operations:

Secondary operations

All items listed under "Primary operations", plus the following:
a) Check the charging system (Chapter 5A).
b) Check the preheating system (Chapter 5C).
c) Check the fuel system (Chapter 4C).

Every 5000 miles (8000 km)

3 Engine oil and filter renewal

1 Frequent oil and filter changes are the most important preventative maintenance procedures which can be undertaken by the DIY owner. As engine oil ages, it becomes diluted and contaminated, which leads to premature engine wear.
2 Before starting this procedure, gather together all the necessary tools and materials. Also make sure that you have plenty of clean rags and newspapers handy, to mop up any spills. Ideally, the engine oil should be warm, as it will drain more easily, and more built-up sludge will be removed with it. Take care not to touch the exhaust or any other hot parts of the engine when working under the vehicle. To avoid any possibility of scalding, and to protect yourself from possible skin irritants and other harmful contaminants in used engine oils, it is advisable to wear gloves when carrying out this work. Access to the underside of the vehicle will be greatly improved if it can be raised on a lift, driven onto ramps, or jacked up and supported on axle stands. Whichever method is chosen, make sure that the vehicle remains level, or if it is at an angle, that the drain plug is at the lowest point. The drain plug is located at the rear of the sump.
3 Remove the oil filler cap.
4 Using a spanner, or preferably a suitable socket and bar, slacken the drain plug about half a turn. Position the draining container under the drain plug, then remove the plug completely (see Haynes Hint).
5 Allow some time for the oil to drain, noting that it may be necessary to reposition the container as the oil flow slows to a trickle.
6 After all the oil has drained, wipe the drain plug and the sealing washer with a clean rag. Examine the condition of the sealing washer, and renew it if it shows signs of scoring or other damage which may prevent an oil-tight seal. Clean the area around the drain plug opening, and refit the plug complete with the washer and tighten it securely.
7 The oil filter is located at the right-hand side of the cylinder block. Note that on models with an oil cooler, the filter is screwed onto the base of the oil cooler which is mounted onto the right-hand side of the engine compartment.
8 Move the container into position under the oil filter.

> **HAYNES HINT**: As the drain plug threads release, move it sharply away so the stream of oil issuing from the sump runs into the container, not up your sleeve!

9 Use an oil filter removal tool to slacken the filter initially, then unscrew it by hand the rest of the way (see illustration). Empty the oil from the old filter into the container.
10 Use a clean rag to remove all oil, dirt and sludge from the filter sealing area on the engine. Check the old filter to make sure that the rubber sealing ring has not stuck to the engine. If it has, carefully remove it.
11 Apply a light coating of clean engine oil to the sealing ring on the new filter, then screw the filter into position on the engine. Tighten the filter firmly by hand only - **do not** use any tools.

Every 5000 miles - diesel models

3.9 Using an oil filter removal tool to slacken the oil filter

12 Remove the old oil and all tools from under the vehicle then, if applicable, lower the vehicle to the ground.
13 Fill the engine through the filler hole, using the correct grade and type of oil (refer to *"Weekly Checks"* for details of topping-up). Pour in half the specified quantity of oil first, then wait a few minutes for the oil to drain into the sump. Continue to add oil, a small quantity at a time, until the level is up to the lower mark on the dipstick. Adding approximately a further 1.0 litre will bring the level up to the upper mark on the dipstick.
14 Start the engine and run it for a few minutes, while checking for leaks around the oil filter seal and the sump drain plug. Note that there may be a delay of a few seconds before the low oil pressure warning light goes out when the engine is first started, as the oil circulates through the new oil filter and the engine oil galleries before the pressure builds up.
15 Stop the engine, and wait a few minutes for the oil to settle in the sump once more. With the new oil circulated and the filter now completely full, recheck the level on the dipstick, and add more oil as necessary.
16 Dispose of the used engine oil safely, with reference to *"General repair procedures"*.

4 Fuel filter water draining

1 There are two possible types of fuel filter arrangement; some early models are equipped with a Bosch fuel filter arrangement and later models are fitted with a Lucas fuel filter arrangement. The Bosch filter is of the spin-on type and looks just like an elongated oil filter, whereas the Lucas filter is sandwiched in-between two aluminium alloy castings. On the Lucas filter the cooling system coolant is circulated around the base of the filter to warm the fuel on cold starts; the base is connected to the engine by two hoses.
2 A water drain plug is provided at the base of the fuel filter/filter housing. On models with a Lucas filter arrangement the water drain screw is on the side of the filter housing base whereas on the Bosch filter arrangement the water drain screw is located on the base of the filter element.
3 Place a suitable container beneath the drain screw **(see illustration)**. To make draining easier, a suitable length of tubing can be attached to the outlet on the screw to direct the fuel flow.
4 Loosen the fuel filter bleed screw, then open the drain screw by turning it anti-clockwise. On models where there is no bleed screw, loosen the fuel inlet union on the filter head.
5 Allow the entire contents of the filter to drain into the container, then securely tighten the drain screw and the bleed screw/fuel filter inlet union (as applicable).
6 Dispose of the drained fuel safely.
7 Prime and bleed the fuel system as described in Chapter 4C.

5 Front brake pad check

1 Apply the handbrake, then jack up the front of the car and support it securely on axle stands (see *"Jacking and vehicle support"*). Remove the front roadwheels.
2 For a comprehensive check, the brake pads should be removed and cleaned. The operation of the caliper can then also be checked, and the condition of the brake disc itself can be fully examined on both sides. Refer to Chapter 9 for further information.
3 On completion refit the roadwheels and lower the car to the ground.

HAYNES HiNT *For a quick check, the thickness of friction material remaining on each brake pad can be measured through the aperture in the caliper body.*

6 Rear brake pad check - models with rear disc brakes

1 Jack up the rear of the car and support it securely on axle stands (see *"Jacking and vehicle support"*). Remove the rear roadwheels.
2 For a comprehensive check, the brake pads should be removed and cleaned. The operation of the caliper can then also be checked, and the condition of the brake disc itself can be fully examined on both sides. Refer to Chapter 9 for further information.
3 On completion refit the roadwheels and lower the car to the ground.

4.3 Water drain screw (arrowed) - Lucas filter (viewed from above)

7 Handbrake check

General

1 The handbrake will normally be kept in adjustment by the action of the drum brakes automatic adjuster, or by the self-adjusting action of the rear disc calipers. Occasionally, the handbrake mechanism may require adjustment to compensate for cable stretch.
2 Chock the front wheels, then jack up the rear of the vehicle, and support securely on axle stands (see *"Jacking and vehicle support"*).
3 Fully release the handbrake and check that the wheels can be rotated easily by hand. The wheels may drag slightly, but there should be no binding.
4 Operate the handbrake, and check that the wheels are fully locked when the handbrake lever is pulled up by 9 notches on models with rear drum brakes, or 12 notches on models with rear disc brakes.
5 If the wheels bind, or if the handbrake lever travel is not as specified, it is likely that the handbrake mechanism is partially seized, or the mechanism is incorrectly adjusted. If the operation of the mechanism is not satisfactory, proceed as follows, according to type.

Models with rear drum brakes

6 Working under the vehicle, remove the cover from the vehicle floor (under the handbrake lever location) to reveal the handbrake adjuster sleeve.
7 Slacken the adjuster locknuts until there is no tension in the handbrake cables **(see illustration)**.
8 Remove the brake drums as described in Chapter 9, and check the operation of the handbrake lever on the trailing shoe **(see illustration)**. The lever should move freely.
9 If necessary, dismantle the components and clean them, then reassemble as described in Chapter 9.
10 On models with Bendix brakes, before refitting the brake drum, place the toothed adjuster segment on the edge of the adjuster lever **(see illustration)**.

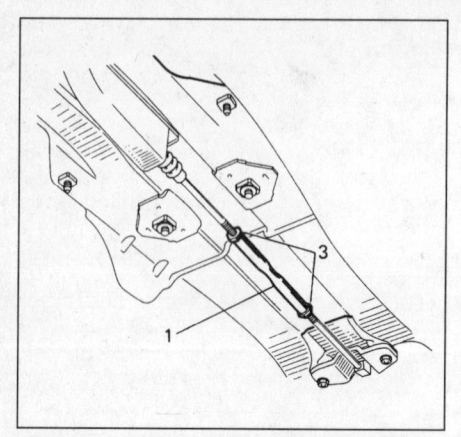

1B•6 Every 5000 miles – diesel models

7.7 Handbrake adjuster sleeve (1) and locknuts (3)

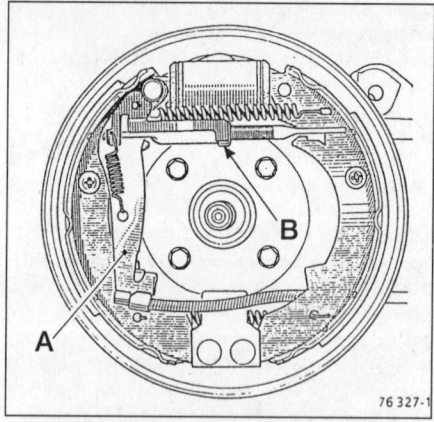

7.8 Handbrake operating lever (A) on trailing shoe, and adjuster star wheel (B) - Girling brakes

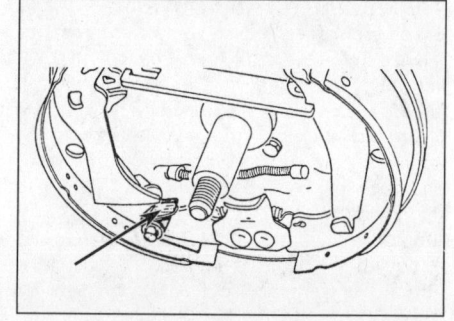

7.10 Place the toothed adjuster segment (arrowed) on the edge of the adjuster lever - Bendix brakes

11 On models with Girling brakes, check that the adjuster star wheel rotates freely in both directions, then back off the star wheel by five or six teeth.
12 Check that the handbrake cables slide freely in their sheaths.
13 Check that the handbrake operating levers are resting correctly on the shoes.
14 Working under the vehicle floor, turn the cable adjuster sleeve until the handbrake operating levers on the brake shoes begin to move when the handbrake lever is pulled beyond the second notch. The operating levers should be free with the handbrake lever on the first and second notches.
15 Tighten the adjuster locknuts.
16 Refit the brake drums as described in Chapter 9, then refit the roadwheels and the adjuster sleeve cover, and lower the vehicle to the ground.
17 With the vehicle resting on its wheels, depress the brake pedal repeatedly to operate the self-adjusting mechanism. It should be possible to hear the adjusting mechanism click as the brake pedal is depressed.

Models with rear disc brakes
18 Proceed as described in paragraphs 6 and 7.
19 Check that the handbrake cables slide freely in their sheaths, and that the operating levers on the calipers move freely.
20 Push the handbrake operating levers on the calipers as far as they will go towards the rear of the vehicle.
21 Working under the vehicle floor, turn the cable adjuster sleeve until the handbrake cable end fittings just contact the operating levers on the calipers, without moving the levers **(see illustration)**.
22 Continue to turn the adjuster sleeve until the handbrake operating levers on the calipers begin to move when the handbrake lever is pulled beyond the second notch. The operating levers should be free with the handbrake lever on the first and second notches.
23 Tighten the adjuster locknuts.

24 Refit the roadwheels and the adjuster sleeve cover, and lower the vehicle to the ground.

8 Clutch check

Check and adjust the clutch cable as described in Chapter 6. Check that the cable moves freely and easily and lubricate its exposed section with multi-purpose grease.

9 Auxiliary drivebelt check and renewal

Checking
1 Vee, multi-toothed, or multi-grooved belts are used to drive the coolant pump, alternator, power steering pump and air conditioning compressor (where fitted) from the crankshaft pulley.
2 Using a socket and extension bar fitted over the crankshaft pulley bolt, rotate the crankshaft so that the entire length of the relevant drivebelt can be examined. Examine the drivebelt for cracks, splitting, fraying or

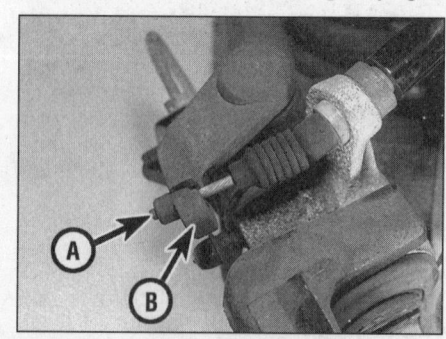

7.21 The handbrake cable end fitting (A) should just contact the operating lever (B) on the caliper

other damage. Check also for signs of glazing (shiny patches) and for separation of the belt plies. Renew the belt if worn or damaged.
3 If the condition of the belt is satisfactory, check the drivebelt tension as described in the following paragraphs.
4 Check the deflection of the belts at the points indicated **(see illustration)**. The belts should be tensioned so that under firm thumb pressure, the deflection of the belt is as specified in the following table. The belts should be tensioned when the engine is cold - if the engine is hot, add 1.0 mm to the figures in the table.

Drivebelt type	Belt deflection
Vee-belt	5.5 to 6.5 mm
Multi-toothed/ multi-grooved belt	4.5 to 5.5 mm

HAYNES HiNT *Correct tensioning of the drivebelt will ensure that it has a long life. A belt which is too slack will slip and squeal. Beware, however, of overtightening, as this can cause wear in the accessory bearings.*

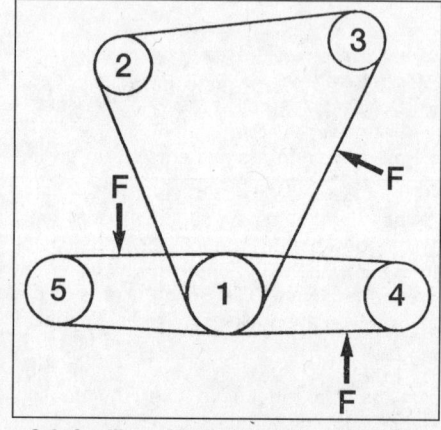

9.4 Auxiliary drivebelt tension checking points (F)

1 Crankshaft
2 Coolant pump
3 Alternator
4 Power steering pump
5 Air conditioning compressor

Every 5000 miles - diesel models 1B•7

Tensioning and removal

5 To tension the crankshaft drivebelts, release the relevant driven accessory adjuster link and mounting bolts, and move the accessory as necessary. Re-tighten the bolts.
6 To remove a drivebelt, slacken the accessory adjuster link and mounting bolts, and push the relevant accessory fully towards the engine. With the drivebelt slack, slip it off the pulleys. If difficulty is experienced removing the belt, turn the crankshaft pulley and at the same time press the belt against the pulley rim, when it will ride up over the rim.
7 It will be obvious that before an inner belt can be removed, the outer belt will have to be removed.
Note: *On certain models with air conditioning, the air conditioning compressor drivebelt is located behind the timing belt. To remove the air conditioning compressor drivebelt, it will therefore be necessary to remove the timing belt as described in Chapter 2.*

Refitting

8 Fit the belt, again turning the crankshaft pulley if the belt is difficult to pass over the pulley rim.
9 Tension the belt in accordance with the figures given in the table at the end of paragraph 4, and tighten the accessory adjuster link and mounting bolts.

10 Idle speed and anti-stall speed check and adjustment

1 The usual type of tachometer (rev counter), which works from ignition system pulses, cannot be used on diesel engines. A diagnostic socket is provided for the use of Renault test equipment, but this will not normally be available to the home mechanic. If it is not felt that adjusting the idle speed 'by ear' is satisfactory, one of the following alternatives may be used.
 a) Purchase or hire of an appropriate tachometer.
 b) Delegation of the job to a Renault dealer or other specialist.
2 Before making adjustments warm up the engine to normal operating temperature. Make sure that the accelerator cable is correctly adjusted (see Chapter 4C).

Idle speed checking and adjustment

3 With the accelerator lever resting against the idle stop, check that the engine idles at the specified speed. If necessary adjust as follows.

Lucas injection pump
4 Loosen the locknut on the idle speed adjustment screw **(see illustrations)**. Turn the screw as required and retighten the locknut.
5 Check the anti-stall adjustment as described later in this Section.

6 Stop the engine and disconnect the tachometer, where applicable.

Early type Bosch injection pump (fast idle thermostatic valve mounted on pump)
7 Loosen the locknut on the idle speed adjustment screw **(see illustration)**. Turn the screw as required and retighten the locknut.

Later type Bosch injection pump (fast idle valve screwed into cylinder head)
8 Loosen the locknut and unscrew the anti-stall adjustment screw until it is clear of the pump accelerator lever **(see illustration)**.
9 Loosen the locknut and turn the idle speed adjustment screw as required, then retighten the locknut.
10 Make the anti-stall adjustment as described later in this Section.
11 Stop the engine and disconnect the tachometer, where applicable.

Anti-stall checking and adjustment

Lucas injection pump
12 Make sure that the engine is at normal operating temperature, and idling at the specified speed, as described previously.
13 Insert a 4 mm shim, such as a selection of feeler blades which add up to the correct thickness, between the pump accelerator lever and the anti-stall adjustment screw.
14 The engine speed should increase to the specified anti-stall speed (see *Specifications*).
15 If adjustment is necessary, loosen the locknut, turn the anti-stall adjustment screw as required, then tighten the locknut.
16 Remove the shim or feeler blade and check the idle speed as described previously.
17 Move the pump accelerator lever to increase the engine speed to approximately 3000 rpm, then quickly release the lever. The deceleration period should be approximately 2.5 to 3.5 seconds, and the engine speed should drop to approximately 50 rpm below idle.
18 If the deceleration is too fast and the engine stalls, unscrew the anti-stall adjustment screw 1/4 turn towards the accelerator lever. If the deceleration is too slow, resulting in poor engine braking, turn the screw 1/4 turn away from the lever.

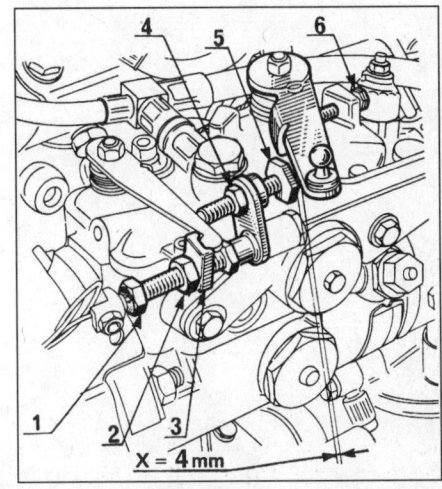

10.4a Adjustment points - early type Lucas pump

1 Idle speed adjusting screw
2 Locknut
3 Locating plate
4 Locknut
5 Anti-stall adjustment screw
6 Maximum speed adjustment screw
X Shim thickness for anti-stall speed adjustment

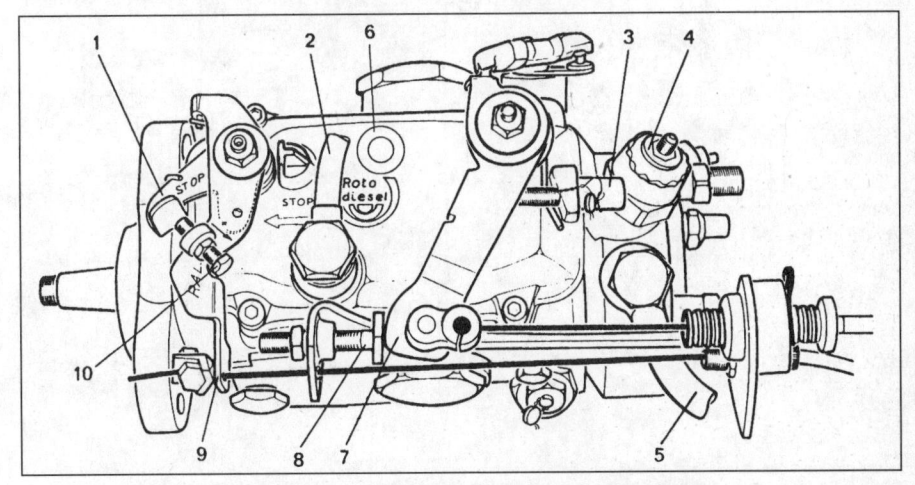

10.4b Adjustment points - later type Lucas pump

1 Manual stop lever
2 Fuel return to the tank
3 Maximum speed adjustment screw
4 Stop solenoid
5 Fuel inlet
6 Injection pump timing access cover
7 Accelerator lever
8 Anti-stall adjustment screw
9 Fast idle lever
10 Idle speed adjustment screw

1B•8 Every 5000 miles - diesel models

19 Retighten the locknut after making an adjustment, then recheck the idle speed and adjust if necessary as described previously.
20 On models with the later type pump, with the engine idling check the operation of the manual stop control by turning the stop lever clockwise (see Chapter 4C). The engine must stop instantly.
21 Where applicable, disconnect the tachometer on completion.

Early type Bosch injection pump (fast idle thermostatic valve mounted on pump)
22 On early type Bosch pumps it is not possible to adjust the anti-stall setting.

Later type Bosch injection pump (fast idle valve screwed into cylinder head)
23 Make sure that the engine is at normal operating temperature, and idling at the specified speed, as described previously.
24 Insert a 1 mm shim or feeler blade between the pump accelerator lever and the anti-stall adjustment screw. The idle speed should rise by approximately 10 to 20 rpm.
25 If adjustment is necessary, loosen the locknut and turn the anti-stall adjustment screw as required. Retighten the locknut.
26 Remove the shim and move the pump accelerator lever to increase the engine speed to approximately 3000 rpm, then quickly release the lever and check that the engine returns to the specified idle speed. Recheck the anti-stall speed setting and readjust, if necessary.
27 With the anti-stall speed correctly set, move the fast idle lever fully towards the flywheel end of the engine and check that the engine speed increases to the specified fast idle speed. If necessary loosen the locknut and turn the fast idle adjusting screw as required, then retighten the locknut.
28 Where applicable, disconnect the tachometer on completion.

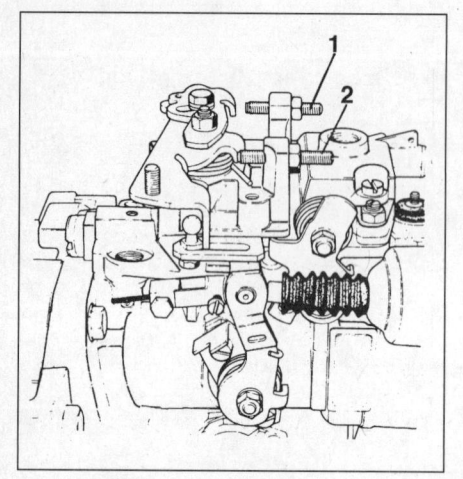

10.7 Adjustment points - early type Bosch injection pump

1 Idle speed adjustment screw
2 Maximum speed adjustment screw

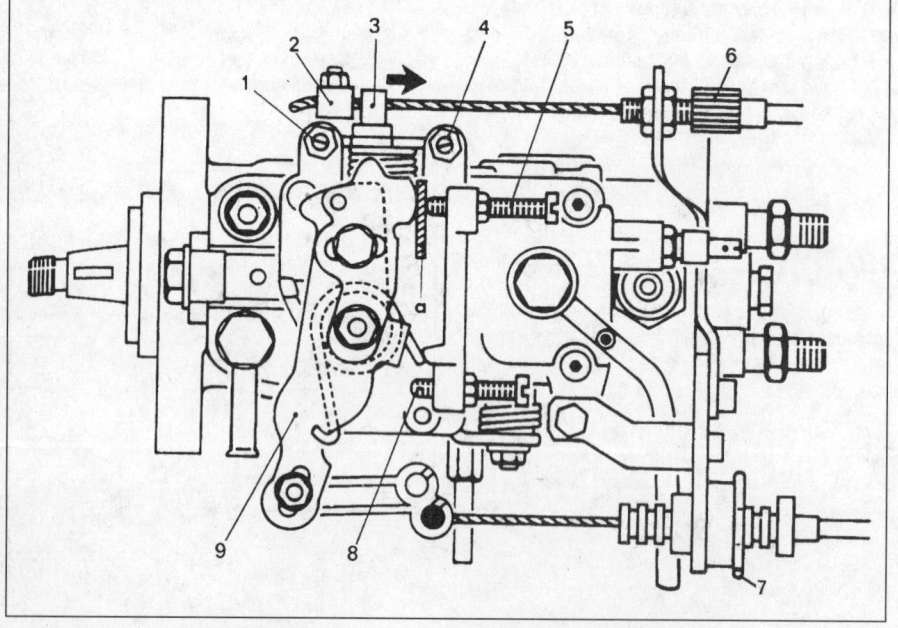

10.8 Adjustment points - later type Bosch injection pump

1 Fast idle adjustment screw
2 Cable end clamp
3 Fast idle lever
4 Idle speed adjustment screw
5 Anti-stall adjustment screw
6 Fast idle cable adjustment screw
7 Accelerator cable spring clip
8 Maximum speed adjustment screw
9 Accelerator lever
a Shim for anti-stall speed adjustment

11 Seat belt check

1 Carefully examine the seat belt webbing for cuts, or any signs of serious fraying or deterioration. If the belt is of the retractable type, pull the belt all the way out of the inertia reel, and examine the full extent of the webbing.
2 Fasten and unfasten the belt, ensuring that the locking mechanism holds securely, and releases properly when intended. If the belt is of the retractable type, check also that the retracting mechanism operates correctly when the belt is released.
3 Check the security of all seat belt mountings and attachments which are accessible without removing any trim or other components.

12 Electrical systems check

1 Check the operation of all electrical equipment, ie, lights, direction indicators, horn, etc. Refer to the appropriate Sections of Chapter 12 for details if any of the circuits are found to be inoperative.
2 Note that stop-light adjustment is described in Chapter 9.
3 Visually check all accessible wiring connectors, harnesses and retaining clips for security, and for signs of chafing or damage. Rectify any faults found.

13 Exhaust system check

1 With the engine cold (at least an hour after the vehicle has been driven), check the complete exhaust system from the engine to the end of the tailpipe. The exhaust system is most easily checked with the vehicle raised on a hoist, or suitably supported on axle stands, so that the exhaust components are readily visible and accessible.
2 Check the exhaust pipes and connections for evidence of leaks, severe corrosion and damage. Make sure that all brackets and mountings are in good condition, and that all relevant nuts and bolts are tight. Leakage at any of the joints or in other parts of the system will usually show up as a black sooty stain in the vicinity of the leak.
3 Rattles and other noises can often be traced to the exhaust system, especially the brackets and mountings. Try to move the pipes and silencers. If the components are able to come into contact with the body or suspension parts, secure the system with new mountings. Otherwise separate the joints (if possible) and twist the pipes as necessary to provide additional clearance.

Every 5000 miles - diesel models

14 Suspension and steering check

Front suspension and steering check

1 Raise the front of the vehicle, and securely support it on axle stands (see *"Jacking and vehicle support"*).
2 Visually inspect the balljoint dust covers and the steering rack-and-pinion gaiters for splits, chafing or deterioration. Any wear of these components will cause loss of lubricant, together with dirt and water entry, resulting in rapid deterioration of the balljoints or steering gear.
3 On vehicles with power steering, check the fluid hoses for chafing or deterioration, and the pipe and hose unions for fluid leaks. Also check for signs of fluid leakage under pressure from the steering gear rubber gaiters, which would indicate failed fluid seals within the steering gear.
4 Grasp the roadwheel at the 12 o'clock and 6 o'clock positions, and try to rock it **(see illustration)**. Very slight free play may be felt, but if the movement is appreciable, further investigation is necessary to determine the source. Continue rocking the wheel while an assistant depresses the footbrake. If the movement is now eliminated or significantly reduced, it is likely that the hub bearings are at fault. If the free play is still evident with the footbrake depressed, then there is wear in the suspension joints or mountings.
5 Now grasp the wheel at the 9 o'clock and 3 o'clock positions, and try to rock it as before. Any movement felt now may again be caused by wear in the hub bearings or the steering track-rod balljoints. If the outer balljoint is worn, the visual movement will be obvious. If the inner joint is suspect, it can be felt by placing a hand over the rack-and-pinion rubber gaiter and gripping the track-rod. If the wheel is now rocked, movement will be felt at the inner joint if wear has taken place.
6 Using a large screwdriver or flat bar, check for wear in the suspension mounting bushes by levering between the relevant suspension component and its attachment point. Some movement is to be expected, as the mountings are made of rubber, but excessive wear should be obvious. Also check the condition of any visible rubber bushes, looking for splits, cracks or contamination of the rubber.
7 With the car standing on its wheels, have an assistant turn the steering wheel back and forth, about an eighth of a turn each way.

14.4 Check for wear in the hub bearings by grasping the wheel and trying to rock it

There should be very little, if any, lost movement between the steering wheel and roadwheels. If this is not the case, closely observe the joints and mountings previously described. In addition, check the steering column universal joints for wear, and also check the rack-and-pinion steering gear itself.

Rear suspension check

8 Chock the front wheels, then jack up the rear of the vehicle and support securely on axle stands (see *"Jacking and vehicle support"*).
9 Working as described previously for the front suspension, check the rear hub bearings, the suspension bushes and the shock absorber mountings for wear.

Shock absorber check

10 Check for any signs of fluid leakage around the shock absorber body, or from the rubber gaiter around the piston rod. Should any fluid be noticed, the shock absorber is defective internally, and should be renewed.
Note: *Shock absorbers should always be renewed in pairs on the same axle.*
11 The efficiency of the shock absorber may be checked by bouncing the vehicle at each corner. Generally speaking, the body will return to its normal position and stop after being depressed. If it rises and returns on a rebound, the shock absorber is probably suspect. Also examine the shock absorber upper and lower mountings for any signs of wear.

15 Roadwheel bolt check

Where applicable, remove the wheel trims, and slacken the roadwheel bolts slightly.
Tighten the bolts to the specified torque, using a torque wrench.

16 Turbocharger boost pressure check - turbo models

Renault specify that the turbocharger boost pressure should be regularly checked at the specified intervals. Checking is a complex procedure, requiring the use of several special tools and adapters and should therefore be entrusted to a Renault dealer.

17 Air conditioning refrigerant check

1 Run the engine, and switch on the air conditioning.
2 After a few minutes, inspect the sight glass (located on top of the drier bottle at the front corner of the engine compartment), and check the fluid flow **(see illustration)**. Clear fluid should be visible - if not, the following will help to diagnose the problem:

a) *Clear fluid flow - the system is functioning correctly.*
b) *No fluid flow - have the system checked for leaks by a Renault dealer or air conditioning specialist.*
c) *Continuous stream of clear air bubbles in fluid - refrigerant level low - have the system recharged by a Renault dealer or air conditioning specialist.*
d) *Milky air bubbles visible - high humidity (have the system checked by a Renault dealer).*

17.2 Air conditioning drier bottle sight glass (arrowed)

1B•10 Every 10 000 miles - diesel models

Every 10 000 miles (16 000 km)

18 Fuel filter renewal

1 There are two possible types of fuel filter arrangement; some early models are equipped with a Bosch fuel filter arrangement and later models are fitted with a Lucas/CAV fuel filter arrangement. The Bosch filter is of the spin-on type and looks just like an elongated oil filter, whereas the Lucas/CAV filter is sandwiched in-between two aluminium alloy castings. On the Lucas/CAV filter the cooling system coolant is circulated around the base of the filter to warm the fuel on cold starts; the base is connected to the engine by two hoses.

Lucas/CAV filter arrangement

2 Drain the contents of the filter as described in Section 4.
3 Support the filter base and undo the through-bolt from the top of the filter head. Withdraw the bolt and recover its washer and sealing ring.
Examine the sealing ring for signs of damage or deterioration and renew if necessary.
4 Lower the filter bowl, taking care not to strain the coolant hoses, and lift out the element. Recover the sealing rings, noting their correct fitted locations, and discard them; new sealing rings are supplied with the filter element.
5 Clean out the filter bowl.
6 Noting that the filter must be installed so that the writing on the filter is the correct way up, fit a large sealing ring to either end of the filter element and position the smaller seal on the top of the element.
7 Manoeuvre the element into position, ensuring that the filter is the correct way up and that all the sealing rings remain correctly seated.
8 Locate the filter base on the element and insert the through-bolt, washer and sealing ring. Check that the sealing rings are correctly positioned and securely tighten the through-bolt.
9 Ensure that the drain screw is securely tightened then bleed the fuel system as described in Chapter 4C. Start the engine and check the filter assembly for signs of leakage.

Bosch filter arrangement

10 Drain the contents of the filter as described in Section 4.
11 Using an oil filter removal tool, slacken the filter initially, then unscrew it by hand the rest of the way.
12 Lower the filter away from the filter head and remove it from the engine compartment. Recover the sealing ring(s) and discard them; new sealing rings are supplied with the filter element. **Note:** *On the original factory-fitted filter, only the large outer sealing ring is fitted and the filter threads are sealed using locking compound. However, all replacement filters should be supplied with two sealing rings, the large outer ring and a smaller inner ring which should be fitted to the centre of the filter.*
13 Fit the new sealing rings to the top of the new filter and lubricate them with a smear of clean fuel. Where necessary, transfer the old drain screw from the base of the original filter onto the new one.
14 Screw on the filter and tighten it firmly by hand only.
15 Ensure that the drain screw is securely tightened then bleed the fuel system as described in Chapter 4C. Start the engine and check the filter assembly for signs of leakage.

19 Manual transmission oil level check

1 Either position the vehicle over an inspection pit, or jack up the front and rear of the vehicle and support it on axle stands. The vehicle must be level for the check to be accurate.
2 Clean the area around the filler/level plug on the left-hand side of the transmission, then slacken and remove the plug from the transmission (see illustration).
3 The transmission oil level should be up to the lower edge of the filler/level plug aperture.
4 If necessary, top-up using the specified type of lubricant until the transmission oil level is correct. Fill the transmission until oil starts to flow out and allow excess oil to drain out.
5 Once the transmission oil level is correct, refit the filler/level plug and tighten it securely.
6 The frequent need for topping-up indicates a leakage, possibly through an oil seal. The cause should be investigated and rectified.

20 Hose and fluid leak check

1 Visually inspect the engine joint faces, gaskets and seals for any signs of water or oil leaks. Pay particular attention to the areas around the cylinder head cover, cylinder head, oil filter and sump joint faces. Bear in mind that, over a period of time, some very slight seepage from these areas is to be expected - what you are really looking for is any indication of a serious leak. Should a leak be found, renew the offending gasket or oil seal by referring to the appropriate Chapters in this manual.
2 Also check the security and condition of all the engine-related pipes and hoses, and all braking system pipes and hoses (see Haynes Hint). Ensure that all cable-ties or securing clips are in place, and in good condition. Clips

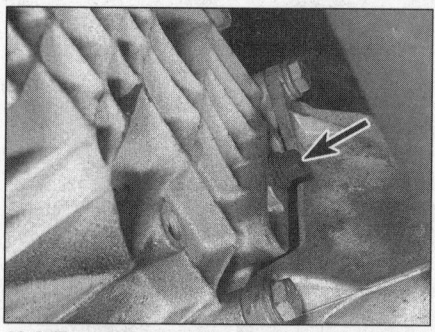

19.2 Transmission filler/level plug (arrowed)

which are broken or missing can lead to chafing of the hoses, pipes or wiring, which could cause more serious problems in the future.
3 Carefully check the radiator hoses and heater hoses along their entire length. Renew any hose which is cracked, swollen or deteriorated. Cracks will show up better if the hose is squeezed. Pay close attention to the hose clips that secure the hoses to the cooling system components. Hose clips can pinch and puncture hoses, resulting in cooling system leaks. If the crimped-type hose clips are used, it may be a good idea to replace them with standard worm-drive clips.
4 Inspect all the cooling system components (hoses, joint faces, etc) for leaks.
5 Where any problems are found on system components, renew the component or gasket with reference to Chapter 3.
6 With the vehicle raised, inspect the fuel tank and filler neck for punctures, cracks and other damage. The connection between the filler neck and tank is especially critical. Sometimes a rubber filler neck or connecting hose will leak due to loose retaining clamps or deteriorated rubber.
7 Carefully check all rubber hoses and metal fuel lines leading away from the fuel tank. Check for loose connections, deteriorated hoses, crimped lines, and other damage. Pay particular attention to the vent pipes and hoses, which often loop up around the filler neck and can become blocked or crimped. Follow the lines to the front of the vehicle, carefully inspecting them all the way. Renew damaged sections as necessary. Similarly, whilst the vehicle is raised, take the opportunity to inspect all underbody brake fluid pipes and hoses.
8 From within the engine compartment, check the security of all fuel, vacuum and brake hose attachments and pipe unions, and inspect all hoses for kinks, chafing and deterioration.
9 Where applicable, check the condition of the power steering fluid pipes and hoses.

> **HAYNES HINT** *A leak in the cooling system will usually show up as white- or rust-coloured deposits on the area adjoining the leak*

Every 15 000 miles - diesel models 1B•11

Every 15 000 miles (24 000 km)

21 Air filter renewal

Models with air cleaner assembly on the right-hand side of the engine compartment

1 Undo the wing nut then lift off the cover and withdraw the filter element **(see illustrations)**.
2 Wipe clean the inside of the housing and insert the new element into the housing. Refit the housing cover and securely tighten the wing nut.

Models with air cleaner in the inlet manifold

3 Slacken the retaining clip and disconnect the inlet duct from the manifold.

4 Undo the retaining screws and remove the filter cover from the inlet manifold. Withdraw the filter element noting which way around it is fitted.
5 Wipe clean the inside of the filter cover and manifold and install the new element, ensuring it is fitted the correct way around.
6 Refit the filter cover and securely tighten its retaining screws. Reconnect the inlet duct and secure in position with its retaining clip.

21.1a Unscrew the wing nut . . .

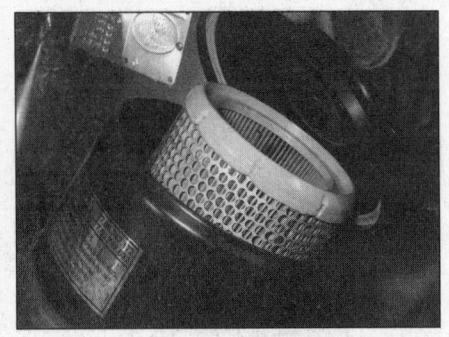

21.1b . . . then lift off the filter cover and remove the filter

Every 40 000 miles (64 000 km)

22 Spare fuse check

Check that spare fuses are in place in the locations provided in the fusebox under the facia (see Chapter 12). It is advisable to carry at least one spare of each rating of fuse fitted. Spare fuses can be obtained from most car accessory shops, or from a Renault dealer.

23 Front wheel alignment check

Refer to the information given in Chapter 10.

24 Rear brake shoe thickness check - models with rear drum brakes

Remove the rear brake drums, and check the brake shoes for signs of wear or contamination. At the same time, also inspect the wheel cylinders for signs of leakage, and the brake drum for signs of wear. Refer to the relevant Sections of Chapter 9 for further information.

25 Manual transmission oil renewal

1 This operation is much quicker and more efficient if the car is first taken on a journey of sufficient length to warm the engine/transmission up to normal operating temperature.
2 Park the car on level ground, switch off the ignition and apply the handbrake firmly. For improved access, jack up the front of the car and support it securely on axle stands. Note that the car must be lowered to the ground and level, to ensure accuracy, when refilling and checking the oil level.
3 Wipe clean the area around the filler/level plug, which is situated on the left-hand side of the transmission. Unscrew the filler/level plug from the transmission.
4 Position a suitable container under the drain plug which is situated on the base of the transmission **(see illustration)**. Note that on some models it will be necessary to remove the protective plastic cover from the base of the transmission to gain access to the drain plug.
5 Unscrew the plug and allow the oil to drain completely into the container. If the oil is hot, take precautions against scalding. Clean both the filler/level and the drain plugs, being especially careful to wipe any metallic particles off the magnetic inserts.
6 When the oil has finished draining, clean the drain plug threads and those of the transmission casing and refit the drain plug, tightening it securely. Where necessary, refit the plastic cover to the transmission. It the car was raised for the draining operation, now lower it to the ground.
7 Refilling the transmission is an extremely awkward operation. Above all, allow plenty of time for the oil level to settle properly before checking it. Note that the car must be parked on flat level ground when checking the oil level.
8 Refill the transmission with the exact amount of the specified type of oil then check the oil level as described in Section 19; if the correct amount was poured into the transmission and a large amount flows out on checking the level, refit the filler/level plug and take the car on a short journey so that the new oil is distributed fully around the transmission components, then check the level again on your return.

25.4 Transmission drain plug

Every 40 000 miles – diesel models

26 Brake fluid renewal

Warning: Brake hydraulic fluid can harm your eyes and damage painted surfaces, so use extreme caution when handling and pouring it. Do not use fluid that has been standing open for some time, as it absorbs moisture from the air. Excess moisture can cause a dangerous loss of braking effectiveness.

1 The procedure is similar to that for the bleeding of the hydraulic system as described in Chapter 9, except that the brake fluid reservoir should be emptied by siphoning, using a clean poultry baster or similar before starting, and allowance should be made for the old fluid to be expelled when bleeding a section of the circuit.
2 Working as described in Chapter 9, open the first bleed screw in the sequence, and pump the brake pedal gently until nearly all the old fluid has been emptied from the master cylinder reservoir. Top-up to the "MAX" level with new fluid, and continue pumping until only the new fluid remains in the reservoir, and new fluid can be seen emerging from the bleed screw. Tighten the screw, and top the reservoir level up to the "MAX" level line.
3 Old hydraulic fluid is invariably much darker in colour than the new, making it easy to distinguish the two.
4 Work through all the remaining bleed screws in the sequence until new fluid can be seen at all of them. Be careful to keep the master cylinder reservoir topped-up to above the "MIN" level at all times, or air may enter the system and greatly increase the length of the task.
5 When the operation is complete, check that all bleed screws are securely tightened, and that their dust caps are refitted. Wash off all traces of spilt fluid, and recheck the master cylinder reservoir fluid level.
6 Check the operation of the brakes before taking the car on the road.

27 Road test

Instruments and electrical equipment

1 Check the operation of all instruments and electrical equipment.
2 Make sure that all instruments read correctly, and switch on all electrical equipment in turn, to check that it functions properly.

Steering and suspension

3 Check for any abnormalities in the steering, suspension, handling or road "feel".
4 Drive the vehicle, and check that there are no unusual vibrations or noises.
5 Check that the steering feels positive, with no excessive "sloppiness", or roughness, and check for any suspension noises when cornering and driving over bumps.

Drivetrain

6 Check the performance of the engine, clutch, transmission and driveshafts.
7 Listen for any unusual noises from the engine, clutch and transmission.
8 Make sure that the engine runs smoothly when idling, and that there is no hesitation when accelerating.
9 Check that the clutch action is smooth and progressive, that the drive is taken up smoothly, and that the pedal travel is not excessive. Also listen for any noises when the clutch pedal is depressed.
10 Check that all gears can be engaged smoothly without noise, and that the gear lever action is not abnormally vague or "notchy".
11 Listen for a metallic clicking sound from the front of the vehicle, as the vehicle is driven slowly in a circle with the steering on full-lock. Carry out this check in both directions. If a clicking noise is heard, this indicates wear in a driveshaft joint (see Chapter 8).

Check the operation and performance of the braking system

12 Make sure that the vehicle does not pull to one side when braking, and that the wheels do not lock prematurely when braking hard.
13 Check that there is no vibration through the steering when braking.
14 Check that the handbrake operates correctly, without excessive movement of the lever, and that it holds the vehicle stationary on a slope.
15 Test the operation of the brake servo unit as follows. Depress the footbrake four or five times to exhaust the vacuum, then start the engine. As the engine starts, there should be a noticeable "give" in the brake pedal as vacuum builds up. Allow the engine to run for at least two minutes, and then switch it off. If the brake pedal is now depressed again, it should be possible to detect a hiss from the servo as the pedal is depressed. After about four or five applications, no further hissing should be heard, and the pedal should feel considerably harder.

Every 70 000 miles (112 000 km)

28 Timing belt renewal

Refer to Chapter 2C.

Every 2 years

29 Coolant renewal

Cooling system draining

Warning: Wait until the engine is cold before starting this procedure. Do not allow antifreeze to come in contact with your skin, or with the painted surfaces of the vehicle. Rinse off spills immediately with plenty of water. Never leave antifreeze lying around in an open container, or in a puddle in the driveway or on the garage floor. Children and pets are attracted by its sweet smell, but antifreeze can be fatal if ingested.

1 With the engine completely cold, remove the expansion tank filler cap. Turn the cap anti-clockwise, wait until any pressure remaining in the system is released, then unscrew it and lift it off.
2 Position a suitable container beneath the radiator bottom hose connection.
3 Slacken the hose clip, pull off the hose and allow the coolant to drain into the container.
4 To assist draining, open the cooling system bleed screws. These are located in various places according to engine and model type. Most models have a bleed screw in the thermostat housing, or in the radiator top hose. Note that some later engines have no bleed screws.
5 When the flow of coolant stops, reposition

the container below the cylinder block drain plug. This is located at the rear of the cylinder block on the right-hand side **(see illustration)**.
6 Remove the drain plug, and allow the coolant to drain into the container.
7 If the coolant has been drained for a reason other than renewal, then provided it is clean and less than two years old, it can be re-used, though this is not recommended.
8 Refit the radiator bottom hose and cylinder block drain plug on completion of draining.

Cooling system flushing

9 If coolant renewal has been neglected, or if the antifreeze mixture has become diluted, then in time, the cooling system may gradually lose efficiency, as the coolant passages become restricted due to rust, scale deposits, and other sediment. The cooling system efficiency can be restored by flushing the system clean.
10 The radiator should be flushed independently of the engine, to avoid unnecessary contamination.

Radiator flushing

11 Disconnect the top and bottom hoses and any other relevant hoses from the radiator, with reference to Chapter 3.
12 Insert a garden hose into the radiator top inlet. Direct a flow of clean water through the radiator, and continue flushing until clean water emerges from the radiator bottom outlet.
13 If after a reasonable period, the water still does not run clear, the radiator can be flushed with a good proprietary cleaning agent. It is important that their manufacturer's instructions are followed carefully. If the contamination is particularly bad, insert the hose in the radiator bottom outlet, and reverse-flush the radiator.

Engine flushing

14 To flush the engine, first refit the cylinder block drain plug, and tighten the cooling system bleed screw(s).
15 Remove the thermostat as described in Chapter 3, then temporarily refit the top hose at its engine connection.

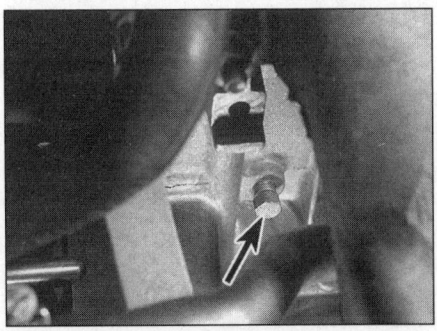

29.5 Cylinder block drain plug (arrowed) - four-cylinder engine

16 With the top and bottom hoses disconnected from the radiator, insert a garden hose into the radiator top hose. Direct a clean flow of water through the engine, and continue flushing until clean water emerges from the radiator bottom hose.
17 On completion of flushing, refit the thermostat and reconnect the hoses with reference to Chapter 3.

Cooling system filling

18 Before attempting to fill the cooling system, make sure that all hoses and clips are in good condition, and that the clips are tight. Note that an antifreeze mixture must be used all year round, to prevent corrosion of the engine components (see Chapter 3). Also check that the cylinder block drain plug is in place and tight.
19 Remove the expansion tank filler cap.
20 Open the cooling system bleed screw(s) (see paragraph 4), where applicable.
21 Place a container under the vehicle, below the expansion tank, to catch any coolant which may be spilt during the topping up procedure. Also place a wad of rags around the expansion tank.
22 Slowly fill the system until the coolant level reaches the top of the expansion tank filler neck.
23 Where applicable, close the bleed screw(s) when coolant free from air bubbles emerges. If more than one bleed screw is fitted, close the screws in sequence, starting with the lowest screw in the system.
24 Start the engine, and run it at a fast idle speed (do not exceed 1500 rpm) for approximately 4 minutes. Keep the level topped up to the top of the expansion tank filler neck.
25 Refit and tighten the expansion tank filler cap.
26 Allow the engine to run for approximately 10 minutes (until the cooling fan cuts in and out on models with an electric fan).
27 Stop the engine and check the coolant level, which should be up to the "MAX" mark on the side of the tank.
28 Allow the engine to cool, then re-check the coolant level with reference to *Weekly checks*. Top-up the level if necessary and refit the expansion tank filler cap.

Antifreeze mixture

29 The antifreeze should always be renewed at the specified intervals. This is necessary not only to maintain the antifreeze properties, but also to prevent corrosion which would otherwise occur as the corrosion inhibitors become progressively less effective.
30 Always use an ethylene-glycol based antifreeze which is suitable for use in mixed-metal cooling systems. The quantity of antifreeze and levels of protection are given in the *Specifications*.
31 Before adding antifreeze, the cooling system should be completely drained, preferably flushed, and all hoses checked for condition and security.
32 After filling with antifreeze, a label should be attached to the expansion tank, stating the type and concentration of antifreeze used, and the date installed. Any subsequent topping-up should be made with the same type and concentration of antifreeze.
33 Do not use engine antifreeze in the windscreen/tailgate washer system, as it will cause damage to the vehicle paintwork. A screenwash additive should be added to the washer system in the quantities stated on the bottle.

Notes

Chapter 2 Part A:
Four cylinder petrol engine

Contents

Camshaft and intermediate shaft oil seals - renewal	9	Engine/transmission mountings - renewal	18
Camshaft and rocker arms - removal, inspection and refitting	10	Flywheel/driveplate - removal, inspection and refitting	16
Camshaft cover - removal and refitting	5	General description	1
Compression test - description and interpretation	2	Initial start-up after major overhaul	29
Crankcase stiffener casting (1995 cc 12-valve and later 2165 cc models) - removal and refitting	15	Intermediate shaft - removal, inspection and refitting	19
		Oil cooler components - removal and refitting	28
Crankshaft - removal, inspection and refitting	20	Oil pump - removal, inspection and refitting	14
Crankshaft oil seals - renewal	17	Operations possible without removing the engine	4
Cylinder block/crankcase - cleaning and inspection	23	Piston rings - removal and refitting	22
Cylinder head - overhaul	12	Piston/connecting rod and cylinder liner assemblies - removal, inspection and refitting	21
Cylinder head - removal and refitting	11		
Engine - removal and refitting, leaving automatic transmission in the car	26	Sump - removal and refitting	13
		Timing belt - removal and refitting	7
Engine - removal and refitting leaving manual transmission in car	24	Timing belt sprockets and tensioner - removal and refitting	8
Engine and automatic transmission - removal and refitting	27	Top dead centre (TDC) for number 1 piston - locating	3
Engine and manual transmission - removal and refitting	25	Valve clearances - adjustment	6

Degrees of difficulty

Easy, suitable for novice with little experience	**Fairly easy,** suitable for beginner with some experience	**Fairly difficult,** suitable for competent DIY mechanic	**Difficult,** suitable for experienced DIY mechanic	**Very difficult,** suitable for expert DIY or professional

Specifications

General
Engine type .... Four-cylinder, in-line mounted longitudinally at front of car. Light alloy construction with wet cylinder liners. Belt-driven overhead camshaft

Engine code:
  1995 cc engine:
    Carburettor models J6R
    Fuel injection models J7R
  2165 cc models J7T
Bore .... 88.0 mm
Stroke:
  1995 cc engine .... 82.0 mm
  2165 cc engine .... 89.0 mm
Firing order .... 1-3-4-2 (No 1 at transmission end)

Timing belt
Timing belt tension deflection (measured using Renault tool Ele 346-04):
  1995 cc 12-valve models .... 10 ± 1 mm
  All other models:
    Timing belt with square profile teeth .... 5.5 to 7.0 mm
    Timing belt with rounded profile teeth .... 6.0 to 8.0 mm

Camshaft
Number of bearings 5
Endfloat:
  1995 cc 12-valve engine .... 0.10 to 0.16 mm
  All other engines .... 0.07 to 0.15 mm

2A•2 Four cylinder petrol engine

Cylinder head
Material .... Light alloy
Overall height:
  1995 cc 12-valve engine .... 90.5 mm
  All other models .... 111.6 mm
Maximum permitted surface distortion .... 0.05 mm

Valves
Valve clearances (engine cold):
  1995 cc 8-valve carburettor models:
    Inlet valves .... 0.10 mm
    Exhaust valves .... 0.25 mm
  1995 cc 8-valve fuel-injected models:
    Inlet valves .... 0.10 to 0.15 mm
    Exhaust valves .... 0.20 to 0.25 mm
  1995 cc 12-valve models:
    Inlet valves .... 0.15 to 0.20 mm
    Exhaust valves .... 0.20 to 0.25 mm
  2165 cc models:
    Inlet valves .... 0.10 to 0.15 mm
    Exhaust valves .... 0.20 to 0.25 mm
Stem diameter:
  1995 cc 12-valve engine .... 7 mm
  All other engines .... 8 mm
Head diameter:
  1995 cc 12-valve engine:
    Small inlet valve .... 30.3 mm
    Large inlet valve .... 36.3 mm
    Exhaust valve .... 40.0 mm
  All other engines:
    Inlet valve .... 44.0 mm
    Exhaust valve .... 38.5 mm
Valve seat angle:
  Inlet valves:
    1995 cc 12-valve engines .... 90°
    All other engines .... 120°
  Exhaust valves .... 90°
Valve seat width:
  1995 cc 12-valve engines .... 1.7 mm
  All other engines:
    Inlet valves .... 1.8 mm
    Exhaust valves .... 1.6 mm
Valve guides:
  Bore:
    1995 cc 12-valve engine .... 7.0 mm
    All other engines .... 8.0 mm
  Outside diameter (standard):
    1995 cc 12-valve engine .... 12.0 mm
    All other engines .... 13.0 mm

Intermediate shaft
Endfloat .... 0.05 to 0.15 mm

Crankshaft
Number of main bearings 5
Main bearing journal diameter:
  Standard .... 62.880 mm
  Undersize .... 62.630 mm
Main bearing journal minimum regrind diameter .... 62.611 mm
Main bearing running clearance* .... 0.025 to 0.050 mm
Crankpin diameter:
  1995 cc engine:
    Standard .... 52.296 mm
    Undersize .... 52.046 mm
  2165 cc engine:
    Standard .... 56.296 mm
    Undersize .... 56.046 mm

Crankpin minimum regrind diameter:
  1995 cc engine .... 52.017 mm
  2165 cc engine .... 56.017 mm
Big-end bearing running clearance* .... 0.025 to 0.050 mm
Crankshaft endfloat:
  1995 cc engine .... 0.07 to 0.25 mm
  2165 cc engine .... 0.13 to 0.30 mm
*These are suggested figures, typical for this type of engine - no exact values are quoted by Renault

Cylinder liners
Height:
  1995 cc engine .... 143.50 mm
  2165 cc engine .... 148.50 mm
Bore .... 88.0 mm
Base locating diameter .... 93.60 mm
Protrusion without O-ring .... 0.07 to 0.13 mm
Maximum difference between any two adjacent liners .... 0.04 mm

Connecting rods and gudgeon pins
Small-end play .... 0.31 to 0.57 mm
Gudgeon pins (press fit in small-end):
  Length:
    1995 cc 12-valve models .... 65 mm
    All other engines .... 75 mm
  Diameter .... 23 mm

Piston rings
Piston ring thickness:
  1995 cc 12-valve engine:
    Top compression .... 1.5 mm
    Second compression .... 1.75 mm
    Oil control .... 3.0 mm
  All other engines:
    Top compression .... 1.75 mm
    Second compression .... 2.0 mm
    Oil control .... 4.0 mm

Lubrication system
Oil pump:
  Type .... Gear, driven from intermediate shaft
  Clearances:
    Gear teeth to body .... 0.05 to 0.12 mm
    Gear endfloat .... 0.02 to 0.10 mm
Oil pressure (at 80°C):
  At idle:
    1995 cc 12-valve models .... 1.25 bar (18 psi)
    All other engines .... 0.8 bar (11 psi)
  At 3000 rpm .... 3.0 bar (43 psi)

Torque wrench settings

	Nm	lbf ft
Big-end cap nuts/bolts:		
1995 cc engine	50	37
2165 cc engine	65	48
Camshaft cover nuts:		
1995 cc 12-valve models	15	11
All other models	6	4
Camshaft sprocket bolt	50	37
Crankcase stiffener casting:		
7 mm bolts	15	11
10 mm bolts	40	30
Crankshaft pulley bolt:		
Bolt 45 mm in length	85	63
Bolt 55 mm or longer	130	97
Cylinder head bolts:		
1995 cc 12-valve models:		
Stage 1	20	15
Stage 2	Angle-tighten 80°	
Wait for three minutes then:		
Stage 3	Angle-tighten a further 80°	

2A•4 Four cylinder petrol engine

Torque wrench settings (continued)	Nm	lbf ft
All other engines:		
Stage 1	20	15
Stage 2	Angle-tighten 105°	
Wait for three minutes then:		
Stage 3	Angle-tighten a further 105°	
Driveplate bolts (automatic transmission models)	70	52
Engine to transmission housing bolts:		
Manual transmission	50	37
Automatic transmission	60	44
Flywheel bolts (manual transmission models)	60	44
Intermediate shaft sprocket bolt	50	37
Main bearing cap bolts	95	70
Oil pump:		
Body bolts	44	32
Cover bolts	12	9
Piston oil jet nozzle bolt - 1995 cc 12-valve models	14	10
Rocker arm adjusting screw locknut	10	7
Rocker arm shaft bolts - 1995 cc 12-valve engine	25	18
Rocker shaft oil filter bolt - all models except 1995 cc 12-valve engine	20	15
Sump retaining bolts:		
Models with a one piece sump	15	11
Models with a two-piece sump (fitted with crankcase stiffener)	10	7
Timing belt tensioner nuts/bolts:		
Spring-loaded tensioner assembly	25	18
Manually-adjusted tensioner:		
Tensioner backplate nut	25	18
Tensioner pulley nut	50	37

1 General description

1 The four-cylinder engine is a in-line, overhead camshaft (OHC) type, mounted at the front of the car **(see illustration)**. The clutch and transmission are attached to the rear of the engine. The Renault 25 range is fitted with both 1995 cc and 2165 cc versions of the engine, the 1995 cc engine being available in both 8-valve and 12-valve forms.

2 The cylinder head is of the crossflow design, having the inlet valves and manifold on the left-hand side of the engine and the exhaust on the right **(see illustrations)**. The inclined valves are operated by a single rocker shaft assembly which is mounted directly above the camshaft. The rocker arms have a screw and locknut type of adjuster for the valve clearances. The camshaft is driven via its sprocket from the timing belt, which in turn is driven by the crankshaft sprocket.

3 The timing belt also drives an intermediate shaft. The intermediate shaft drives the oil pump by means of a short driveshaft geared to it.

4 The distributor is driven from the rear end of the camshaft by means of an offset dog.

5 A spring-loaded tensioner assembly provides the timing belt tension adjustment on early models, on later models the timing belt is tensioned by a manually adjusted tensioner. A single, twin or triple pulley is mounted on the front of the crankshaft and this drives the alternator/water pump drivebelt, the power

1.1 Cutaway view of four-cylinder engine

Four cylinder petrol engine 2A•5

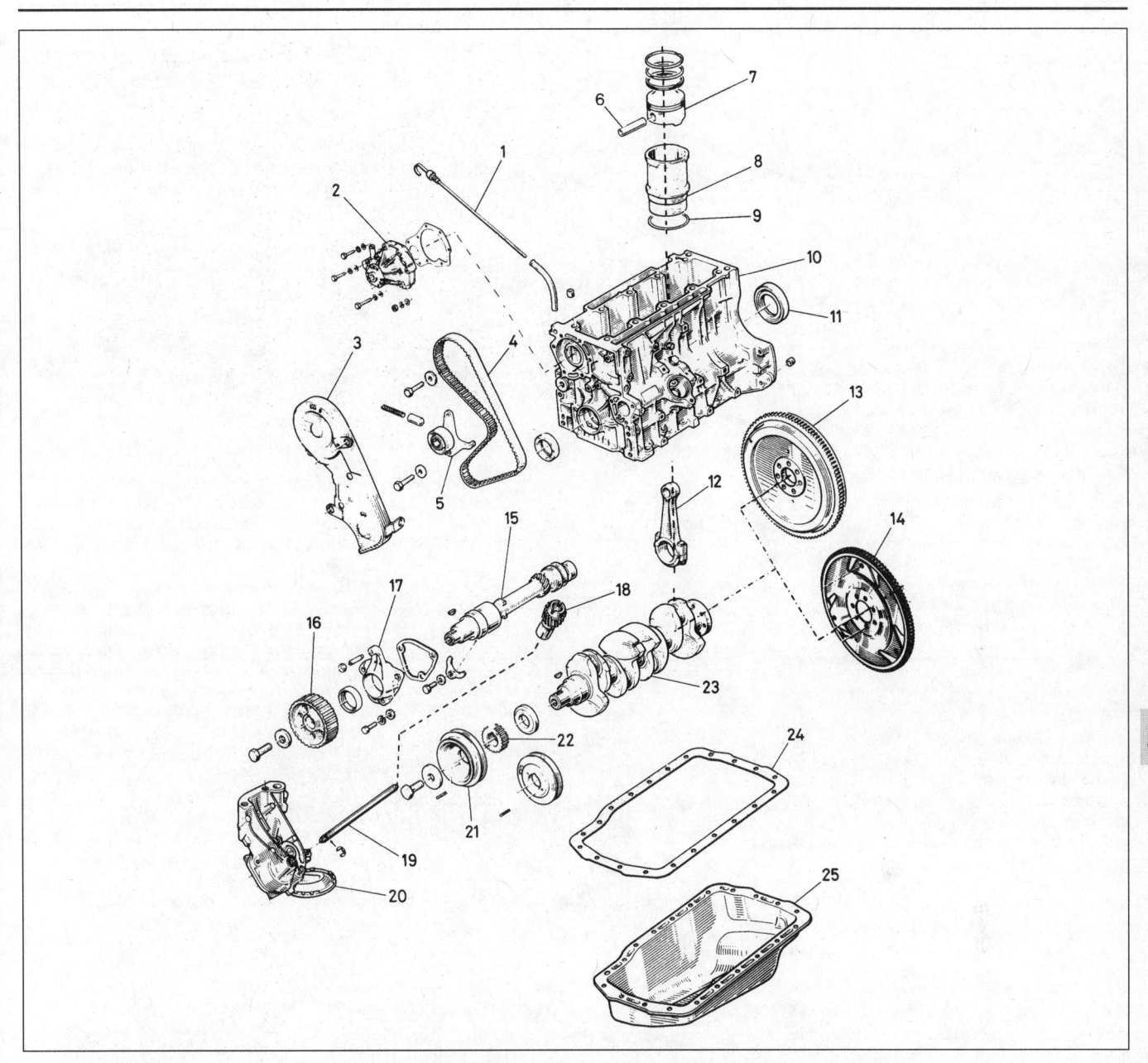

1.2a Cylinder block and associated components (early model shown)

1 Dipstick
2 Water pump
3 Timing belt cover
4 Timing belt
5 Timing belt tensioner
6 Gudgeon pin
7 Piston
8 Cylinder liner
9 Cylinder liner sealing ring
10 Cylinder block
11 Crankshaft rear oil seal
12 Connecting rod
13 Flywheel (manual transmission)
14 Driveplate (automatic transmission)
15 Intermediate shaft
16 Intermediate shaft sprocket
17 Intermediate shaft cover
18 Oil pump drive gear
19 Oil pump driveshaft
20 Oil pump
21 Crankshaft pulley
22 Crankshaft sprocket
23 Crankshaft
24 Sump gasket
25 Sump

steering pump drivebelt and the air conditioning compressor drivebelt, as applicable. The crankshaft runs in five main bearings. The crankshaft endfloat is taken up by thrustwashers which are fitted the upper half of number 2 main bearing.

6 Aluminium pistons are employed, the gudgeon pins being a press fit in the connecting rod small-ends and a sliding fit in the pistons. The No 1 piston is located at the transmission end of the engine (at the rear). As is typical with Renault engines, removable wet cylinder liners are employed, each being sealed in the crankcase by a flange and O-ring. The liner protrusion above the top surface of the crankcase is crucial; when the cylinder head and gasket are tightened down they compress the liners to provide the upper and lower seal of the engine coolant circuit within the engine. The cylinder head and crankcase are manufactured in light alloy.

2A•6 Four cylinder petrol engine

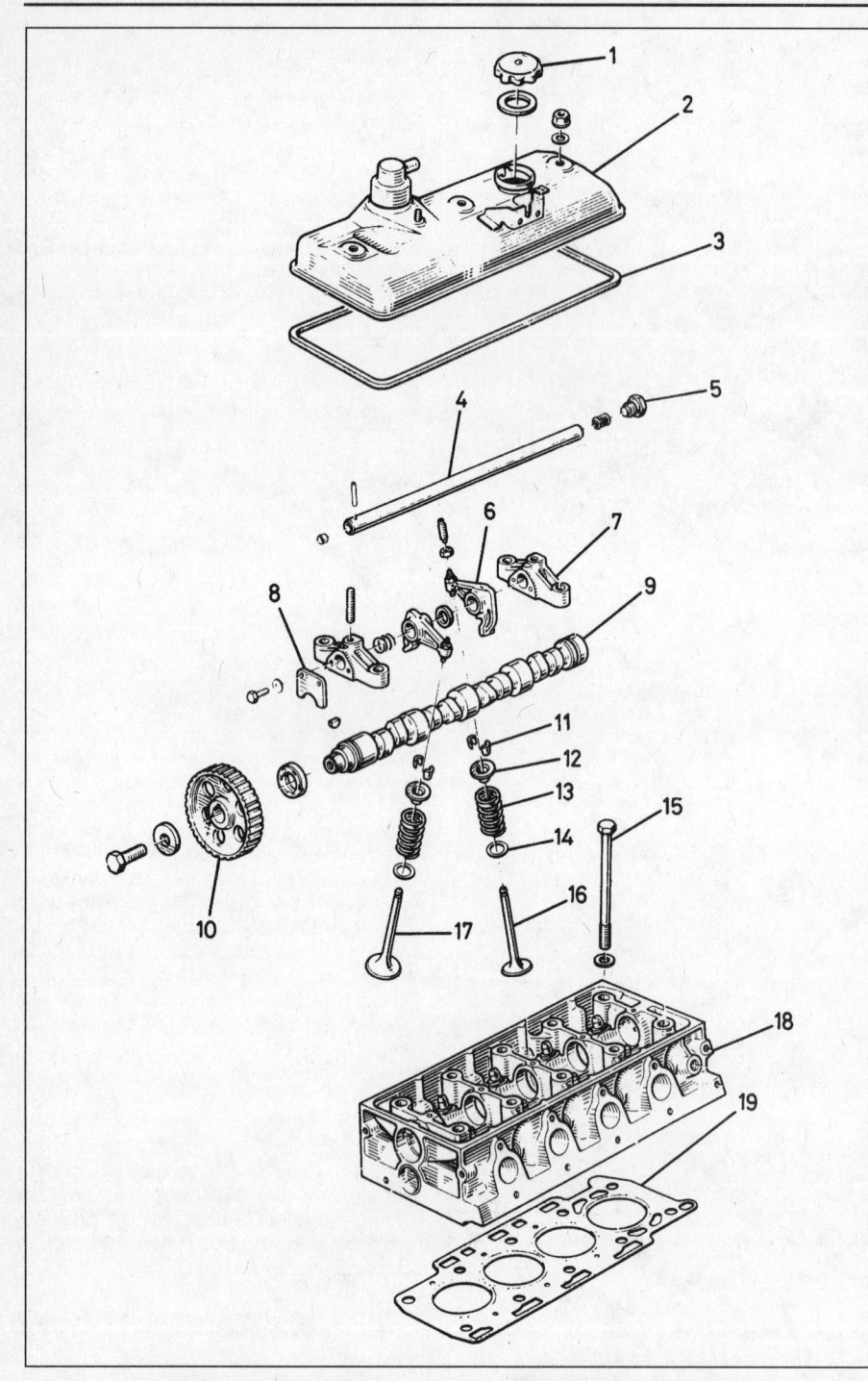

1.2b Cylinder head and associated components - 8-valve models

1 Oil filler cap
2 Camshaft cover
3 Seal
4 Rocker shaft
5 Rocker shaft end bolt and oil filter
6 Rocker arm
7 Rocker shaft pedestal
8 Camshaft thrust plate
9 Camshaft
10 Camshaft sprocket
11 Collets
12 Valve spring retainer
13 Valve spring
14 Valve spring seat
15 Cylinder head bolt
16 Exhaust valve
17 Inlet valve
18 Cylinder head
19 Gasket

2 Compression test - description and interpretation

1 When engine performance is down, or if misfiring occurs which cannot be attributed to the ignition or fuel systems, a compression test can provide diagnostic clues as to the engine's condition. If the test is performed regularly, it can give warning of trouble before any other symptoms become apparent.

2 The engine must be fully warmed-up to normal operating temperature, the battery must be fully charged, and all the spark plugs must be removed (Chapter 1A). The aid of an assistant will also be required.

3 Disable the ignition system by disconnecting the ignition HT coil lead from the distributor cap and earthing it on the cylinder block. Use a jumper lead or similar wire to make a good connection.

4 Fit a compression tester to the No 1 cylinder spark plug hole - the type of tester which screws into the plug thread is to be preferred.

5 Have the assistant hold the throttle wide open, and crank the engine on the starter motor; after one or two revolutions, the compression pressure should build up to a maximum figure, and then stabilise. Record the highest reading obtained.

6 Repeat the test on the remaining cylinders, recording the pressure in each.

7 All cylinders should produce very similar pressures; a difference of more than 2 bars between any two cylinders indicates a fault. Note that the compression should build up quickly in a healthy engine; low compression on the first stroke, followed by gradually-increasing pressure on successive strokes, indicates worn piston rings. A low compression reading on the first stroke, which does not build up during successive strokes, indicates leaking valves or a blown head gasket (a cracked head could also be the cause). Deposits on the undersides of the valve heads can also cause low compression.

8 Although Renault do not specify exact compression pressures, as a guide, any cylinder pressure of below 10 bars can be considered as less than healthy. Refer to a Renault dealer or other specialist if in doubt as to whether a particular pressure reading is acceptable.

9 If the pressure in any cylinder is low, carry out the following test to isolate the cause. Introduce a teaspoonful of clean oil into that cylinder through its spark plug hole, and repeat the test.

10 If the addition of oil temporarily improves the compression pressure, this indicates that bore or piston wear is responsible for the pressure loss. No improvement suggests that leaking or burnt valves, or a blown head gasket, may be to blame.

11 A low reading from two adjacent cylinders is almost certainly due to the head gasket

Four cylinder petrol engine 2A•7

having blown between them; the presence of coolant in the engine oil will confirm this.

12 If one cylinder is about 20 percent lower than the others and the engine has a slightly rough idle, a worn camshaft lobe could be the cause.

13 If the compression reading is unusually high, the combustion chambers are probably coated with carbon deposits. If this is the case, the cylinder head should be removed and decarbonised.

14 On completion of the test, refit the spark plugs and reconnect the ignition system.

3 Top dead centre (TDC) for number 1 piston - locating

1 Top dead centre (TDC) is the highest point in the cylinder that each piston reaches as the crankshaft turns. Each piston reaches TDC at the end of the compression stroke and again at the end of the exhaust stroke; however, for the purpose of timing the engine, TDC refers to the position of No1 piston at the end of its compression stroke. On all engines, No1 piston (and cylinder) is at the transmission (rear) end of the engine.

2 When No 1 piston is at TDC, the camshaft sprocket timing mark should be aligned with the pointer in the timing belt cover window and the timing mark on the flywheel/driveplate should be aligned with the TDC (0°) mark on the transmission bellhousing.

3 To align the timing marks, the crankshaft must be turned. This should be done by using a spanner/socket and extension bar on the crankshaft pulley bolt. If necessary to improve access to the bolt, firmly apply the handbrake then jack up the front of the vehicle and support it on axle stands (see "*Jacking and vehicle support*").

4 Turn the crankshaft in the normal direction of rotation (clockwise) whilst keeping an eye on the timing belt cover window. Align the timing mark with cover pointer then check that the flywheel/driveplate mark is correctly aligned with the TDC (0°) mark on the bellhousing **(see illustration)**. The engine is now positioned with No1 piston at TDC on its compression stroke.

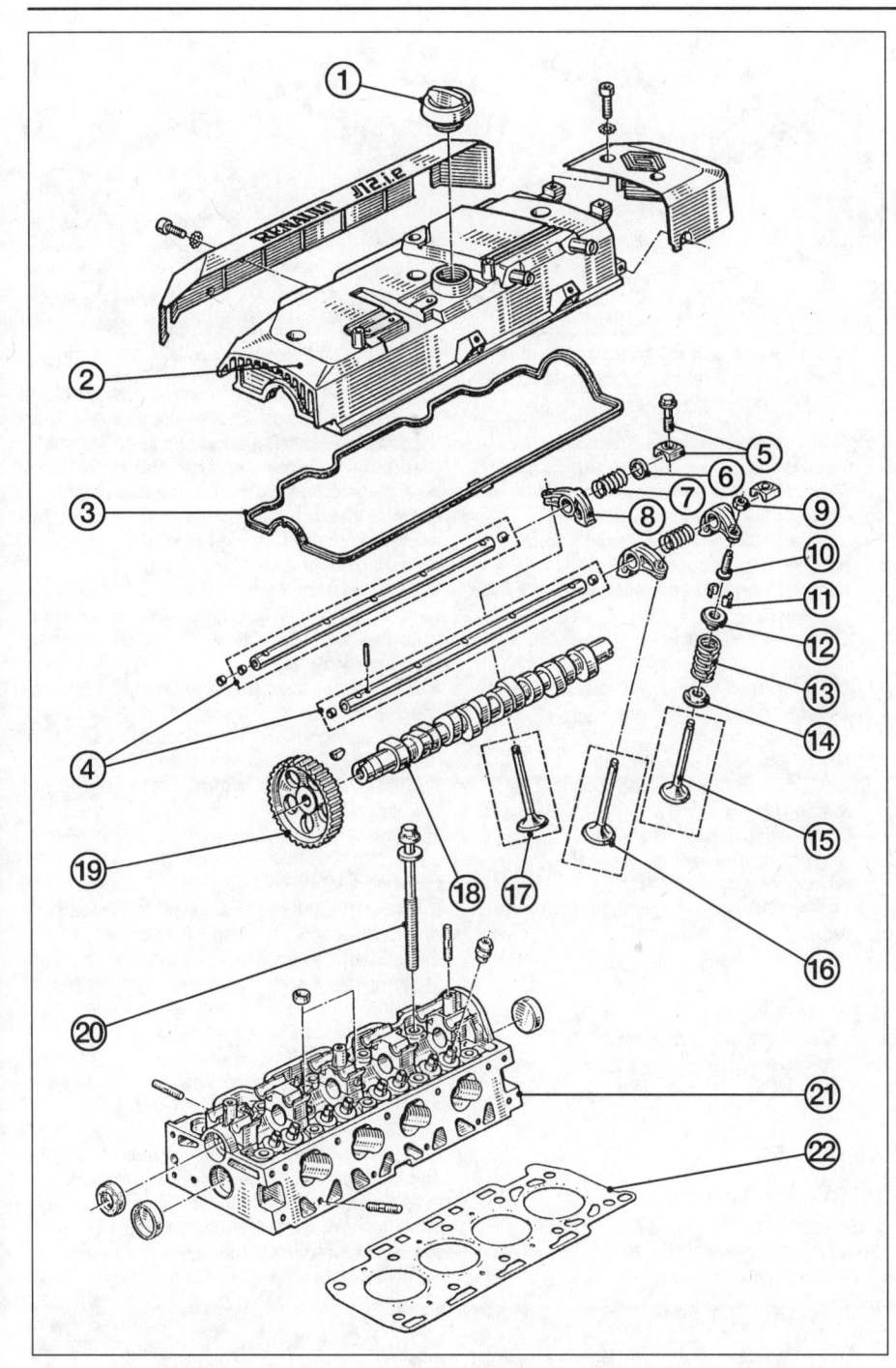

1.2c Cylinder head and associated components - 1995 cc 12-valve models

1 Oil filler cap
2 Camshaft cover
3 Seal
4 Rocker shaft
5 Rocker shaft retaining bolt and plate
6 Spacer
7 Spring
8 Rocker arm
9 Rocker arm adjusting screw locknut
10 Rocker arm adjusting screw
11 Collets
12 Valve spring retainer
13 Valve spring
14 Valve spring seat
15 Small inlet valve
16 Large inlet valve
17 Exhaust valve
18 Camshaft
19 Camshaft sprocket
20 Cylinder head bolt
21 Cylinder head
22 Gasket

3.4 Flywheel/driveplate timing mark at TDC position

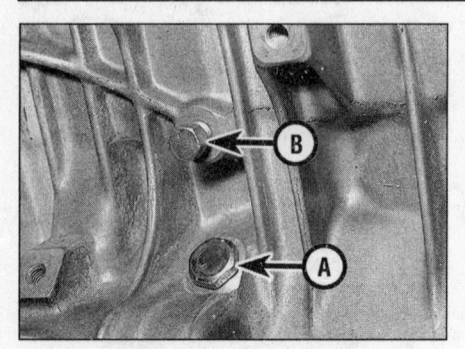

2A•8 Four cylinder petrol engine

3.5a Be sure to unscrew the timing rod access bolt (A) and not the coolant drain plug (B)

3.5b Insert a suitable diameter rod in through the cylinder block hole . . .

3.5c . . . until it engages with the timing slot in the crankshaft web

5 If necessary, the crankshaft can be locked in position to prevent unnecessary rotation. To do this, unscrew the access bolt from the left-hand side of the cylinder block, located just in front of the starter motor, and insert a dowel rod of suitable diameter to be a snug fit in the hole. Engage the rod in the timing slot provided for this purpose in the crankshaft, noting that it may be necessary to rotate the crankshaft slightly to do this **(see illustrations)**. Once in place it should be impossible to turn the crankshaft. If the crankshaft will still move to and fro slightly, then the dowel rod has entered a balance hole in the crankshaft, instead of the timing slot.

Note: *Do not attempt to rotate the engine whilst the crankshaft is locked in position. If the engine is to be left in this state for a long period of time, it is a good idea to place warning notices inside the vehicle, and in the engine compartment. This will reduce the possibility of the engine being accidentally cranked on the starter motor, which is likely to cause damage with the locking rod in place.*

4 Operations possible without removing the engine

1 The following operations are possible without the need to remove the engine from the car.

a) Compression pressure - testing.
b) Valve clearances - checking and adjusting
c) Camshaft cover - removal and refitting.
d) Timing belt - removal and refitting.
e) Timing belt tensioner and sprockets - removal and refitting.
f) Camshaft and intermediate shaft oil seals - renewal.
g) Camshaft and rocker arms - removal and refitting.
h) Cylinder head - removal and refitting.
i) Cylinder head and pistons - decarbonising.
j) Sump - removal and refitting.
k) Oil pump - removal, inspection and refitting.
l) Crankshaft oil seals - renewal.
m) Engine/transmission mountings - inspection and renewal.
n) Flywheel/driveplate - removal and refitting.

5 Camshaft cover - removal and refitting

Removal

1995 cc 12-valve models

1 Undo the retaining screws and remove the three plastic covers from the camshaft cover **(see illustrations)**.

2 Undo the retaining bolts and release the HT lead retaining bracket from the right-hand side of the camshaft cover **(see illustration)**.
3 Undo the retaining bolts and release the accelerator cable bracket from the top of the camshaft cover **(see illustration)**.
4 Slacken and remove the two nuts, and remove both the fuel hose retaining clamps from the left hand side of the camshaft cover **(see illustrations)**.
5 Disconnect the breather hoses, then undo the retaining nuts and lift off the camshaft cover along with its rubber seal **(see illustrations)**. Examine the seal for signs of damage or deterioration, and renew if necessary.

All other models

6 Where the throttle linkage pivot is mounted onto the top of the camshaft cover, disconnect the accelerator cable and link rod (as applicable) so the pivot assembly is free to be removed with the camshaft cover (see relevant part of Chapter 4).
7 Detach the HT lead support bracket, crankcase ventilation hoses and (where applicable) the hot air inlet duct to the carburettor.
8 Undo the nuts and washers and lift off the camshaft cover and seal **(see illustrations)**.
9 Remove the gasket, if necessary, and thoroughly clean the cover. Examine the seal for signs of damage or deterioration, and renew if necessary.

5.1a On 1995 cc 12-valve engines, undo the retaining screws (arrowed) . . .

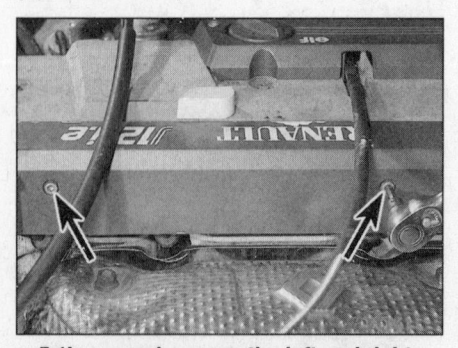

5.1b . . . and remove the left and right-hand plastic covers from the camshaft cover

5.1c Undo the two screws and remove the rear cover from the camshaft cover

Four cylinder petrol engine 2A•9

5.2 Undo the screws (arrowed) and free the HT lead bracket from the cover

5.3 Unbolt the accelerator cable bracket from the top of the camshaft cover

5.4a Undo the two nuts (arrowed) . . .

5.4b . . . and remove the fuel hose clamps

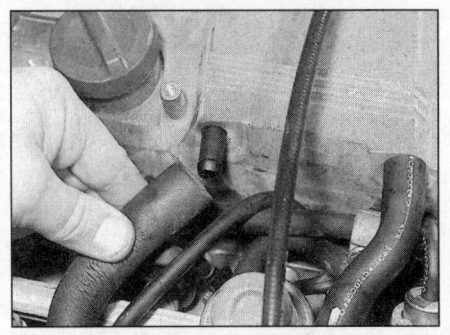

5.5a Disconnect the breather hoses . . .

5.5b . . . and remove the camshaft cover and seal from the engine

5.8a On 8-valve engines, remove the retaining nuts and washers . . .

5.8b . . . and lift off the camshaft cover and seal

Refitting

10 Refitting is a reverse of removal, making sure the seal remains correctly seated in the cover. Tighten the cover retaining nuts to the specified torque.

6 Valve clearances - adjustment

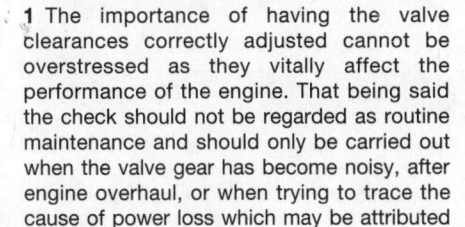

1 The importance of having the valve clearances correctly adjusted cannot be overstressed as they vitally affect the performance of the engine. That being said the check should not be regarded as routine maintenance and should only be carried out when the valve gear has become noisy, after engine overhaul, or when trying to trace the cause of power loss which may be attributed to the valve. The clearances are checked as follows noting that the engine must be cold for the check to be accurate.
2 Remove the camshaft cover as described in Section 5.
3 Position the engine with No 1 piston at TDC on its compression stroke as described in Section 3 and proceed as described under the relevant sub-heading.

1995 cc carburettor engines

4 Turn the crankshaft clockwise (viewed from the front of the vehicle) by 90°. Looking through the viewing aperture on the timing belt rear cover (cylinder head side) check to see whether there is a mark on the camshaft sprocket rim at this point **(see illustration 6.18)**. If no mark is visible adjust the valve clearances as follows. If a mark is visible, use the procedure described in paragraphs 16 to 23.

5 It is important that the clearance of the relevant valve is checked and adjusted when the valve is fully closed and the rocker arm rests on the heel of the cam (directly opposite the peak). This can be ensured by carrying out the adjustments in the following sequence, remembering that No 1 cylinder is at the transmission end of the engine. Refer to the *Specifications* for the correct clearances, the valve locations can be determined from the position of the manifolds.

Valve fully open	Adjust valves
No 1 cyl exhaust	No 3 cyl inlet and No 4 cyl exhaust
No 3 cyl exhaust	No 4 cyl inlet and No 2 cyl exhaust
No 4 cyl exhaust	No 2 cyl inlet and No 1 cyl exhaust
No 2 cyl exhaust	No 1 cyl inlet and No 3 cyl exhaust

6 With the relevant valve fully open, check the clearances of the two valves specified. Clearances are checked by inserting a feeler blade of the correct thickness between the valve stem and the rocker arm adjusting screw, the feeler blade should be a light, sliding fit. If adjustment is necessary, slacken the adjusting screw locknut and turn the screw as necessary until the feeler blade is a light sliding fit. Once the correct clearance is obtained hold the adjusting screw and securely tighten the locknut **(see illustration)**. Recheck the valve clearance and adjust again if necessary.

2A•10 Four cylinder petrol engine

6.6 Adjusting a valve clearance

6.10 On 1995 cc 12-valve models, align the camshaft sprocket timing marks (arrowed) with the cylinder head surface

6.12 Adjusting an exhaust valve clearance - 1995 cc 12-valve engine

7 Rotate the crankshaft until the next valve in the sequence is fully open and check the clearances of the next two specified valves.
8 Repeat the procedure until all eight valve clearances have been checked and, if necessary, adjusted.
9 Refit the camshaft cover as described in Section 5.

1995 cc 12-valve models

10 Turn the crankshaft in a clockwise direction to align the first set of timing marks on the rear of the camshaft sprocket, one on either side of the rim, parallel with the cylinder head surface (see illustration).
11 In this position, check the first three valves given in the table below. The valve clearances are given in the *Specifications* at the start of this Chapter. The valve locations can be determined from the position of the manifolds, remembering No 1 cylinder is at the transmission end of the engine.

Camshaft sprocket timing marks	Adjust valves
1st marks	No 2 cyl inlets and No 4 cyl exhaust
2nd marks	No 1 cyl inlets and No 2 cyl exhaust
3rd marks	No 3 cyl inlets and No 1 cyl exhaust
4th marks	No 4 cyl inlets and No 3 cyl exhaust

12 Check and, if necessary adjust, the clearances of the three valves specified as described in paragraph 6 (see illustration).
13 Rotate the engine so that the camshaft sprocket turns through 90° and the second set of sprocket timing marks are aligned with the cylinder head surface. Check the clearances of the next three specified valves.
14 Repeat the procedure for the third and fourth set of camshaft sprocket marks until all twelve valve clearances have been checked and, if necessary, adjusted.
15 Refit the camshaft cover as described in Section 5.

All other models

16 Whilst positioning No 1 cylinder at TDC (see Section 3) check whether there are timing marks on the rear face of the camshaft sprocket, the marks will be visible through the aperture in the timing belt rear cover.
17 With No 1 cylinder positioned at TDC on its compression stroke, continue as described under the relevant heading.

Models where the camshaft sprocket is equipped with rear timing marks

18 Turn the crankshaft clockwise (viewed from the front of the vehicle) by 90°. Looking through the viewing aperture on the timing belt rear cover (cylinder head side) align the mark on the rear face of the camshaft sprocket rim with the pointer in the rear cover (see illustration).
19 In this position, check the first two valves given in the table below. The valve clearances are given in the *Specifications* at the start of this Chapter. The valve locations can be determined from the position of the manifolds, remembering No1 cylinder is the transmission end of the engine.

Camshaft sprocket timing marks	Adjust valves
1st mark	No 2 cyl inlet and No 4 cyl exhaust
2nd mark	No 1 cyl inlet and No 2 cyl exhaust
3rd mark	No 3 cyl inlet and No 1 cyl exhaust
4th mark	No 4 cyl inlet and No 3 cyl exhaust

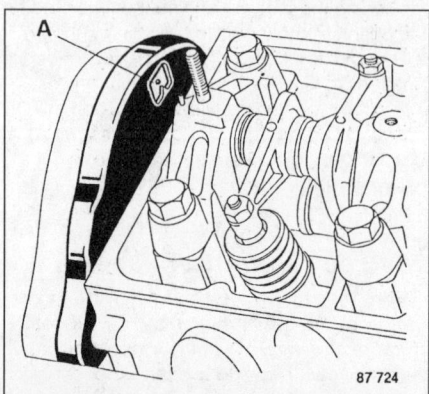

6.18 Align the camshaft sprocket rear timing mark (A) with the pointer on the timing belt rear cover

20 Check and, if necessary, adjust the clearances of the valves specified as described in paragraph 6.
21 Rotate the engine so that the camshaft sprocket turns through 90° and the second mark on the sprocket is aligned with the rear cover pointer. Check the clearances of the next two specified valves.
22 Repeat the procedure for the third and fourth camshaft sprocket marks until all eight valve clearances have been checked and, if necessary, adjusted.
23 Refit the camshaft cover as described in Section 5.

Models where the camshaft sprocket has no rear timing marks

24 Referring to illustration 6.24, make up the template shown out of cardboard.
25 With No 1 cylinder at TDC on its compression stroke, stick the template securely to the crankshaft pulley. Using white paint, make an alignment mark on the timing belt cover alongside the "A" mark of the template.
26 Turn the crankshaft clockwise (viewed from the front of the vehicle) until the template "B" mark is aligned with the mark made on the

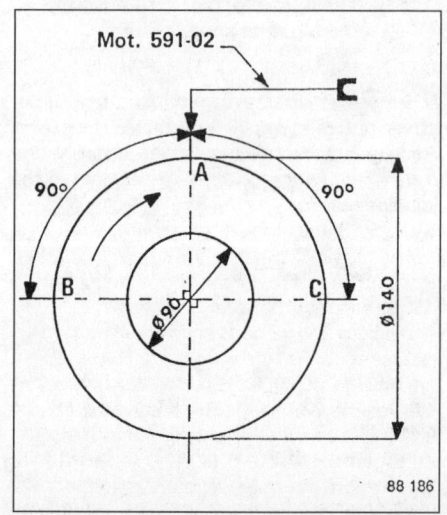

6.24 Valve clearance adjustment template for engines without camshaft sprocket rear timing marks - dimensions in mm

Four cylinder petrol engine 2A•11

timing belt cover. In this position, noting that No 1 cylinder is at the transmission end of the engine, check and adjust the clearance of No 2 cylinder inlet valve and No 4 cylinder exhaust valve (see paragraph 6).

27 Rotate the crankshaft sprocket through 180° so the template "C" mark is aligned with the mark made on the timing belt cover. Check and, if necessary, adjust the clearances of No 1 cylinder inlet valve and No 2 cylinder exhaust valve.

28 Rotate the crankshaft sprocket through another 180° until the template "B" mark is realigned with the timing belt cover mark. Check and, if necessary, adjust the clearances of No 3 cylinder inlet valve and No 1 cylinder exhaust valve.

29 Rotate the crankshaft sprocket through another 180° until the template "C" mark is realigned with the timing belt cover mark. Check and, if necessary, adjust the clearances of No 4 cylinder inlet valve and No 3 cylinder exhaust valve.

30 Once all eight valve clearances have been checked and adjusted, refit the camshaft cover as described in Section 5 and remove the template from the end of the crankshaft pulley.

7 Timing belt - removal and refitting

Removal

1995 cc 12-valve models

1 Disconnect the battery negative terminal. To improve access to the timing belt, remove the radiator as described in Chapter 3.

2 Firmly apply the handbrake then jack up the front of the vehicle and support it on axle stands (see *"Jacking and vehicle support"*).

3 Unbolt and remove the engine undertray (where fitted).

4 Remove the auxiliary drivebelt(s) as described in Chapter 1A.

5 Slacken and remove the timing belt upper cover retaining nuts and remove the cover. Recover the rubber mountings and spacers from the cover noting their correct fitted locations.

6 Slacken and remove the timing belt lower cover nut and bolt and remove the cover. Recover the rubber mountings and spacers noting their correct fitted locations.

7 Slacken and remove the crankshaft pulley retaining bolt and washer. To prevent crankshaft rotation whilst the retaining bolt is slackened, remove the lower cover plate and lock the flywheel/driveplate ring gear, using an arrangement similar to that shown **(see illustration)**. On manual transmission models, if the engine is in the vehicle the crankshaft can be retained by select top gear and applying the brakes firmly. *Do not be tempted to use the crankshaft locking pin to prevent*

the crankshaft from rotating (see Section 3). Remove the crankshaft pulley then temporarily refit the pulley bolt to the end of the crankshaft.

8 Rotate the crankshaft in a clockwise direction until the crankshaft sprocket Woodruff key, the camshaft sprocket and the intermediate shaft sprocket timing marks are positioned as shown **(see illustration)**. In this position, all four pistons are at mid-stroke.

9 Slacken the tensioner pulley backplate retaining nut and the large pulley retaining nut, and pivot the pulley and backplate away from the belt to release the belt tension.

10 If the timing belt is to be re-used, use white paint or similar to mark the direction of rotation on the belt (if markings do not already exist), then slip the belt off the sprockets.

11 Check the timing belt carefully for any signs of uneven wear, splitting or oil contamination; renew it if there is the slightest doubt about its condition. If signs of oil contamination are found, trace the source of the oil leak and rectify it, then wash down the engine timing belt area and all related components to remove all traces of oil.

All other models

Note: There are two possible types of timing belt and sprockets fitted. Some early models may be fitted with sprockets and a belt where the teeth have a square profile whereas on later models the belt and sprocket teeth have a rounded profile. If a new timing belt is to be fitted ensure its teeth have the correct profile to match the sprockets **(see illustration)**.

12 Carry out the operations described in paragraphs 1 to 4.

13 Position No 1 cylinder at TDC on its compression stroke as described in Section 3 and lock the crankshaft in position.

14 Slacken and remove the timing belt cover retaining nuts and bolts and remove the cover. Recover the collars, rubber mountings and spacers from the cover noting their correct fitted locations.

15 Check that the sprocket timing marks are positioned as shown **(see illustrations)**.

16 Remove the crankshaft locking pin then remove the crankshaft pulley as described in paragraph 7. Refit the bolt to the end of the

7.7 Use a fabricated tool like this one to lock the flywheel/driveplate and prevent crankshaft rotation

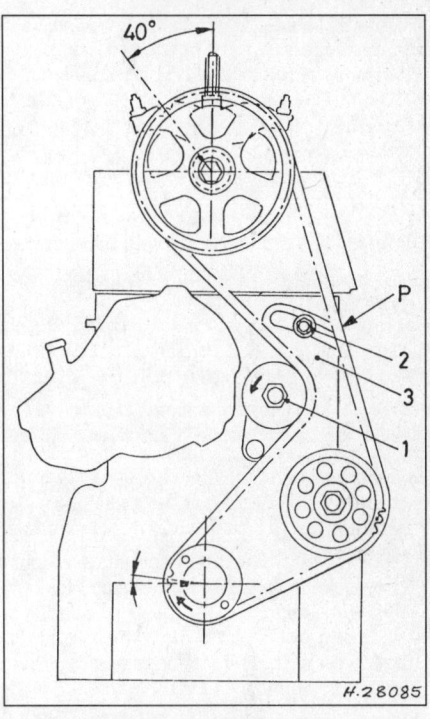

7.8 Timing belt and sprocket timing marks - 1995 cc 12-valve engine

1 Tensioner pulley retaining nut
2 Tensioner pulley backplate retaining nut
3 Tensioner pulley backplate
P Timing belt deflection measuring point

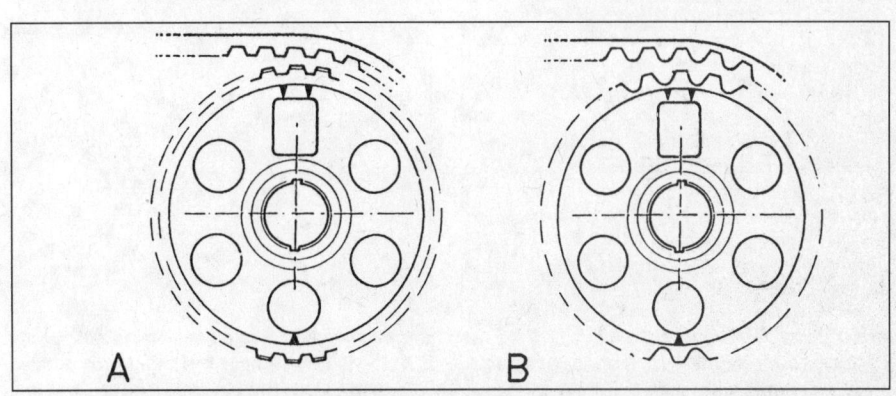

7.12 Early (A) and later (B) types of sprocket and timing belt teeth profiles

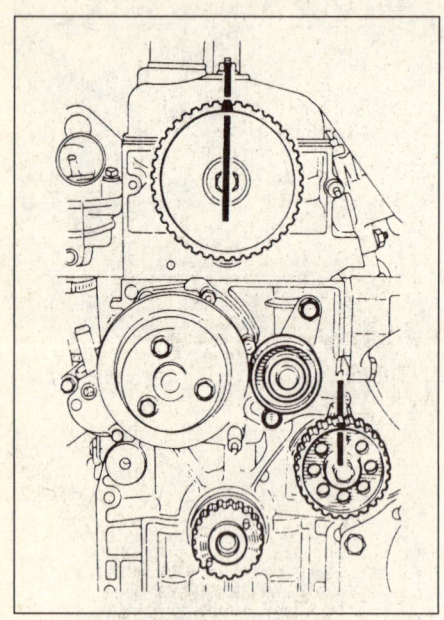

7.15a Timing sprocket positions with No 1 cylinder at TDC on its compression stroke

7.15b The camshaft sprocket timing mark should be aligned with the camshaft cover stud . . .

7.15c . . . and the crankshaft sprocket should have its Woodruff key uppermost

crankshaft and lock the crankshaft in position again with the pin.

17 On early models with a spring-loaded timing belt tensioner mechanism, loosen the tensioner retaining nut and bolt. Pivot the tensioner away from the timing belt and hold it in position by tightening the retaining nut and bolt **(see illustrations)**.

18 On later models with a manually adjusted tensioner, slacken the tensioner pulley backplate retaining nut and the large pulley retaining nut. Pivot the pulley and backplate away from the belt to release the belt tension.

19 If the timing belt is to be re-used, use white paint or similar to mark the direction of rotation on the belt (if markings do not already exist), then slip the belt off the sprockets. **Do not** rotate the camshaft whilst the belt is removed as there is a risk of the valves contacting the pistons.

20 Check the timing belt carefully for any signs of uneven wear, splitting or oil contamination; renew it if there is the slightest doubt about its condition. If signs of oil contamination are found, trace the source of the oil leak and rectify it, then wash down the engine timing belt area and all related components to remove all traces of oil.

Refitting

1995 cc 12-valve models

21 Before refitting, thoroughly clean the timing belt sprockets, and check that the tensioner pulley rotates freely, without any sign of roughness. If there is any sign of roughness in the pulley bearing, then the tensioner assembly must be renewed (see Section 8).

22 Ensure all the sprocket timing marks are still positioned as shown in **illustration 7.8**. Manoeuvre the timing belt into position, ensuring that the arrows on the belt are pointing in the direction of rotation (clockwise when viewed from the front end of the engine).

23 On the outside of the timing belt, there are three timing marks in the form of straight lines **(see illustration)**. These timing marks are painted on the belt to ensure that the valve timing is correctly set.

24 Starting at the camshaft sprocket, align timing mark number 1 with the mark on the sprocket rim. Keeping the belt taut along its front run, so that all slack is on the tensioner pulley side of the belt, fit the belt over the intermediate shaft sprocket and crankshaft sprocket. Do not twist the belt sharply while refitting it, and ensure that the belt teeth are correctly seated on the sprockets.

25 If all is well, timing mark number 2 should align with the mark on the intermediate shaft sprocket, and timing mark number 3 should align with the mark on the crankshaft sprocket, indicating that the valve timing is correctly set. If not, disengage the belt and adjust the sprocket positions as required until all timing belt and sprocket marks are correctly aligned.

26 With the valve timing correctly set, pivot the tensioner backplate towards the belt until the upper retaining nut is located at the mid-point of the backplate slot. Hold the backplate in this position, and tighten its retaining nut to the specified torque.

27 The timing belt tension is then adjusted by rotating the eccentric tensioner pulley using a peg spanner fitted to the two holes in the pulley centre. In the absence of the special Renault peg spanner (Mot. 1135) a home-made peg spanner can be used. Alternatively, obtain two suitable diameter bolts and insert these in the pulley holes; the pulley can then be rotated using a screwdriver or bar inserted between the bolts.

28 Using one of the tools described above, rotate the tensioner pulley towards the belt until the deflection at the mid-point of the front run of the belt (between the camshaft sprocket and intermediate shaft sprocket, **Point P in illustration 7.8**) is as given in the *Specifications*. Renault dealers use the special belt-tensioning tool shown to measure the belt deflection. In the absence of this tool,

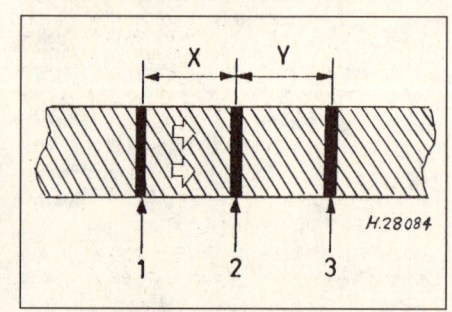

7.17a On models with a spring-loaded tensioner, slacken the retaining nut and bolt . . .

7.17b . . . then pivot the tensioner pulley away from the bolt and secure it in position by retightening the nut and bolt

7.23 Timing belt markings - 1995 cc 12-valve engine

Distance X (between marks 1 and 2) - 44 teeth
Distance Y (between marks 2 and 3) - 23 teeth

Four cylinder petrol engine 2A•13

an approximate setting may be achieved by measuring the belt deflection under firm thumb pressure **(see illustration)**.

29 With the belt correctly tensioned, hold the tensioner pulley stationary, and tighten the large pulley retaining nut to the specified torque setting.

30 Rotate the crankshaft through two complete turns in a clockwise direction and check that the sprocket timing marks are still correctly positioned. If all is well, recheck the belt deflection and adjust as necessary. If the belt deflection is being checked without the special Renault tool, the belt tension must be checked by a Renault dealer at the earliest possible opportunity.

31 With the timing belt correctly tensioned, refit the timing belt covers making sure the rubber mountings and spacers are correctly positioned. Refit the washers and retaining nuts and bolts and tighten them securely.

32 Remove the crankshaft pulley bolt then refit the pulley, making sure it is correctly engaged with the sprocket pins **(see illustration)**. Apply a few drops of locking compound (Renault recommend the use of Loctite Frenbloc) to the threads of the pulley retaining bolt then fit the bolt and tighten the it to the specified torque setting, using the method employed on removal to prevent crankshaft rotation.

33 Refit the auxiliary drivebelt(s) as described in Chapter 1A and refit the undertray (where fitted).

34 Lower the vehicle to the ground and reconnect the battery. Where necessary, refit the radiator as described in Chapter 3.

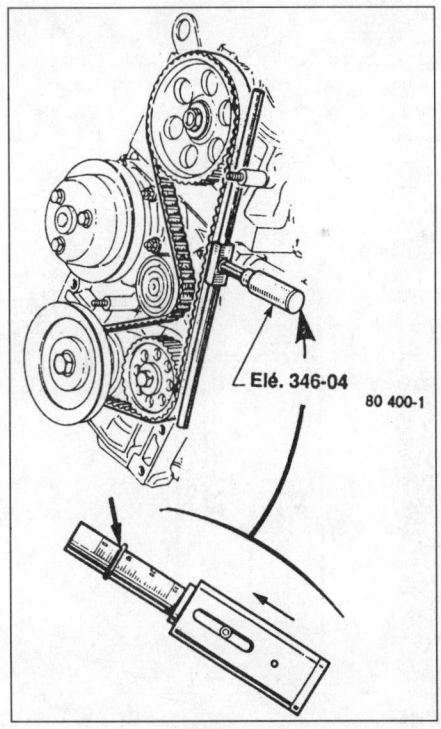

All other models

35 Before refitting, thoroughly clean the timing belt sprockets, and check that the tensioner pulley rotates freely, without any sign of roughness. If there is any sign of roughness in the pulley bearing, then the tensioner assembly must be renewed (see Section 8). Check that the camshaft and intermediate shaft sprocket timing marks are still correctly positioned (see paragraph 15).

36 Starting at the camshaft sprocket, keeping the belt taut along its front run, so that all slack is on the tensioner pulley side of the belt, fit the belt over the intermediate shaft sprocket and crankshaft sprocket. Do not twist the belt sharply while refitting it, and ensure that the belt teeth are correctly seated on the sprockets.

37 On early models with a spring-loaded tensioner, check the sprocket timing marks are correctly positioned then loosen the tensioner nut and bolt. Check that the tensioner is forced against the timing belt by its spring then tighten its retaining bolt and nut to the specified torque setting.

38 On later models, tension the timing belt as described in paragraphs 26 to 29.

39 On all models, remove the locking pin from the crankshaft and rotate the crankshaft through two complete turns in a clockwise direction. Refit the crankshaft locking pin, check that the sprocket timing marks are still correctly positioned then check the belt deflection as described in paragraph 28. If the belt deflection is being checked without the special Renault tool, the belt tension must be checked by a Renault dealer at the earliest possible opportunity.

40 If necessary, adjust the timing belt by slackening the tensioner retaining nut and bolt (early models with spring-loaded tensioner) or by slackening the tensioner pulley nut and rotating the pulley (later models with manually-adjusted tensioner) as described previously. Tighten the tensioner bolt and nut/nut (as applicable) to the specified torque setting.

41 Once the timing belt is correctly tensioned, refit all the remaining components as described in paragraphs 31 to 34.

8 Timing belt sprockets and tensioner - removal and refitting

1 Remove the timing belt as described in Section 7 and proceed as follows under the relevant sub-heading.

Camshaft sprocket - 1995 cc 12-valve models

Removal

2 Slacken the sprocket retaining bolt and remove it. To prevent rotation as the bolt is slackened, a sprocket-holding tool will be required. In the absence of the special Renault tool, an acceptable substitute can be fabricated as follows. Use two lengths of steel strip (one long, the other short), and three nuts and bolts; one nut and bolt forms the pivot of a forked tool, with the remaining two nuts and bolts at the tips of the "forks" to engage with the sprocket spokes as shown **(see illustration)**.

3 With the retaining bolt removed, slide off the sprocket. Examine the oil seal for signs of oil leakage and, if necessary, renew it as described in Section 9.

Refitting

4 Prior to refitting, clean the threads of the camshaft and sprocket retaining bolt.

5 Align the locating key with the camshaft groove and refit the sprocket to the camshaft.

6 Apply fresh locking compound (Renault recommend the use of Loctite Frenbloc) to the retaining bolt threads then fit the bolt and tighten it to the specified torque setting, using the holding tool to prevent rotation.

7 Refit the timing belt as described in Section 7.

Camshaft sprocket - all models except 1995 cc 12-valve models

Removal

8 Remove the sprocket as described in paragraphs 2 and 3. If the sprocket Woodruff key is a loose fit in the camshaft, remove it and store it with the sprocket for safe-keeping **(see illustration)**.

7.28 Using the special Renault tool to check timing belt deflection

7.32 Ensure the crankshaft pulley is correctly located on the sprocket pins on refitting

8.2 Using a home-made tool to retain the camshaft sprocket whilst the retaining bolt is slackened

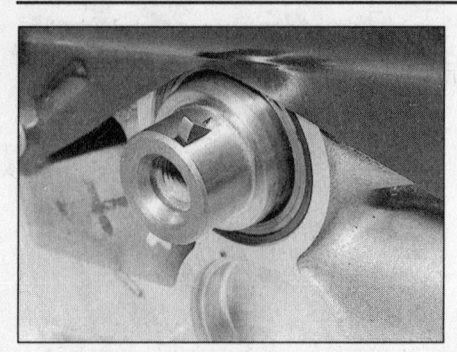

2A•14 Four cylinder petrol engine

8.8 On 8-valve engines, remove the Woodruff key from camshaft end

Refitting

9 Where removed, refit the Woodruff key to the camshaft groove.
10 Ensuring the sprocket is fitted the correct way around, locate the sprocket on the camshaft end, aligning its slot with the Woodruff key (see illustration).
11 Apply fresh locking compound (Renault recommend the use of Loctite Frenbloc) to the retaining bolt threads then fit the bolt and tighten it to the specified torque setting, using the holding tool to prevent rotation.
12 Refit the timing belt as described in Section 7.

Intermediate shaft sprocket

Removal

13 Remove the sprocket as described in paragraphs 2 and 3. If the sprocket Woodruff key is a loose fit in the camshaft, remove it and store it with the sprocket for safe-keeping (see illustrations).

Refitting

14 Where removed, refit the Woodruff key to the intermediate shaft groove.
15 Ensuring the sprocket is fitted the correct way around, locate the sprocket on the camshaft end, aligning its slot with the Woodruff key (see illustration).
16 Apply fresh locking compound (Renault recommend the use of Loctite Frenbloc) to the retaining bolt threads then fit the bolt and

tighten it to the specified torque setting, using the holding tool to prevent rotation.
17 Refit the timing belt as described in Section 7.

Crankshaft sprocket

Removal

18 Unscrew the crankshaft pulley bolt and slide the sprocket off from the end of the crankshaft (see illustration).
19 Remove the Woodruff key from the crankshaft and slide off the large spacer (see illustrations). Examine the crankshaft oil seal for signs of oil leakage and, if necessary, renew it as described in Section 17. **Note:** *On some models equipped with air conditioning, the spacer is replaced by the air conditioning drivebelt pulley, the pulley is driven from the sprocket by two roll pins.*

Refitting

20 Slide the spacer onto the end of the crankshaft and refit the Woodruff key.
21 Slide on the sprocket, making sure the crankshaft pulley pins facing outwards, then refit the timing belt as described in Section 7.

Timing belt tensioner - early models with spring-loaded tensioner

Removal

22 Slacken and remove the tensioner retaining nut and bolt along with their washers whilst firmly holding the tensioner pulley. With the nut and bolt removed slowly pivot the pulley assembly away from the spring, until all tension is relieved, then remove the tensioner pulley assembly and withdraw the plunger and spring from the coolant pump housing (see illustrations).

Refitting

23 Fit the spring to the inside of the plunger and refit the plunger assembly to the coolant pump housing.
24 Compress the plunger and manoeuvre the tensioner pulley assembly into position. Refit the tensioner retaining nut and bolt and check that the tensioner assembly pivots smoothly and is forced towards the timing belt by the tensioner spring.

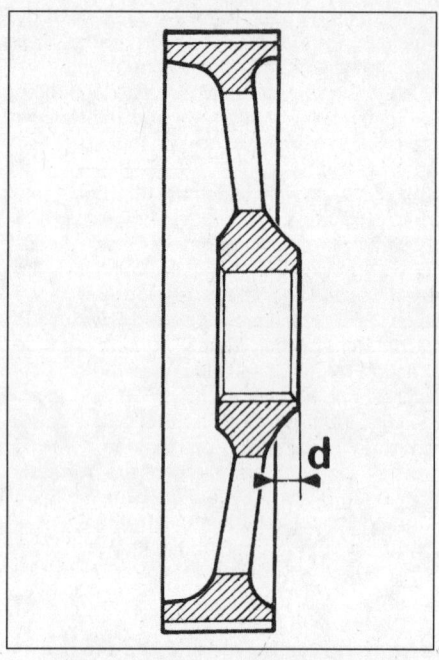

8.10 On 8-valve engines ensure the camshaft sprocket is fitted with its offset (d) towards the cylinder head

25 Pivot the pulley fully away from the timing belt and securely tighten its retaining nut and bolt to keep it there. Refit the timing belt as described in Section 7.

Timing belt tensioner - later models with manually-adjusted tensioner

Removal

26 Slacken and remove the tensioner backplate retaining bolt and nut and remove the tensioner assembly.

8.13a Slacken and remove the retaining bolt and washer . . .

8.13b . . . and remove the intermediate shaft sprocket

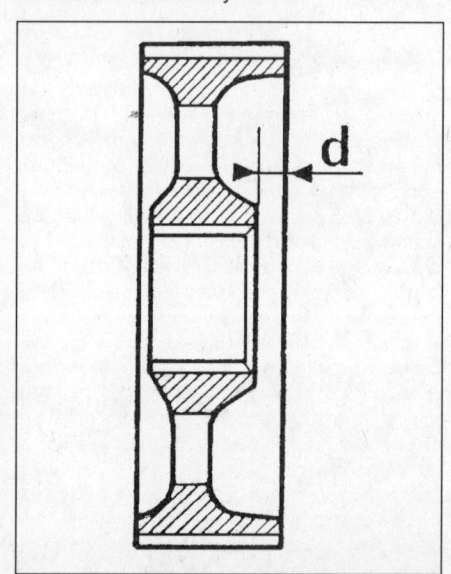

8.15 On refitting, ensure the intermediate shaft sprocket is fitted with its offset (d) towards the cylinder block

Four cylinder petrol engine 2A•15

8.18 Slide off the crankshaft sprocket...

8.19a ...then remove the Woodruff key...

8.19b ...and spacer from the crankshaft end

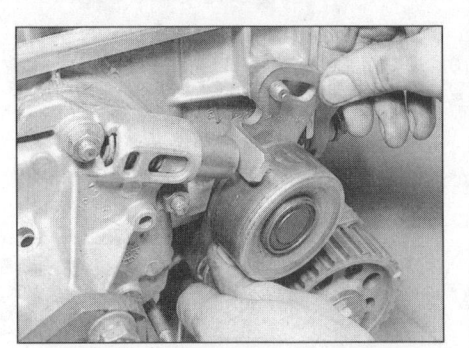

8.22a On models with a spring-loaded tensioner, remove the tensioner pulley assembly...

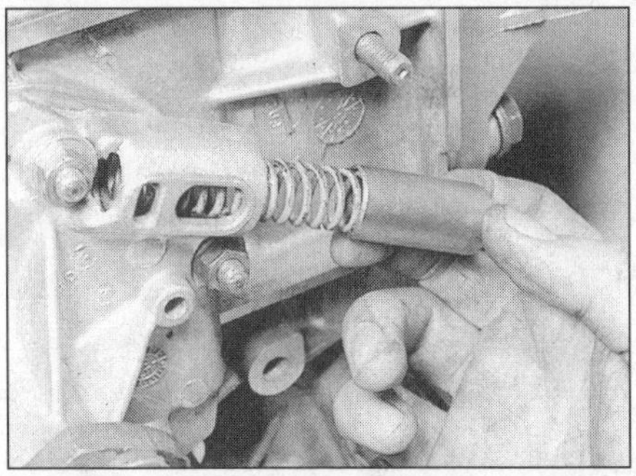

8.22b ...and recover the plunger and spring from the coolant pump housing

Refitting

27 Fit the new tensioner assembly and refit the timing belt as described in Section 7.

9 Camshaft and intermediate shaft oil seals - renewal

Camshaft oil seal

1 Remove the camshaft sprocket as described in Section 8.
2 Make a note of the correct fitted depth of the seal then punch or drill two small holes opposite each other in the oil seal. Screw a self-tapping screw into each and pull on the screws with pliers to extract the seal.
3 Clean the seal housing and polish off any burrs or raised edges which may have caused the seal to fail in the first place.
4 Lubricate the lips of the new seal with clean engine oil and ease it into position on the end of the shaft. Press the seal into its housing until it is positioned at the same depth as the original was prior to removal. If necessary, a suitable tubular drift, such as a socket, which bears only on the hard outer edge of the seal can be used to tap the seal into position. Take great care not to damage the seal lips during fitting and ensure that the seal lips face inwards. Note that if the surface of the shaft was noted to be badly scored, press the new seal slightly further into its housing so that its lip is running on an unmarked area of the shaft.
5 Refit the camshaft sprocket as described in Section 8.

Intermediate shaft oil seal

6 Remove the intermediate shaft sprocket as described in Section 8.
7 Undo the retaining bolts and slide the intermediate shaft cover off the end of the shaft. Recover the gasket and discard it, a new one should be used on refitting.
8 Using a suitable flat-bladed screwdriver, lever the seal out of the cover whilst taking great care not to mark the cover.
9 Position the new seal in the cover and tap it into position using a suitable tubular drift, such as a socket, which bears only on the hard outer edge of the seal. Ensure that the seal lip is facing inwards.
10 Remove all traces of gasket from the cover and cylinder block/crankcase mating surfaces and position a new gasket on the cylinder block.
11 Slide the cover into position carefully easing the seal over the end of the intermediate shaft.
12 To prevent the possibility of oil leakage smear the shafts of the cover retaining bolts with a suitable sealant (Renault recommend the use of Loctite Frenetanch).
13 Ensure the gasket is correctly positioned then refit the cover retaining bolts and tighten them securely.
14 Refit the intermediate shaft sprocket as described in Section 8.

10 Camshaft and rocker arms - removal, inspection and refitting

Removal

1995 cc 12-valve models

1 Remove the camshaft cover and timing belt as described in Sections 5 and 7.
2 Slacken all the rocker arm adjusting screw locknuts, and unscrew the adjusting screws until valve spring pressure is relieved from the rocker arms.
3 Fully unscrew the rocker arm shaft retaining bolts, and lift the rocker arm assemblies away from the cylinder head, noting the correct fitted positions of the shaft locating pins. If the pins are a loose fit, remove them and store them safely. **Do not** under any circumstances remove the bolts from the rocker shaft.
4 To remove the camshaft, remove the camshaft sprocket as described in Section 8.
5 The camshaft bearing caps should be marked 1 to 5, number 1 being at the timing belt end of the engine **(see illustration)**. If identification marks do not already exist, make some using white paint or a marker pen. Also mark each cap to indicate which way round it is fitted. This will avoid the possibility of refitting the caps the wrong way round.
6 Lift off the bearing caps. Remove the camshaft from the cylinder head and discard its oil seal; a new seal must be used on refitting.

10.5 On 1995 cc 12-valve engines identification numbers are cast into the top of each camshaft bearing cap

2A•16 Four cylinder petrol engine

All models except 1995 cc 12-valve models

Note: *The rocker arm assembly is secured to the top of the cylinder head by the cylinder head bolts. Although in theory it is possible to undo the head bolts and remove the rocker arm assembly without removing the head, in practice, this is not recommended. Once the bolts have been removed, the head gasket will be disturbed, and the gasket will almost certainly leak or blow after refitting. For this reason, removal of the rocker arm assembly and camshaft should not be done without removing the cylinder head and renewing the head gasket.*

7 Remove the cylinder head as described in Section 11.
8 Remove the camshaft sprocket as described in Section 8.
9 Undo the retaining bolts and remove the timing belt rear cover from the cylinder head **(see illustration)**.
10 Carefully slide the camshaft out of the front of the cylinder head along with its oil seal. Take care during its removal not to snag any of the lobe corners on the bearings as they are passed through the cylinder head **(see illustration)**.

Inspection

1995 cc 12-valve models

11 Examine the camshaft bearing surfaces and cam lobes for signs of wear ridges and scoring. Renew the camshaft if any of these conditions are apparent. Examine the condition of the bearing surfaces both on the camshaft journals and in the cylinder head and bearing caps. If the head bearing surfaces are worn excessively, it may be necessary to renew the cylinder head. Consult a Renault dealer or engine reconditioning specialist.

10.9 On 8-valve engines unbolt the timing belt rear cover . . .

12 Each rocker arm assembly can be dismantled by removing the retaining bolts one at a time and sliding the various components off the end of the shafts. Keep all components in their correct fitted order, making a note as each is removed, to ensure correct reassembly.
13 Examine the rocker arm surfaces which contact the camshaft lobes for wear ridges and scoring. Renew any rocker arms on which these conditions are apparent. Also ensure that the rocker arm and shaft oilways are clear and free of blockages.
14 Examine the rocker arm and shaft bearing surfaces for wear ridges and scoring. If there are obvious signs of wear, the rocker arm and/or shaft must be renewed.

All models except 1995 cc 12-valve models

15 Inspect the camshaft and head bearing surfaces as described in paragraph 11.
16 To dismantle the rocker arm assembly, unscrew the bolt from the rear end of the rocker shaft whilst retaining the rocker pedestal to prevent it be sprung off the end of the shaft. Recover the oil filter from inside the shaft then slide off the various components, keeping all components in their correct fitted

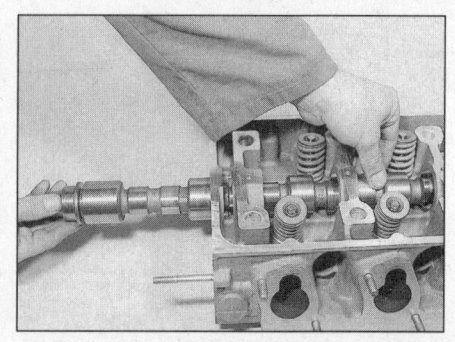

10.10 . . . then slide the camshaft out from the cylinder head

order **(see illustration)**. Make a note of each components correct fitted position as it is removed, to ensure it is positioned correctly on reassembly. To separate the front pedestal and shaft, tap out the roll pin. If necessary the camshaft thrust plate can also be unbolted from the front of the pedestal.
17 Examine the rocker arm surfaces which contact the camshaft lobes for wear ridges and scoring. Renew any rocker arms on which these conditions are apparent. Also ensure that the rocker arm and shaft oilways are clear and free of blockages.
18 Examine the rocker arm and shaft bearing surfaces for wear ridges and scoring. If there are obvious signs of wear, the rocker arm and/or shaft must be renewed.

Refitting

1995 cc 12-valve models

19 Where necessary, reassemble the rocker arms by inserting the first bolts and locating plates. Apply a smear of clean engine oil to the shaft then slide on all removed components, ensuring each is correctly fitted in its original position, securing them in position with the end bolts and locating plates.

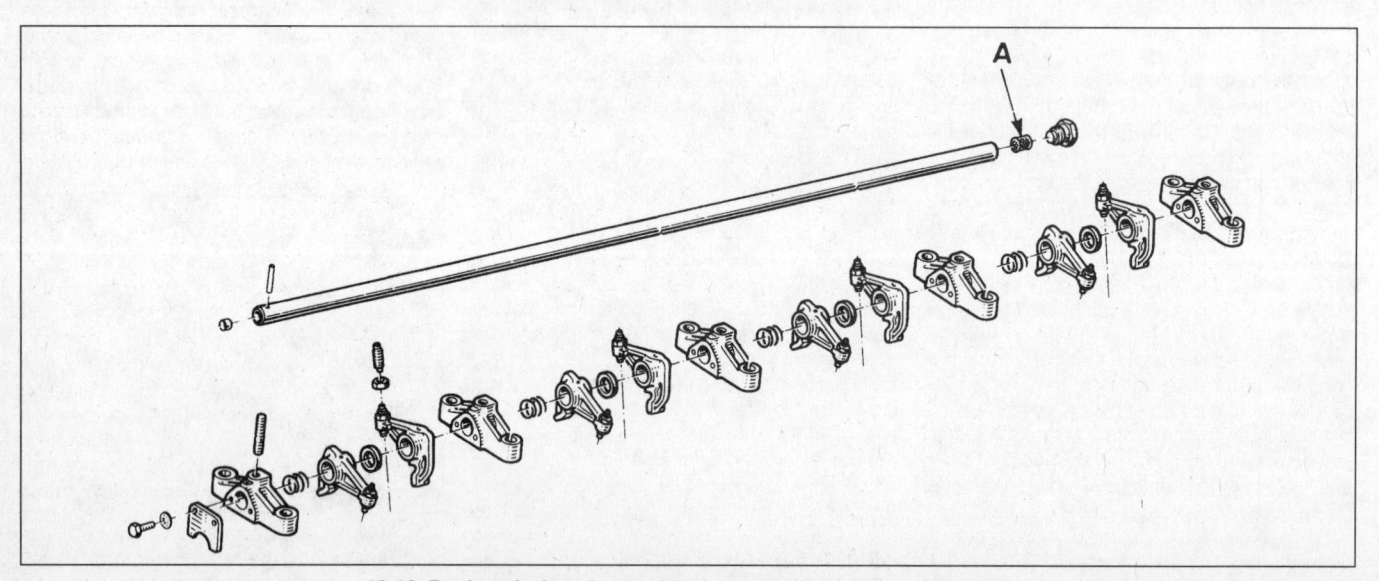

10.16 Rocker shaft and associated components. Note oil filter (A)

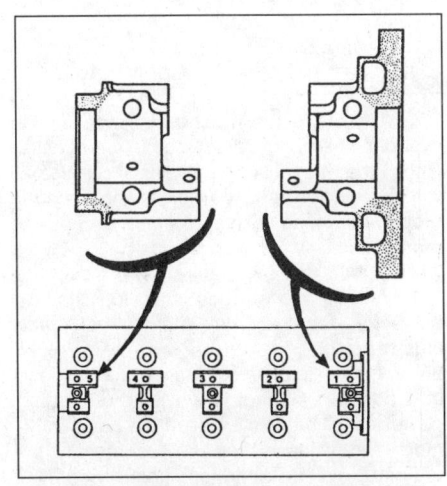

Four cylinder petrol engine 2A•17

10.21 On 1995 cc 12-valve engines, apply sealant to the shaded areas of numbers 1 and 5 camshaft bearing caps

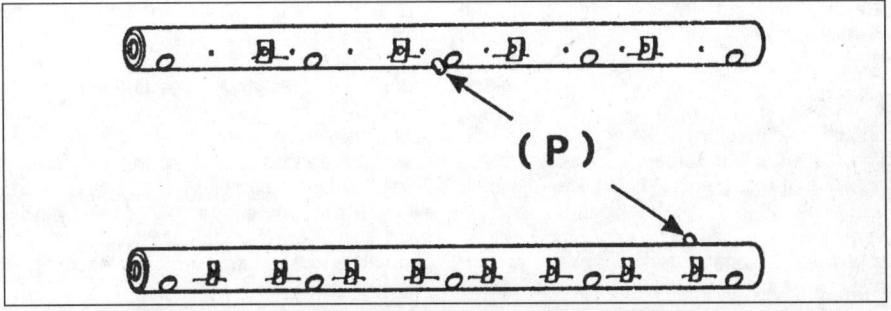

10.23 On 1995 cc 12-valve engines ensure the locating pins (P) are in position in the exhaust (upper) and inlet (lower) rocker shaft prior to refitting

20 Lubricate the bearings in the cylinder head with clean engine oil, and refit the camshaft. Temporarily fit the camshaft sprocket and position it as shown in **illustration 7.8**.
21 Ensure the bearing cap and cylinder head mating surfaces are clean and dry, then apply a smear of sealant (Renault recommend the use of Loctite 518) to the areas of bearing caps numbers 1 and 5 as shown **(see illustration)**.
22 Refit the bearing caps to the cylinder head, using the identification marks to ensure that each is fitted in its original location.
23 Ensure that the rocker shaft locating pins are in position in each shaft. The exhaust shaft pin is located in the centre of the shaft at the bottom, and the inlet shaft pin is fitted to the top of the shaft on the timing belt end **(see illustration)**.
24 Refit both rocker shafts to the camshaft bearing caps, and tighten the retaining bolts evenly and progressively to the specified torque. Before tightening the bolts, ensure that the rocker arms, springs and collars are correctly positioned against the sides of each bearing cap, and not trapped between the

shaft and caps **(see illustration)**. Note that the rocker arms must be fitted and the retaining bolts tightened as quickly as possible after fitting the bearing caps, before the sealant hardens.
25 Fit a new camshaft oil seal then refit the camshaft sprocket as described in Section 8 and 9.
26 Adjust the valve clearances as described in Section 6 then refit the camshaft cover as described in Section 5.

All models except 1995 cc 12-valve models

27 If the rocker arm assembly was dismantled, refit the rocker shaft to the front pedestal, aligning its locating hole with the pedestal threaded hole, and refit the grub screw tightening it securely. Apply a smear of clean engine oil to the shaft then slide on all removed components, ensuring each is correctly fitted in its original position. Note that the pedestals are not symmetrical and must be fitted the correct way around.
28 Once all components are in position on the shaft, compress the rear pedestal. Install the new oil filter in the shaft bore, then refit the bolt and tighten it to the specified torque setting.
29 Insert the camshaft carefully into the cylinder head, guiding the cam sections through the bearing apertures so as not to score the bearing surfaces.
30 With the camshaft in position, fit a new oil seal as described in Section 9.
31 Refit the timing belt rear cover to the head and securely tighten its retaining bolts.
32 Refit the camshaft sprocket then fit the cylinder head as described in Sections 8 and 11.

11 Cylinder head - removal and refitting

Note: *New cylinder head bolts will be required on refitting.*

Removal

1995 cc 12-valve models
1 Disconnect the battery negative terminal.

2 Drain the cooling system as described in Chapter 1A.
3 Remove the inlet and exhaust manifolds as described in Chapter 4A or 4B.
4 Remove the distributor as described in Chapter 5B.
5 Remove the camshaft cover as described in Section 5.
6 Remove the timing belt as described in Section 7.
7 Slacken the retaining clips and disconnect the relevant coolant hose(s) from the cylinder head. Where necessary, also unbolt the dipstick tube from the head.
8 Disconnect the connectors from the cylinder head switches and sensors. Where necessary, detach the diagnostic socket and tie it to one side.
9 Working in the **reverse** of the sequence shown in **illustration 11.28**, progressively slacken the cylinder head bolts by half a turn at a time until all nuts and bolts can be unscrewed by hand.
10 With the exception of the front right-hand bolt, remove each bolt and washer in turn.
11 With the front right-hand bolt (bolt number 10 in the tightening sequence) still in position, the joint between the cylinder head and gasket and the cylinder block/crankcase must now be broken without disturbing the wet liners; although these liners are better located and sealed than some wet liner engines, there is still a risk of coolant and foreign matter leaking into the sump if the cylinder head is lifted carelessly. If care is not taken and the liners are moved, there is also a possibility of the bottom seals being disturbed, causing leakage after refitting the head.
12 To break the joint, strike the side of the head firmly using a soft-faced mallet or hammer and block of wood.
13 Once the gasket seal is broken, remove the right-hand front bolt and washer from the cylinder head. On all models, except 1995 cc 12-valve models, the rocker shaft assembly can then be removed from the cylinder head.
14 The cylinder head can then be lifted out of the engine compartment; use assistance if possible as it is a heavy assembly. Remove the gasket from the top of the block, noting the position of the locating dowel. If the dowel is a loose fit, remove it and store it with the head for safe-keeping.

10.24 Prior to tightening the retaining bolts make sure the rocker arms and collars are correctly positioned against the sides of the bearing caps

2A•18 Four cylinder petrol engine

15 Note **do not** attempt to rotate the crankshaft with the cylinder head removed, otherwise the wet liners may be displaced. Operations that require the rotation of the crankshaft (eg cleaning the piston crowns), should only be carried out once the cylinder liners are firmly clamped in position. In the absence of the special Renault liner clamping tool (Mot. 521-01), the liners can be clamped in position using large flat washers positioned underneath suitable-length bolts, or the original head bolts with suitable spacers fitted to their shanks **(see illustration)**.

Preparation for refitting

16 The mating faces of the cylinder head and cylinder block/crankcase must be perfectly clean before refitting the head. Use a hard plastic or wood scraper to remove all traces of gasket and carbon; also clean the piston crowns. Take particular care as the soft aluminium alloy is damaged easily. Also, make sure that the carbon is not allowed to enter the oil and water passages - this is particularly important for the lubrication system, as carbon could block the oil supply to any of the engine's components. Using adhesive tape and paper, seal the water, oil and bolt holes in the cylinder block/crankcase. To prevent carbon entering the gap between the pistons and bores, smear a little grease in the gap. After cleaning each piston, use a small brush to remove all traces of grease and carbon from the gap, then wipe away the remainder with a clean rag. Clean all the pistons in the same way.

17 Check the mating surfaces of the cylinder block/crankcase and the cylinder head for nicks, deep scratches and other damage. If slight, they may be removed carefully with a file, but if excessive, machining may be the only alternative to renewal.

18 If warpage is suspected of the cylinder head gasket surface, use a straight-edge to check it for distortion. Refer to Section 12 if necessary.

19 The cylinder head bolts must be discarded and renewed, regardless of their apparent condition.

20 Prior to refitting the cylinder head, check the cylinder liner protrusion as described in Section 23. Although the measuring procedure specifies that the liners should be fitted without seals, the seals have no effect on liner protrusion. If the liner protrusion exceeds the specified limits, seek the advice of a Renault dealer prior to refitting the head.

21 Ensure that the cylinder head bolt holes in the crankcase are clean and free of oil. Syringe or soak up any oil left in the bolt holes, and in the oil feed hole on the rear left-hand corner of the block (all models except 1995 cc 12-valve). This is most important in order that the correct bolt tightening torque can be applied.

22 On all models except 1995 cc 12-valve models, slacken the rocker arm adjusting screw locknuts and fully unscrew the adjusting screws. This will prevent the valve spring pressure affecting the cylinder head bolt tightness.

Refitting

23 Wipe clean the mating surfaces of the cylinder head and cylinder block/crankcase. Check that the locating dowel is in position at the front right-hand corner of the cylinder block/crankcase surface. Where necessary, remove the cylinder liner clamps.

24 Position the new gasket on the cylinder block/crankcase surface ensuring that its correctly located on the dowel **(see illustration)**.

25 With the aid of an assistant lower the head into position, engaging it with the locating dowel. On all models except 1995 cc 12-valve models, refit the rocker arm assembly to the top of the cylinder head **(see illustrations)**.

26 Lightly oil under the head and on the threads of each new cylinder head bolt.

27 Carefully enter each bolt into its hole, screwing it in by hand only until finger-tight.

28 Working progressively and in the sequence shown, tighten the cylinder head bolts to their Stage 1 torque setting, using a torque wrench and suitable socket **(see illustration)**.

29 Once all the bolts have been tightened to their Stage 1 setting, working again in the given sequence, angle-tighten the bolts through the specified Stage 2 angle, using a socket and extension bar. It is recommended that an angle-measuring gauge is used during this stage of the tightening, to ensure accuracy.

30 Wait for three minutes and then work around in the specified sequence and tighten the bolts through the specified Stage 3 angle setting. Note that no further re-tightening of the bolts will be required.

31 Reconnect the wiring connectors to the cylinder head switches and sensors.

32 Reconnect the coolant hoses and securely tighten the retaining clips.

33 Refit the timing belt as described in Section 7.

34 On all models except 1995 cc 12-valve models, adjust the valve clearances as described in Section 6.

35 Refit the camshaft cover as described in Section 5.

36 Refit the distributor as described in Chapter 5B.

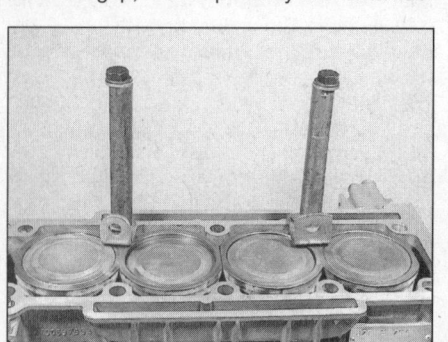

11.15 Cylinder liner clamps can be improvised using the cylinder head bolts and suitable spacers

11.24 Fit the new head gasket . . .

11.25a . . . and lower the head onto the block

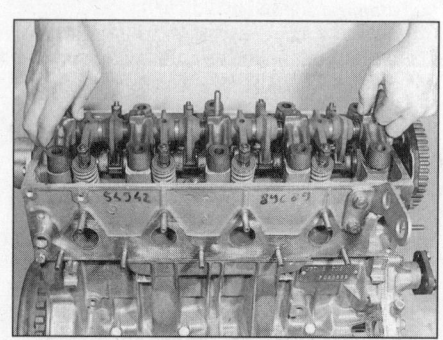

11.25b On all 8-valve models refit the rocker shaft assembly then fit the cylinder head bolts

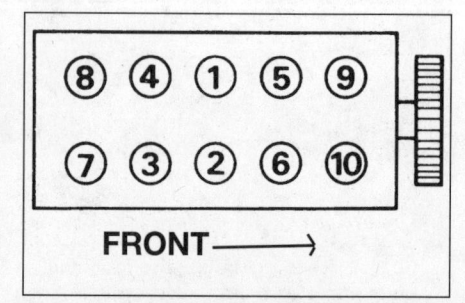

11.28 Cylinder head bolt tightening sequence

Four cylinder petrol engine 2A•19

12.3a Compress the valve spring and withdraw the collets . . .

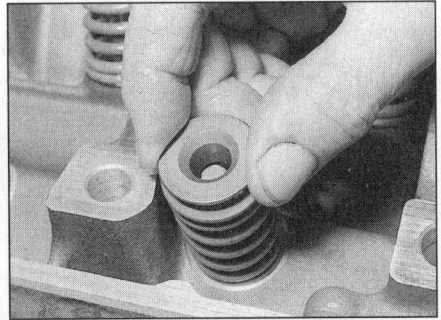

12.3b . . . then lift off the spring retainer . . .

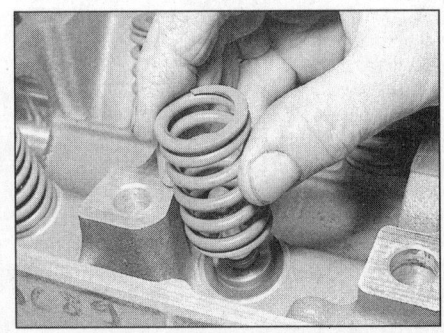

12.3c . . . and the valve spring

12.4 Slide the valve out of position . . .

12.5 . . . then remove the spring seat and valve guide oil seal

37 Refit the inlet and exhaust manifolds as described in Chapter 4A or 4B.
38 Refill the cooling system as described in Chapter 1A.
39 On completion, check the oil level (see *Weekly checks*).

12 Cylinder head - overhaul

1 Remove the cylinder head as described in Section 11 and clean its external surfaces.
2 On 1995 cc 12-valve models remove the camshaft and rocker arm assemblies as described in Section 10.
3 Remove the valves by compressing the valve spring with a suitable valve spring compressor and lifting out the collets. Release the compressor and remove the valve spring retainer and spring **(see illustrations)**.
4 Remove the valves, keeping them identified for location to ensure correct refitting **(see illustration)**.
5 Lever the valve stem oil seals from the valve guides and remove the spring seat **(see illustration)**.
6 With the valves removed, clean the carbon from them and from the combustion chambers and ports. The piston crowns can also be cleaned at this stage but take care not to allow carbon to drop between the pistons and bores. To prevent this, clean two pistons at a time with them at the top of their bores and press a little grease between the pistons and bores. Seal off the remaining cylinders, and oil and water channels with paper. After cleaning, move the pistons down the bore and wipe out the grease which will contain the particles of carbon.
7 Examine the heads of the valves and the valve seats for pitting and burning. If the pitting on valve and seat is slight it can be removed by grinding the valves and seats together with coarse, and then fine, valve grinding paste. If the pitting is deep, the valves will have to be reground on a valve grinding machine and the seats will have to be recut with a valve seat cutter. Both these operations are a job for your Renault dealer or motor engineering specialist.
8 Check the valve guides for wear by inserting the valve in the guide and attempting to move the valve from side to side. If there is excess movement the valve guides should be renewed by a Renault dealer or motor engineering specialist.
9 When grinding slightly-pitted valves and valve seats with carborundum paste, continue as follows. Apply a little coarse grinding paste to the valve head, and using a suction-type valve grinding tool, grind the valve into its seat with a rotary movement, lifting the valve and turning it from time to time. A light spring under the valve head will assist in this operation. When a dull matt, even surface finish appears on both the valve and the valve seat, clean off the coarse paste. Repeat the grinding operation with a fine grinding paste until a continuous ring of light grey matt finish appears on both valve and valve seat. Carefully clean off all traces of grinding paste.
10 Check the valve springs for damage and if possible compare their length with that of a new spring. Renew them if necessary.
11 Using a straight edge and feeler blade check the joint face of the cylinder head for distortion. If greater than the maximum amount given in the *Specifications* it may be possible to have the head machined flat. Consult a Renault dealer if necessary.
12 Begin reassembly by refitting the spring seats.
13 Press the valve stem oil seals onto the guides.
14 To fit the valves, lubricate the valve stem with engine oil and insert it in the valve guide.
15 Refit the spring and spring retainer, then use the valve spring compressor to compress the spring until the collets can be fitted in position in the slots in the valve stem. Release the compressor slowly, and check that the collets are seated correctly.
16 After fitting all the valves, tap the tops of the springs lightly to ensure correct seating of the collets.

13 Sump - removal and refitting

Removal

Note: *On 1995 cc 12-valve models and later 2165 cc models a modified sump arrangement is fitted to the engine. In place of the standard one-piece sump, a two piece arrangement is fitted; a small pressed steel sump being bolted to the base of a large alloy crankcase stiffener (see Section 15).*

1 Drain the engine oil (see Chapter 1A).
2 Unbolt and remove the lower cover plate from the base of the transmission. On models with a one-piece aluminium sump undo the bolts securing the transmission to the sump.
3 Where necessary, remove the engine oil sensor (see Chapter 5A).
4 To improve access, unbolt and remove the wiring harness duct (where fitted) which runs under and across the front of the sump pan.
5 On models with a one-piece sump, raise the front of the car, support it securely so that the front roadwheels hang free. This will have the effect of moving the anti-roll bar towards

2A

2A•20 Four cylinder petrol engine

the front of the car to make extraction of the sump rear screws easier.
6 Unscrew the sump retaining screws and remove the sump pan along with its gasket. If it is stuck tight, tap it gently with a wooden or plastic-faced hammer. Discard the gasket, a new one should be used on refitting. **Note:** *On some models with a one-piece sump, it may be necessary to unbolt the engine mountings and raise the engine slightly to gain the necessary clearance required to manoeuvre the sump out of position.*

Refitting

7 Refitting is the reverse of removal noting the following.
  a) Ensuring the mating surfaces are clean and dry and using a new gasket.
  b) On models with a one-piece aluminium sump ensure that the sump is correctly mated to the transmission before tighten its retaining bolts; failure to do so could lead to the sump cracking when the transmission to engine bolts are tightened.
  c) On completion refill the engine with oil as described in Chapter 1.

14 Oil pump - removal, inspection and refitting

Removal

1 Remove the sump as described in Section 13 then proceed as described under the relevant sub-heading.

1995 cc 12-valve models and later 2165 cc models with a two-piece sump arrangement

2 Remove the four bolts securing the oil pump strainer cover to the pump body. Lower the cover and recover the oil pump gears.
3 Undo the two bolts securing the pump body to the base of the crankcase, and remove the body from the crankcase along with its driveshaft.

All other models with a one-piece sump

4 Undo the retaining bolts and remove the oil

14.4 Undo the mounting bolts and remove the oil pump from the engine unit

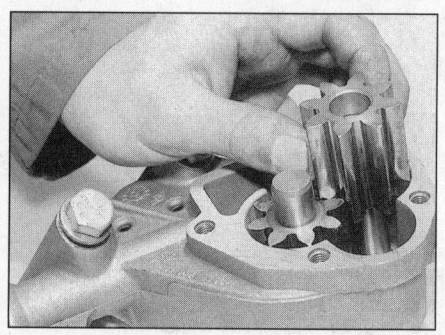

pump from the base of the crankcase along with its driveshaft **(see illustration)**.

Inspection

5 If not already having done so, undo the retaining bolts then lift off the cover and extract the gears from the pump **(see illustration)**.
6 Inspect for any signs of damage or excessive wear. Use a feeler blade and check the clearance between the rotor (gear) tips and the inner housing **(see illustration)**.
7 Also, check the gear endfloat using a straight-edge rule laid across the body of the pump and feeler blade inserted between the rule and gears **(see illustration)**.
8 Do not overlook the relief valve assembly. To extract it, remove the split pin and withdraw the cup, spring, guide and piston **(see illustration)**. Again, look for signs of excessive wear or damage.
9 Compare the clearances with the allowable tolerances given in the *Specifications* at the start of this Chapter. If any component is worn, renew the pump as a complete unit.
10 Check the pump driveshaft for signs of wear or distortion and renew if necessary. Make sure that the circlip is securely located in the driveshaft groove.

14.5 Removing the oil pump gears

Refitting

1995 cc 12-valve models and later 2165 cc models with a two-piece sump arrangement

11 Fit the driveshaft to the pump body making sure its circlip is at the lower end.
12 Manoeuvre the pump body into position, aligning the driveshaft with its drive gear, and tighten the pump mounting bolts to the specified torque.
13 Lubricate the pump gears with clean engine oil and fit them to the body. Refit the pump cover and tighten its retaining bolts to the specified torque.
14 Refit the sump as described in Section 13.

All other models with a one-piece sump

15 Where necessary, lubricate the gears with clean engine oil and fit them to the pump body. Refit the pump cover and tighten its retaining bolts to the specified torque.
16 Refit the pump and driveshaft as described in paragraphs 11 and 12 then refit the sump as described in Section 13.

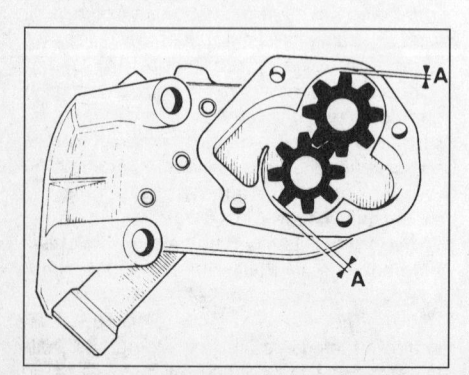

14.6 Oil pump gear to body clearance

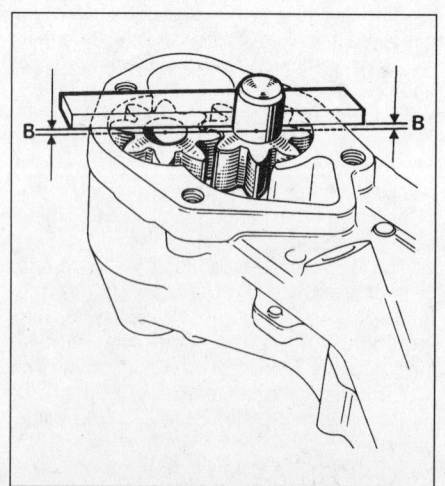

14.7 Oil pump gear endfloat

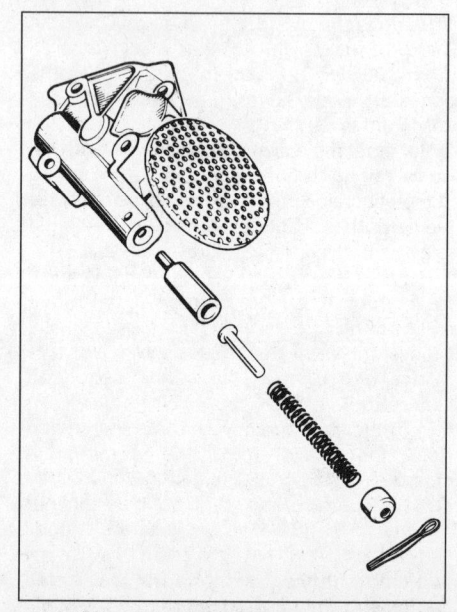

14.8 Oil pressure relief valve components

Four cylinder petrol engine 2A•21

15 Crankcase stiffener casting (1995cc 12-valve and later 2165cc) - removal and refitting

Removal

1 Remove the oil pump as described in Section 14.
2 Withdraw the dipstick from its tube.
3 Disconnect the wiring from the oil level sensor (where fitted) and remove the sensor from the side of the crankcase.
4 Where necessary, slacken and remove the bolt(s) securing the inlet manifold support stay(s) to the side of the stiffener casting.
5 Undo all the bolts securing the crankcase stiffener casting to the base of the crankcase (see illustration). Note that due to a lack of clearance, it may not be possible to withdraw all of the bolts; some of them will have to be removed with the stiffener.
6 With all the bolts slackened, release the stiffener casting from the base of the crankcase, and manoeuvre it out from underneath the vehicle. Withdraw the remaining retaining bolts from the stiffener, noting their locations. Discard the gasket; a new one must be used on refitting.

Refitting

7 Remove all traces of gasket from the stiffener casting and crankcase mating surfaces. Ensure the mating surfaces are clean and dry, and fit a new gasket to the top of the stiffener, using a few drops of sealing compound to hold the gasket in position.
8 Insert those bolts which could not be withdrawn due to a lack of clearance into position in the stiffener casting. This is important, since it will not be possible to insert them once the casting is manoeuvred into position.
9 Manoeuvre the casting assembly into position and, referring to **illustration 15.5**, insert the remaining bolts into their correct locations.
10 Tighten all the bolts lightly at first, then work around the casting and tighten each bolt to its correct torque setting.
11 Refit the oil pump and sump as described earlier in this Chapter.

16 Flywheel/driveplate - removal, inspection and refitting

Removal

1 Remove the transmission as described in Chapter 7A or 7B (as applicable). On manual transmission models remove the clutch assembly as described in Chapter 6.
2 Prevent the flywheel/driveplate from turning by locking the ring gear teeth with a similar arrangement to that shown in **illustration 7.7**.

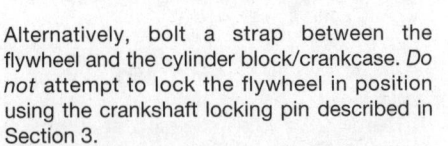

15.5 Crankcase stiffener casting bolt locations - 1995 cc 12-valve models and later 2165 cc models

White circle - 6 mm sump bolts
Black circle - 7 mm bolts
Triangles - 10 mm bolts

Alternatively, bolt a strap between the flywheel and the cylinder block/crankcase. *Do not* attempt to lock the flywheel in position using the crankshaft locking pin described in Section 3.
3 Slacken and remove the flywheel/driveplate retaining bolts, and discard them; they must be renewed whenever they are disturbed.
4 Remove the flywheel/driveplate. Do not drop it, as it is very heavy.

Inspection

5 If the ring gear is badly worn or has missing teeth, renew the flywheel/driveplate.
6 On manual transmission models, if the flywheel's clutch mating surface is deeply scored, cracked or otherwise damaged, the flywheel must be renewed. However, it may be possible to have it surface-ground; seek the advice of a Renault dealer or engine reconditioning specialist.

Refitting

7 Clean the mating surfaces of the flywheel/driveplate and crankshaft. Remove any remaining locking compound from the threads of the crankshaft holes, using the correct-size tap, if available.

> **HAYNES HINT**: *If a suitable tap is not available, cut two slots into the threads of one of the old flywheel bolts and use the bolt to remove the locking compound from the threads.*

8 If the new flywheel retaining bolts are not supplied with their threads already pre-coated, apply a suitable thread-locking compound (Renault recommend the use of Loctite Frenatanch) to the threads of each bolt (see illustration).
9 Offer up the flywheel/driveplate to the flywheel and align it with the retaining bolt holes.
10 Fit the new retaining bolts, then lock the flywheel/driveplate using the method employed on dismantling, and tighten the retaining bolts to the specified torque.
11 Refit the clutch (where necessary) as described in Chapter 6. Remove the locking tool, and refit the transmission as described in Chapter 7A or 7B.

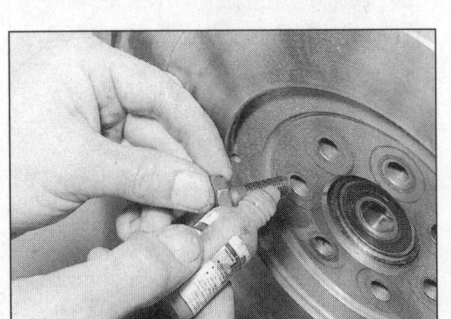

16.8 Apply thread locking compound to the flywheel/driveplate bolts before installing them

2A•22 Four cylinder petrol engine

17 Crankshaft oil seals - renewal

Front (timing belt end) oil seal

1 Remove the crankshaft sprocket and spacer as described in Section 8.
2 Make a note of the correct fitted depth of the seal in its housing then punch or drill two small holes opposite each other in the seal. Screw a self-tapping screw into each, and pull on the screws with pliers to extract the seal. Alternatively, the seal can be levered out of position using a suitable flat-bladed screwdriver, taking great care not to damage the crankshaft shoulder or seal housing.
3 Clean the seal housing, and polish off any burrs or raised edges, which may have caused the seal to fail in the first place.
4 Lubricate the lips of the new seal with clean engine oil, and carefully locate the seal on the end of crankshaft (see illustration). Note that its sealing lip must face inwards. Take care not to damage the seal lips during fitting.
5 Using a suitable tubular drift (such as a socket) which bears only on the hard outer edge of the seal, tap the seal into position, to the same depth in the housing as the original was prior to removal. The inner face of the seal must end up flush with the inner wall of the crankcase.
6 Wash off any traces of oil, then refit the crankshaft sprocket as described in Section 8.

Rear (flywheel/driveplate) oil seal

7 Remove the flywheel/driveplate as described in Section 16.
8 Make a note of the correct fitted depth of the seal in its housing. Punch or drill two small holes opposite each other in the seal. Screw a self-tapping screw into each, and pull on the screws with pliers to extract the seal.
9 Clean the seal housing, and polish off any burrs or raised edges, which may have caused the seal to fail in the first place.
10 Lubricate the lips of the new seal with clean engine oil, and carefully locate the seal on the end of the crankshaft (see illustration).
11 Using a suitable tubular drift, which bears only on the hard outer edge of the seal, drive the seal into position, to the same depth in the housing as the original was prior to removal.
12 Wash off any traces of oil, then refit the flywheel/driveplate as described in Section 16.

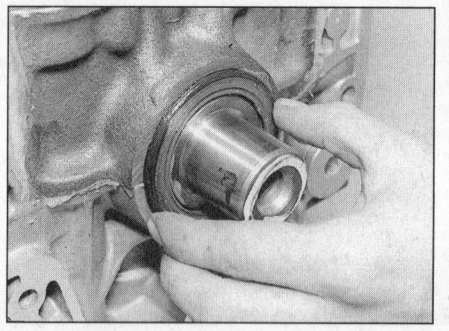

17.4 Fitting the crankshaft front oil seal

17.10 Fitting the crankshaft rear oil seal

18 Engine/transmission mountings - renewal

1 The engine and transmission flexible mountings may be renewed with the units in the car.
2 Support the engine or transmission securely on a jack using a block of wood as an insulator.
3 Renew only one mounting at a time by unscrewing the through-bolt nuts and unbolting the mounting bracket from the crankcase, transmission casing or body member (see illustration).

19 Intermediate shaft - removal, inspection and refitting

Removal

1 In order to gain the clearance necessary to withdraw the shaft from the front of the cylinder block whilst the engine is in the vehicle, remove the radiator as described in Chapter 3 and unbolt the grille panel as described in Chapter 11.
2 On carburettor models, remove the fuel pump as described in Chapter 4A.
3 Remove the intermediate shaft sprocket as described in Section 8.
4 Undo the retaining bolts then slide the cover off the end of the intermediate shaft and recover the gasket (see illustrations).
5 Undo the retaining bolts and remove the oil pump drive gear cover from the left-hand side of the cylinder block (see illustration).
6 Carefully withdraw the oil pump drivegear, taking great care not to dislodge the oil pump driveshaft (see illustration). **Note:** *Great care must be taken to ensure that the driveshaft is not dislodged from the oil pump. If the driveshaft is dislodged it will drop down into the bottom sump. If this happens, the sump will have to be removed in order to recover the driveshaft.*
7 Unscrew the retaining bolt and washer then slide out the intermediate shaft thrust plate and withdraw the intermediate shaft (see illustrations).

Inspection

8 Examine the intermediate shaft and oil pump driveshaft for pitting, scoring or wear ridges on the bearing journals, and for chipping or wear of the gear teeth. Renew as necessary. Check the intermediate shaft bearings in the cylinder block for wear and, if worn, have these renewed by your Renault dealer or suitably equipped engineering works. Wipe them clean if they are still serviceable.

18.3 Engine rubber mounting

19.4a Slacken and remove the retaining bolts . . .

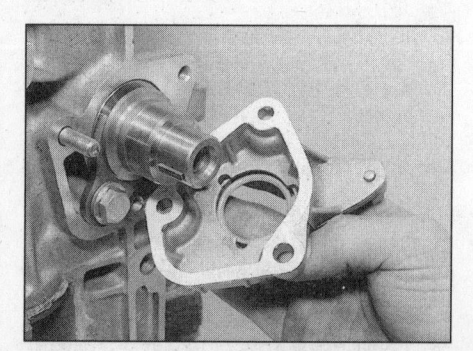

19.4b . . . and remove the intermediate shaft cover and gasket

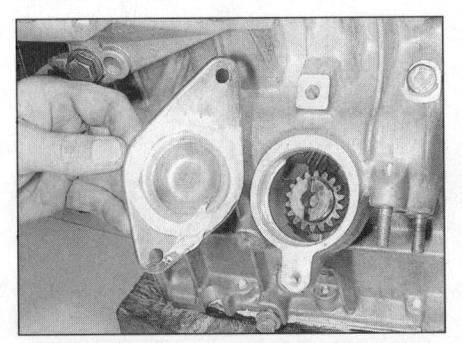

Four cylinder petrol engine 2A•23

19.5 Unbolt the cover from the left-hand side of the cylinder block...

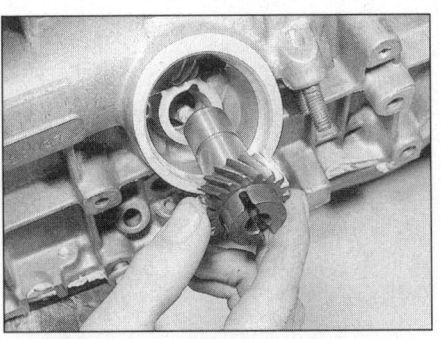

19.6 ...and carefully remove the oil pump drive gear

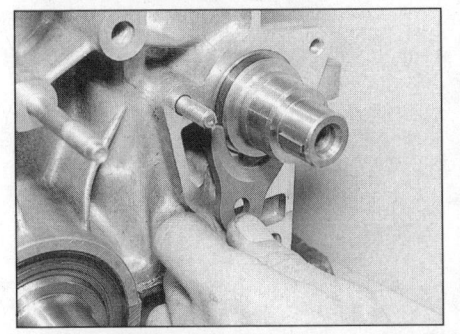

19.7a Unbolt the intermediate shaft thrust plate...

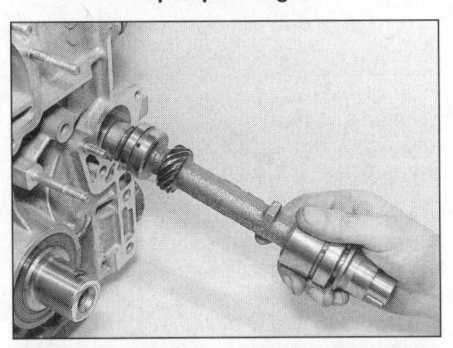

19.7b ...and slide the intermediate shaft out of position

9 Temporarily fit the thrust plate to its position on the intermediate shaft, and use a feeler blade to check that the endfloat is as given in the *Specifications*. If it is greater than the upper tolerance, a new thrust plate should be obtained, but first check the thrust surfaces on the shaft to ascertain if wear has occurred here.

Refitting

10 Clean off all traces of the old gasket or sealant from the intermediate shaft cover. Lever the old oil seal out of the cover and position a new seal in the cover and tap it into position using a suitable tubular drift, such as a socket, which bears only on the hard outer edge of the seal. Ensure that the seal lip is facing inwards.
11 Liberally lubricate the intermediate shaft, and slide it into its bearings.
12 Slide the thrust plate in position and securely tighten its retaining bolt.
13 Place a new gasket in position on the cylinder block/crankcase and carefully ease the cover over the end of the shaft and slide it into position.
14 To prevent the possibility of oil leakage smear the shafts of the cover retaining bolts with a suitable sealant (Renault recommend the use of Loctite Frenetanch). Ensure the gasket is correctly positioned then refit the cover retaining bolts, tightening them securely.
15 Ensure the driveshaft is correctly engaged with the oil pump then refit the oil pump drive gear, engaging it with driveshaft.
16 Apply a bead of sealant to oil pump drivegear cover (Renault recommend CAF 4/60 Thixo) then refit the cover and securely tighten its retaining bolts.
17 Refit the intermediate shaft sprocket and install the timing belt as described in Sections 7 and 8.
18 On carburettor models refit the fuel pump as described in Chapter 4A.
19 In the engine is in the vehicle refit the front grille panel (Chapter 11) then refit the radiator (Chapter 3).

20 Crankshaft - removal, inspection and refitting

Removal

1 Remove the crankshaft sprocket, flywheel/driveplate, the oil pump and (where necessary) the crankcase stiffener casting as described in the relevant Sections of this Chapter (as applicable).
2 Remove the liner/piston and connecting rod assemblies as described in Section 21. If no work is to be done on the pistons and connecting rods, there is no need to remove the cylinder head and withdraw the liner assemblies from the cylinder block/crankcase. Instead unbolt the big-end bearing caps and push the pistons sufficiently up the bores so that the connecting rods are positioned clear of the crankshaft journals.
3 Before proceeding further, check the crankshaft endfloat using a dial gauge in contact with the end of the crankshaft. Push the crankshaft fully one way and then zero the gauge. Push the crankshaft fully the other way and check the endfloat. The result can be compared with the specified amount and will give an indication as to whether new thrustwashers are required. Note that thrustwashers are available in four different thickness. If a dial gauge is not available, feeler blades can be used. First push the crankshaft fully towards the flywheel end of the engine, then use feeler blades to measure the gap between the web of number 2 crankpin and the thrustwasher **(see illustration)**.
4 The main bearing caps should be numbered 1 to 5 from the transmission end of the engine **(see illustration)**. If not, mark them accordingly using a centre punch. Also note the correct fitted depth of both the front and rear crankshaft oil seals in the bearing caps.
5 Slacken and remove the main bearing cap retaining bolts and lift off each bearing cap, noting the locating dowels **(see illustration)**. Recover the lower bearing shells and tape them to their respective caps for safe-keeping. Remove the rubber sealing strips (where fitted) from the sides of number 1 and/or 5 main bearing caps and discard them (as applicable).
6 Lift out the crankshaft and discard the oil seals.
7 Recover the upper bearing shells from the cylinder block and tape them to their respective caps for safe keeping. Remove the thrustwasher halves from the side of number 2 main bearing and store them with the bearing caps **(see illustration)**.

20.3 Checking crankshaft endfloat using a feeler blade

20.4 Main bearing caps have identification numbers cast into them

2A•24 Four cylinder petrol engine

20.5 Lift off the main bearing cap noting the locating dowels

20.7 Remove the thrustwashers from number 2 main bearing

20.14 Measuring a crankshaft main bearing journal

Inspection

8 Clean the crankshaft using paraffin or a suitable solvent, and dry it, preferably with compressed air if available. Be sure to clean the oil holes with a pipe cleaner or similar probe, to ensure that they are not obstructed.

⚠ *Warning: Wear eye protection when using compressed air*

9 Check the main and big-end bearing journals for uneven wear, scoring, pitting and cracking.
10 Big-end bearing wear is accompanied by distinct metallic knocking when the engine is running, particularly noticeable when the engine is pulling from low revs, and some loss of oil pressure.
11 Main bearing wear is accompanied by severe engine vibration and rumble - getting progressively worse as engine revs increase - and again by loss of oil pressure.
12 Check the bearing journal for roughness by running a finger lightly over the bearing surface. Any roughness (which will be accompanied by obvious bearing wear) indicates the that the crankshaft requires regrinding (where possible) or renewal.
13 If the crankshaft has been reground, check for burrs around the crankshaft oil holes (the holes are usually chamfered, so burrs should not be a problem unless regrinding has been carried out carelessly). Remove any burrs with a fine file or scraper, and thoroughly clean the oil holes as described previously.
14 Using a micrometer, measure the diameter of the main and big-end bearing journals and compare the results with the *Specifications* **(see illustration)**. By measuring the diameter at a number of points around each journal's circumference, you will be able to determine whether or not the journal is out-of-round. Take the measurement at each end of the journal, near the webs, to determine if the journal is tapered. Compare the results obtained to those given in the *Specifications*.
15 Check the oil seal contact surfaces at each end of the crankshaft for wear and damage. If the seal has worn an excessive groove in the surface of the crankshaft, consult an engine overhaul specialist who will be able to advise whether a repair is possible or whether a new crankshaft is necessary.
16 If the crankshaft has worn beyond the specified limits, it must be renewed. Note that Renault produce a set of oversize bearing shells for both the main bearings and big-end bearings so, if the crankshaft journals have not already been reground, it may be possible to have the crankshaft reconditioned and fit the oversize shells instead of renewing the crankshaft. Seek the advice of your Renault dealer or suitable engine specialist on the best course of action.

Refitting

Note: It is recommended that new main bearing shells are fitted regardless of the condition of the original ones.

Selection of bearing shells

17 There are two different sizes of bearing shell available; the standard size shell for use with an original crankshaft and an oversize shell for use once the crankshaft has been reground.
18 The relevant set of bearing shells required can be obtained by measuring the diameter of the crankshaft main bearing journals (see paragraph 14). This will show if the crankshaft is original or whether its journals have been reground, identifying if either standard or oversize bearing shells are required.
19 If the access to the necessary measuring equipment cannot be gained, the size of the bearing shells can be identified by the markings stamped on the rear of each shell. Details of these markings should be supplied to your Renault dealer who will then be able to identify the size of shell fitted.
20 Whether the original shells or new shells are being fitted, it is recommended that the running clearance is checked as follows prior to installation.

Main bearing running clearance check

21 Clean the backs of the bearing shells and the bearing locations in both the cylinder block/crankcase and the main bearing caps.
22 Press the bearing shells into their locations, ensuring that the tab on each shell engages in the notch in the cylinder block/crankcase or main bearing cap, and taking care not to touch any shell's bearing surface with your fingers. If the original bearing shells are being used for the check ensure that they are refitted in their original locations. The clearance can be checked in either of two ways.
23 One method (which will be difficult to achieve without a range of internal micrometers or internal/external expanding calipers) is to refit the main bearing caps to the cylinder block/crankcase, with bearing shells in place. With the cap retaining bolts correctly tightened, measure the internal diameter of each assembled pair of bearing shells. If the diameter of each corresponding crankshaft journal is measured and then subtracted from the bearing internal diameter, the result will be the main bearing running clearance.
24 The second (and more accurate) method is to use an American product known as Plastigage. This consists of a fine thread of perfectly round plastic which is compressed between the bearing shell and the journal. When the shell is removed, the plastic is deformed and can be measured with a special card gauge supplied with the kit. The running clearance is determined from this gauge. Plastigage is sometimes difficult to obtain but enquiries at one of the larger specialist quality motor factors should produce the name of a stockist in your area. The procedure for using Plastigage is as follows.
25 With the main bearing upper shells in place, carefully lay the crankshaft in position. Do not use any lubricant; the crankshaft journals and bearing shells must be perfectly clean and dry.
26 Cut several lengths of the appropriate size Plastigage (they should be slightly shorter than the width of the main bearings) and place one length on each crankshaft journal axis **(see illustration)**.
27 With the main bearing lower shells in position, refit the main bearing caps, tightening their retaining bolts to the specified torque. Take care not to disturb the Plastigage and **do not** rotate the crankshaft at any time during this operation.
28 Remove the main bearing caps again taking great care not to disturb the Plastigage or rotate the crankshaft.

Four cylinder petrol engine 2A•25

20.26 Plastigage in place on crankshaft journal

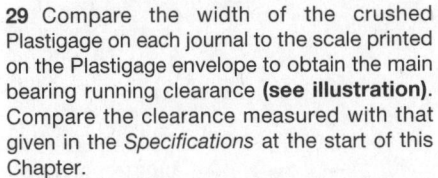

20.29 Measure the width of the crushed Plastigage using the scale on the card provided

20.36 Fit the main bearing shells . . .

29 Compare the width of the crushed Plastigage on each journal to the scale printed on the Plastigage envelope to obtain the main bearing running clearance **(see illustration)**. Compare the clearance measured with that given in the *Specifications* at the start of this Chapter.

30 If the clearance is significantly different from that expected, the bearing shells may be the wrong size (or excessively worn if the original shells are being re-used). Before deciding that the crankshaft is worn, make sure that no dirt or oil was trapped between the bearing shells and the caps or block when the clearance was measured. If the Plastigage was wider at one end than at the other, the crankshaft journal may be tapered.

31 Note that Renault do not specify a running clearance for these engines. The figure given in the *Specifications* is a guide figure which is typical for this type of engine. Before condemning the components concerned, seek the advice of your Renault dealer or suitable engine repair specialist. They will also be able to inform as to the best course of action and whether it is possible to have the crankshaft journals reground or whether renewal will be necessary.

32 Where necessary, obtain the correct size of bearing shell and repeat the running clearance checking procedure as described above.

33 On completion, carefully scrape away all traces of the Plastigage material from the crankshaft and bearing shells using a fingernail or other object which is unlikely to score the bearing surfaces.

Final crankshaft refitting

34 Carefully lift the crankshaft out of the cylinder block once more.

35 Using a little grease, stick the thrustwashers to each side of the number 2 main bearing upper location; ensure that the oilway grooves on each thrustwasher face outwards (away from the cylinder block).

36 Place the bearing shells in their locations as described above in paragraphs 21 and 22 **(see illustration)**. If new shells are being fitted, ensure that all traces of the protective grease are cleaned off using paraffin. Wipe dry the shells and caps with a lint-free cloth.

37 Liberally lubricate each bearing shell in the cylinder block/crankcase with clean engine oil then lower the crankshaft into position ensuring that the bearing shells and thrustwashers remain correctly seated **(see illustrations)**.

38 Check the crankshaft endfloat as described in paragraph 3.

39 Ensure the cap locating dowels are in position and fit the main bearing caps numbers 2 to 4. Ensure the caps are fitted in their correct locations and the correct way round. Insert the bearing cap bolts and tighten them to the specified torque setting **(see illustration)**.

40 There are two possible ways of sealing the end main bearing cap(s) (as applicable). The first is by fitting rubber sealing strips to the cap grooves, and the second is by filling the grooves with a special sealant kit available from your Renault dealer **(see illustration)**. The second method using the sealant is extremely messy and if carried out carelessly can lead to the oilways being blocked. It is therefore recommended that the sealing strips are used as follows.

41 Two different thickness of sealing strip are available and it is first necessary to decide which size is needed. The thinner (5.1 mm thick) sealing strip is unmarked whereas the thicker (5.4 mm thick) sealing strip has a colour marking on it.

42 To select the correct size of sealing strip, fit the main bearing end cap(s) in position and lightly tighten the retaining bolts. Using a suitable twist drill, measure the gap between the inner edge of each main bearing cap seal

20.37a . . . then lubricate them with clean engine oil . . .

20.37b . . . and refit the crankshaft

20.39 Fit the centre main bearing caps and tighten the retaining bolts to the specified torque

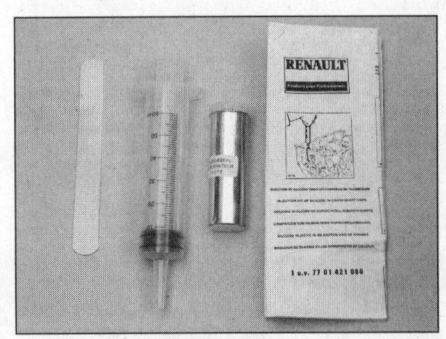

20.40 Renault sealing kit for main bearing cap grooves. Full instructions are supplied with the kit

2A•26 Four cylinder petrol engine

20.42 Using a twist drill to measure main bearing cap groove to cylinder block clearance

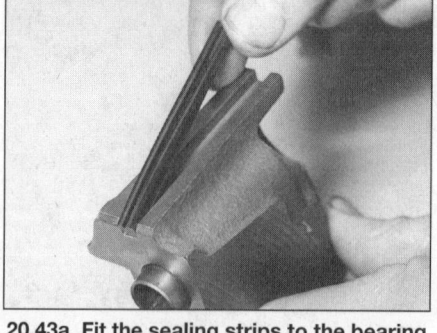

20.43a Fit the sealing strips to the bearing cap so that its groove is facing away from the cap...

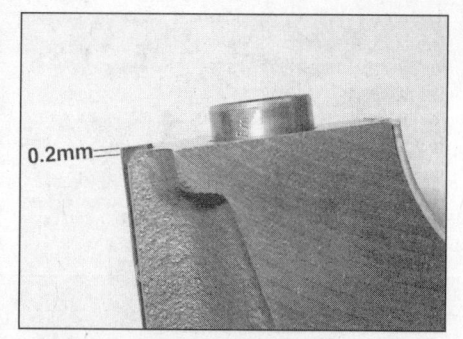

20.43b ...and position the strip so that it protrudes above the cap mating surface by approximately 0.2 mm

20.45 Apply a smear of suitable sealant to the cylinder block/crankcase...

20.46 ...then slide the bearing cap assembly into position using the metal strips to ensure the sealing strips are not displaced

47 With the cap in position tighten the retaining bolts to the specified torque setting then carefully pull out the metal strips with a pair of pliers in a horizontal direction. Using a sharp knife, trim the lower end of each sealing strip so that the strips protrude above the sump mating surface by approximately 0.5 mm **(see illustrations)**.

48 Repeat the procedure for the opposite bearing cap.

49 Refit/reconnect the piston connecting rod assemblies to the crankshaft as described in Section 21.

50 Refit the crankshaft oil seals, crankshaft sprocket, flywheel/driveplate, crankcase stiffener (where fitted) and the oil pump as described in the relevant Sections of this Chapter (as applicable).

groove and the cylinder block/crankcase **(see illustration)**. If this dimension is less than 5 mm then the thinner sealing strips will be required. If the dimension is greater than or equal to 5 mm then the thicker sealing strips will be required. Unbolt the bearing cap(s) and remove.

43 Fit the correct rubber sealing strips to each groove in the bearing cap ensuring that the groove in each strip is facing outwards. Position each sealing strip so that is protrudes approximately 0.2 mm above the upper (cylinder block/crankcase) mating surface of the bearing cap **(see illustrations)**.

44 To ease installation, obtain two thin metal strips of 0.25 mm thickness or less. These can then be used to prevent the strips moving as the cap is being fitted. Old feeler blades are ideal for this purpose, provided all burrs which may damage the sealing strips are first removed.

45 Apply a thin coating of sealant (Renault recommend the use of CAF 4/60 THIXO) to the mating surface of the cylinder block/crankcase, taking great care not to block the oil return holes **(see illustration)**.

46 Oil both sides of the metal strips, and hold them on the sealing strips. Ease the main bearing cap into position and insert the bolts loosely **(see illustration)**. Just before the cap touches the cylinder block/mating surface, check that the sealing strips are still protruding from the cap. If not remove the cap and repeat the fitting procedure.

21 Piston/connecting rod and cylinder liner assemblies - removal, inspection, refitting

Removal

Note: *It is possible to remove the piston/connecting rod and liner assemblies without removing the engine from the vehicle although this is not recommended. To ensure absolute cleanliness it is recommended that the engine is removed from the vehicle and the cylinder liner assemblies removed with the engine on a bench.*

20.47a Tighten the bearing cap retaining bolts to the specified torque...

20.47b ...then carefully slide out the metal strips

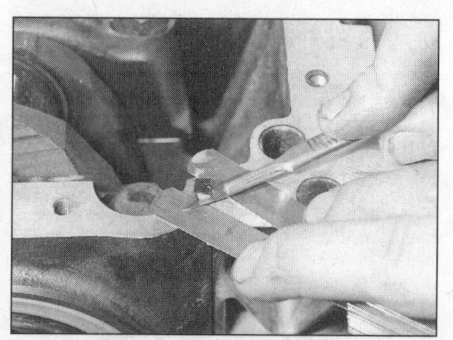

20.47c Trim off the ends of the sealing strips so that each one protrudes from the bearing cap by approximately 0.5 mm

Four cylinder petrol engine 2A•27

1 Remove the cylinder head as described in Section 11.
2 Remove the oil pump as described in Section 14.
3 On 1995 cc 12-valve and later 2165 cc models also remove the crankcase stiffener casting as described in Section 15.
4 Using a hammer and centre-punch, paint or similar, mark each connecting rod big-end bearing cap with its respective cylinder number on the flat machined surface provided; if the engine has been dismantled before, note carefully any identifying marks made previously. Note that No 1 cylinder is at the transmission end of the engine.
5 Turn the crankshaft to bring pistons 1 and 4 to BDC (bottom dead centre).
6 Unscrew the nuts/bolts from number 1 piston big-end bearing cap, then take off the cap and recover the bottom half bearing shell **(see illustration)**. If the bearing shells are to be re-used, tape the cap and the shell together. If the bearing cap locating dowels/pins (as applicable) are a loose fit, remove them and store them with the cap for safe-keeping.
7 Push the piston up the bore to disengage the connecting rod from the crankshaft. Recover the upper bearing shell, and tape it to the connecting rod for safe keeping.
8 Using a suitable piece of wood, such as the handle of a hammer, tap the liner upwards and out of position from the base of the cylinder block/crankcase **(see illustration)**.
9 Withdraw the liner/piston and connecting rod assembly from the top of the cylinder block/crankcase. Mark the liner in some way to indicate its correct fitted direction then remove the sealing ring from the base of liner and discard.
10 Refit the big-end cap to the connecting rod and loosely tighten its retaining nuts/bolts. This will help to keep the components in their correct order.
11 Repeat the procedure and remove number 4 liner/piston and connecting rod assembly in the same way.
12 Turn the crankshaft through 180° to bring pistons 2 and 3 to BDC (bottom dead centre) and remove them in the same way.
13 If necessary, mark the relevant cylinder number on the liner and withdraw the piston/connecting rod assembly from the base of each liner.

Inspection

Piston and connecting rod assembly

14 Before the inspection process can begin, the piston/connecting rod assemblies must be cleaned, and the original piston rings removed as described in Section 22.
15 Scrape away all traces of carbon from the top of the piston. A hand-held wire brush (or a piece of fine emery cloth) can be used, once the majority of the deposits have been scraped away.
16 Remove the carbon from the ring grooves in the piston, using an old ring. Break the ring in half to do this (be careful not to cut your fingers - piston rings are sharp). Be careful to remove only the carbon deposits - do not remove any metal, and do not nick or scratch the sides of the ring grooves.
17 Once the deposits have been removed, clean the piston/connecting rod assembly with paraffin or a suitable solvent, and dry thoroughly. Make sure that the oil return holes in the ring grooves are clear.
18 If the pistons and cylinder liners are not damaged or worn excessively, the original pistons can be refitted. Normal piston wear shows up as even vertical wear on the piston thrust surfaces, and slight looseness of the top ring in its groove. New piston rings should always be used when the engine is reassembled.
19 Carefully inspect each piston for cracks around the skirt, around the gudgeon pin holes, and at the piston ring "lands" (between the ring grooves).
20 Look for scoring and scuffing on the piston skirt, holes in the piston crown, and burned areas at the edge of the crown. If the skirt is scored or scuffed, the engine may have been suffering from overheating, and/or abnormal combustion which caused excessively high operating temperatures. The cooling and lubrication systems should be checked thoroughly. Scorch marks on the sides of the pistons show that blow-by has occurred. A hole in the piston crown, or burned areas at the edge of the piston crown, indicates that abnormal combustion (pre-ignition, knocking, or detonation) has been occurring. If any of the above problems exist, the causes must be investigated and corrected, or the damage will occur again. The causes may include incorrect ignition timing or a faulty injector (as applicable).
21 Corrosion of the piston, in the form of pitting, indicates that coolant has been leaking into the combustion chamber and/or the crankcase. Again, the cause must be corrected, or the problem may persist in the rebuilt engine.
22 It is not possible to renew the pistons separately; pistons are only supplied with piston rings and a liner, as a part of a matched assembly.
23 Examine each connecting rod carefully for signs of damage, such as cracks around the big-end and small-end bearings. Check that the rod is not bent or distorted. Damage is highly unlikely, unless the engine has been seized or badly overheated. Detailed checking of the connecting rod assembly can only be carried out by a Renault dealer or engine repair specialist with the necessary equipment.
24 The gudgeon pins are an interference fit in the connecting rod small-end bearing. Therefore, piston and/or connecting rod renewal should be entrusted to a Renault dealer or engine repair specialist, who will have the necessary tooling to remove and install the gudgeon pins.

Cylinder liner

25 Refer to Section 23.

Refitting

Note: *It is recommended that new piston rings and big-end bearing shells are fitted regardless of the condition of the original ones.*

Selection of bearing shells

26 See Section 20.

Big-end bearing running clearance check

27 Clean the backs of the bearing shells and the bearing locations in both the connecting rod and bearing cap.
28 Press the bearing shells into their locations, ensuring that the tab on each shell engages in the notch in the connecting rod and cap and taking care not to touch any shell's bearing surface with your fingers.
29 The upper shells, which are fitted to the connecting rods may or may not (depending on model) have an oil hole in them, the lower shells which are fitted to the bearing caps, are plain. If the original bearing shells are being used for the check ensure that they are refitted in their original locations. The clearance can be checked in either of two ways.
30 One method is to refit the big-end bearing cap to the connecting rod, with bearing shells in place. With the cap retaining nuts/bolts tightened to the specified torque, use an internal micrometer or vernier caliper to measure the internal diameter of each

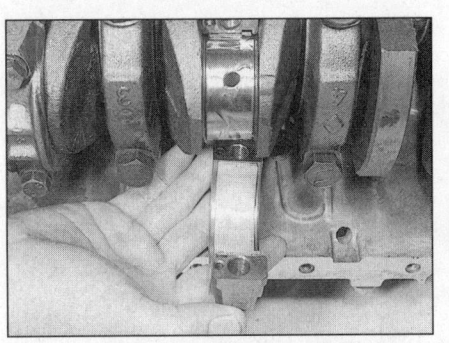

21.6 Removing a big-end bearing cap

21.8 Removing a cylinder liner assembly

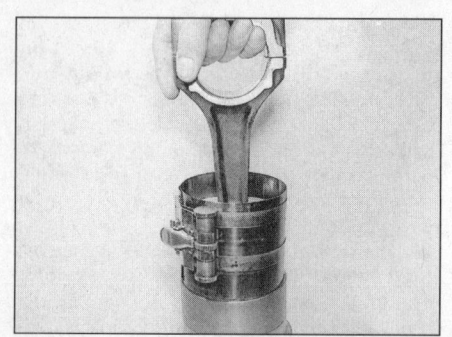

2A•28 Four cylinder petrol engine

21.36 Using a piston ring compressor to fit the piston/connecting rod assembly to the cylinder liner

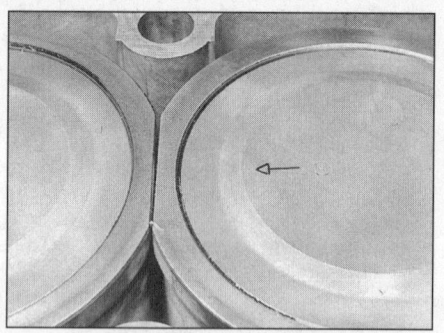

21.39 Ensure the piston arrow is pointing towards the transmission end of the engine

21.40 Tighten the big-end bearing cap bolts/nuts to the specified torque setting

assembled pair of bearing shells. If the diameter of each corresponding crankshaft journal is measured and then subtracted from the bearing internal diameter, the result will be the big-end bearing running clearance.

31 The second method is to use Plastigage as described in Section 20, paragraphs 24 to 33. Place a strand of Plastigage on each (cleaned) crankpin journal and refit the (clean) piston/connecting rod assemblies, shells and big-end bearing caps, tightening the nuts/bolts to the specified torque wrench settings. Take care not to disturb the Plastigage. Dismantle the assemblies without rotating the crankshaft and use the scale printed on the Plastigage envelope to obtain the big-end bearing running clearance. On completion of the measurement, carefully scrape off all traces of Plastigage from the journal and shells using a fingernail or other object which will not score the components.

Final piston/connecting rod assembly refitting

32 Check the liner protrusions as described in Section 23. If the original liner/piston and connecting rod assemblies are being refitted then ensure that they are refitted in their original positions. If new liner/piston and connecting rod assemblies are being installed, position and number the assemblies as described in Section 23.

33 Ensure the bearing shells are correctly refitted as described above in paragraphs 27 to 29. If new shells are being fitted, ensure that all traces of the protective grease are cleaned off using paraffin. Wipe dry the shells and connecting rods with a lint-free cloth.

34 Lubricate the liner bores, the pistons and piston rings then lay out each piston/connecting rod assembly with its liner.

35 Starting with assembly number 1, make sure that the piston rings are still spaced as described in Section 22, then clamp them in position with a piston ring compressor.

36 Insert the piston/connecting rod assembly into the bottom of number 1 liner and tap into position **(see illustration)**. Once the piston rings have entered the liner, remove the ring compressor, then slide the piston into position until the piston crown is flush with the top of the liner.

37 Fit a new sealing ring to the liner seat. Apply a smear of oil to the sealing ring to ease installation.

38 Ensure that the bearing shell is still correctly installed and carefully install the liner/piston and connecting rod assembly in the cylinder block, making sure it is fitted the correct way around.

39 Ensure the piston/connecting rod assembly is correctly positioned, with the arrow on the piston crown pointing towards the flywheel/driveplate end of the engine **(see illustration)**.

40 Taking care not to mark the liner bores, liberally lubricate the crankpin and both bearing shells, then pull the piston/connecting rod assembly down the liner and onto the crankpin. Refit the big-end bearing cap, aligning the cap with its locating pins/dowels (which means that the bearing shell locating tabs abut each other), and tighten its retaining nuts/bolts to the specified torque setting **(see illustration)**.

41 Refit the remaining three liner/piston and connecting rod assemblies in the same way.

42 Once all assemblies are correctly installed, clamp the liners in position as described in Section 11.

43 Rotate the crankshaft, and check that it turns freely, with no signs of binding or tight spots.

44 Refit the cylinder head, crankcase stiffener (where fitted) and oil pump as described in the relevant Sections of this Chapter.

22 Piston rings - removal and refitting

Removal

1 With the piston removed from the liner, carefully expand the old rings over the top of the pistons. The use of two or three old feeler blades will be helpful in preventing the rings dropping into empty grooves. Be careful not to scratch the piston with the ends of the ring. The rings are brittle, and will snap if they are spread too far. They're also very sharp - protect your hands and fingers. Note that the third ring incorporates an expander. Always remove the rings from the top of the piston. Keep each set of rings with its piston if the old rings are to be re-used.

Refitting

2 Install the new rings by fitting by fitting them over the top of the piston starting with the oil control ring. Ensure that the second compression is correctly installed with its stepped surface at the bottom **(see illustration)**.

3 With the piston rings correctly installed, check that each ring is free to rotate easily in its groove. Position the ring end gaps so that are spaced at 120° intervals.

23 Cylinder block/crankcase - cleaning and inspection

Cleaning

1 Remove all external components and electrical switches/sensors from the block. For complete cleaning, ideally the core plugs should be removed. Drill a small hole in the plugs, then insert a self-tapping screw and

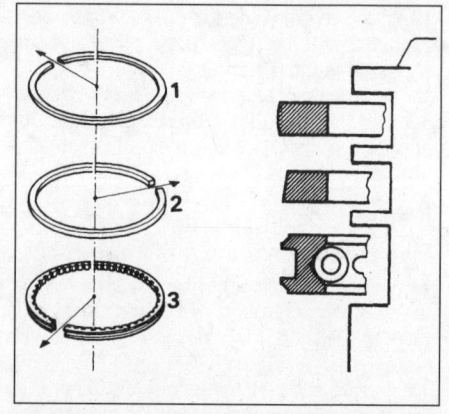

22.2 Piston ring fitting details. Position the end gaps as shown

1 Top ring
2 Second ring
3 Oil control ring

Four cylinder petrol engine 2A•29

pull out the plugs using a pair of grips or a slide hammer.

2 On 1995 cc 12-valve models, undo the retaining bolts and washers and remove the piston oil jet nozzles from the inside the cylinder block. Note that the nozzles are different and are not interchangeable; note the correct fitted position of each nozzle as it is removed and recover the O-ring from the back of each nozzle.

3 Scrape all traces of gasket from the cylinder block/crankcase, taking care not to damage the gasket/sealing surfaces.

4 Remove all oil gallery plugs. The plugs are usually very tight - they may have to be drilled out and the holes re-tapped. Use new plugs when the engine is reassembled.

5 If any of the castings are extremely dirty, all should be steam cleaned.

6 After the castings are returned, clean all oil holes and oil galleries one more time. Flush all internal passages with warm water until the water runs clear, then dry thoroughly and apply a light film of oil to all mating surfaces to prevent rusting. If you have access to compressed air, use it to speed up the drying process and to blow out all the oil holes and galleries.

> ⚠ **Warning: Wear eye protection when using compressed air!**

7 If the castings are not very dirty, you can do an adequate cleaning job with hot (as hot as you can stand!), soapy water and a stiff brush. Take plenty of time and do a thorough job. Regardless of the cleaning method used, be sure to clean all oil holes and galleries very thoroughly and to dry all components well; protect all mating surfaces as described above to prevent rusting.

8 All threaded holes must be clean to ensure accurate torque readings during reassembly.

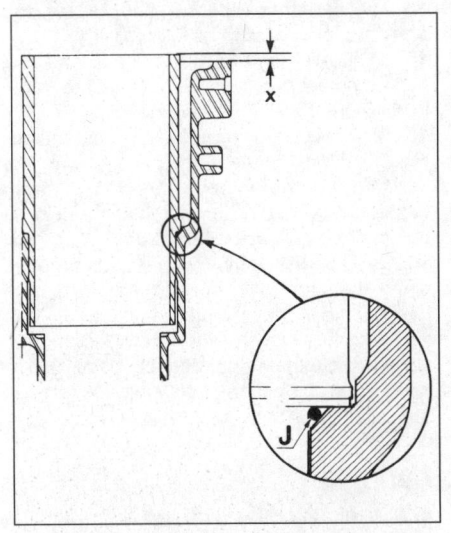

23.15 Cylinder liner cross-sectional view

J Sealing ring
X Protrusion above block surface

To clean all threads, run the proper size tap into each of the holes to remove rust, corrosion, thread sealant or sludge and to restore damaged threads. If possible, use compressed air to clear the holes of debris produced by this operation; a good alternative is to inject aerosol-applied water-dispersant lubricant into each hole, using the long spout usually supplied.

9 Apply suitable sealant to the new oil gallery plugs and insert them into the holes in the block. Tighten them securely.

> ⚠ **Warning: Wear eye protection when cleaning out these holes in this way!**

10 On 1995 cc 12-valve model, fit a new O-ring to rear of each piston oil jet nozzles. Install the nozzles in their original positions inside the cylinder block and refit the retaining bolts and washers, tightening then to the specified torque setting.

11 If the engine is not going to be reassembled right away, cover it with a large plastic bag to keep it clean; protect the all mating surfaces as described above to prevent rusting.

Inspection

12 Check the each cylinder liner for scuffing and scoring. Check for signs of a wear ridge at the top of the cylinder indicating that the liner is excessively worn.

13 Since Renault do not state any specific wear limits for the cylinder liners or pistons, it is not possible to assess the amount of wear by direct measurement. If there is any doubt about the condition of the cylinder liner seek the advice of a Renault dealer or suitable engine reconditioning specialist.

14 If renewal is necessary, new liners complete with piston and piston rings can be purchased from a Renault dealer. Note that it is not possible to renew the liners individually, as they are supplied as a matched assembly complete with piston and rings.

15 With the sealing rings removed from each liner, install the liners in their original position in the cylinder block. Using a dial gauge or a straight-edge and feeler blade, check that the protrusion of each liner above the upper surface of the cylinder block is within the limits given in the *Specifications*, and that the maximum difference between any two liners is not exceeded **(see illustration)**. If this is not the case seek the advice of a Renault dealer before proceeding with the engine rebuild.

16 Note that if new liner assemblies are being installed the liners must be arranged so that the liner protrusions gradually decrease/increase from one end of the cylinder block to the other, ie. the largest protrusion is at one end with the smallest at the opposite end **(see illustration)**. Use the liner protrusion measurements to arrange the liners as required then number each one with its respective cylinder number, remembering No.1 cylinder is at the transmission end of the engine.

24 Engine - removal and refitting leaving manual transmission in car

Removal

1 Disconnect the battery negative terminal.
2 Remove the bonnet as described in Chapter 11.
3 Drain the engine oil and cooling system as described in Chapter 1A
4 Remove the radiator and ducts as described in Chapter 3.
5 Remove the front grille panel as described in Chapter 11.
6 Remove the air cleaner and disconnect the accelerator cable from the throttle linkage as described in Chapter 4A or 4B.
7 Remove the exhaust front pipe as described in Chapter 4A or 4B.
8 On models with an oil cooler, unbolt the oil cooler mounting bracket so the oil cooler is free to be removed with the engine.
9 On models with power steering, unbolt the pump from the side of the engine and position it clear (see Chapter 1A0). Note that there is no need to disconnect the pipes from the pump, simply release them from all the relevant retaining clips.
10 On fuel injection models, remove the crankshaft sensor as described in Chapter 4B.
11 Disconnect all vacuum hoses including the one for the brake servo. Identify each hose with masking tape using a different number or letter if there is likely to be any doubt as to its reconnection.

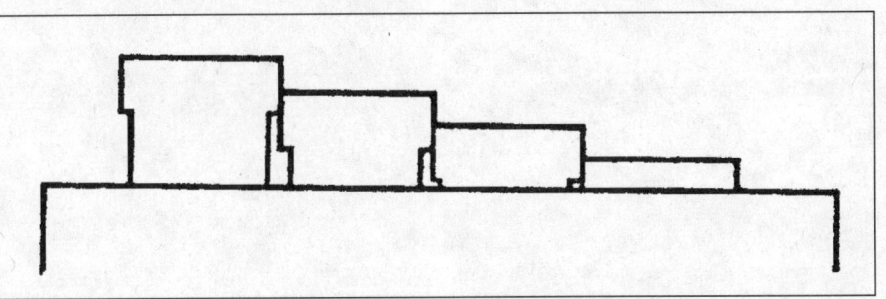

23.16 If new cylinder liners are being fitted, arrange them so the protrusion gradually decrease/increase from one end of the block to the other

2A•30 Four cylinder petrol engine

12 Disconnect the coolant hoses, including those for the heater at the engine compartment rear bulkhead.
13 On 1995 cc 12-valve and 2165 cc engines, unclip the electronic control unit box from the left-hand wing valance (near the rear of the headlamp) and rest it on top of the engine. The wiring plugs will have to be disconnected before this can be done. Disconnect all engine electrical and earth leads (see illustration), identifying them with numbered masking tape if necessary.
14 Disconnect the HT lead from the ignition coil.
15 Unbolt the loom protector from under the front end of the sump pan and place it to one side (see illustration).
16 Disconnect the fuel lines (see Chapter 4A or 4B).
17 Remove the starter motor (see Chapter 5A).
18 Disconnect the earth strap which runs between the clutch bellhousing and the body member (see illustration).
19 Attach a suitable hoist to the engine lifting lugs and just take its weight.
20 Unscrew the engine-to-transmission connecting bolts.
21 Unbolt and remove the reinforcement brackets at the base of the bellhousing, then unbolt and remove the bellhousing lower cover plate (see illustration).
22 Place a jack under the transmission.
23 Unscrew the engine mounting through-bolt nuts.
24 Make a final check to ensure that any components, hoses etc affecting engine removal have been removed or disconnected as applicable.
25 Raise the engine to clear the mounting brackets, then withdraw it forward to disconnect it from the transmission and lift it up and out of the engine compartment.

Refitting

26 Refitting is the reverse of removal noting the following.
 a) Apply a smear of high-melting-point grease (Renault recommend the use of Molykote BR2) to the splines of the transmission input shaft. Do not apply too much, otherwise there is a possibility of the grease contaminating the clutch friction plate.
 b) Ensure that the locating dowels are correctly positioned in the engine or transmission before lowering the engine into place.
 c) Ensure that the wiring loom is correctly routed, and retained by all the relevant retaining clips; all connectors should be correctly and securely reconnected.
 d) Ensure that all coolant hoses are correctly reconnected, and securely retained by their retaining clips.
 e) Adjust the accelerator cable as described in Chapter 4.
 f) Refill the engine with correct quantity and type of oil, as described in Chapter 1A.
 g) Refill the cooling system as described in Chapter 1A.

25 Engine and manual transmission - removal and refitting

Removal

1 The operations include those described in paragraphs 1 to 19 of the preceding Section plus the following additional work.
2 Drain the transmission oil (see Chapter 1A).
3 Disconnect the gearchange link rods and the reverse interlock operating cable as described in Chapter 7A.
4 Disconnect the reversing lamp switch leads.
5 Disconnect the speedometer sensor.
6 On models with a cable-operated clutch, disconnect the clutch cable from the release arm as described in Chapter 6.
7 On models with a hydraulically-operated clutch, remove the shield and unbolt the clutch slave cylinder (see Chapter 6). Tie it up out of the way without disconnecting the fluid hose.
8 Raise the front of the car so that the roadwheels are clear of the ground, support securely on stands and remove the roadwheels.
9 Remove the driveshafts as described in Chapter 8.
10 With the hoist attached to the engine lifting lugs, and taking the weight of the engine and transmission, unscrew the engine mounting through-bolts and those for the transmission. To provide clearance during removal, unbolt and remove the transmission mounting brackets (see illustrations).
11 Withdraw the engine/transmission forward and upwards from the engine compartment (see illustration).
12 Clean away external dirt using paraffin and a stiff brush or a water soluble solvent.
13 To separate the engine from the transmission, proceed as follows.
14 Unscrew the bolts which hold the reinforcement brackets and cover plate to the lower face of the clutch bellhousing.
15 Unscrew and remove the engine-to-transmission connecting bolts. Note the location of the various brackets, wiring harness clips and other attachments held by some of the bolts (see illustration).
16 Support the engine with wooden blocks under the sump pan so that the transmission has a clear gap underneath, and then withdraw the transmission from the engine. Do not allow the weight of the transmission to hang upon the clutch shaft while the latter is still engaged with the clutch driven plate.

Refitting

17 Check that the clutch driven plate is centralised, and then apply a smear of high-temperature grease (Renault recommend the use of Molykote BR2) to the clutch input shaft splines.

24.13 Engine earth lead

24.15 Wiring loom protector bolt

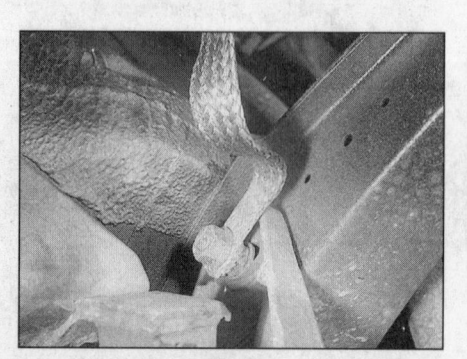

24.18 Transmission earth strap

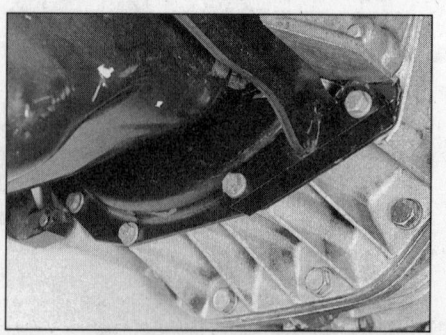

24.21 Bellhousing lower cover plate and reinforcement bracket

Four cylinder petrol engine 2A•31

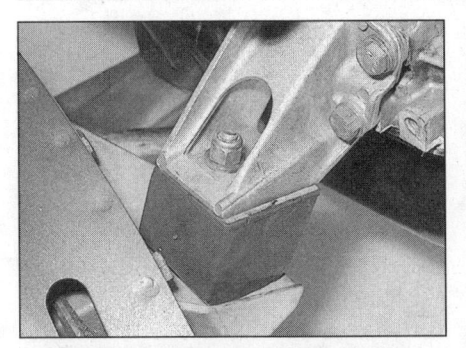

25.10a Right-hand engine mounting

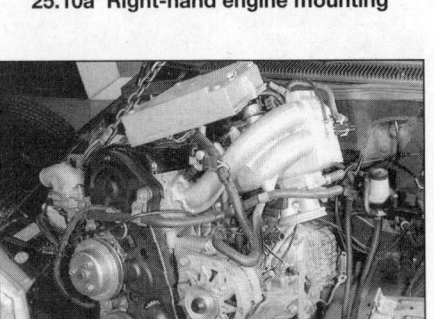

25.10b Transmission mounting

25.11 Removing the engine/transmission unit

25.15 Unscrewing engine/transmission bolt

18 Offer the transmission to the engine and locate it on the positioning dowels.
19 Screw in the connecting bolts, noting the locations of the lifting lugs and clips.
20 Refit the cover plate to the front face of the clutch bellhousing and screw in the reinforcement bracket bolts.
21 The remaining operations are a reversal of removal but observe the following points.

a) Arrange the lifting gear so that the transmission is slanting downwards at a steep angle and the engine/transmission is canted over to the left. It will be found easier to install the engine if the right-hand mounting bracket bolts are loosened.
b) Refit the driveshafts as described in Chapter 8.
c) Ensure that the wiring loom is correctly routed, and retained by all the relevant retaining clips; all connectors should be correctly and securely reconnected.
d) Ensure that all coolant hoses are correctly reconnected, and securely retained by their retaining clips.
e) On models with a cable-operated clutch, adjust the cable as described in Chapter 6.
f) Adjust the accelerator cable as described in Chapter 4.
g) Refill the engine and transmission with correct quantity and type of oil, as described in Chapter 1A.
h) Refill the cooling system as described in Chapter 1A.

26 Engine - removal and refitting leaving automatic transmission in the car

Removal

1 Carry out the operations described in paragraphs 1 to 18 of Section 24.
2 Working through the starter motor aperture, lock the teeth of the ring gear using a suitable tool (see illustration 7.7) and unscrew the driveplate-to-torque converter fixing bolts.
3 In order to bring each bolt into view, the crankshaft will have to be turned by means of a socket on its pulley bolt while the ring gear locking tool is temporarily removed.
4 Unscrew and remove the bolts which hold the torque converter lower cover plate. The cover plate will drop off its dowels only after the engine and transmission are separated.
5 Unscrew and remove the engine-to-transmission bolts.
6 Attach a hoist to the engine lifting lugs and just take the weight of the engine.
7 Place a jack under the transmission.
8 Unscrew the engine mounting through-bolt nuts.
9 Raise the engine to clear the mounting brackets, then withdraw it forward to disconnect it from the transmission and lift it up and out of the engine compartment.
10 Bolt a retaining bar to one of the torque converter housing flange bolts in order to retain the converter in full engagement with the transmission and to prevent damage to the oil seal.

Refitting

11 Apply a smear of high temperature grease (Renault recommend the use of Molykote BR2) grease to the locating boss on the torque converter.
12 Remove the temporary holding bar from the torque converter.
13 Lower the engine into the car, making sure that the torque converter does not move forward as the engine is connected and bolted to the transmission. Note that the upper bellhousing bolts must be in position before the engine is connected to the transmission as they cannot be fitted afterwards.
14 The bellhousing lower cover must be located on the dowels before the engine and transmission are brought together, and the alignment mark on the driveplate must be between the two marks on the torque converter (see Chapter 7B).
15 The remaining reconnection and refitting operations are reversals of removal, but observe the following points.

a) Ensure that the wiring loom is correctly routed, and retained by all the relevant retaining clips; all connectors should be correctly and securely reconnected.
b) Ensure that all coolant hoses are correctly reconnected, and securely retained by their retaining clips.
c) Adjust the accelerator cable as described in Chapter 4A or 4B.
d) Adjust the selector cable as described in Chapter 7B.
e) Refill the engine with correct quantity and type of oil, as described in Chapter 1A. Also check the transmission fluid level.
f) Refill the cooling system as described in Chapter 1A.

27 Engine and automatic transmission - removal and refitting

Removal

1 Carry out the operations described in paragraphs 1 to 18 of Section 24.
2 Disconnect the earth straps.
3 Remove the transmission dipstick tube.
4 Disconnect the fluid hoses from the transmission fluid cooler and plug them (see Chapter 7B).
5 Disconnect the TDC sensor and the speedometer sensor, also the reversing lamp switch leads.
6 Remove the driveshafts as described in Chapter 8.
7 Disconnect the selector control linkage or cable, governor cable (where fitted) and the vacuum pipe and computer wiring plugs using the information in Chapter 7B (as applicable).
8 With the hoist attached to the lifting lugs and taking the weight of the engine and transmission, remove the transmission

2A

2A•32 Four cylinder petrol engine

mountings and unscrew the nuts from the engine mounting through-bolts. Prepare to guide the computer out with the transmission.
9 Withdraw the engine/transmission forward and upwards out of the engine compartment.
10 Clean away external dirt using paraffin and a stiff brush or a water soluble solvent.
11 Separate the transmission from the engine as described in paragraphs 5 to 7. Pull the transmission from the engine and then fit a torque converter retaining bar as described in paragraph 10.

Refitting
12 Refer to Section 26.

28 Oil cooler components - removal and refitting

Removal
1 Disconnect the battery negative lead and proceed as described under the relevant sub-heading.

Oil cooler
2 If necessary to improve access to the oil cooler, firmly apply the handbrake then jack up the front of the vehicle and support it on axle stands (see *Jacking and vehicle support*").
3 Position a suitable container beneath the oil filter, then unscrew the filter using an oil filter removal tool, and drain the oil into the container. If the oil filter is damaged or disfigured during removal, it must be renewed and the engine should be filled with clean oil on refitting. Disconnect the oil hoses from the cooler as described in paragraph 6.
4 Release the hose clips and disconnect the coolant hoses from the oil cooler. To minimise coolant loss, clamp the coolant hoses using a suitable brake hose or G-clamp prior to disconnecting them.
5 Unscrew the oil cooler/oil filter mounting bolt from the base of the mounting plate, and withdraw the cooler. Discard the oil cooler sealing ring, a new one must be used on refitting.

Oil hoses
6 To remove an oil cooler hose, position a suitable hose beneath the oil cooler then slacken the union nut securing the hose to the cooler mounting plate. Disconnect the hose and allow the oil to drain into the container then unscrew the hose from the adapter plate on the cylinder block/crankcase and remove it from the engine compartment. Recover the sealing ring from each hose union and discard them; new ones should be used on refitting.
7 If necessary, remove the second hose in the same way. If both hoses are to be removed or disconnected at the same time, make alignment marks between the hose unions and adapter plate/cooler mounting plate (as applicable) to remove the possibility of the hoses being wrongly reconnected on refitting.

Cylinder block/crankcase adapter plate
8 Remove both the oil hoses as described above.
9 There are two possible types of adapter plate fitted; the first is secured to the cylinder block by three bolts and the second by a large centre bolt.
10 To remove the first type, slacken and remove the three bolts and washers securing the plate to the side of the block then remove the plate and recover both its sealing rings. Discard both sealing rings; new ones should be used on refitting.
11 To remove the second type, unscrew the large centre bolt and washer then slide the adapter plate off its mounting stud. Recover the large sealing ring from the back of the adapter plate and discard it; a new one should be used on refitting.

Refitting

Oil cooler
12 Fit a new sealing ring to the recess in the top of the oil cooler and locate the cooler on mounting plate. Ensure that the cooler coolant hose unions are facing towards the rear of the vehicle then refit the oil cooler/oil filter mounting bolt and tighten it securely.
13 Apply a smear of clean engine oil to the oil filter sealing ring and screw the filter securely onto the cooler, tightening it only by hand.
14 Reconnect the coolant hoses to the oil cooler, securing them in position with the retaining clips, and remove the clamp.
15 Lower the vehicle to the ground and top-up or refill the engine oil (as applicable).
16 Check the coolant level as described in Chapter 2, then start the engine and check the oil cooler assembly for signs of leakage.

Oil hoses
17 Fit a new sealing ring to the hose union then reconnect the hose to the adapter, tightening it securely. If both hoses have been disconnected or removed use the marks made on removal to ensure that the hose is reconnected to the correct union.
18 Fit a new sealing ring to the cooler end of the hose then reconnect the hose to the cooler mounting plate and securely tighten its union nut. Again, if both hoses have been disconnected use the marks made on removal to ensure that the hose is connected to the correct union.
19 Where necessary fit the second hose in the same way.
20 Reconnect the battery then start up the engine and check the hose unions for signs of leakage.

Cylinder block/crankcase adapter plate
21 Where the first type of adapter plate is fitted, locate the new sealing rings in their recesses in the rear of the adapter plate. Manoeuvre the plate into position on the block, taking great care not displace either sealing ring, then refit the three retaining bolts and washers and tighten them securely.
22 Where the second type of adapter plate is fitted, locate the new sealing ring in the recess in the rear of the adapter plate. Locate the adapter plate on its retaining stud, ensuring that the sealing ring is not displaced, noting that the locating lug should be positioned at the top of the plate. With the plate correctly positioned, refit the centre bolt and washer and tighten it securely.
23 Refit the oil hoses as described above.

29 Initial start-up after major overhaul

1 With the engine refitted in the vehicle, double-check the engine oil and coolant levels. Make a final check that everything has been reconnected, and that there are no tools or rags left in the engine compartment.
2 Remove the spark plugs. Disable the ignition system by disconnecting the ignition HT coil lead from the distributor cap, and earthing it on the cylinder block. Use a jumper lead or similar wire to make a good connection.
3 Turn the engine on the starter until the oil pressure warning light goes out. Refit the spark plugs, and reconnect the spark plug (HT) leads, referring to Chapter 1A for further information. Reconnect the HT lead to the distributor.
4 Start the engine, noting that this may take a little longer than usual, due to the fuel system components having been disturbed.
5 While the engine is idling, check for fuel, water and oil leaks. Don't be alarmed if there are some odd smells and smoke from parts getting hot and burning off oil deposits.
6 Assuming all is well, keep the engine idling until hot water is felt circulating through the top hose, then switch off the engine.
7 Check the ignition timing and the idle speed settings (as appropriate), then switch the engine off.
8 After a few minutes, recheck the oil and coolant levels as described in *Weekly Checks*, and top-up as necessary.
9 If they were tightened as described, there is no need to re-tighten the cylinder head bolts once the engine has first run after reassembly.
10 If new pistons, rings or crankshaft bearings have been fitted, the engine must be treated as new, and run-in for the first 500 miles (800 km). *Do not* operate the engine at full-throttle, or allow it to labour at low engine speeds in any gear. It is recommended that the oil and filter be changed at the end of this period.

Chapter 2 Part B:
Six cylinder petrol engine

Contents

Camshaft and rocker arms - removal, inspection and refitting	10
Camshaft oil seals - renewal	14
Compression test - description and interpretation	2
Crankshaft - removal, inspection and refitting	18
Crankshaft oil seals - renewal	15
Cylinder block/crankcase - cleaning and inspection	21
Cylinder head - overhaul	12
Cylinder heads - removal and refitting	11
Engine - removal and refitting leaving automatic transmission in car	24
Engine - removal and refitting leaving manual transmission in car	22
Engine and automatic transmission - removal and refitting	25
Engine and manual transmission - removal and refitting	23
Engine/transmission mountings - renewal	17
Flywheel/driveplate - removal, inspection and refitting	16
General description	1
Initial start-up after major overhaul	27
Oil cooler components - removal and refitting	26
Oil pump - removal, inspection and refitting	8
Operations possible without removing the engine	4
Piston rings - removal and refitting	20
Piston/connecting rod and cylinder liner assemblies - removal, inspection and refitting	19
Rocker covers - removal and refitting	5
Sump - removal and refitting	13
Timing chains and sprockets - removal, inspection and refitting	9
Timing cover - removal and refitting	7
Top dead centre (TDC) for number 1 piston - locating	3
Valve clearances - adjustment	6

Degrees of difficulty

Easy, suitable for novice with little experience

Fairly easy, suitable for beginner with some experience

Fairly difficult, suitable for competent DIY mechanic

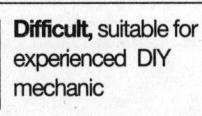

Difficult, suitable for experienced DIY mechanic

Very difficult, suitable for expert DIY or professional

Specifications

General
Engine type .... Six-cylinder, 90° vee configuration mounted longitudinally at front of car. Light alloy construction with wet cylinder liners. Chain-driven overhead camshafts

Engine code:
   2458 cc turbo engine Z7U
   2664 cc engine Z7V
   2849 cc engine Z7W

Bore:
   2458 cc turbo and 2849 cc engine .... 91.0 mm
   2664 cc engine .... 88.0 mm

Stroke:
   2458 cc turbo engine .... 63.0 mm
   2664 cc and 2849 cc engines .... 73.0 mm

Firing order .... 1-6-3-5-2-4 (No 1 at transmission end of left-hand cylinder bank)

Camshaft
Number of bearings 4
Endfloat .... 0.07 to 0.14 mm

Cylinder head
Overall height:
   2458 cc turbo and 2664 cc engines .... 111.07 ± 0.15 mm
   2849 cc engine .... 110.83 mm
Maximum permitted surface distortion .... 0.05 mm

Valves
Valve clearances (engine cold):
   Inlet .... 0.10 mm
   Exhaust .... 0.25 mm
Stem diameter .... 8.0 mm

2B•2 Six cylinder petrol engine

Valves (continued)
Head diameter:
  Inlet:
    2458 cc turbo engine .... 43.8 mm
    2664 cc engine .... 44.0 mm
    2849 cc engine .... 45.3 mm
  Exhaust:
    2458 cc turbo and 2664 cc engine .... 37.0 mm
    2849 cc engine .... 38.5 mm
Valve seat angle:
  Inlet:
    2458 cc turbo and 2849 cc engine .... 120°
    2849 cc engine .... 90°
  Exhaust .... 90°
Valve seat width:
  Inlet:
    2458 cc turbo engine .... 1.3 to 1.7 mm
    2664 cc and 2849 cc engines .... 1.7 to 2.1 mm
  Exhaust .... 2.0 to 2.4 mm
Valve guides:
  Bore .... 8.0 mm
  Outside diameter (nominal) .... 13.0 mm
  Repair oversizes:
    With one groove .... 13.10 mm
    With two grooves .... 13.25 mm
Valve spring free length .... 47.2 mm

Crankshaft
Number of main bearings 4
Main bearing journal diameter .... 70.062 mm
Main bearing journal minimum regrind diameter - 2664 cc engine only . 69.762 mm
Main bearing running clearance* .... 0.025 to 0.050 mm
Crankpin diameter:
  2458 cc turbo and 2849 cc engines .... 60.000 mm
  2664 cc engine .... 52.290 mm
Crankpin minimum regrind diameter - 2664 cc engine only .... 51.996 mm
Big-end bearing running clearance* .... 0.025 to 0.050 mm
Crankshaft endfloat .... 0.07 to 0.27 mm
*These are suggested figures, typical for this type of engine - no exact values are quoted by Renault

Cylinder liners
Bore:
  2458 cc turbo and 2849 cc engines .... 91.0 mm
  2664 cc engine .... 88.0 mm
Base locating diameter:
  2458 cc turbo engine .... 96.48 mm
  2664 cc engine .... 93.48 mm
  2849 cc engine .... 97.68 mm
Liner protrusion:
  2458 cc turbo and 2664 cc engines .... 0.16 to 0.23 mm (fitted with paper type seals)
  2849 cc engine .... 0.13 to 0.20 mm (fitted with lacquered type seals)
Maximum difference in protrusion between two adjacent liners ...... 0.04 mm
Liner base seal thickness:
  Paper seals:
    Blue .... 0.087 mm
    White .... 0.102 mm
    Red .... 0.122 mm
    Yellow .... 0.147 mm
  Lacquered seals:
    Red .... 0.116 mm
    Unmarked .... 0.136 mm
    Blue .... 0.166 mm

Connecting rods and gudgeon pins
Gudgeon pins:
  Length:
    2849 cc engine .... 60.0 mm
    2458 cc turbo and 2664 cc engine .... 72.0 mm

Six cylinder petrol engine

Gudgeon pins (continued)
Outside diameter:
- 2664 cc engine .... 23.5 mm
- 2458 cc turbo and 2849 cc engine .... 25.0 mm

Piston rings
Piston ring thickness:
- Top compression:
  - 2458 cc turbo engine .... 1.75 mm
  - 2664 cc and 2849 cc engine .... 1.5 mm
- Second compression .... 2.0 mm
- Oil control .... 4.0 mm

Lubrication system
Oil pump: .... Gear-driven by chain from crankshaft pulley
Oil pressure (at 80°C):
- 2849 cc engine:
  - At idle .... 1.0 bar (14 psi)
  - At 5500 rpm .... 4.0 bar (58 psi)
- 2458 cc turbo and 2664 cc engines:
  - At 900 rpm .... 2.2 bar (32 psi)
  - At 4000 rpm .... 4.4 bar (64 psi)

Torque wrench settings

	Nm	lbf ft
Big-end bearing cap nuts	45	33
Camshaft sprocket bolt:		
2458 cc turbo and 2849 cc engine	80	59
2664 cc engine	75	55
Crankshaft pulley nut	180	132
Cylinder head bolts:		
2458 cc turbo engine:		
Stage 1	60	44
Fully slacken all bolts then:		
Stage 2	20	15
Stage 3	Angle-tighten through 127°	
Run the engine for 15 minutes, leave to cool (6 hours min) then:		
Stage 4	Angle-tighten a further 25°	
2664 cc engine:		
Stage 1	60	44
Fully slacken all bolts then:		
Stage 2	20	15
Stage 3	Angle-tighten through 115°	
Run the engine for 15 minutes, leave to cool (6 hours min) then fully slacken all bolts and:		
Stage 4	20	15
Stage 5	Angle-tighten a further 115°	
2849 cc turbo engine:		
Stage 1	60	44
Fully slacken all bolts then:		
Stage 2	20	15
Stage 3	Angle-tighten through 106°	
Run the engine for 15 minutes and leaver to cool for at least 6 hours:		
Stage 4	Angle-tighten a further 45°	
Main bearing cap nuts:		
Stage 1	30	23
Stage 2	Angle-tighten through 75°	
Driveplate bolts - automatic transmission models	65	48
Engine to transmission bolts:		
Manual transmission	50	37
Automatic transmission	60	44
Flywheel bolts - manual transmission models	45	33
Main bearing casting to cylinder block bolts	20	15
Oil pump cover bolts	15	11
Oil pump sprocket bolts	7	5
Sump bolts	20	15
Steering pump drivebelt camshaft pulley bolt - 2664 cc engine	100	73
Timing cover bolts	15	11

2B•4 Six cylinder petrol engine

1 General description

1 The six cylinder engine fitted to the Renault is of the V6, overhead camshaft (OHC) type. The engine is available in three forms; 2458 cc Turbo, 2664 cc and 2849 cc versions **(see illustration)**. Its basic design is as follows.
2 The cylinder heads are of the crossflow type: that is to say, the inlet ports are on one side of the head and the exhaust ports on the other. There are separate alloy-cast iron valve seats fitted and the cylinder head bolts also retain the rocker arm mechanism.
3 The camshafts, journalled in four bearings of different sizes, are chain-driven from the front of the crankshaft; each chain being tensioned separately. The oil pump too is chain-driven off the crankshaft at the front end, and is located in the front of the block. The whole driving system is covered by a single section light alloy cover plate.
4 At the end of each camshaft there is an extra driving arrangement: on 2664 cc engines, the distributor is worm driven from the rear right-hand side, while on 2458 cc Turbo and 2849 cc engines, the distributor is driven from the front left-hand side. The power steering pump unit is also driven from the rear end of the left-hand camshaft, by means of an external pulley.
5 The cylinder block has wet replaceable cast iron liners, which are class matched to the very light aluminium alloy pistons. These have two compression rings and one oil scraper ring.
6 The crankshaft is short, strong and has ground, surface-hardened bearing journals; it runs in four well-proportioned main bearings in the cylinder block. The main bearing caps and lower half of the crankcase are separate units, but the nuts for the main bearing caps are underneath (outside) the lower crankcase. This helps give added torsion strength to the light alloy engine construction. There are only three big-end crankpins, two connecting rods being mounted on each. The big-end bearings and main bearings run in renewable shells and all the crankpins can be ground to a limited undersize. The endfloat on the crankshaft is governed by variable size thrustwashers.
7 The connecting rods are made of drop-forged steel; the gudgeon pins are pressed into them and journalled in the pistons.
8 On Turbo versions, although the specified engine overhaul and servicing details generally apply, special reference should be made to the *Specifications* at the beginning of this Chapter and to relevant paragraphs detailing Turbo differences in some Sections of this Chapter.
9 The engine lubrication system relies on a crankshaft driven oil pump and incorporates a full-flow filter.
10 On Turbo models, the system incorporates an oil cooler and supplies oil to the turbocharger.

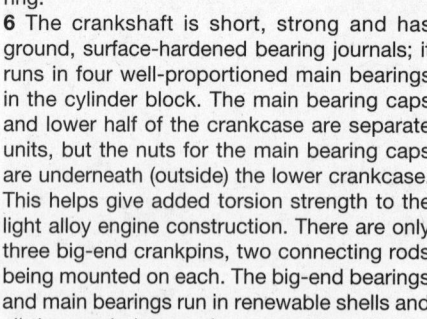

2 Compression test - description and interpretation

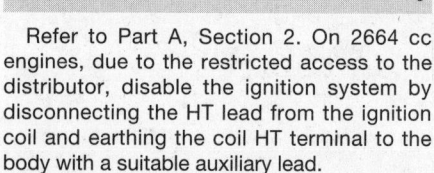

Refer to Part A, Section 2. On 2664 cc engines, due to the restricted access to the distributor, disable the ignition system by disconnecting the HT lead from the ignition coil and earthing the coil HT terminal to the body with a suitable auxiliary lead.

3 Top dead centre (TDC) for number 1 piston - locating

1 Top dead centre (TDC) is the highest point in the cylinder that each piston reaches as the crankshaft turns. Each piston reaches TDC at the end of the compression stroke and again at the end of the exhaust stroke; however, for the purpose of timing the engine, TDC refers to the position of No1 piston at the end of its compression stroke. On all engines, No1 piston (and cylinder) is at the transmission (rear) end of the left-hand cylinder bank.
2 To determine when No 1 piston is on its compression stroke, it will be necessary to remove one of the rocker covers. The movement of the rocker arms can then be used to identify when No 1 cylinder is on its compression stroke.
3 With the rocker cover removed, rotate the crankshaft using a spanner/socket and extension bar on the crankshaft pulley bolt. If necessary to improve access to the bolt, firmly apply the handbrake then jack up the front of the vehicle and support it on axle stands (see "*Jacking and vehicle support*").
4 Turn the crankshaft in the normal direction of rotation (clockwise) whilst keeping an eye on the transmission housing aperture. Align the flywheel/driveplate mark with the TDC (0°) mark on the bellhousing **(see illustration)**.
5 Rotate the crankshaft back and forth slightly whilst observing the movement of the rocker arms. If No 1 cylinder is at TDC on its compression stroke, the rocker arms of No 5 cylinder (centre cylinder on the right-hand bank) will be rocking (ie. exhaust valve closing, inlet valve opening). If the rocker arms of No 1 cylinder are rocking, rotate the crankshaft through one complete turn in the correct direction of rotation.
6 Realign the flywheel/driveplate marking with the TDC (0°) mark; No 1 cylinder is now positioned at TDC on its compression stroke.

4 Operations possible without removing the engine

The following operations are possible without the need to remove the engine from the car.
a) *Compression pressure - testing.*

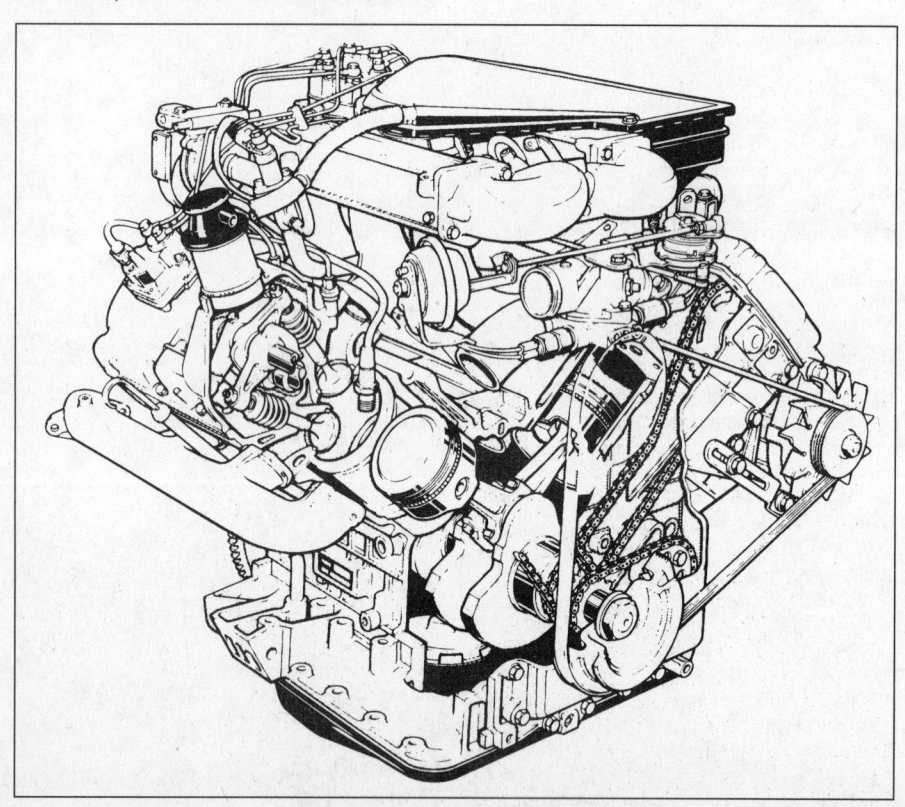

1.1 Cutaway view of 2664 cc engine

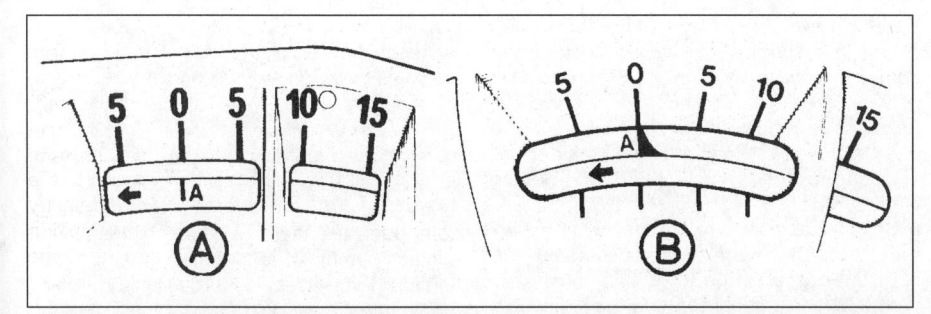

Six cylinder petrol engine 2B•5

3.4 Flywheel/driveplate timing mark aligned with transmission housing 0° timing mark
A Manual transmission *B Automatic transmission*

5.3a Slacken and remove the retaining screws ...

5.3b ... and lift off the rocker cover (2664 cc engine shown)

b) Valve clearances - checking and adjusting
c) Rocker covers - removal and refitting.
d) Oil pump - removal, inspection and refitting.
e) Timing chains and sprockets - removal and refitting.
f) Camshaft and intermediate shaft oil seals - renewal.
g) Camshaft and rocker arms - removal and refitting.
h) Cylinder head - removal and refitting.
i) Cylinder head and pistons - decarbonising.
j) Sump - removal and refitting.
k) Crankshaft oil seals - renewal.
l) Engine/transmission mountings - inspection and renewal.
m) Flywheel/driveplate - removal and refitting.

5 Rocker covers - removal and refitting

Removal

1 To gain access to the rocker covers, remove the air cleaner and/or necessary inlet ducts as described in the relevant part of Chapter 4.
2 Remove the oil filler cap then slacken the retaining clips and disconnect the breather hose(s) from the rocker cover(s) (as applicable)
3 Position the relevant hoses clear of the rocker cover(s) then slacken and remove the retaining bolts and washers and lift off the cover(s) **(see illustrations)**. Remove the gasket(s) and discard them, new one(s) should used on refitting.

Refitting

4 Refitting is the reverse of removal using new gasket(s). To prevent possible oil leakage, prior to fitting the gaskets, apply a smear of sealant (Renault recommend CAF 4/60 Thixo) to the areas where the timing cover meets the cylinder head.

6 Valve clearances - adjustment

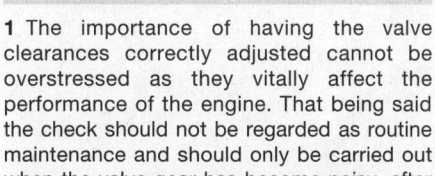

1 The importance of having the valve clearances correctly adjusted cannot be overstressed as they vitally affect the performance of the engine. That being said the check should not be regarded as routine maintenance and should only be carried out when the valve gear has become noisy, after engine overhaul, or when trying to trace the cause of power loss which may be attributed to the valve. The clearances are checked as follows noting that the engine must be cold for the check to be accurate.
2 Remove both rocker covers (see Section 5).
3 Position No 1 cylinder at TDC on its compression stroke as described in Section 3.
4 Check and adjust as necessary the following valve clearances. Refer to the *Specifications* for the correct clearances, the valve locations can be determined from the position of the manifolds **(see illustration)**.
 No 1 cylinder inlet valve
 No 2 cylinder inlet valve
 No 4 cylinder inlet valve
 No 1 cylinder exhaust valve
 No 3 cylinder exhaust valve
 No 6 cylinder exhaust valve

5 Clearances are checked by inserting a feeler blade of the correct thickness between the valve stem and the rocker arm adjusting screw, the feeler blade should be a light, sliding fit. If adjustment is necessary, slacken the adjusting screw locknut and turn the screw as necessary until the feeler blade is a light sliding fit. Once the correct clearance is obtained hold the adjusting screw and securely tighten the locknut **(see illustration)**. Recheck the valve clearance and adjust again if necessary.
6 Once all six specified valves have been checked, rotate the crankshaft through one complete turn in the correct direction of rotation and realign the flywheel/driveplate mark with the TDC (0°) mark on the bellhousing. Check that No 1 cylinder rocker arms are "rocking" and then check/adjust the following valves:
 No 3 cylinder inlet valve
 No 5 cylinder inlet valve
 No 6 cylinder inlet valve
 No 2 cylinder exhaust valve
 No 4 cylinder exhaust valve
 No 5 cylinder exhaust valve

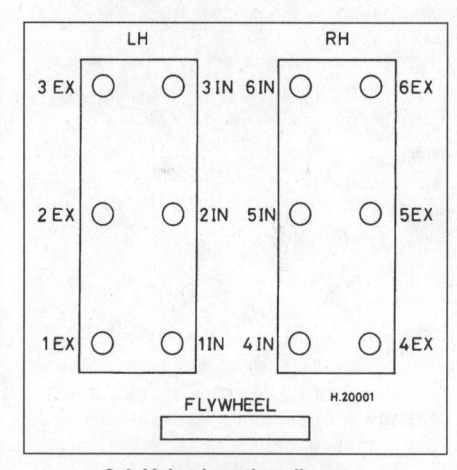

6.4 Valve location diagram

6.5 Adjusting a valve clearance

2B•6 Six cylinder petrol engine

7 When all twelve valve clearances have been checked and where necessary adjusted, refit the rocker covers as described in Section 5.

7 Timing cover - removal and refitting

Removal

1 Drain the engine oil as described in Chapter 1A.
2 Remove the radiator and coolant pump as described in Chapter 3.
3 Remove the auxiliary drivebelt(s) as described in Chapter 1A and unbolt the alternator adjusting strap from the engine.
4 Remove the rocker covers as described in Section 5.
5 Unbolt and remove the engine undertray.
6 On 2458 cc turbo and 2849 cc engines, remove the distributor cap and rotor arm as described in Chapter 5B. Slacken and remove the camshaft sprocket bolt (see paragraph 8), remove the rotor drive flange and locating pin then refit the bolt and tighten securely,
7 On models with air conditioning, remove the cooling fan motor(s) and unbolt the air conditioning compressor and position it clear of the engine. **Do not** disconnect the refrigerant pipes from the compressor (see *Warnings* in Chapter 3).
8 To prevent crankshaft rotation whilst the pulley retaining nut is slackened, unbolt the cover plate from the right-hand side of the engine and lock the flywheel/driveplate ring gear, using an arrangement similar to that shown **(see illustration)**. On manual transmission models, if the engine is still in the vehicle the crankshaft can be retained by selecting top gear and applying the brakes firmly.
9 Unscrew and remove the crankshaft pulley nut. When the nut is removed, release the crankshaft and rotate it so that the pulley Woodruff key is at the top and then withdraw the crankshaft pulley **(see illustration)**.
10 Remove the retaining bolts from the timing cover, noting their varying lengths and locations as they are removed. When all the bolts are removed withdraw the timing case. It may be stuck in position, in which case some careful light taps with a soft-headed mallet from the rear will help to unseal it. Remove the gaskets and discard them and recover the cover locating dowels.

Refitting

11 Ensure that the timing cover and cylinder block mating surfaces are clean and dry. Ensure that the locating dowels are in position then fit the new gaskets and cover **(see illustrations)**. On 2458 cc turbo and 2849 cc engines take care not to damage the camshaft oil seal as the cover is fitted.
12 Fit the timing cover retaining bolts in their original locations, noting that the lower retaining bolt threads should be smeared with sealant (Renault recommend the use of Loctite Frenatanch) to prevent possible oil leakage. Tighten all the cover bolts to the specified torque **(see illustration)**.
13 Apply a smear of oil to the crankshaft pulley oil seal surface and carefully slide the pulley into position, aligning it with the Woodruff key.
14 Apply a few drops of thread locking compound (Renault recommend the use of Loctite Frenetanch) to the pulley retaining and tighten it to the specified torque whilst prevent crankshaft rotation using the method employed on removal.
15 Trim off the top edges of the timing cover gasket which protrude above the cylinder head surface then refit the rocket covers.
16 On 2458 cc and 2849 cc engines, remove the left-hand camshaft sprocket bolt and refit the rotor drive flange pin to the spacer hole. Refit the rotor flange, making sure it is correctly engaged with the pin and tighten the camshaft sprocket bolt to the specified torque. Refit the rotor arm and distributor cap as described in Chapter 5B.
17 The remainder of refitting is the reverse of removal referring to the relevant Chapters for further information. On completion refill the engine with oil and the cooling system as described in Chapter 1A.

8 Oil pump - removal, inspection and refitting

Removal

1 The oil pump is located in the front face of the cylinder block. Access to it is gained by removing the timing cover. The pump is driven by a chain from the crankshaft.
2 Remove the timing cover as described in Section 7.
3 Unscrew and remove the three oil pump sprocket securing bolts and then withdraw the oil pump driven sprocket and chain. If necessary slide the drive sprocket off the end of the crankshaft, noting which way around it is fitted, then remove the Woodruff key and slide off the spacer **(see illustrations)**.
4 Unscrew the oil pump unit retaining bolts

7.8 Use a fabricated tool like this one to lock the flywheel/driveplate ring gear and prevent crankshaft rotation

7.9 Removing the crankshaft pulley

7.11a Fit the new gaskets to the cylinder block . . .

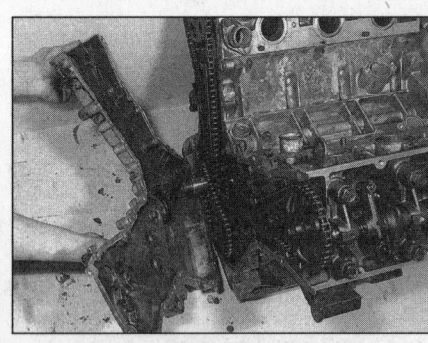

7.11b . . . and refit the timing cover

7.12 Refit the timing cover bolts and tighten them to the specified torque setting

Six cylinder petrol engine 2B•7

8.3a Undo the retaining bolts and remove the oil pump sprocket and drive chain . . .

8.3b . . . then slide off the drive sprocket . . .

8.3c . . . and spacer from the crankshaft

8.4 Undo the retaining bolts and remove the oil pump from the cylinder block

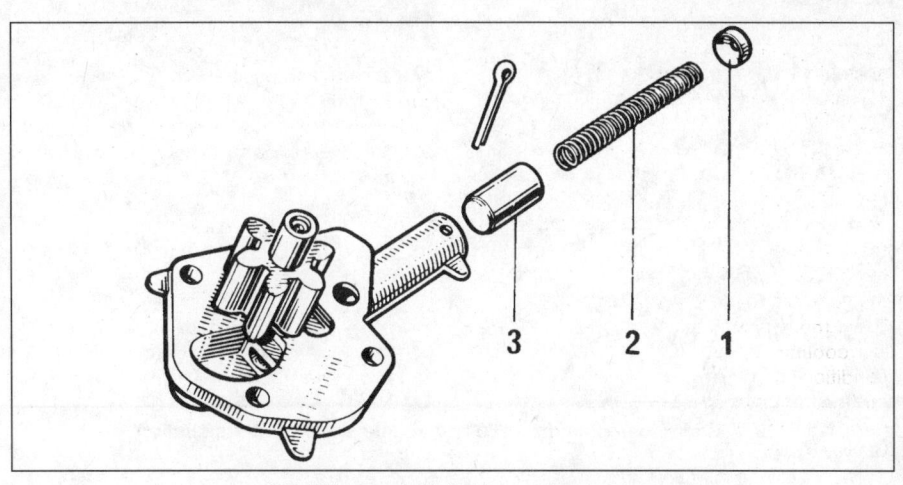

8.7 Oil pump relief valve components

1 Spring seat   2 Spring   3 Plunger

and withdraw the pump unit from the front end of the cylinder block. It may be tight, but do not attempt to free it by hammering or levering on the relief valve boss (with the split pin through it) **(see illustration)**.
5 Remove the oil pump gear and shaft.

Inspection

6 Check the pump rotor and its housing for any signs of wear or damage. The complete pump assembly must be renewed if found to be defective. If the pump housing in the cylinder block is excessively worn seek the advice of a Renault dealer or other suitable engine repair specialist.
7 To inspect the oil pressure relief valve, extract the split pin and withdraw the spring seat, spring and release valve plunger **(see illustration)**. Check the end face of the plunger and its contact face within the body

8.15 Prime the oil pump by injecting oil in through the hole behind the oil filter, with the filter removed

for signs of scoring or damage, and renew if necessary.
8 The drive chain and sprockets should also be checked for wear and renewed if worn.
9 Reassembly of the pump unit is a reversal of the dismantling procedure, making sure the relief valve plunger is fitted the correct way round.

Refitting

10 Lubricate the pump gear and shaft with clean engine oil then fit the pump assembly to the cylinder block. Tighten the retaining bolts to the specified torque then rotate the sprocket flange to ensure that the pump gear rotates freely.
11 Where removed, refit the spacer to the end of the crankshaft then fit the sprocket Woodruff key and slide on the oil pump drive sprocket, making sure it is fitted the correct way around.
12 Wipe clean the threads of the pump sprocket bolts and apply a few drops of thread locking compound to each of them (Renault recommend the use of Loctite Frenetanch).
13 Fit the pump driven sprocket and chain then fit the sprocket retaining bolts and tighten them to the specified torque.
14 Refit the timing cover as described in Section 7.

15 Prior to starting the engine, it is recommended that the oil pump is primed with clean engine oil by injecting it in through the hole behind the oil filter **(see illustration)**. If not disable the ignition system (see Section 2) and turn the engine over on the starter motor until the oil pressure warning light goes out. Then reconnect the ignition system and start the engine.

9 Timing chains and sprockets - removal, inspection and refitting

Removal

1 Remove the timing cover as described in Section 7 ensuring that the crankshaft is positioned with its Woodruff key is uppermost. This will set No 1 piston 15.0 mm down its bore BTDC to prevent piston-to-valve contact **(see illustration)**.
2 Unscrew and remove the three oil pump sprocket securing bolts and then withdraw the oil pump driven sprocket and chain. If necessary slide the drive sprocket off the end of the crankshaft, noting which way around it is fitted, then remove the Woodruff key and slide off the spacer (see Section 8).

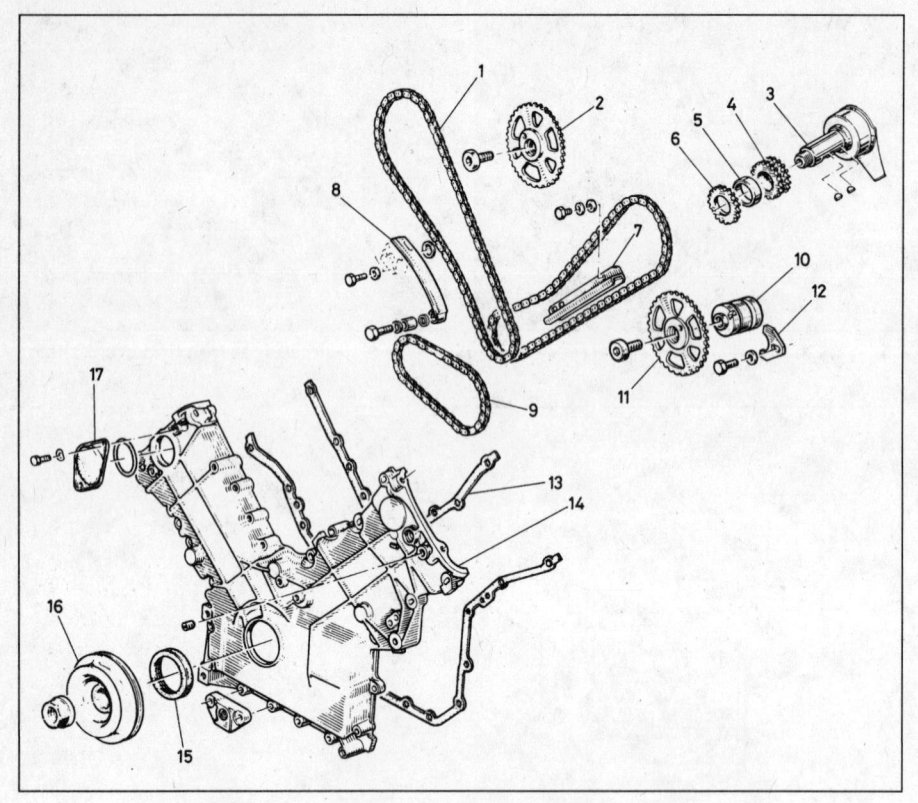

2B•8 Six cylinder petrol engine

9.1 Timing components - 2664 cc engine shown (others similar)

1 Timing chain
2 Camshaft sprocket
3 Crankshaft
4 Crankshaft sprocket
5 Spacer
6 Oil pump drive sprocket
7 Chain slipper (fixed)
8 Chain slipper moveable
9 Oil pump drive chain
10 Camshaft
11 Camshaft sprocket
12 Camshaft thrust plate
13 Gasket
14 Timing cover
15 Timing cover oil seal
16 Crankshaft pulley
17 Camshaft sprocket cover plate

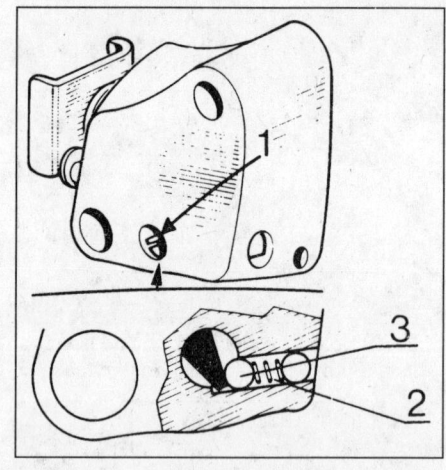

9.5a Chain tensioner retaining lock details

1 Retaining lock   2 Spring   3 Ball

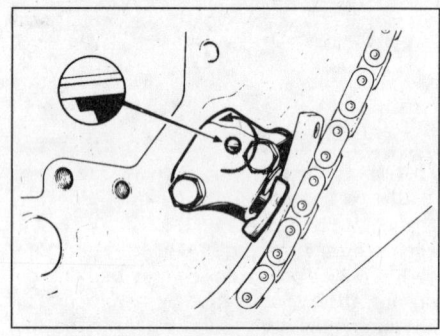

9.5b Release the chain tensioner by turning the ratchet mechanism anti-clockwise

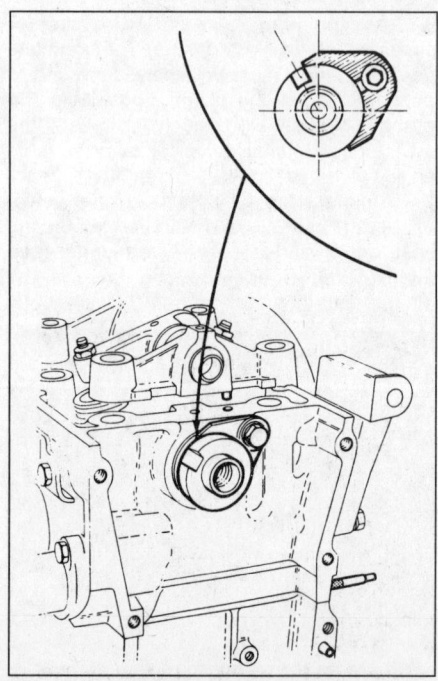

9.11 Align the left-hand camshaft slot with the thrust plate as shown

3 Slacken and remove the camshaft sprocket bolts and spacers.

4 Using quick drying paint, mark all timing components LH or RH if they are to be used again, so that they can be refitted in their original positions.

5 Do not disturb the chain tensioner retaining lock, but retract both chain tensioners by turning the ratchet mechanism anti-clockwise using a screwdriver **(see illustrations)**.

6 Remove the right-hand and then the left-hand timing chains complete with camshaft sprockets. If the sprocket locating pins are a loose fit, remove them and store them with the sprockets.

7 Remove the chain tensioners, the chain slippers and filters. If necessary, slide the sprocket off the crankshaft, noting which way around it is fitted, and remove the Woodruff key.

Inspection

8 After a considerable mileage the timing chains and sprockets will have worn and possibly be in need of renewal.

9 Clean and examine the camshaft sprockets and also the crankshaft and oil pump drive sprockets. If they are worn, indicated by hooked teeth, then they must be renewed. It is possible to renew the chains at time of major overhaul; if new sprockets are being fitted, new chains must be used.

10 Check the respective tensioner blades and pivots for wear. The tensioner block itself should not be dismantled; if suspect, renew it. If the retaining lock in the tensioner is accidentally removed then it must be renewed since there is no way of checking the lock finger position relative to the thrust ball during assembly, and it is possible to jam the tip of the finger with the spring.

Refitting

11 Commence refitting by turning the left-hand camshaft until the slot is aligned with the thrust plate as shown **(see illustration)**. In this position the valves of No 1 cylinder will be "rocking" (one valve seating as the other one starts to rise).

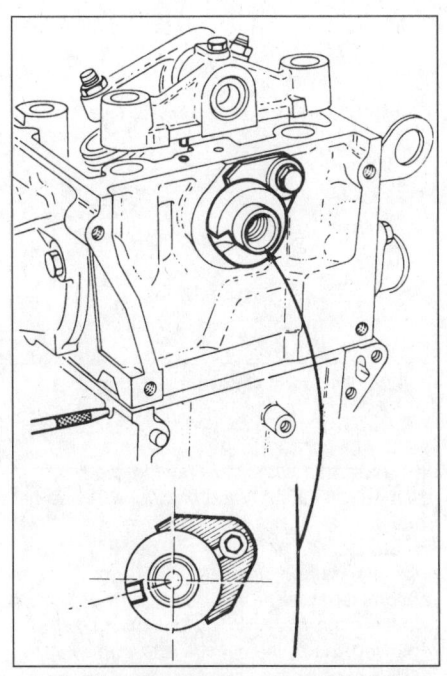

Six cylinder petrol engine 2B•9

9.12 Align the right-hand camshaft with the thrust plate as shown

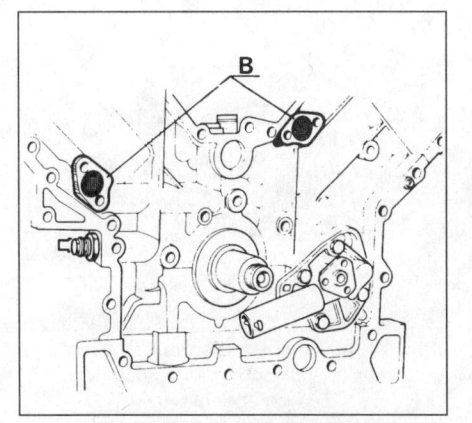

9.13 Ensure the chain tensioner oil filters (B) are clean and undamaged

12 Now turn the right-hand camshaft until the slot is aligned with the thrust plate as shown **(see illustration)**. In this position, the valves of No 6 cylinder will be `rocking'.

13 Clean the tensioner oil filters and fit them **(see illustration)**.

14 Refit the chain tensioners, the chain slippers and guides.

15 Refit the Woodruff key to the crankshaft then slide on the sprocket so that the timing mark is visible **(see illustration)**.

16 The left-hand side timing chain assembly is fitted first. Rotate the crankshaft so that its keyway is facing upwards to the centre-line of the left-hand cylinder bank.

17 Fit the left-hand camshaft sprocket into the chain so that the timing mark on the sprocket front face is between the two marked links of the chain. Supporting the chain and sprocket in this position, fit the chain lower end over the crankshaft sprocket and align the single marking on the chain link with the timing mark on the front face of the sprocket. The chain is fitted over the rear teeth of the double timing sprocket **(see illustrations)**.

18 As the cylinder head and camshaft are in position, then the camshaft sprocket can be fitted to the camshaft so that the driving pin engages its slot in the shaft flange. Locate the camshaft retaining bolt and tighten it to the specified torque whilst preventing the crankshaft from turning.

19 To fit the right-hand timing chain assembly on 2664 cc engines, first turn the crankshaft through 150° so that the timing mark on the crankshaft sprocket aligns with the oil pump cover lower retaining bolt **(see illustration)**.

20 On 2458 cc turbo and 2849 cc engines, turn the crankshaft through 120° so that the timing mark on the crankshaft sprocket aligns with the edge of the left-hand fixed chain slipper **(see illustration)**.

21 Now fit the right-hand camshaft sprocket into its chain so that its timing mark is between the twin markings on the chain links. Supporting the chain and sprocket in this position, fit the chain over the crankshaft sprocket so that the single marking on the chain link aligns with the crankshaft sprocket timing mark. This chain is fitted over the front teeth of the crankshaft sprocket **(see illustration)**.

22 The camshaft sprocket can be fitted to the camshaft, engaging its drive pin into the flange slot. The camshaft sprocket retaining bolts should then be refitted. Prevent the crankshaft from turning and tighten the retaining bolt to the specified torque setting.

23 The timing chain tensioners can be reset when the respective camshaft sprockets are bolted in position on their camshafts and the chains fitted. Press the tensioner shoe in so that it touches the tensioner body, then release the shoe but do not assist the spring action.

24 If the timing chains and sprockets have been correctly assembled, then the timing marks on the crankshaft and camshaft

9.15 Refit the crankshaft sprocket making sure its timing mark is facing outwards

9.17a Position the camshaft sprocket timing mark between the two marked timing chain links . . .

9.17b . . . and align the crankshaft sprocket mark with the other marked chain link

9.17c Left-hand timing chain and sprockets show correct alignment of timing marks

9.19 On 2664 cc engines align the crankshaft sprocket timing mark with the oil pump cover lower retaining bolt hole (shown with pump removed)

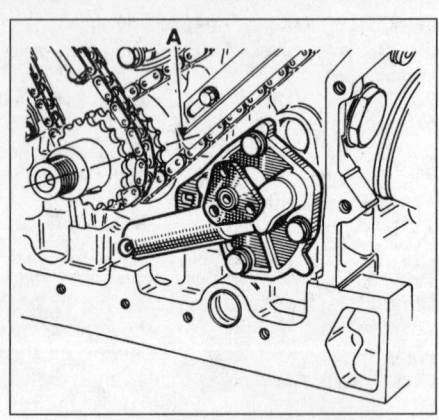

2B•10 Six cylinder petrol engine

9.20 On 2458 cc turbo and 2849 cc engines align the crankshaft sprocket timing mark with the edge of the timing chain slipper (A)

9.21 Timing chain and sprocket alignment marks with both chains installed (2664 cc engine shown)

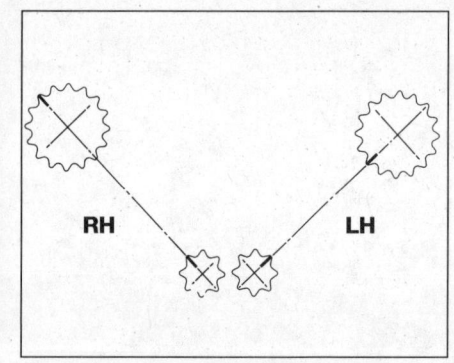

9.24 Sprocket timing mark alignment

sprockets will be aligned as shown **(see illustration)**. Turn the crankshaft only clockwise when checking.
25 Refit the oil pump sprockets and drive chain as described in paragraphs 11 to 13 of Section 8.
26 Refit the timing cover as described in Section 7.

10 Camshafts and rocker arms - removal, inspection and refitting

Removal

Note: *The rocker arm assembly is secured to the top of the cylinder head by the cylinder head bolts. Although in theory it is possible to undo the head bolts and remove the rocker arm assembly without removing the head, in practice, this is not recommended. Once the bolts have been removed, the head gasket will be disturbed, and the gasket will almost certainly leak or blow after refitting. For this reason, removal of the rocker arm assembly and camshaft should not be done without removing the cylinder head and renewing the head gasket.*

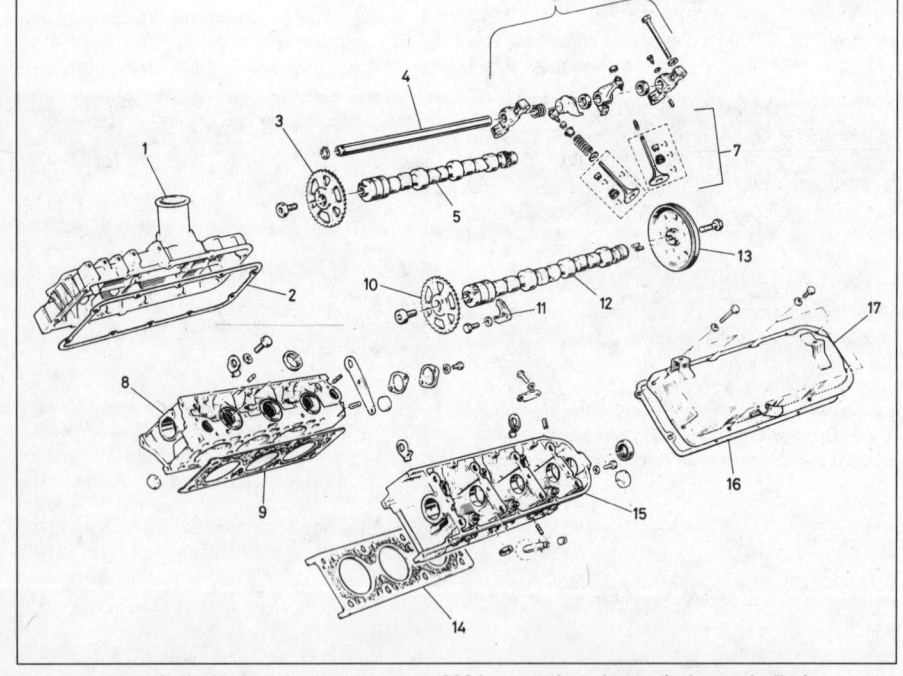

1 Remove the cylinder head and rocker arm assembly as described in Section 11. If the camshaft is to be removed, proceed as follows **(see illustration)**.
2 On the left-hand cylinder head on 2664 cc engines where the power steering pump is driven by the camshaft, slacken and remove the power steering pump pulley bolt and remove the pulley. To prevent rotation as the bolt is slackened, use an old drivebelt and a pair of grips to retain the pulley. If the pulley locating pins are a loose fit remove them and store them with the pulley for safe-keeping. Using a flat bladed screwdriver, carefully lever out the camshaft oil seal.
3 On all other models, undo the retaining bolts and remove the end cover and gasket from the rear of the cylinder head (where fitted).
4 Remove the retaining bolt and thrust plate and slide the camshaft out of the cylinder head **(see illustration)**. Take care during its removal not to snag any of the lobe corners on the bearings as they are passed through the cylinder head

10.1 Cylinder head components - 2664 cc engine shown (others similar)

1 Rocker cover
2 Gasket
3 Right-hand camshaft sprocket
4 Rocker shaft
5 Right-hand camshaft
6 Rocker arms and associated components
7 Valve components
8 Right-hand cylinder head
9 Gasket
10 Left-hand camshaft sprocket
11 Camshaft thrust plate
12 Left-hand camshaft
13 Power steering pump drive pulley
14 Gasket
15 Left-hand cylinder head
16 Gasket
17 Rocket cover

10.4 Unbolting the camshaft thrust plate

Six cylinder petrol engine 2B•11

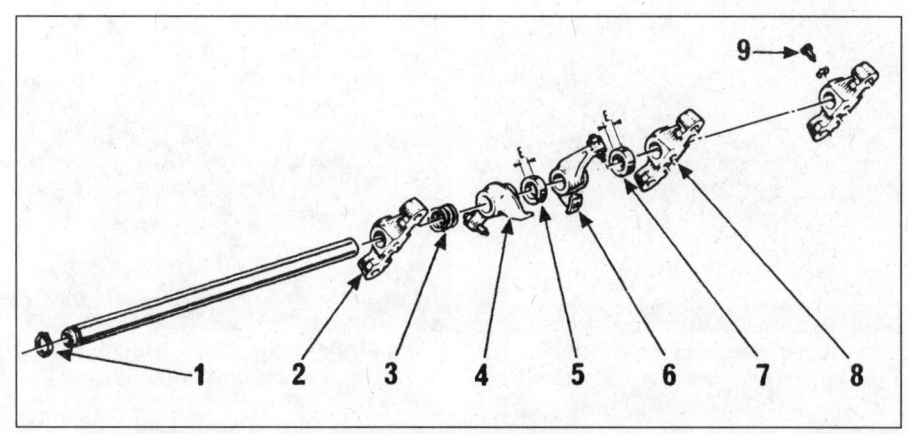

10.5 Rocker shaft components

1 Circlip
2 Pedestal
3 Spring
4 Rocker arm
5 5.35 mm spacer
6 Rocker arm
7 8.2 mm spacer
8 Pedestal
9 End pedestal bolt

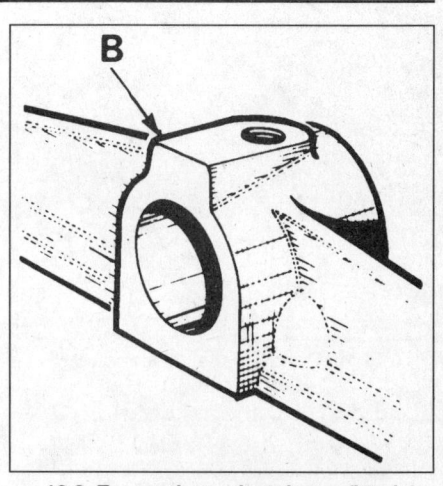

10.8 Ensure the pedestals are fitted with the boss (B) towards the rocker shaft circlip

Inspection

Rocker arms

Note: *Although the right-hand and left-hand bank rocker assemblies are identical, they should be dismantled separately to ensure components are not interchanged.*

5 Unscrew the bolt from the end bearing pedestal and slide off the pedestals, rocker arms, spacers and springs **(see illustration)**. Keep all components in their correct fitted order, making a note as each is removed, to ensure correct reassembly. Note the direction of fitting of each rocker shaft to ensure correct reassembly. If refitted incorrectly, the oil feed holes will be blocked.

6 Check the rocker arm wear faces (where they contact the camshaft lobes). Where they are only slightly worn they can be refaced. More excessive wear will necessitate their renewal. Inspect the rocker shafts for wear and renew if necessary. Check the oil feed holes and clear them out if blocked. **Note:** *The oilway plugs at the end of each rocker shaft are press fitted and should not be removed.* Also check the spacers; there are two different widths of spacer, these being 5.35 mm and 8.2 mm wide respectively. Renew any component which is badly worn.

10.11 Checking camshaft endfloat

7 To reassemble each rocker shaft assembly ready for refitting, first ensure that all parts are clean and then as they are assembled lubricate them with clean engine oil.

8 If not already in position, locate the circlip into the end groove, then slide the bearing pedestal down the shaft with the flat on the boss facing the circlip **(see illustration)**. Locate a spacer spring, followed by a rocker arm located with its adjuster screw on the left (from the circlip end). Fit the thinner spacer into position followed by the next rocker arm, fitted with its adjuster screw on the right then locate the thicker spacer collar, followed by the bearing pedestal with the flat on its boss towards the snap-ring end.

9 Repeat the assembly sequence in paragraph 8 for the remaining pairs of rocker arms, spacers and springs. Ensure that the respective rocker components are correctly fitted and that the shaft oil holes face downwards. Align the end pedestal retaining bolt hole with the rocker shaft hole then refit the retaining bolt and washer and tighten securely.

Camshaft

10 Examine the camshaft bearing surfaces and cam lobes for signs of wear ridges and scoring. Renew the camshaft if any of these conditions are apparent. Examine the condition of the bearing surfaces both on the camshaft journals and in the cylinder head. Renew the camshaft if worn. If the head bearing surfaces are worn excessively, it may be necessary to renew the cylinder head. Consult a Renault dealer or engine reconditioning specialist.

11 With the camshaft in position, refit the thrust plate and securely tighten its retaining bolt. Check the endfloat using a feeler blade and renew the thrust plate if the end float exceeds the specified limit **(see illustration)**.

Refitting

12 Lubricate the camshaft bearings with clean engine oil and slide the camshaft into position. Engage the thrust plate with the camshaft groove and refit its retaining bolt.

13 Where necessary, fit the end cover using a new gasket and securely tighten its retaining bolts.

14 On the left-hand cylinder head on 2664 cc engines, carefully ease a new oil seal over the camshaft end, ensuring its sealing lip is facing inwards. Press the seal squarely into the cylinder head then refit the locating pins to the camshaft end. Refit the drivebelt pulley, engaging it with the pins and tighten its retaining bolt to the specified torque, using the method described in paragraph 2 to prevent rotation.

15 Refit the rocker shaft and cylinder head as described in Section 11.

11 Cylinder heads - removal and refitting

Note: *If one or both cylinder heads are to be removed with the engine in the vehicle or where the rest of the engine is not being dismantled, you will need to keep the timing chain taut throughout the operation. If the chain tension is slackened for any reason you will have to remove the timing cover in order to reset the chain tensioner(s). A separate chain is used to drive each camshaft. To keep the chains taut you will need to use special Renault tool (Mot. 589) or be able to fabricate a similar support for the sprocket(s)* **(see illustrations)**. *The dummy bearing bracket (F) will only be required if the crankshaft has to be turned for any reason whilst the head(s) are removed, such as for pistons/connecting rods/liners removal, in order to maintain the valve timing.*

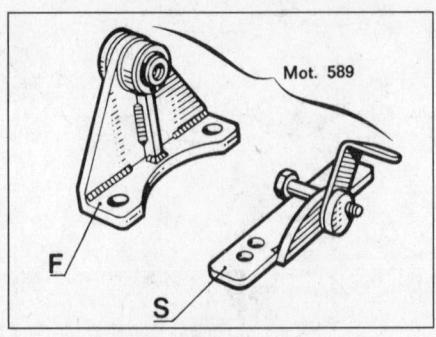

2B•12 Six cylinder petrol engine

11.1a Renault special sprocket support tools

F Dummy support bracket (needed if the crankshaft is to be rotated)
S Sprocket support bracket

11.1b Fabricated sprocket support bracket

11.6a Unscrew the access plug from the timing cover . . .

Removal

1 Proceed as follows for each cylinder head in turn noting the differences for left- and right-hand sides.
2 Where the engine is still in the vehicle, carry out the following preliminary operations:
a) Drain the cooling system and disconnect the appropriate cooling and heating hoses as necessary, referring to Chapter 3 if required for further details. Note their respective connections
b) Disconnect the battery earth lead
c) Disconnect the ignition leads and other associated electrical wires from the cylinder head(s), noting their respective positions. Also detach the diagnostic socket and bracket for removal of the right-hand cylinder head
d) Remove the alternator (left-hand cylinder head)
e) On Turbo engines, refer to Chapter 4B for details of disconnection of turbocharger and its cooling system components
f) Detach the exhaust front pipe(s) from the manifold connection(s)
g) On the left-hand cylinder head on 2664 cc engines, remove the power steering drivebelt and pump unit, but do not disconnect the hoses. Rest the pump unit on the bulkhead out of the way so that it does not leak (see Chapter 1A).

3 Refer to Chapter 4B and remove the inlet manifold.
4 Remove the dipstick tube clip on the right-hand cylinder head.
5 Unbolt and remove the rocker cover(s).
6 To gain access to the camshaft sprocket bolt, unscrew and remove the camshaft sprocket bolt access plug from the timing cover or unbolt and remove the cover plate from the front face of the timing cover **(see illustrations)**.
7 On 2458 cc turbo and 2849 cc engines, if the left-hand head is being removed, remove the distributor cap and rotor arm as described in Chapter 5B. Slacken and remove the camshaft sprocket bolt, remove the rotor drive flange and locating pin then refit the bolt and tighten securely. To gain additional clearance, unbolt the distributor body and carefully lever out the camshaft oil seal.
8 On 2664 cc engines remove the distributor if the right-hand head is to be removed (see Chapter 5B)
9 Turn the engine (using a spanner on the crankshaft pulley nut) so that the left-hand camshaft sprocket drive pin is at the top as shown **(see illustration)**.
10 Locate the camshaft sprocket support (Renault tool Mot. 589) in position on the top face of the timing case, with the bolts holding it in position moderately tightened in the two rocker cover front fixing holes. In the event of this tool not being available, fabricate a suitable support from a piece of angled steel as shown **(see illustration)**. Cut it to size and drill the necessary holes for the retaining bolts.
11 With the camshaft sprocket(s) suitably supported, loosen but do not remove the camshaft sprocket retaining bolt.

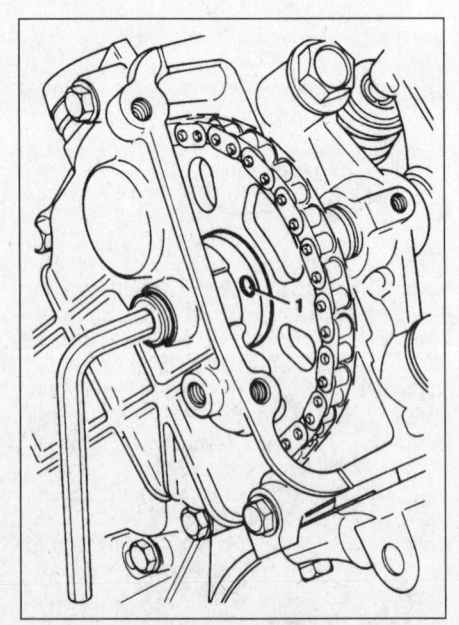

11.9 Position the crankshaft so that the left-hand camshaft sprocket drive pin 1 is positioned as shown

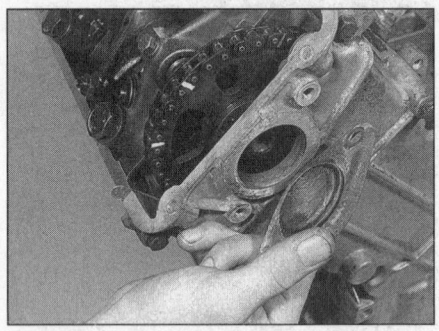

11.6b . . . or remove the cover plate (as applicable) to gain access to the camshaft sprocket bolt

12 Progressively loosen and remove the cylinder head retaining bolts in the reverse of the sequence shown in **illustration 11.36**, then lift clear the rocker assembly. Mark them accordingly to avoid confusion (right and left).
13 Unscrew the camshaft thrust plate bolt and disengage the thrust plate from its camshaft recess.
14 Slowly loosen off the camshaft sprocket bolt until the camshaft is disengaged from its sprocket.
15 Using a suitable rod, press the cylinder head locating dowels down in the block - the positions of the dowels are shown in **illustration 11.29a (see illustration)**.
16 Unscrew and remove the timing cover-to-cylinder head retaining bolts. There are four bolts to each cylinder head.
17 The cylinder head(s) are now ready for removal. The joint between the cylinder head

11.10 Fabricated sprocket support tool in position on the timing cover

Six cylinder petrol engine 2B•13

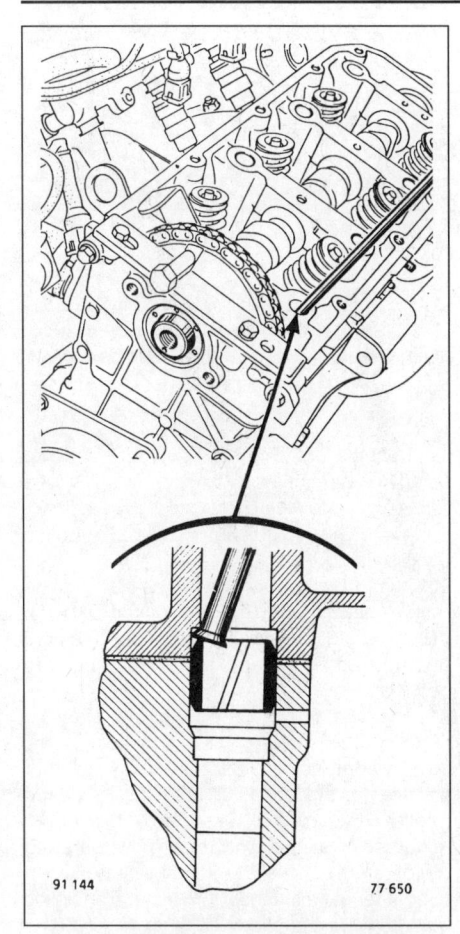

11.15 Using a suitable rod, press the locating dowels down into the cylinder block to allow the head to move sideways

and gasket and the cylinder block/crankcase must now be broken without disturbing the wet liners; although these liners are better located and sealed than some wet liner engines, there is still a risk of coolant and foreign matter leaking into the sump if the cylinder head is lifted carelessly. If care is not taken and the liners are moved, there is also a possibility of the bottom seals being disturbed, causing leakage after refitting the head.

18 To break the joint, strike the side of the head firmly using a soft-faced mallet or hammer and block of wood. Once the gasket seal is broken, lift the cylinder head upwards and away from the engine and remove the cylinder head gasket. If the camshaft sprocket pin is a loose fit, take care not to drop it down into the engine and remove it from the camshaft for safe keeping.

19 As soon as the cylinder head is removed, clamp the cylinder liners in position **(see illustration)**. In the absence of the special Renault liner clamping tool shown, the liners can be clamped in position using large flat washers positioned underneath suitable-length bolts, or the original head bolts with suitable spacers fitted to their shanks

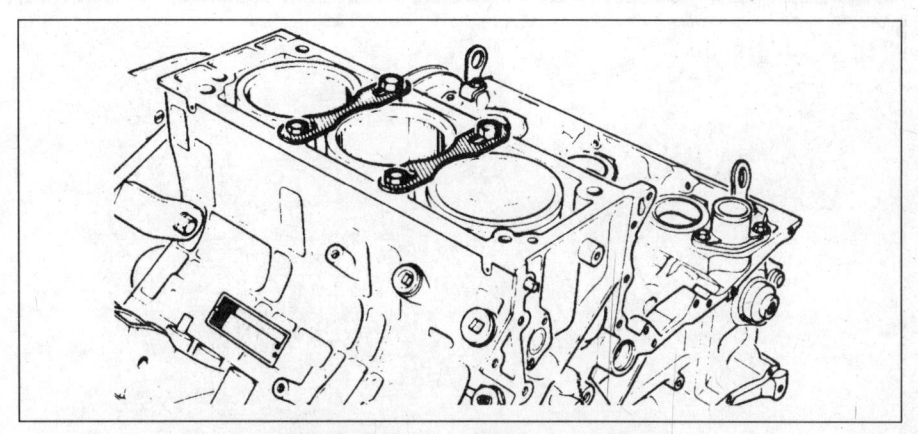

11.19 Cylinder liner clamps in position

⚠ *Warning: Whilst the cylinder head(s) are removed from the engine it is important not to turn the engine unless the liners are clamped securely in position and camshaft sprocket bearing bracket(s) are fitted (see Note at the start of Section).*

Preparation for refitting

20 The mating faces of the cylinder head and cylinder block/crankcase must be perfectly clean before refitting the head. Use a hard plastic or wood scraper to remove all traces of gasket and carbon; also clean the piston crowns. Take particular care as the soft aluminium alloy is damaged easily. Also, make sure that the carbon is not allowed to enter the oil and water passages - this is particularly important for the lubrication system, as carbon could block the oil supply to any of the engine's components. Using adhesive tape and paper, seal the water, oil and bolt holes in the cylinder block/crankcase. To prevent carbon entering the gap between the pistons and bores, smear a little grease in the gap. After cleaning each piston, use a small brush to remove all traces of grease and carbon from the gap, then wipe away the remainder with a clean rag. Clean all the pistons in the same way.

21 Check the mating surfaces of the cylinder block/crankcase and the cylinder head for nicks, deep scratches and other damage. If slight, they may be removed carefully with a file, but if excessive, machining may be the only alternative to renewal.

22 If warpage is suspected of the cylinder head gasket surface, use a straight-edge to check it for distortion. Refer to Section 12 if necessary.

23 The cylinder head bolts must be discarded and renewed, regardless of their apparent condition.

24 Prior to refitting the cylinder head, check the cylinder liner protrusion as described in Section 21. If the liner protrusion exceeds the specified limits, seek the advice of a Renault dealer prior to refitting the head.

25 Ensure that the cylinder head bolt holes in the crankcase are clean and free of oil.

Syringe or soak up any oil left in the bolt holes, and in the oil feed holes. This is most important in order that the correct bolt tightening torque can be applied.

26 Slacken the rocker arm adjusting screw locknuts and fully unscrew the adjusting screws. This will prevent the valve spring pressure affecting the cylinder head bolt tightness.

Refitting

27 Refit one cylinder head at a time in the following manner, taking care not to allow the timing chain to become slack at any time during assembly. If it does, you will have to remove the timing cover after the head has been fitted in order to reset the chain tension and check the valve timing - see Section 8. A separate tensioner is fitted to each timing chain.

28 New timing cover to cylinder head gasket sections must be used. Clean off any of the old gasket still remaining and then cut to size and locate new gasket sections as shown **(see illustration)**. Stick them in position by smearing with a thin layer of RTV sealant.

29 Wipe clean the mating surfaces of the cylinder head and cylinder block/crankcase.

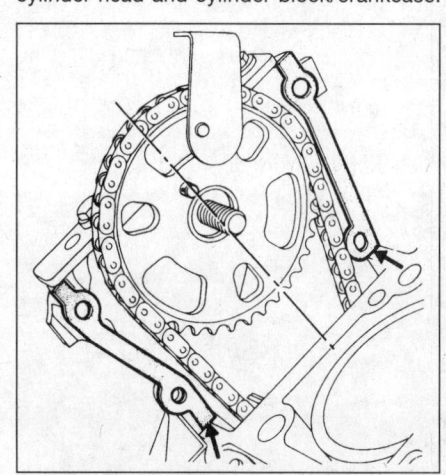

11.28 Trim off the sections of the new timing cover gaskets and fit them to the timing cover as shown

2B•14 Six cylinder petrol engine

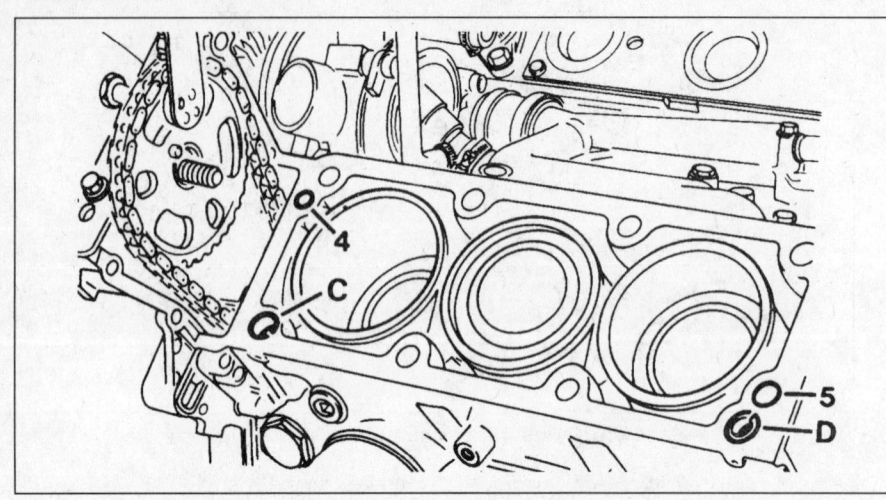

11.29a Cylinder head locating dowels (C and D) and oil feed holes (4 and 5)

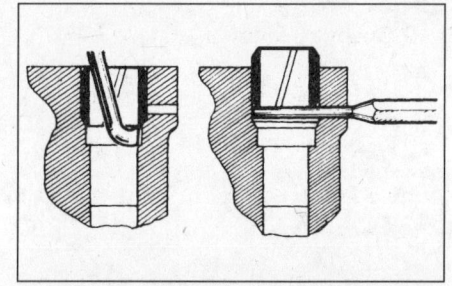

11.29b Lift the dowels back out of the cylinder block use suitable tools as shown

Check that the locating dowels are in position and are protruding above the cylinder block surface then, where necessary, remove the cylinder liner clamps **(see illustrations)**.

30 Position the new gasket on the cylinder block/crankcase surface ensuring that its correctly located on the dowels **(see illustration)**.

31 With the aid of an assistant lower the head into position, engaging it with the locating dowels. Refit the rocker arm assembly to the top of the cylinder head **(see illustration)**.

32 With the head in position, insert the bolts securing the timing cover to cylinder head, tightening them lightly only at this stage.

33 Align the camshaft drive pin with its hole and mate the camshaft with the sprocket. Slide the camshaft thrust plate back into position in the camshaft groove and securely tighten the retaining bolt. Refit the camshaft sprocket retaining bolt and spacer and tighten it securely.

34 Lightly oil under the head and on the threads of each new cylinder head bolt.

35 Carefully enter each bolt into its hole, screwing it in by hand only until finger-tight.

36 Working progressively and in the sequence shown, tighten the cylinder head bolts to their Stage 1 torque setting, using a torque wrench and suitable socket **(see illustration)**. Work in the reverse of the specified sequence and fully slacken all the bolts then go around again in sequence and retighten the bolts to the specified Stage 2 torque setting.

37 Once all the bolts have been tightened to their Stage 2 setting, working again in the given sequence, angle-tighten the bolts through the specified Stage 3 angle, using a socket and extension bar. It is recommended that an angle-measuring gauge is used during this stage of the tightening, to ensure accuracy. Note that further retightening of the bolts will be required once the engine has been warmed up (see paragraph 45).

38 Check and adjust the valve clearances as described in Section 6.

39 On 2664 cc engines refit the distributor as described in Chapter 5B (where removed).

40 On 2458 cc and 2849 cc engines, where the left-hand head has been removed, press a new oil seal into the timing cover, taking care not to damage its sealing lip then refit the distributor body. Remove the camshaft sprocket bolt then refit the rotor drive flange and retaining pin and refit the sprocket bolt. Tighten the camshaft sprocket bolt to the specified torque then refit the rotor arm and distributor cap as described in Chapter 5B.

41 Tighten the camshaft sprocket retaining bolt(s) to the specified torque setting (if not already having done so). Also tighten the bolts securing the timing cover to the cylinder head to the specified torque.

42 Refit the camshaft sprocket access cover using a new gasket and securely tighten its retaining bolts. Where an access plug is fitted, refit the plug to the timing cover and tighten it securely.

43 Trim off the protruding edges of the timing cover gaskets and refit the rocker cover(s), using the old gaskets; the covers will have to be removed again to allow the head bolts to be retightened.

44 The remainder of refitting is a direct reversal of removal.

45 On completion, refill the cooling system and top-up the engine oil level then start the engine and run it for approximately 15 minutes to ensure it reaches full operating temperature. Turn the engine off and leave it to cool for at least 6 hours. When the engine is cold, remove the rocker cover(s). On 2458 cc turbo and 2849 cc engines, tighten the cylinder head bolts through the specified Stage 4 angle, working in the sequence shown in **illustration 11.36**. On 2664 cc engines, working in the reverse of the sequence shown fully slacken all the cylinder head bolts, then working in the specified sequence, tighten the bolts first to the Stage 4 specified torque setting then go around again in the specified sequence and tighten them through the specified Stage 5 angle. On all models, it is recommended that an angle-measuring gauge is used during this stage of the tightening, to ensure accuracy. Once the cylinder head bolts have been tightened, refit the rocker cover(s) using the new gasket(s).

11.30 Locate the new gasket on the dowels . . .

11.31 . . . and lower the cylinder head into position

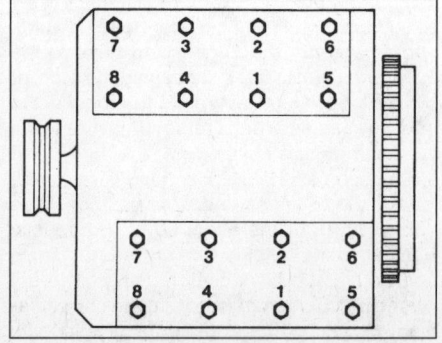

11.36 Cylinder bolt tightening sequence

Six cylinder petrol engine 2B•15

12 Cylinder head - overhaul

The cylinder head can be overhauled as described in Chapter 2A, Section 12 once the camshaft has been removed as described in Section 10.

13 Sump - removal and refitting

Removal

1 Drain the engine oil (see Chapter 1A).
2 Raise the front of the car so that the front roadwheels hang free. This will cause the anti-roll bar to move forward sufficiently far to enable the rear sump pan screws to be extracted (see illustration).
3 Unscrew and remove the screws and remove the sump pan. If it is stuck tight, cut around the gasket using a sharp knife.
4 Unscrew the three bolts and remove the oil pick-up pipe and sealing ring.
5 Unscrew the six bolts and remove the baffle plate from the base of the main bearing cap casting.

Refitting

6 Relocate the baffle plate and secure with bolts and washers (see illustration).
7 Locate a new sealing ring onto the bottom end of the oil pick-up pipe, then refit the pipe assembly into position (see illustrations).
8 Fit the new sump gasket into position and then refit the sump pan. Locate and tighten its retaining bolts to the specified torque. Make sure that the sump drain plug is tight.
9 Lower the car to the ground and fill the engine with oil.

14 Camshaft oil seals - renewal

Left-hand rear oil seal - 2664 cc engine

1 Slacken the camshaft pulley bolt then loosen the power steering pump and remove its drivebelt (see Chapter 1A).
2 Remove the camshaft pulley retaining bolt and then withdraw the pulley (see illustration). Take care not to lose the pulley locating pins.
3 The oil seal can now be carefully eased out of its housing using a screwdriver.
4 Lubricate the new seal before fitting and then insert it into its housing, using a suitable diameter tube drift to ensure that it is fitted correctly and without distortion. When in position the new seal should be flush with the cylinder head rear face.

13.2 Slacken and remove the sump retaining screws

13.7a Fit a new sealing ring to the oil pick-up pipe . . .

5 Relocate the pulley, making sure it is engaged with the locating pins and refit its retaining bolt.
6 Refit and tension the drivebelt (see Chapter 1A) then tighten the pulley retaining bolt to the specified torque.

Left-hand front oil seal - 2458 cc turbo and 2849 cc engines

7 Remove the distributor cap and rotor arm as described in Chapter 5B (see illustration).
8 Slacken and remove the camshaft sprocket bolt, remove the rotor drive flange and locating pin. Temporarily refit the sprocket bolt and tighten it securely.

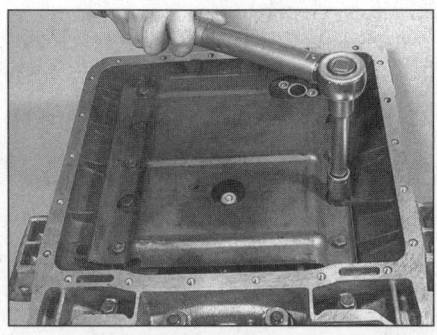

13.6 Refit the baffle plate and securely tighten its retaining screws

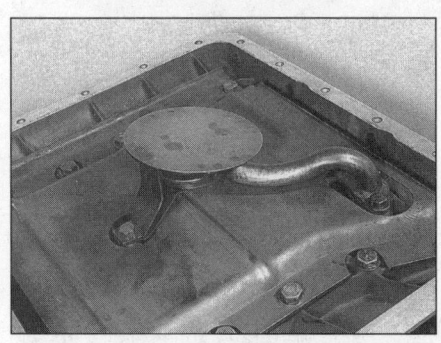

13.7b . . . then refit the pipe and securely tighten its retaining bolts

14.2 Removing the power steering pump drivebelt pulley - 2664 cc engine

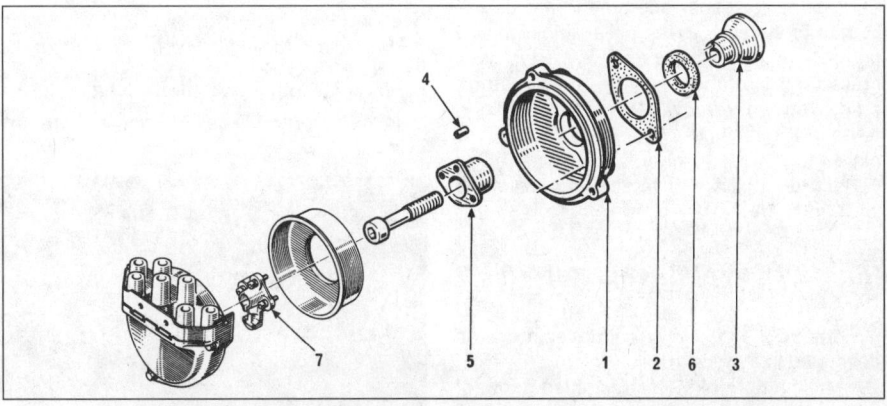

14.7 Distributor drive components - 2458 cc turbo and 2849 cc engines

1 Distributor body
2 Gasket
3 Spacer
4 Locating pin
5 Rotor drive flange
6 Oil seal
7 Rotor arm

2B•16 Six cylinder petrol engine

9 Unbolt the distributor body and gasket from the timing cover then carefully lever the oil seal out of position using a flat-bladed screwdriver.
10 Press the new seal into position taking care not to damage its sealing lip.
11 Refit the distributor body and gasket and securely tighten its retaining bolts.
12 Remove the sprocket bolt then refit the rotor drive flange and locating pin and tighten the sprocket bolt to the specified torque.
13 Refit the rotor arm and distributor (see Chapter 5B).

15 Crankshaft oil seals - renewal

Front (timing cover) seal

1 Remove the auxiliary drivebelt(s) as described in Chapter 1A.
2 To prevent crankshaft rotation whilst the pulley retaining nut is slackened, unbolt the cover plate from the right-hand side of the engine and lock the flywheel/driveplate ring gear, using an arrangement similar to that shown (see illustration 7.8). On manual transmission models, if the engine is still in the vehicle the crankshaft can be retained by selecting top gear and applying the brakes firmly.
3 Unscrew and remove the crankshaft pulley nut. When the nut is removed, release the crankshaft and rotate it so that the pulley Woodruff key is at the top and then withdraw the crankshaft pulley.
4 Carefully lever out the oil seal from the timing cover using a flat-bladed screwdriver.
5 Press the new seal into position making sure its sealing lip is facing inwards.
6 Apply a smear of oil to the crankshaft pulley oil seal surface and carefully slide the pulley into position, aligning it with the Woodruff key.
7 Apply a few drops of thread locking compound (Renault recommend the use of Loctite Frenetanch to the pulley retaining and tighten it to the specified torque whilst prevent crankshaft rotation using the method employed on removal. Where necessary, remove the locking tool and refit the cover plate.
8 Refit the auxiliary drivebelt(s) as described in Chapter 1A.

Rear (flywheel/driveplate) oil seal

9 Remove the flywheel/driveplate as described in Section 16.
10 Make a note of the correct fitted depth of the seal in its housing. Punch or drill two small holes opposite each other in the seal. Screw a self-tapping screw into each, and pull on the screws with pliers to extract the seal.
11 Clean the seal housing, and polish off any burrs or raised edges, which may have caused the seal to fail in the first place.
12 Lubricate the lips of the new seal with clean engine oil, and carefully locate the seal on the end of the crankshaft.
13 Using a suitable tubular drift, which bears only on the hard outer edge of the seal, drive the seal into position, to the same depth in the housing as the original was prior to removal.
14 Wash off any traces of oil, then refit the flywheel/driveplate as described in Section 16.

16 Flywheel/driveplate - removal, inspection and refitting

Refer to Chapter 2A, Section 16, note that on some models a spacer may be fitted on either side of the driveplate (see illustration).

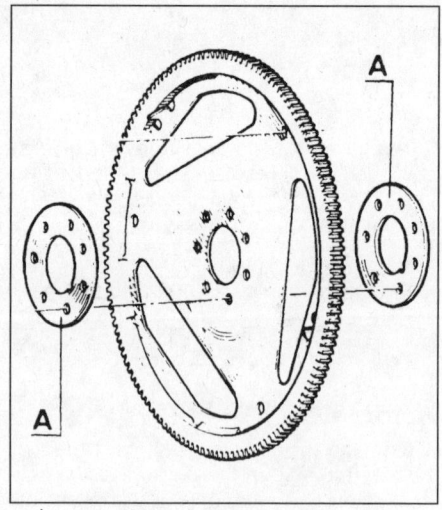

17 Engine/transmission mountings - renewal

Refer to Chapter 2A, Section 18 noting the different types of mounting shown (see illustrations).

18 Crankshaft - removal, inspection and refitting

Removal

1 Remove the timing chains, flywheel/driveplate and the sump (including the oil pick-up pipe and baffle plate) as described in the relevant Sections of this Chapter (see illustration).
2 If the cylinder liner/piston and connecting rod assemblies are to be removed, remove the cylinder head(s) as described in Section 11.
3 Undo the retaining bolts, not forgetting the two bolts securing the base of the housing to the main bearing cap casting, and remove the crankshaft rear oil seal housing from the cylinder block.

16.1 Driveplate and spacers (A)

4 Before proceeding further, check the crankshaft endfloat using a dial gauge in contact with the end of the crankshaft. Push the crankshaft fully one way and then zero the gauge. Push the crankshaft fully the other way and check the endfloat. The result can be compared with the specified amount and will give an indication as to whether new thrustwashers are required. Note that thrustwashers are available in four different thickness. If a dial gauge is not available, feeler blades can be used. First push the crankshaft fully towards the flywheel/driveplate end of the engine, then use feeler blades to measure the gap between the crankshaft rear flange and the thrustwasher (see illustration).
5 Turn the engine upside down then slacken and remove the outer bolts securing the main bearing cap casting to the base of the cylinder block.
6 Working in the reverse of the sequence shown in illustration 18.38a, evenly and progressively slacken and remove the main bearing cap nuts and washers then lift the main bearing cap casting away from the block. Remove the sealing ring from the oil pipe and discard it. It is recommended that the main bearing nuts are discarded and new ones used on refitting.

17.1a Engine rubber mounting

17.1b Transmission rubber mounting

Six cylinder petrol engine 2B•17

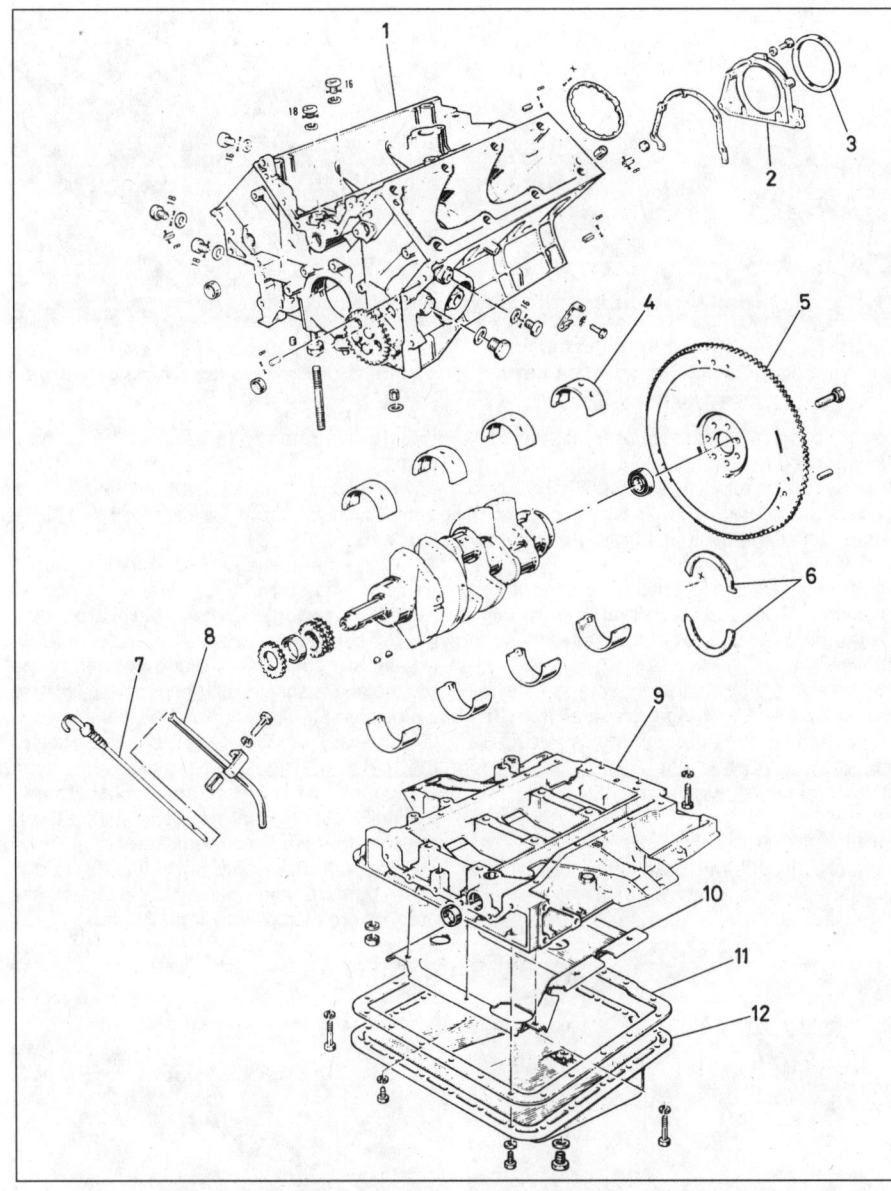

18.1 Cylinder block/crankcase and associated components

1 Cylinder block/crankcase
2 Crankshaft rear oil seal housing
3 Oil seal
4 Main bearing shells
5 Flywheel/driveplate
6 Crankshaft thrust washers
7 Dipstick
8 Dipstick tube
9 Main bearing casting
10 Baffle plate
11 Gasket
12 Sump

18.4 Checking crankshaft endfloat

7 Working as described in Section 19, unbolt the big-end bearing caps and push the pistons sufficiently up the bores so that the connecting rods are positioned clear of the crankshaft journals.

8 The main bearing caps should be numbered 1 to 4 from the transmission end of the engine. If not, mark them accordingly using a centre punch to indicate their correct fitted locations and which way around they are fitted.

9 Lift off each bearing cap and recover the lower bearing shells. If necessary, tape them to their respective caps for safe-keeping. Also recover the lower half of the thrustwasher from the rear bearing cap.

10 Lift out the crankshaft and recover the upper bearing shells from the cylinder block and tape them to their respective caps for safe keeping. Remove the upper thrustwasher half from the side of the rear main bearing.

Inspection

11 Refer to Chapter 2A, Section 20 noting that Renault oversize main bearing and big-end bearing shells are only available for 2664 cc engines.

Refitting

Note: *It is recommended that new main bearing shells are fitted regardless of the condition of the original ones.*

Selection of bearing shells

12 On 2458 cc turbo and 2849 cc engines, Renault produce only standard size bearings, no oversizes being available. On 2664 cc engines, there are two different sizes of bearing shell available; the standard size shell for use with an original crankshaft and an oversize shell for use once the crankshaft has been reground (see Chapter 2A, Section 20). Whether the original shells or new shells are being fitted, it is recommended that the running clearance is checked as follows prior to installation.

Main bearing running clearance check

13 Clean the backs of the bearing shells and the bearing locations in both the cylinder block/crankcase and the main bearing caps.

14 Press the bearing shells into their locations, noting that the upper shells which are fitted to the crankcase are grooved and the lower shells are plain. Ensure that the tab on each shell engages in the notch in the cylinder block/crankcase or main bearing cap, and take care not to touch any shell's bearing surface with your fingers. If the original bearing shells are being used for the check ensure that they are refitted in their original locations. The clearance can be checked in either of two ways.

15 One method (which will be difficult to achieve without a range of internal micrometers or internal/external expanding calipers) is to refit the main bearing caps and casting to the cylinder block/crankcase, with bearing shells in place. With the main bearing nuts correctly tightened (see paragraphs 38 and 39), measure the internal diameter of each assembled pair of bearing shells. If the diameter of each corresponding crankshaft journal is measured and then subtracted from the bearing internal diameter, the result will be the main bearing running clearance.

16 The second (and more accurate) method is to use an American product known as Plastigage. This consists of a fine thread of perfectly round plastic which is compressed between the bearing shell and the journal.

2B•18 Six cylinder petrol engine

18.18 Plastigage in place on a crankshaft journal

18.21 Measure the width of the deformed Plastigage using the scale on the card supplied

18.27 Fit the thrustwasher half making sure its oil grooves are facing outwards

When the shell is removed, the plastic is deformed and can be measured with a special card gauge supplied with the kit. The running clearance is determined from this gauge. Plastigage is sometimes difficult to obtain but enquiries at one of the larger specialist quality motor factors should produce the name of a stockist in your area. The procedure for using Plastigage is as follows.

17 With the main bearing upper shells in place, carefully lay the crankshaft in position. Do not use any lubricant; the crankshaft journals and bearing shells must be perfectly clean and dry.

18 Cut several lengths of the appropriate size Plastigage (they should be slightly shorter than the width of the main bearings) and place one length on each crankshaft journal axis **(see illustration)**.

19 With the main bearing lower shells in position, refit the main bearing caps in their correct positions. Refit the main bearing casting and tighten the main bearing nuts as described in paragraphs 38 and 39. Take care not to disturb the Plastigage and **do not** rotate the crankshaft at any time during this operation.

20 Remove the main bearing casting and caps again taking great care not to disturb the Plastigage or rotate the crankshaft.

21 Compare the width of the crushed Plastigage on each journal to the scale printed on the Plastigage envelope to obtain the main bearing running clearance **(see illustration)**. Compare the clearance measured with that given in the *Specifications* at the start of this Chapter.

22 If the clearance is significantly different from that expected, the bearing shells may be the wrong size (or excessively worn if the original shells are being re-used). Before deciding that the crankshaft is worn, make sure that no dirt or oil was trapped between the bearing shells and the caps or block when the clearance was measured. If the Plastigage was wider at one end than at the other, the crankshaft journal may be tapered.

23 Note that Renault do not specify a running clearance for these engines. The figure given in *Specifications* is a guide figure which is typical for this type of engine. Before condemning the components concerned, seek the advice of your Renault dealer or suitable engine repair specialist. They will also be able to inform as to the best course of action on any possible repairs or whether renewal will be necessary.

24 Where necessary, obtain the correct size of bearing shell and repeat the running clearance checking procedure as described above.

25 On completion, carefully scrape away all traces of the Plastigage material from the crankshaft and bearing shells using a fingernail or other object which is unlikely to score the bearing surfaces.

Final crankshaft refitting

26 Carefully lift the crankshaft out of the cylinder block once more.

27 Using a little grease, stick the upper thrustwasher half to the side of the cylinder block rear main bearing; ensure that the oilway grooves on each thrustwasher face outwards (away from the cylinder block) **(see illustration)**.

28 Place the bearing shells in their locations as described above in paragraphs 13 and 14 **(see illustration)**. If new shells are being fitted, ensure that all traces of the protective grease are cleaned off using paraffin. Wipe dry the shells and bearing caps with a lint-free cloth.

29 Liberally lubricate each bearing shell in the cylinder block/crankcase with clean engine oil then lower the crankshaft into position ensuring that the bearing shells and thrustwasher remain correctly seated.

30 Fit the thrustwasher lower half to the rear main bearing cap making sure its oilway grooves are facing away from the cap.

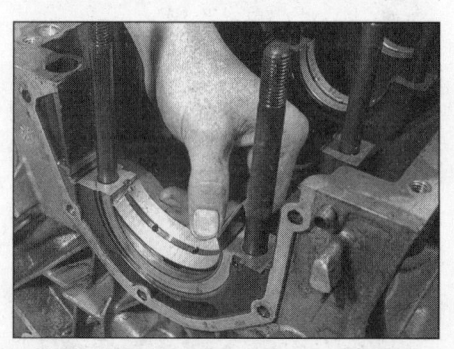

18.28 Make sure the main bearing shell tab is correctly located in the block cutout

18.31 Refitting a main bearing cap

18.33a Refit the crankshaft rear oil seal housing ...

18.33b ... and lightly tighten its retaining bolts

Six cylinder petrol engine 2B•19

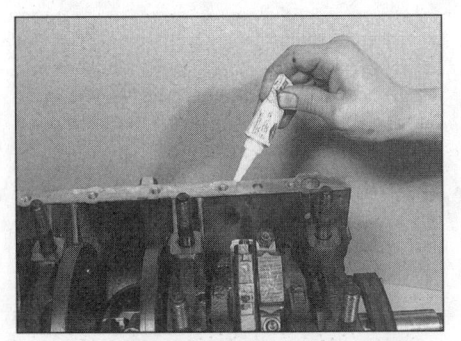

18.34 Coat the cylinder block mating surface with sealant

31 Fit the main bearing caps to the cylinder block using the identification marks to ensure that they are fitted the correct way around and in their original locations (the bosses on the base of each cap should be facing towards the timing chain end of the block - see **illustration 18.35a) (see illustration)**. As the rear cap is fitted, ensure that the thrustwasher half remains correctly seated.

32 Pull the connecting rods down into position and refit the big-end bearing caps as described in Section 19.

33 Lever the oil seal out from the rear oil seal housing and press a new one into position making sure its sealing lip is facing inwards. Fit a new gasket to the rear of the cylinder block and ease the housing into position taking care not to damage the oil seal. Fit the housing bolts, tightening them lightly only,

18.36 ... and fit the main bearing nuts and washers

18.37 Go around and with a straight edge and ensure the main bearing casting is correctly aligned with the cylinder block

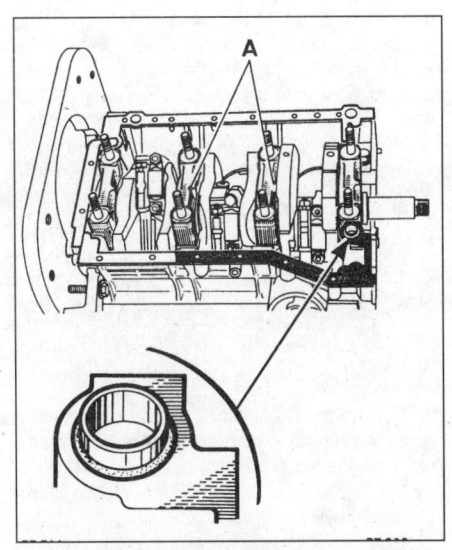

18.35a Fit a new sealing ring to the oil pipe. Make sure the main bearing caps are fitted the correct way around with the bosses (A) facing towards the timing chain end of the engine

and trim off the protruding edges of the gasket **(see illustrations)**.

34 Ensure that the cylinder block and main bearing cap casting mating surfaces are clean and dry and apply a thin coat of sealant (Renault recommend the use of CAF 4/60 Thixo) to the mating surface of the cylinder block **(see illustration)**.

35 Fit a new sealing ring to the oil pipe and carefully lower the main bearing cap casting into position on the cylinder block **(see illustrations)**.

36 Fit the main bearing washers and nuts and screw in the bolts securing the casting to the base of cylinder block. **Do not** tighten any of the nuts and bolts yet **(see illustration)**.

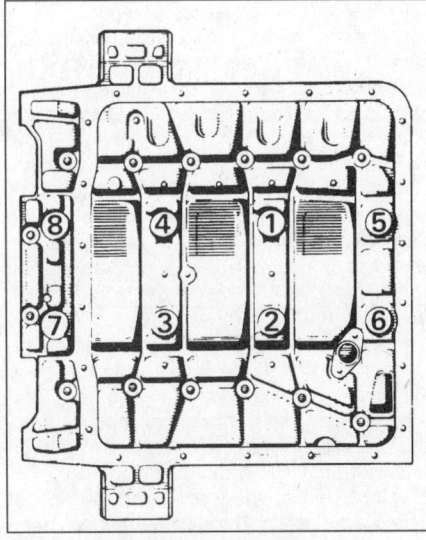

18.38a Main bearing cap nut tightening sequence

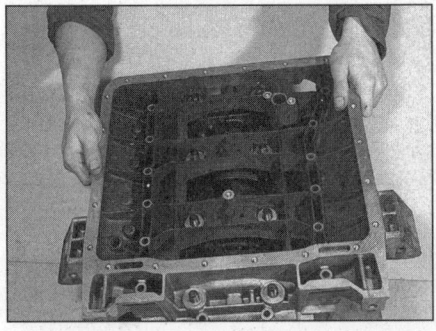

18.35b Lower the main bearing casting onto the cylinder block ...

37 Using two straight edges, check all around the outer edge of the main bearing casting to ensure that it is correctly aligned with the cylinder block. This is especially important on the transmission end of the engine **(see illustration)**.

⚠ **Warning: If the main bearing cap casting is not correctly aligned with the cylinder block, there is a risk of the casting distorting when the transmission is bolted to the engine.**

38 Once the main bearing cap casting is correctly aligned, working in the sequence shown, tighten the main bearing cap nuts to the specified stage 1 torque setting **(see illustrations)**.

39 Once all the nuts have been tightened to their Stage 1 setting, working again in the given sequence, angle-tighten the bolts through the specified Stage 2 angle, using a socket and extension bar. It is recommended that an angle-measuring gauge is used during this stage of the tightening, to ensure accuracy.

40 Check that the crankshaft rotates freely then work around the outer edge of the main bearing casting and tighten all the retaining bolts securing it to the base of the cylinder block/oil seal housing to the specified torque.

41 Securely tighten the remaining rear oil seal housing retaining bolts.

42 Refit the sump, flywheel/driveplate, cylinder heads and timing chains (as applicable) as described in the relevant Sections of this Chapter.

18.38b Tighten the main bearing cap nuts first to the specified Stage 1 torque and then through the specified stage 2 angle

19 Piston/connecting rod and cylinder liner assemblies - removal, inspection & refitting

Removal

1 Remove the timing chains, cylinder head(s), flywheel/driveplate and the sump (including the oil pick-up pipe and baffle plate) as described in the relevant Sections of this Chapter.
2 Unbolt and remove the main bearing cap casting as described in paragraphs 5 and 6 of Section 18.
3 Using a hammer and centre-punch, paint or similar, mark each connecting rod big-end bearing cap with its respective cylinder number on the flat machined surface provided; if the engine has been dismantled before, note carefully any identifying marks made previously. Note that No 1 cylinder is at the transmission end of the left-hand cylinder bank.
4 Turn the crankshaft to bring No 1 and 4 pistons to BDC (bottom dead centre).
5 Unscrew the nuts from number 1 piston big-end bearing cap, then take off the cap and recover the bottom half bearing shell (see illustration). If the bearing shells are to be re-used, tape the cap and the shell together.
6 Push the piston up the bore to disengage the connecting rod from the crankshaft. Recover the upper bearing shell, and tape it to the connecting rod for safe keeping.
7 Using a suitable piece of wood, such as the handle of a hammer, tap the liner upwards and out of position from the base of the cylinder block/crankcase.
8 Withdraw the liner/piston and connecting rod assembly from the top of the cylinder block/crankcase. Mark the liner in some way to indicate its correct fitted direction then remove the seal from the base of liner and discard.
9 Refit the big-end cap to the connecting rod and loosely tighten its retaining nuts/bolts. This will help to keep the components in their correct order.
10 Repeat the procedure and remove number 4 liner/piston and connecting rod assembly in the same way.
11 Turn the crankshaft through 120° to the next to pistons to BDC (bottom dead centre) and remove them in the same way. Rotate the crankshaft through a further 120° and remove the remaining pistons in the same way.
12 If necessary, mark the relevant cylinder number on the liner and withdraw the piston/connecting rod assembly from the base of each liner.

Inspection

Piston and connecting rod assembly

13 Before the inspection process can begin, the piston/connecting rod assemblies must be cleaned, and the original piston rings removed as described in Section 20.
14 Scrape away all traces of carbon from the top of the piston. A hand-held wire brush (or a piece of fine emery cloth) can be used, once the majority of the deposits have been scraped away.
15 Remove the carbon from the ring grooves in the piston, using an old ring. Break the ring in half to do this (be careful not to cut your fingers - piston rings are sharp). Be careful to remove only the carbon deposits - do not remove any metal, and do not nick or scratch the sides of the ring grooves.
16 Once the deposits have been removed, clean the piston/connecting rod assembly with paraffin or a suitable solvent, and dry thoroughly. Make sure that the oil return holes in the ring grooves are clear.
17 If the pistons and cylinder liners are not damaged or worn excessively, the original pistons can be refitted. Normal piston wear shows up as even vertical wear on the piston thrust surfaces, and slight looseness of the top ring in its groove. New piston rings should always be used when the engine is reassembled.
18 Carefully inspect each piston for cracks around the skirt, around the gudgeon pin holes, and at the piston ring "lands" (between the ring grooves).
19 Look for scoring and scuffing on the piston skirt, holes in the piston crown, and burned areas at the edge of the crown. If the skirt is scored or scuffed, the engine may have been suffering from overheating, and/or abnormal combustion which caused excessively high operating temperatures. The cooling and lubrication systems should be checked thoroughly. Scorch marks on the sides of the pistons show that blow-by has occurred. A hole in the piston crown, or burned areas at the edge of the piston crown, indicates that abnormal combustion (pre-ignition, knocking, or detonation) has been occurring. If any of the above problems exist, the causes must be investigated and corrected, or the damage will occur again. The causes may include incorrect ignition timing or a faulty injector (as applicable).
20 Corrosion of the piston, in the form of pitting, indicates that coolant has been leaking into the combustion chamber and/or the crankcase. Again, the cause must be corrected, or the problem may persist in the rebuilt engine.
21 It is not possible to renew the pistons separately; pistons are only supplied with piston rings and a liner, as a part of a matched assembly.
22 Examine each connecting rod carefully for signs of damage, such as cracks around the big-end and small-end bearings. Check that the rod is not bent or distorted. Damage is highly unlikely, unless the engine has been seized or badly overheated. Detailed checking of the connecting rod assembly can only be carried out by a Renault dealer or engine repair specialist with the necessary equipment.
23 On 2664 cc engines, the gudgeon pins are an interference fit in the connecting rod small-end bearing. Therefore, piston and/or connecting rod renewal should be entrusted to a Renault dealer or engine repair specialist, who will have the necessary tooling to remove and install the gudgeon pins.
24 On 2458 cc turbo and 2849 cc engines, the gudgeon pins are of the floating type, secured in position by two circlips. On these engines, the pistons and connecting rods can be separated as follows.
25 Using a small flat-bladed screwdriver, prise out the circlips, and push out the gudgeon pin. Hand pressure should be sufficient to remove the pin, if not warm the piston gently (by soaking it in hot water) to ease removal. Identify the piston and rod to ensure correct reassembly. Discard the circlips - new ones *must* be used on refitting.
26 Examine the gudgeon pin and connecting rod small-end bearing for signs of wear or damage. The connecting rods themselves should not be in need of renewal, unless seizure or some other major mechanical failure has occurred. Check the alignment of the connecting rods visually, and if the rods are not straight, take them to an engine overhaul specialist for a more detailed check.
27 Examine all components, and obtain any new parts from your Peugeot dealer. If new pistons are purchased, they will be supplied complete with gudgeon pins and circlips. Circlips can also be purchased individually.
28 Position the piston so that the arrow on the piston crown is positioned as shown in relation to the connecting rod big-end bearing shoulder; On the left-hand cylinder bank (cylinder No 1 to 3) the arrow on the piston crown should point away from the connecting rod shoulder, and on the right-hand bank (cylinder No 4 to 6) the arrow on the piston should point towards the side of the connecting rod shoulder (see illustrations).
29 With the piston and connecting rod correctly mated, apply a smear of clean engine oil to the gudgeon pin. Slide it into the piston and through the connecting rod small-end. Check that the piston pivots freely on the rod, then secure the gudgeon pin in position with two new circlips. Ensure that each circlip is correctly located in its groove in the piston.

19.5 Fitting a big-end bearing cap

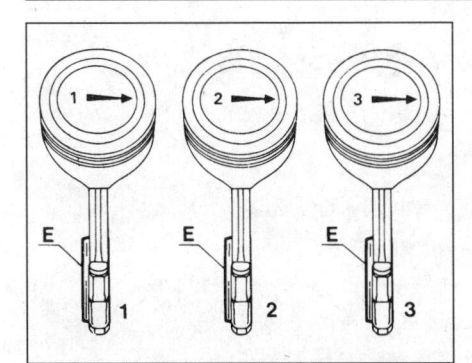

Six cylinder petrol engine 2B•21

19.28a On the left-hand cylinder bank, ensure the piston arrows point away from the connecting rod shoulders (E)

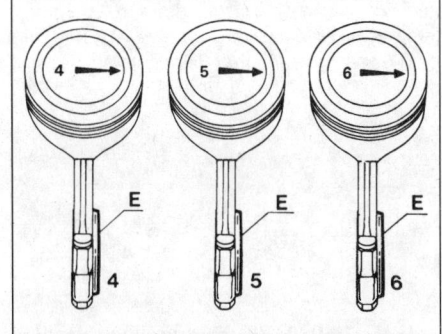

19.28b On the right-hand cylinder bank, ensure the piston arrows face towards the side of the connecting rod shoulders (E)

Cylinder liner

30 Refer to Section 21.

Refitting

Note: *It is recommended that new piston rings and big-end bearing shells are fitted regardless of the condition of the original ones.*

Selection of bearing shells

31 On 2458 cc turbo and 2849 cc engines, Renault produce only standard size bearings, no oversizes being available. On 2664 cc engines, there are two different sizes of bearing shell available; the standard size shell for use with an original crankshaft and an oversize shell for use once the crankshaft has been reground (see Chapter 2A, Section 20). Whether the original shells or new shells are being fitted, it is recommended that the running clearance is checked as follows prior to installation.

Big-end bearing running clearance check

32 Clean the backs of the bearing shells and the bearing locations in both the connecting rod and bearing cap.
33 Press the bearing shells into their locations, ensuring that the tab on each shell engages in the notch in the connecting rod and cap and taking care not to touch any shell's bearing surface with your fingers **(see illustration)**.
34 The upper shells, which are fitted to the connecting rods may or may not (depending on model) have an oil hole in them, the lower shells which are fitted to the bearing caps, are plain. If the original bearing shells are being used for the check ensure that they are refitted in their original locations. The clearance can be checked in either of two ways.
35 One method is to refit the big-end bearing cap to the connecting rod, with bearing shells in place. With the cap retaining nuts tightened to the specified torque, use an internal micrometer or vernier caliper to measure the internal diameter of each assembled pair of bearing shells. If the diameter of each corresponding crankshaft journal is measured and then subtracted from the bearing internal diameter, the result will be the big-end bearing running clearance.
36 The second method is to use Plastigage as described in Section 18, paragraphs 16 to 25. Place a strand of Plastigage on each (cleaned) crankpin journal and refit the (clean) piston/connecting rod assemblies, shells and big-end bearing caps, tightening the nuts to the specified torque wrench settings. Take care not to disturb the Plastigage. Dismantle the assemblies without rotating the crankshaft and use the scale printed on the Plastigage envelope to obtain the big-end bearing running clearance. On completion of the measurement, carefully scrape off all traces of Plastigage from the journal and shells using a fingernail or other object which will not score the components.

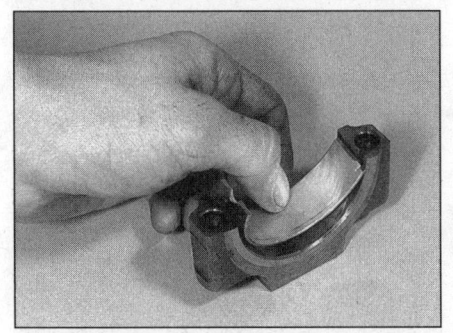

Final piston/connecting rod assembly refitting

37 Check the liner protrusions as described in Section 21. If the original liner/piston and connecting rod assemblies are being refitted then ensure that they are refitted in their original positions. If new liner/piston and connecting rod assemblies are being installed, position and number the assemblies as described in Section 21.
38 Ensure that the bearing shells are correctly refitted as described above in paragraphs 32 to 34. If new shells are being fitted, ensure that all traces of the protective grease are cleaned off using paraffin. Wipe dry the shells and connecting rods with a lint-free cloth.
39 Lubricate the liner bores, the pistons and piston rings then lay out each piston/connecting rod assembly with its liner.
40 Starting with assembly number 1, make sure that the piston rings are still spaced as described in Section 20, then clamp them in position with a piston ring compressor.
41 Insert the piston/connecting rod assembly into the bottom of number 1 liner and tap into position. Once the piston rings have entered the liner, remove the ring compressor, then slide the piston into position until the piston crown is flush with the top of the liner.
42 Fit a new base seal to the liner seat, making sure the seal teeth are correctly engaged with liner groove **(see illustration)**.
43 Ensure that the bearing shell is still correctly installed and carefully install the liner/piston and connecting rod assembly in the cylinder block, making sure it is fitted the correct way around.

19.33 Fit the bearing shells making sure their tabs locate correctly in the cutouts

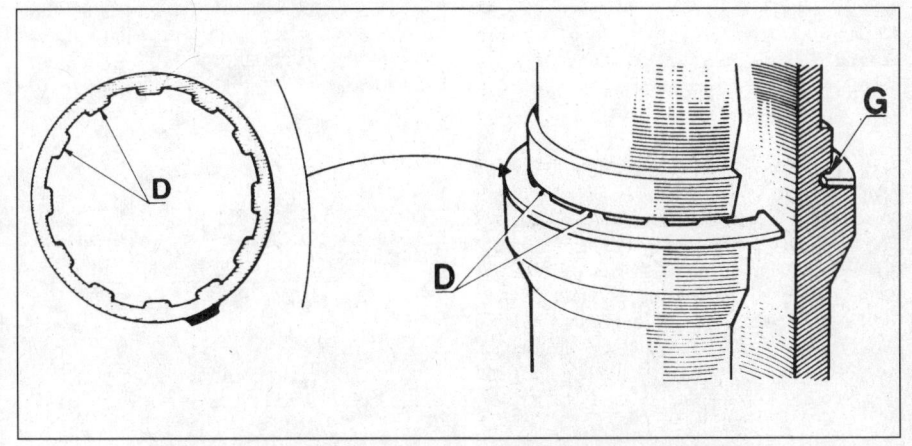

19.42 Fit a new base seal to each liner making sure its teeth (D) engage correctly in the liner groove (G)

2B•22 Six cylinder petrol engine

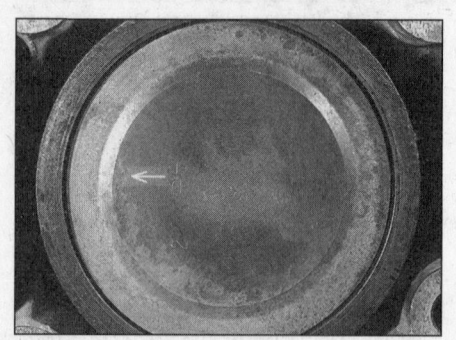

19.44 Fit the piston and liner assemblies making sure the arrow on each piston crown points towards the timing chain end of the engine

19.45 Tighten the big-end bearing cap nuts to the specified torque setting

44 Ensure that the piston/connecting rod assembly is correctly positioned, with the arrow on the piston crown pointing towards the timing chain end of the engine **(see illustration)**.
45 Taking care not to mark the liner bores, liberally lubricate the crankpin and both bearing shells, then pull the piston/connecting rod assembly down the liner and onto the crankpin. Refit the big-end bearing cap and tighten its retaining nuts to the specified torque setting **(see illustration)**.
46 Refit the remaining liner/piston and connecting rod assemblies in the same way.
47 Once all assemblies are correctly installed, clamp the liners in position as described in Section 11.
48 Rotate the crankshaft, and check that it turns freely, with no signs of binding or tight spots.
49 Refit the main bearing cap casting as described in paragraphs 34 to 40 of Section 18.
50 Refit the cylinder head(s), timing chains, flywheel/driveplate and sump as described in the relevant Sections of this Chapter.

20 Piston rings - removal and refitting

Removal

With the piston removed from the liner, carefully expand the old rings over the top of the pistons. The use of two or three old feeler blades will be helpful in preventing the rings dropping into empty grooves. Be careful not to scratch the piston with the ends of the ring. The rings are brittle, and will snap if they are spread too far. They're also very sharp - protect your hands and fingers. Note that the third ring incorporates an expander. Always remove the rings from the top of the piston. Keep each set of rings with its piston if the old rings are to be re-used.

Refitting

Install the new rings by fitting by fitting them over the top of the piston starting with the oil control expander ring. Ensure that the second compression is correctly installed the correct way around and position the piston ring end gaps as shown in relation to the gudgeon pin axis **(see illustration)**. Note that certain models maybe fitted with a one-piece oil control ring.

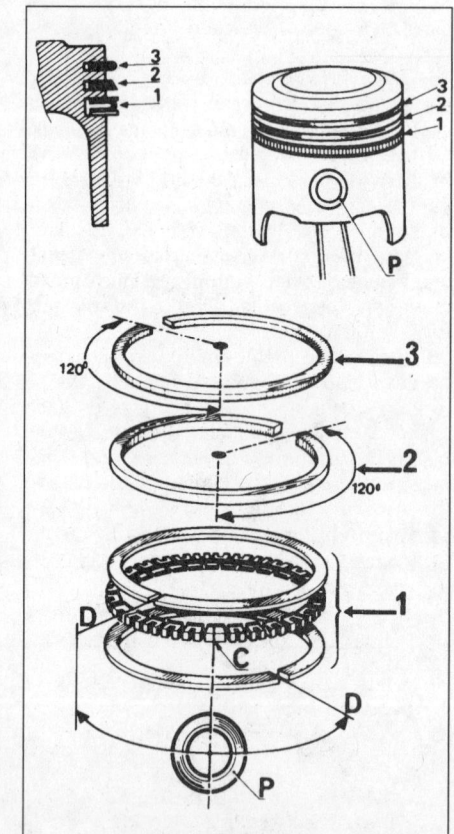

20.2 Piston ring fitting details

1 Oil control ring
2 Second compression ring
3 Top compression ring
C End gap
D End gap
P Gudgeon pin

21 Cylinder block/crankcase - cleaning and inspection

Cleaning

1 Refer to Chapter 2A, Section 23.

Inspection

2 Check the each cylinder liner for scuffing and scoring. Check for signs of a wear ridge at the top of the cylinder indicating that the liner is excessively worn.
3 Since Renault do not state any specific wear limits for the cylinder liners or pistons, it is not possible to assess the amount of wear by direct measurement. If there is any doubt about the condition of the cylinder liner seek the advice of a Renault dealer or suitable engine reconditioning specialist.
4 If renewal is necessary, new liners complete with piston and piston rings can be purchased from a Renault dealer. Note that it is not possible to renew the liners individually, as they are supplied as a matched assembly complete with piston and rings.
5 Make sure the base seals are correctly fitted, with their teeth engaged in the liner groove, and install the liners in their original position in the cylinder block. Using a dial gauge or a straight-edge and feeler blade, check that the protrusion of each liner above the upper surface of the cylinder block is within the limits given in the *Specifications*, and that the maximum difference between any two liners is not exceeded **(see illustration)**.
6 Varying thicknesses of liner base seals are available in order to obtain the correct liner protrusion above the top face of the cylinder block faces on each bank. The liner base seals are available in various thicknesses (see *Specifications*) and are colour coded for identification. New base seals should be fitted whenever the liners are disturbed, they are selected as follows.
7 Fit the liners into their correct locations in the cylinder blocks having first removed their base seals. Lay a straight-edge across the top face of a liner and measure the gap to the top face of the cylinder block with feeler blades. Repeat this with the two other liners in the

21.5 Checking cylinder liner protrusion

Six cylinder petrol engine 2B•23

bank concerned and make a note of the respective protrusions.

8 Noting that the protrusion should be as close to the maximum allowable as possible, now subtract the measurement reading for each liner taken from maximum protrusion allowable (see *Specifications*) to determine the seal thickness required.

9 Seal thickness equal to, or just less than, the calculated thickness should then be selected. As an example on a 2664 cc engine, if the protrusion measurement is 0.10 mm, subtract this from 0.23 mm to give 0.13 mm. In this instance a red coded base seal is required, which has a thickness of 0.122 mm.

10 Repeat this procedure until the correct base seal for each liner has been selected, ensuring that the maximum allowable difference between any two liners is not exceeded.

11 Fit the new base seals, making sure their teeth are correctly engaged with the liner groove, then fit the liners to the block and check the protrusions as described above to ensure that the correct seals have been selected.

12 Note that if new liner assemblies are being installed the liners must be arranged so that the liner protrusions gradually decrease/increase from one end of the cylinder block to the other, ie. the largest protrusion is at one end with the smallest at the opposite end **(see illustration)**. Use the liner protrusion measurements to arrange the liners as required then number each one with its respective cylinder number, remembering No.1 cylinder is at the transmission end of the left-hand cylinder bank.

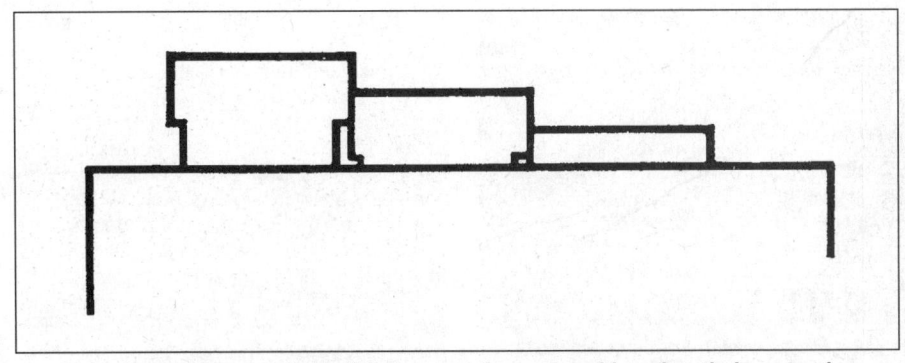

22 Engine - removal and refitting leaving manual transmission in car

Removal

1 Remove the bonnet as described in Chapter 1A.

21.12 If new liners are being fitted ensure they are positioned so their protrusion gradually decrease/increase from one end of the block to the other

2 Disconnect the battery.
3 Remove the air cleaner as described in Chapter 4B.
4 Drain the engine oil and cooling system as described in Chapter 1A.
5 Remove the radiator grille and headlamp wipers (if fitted).
6 Unbolt the radiator top rail and remove it with the side deflectors.
7 Remove the radiator assembly as described in Chapter 3.
8 On cars equipped with air conditioning, unbolt the compressor and position it clear of the engine. Do not disconnect the refrigerant lines from the compressor (see **Warnings** in Chapter 3).
9 Unbolt the ignition electronic control box and place it on top of the engine.
10 Disconnect the accelerator cable.
11 Unbolt and remove the exhaust front pipe(s) as described in Chapter 4B.
12 Disconnect all vacuum hoses including the one for the brake servo. Identify each hose with numbered tape so there will be no confusion at reconnection.
13 Disconnect the coolant hoses including those for the heater at the engine compartment rear bulkhead.
14 Disconnect all engine electrical leads, identifying them with numbered tape if necessary.
15 Disconnect the HT lead from the ignition coil, and the alternator leads.
16 Unbolt the loom protector from the front end of the sump pan and place it to one side.

17 Disconnect the fuel lines as described in Chapter 4B.
18 Remove the starter motor as described in Chapter 5A.
19 On Turbo models, remove the alternator, the turbocharger heat shield and the inlet scoop as described in Chapter 4B.
20 Release the power steering pump fluid hoses from their clips, but do not disconnect them. Unbolt the pump and tie it to one side of the engine compartment (see Chapter 1A).
21 Unbolt the engine to body earth strap.
22 On Turbo models, disconnect the oil pipe bracket.
23 Remove the clutch slave cylinder heat shield, the TDC sensor and cover plate.
24 On Turbo models, disconnect the cylinder block oil hoses, the absolute pressure sensor pipes and the pressure indicator connection at the front strut turret **(see illustrations)**.
25 Remove the left-hand engine damper and its bracket **(see illustration)**.
26 Disconnect the right-hand engine damper bottom mounting.
27 Release the cruise control diaphragm unit.
28 Unbolt the auxiliary air valve.
29 Remove the cover plate from the lower part of the flywheel housing.
30 Attach a hoist to the engine lifting lugs and just take its weight.
31 Unscrew the engine-to-transmission connecting bolts.
32 Place a jack under the transmission.
33 Unscrew the engine mounting through-bolt nuts and disconnect the mountings.

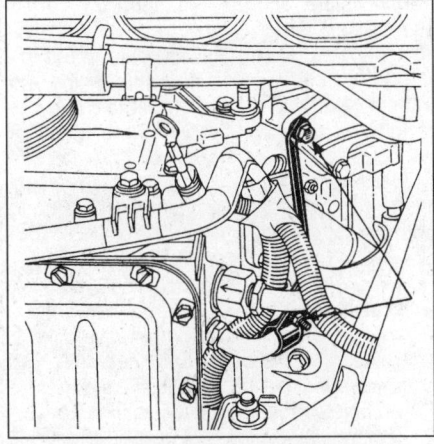

22.24a Oil pipe bracket fixings - 2458 cc turbo engine

22.24b Pressure indicator hose connection (E) - 2458 cc turbo engine

22.25 Engine damper - viewed from underneath

2B•24 Six cylinder petrol engine

34 Pull the engine forwards until it is clear of the clutch shaft and then hoist it up out of the engine compartment.

Refitting

35 Refitting is the reverse of removal noting the following.
 a) Apply a smear of high-melting-point grease (Renault recommend the use of Molykote BR2) to the splines of the transmission input shaft. Do not apply too much, otherwise there is a possibility of the grease contaminating the clutch friction plate.
 b) Ensure that the locating dowels are correctly positioned in the engine or transmission before lowering the engine into place.
 c) Ensure that the wiring loom is correctly routed, and retained by all the relevant retaining clips; all connectors should be correctly and securely reconnected.
 d) Ensure that all coolant hoses are correctly reconnected, and securely retained by their retaining clips.
 e) Adjust the accelerator cable as described in Chapter 4B.
 f) Refill the engine with correct quantity and type of oil, as described in Chapter 1A.
 g) Refill the cooling system as described in Chapter 1A.

23 Engine and manual transmission - removal and refitting

Removal

1 The operations include those described in the preceding Section plus the following additional work. Do not remove the engine-to-transmission connecting bolts nor the bellhousing lower cover plate prior to removal of the engine/transmission.
2 Drain the transmission oil as described in Chapter 1A.
3 Disconnect the gearchange link rods and the reverse interlock operating cable as described in Chapter 7A.
4 Disconnect the reversing lamp switch leads.
5 Disconnect the speedometer sensor.
6 Remove the heat shield, unbolt the clutch slave cylinder and tie it up out of the way without disconnecting the fluid hose.
7 Remove the driveshafts as described in Chapter 8.
8 With the hoist taking the weight of the engine and transmission, unscrew the engine mounting through-bolts and unbolt the transmission mounting brackets.
9 Withdraw the engine/transmission forwards and upwards from the engine compartment.
10 Clean away external dirt using paraffin and a stiff brush or a water soluble solvent.

11 Remove the bellhousing lower cover plate.
12 Unbolt and remove the starter motor.
13 Unscrew and remove the engine-to-transmission connecting bolts. Note the location of the various brackets, wiring harness clips and other attachments held by some of the bolts.
14 Support the engine on wooden blocks placed under the sump pan so that the transmission has a clear gap underneath it, and then withdraw the transmission from the engine. Do not allow the weight of the transmission to hang upon the clutch shaft, while the latter is still engaged with the clutch driven plate.

Refitting

15 Check that the clutch driven plate is centralised, and then apply a smear of high-temperature grease (Renault recommend the use of Molykote BR2) to the clutch input shaft splines.
16 Offer the transmission to the engine and locate it on the positioning dowels.
17 Screw in the connecting bolts, noting the locations of the lifting lugs and clips.
18 Refit the cover plate to the front face of the clutch bellhousing and screw in the reinforcement bracket bolts.
19 Refit the starter motor and heatshield and securely tighten its bolts.
20 The remaining operations are a reversal of removal but observe the following points.
 a) Arrange the lifting gear so that the transmission is slanting downwards at a steep angle and the engine/transmission is canted over to the left. It will be found easier to install the engine if the right-hand mounting bracket bolts are loosened.
 b) Refit the driveshafts as described in Chapter 8.
 c) Ensure that the wiring loom is correctly routed, and retained by all the relevant retaining clips; all connectors should be correctly and securely reconnected.
 d) Ensure that all coolant hoses are correctly reconnected, and securely retained by their retaining clips.
 e) Adjust the accelerator cable as described in Chapter 4B.
 f) Refill the engine and transmission with correct quantity and type of oil, as described in Chapter 1A.
 g) Refill the cooling system as described in Chapter 1A.

24 Engine - removal and refitting leaving automatic transmission in car

Removal

1 Carry out the operations described in Section 22 excluding those applicable to Turbo models.

2 The following additional tasks must also be carried out.
3 Disconnect the earth straps between transmission and body.
4 Unbolt and remove the starter motor.
5 Disconnect the vacuum pipe from the inlet manifold.
6 Disconnect the governor cable at the throttle housing (where necessary).
7 Remove the lower cover plate from the torque converter housing.
8 Lock the teeth of the driveplate starter ring gear using a suitable tool and, then unscrew the driveplate-to-torque converter fixing bolts. In order to bring each bolt into view, the crankshaft will have to be turned by means of its pulley bolt while the ring gear locking tool is temporarily removed.
9 Unscrew and remove the transmission bellhousing-to-engine connecting bolts.
10 Attach a hoist to the engine lifting lugs and just take its weight.
11 Place a jack under the transmission.
12 Remove the engine mounting through-bolts.
13 Raise the engine to clear the mounting brackets and then withdraw it forward to disconnect it from the transmission, and lift it up and out of the engine compartment.
14 Bolt a retaining bar to one of the torque converter housing flange bolts in order to retain the converter in full engagement with the transmission, and to prevent damage to the oil seal and loss of fluid (see Chapter 7B).

Refitting

15 Apply a smear of high temperature grease (Renault recommend the use of Molykote BR2) grease to the locating boss on the torque converter.
16 Remove the temporary holding bar from the torque converter.
17 Lower the engine into the car, making sure that the torque converter does not move forward as the engine is connected and bolted to the transmission. Note that the upper bellhousing bolts must be in position before the engine is connected to the transmission as they cannot be fitted afterwards.
18 The bellhousing lower cover must be located on the dowels before the engine and transmission are brought together, and the alignment mark on the driveplate must be between the two marks on the torque converter (see Chapter 7B).
19 The remaining reconnection and refitting operations are reversals of removal, but observe the following points.
 a) Ensure that the wiring loom is correctly routed, and retained by all the relevant retaining clips; all connectors should be correctly and securely reconnected.
 b) Ensure that all coolant hoses are correctly reconnected, and securely retained by their retaining clips.
 c) Adjust the accelerator cable as described in Chapter 4A or 4B.

Six cylinder petrol engine 2B•25

d) *Adjust the selector cable as described in Chapter 7B.*
e) *Refill the engine with correct quantity and type of oil, as described in Chapter 1A. Also check the transmission fluid level.*
f) *Refill the cooling system as described in Chapter 1A.*

25 Engine and automatic transmission - removal and refitting

Removal

1 Carry out the operations described in paragraphs 1 to 28 of Section 22 excluding those applicable to Turbo models.
2 Drain the final drive lubricant as described in Chapter 7B.
3 Disconnect the selector cable as described in Chapter 7B.
4 Remove the driveshafts (see Chapter 8).
5 Remove the left-hand engine damper and bracket.
6 Take the weight of the engine and transmission and disconnect the engine and transmission mountings.
7 Lift the engine/transmission forward and then up and out of the engine compartment.
8 Clean away external dirt using paraffin and a stiff brush or a water soluble solvent.
9 Separate the transmission from the engine, described in paragraphs 7 to 9 of Section 24. Pull the transmission from the engine and then fit a torque converter retaining bar as described in Section 24, paragraph 14.

Refitting

10 Refer to Section 24.

26 Oil cooler components - removal and refitting

Refer to Chapter 2A, Section 28.

27 Initial start-up after major overhaul

Refer to Chapter 2A, Section 29. Note that the cylinder head bolts will require retightening after the engine has been warmed up if the cylinder head(s) has been removed (see Section 11).

Notes

Chapter 2 Part C:
Diesel engine

Contents

Camshaft and intermediate shaft oil seals - renewal	11
Camshaft and rocker arms - removal, inspection and refitting	12
Compression and leakdown tests - description and interpretation	2
Crankcase stiffener casting (later J8S 708 turbo engines) - removal and refitting	17
Crankshaft - removal, inspection and refitting	23
Crankshaft oil seals - renewal	19
Crankshaft pulley - removal and refitting	7
Cylinder block/crankcase - cleaning and inspection	26
Cylinder head - overhaul	14
Cylinder head - removal and refitting	13
Cylinder head cover - removal and refitting	5
Engine - removal and refitting leaving transmission in car	28
Engine and transmission - removal and refitting	27
Engine/transmission mountings - renewal	21
Flywheel - removal, inspection and refitting	20
General description	1
Initial start-up after major overhaul	29
Intermediate shaft - removal, inspection and refitting	22
Oil cooler components - removal and refitting	18
Oil pump - removal, inspection and refitting	16
Operations possible without removing the engine	4
Piston rings - removal and refitting	25
Piston/connecting rod and cylinder liner assemblies - removal, inspection and refitting	24
Sump - removal and refitting	15
Timing belt - removal, inspection and refitting	9
Timing belt cover - removal and refitting	8
Timing belt tensioner and sprockets - removal, inspection and refitting	10
Top dead centre (TDC) for number 1 piston - locating	3
Valve clearances - adjustment	6

Degrees of difficulty

Easy, suitable for novice with little experience
Fairly easy, suitable for beginner with some experience
Fairly difficult, suitable for competent DIY mechanic
Difficult, suitable for experienced DIY mechanic
Very difficult, suitable for expert DIY or professional

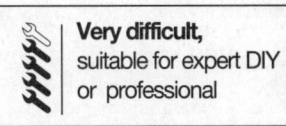

Specifications

Engine type	Four-cylinder, in-line mounted longitudinally at front of car. Light alloy construction with wet cylinder liners. Belt-driven overhead camshaft
Engine code	J8S
Bore	86.0 mm
Stroke	89.0 mm

Timing belt

Timing belt tension deflection (measured using Renault tool Ele 346-04)	See Text

Camshaft

Number of bearings	5
Endfloat	0.05 to 0.15 mm

Cylinder head

Material	Light alloy
Overall height	104.5 mm
Maximum permitted surface distortion	0.05 mm
Swirl chamber protrusion from gasket face	0.01 to 0.04 mm
Cylinder head to piston clearance	See Text

Valves

Valve clearances (engine cold):	
Inlet valves	0.20 mm
Exhaust valves	0.25 mm
Stem diameter	8 mm
Head diameter:	
Inlet valve	40.2 mm
Exhaust valve	33.2 mm
Valve head recess below the gasket face	0.80 to 1.15 mm
Valve seat angle	90°
Valve seat width	1.6 to 1.9 mm

Valves (continued)
Valve guides:
- Bore .... 8.0 mm
- Outside diameter:
  - Standard .... 13.1 mm
  - Oversize .... 13.3 mm

Intermediate shaft
Endfloat .... 0.05 to 0.15 mm

Crankshaft
Number of main bearings 5
Main bearing journal diameter:
- Standard .... 62.880 mm
- Undersize .... 62.630 mm

Main bearing journal minimum regrind diameter .... 62.611 mm
Main bearing running clearance .... 0.025 to 0.050 mm
Crankpin diameter:
- Standard .... 52.296 mm
- Undersize .... 52.046 mm

Crankpin minimum regrind diameter .... 52.017 mm
Big-end bearing running clearance .... 0.025 to 0.050 mm
Crankshaft endfloat .... 0.20 to 0.30 mm

*These are suggested figures, typical for this type of engine - no exact values are quoted by Renault

Cylinder liners
Bore .... 86.0 mm
Base locating diameter .... 93.6 mm
Protrusion without O-ring .... 0.05 to 0.12 mm
Maximum difference between any two adjacent liners ...... 0.04 mm

Connecting rods and gudgeon pins
Connecting rod big-end endfloat:
- Normally-aspirated engine .... 0.31 to 0.57 mm
- Turbo engine .... 0.31 to 0.50 mm

Piston rings
Piston ring thickness:
- Top and second compression:
  - Normally-aspirated engine .... 2.0 mm
  - Turbo engine .... 2.5 mm
- Oil control .... 4.0 mm

Lubrication system
Oil pump:
- Type .... Gear, driven from intermediate shaft
- Clearances:
  - Gear teeth to body .... 0.10 to 0.24 mm
  - Gear endfloat .... 0.020 to 0.085 mm

Oil pressure (at 80°C):
- At idle .... 0.8 bar (11 psi)
- At 3000 rpm:
  - Normally-aspirated engine .... 3.0 bar (43 psi)
  - Turbo engine .... 3.5 bar (50 psi)

Torque wrench settings

	Nm	lbf ft
Big-end cap bolts	65	48
Camshaft sprocket bolt	50	37
Crankcase stiffener casting - later J8S 708 engines:		
7 mm bolts	15	11
10 mm bolts	40	30
Crankshaft pulley bolt	135	101
Engine to transmission housing bolts	50	37
Flywheel bolts	60	44
Injection pump sprocket nut	50	37
Intermediate shaft sprocket bolt	50	37
Main bearing cap bolts	95	70
Oil pump:		
Body bolts	44	32
Cover bolts	12	9

Diesel engine 2C•3

Torque wrench settings (continued)

	Nm	lbf ft
Cylinder head bolts (see Text):		
Stage 1	30	22
Stage 2	50	37
Working on each bolt in the specified sequence, slacken the bolt then:		
Stage 3	20	15
Stage 4 (refer to illustration 13.41):		
Nuts number 1, 8 and 9 in tightening sequence	Angle-tighten a further 120°	
Bolts number 2, 3, 6, 8, 10, 11, 14 and 15	Angle-tighten a further 60°	
Bolts number 4, 5, 12 and 13	Angle-tighten a further 70°	
Bolts number 16 and 17	Angle-tighten a further 80°	
Go around in specified sequence and:		
Stage 5 (refer to illustration 13.41):		
Nuts number 1, 8 and 9 in tightening sequence	Angle-tighten a further 120°	
Bolts number 2, 3, 6, 7, 10, 11, 14 and 15	Angle-tighten a further 60°	
Bolts number 4, 5, 12 and 13	Angle-tighten a further 70°	
Bolts number 16 and 17	Angle-tighten a further 80°	
Piston oil jet nozzle bolt - later turbo models	14	10
Rocker shaft:		
Shaft end bolt	20	15
Rocker arm pedestal bolts	30	22
Sump retaining bolts:		
Models with a one piece sump	15	11
Models with a two-piece sump (fitted with crankcase stiffener)	10	7

1 General description

1 The diesel engine is of four-cylinder overhead camshaft design, mounted longitudinally at the front of the vehicle, with the transmission mounted on the rear end of the engine **(see illustration)**.

2 The camshaft is driven by a toothed timing belt. The camshaft operates the eight valves via rocker arms which are mounted on a shaft which is positioned directly above the camshaft. Valve clearances are adjusted via the screw and locknut arrangement fitted to each rocker arm. The inlet and exhaust valves are each closed by coil springs and operate in guides pressed into the cylinder head.

3 The toothed timing belt also drives the fuel injection pump and intermediate shaft. The coolant pump is driven by the same drivebelt as the alternator.

4 The crankshaft runs in five main bearings of the usual shell type. Endfloat is controlled by thrustwashers either side of number 2 main bearing.

5 The connecting rods rotate on horizontally-split bearing shells at their big-ends. The pistons are attached to the connecting rods by fully floating gudgeon pins which are retained by circlips. The aluminium alloy pistons are fitted with three piston rings, comprising two compression rings and an oil control ring.

6 The cylinder bores are formed by replaceable wet liners that are located from their bottom ends; sealing rings are fitted at the base of each liner to prevent the escape of coolant into the sump.

7 Lubrication is by means of a gear type oil pump which is driven by the intermediate shaft via a worm gear. It draws oil through a strainer located in the sump and then forces it through a full-flow cartridge-type filter into galleries in the cylinder block/crankcase, from where it is distributed to the crankshaft (main bearings) and camshaft. The big-end bearings are supplied with oil via internal drillings in the crankshaft, while the camshaft bearings also receive a pressurised supply. The camshaft lobes and valves are lubricated by splash, as are all other engine components.

8 On Turbo models the turbocharger bearings also receive a pressurised oil supply, the turbocharger being linked to the cylinder block oil galleries by a feed and return hose. The pistons are also cooled by oil sprayed

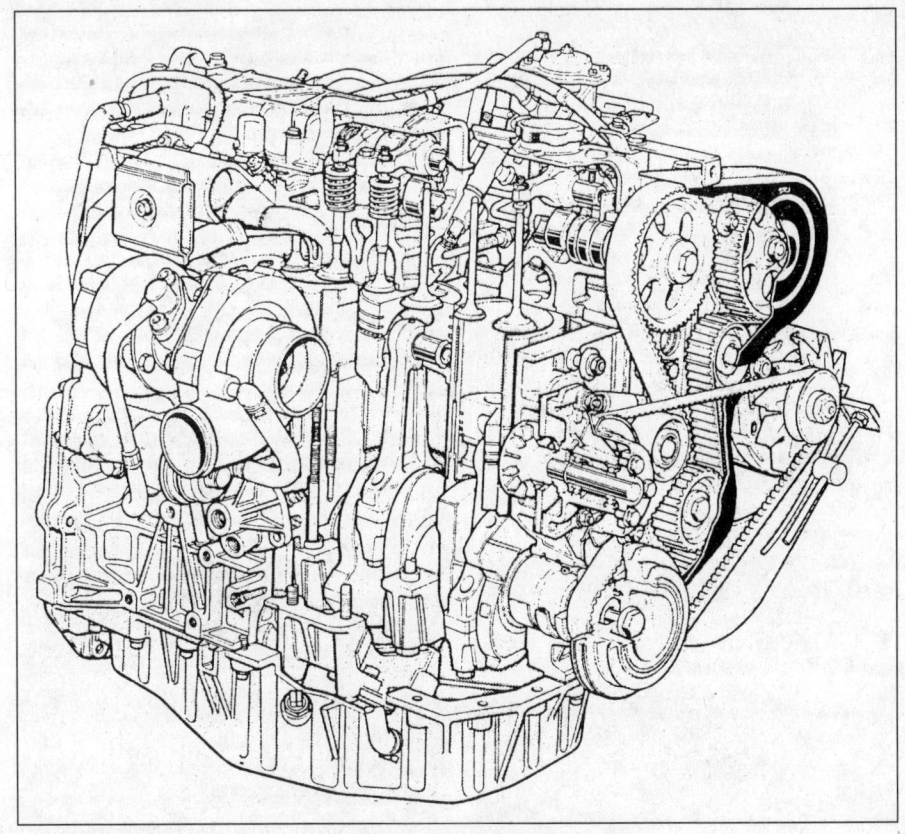

1.1 Cutaway view of the turbo engine

2C•4 Diesel engine

onto the underside of each assembly by the jets (one for each cylinder) mounted on the cylinder block.

9 On Turbo models an oil cooler is fitted to keep the oil temperature constant under extreme use. The cooler is mounted on the right-hand wing valance and is linked to the right-hand side of the cylinder block/crankcase by a feed and return hose. The oil is forced from the pump directly to the oil cooler before returning to the cylinder block to circulate around the engine components as described above. Where an oil cooler is fitted the oil filter is screwed onto the underside of the cooler rather than onto the side of the cylinder block.

2 Compression and leakdown tests - description and interpretation

Compression test

Note: *A compression tester specifically designed for diesel engines must be used for this test.*

1 When engine performance is down, or if misfiring occurs which cannot be attributed to the ignition or fuel systems, a compression test can provide diagnostic clues as to the engine's condition. If the test is performed regularly it can give warning of trouble before any other symptoms become apparent.

2 A compression tester specifically intended for diesel engines must be used, because of the higher pressures involved. The tester is connected to an adapter which screws into the glow plug or injector hole. An adapter suitable for use in the injector hole will be preferable, due to the limited access to the glow plug holes. It is unlikely to be worthwhile buying such a tester for occasional use, but it may be possible to borrow or hire one - if not, have the test performed by a garage.

3 Unless specific instructions to the contrary are supplied with the tester, observe the following points.
 a) *The battery must be in a good state of charge, the air filter must be clean and the engine should be at normal operating temperature.*
 b) *All the injectors or glow plugs should be removed before starting the test. If removing the injectors, also remove the flame shield washers, otherwise they may be blown out.*
 c) *The stop solenoid must be disconnected to prevent fuel from being discharged. To disconnect the solenoid undo the retaining nut and disconnect the wire from the solenoid which is screwed into the top of the injection pump.*

4 There is no need to hold the accelerator pedal down during the test because the diesel engine air inlet is not throttled.

5 The actual compression pressures measured are not so important as the balance between cylinders.

6 The cause of poor compression is less easy to establish on a diesel engine than on a petrol one. The effect of introducing oil into the cylinders ('wet' testing) is not conclusive, because there is a risk that the oil will sit in the swirl chamber or in the recess on the piston crown instead of passing to the rings. However, the following can be used as a rough guide to diagnosis.

7 All cylinders should produce very similar pressures; a difference greater than 2 bars between any two cylinders indicates the existence of a fault. Note that the compression should build up quickly in a healthy engine; low compression on the first stroke, followed by gradually increasing pressure on successive strokes, indicates worn piston rings. A low compression reading on the first stroke, which does not build up during successive strokes, indicates leaking valves or a blown head gasket (a cracked head could also be the cause). Deposits on the undersides of the valve heads can also cause low compression.

8 A low reading from two adjacent cylinders is almost certainly due to the head gasket having blown between them; the presence of coolant in the engine oil will confirm this.

9 Although Renault do not specify any exact compression pressures, as a guide, any cylinder pressure less than 20 bars can be considered as less than ideal. Refer to your Renault dealer for the exact specified compression pressure for your vehicle.

10 If the compression reading is unusually high, the cylinder head surfaces, valves and pistons are probably coated with carbon deposits. If this is the case, the cylinder head should be removed and decarbonised.

Leakdown test

11 A leakdown test measures the rate at which compressed air fed into the cylinder is lost. It is an alternative to a compression test and in many ways it is better, since the escaping air provides easy identification of where pressure loss is occurring (piston rings, valves or head gasket).

12 The equipment needed for leakdown testing is unlikely to be available to the home mechanic. If poor compression is suspected, have the test performed by a suitably equipped garage.

3 Top Dead Centre (TDC) for number one piston - locating

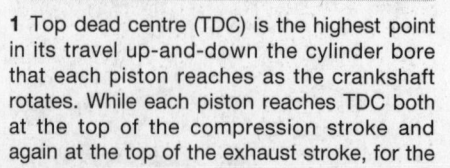

1 Top dead centre (TDC) is the highest point in its travel up-and-down the cylinder bore that each piston reaches as the crankshaft rotates. While each piston reaches TDC both at the top of the compression stroke and again at the top of the exhaust stroke, for the purpose of timing the engine, TDC refers to the position of No. 1 piston at the end of its compression stroke. No. 1 piston (and cylinder) is at the flywheel end of the engine. Note that the crankshaft rotates clockwise when viewed from the front of the car.

2 Disconnect the battery negative lead.

3 Apply the handbrake and ensure that the transmission is in neutral, then jack up the front of the vehicle and support it on axle stands (see "*Jacking and vehicle support*"). To improve access to the pulley, work around the plastic undershield, removing its retaining screws and prising out its retaining clips, and remove the undershield from the underneath the vehicle.

4 The crankshaft can rotated using a spanner, or socket and extension bar, applied to the crankshaft pulley bolt. Rotate the crankshaft clockwise until the timing marks on the camshaft sprocket and injection pump sprocket rims appear in the windows in the top of timing belt outer cover and align the mark with the pointer on the base of each window.

5 With the timing marks and pointers aligned, No. 1 piston is situated at TDC at the end of its compression stroke **(see illustration)**.

6 If required, the crankshaft can be locked in position to prevent engine rotation. To do this, remove the plug from the left-hand side of cylinder block/crankcase. In the absence of the special Renault locking pin (Mot. 861), it will be necessary to obtain a rod which is 8 mm in diameter and at least 100 mm in length. Ensure that the camshaft sprocket timing mark is correctly positioned as described above then insert the locking pin through the cylinder block/crankcase hole and locate it in the crankshaft web slot. If the camshaft sprocket timing mark is not accurately aligned, the crankshaft will not be correctly positioned and the locking pin may engage with one of the balance holes in the web instead of the TDC slot **(see illustrations)**.

Note: *Do not attempt to rotate the engine whilst the crankshaft is locked in position. If the engine is to be left in this state for a long period of time, it is a good idea to place*

3.5 With No 1 cylinder at TDC on its compression stroke both timing mark pointers on the cover should be aligned with the marks on the camshaft and injection pump sprockets

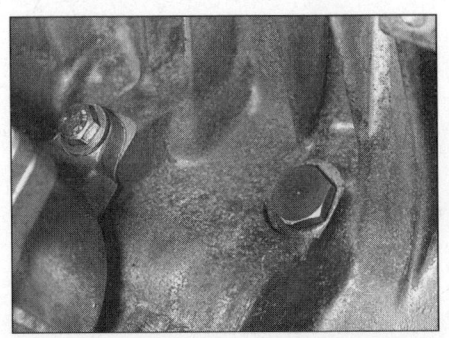

Diesel engine 2C•5

3.6a To lock the crankshaft in position, unscrew the plug from the cylinder block . . .

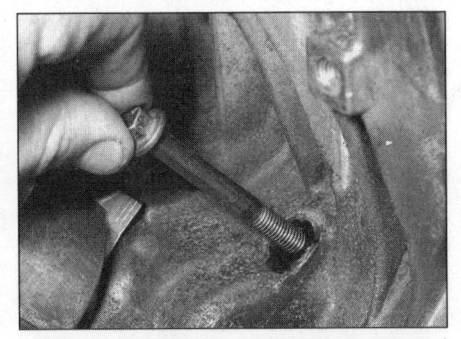

3.6b . . . and insert a pin of the correct dimensions such as a bolt

suitable warning notices inside the vehicle, and in the engine compartment. This will reduce the possibility of the engine being accidentally cranked on the starter motor, which is likely to cause damage with the locking pin in place.

4 Operations possible without removing the engine

The following work can be carried out with the engine in the car

- a) Compression and leakdown pressure - testing.
- b) Cylinder head cover - removal and refitting.
- c) Crankshaft pulley - removal and refitting.
- d) Timing belt covers - removal and refitting.
- e) Timing belt - removal and refitting.
- f) Timing belt tensioner and sprockets - removal and refitting.
- g) Camshaft and intermediate shaft oil seals - renewal.
- h) Camshaft and rocker arms - removal and refitting.
- i) Cylinder head - removal and refitting.
- j) Cylinder head and pistons - decarbonising.
- k) Sump - removal and refitting.
- l) Oil pump - removal, overhaul and refitting.
- m) Crankshaft oil seals - renewal.
- n) Engine/transmission mountings - inspection and renewal.
- o) Flywheel - removal and refitting.

5 Cylinder head cover - removal and refitting

Removal

1 Disconnect the battery negative lead.
2 Slacken the retaining clip (where fitted) and disconnect the breather hose from the rear of the cover.
3 Slacken and remove the cylinder head cover retaining bolts, noting the correct fitted positions of all brackets retained by the bolts, then lift off the cover, complete with gasket.

Remove the cover from the engine compartment.
4 Remove the gasket from the cover and examine it for signs damage or deterioration, and renew if necessary.

Refitting

5 On reassembly, carefully clean the cylinder head mating surfaces and the cover gasket's groove and remove all traces of oil. Seat the gasket in its groove in the cover noting that the gasket is not symmetrical and can only be fitted one way. On later models the gasket is equipped with a locating lug which should be positioned at the rear of the cover, and on early models the gasket must be fitted with its graphite surface facing the cylinder head cover.
6 Ensure that the gasket is correctly fitted then manoeuvre the cover into position on the cylinder head **(see illustration)**.
7 Refit the cover retaining bolts and tighten them securely, ensuring all the relevant brackets are correctly positioned.
8 Reconnect the breather hose to the cylinder head cover and reconnect the battery negative lead.

6 Valve clearances - adjustment

1 The importance of having the valve clearances correctly adjusted cannot be overstressed as they vitally affect the performance of the engine. That being said

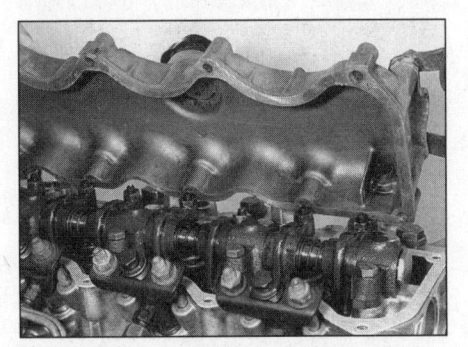

5.6 Position a new gasket on the cylinder head and refit the head cover

the check should not be regarded as routine maintenance and should only be carried out when the valve gear has become noisy, after engine overhaul, or when trying to trace the cause of power loss which may be attributed to the valve. The clearances are checked as follows noting that the engine must be cold (engine not having been started for at least two and a half hours) for the check to be accurate.
2 Apply the handbrake then jack up the front of the car and support it on axle stands to improve access to the crankshaft pulley bolt. The engine can then be turned over using a suitable socket and extension bar fitted to the bolt.
3 Remove the cylinder head cover as described in Section 5.
4 It is important that the clearance of the relevant valve is checked and adjusted when the valve is fully closed and the rocker arm rests on the heel of the cam (directly opposite the peak). This can be ensured by carrying out the adjustments in the following sequence, noting that number 1 cylinder is at the transmission end of the engine. The valve locations are as shown and the correct valve clearances are given in the Specifications at the start of this Chapter, the valve locations are as shown **(see illustration)**.

Valve fully open	Adjust valves
No 1 cyl exhaust	No 3 cyl inlet and No 4 cyl exhaust
No 3 cyl exhaust	No 4 cyl inlet and No 2 cyl exhaust
No 4 cyl exhaust	No 2 cyl inlet and No 1 cyl exhaust
No 2 cyl exhaust	No 1 cyl inlet and No 3 cyl exhaust

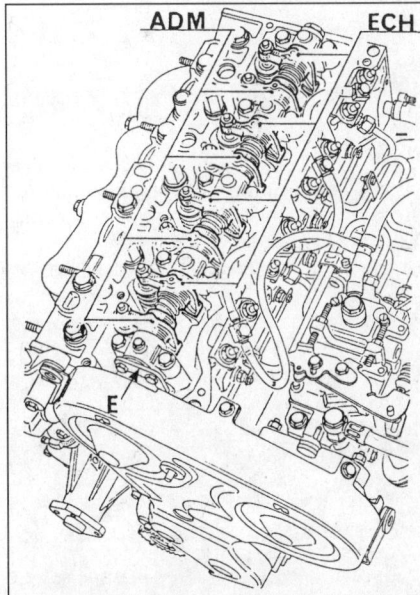

6.4 Valve locations

ADM = Inlet valves
ECH = Exhaust valves
E shows the front end of the camshaft

2C•6 Diesel engine

6.5a With the relevant valve fully open (rocker arm pad on the peak of the cam) . . .

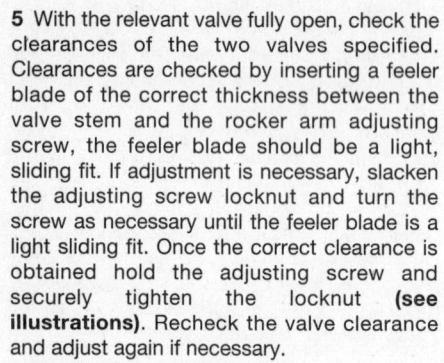

6.5b . . . check and adjust the clearances on the two valves specified

7.2 Use a fabricated tool like this one to lock the flywheel ring gear and prevent crankshaft rotation

5 With the relevant valve fully open, check the clearances of the two valves specified. Clearances are checked by inserting a feeler blade of the correct thickness between the valve stem and the rocker arm adjusting screw, the feeler blade should be a light, sliding fit. If adjustment is necessary, slacken the adjusting screw locknut and turn the screw as necessary until the feeler blade is a light sliding fit. Once the correct clearance is obtained hold the adjusting screw and securely tighten the locknut **(see illustrations)**. Recheck the valve clearance and adjust again if necessary.

6 Rotate the crankshaft until the next valve in the sequence is fully open and check the clearances of the next two specified valves.

7 Repeat the procedure until all eight valve clearances have been checked and if necessary, adjusted, then refit the cylinder head cover as described in Section 5.

7 Crankshaft pulley - removal and refitting

Removal

1 Remove the power steering pump and/or alternator auxiliary drivebelt(s) (as applicable) as described in Chapter 1B.

2 To prevent crankshaft rotation whilst the pulley retaining bolt is slackened, select top gear and have an assistant apply the brakes firmly. If the engine has been removed from the vehicle, lock the flywheel ring gear using the arrangement shown **(see illustration)**. **Do not** attempt to lock the crankshaft by inserting the locking pin through the cylinder block and into the crankshaft TDC slot.

3 Unscrew the retaining bolt and remove the pulley from the end of the crankshaft.

Refitting

4 Locate the pulley on the end of the crankshaft, aligning its locating holes with the sprocket roll pins, and refit the retaining bolt **(see illustration)**.

5 Lock the crankshaft by the method used on removal and tighten the pulley retaining bolt to the specified torque setting.

6 Refit the auxiliary drivebelt(s) as described in Chapter 1B.

8 Timing belt cover - removal and refitting

Removal

1 Access to the timing belt cover is very limited when the engine is fitted to the vehicle. To improve access, remove the upper crossmember from the front of the vehicle then lift the radiator out from its lower mountings and position as far in front of the timing belt cover as possible, without disconnecting the radiator hoses (see Chapter 3). Also remove the cooling fan.

2 To improve access to the lower timing belt cover retaining bolts and nuts, firmly apply the handbrake then jack up the front of the vehicle and support it on axle stands (see "Jacking and vehicle support"). Work around the plastic undershield, removing its retaining screws and prising out its retaining clips, and remove the undershield from the underneath the vehicle.

3 Remove the auxiliary drivebelt(s) as described in Chapter 1B.

4 Undo the nuts and free the cable guide from the front of the timing belt cover.

5 Work around the cover and remove all the nuts and bolts along with their washers.

6 Carefully withdraw the cover from the engine compartment and recover the collars from the centre of the each cover rubber mounting.

7 With the cover removed, slide the three spacers off the cover mounting studs, noting the correct location of each spacer as it is removed.

8 Where necessary, unbolt and remove the rear timing belt cover from the cylinder head.

9 Examine the cover mounting rubbers for signs of damage or deterioration and renew as necessary.

Refitting

10 Refit the rear timing belt cover (where removed) to the cylinder head and securely tighten its retaining bolts.

11 Noting that all three spacers are a different length, refit each spacer to its respective cover mounting stud **(see illustration)**.

12 Ensure that all the cover mounting rubbers are correctly fitted and fit the collars into position in the centre of each rubber.

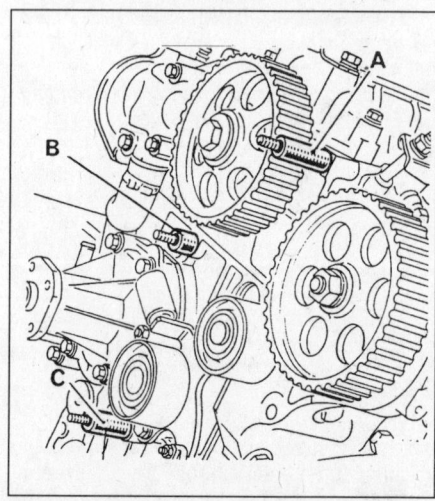

7.4 Align the pulley holes with the sprocket roll pins (arrowed) when refitting the crankshaft pulley

8.11 Timing belt cover spacer locations

A Long spacer
B Short spacer
C Medium spacer

13 Refit the timing belt cover to the engine, taking great care not to displace the mounting rubbers or collars. Refit the washers and cover retaining nuts and bolts, tightening them securely (see illustration).
14 Refit the cable guide and securely tighten its retaining nuts.
15 Refit the auxiliary drivebelt(s) as described in Chapter 1B.
16 Refit the undershield ensuring that it is securely retained by all the necessary clips and that its retaining screws are securely tightened.
17 Lower the vehicle to the ground then refit the cooling fan and radiator as described in Chapter 3.

9 Timing belt - removal, inspection and refitting

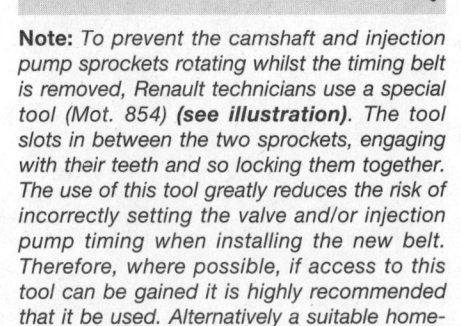

Note: *To prevent the camshaft and injection pump sprockets rotating whilst the timing belt is removed, Renault technicians use a special tool (Mot. 854) (see illustration). The tool slots in between the two sprockets, engaging with their teeth and so locking them together. The use of this tool greatly reduces the risk of incorrectly setting the valve and/or injection pump timing when installing the new belt. Therefore, where possible, if access to this tool can be gained it is highly recommended that it be used. Alternatively a suitable home-made substitute can be fabricated to lock the sprockets together.*

Removal

1 Disconnect the battery negative lead.
2 Position No. 1 piston at TDC as described in Section 3 and lock the crankshaft in position with the locking pin.
3 Remove the crankshaft pulley as described in Section 7. To prevent the possibility of the locking pin being damaged, remove the pin temporarily as the pulley retaining bolt is slackened and refit it once the bolt is slack.
4 Remove the timing belt cover as described in Section 8.

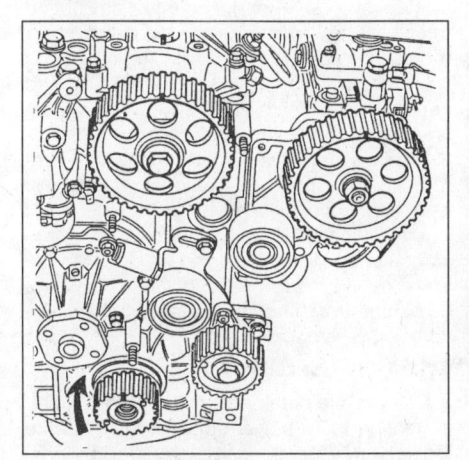

9.5a Sprocket timing mark locations

8.13 Do not omit the collars when refitting the timing cover retaining bolts

5 Check that the camshaft, injection pump and crankshaft sprockets are positioned as shown. The camshaft sprocket timing mark should be aligned with boss on the centre of the cylinder head cover and the injection pump sprocket mark should be aligned with the centre of the boss on the injection pump. If alignment marks do not already exist, using white paint or similar to mark the cylinder head cover and injection pump; the marks can then be used on refitting to ensure that the valve and pump timing is correctly set (see illustrations).
6 Insert the sprocket locking tool (where available) in between the camshaft and injection pump sprockets.
7 If the timing belt is to be re-used and the original direction of rotation arrows are not visible, use white paint or similar to mark the direction of rotation on the belt.
8 Slacken both the tensioner pulley retaining nut and bolt and pivot the pulley fully away from the timing belt. Hold the tensioner in this position and securely tighten the retaining nut and bolt to keep it there.
9 Slip the belt off the sprockets and remove it from the engine. If the camshaft and injection pump sprockets were not locked in position, they must not be rotated whilst the belt is removed.
10 Examine the air conditioning compressor drivebelt for signs of damage or deterioration and renew if necessary. Due to the amount of work required to renew the belt, it is

9.5b Make alignment marks on the cylinder head and injection pump for the sprocket timing marks prior to removing the timing belt

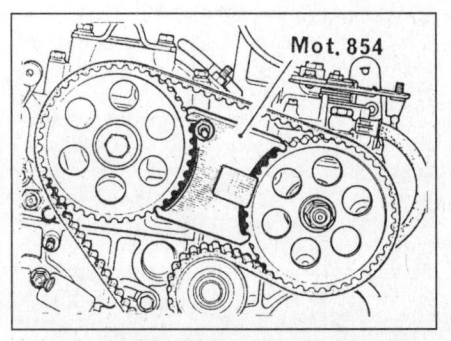

9.1 Special Renault tool for locking the injection pump and camshaft sprockets in position

recommended that it should be renewed regardless of its apparent condition. Refer to Chapter 1B for further information.

Inspection

11 Check the timing belt carefully for any signs of uneven wear, splitting or oil contamination and renew it if there is the slightest doubt about its condition. If the engine is undergoing an overhaul and has covered more than or close to 70 000 miles (115 000 km) since the original belt was fitted, renew the belt as a matter of course. If signs of oil contamination are found, trace the source of the oil leak and rectify it, then wash down the engine timing belt area and all related components to remove all traces of oil. Whilst the belt is removed examine the tensioner and idler pulley bearings as described in Section 10.

Refitting

12 Prior to refitting the timing belt, check the clearance between the tensioner pulley backplate and the adjusting screw on the intermediate shaft cover is 0.1 mm. This is best done using feeler blades, a 0.1 mm feeler blade should be a light, sliding fit between the two components. If not, slacken the locknut and adjust the screw as required (see illustration). Once the clearance is correctly adjusted, hold the screw stationary and securely tighten its locknut.

9.12 Adjusting the tensioner pulley backplate clearance

2C•8 Diesel engine

13 On reassembly, thoroughly clean the timing belt sprockets and check that they are positioned (see paragraph 5) so that the camshaft and injection pump sprockets are correctly aligned with the marks made or noted prior to removal. If the sprockets were locked in position they should not have moved.

14 Offer up the timing belt, observing any marks indicating the direction of rotation and, starting at the crankshaft sprocket and working in an anti-clockwise direction, align the lines on the belt with the timing marks on each sprocket and engage the belt with the intermediate shaft, injection pump and camshaft sprockets **(see illustrations)**. Ensure that the belt front run and top run is taut ie, all slack is on the tensioner pulley side of the belt. Do not twist the belt sharply while refitting it, ensuring that the belt teeth are correctly seated centrally in the sprockets and that the timing marks remain in alignment. If a used belt is being refitted, ensure that the arrow mark made on removal points in the normal direction of rotation as before. Note that if the belt is correctly installed, there should be a total of twenty tooth troughs (on the timing belt) between the camshaft and injection pump sprocket timing marks.

15 Remove the sprocket locking tool (where fitted) and slacken the tensioner pulley retaining nut and bolt. Check that the tensioner pulley is forced against the timing belt under spring pressure then securely tighten the pulley retaining nut and bolt.

16 Temporarily refit the timing belt cover and check that the camshaft sprocket and injection pump timing marks are aligned with their respective pointers. If not, the belt will have to be removed and the refitting procedure repeated.

17 Refit the crankshaft pulley, tightening its retaining bolt loosely only at this stage, and remove the crankshaft locking pin. Rotate the crankshaft through two complete turns in a clockwise direction until both the sprocket timing marks are realigned with their pointers. **Do not** under any circumstances rotate the crankshaft anti-clockwise.

18 Remove the timing belt cover then slacken the tensioner retaining nut and bolt by half a turn and retighten them securely.

19 The belt tension should now ideally be checked using the Renault service tool Ele. 346. If the tool is available, use it to check that the belt deflection is 3 to 5 mm midway between the camshaft and injection pump sprockets.

20 If the special tool is not available, an approximate check of the belt tension can be made using a spring balance and steel rule. At the mid-point between the camshaft and injection pump sprockets the belt deflection should be 3 to 5 mm under a force of 30 N **(see illustration)**. It must be stressed however that this is only an approximate check, the tension can only be accurately checked using the Renault tool.

9.14a On fitting align the lines on the timing belt with the timing marks on the crankshaft sprocket . . .

9.14b . . . camshaft sprocket . . .

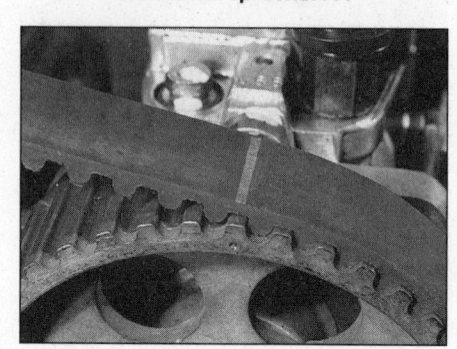

9.14c . . . and injection pump sprocket

9.20 Using a spring balance and ruler to check timing belt tension

21 If the belt tension is not correctly set, repeat the operations described in paragraphs 17 to 20 (as applicable) until the correct tension is achieved.

22 Once the belt is correctly tensioned refit the timing belt cover as described in Section 8.

23 Refit the plug to the cylinder block/crankcase locking pin hole and tighten the crankshaft pulley retaining bolt to the specified torque setting.

24 Refit and tension the auxiliary drivebelt(s) as described in Chapter 1B.

10 Timing belt tensioner and sprockets - removal, inspection and refitting

Note: *This Section describes as individual operations the removal and refitting of the components concerned - if more than one of them are to be removed at the same time, start by removing the timing belt as described in Section 9, then remove the actual component as described below, ignoring the preliminary dismantling steps.*

Removal

1 Disconnect the battery negative lead.
2 Position No. 1 piston at TDC as described in Section 3 and lock the crankshaft in position with the locking pin.
3 Remove the timing belt cover as described in Section 8 and proceed as described under the relevant sub-heading.

Camshaft sprocket

4 Carry out the operation described in paragraph 5 of Section 9.
5 Slacken both the tensioner pulley retaining nut and bolt and pivot the pulley fully away from the timing belt. Hold the tensioner in this position and securely tighten the retaining nut and bolt to keep it there.
6 Remove the belt from the camshaft sprocket taking care not to twist it too sharply; use the fingers only to handle the belt.
7 Slacken the camshaft sprocket retaining bolt whilst holding the sprocket stationary with a suitable peg spanner which engages with the sprocket holes. A suitable home-made tool can be fabricated from two lengths of steel strip (one long, the other short) and three nuts and bolts; one nut and bolt forming the pivot of a forked tool with the remaining two nuts and bolts at the tips of the 'forks' to engage with the sprocket spokes as shown **(see illustration)**.
8 Unscrew the retaining bolt and washer and remove the sprocket from the end of the camshaft, noting which way around it is fitted. If the Woodruff key is a loose fit in the camshaft end, remove it and store it with the sprocket for safe-keeping. Note that the camshaft must not be rotated whilst the sprocket is removed.

Crankshaft sprocket

9 Remove the crankshaft pulley as described in Section 7. To prevent the possibility of the locking pin being damaged, temporarily remove the pin from the crankshaft as the bolt

Diesel engine 2C•9

10.7 Using the fabricated home-made tool to retain the camshaft sprocket whilst the bolt is slackened

10.13a Slide off the crankshaft sprocket . . .

10.13b . . . then remove the Woodruff key

is slackened and refit it once the bolt is slack.
10 Carry out the operation described in paragraph 5 of Section 9. If the sprocket locking tool is available, insert it in between the camshaft and injection pump sprockets to lock the sprockets together.
11 Slacken both the tensioner pulley retaining nut and bolt and pivot the pulley fully away from the timing belt. Hold the tensioner in this position and securely tighten the retaining nut and bolt to keep it there.
12 Work the belt clear of the crankshaft sprocket taking care not to twist it too sharply; use the fingers only to handle the belt.
13 Slide the sprocket off the end of the crankshaft. If the Woodruff key is a loose fit in the crankshaft end, remove it and store it with the sprocket for safe-keeping **(see illustrations)**.
14 On models not equipped with air conditioning, slide the spacer off the end of the crankshaft and store it with the sprocket **(see illustration)**.
15 On models with air conditioning, the air conditioning compressor drivebelt drive pulley is situated behind the crankshaft sprocket. Where necessary, remove the drivebelt as described in Chapter 1B and slide off the drive pulley.

Injection pump sprocket

16 Carry out the operation described in paragraph 5 of Section 9.
17 Slacken both the tensioner pulley retaining nut and bolt and pivot the pulley fully away from the timing belt. Hold the tensioner

in this position and securely tighten the retaining nut and bolt to keep it there.
18 Remove the belt from the injection pump sprocket taking care not to twist it too sharply; use the fingers only to handle the belt.
19 Hold the sprocket stationary using the tool described in paragraph 7 and slacken the sprocket retaining nut **(see illustration)**. Unscrew the nut and position it so that it is flush with the end of the pump shaft; the nut will protect shaft threads during the following operation.
20 A suitable puller will then be needed to free the sprocket from its taper on the pump shaft. The puller should be inserted through the holes in the sprocket so that its legs bear against the back of the sprocket and not positioned so that they bear against the sprocket teeth. Screw in the puller centre bolt

10.14 On models not equipped with air conditioning slide the spacer off the crankshaft

until it contacts the pulley shaft and draw the sprocket off the pump shaft taper **(see illustration)**. **Do not** be tempted to strike the pump with a hammer in an attempt to free the sprocket as the pump internals will almost certainly be damaged.
21 Remove the puller then remove the sprocket retaining nut and washer and slide off the sprocket. If the Woodruff key is a loose fit in the pump shaft, remove it and store it with the sprocket for safe-keeping **(see illustrations)**.

Intermediate shaft sprocket

22 Carry out the operation described in paragraph 5 of Section 9. If the sprocket locking tool is available, insert it in between the camshaft and injection pump sprockets to lock the sprockets together.

2C

10.19 Using the fabricated home-made tool to retain the injection pump sprocket whilst the bolt is slackened

10.20 Using a puller to draw the sprocket off the injection pump shaft taper

10.21a Remove the nut and washer . . .

10.21b . . . then remove the sprocket and recover the Woodruff key (arrowed)

2C•10 Diesel engine

23 Slacken both the tensioner pulley retaining nut and bolt and pivot the pulley fully away from the timing belt. Hold the tensioner in this position and securely tighten the retaining nut and bolt to keep it there.
24 Remove the belt from the intermediate shaft sprocket taking care not to twist it too sharply; use the fingers only to handle the belt.
25 In the absence of the special Renault sprocket holding tool, Mot. 855, a length of old timing belt will be required to prevent the intermediate shaft from rotating as the bolt is slackened. Wrap the timing belt around the sprocket and clamp it firmly with a pair of grips. Another possible way of retaining the sprocket is to jam a large flat-bladed screwdriver in the sprocket teeth but this is not recommended due to the risk of damaging the sprocket.
26 Slacken and remove the retaining bolt and washer and slide the sprocket off the end of the shaft. If the Woodruff key is a loose fit in the shaft, remove it and store it with the sprocket for safe-keeping **(see illustration)**.

Tensioner assembly

27 Carry out the operation described in paragraph 5 of Section 9. If the sprocket locking tool is available, insert it in between the camshaft and injection pump sprockets to lock the sprockets together.
28 Slacken both the tensioner pulley retaining nut and bolt and pivot the pulley fully away from the timing belt. Hold the tensioner in this position and securely tighten the retaining nut and bolt to keep it there.
29 Position the belt clear of the tensioner assembly, taking care not to twist it too sharply; use the fingers only to handle the belt.
30 Slacken and remove the tensioner retaining nut and bolt along with their washers whilst firmly holding the tensioner pulley. With the nut and bolt removed slowly pivot the pulley assembly away from the spring, until all tension is relieved, then remove the tensioner pulley assembly and withdraw the plunger and spring from the coolant pump housing.

Tensioner idler pulley

Note: *The tensioner idler pulley is an integral part of the injection pump front mounting bracket and is not available separately.*
31 Remove the injection pump sprocket as described in paragraphs 16 to 21 of this Section.
32 Slacken and remove the nuts and washers securing the injection pump to the front mounting bracket studs **(see illustration)**.
33 Undo the bolts securing the front pump mounting bracket to the side of the cylinder block/crankcase and remove the bracket from the engine **(see illustrations)**.

Inspection

34 Thoroughly clean the relevant sprocket(s)

10.26 Remove the auxiliary shaft and recover the Woodruff key (arrowed)

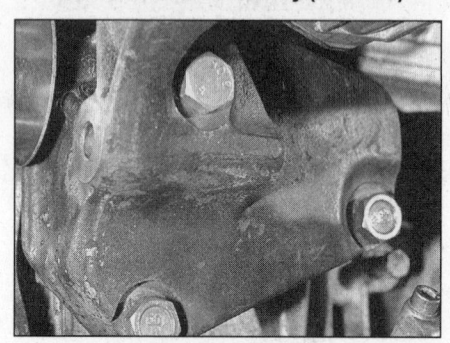

10.33a ... then undo the bolts securing the front mounting bracket to the cylinder block ...

and renew any that show signs of wear, damage or cracks.
35 Clean the idler and tensioner pulleys but do not use any strong solvent which may enter the pulley bearings. Check that each pulley rotates freely on the backplate, with no sign of stiffness or of free play. Renew the

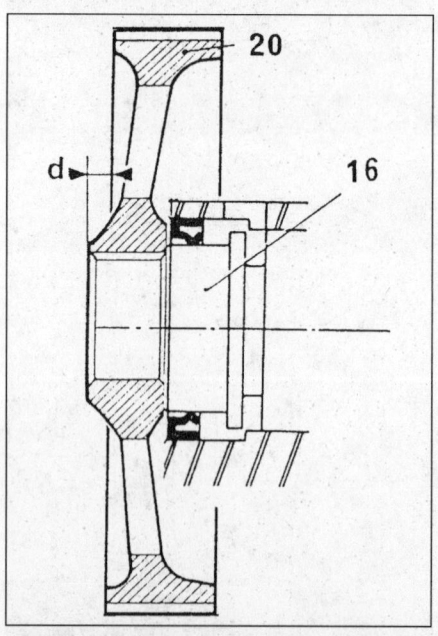

10.36a Ensure the camshaft sprocket (20) is fitted on the camshaft (16) so that the sprocket hub protrudes out in front of the timing belt run (d)

10.32 Unscrew the injection pump front mounting nuts ...

10.33b ... and remove the mounting bracket and idler pulley assembly

assembly if there is any doubt about its condition or if there are any obvious signs of wear or damage. It is recommended that the tensioner spring is renewed regardless of its apparent condition, since its condition is critical.

Refitting

Camshaft sprocket

36 Refit the Woodruff key (where removed) to the slot in the camshaft end and slide on the sprocket, ensuring it is fitted the correct way around **(see illustrations)**. If the sprocket is installed incorrectly it will not be aligned centrally with the timing belt.
37 Refit the sprocket retaining bolt and washer and tighten it to the specified torque setting whilst preventing rotation using the method employed on removal **(see illustrations)**.

10.36b Refit the camshaft sprocket, ensuring it is fitted the correct way around ...

Diesel engine 2C•11

10.37a ... then refit the retaining bolt and washer ...

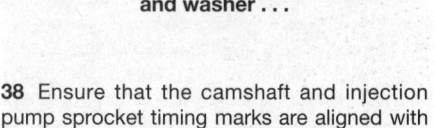

10.37b ... and tighten it to the specified torque setting

10.39 Engage the timing belt with the injection pump sprocket ensuring the sprocket timing marks are correctly positioned

38 Ensure that the camshaft and injection pump sprocket timing marks are aligned with the marks made or noted prior to removal.

39 Ensuring that the belt front run and top run is taut ie, all slack is on the tensioner pulley side of the belt, engage the belt with injection pump and camshaft sprockets. Do not twist the belt sharply while refitting it and ensure that the belt teeth are correctly seated centrally in the sprockets and that the timing marks remain in alignment **(see illustration)**.

40 Slacken the tensioner pulley retaining nut and bolt. Check that the tensioner pulley is forced against the timing belt under spring pressure then securely tighten the pulley retaining nut and bolt.

41 Temporarily refit the timing belt cover and check that the camshaft sprocket and injection pump timing marks are aligned with their respective pointers. If not, the tensioner will have to be released and the belt relocated on the sprockets.

42 Remove the crankshaft locking pin and rotate the crankshaft through two complete turns in a clockwise direction until both the sprocket timing marks are realigned with their pointers. **Do not** under any circumstances rotate the crankshaft anti-clockwise.

43 Remove the timing belt cover then slacken the tensioner retaining nut and bolt by half a turn each then retighten them securely. Check the timing belt tension as described in paragraphs 19 to 21 of Section 9.

44 Once the timing belt is correctly tensioned, refit the timing belt cover as described in Section 8.

45 Refit the plug to the cylinder block/crankcase timing hole and reconnect the battery.

Crankshaft sprocket

46 On models not equipped with air conditioning, slide the spacer onto the end of the crankshaft.

47 On models with air conditioning, where necessary, slide the drive pulley onto the crankshaft and refit the air conditioning compressor drivebelt as described in Chapter 1B.

48 Refit the Woodruff key (where removed) to the slot in the crankshaft, and slide on the sprocket ensuring it is fitted the correct way around. On models equipped with air conditioning ensure that the sprocket roll pins engage correctly with the holes in the drive pulley.

49 Ensure that the camshaft and injection pump sprockets are correctly aligned with the marks made or noted on removal. If the locking tool has been used the sprockets will not have moved.

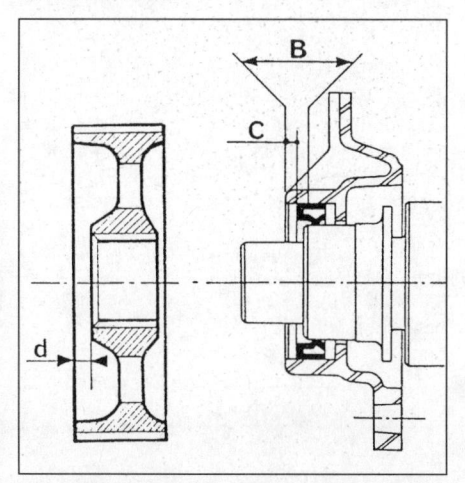

50 Ensuring that the belt front run and top run is taut ie, all slack is on the tensioner pulley side of the belt, engage the belt with crankshaft. Do not twist the belt sharply while refitting it and ensure that the belt teeth are correctly seated centrally in the sprockets and that the timing marks remain in alignment.

51 Adjust the timing belt tension as described in paragraphs 17 to 24 of Section 9.

Injection pump sprocket

52 Ensure that the sprocket and pump shaft are clean and dry and (where necessary) refit the Woodruff key to the pump shaft.

53 Slide the sprocket onto the shaft and refit the washer and retaining nut.

54 Hold the sprocket stationary using the method employed on removal and tighten the sprocket retaining nut to the specified torque setting **(see illustration)**.

55 Carry out the operations described in paragraphs 38 to 45 of this Section.

Intermediate shaft sprocket

56 Refit the Woodruff key to the intermediate shaft slot and slide on the sprocket, ensuring it is fitted the correct way around **(see illustrations)**. If the sprocket is installed incorrectly it will not be aligned centrally with the timing belt.

10.54 Tighten the injection pump sprocket retaining nut to the specified torque setting

10.56a Ensure the auxiliary shaft sprocket is installed so that the side on which the sprocket hub is recessed the most (d) faces away from the cylinder block

10.56b Measure the auxiliary shaft sprocket hub recess before refitting to ensure it is installed the correct way around

2C•12 Diesel engine

10.57 Using a length of old timing belt and a pair of grips to retain the auxiliary shaft sprocket whilst its bolt is tightened

10.61 Refit the plunger and spring to the coolant pump housing . . .

10.62 . . . then install the timing belt tensioner pulley and check that it is free to pivot smoothly

10.63 Force the plunger fully into the coolant pump housing and securely tighten the tensioner retaining nut

57 Refit the retaining bolt and washer and tighten it to the specified torque setting, using the method employed on removal to prevent rotation **(see illustration)**.
58 Ensure that the camshaft and injection pump sprockets are correctly aligned with the marks made or noted on removal. If the sprockets were locked in position they will not have moved.
59 Ensuring that the belt front run and top run is taut ie, all slack is on the tensioner pulley side of the belt, engage the belt with intermediate shaft sprocket. Do not twist the belt sharply while refitting it and ensure that the belt teeth are correctly seated centrally in the sprockets and that the timing marks remain in alignment.
60 Carry out the operations described in paragraphs 40 to 45 of this Section.

Tensioner assembly

61 Fit the spring to the inside of the plunger and refit the plunger assembly to the coolant pump housing **(see illustration)**.
62 Compress the plunger and manoeuvre the tensioner pulley assembly into position. Refit the tensioner retaining nut and bolt and check that the tensioner assembly pivots smoothly and is forced towards the timing belt by the tensioner spring **(see illustration)**.
63 Pivot the pulley fully away from the timing belt and securely tighten its retaining nut and bolt to keep it there **(see illustration)**.
64 Ensure that the camshaft and injection pump sprockets are correctly aligned with the marks made or noted on removal. If the sprockets were locked in position they will not have moved.
65 Locate the belt in front of the tensioner pulley ensuring that the belt front run and top run is taut and all slack is on the tensioner pulley side of the belt. Do not twist the belt sharply while refitting it and ensure that the belt teeth are correctly seated centrally in the sprockets and that the timing marks remain in alignment.
66 Carry out the operations described in paragraphs 40 to 45 of this Section.

Tensioner idler pulley

67 Manoeuvre the injection pump mounting bracket into position, engaging its studs with the front of the injection pump. Refit the bolts which secure the bracket to the cylinder block/crankcase and tighten them securely.
68 Refit the washers and nuts securing the bracket to the injection pump and tighten them securely.
69 Refit the injection pump sprocket as described in paragraphs 52 to 55 of this Section.

11 Camshaft and intermediate shaft oil seals - renewal

Note: *If either the camshaft or intermediate shaft oil seal is to be renewed with the timing belt still in place, check that the belt is free from oil contamination (renew the belt as a matter of course if signs of oil contamination are found; see Section 25), then cover the belt to protect it from contamination by oil while work is in progress and ensure that all traces of oil are removed from the area before the belt is refitted.*

Camshaft oil seal

1 Remove the camshaft sprocket as described in Section 10.
2 Make a note of the correct fitted depth of the seal then punch or drill two small holes opposite each other in the oil seal. Screw a self-tapping screw into each and pull on the screws with pliers to extract the seal.
3 Clean the seal housing and polish off any burrs or raised edges which may have caused the seal to fail in the first place.
4 Lubricate the lips of the new seal with clean engine oil and ease it into position on the end of the shaft. Press the seal into its housing until it is positioned at the same depth as the original was prior to removal. If necessary, a suitable tubular drift, such as a socket, which bears only on the hard outer edge of the seal can be used to tap the seal into position. Take great care not to damage the seal lips during fitting and ensure that the seal lips face inwards. Note that if the surface of the shaft was noted to be badly scored, press the new seal slightly further into its housing so that its lip is running on an unmarked area of the shaft.
5 Refit the camshaft sprocket as described in Section 10.

Intermediate shaft oil seal

6 Remove the intermediate shaft sprocket as described in Section 10. Secure the timing belt clear of the working area so that it is not contaminated with oil during the following procedure.
7 Undo the retaining bolts and slide the intermediate shaft cover off the end of the shaft. Recover the gasket and discard it, a new one should be used on refitting.
8 Using a suitable flat-bladed screwdriver, lever the seal out of the cover whilst taking great care not to mark the cover **(see illustration)**.

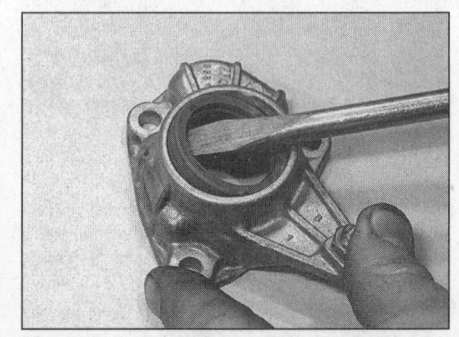

11.8 Lever out the auxiliary shaft oil seal with a suitable flat-bladed screwdriver . . .

Diesel engine 2C•13

9 Position the new seal in the cover and tap it into position using a suitable tubular drift, such as a socket, which bears only on the hard outer edge of the seal **(see illustration)**. Ensure that the seal lip is facing inwards.
10 Remove all traces of gasket from the cover and cylinder block/crankcase mating surfaces and position a new gasket on the cylinder block.
11 Slide the cover into position carefully easing the seal over the end of the intermediate shaft.
12 To prevent the possibility of oil leakage smear the shafts of the cover retaining bolts with a suitable sealant. Renault recommend the use of Loctite Frenetanch (available from your Renault dealer); in the absence of this ensure a good quality sealant is used.
13 Ensure that the gasket is correctly positioned then refit the cover retaining bolts and tighten them securely.
14 Refit the intermediate shaft sprocket as described in Section 10.

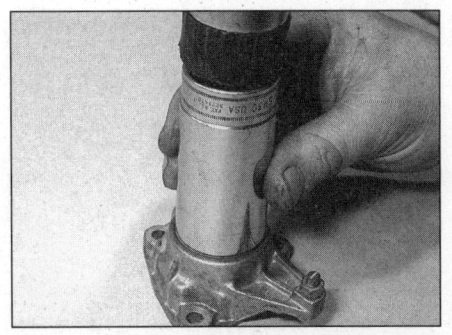

12 Camshaft and rocker arms - removal, inspection and refitting

Removal

Rocker arm assembly

1 Position No. 1 piston at TDC as described in Section 3 and lock the crankshaft in position with the locking pin.
2 Remove the cylinder head cover as described in Section 5.
3 Working in a diagonal sequence, evenly and progressively slacken the rocker arm pedestal retaining bolts by half a turn at a time until all valve spring pressure is relieved from the arms. The bolts can then be unscrewed and removed from the cylinder head.
4 Lift the rocker arm assembly away from the cylinder head noting the correct locations of the pedestal locating dowels. If the dowels are a loose fit, remove them and store them with the rocker shaft for safe-keeping **(see illustration)**.
5 To dismantle the rocker arm assembly,

11.9 ... and tap the new seal into position using a suitable tubular drift

unscrew the bolt from the rear end of the rocker shaft whilst retaining the rocker pedestal to prevent it be sprung off the end of the shaft. Recover the oil filter from inside the shaft then slide off the various components, keeping all components in their correct fitted order **(see illustrations)**. Make a note of each components correct fitted position as it is removed, to ensure it is positioned correctly on reassembly.
6 To separate the front pedestal and shaft, unscrew the grub screw from the top of the pedestal. If necessary, the camshaft thrust plate can also be unbolted from the front of the pedestal **(see illustration)**.

Camshaft

Note: The camshaft is slid out the front of the cylinder head and therefore cannot be removed without first removing the cylinder head from the vehicle.

7 Remove the cylinder head as described in Section 13.
8 With the head on the bench, temporarily refit the rocker shaft assembly making sure the thrust plate is correctly engaged with the camshaft slot. Set up a dial gauge on the front end of the camshaft and measure the camshaft endfloat whilst moving the camshaft to and fro. If the endfloat exceeds the specified service limit the camshaft thrust plate should be renewed on refitting.
9 Remove the camshaft sprocket as described in paragraphs 7 and 8 of Section 10.

12.4 Remove the rocker arm assembly noting the correct fitted locations of the locating dowels (arrowed)

10 Make a note of the correct fitted depth of the seal then punch or drill two small holes opposite each other in the oil seal. Screw a self-tapping screw into each and pull on the screws with pliers to extract the seal.
11 Slide the camshaft out of the front of the cylinder head.

Inspection

Rocker arm assembly

12 Examine the rocker arm bearing surfaces which contact the camshaft lobes for wear ridges and scoring. Renew any rocker arms on which these conditions are apparent. If a rocker arm bearing surface is badly scored also examine the corresponding lobe on the camshaft for wear as it is likely that both will be worn. Renew worn components as necessary, the rocker arm assembly can be dismantled as in paragraphs 4 and 5. Renew the rocker shaft oil filter regardless of its apparent condition.
13 Inspect the ends of the adjusting screws for signs of wear or damage and renew as required.
14 If the rocker arm assembly has been dismantled, examine the rocker arm and shaft bearing surfaces for wear ridges and scoring. If there are obvious signs of wear, the relevant rocker arm(s) and/or shaft must be renewed. Also check that the rocker arm oil holes are clear by passing a piece of wire through each one **(see illustration)**.

12.5a Unscrew the bolt from the end of the rocker shaft ...

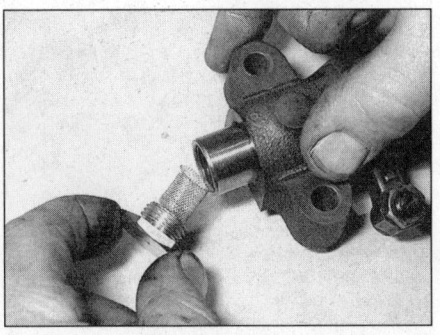

12.5b ... remove the oil filter and slide off the rocker arm components

12.6 The camshaft thrust plate is secured to the front rocker arm pedestal by two bolts

2C•14 Diesel engine

12.14 Use a length of wire to check each rocker arm oil hole is unblocked

12.17a Refit the front rocker arm pedestal, aligning it with the rocker shaft hole (arrowed) . . .

12.17b . . . refit the grub screw and tighten it securely . . .

12.17c . . . then check that the rocker shaft oilways are facing downwards

12.17d Slide on all rocker shaft components in their original fitted positions . . .

12.17e . . . making sure the protruding side of each pedestal is facing the rear of the shaft

Camshaft

15 Examine the camshaft bearing surfaces and cam lobes for signs of wear ridges and scoring. Renew the camshaft if any of these conditions are apparent. Examine the condition of the bearing surfaces both on the camshaft journals and in the cylinder head. If the head bearing surfaces are worn excessively, the cylinder head will need to be renewed.

16 Examine the camshaft thrust plate for signs of wear or scoring and renew if necessary; the plate should also be renewed if the endfloat was checked prior to removal and was found to be excessive. The thrust plate is secured to the front of the front rocker arm pedestal by two bolts.

Refitting

Rocker arm assembly

17 If the rocker arm assembly was dismantled, refit the rocker shaft to the front pedestal, aligning its locating hole with the pedestal threaded hole, and refit the grub screw tightening it securely. Apply a smear of clean engine oil to the shaft then slide on all removed components, ensuring each is correctly fitted in its original position. Note that the pedestals are not symmetrical and must be fitted the correct way around with the protruding side of the pedestal facing the rear end of the rocker shaft **(see illustrations)**.

18 Once all components are in position on the shaft, compress the rear pedestal. Install the new oil filter in the shaft bore, then refit the bolt and tighten it to the specified torque setting.

19 Ensure that the pedestal locating dowels are correctly positioned then lower the rocker arm assembly into position on the cylinder head, ensuring that the thrust plate engages correctly with the camshaft slot **(see illustrations)**.

20 Refit the retaining bolts and, working in a diagonal sequence, evenly and progressively tighten them until all the pedestals are contacting the cylinder head. As the bolts are tightened ensure that the thrust plate remains correctly aligned with the groove in the camshaft. With all the pedestals in contact with the head, work around the retaining bolts and tighten them to the specified torque setting **(see illustration)**.

12.19a Ensure the locating dowel are in position . . .

12.19b . . . and lower the rocker shaft into position aligning the thrust plate with the camshaft groove

12.20 Tighten the rocker shaft retaining bolts to the specified torque setting

Diesel engine 2C•15

21 Check and, if necessary, adjust the valve clearances as described in Section 6 then refit the cylinder head cover as described in Section 6.

Camshaft

22 Ensure that the cylinder head and camshaft bearing surfaces are clean then liberally oil the camshaft bearings and lobes and slide the camshaft back into position in the cylinder head.

23 Lubricate the lips of the new seal with clean engine oil and ease it into position on the end of the shaft. Press the seal into the cylinder head until it is positioned at the same depth as the original was prior to removal. If necessary, a suitable tubular drift, such as a socket, which bears only on the hard outer edge of the seal can be used to tap the seal into position **(see illustrations)**. Take great care not to damage the seal lips during fitting and ensure that the seal lips face inwards.

24 Check and, if necessary, adjust the valve clearances as described in Section 6.

25 Refit the camshaft sprocket as described in paragraphs 36 and 37 of Section 10.

26 Refit the cylinder head as described in Section 13.

12.23a Slide the new camshaft oil seal carefully over the camshaft . . .

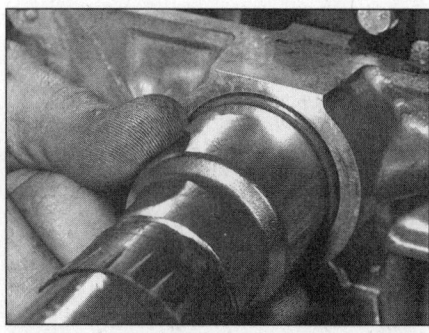

12.23b . . . and tap it into position using a suitable tubular drift

10 Undo the retaining nuts and disconnect the wiring from each of the glow plugs. Also disconnect the wiring connector from the injection pump stop solenoid and the coolant temperature sender unit(s) which is/are screwed into the front of the cylinder head.

11 On models with clamp-type injectors, wipe clean the injectors then undo the union bolt securing the fuel return hose to the top of each injector and recover the sealing washer from each side of the hose union and, where necessary, the filter. With all four union bolts removed position the return hose clear of the cylinder head. Cover the hose and injector unions to prevent the ingress of dirt into the fuel system. Undo the union nuts and free the

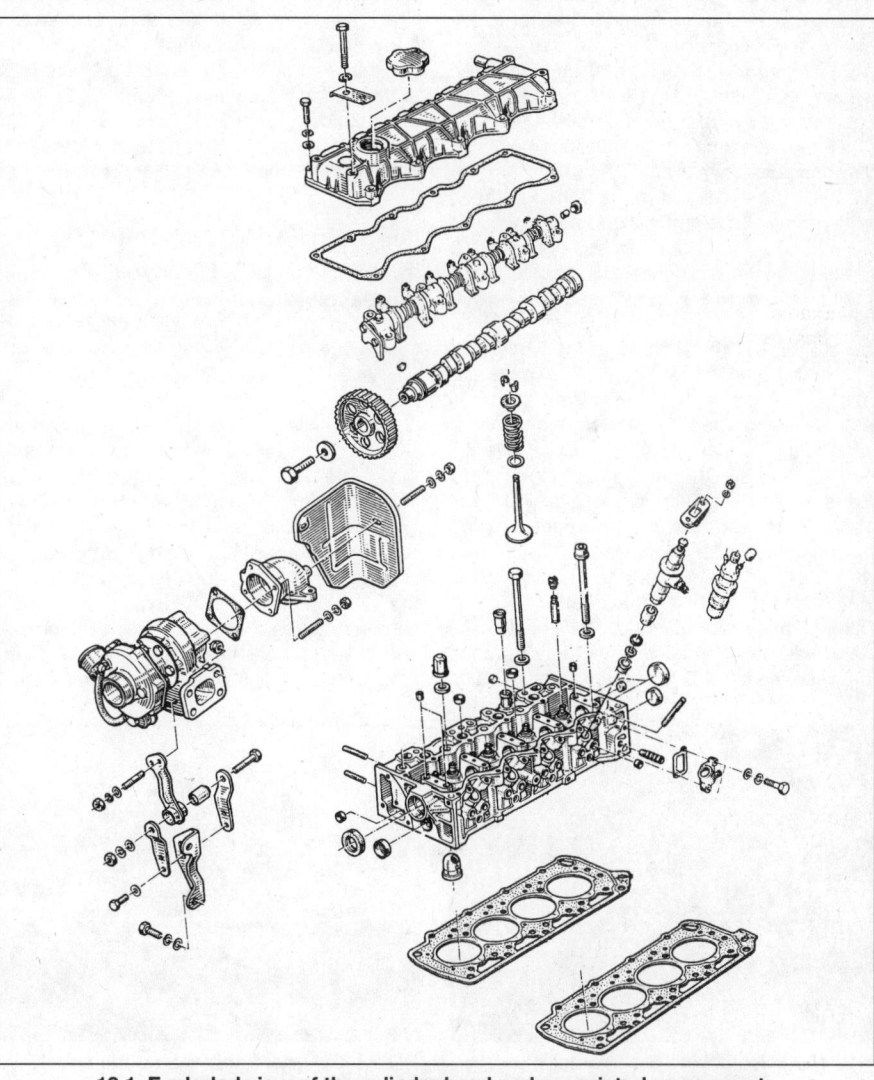

13 Cylinder head - removal and refitting

Note: *Great care must be taken not to allow dirt to enter the fuel system during the following procedure.*

Removal

1 Disconnect the battery negative lead.

2 Position No. 1 piston at TDC as described in Section 3 and lock the crankshaft in position with the locking pin.

3 Remove the timing belt cover as described in Section 8 and carry out the operation described in paragraph 5 of Section 9.

4 Slacken both the tensioner pulley retaining nut and bolt and pivot the pulley fully away from the timing belt. Hold the tensioner in this position and securely tighten the retaining nut and bolt to keep it there. Disengage the belt from the camshaft sprocket taking care not to twist it too sharply; use the fingers only to handle the belt.

5 Remove the cylinder head cover as described in Section 5.

6 Remove the rocker shaft assembly as described in Section 12.

7 Drain the cooling system as described in Chapter 1.

8 Remove the inlet and exhaust manifolds as described in Chapter 4C.

9 On models where the thermostatic fast idle valve is mounted onto the injection pump, slacken the retaining clips and disconnect the coolant hoses from the valve.

13.1 Exploded view of the cylinder head and associated components (turbo model shown)

2C•16 Diesel engine

injector pipes from the four injectors noting there is no need to remove the injector pipes completely.

12 On models with screw-type injectors, wipe clean the pipe unions then slacken the union nut securing the injector pipes to the top of each injector and the four union nuts securing the pipes to the rear of the injection pump; as each pump union nut is slackened, retain the adapter with a suitable open-ended spanner to prevent it being unscrewed from the pump. With all the union nuts undone remove the injector pipe assembly from the engine **(see illustration)**. Disconnect the fuel return pipe from rear injector then cover the pump and injector unions to prevent the ingress of dirt into the fuel system.

13 Where necessary, undo the retaining bolts and free the oil separator chamber from the rear of the cylinder head.

14 Slacken the retaining clips and disconnect all the relevant coolant hoses from thermostat housing and cylinder head.

15 Slacken and remove the bolt securing the rear injection pump mounting bracket to the side of the cylinder head.

16 Working in the reverse of the sequence shown in **illustration 13.41**, progressively slacken the cylinder head nuts and bolts by half a turn at a time until all nuts and bolts can be unscrewed by hand.

17 With the exception of the front right-hand bolt, remove each nut/bolt and washer.

18 With the front right-hand bolt (number 16 in the tightening sequence) still in position, the joint between the cylinder head and gasket and the cylinder block/crankcase must now be broken without disturbing the wet liners; although these liners are better located and sealed than some wet liner engines, there is still a risk of coolant and foreign matter leaking into the sump if the cylinder head is lifted carelessly. If care is not taken and the liners are moved, there is also a possibility of the bottom seals being disturbed, causing leakage after refitting the head.

19 To break the joint, strike the side of the head firmly using a soft-faced mallet or hammer and block of wood. Movement of the head will be minimal due to the clearance between the head and its three locating studs but it should be sufficient to free the head from the gasket.

20 Once the gasket seal is broken, remove the right-hand front bolt and washer. Discard all the nuts and bolts, new ones should be used on refitting.

21 The cylinder head can then be lifted out of the engine compartment; use assistance if possible as it is a heavy assembly. Remove the gasket from the top of the block, noting the position of the locating dowel. If the dowel is a loose fit, remove it and store it with the head for safe-keeping. Do not discard the gasket, it will be needed for identification purposes (see paragraphs 30 and 31).

22 Note **do not** attempt to rotate the crankshaft with the cylinder head removed, otherwise the wet liners may be displaced. Operations that require the rotation of the crankshaft (eg cleaning the piston crowns), should only be carried out once the cylinder liners are firmly clamped in position. In the absence of the special Renault liner clamping tool (Mot. 521-01), the liners can be clamped in position using large flat washers positioned underneath suitable-length bolts, or the original head bolts with suitable spacers fitted to their shanks **(see illustration)**.

23 If the cylinder head is to be dismantled for overhaul, remove the camshaft as described in Section 12, then refer to Section 14.

Preparation for refitting

24 The mating faces of the cylinder head and cylinder block/crankcase must be perfectly clean before refitting the head. Use a hard plastic or wood scraper to remove all traces of gasket and carbon; also clean the piston crowns. Take particular care as the soft aluminium alloy is damaged easily. Also, make sure that the carbon is not allowed to enter the oil and water passages - this is particularly important for the lubrication system, as carbon could block the oil supply to any of the engine's components. Using adhesive tape and paper, seal the water, oil and bolt holes in the cylinder block/crankcase. To prevent carbon entering the gap between the pistons and bores, smear a little grease in the gap. After cleaning each piston, use a small brush to remove all traces of grease and carbon from the gap, then wipe away the remainder with a clean rag. Clean all the pistons in the same way.

25 Check the mating surfaces of the cylinder block/crankcase and the cylinder head for nicks, deep scratches and other damage. If slight, they may be removed carefully with a file, but if excessive, machining may be the only alternative to renewal.

26 If warpage is suspected of the cylinder head gasket surface, use a straight-edge to check it for distortion. Refer to Section 14.

27 Renew the cylinder nuts and bolts, regardless of their apparent condition.

28 Prior to refitting the cylinder head, check the cylinder liner protrusion as described in Chapter 2A, Section 20. Although the measuring procedure specifies that the liners should be fitted without seals, the seals have no effect on liner protrusion. If the liner protrusion exceeds the specified limits, seek the advice of a Renault dealer prior to refitting the head. If the liner protrusion is within the specified limits the new cylinder head gasket should be selected as follows.

29 When purchasing a new cylinder head gasket, it is essential that a gasket of the correct thickness is obtained. There are three different thicknesses available; 1.6 mm, 1.7 mm and 1.8 mm.

30 On early models the gasket is identified as follows; the 1.6 mm gasket has `1.6' stamped on the gasket surface, the 1.7 mm gasket has no marking and the 1.8 mm gasket has `1.8' stamped on the gasket surface. It is unlikely, however, that this marking will still be visible on the original gasket.

31 On later models the gasket thickness is indicated by the hole(s) punched in the tab situated at the rear end of the gasket **(see illustrations)**; this tab is visible even when the head is still fitted to the block. The 1.6 mm gasket has 2 holes punched in the tab, the 1.7 mm gasket has 1 hole punched in the tab and the 1.8 mm gasket 3 holes punched in the tab.

32 If the gasket thickness can be determined from its original markings, and no work which could affect the piston protrusion has been carried out, then a gasket of the same thickness as the original can be fitted. That being said, it is still advisable that the new gasket should be selected by direct measurement as described in the following paragraphs.

33 If the thickness of the original gasket is unknown, or if work has been carried out which affects piston protrusion, then the correct gasket must be selected by using the following procedure. Note that all gaskets now supplied by Renault have the later type of identification marking.

34 Ensure that the liners are clamped squarely and securely in position so that are correctly seated in the cylinder block then remove all traces of carbon from the piston crowns as described above.

13.12 Removing the injector pipes

13.22 Cylinder liners clamped in position using sockets positioned over the cylinder head studs

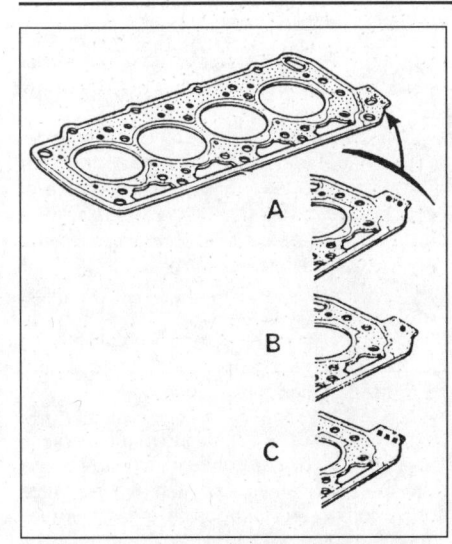

Diesel engine 2C•17

13.31a Cylinder head gasket thickness identification hole(s)

A 1.6 mm thick gasket
B 1.7 mm thick gasket
C 1.8 mm thick gasket

13.31b Cylinder head thickness identification holes - 1.8 mm thick gasket shown

13.35 Measuring piston protrusion

35 Using a dial gauge, establish were TDC is then measure the protrusion of each piston above the cylinder block gasket surface. Take two measurements from each piston, one on each side of the piston and take the average of the two as the correct piston protrusion **(see illustration)**. Note that the piston must not move during the measuring procedure, do not exert any force on the piston crown as this will cause the piston to tip in the bore; if the piston moves repeat both measurements. The correct gasket is selected using the **largest** piston protrusion measurement obtained; the required gasket thickness is as follows.

Piston protrusion	Gasket thickness required
Less than 0.96 mm	1.6 mm gasket
0.96 to 1.04 mm	1.7 mm gasket
Greater than 1.04 mm	1.8 mm gasket

36 Remove all traces of carbon from the valve heads and cylinder head combustion chamber surface. Using a dial gauge, measure the amount by which each valve head is recessed below the cylinder head combustion chamber surface **(see illustration)**. Note down these measurements and check that they are all within the specified limits given in the *Specifications* at the start of this Chapter. If not the cylinder head should be dismantled and overhauled as described in Section 14. Make a special note of the valve with the smallest recess on the cylinder with the greatest piston protrusion (ie. the cylinder which was used to calculate the thickness of the head gasket required) as this valve is the one on which the piston-to-cylinder head clearance will be checked once the head has been refitted (see paragraphs 45 to 50).

Refitting

37 Wipe clean the mating surfaces of the cylinder head and cylinder block/crankcase. Check that the locating dowel is in position at the front right-hand corner of the cylinder block/crankcase surface **(see illustration)**. Where necessary, remove the cylinder liner clamps.

38 Position the new gasket on the cylinder block/crankcase surface ensuring that its identification tab is situated at the rear end of the gasket **(see illustration)**.

39 With the aid of an assistant lower the head into position, engaging it with the locating dowel **(see illustration)**.

40 Keeping all the cylinder head bolts in their correct fitted order, lightly oil under the head and on the threads of each bolt. Carefully enter each bolt into its hole, screwing it in by hand only until finger-tight, then fit the three washers and nuts to the head studs **(see illustration)**.

13.36 Measuring valve recess in the cylinder head

13.37 Ensure the locating dowel is in position . . .

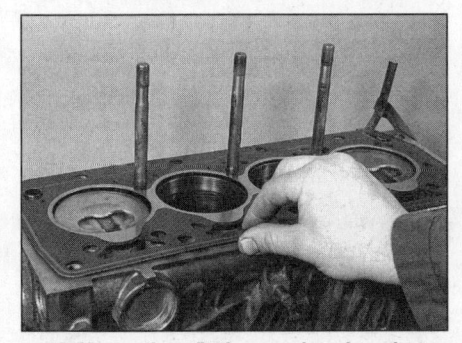

13.38 . . . then fit the new head gasket

13.39 Lower the head into position . . .

13.40 . . . and fit the cylinder head bolts and nuts

2C•18 Diesel engine

41 Working in the sequence shown, tighten all the head nuts and bolts first to their stage 1 torque setting, then working again in the specified sequence, tighten them to the specified stage 2 torque setting **(see illustration)**.

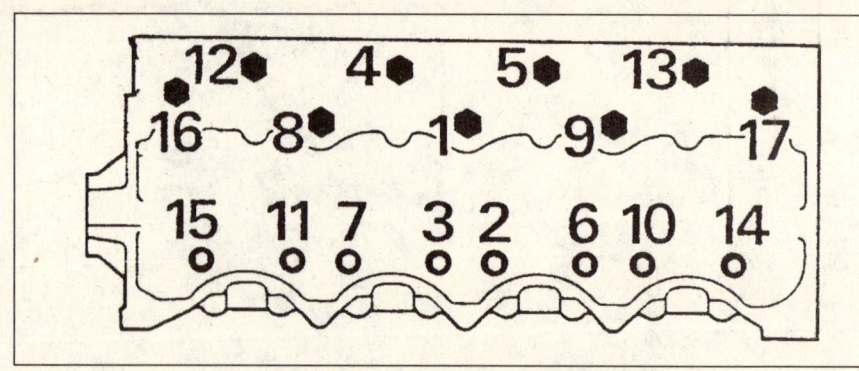

42 Once all nuts/bolts are tightened to the stage 2 specified torque setting, fully slacken the first nut (number 1) in the tightening sequence. Tighten the nut to the specified stage 3 torque setting then angle-tighten it through its specified Stage 4 angle, using a socket and extension bar. It is recommended that an angle-measuring gauge is used during this stage of the tightening, to ensure accuracy. **Note:** *The Stage 4 (and 5) tightening angles differ depending on the bolt/nut location.* Repeat this operation on the remaining nuts/bolts, working through them in the specified sequence.

43 When all nuts/bolts have been tightened through their Stage 3 and 4 settings, work around in the specified sequence and tighten them through the specified Stage 5 angle, again using an angle-tightening gauge to ensure accuracy. Note that no further retightening of the nut bolts will be required once the engine has been warmed up.

44 With the cylinder head in position, the piston-to-cylinder head clearance should now be checked. If no work affecting the clearance has been carried out and a cylinder head gasket of the same thickness as the original is being fitted, this operation is not strictly necessary. However, it is still recommended that the clearance is checked as a precaution. Proceed as described in paragraphs 45 to 50 to check the piston-to-cylinder head clearance. If it is not wished to check the clearance proceed as described in paragraph 51 onwards.

45 The piston-to-cylinder head clearance is checked on the cylinder with the largest piston protrusion, by measuring the clearance between the valve with the smallest recess and the piston crown when the piston is at TDC (see paragraphs 35 and 36).

46 Rotate the crankshaft so that the required piston is at TDC then remove the valve spring from the valve with the smallest recess. This is best achieved using a bar type valve spring compressor which is bolted to the cylinder

13.41 Cylinder head nut/bolt tightening sequence

head. Compress the valve spring and withdraw the collets then release the compressor and lift off the spring retainer and valve spring **(see illustration)**.

47 Drop the valve down so that it rests on the piston crown and set up a dial gauge on the tip of the valve stem. Rotate the crankshaft back and forth slightly whilst using the dial to establish when the piston is at TDC. With TDC established, zero the dial gauge then lift the valve whilst measuring the piston-to-valve clearance (ie. the amount of movement of the valve) **(see illustration)**.

48 The piston-to-cylinder head clearance is then calculated by subtracting the valve head recess measurement from the piston-to-valve clearance. For example, if the piston-to-valve clearance was 1.57 mm and the valve recess measurement was 0.87 mm, the piston-to-cylinder head clearance would be 0.7 mm (1.57 mm - 0.87 mm = 0.7 mm).

49 The piston-to-cylinder head clearance must be at least 0.6 mm or greater. If this is not the case then a cylinder head gasket of the wrong thickness has been fitted. **Note:** *The engine **must not** be run if the clearance is less than 0.6 mm as serious engine damage is likely to occur. The cylinder head must be removed and a thicker gasket fitted.* If the thickest head gasket is already fitted then some other problem exists and the advice of your Renault dealer must be sought about the best course of action.

50 When the piston-to-cylinder head clearance is known to be correct, refit the valve spring and retainer to the valve.

Compress the valve spring and refit the collets ensuring that they are correctly located in the recess on the valve stem. Remove the valve spring compressor and, using a hammer and interposed block of wood, tap the end of the valve stem to settle the components in position.

51 With the head correctly installed, refit the injection pump rear bracket retaining bolt and tighten it securely.

52 Connect all the relevant coolant hoses to the cylinder head and securely tighten their retaining clips.

53 Where necessary, refit the oil separator to the rear of the cylinder head tightening its retaining bolts securely.

54 On models with screw-type injectors, remove the plugs from the fuel unions and manoeuvre the injector pipe assembly back into position. Align the pipe ends with the injectors and injection pump unions and tighten the pipe union nuts to the specified torque setting. Reconnect the return pipe to the rear injector.

55 On models with clamp-type injectors, align the injector pipes with their unions on the injectors and tighten the pipe union nuts to the specified torque setting. Position a new sealing washer on both sides of each return pipe union and, where necessary, refit the filter to the return hose union. Refit both union bolts, tightening them to the specified torque setting.

56 Reconnect the wiring to the glow plugs and securely tighten the retaining nuts. Where necessary, reconnect the wiring connector to the coolant temperature sender unit.

57 Where removed, refit the thermostatic fast idle valve as described in Chapter 4C.

58 Refit the inlet and exhaust manifolds as described in Chapter 4C.

59 With No. 1 cylinder at TDC and the crankshaft locked in position with the locking pin, position the camshaft and injection pump sprockets so that their timing marks are aligned with marks made or noted on removal.

60 Refit the rocker shaft assembly as described in Section 12.

61 With the shaft in position, carry out the operations described in paragraphs 38 to 44 of Section 10.

62 With the timing belt correctly fitted and

13.46 Using a bar type valve spring compressor to compress the valve spring

13.47 Measuring piston to valve clearance

the cover installed, check and, if necessary, adjust the valve clearances as described in Section 6.
63 Refit the cylinder head cover as described in Section 5.
64 Refill the cooling system as described in Chapter 3.
65 Refit the plug to the cylinder block/crankcase timing hole and reconnect the battery. Ensure that the oil and coolant levels are correct before starting the engine.

14 Cylinder head - overhaul

1 Refer to Chapter 2A, Section 12 for information on the cylinder head and valves. Check the swirl chambers as follows.
2 Inspect the swirl chambers for burning or damage such as cracking (see illustration). Small cracks in the chambers are acceptable; renewal of the chambers will only be required if chamber tracts are badly burned and disfigured or if they are no longer a tight fit in the cylinder head. If there is any doubt as to the swirl chamber condition, seek the advice of a Renault dealer or a suitable repairer who specialises in diesel engines. Swirl chamber renewal should be entrusted to a specialist.
3 Using a dial test indicator check that the swirl chamber protrusion is within the limits given in the *Specifications*. Zero the dial test indicator on the gasket surface of the cylinder head, then measure the protrusion of the swirl chamber (see illustration). If the protrusion is not within the specified limits the advice of a Renault dealer or suitable repairer who specialises in diesel engines should be sought.

15 Sump - removal and refitting

Refer to Chapter 2A, Section 13 noting that most models are fitted with a one-piece sump but some later (J8S 708) turbo engines have the two-piece arrangement.

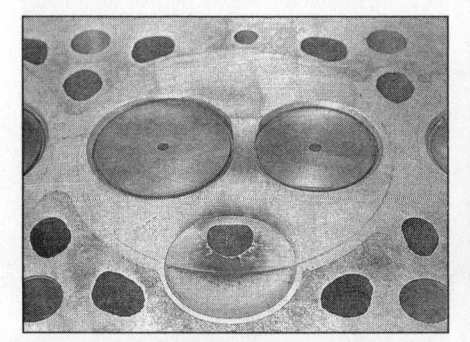

14.2 This swirl chamber is showing initial signs of cracking

16 Oil pump - removal, inspection and refitting

Refer to Chapter 2A, Section 14 noting that most models are fitted with a one-piece sump but some later (J8S 708) turbo engines have the two-piece arrangement.

17 Crankcase stiffener casting (later J8S 708 turbo engines) - removal and refitting

Refer to Chapter 2A, Section 15.

18 Oil cooler components - removal and refitting

Refer to Chapter 2A, Section 28.

19 Crankshaft oil seals - renewal

Front (timing belt) oil seal

1 Remove the crankshaft sprocket and spacer/air conditioning drive pulley (as applicable) as described in Section 10. Secure the timing belt clear of the working area so that it cannot be contaminated with oil. Make a note of the correct fitted depth of the seal in its housing.
2 Punch or drill two small holes opposite each other in the seal. Screw a self-tapping screw into each and pull on the screws with pliers to extract the seal. Alternatively, the seal can be levered out of position using a suitable flat-bladed screwdriver taking great care not to damage the crankshaft shoulder or seal housing.
3 Clean the seal housing and polish off any burrs or raised edges which may have caused the seal to fail in the first place.
4 Lubricate the lips of the new seal with clean engine oil and carefully locate the seal on the end of crankshaft noting that its sealing lip

14.3 Measuring swirl chamber protrusion

must be facing inwards. Take care not to damage the seal lips during fitting.
5 Press the seal into its housing until it is positioned at the same depth as the original was prior to removal. If necessary, a suitable tubular drift, such as a socket, which bears only on the hard outer edge of the seal can be used to tap the seal into position. Take great care not to damage the seal lips during fitting and ensure that the seal lips face inwards. Note that if the surface of the crankshaft was noted to be badly scored, press the new seal slightly further into its housing so that its lip is running on an unmarked area of the crank.
6 Wash off any traces of oil, then refit the crankshaft sprocket and associated components as described in Section 10.

Rear (flywheel) oil seal

7 Remove the flywheel as described in Section 20. Make a note of the correct fitted depth of the seal in its housing.
8 Renew the seal as described above in paragraphs 2 to 5.
9 Wash off any traces of oil, then refit the flywheel as described in Section 20.

20 Flywheel - removal and refitting

Refer to Chapter 2A, Section 16.

21 Engine/transmission mountings - renewal

Refer to Chapter 2A, Section 18. An engine movement damper is fitted to most vehicles. To remove the damper, simply undo the two retaining bolts and manoeuvre the unit out from the engine compartment. If necessary, the damper mounting brackets can then be removed once their retaining bolts have been undone. Refitting is the reverse of removal ensuring all the retaining bolts are securely tightened.

22 Intermediate shaft - removal, inspection and refitting

Removal

1 In order to gain the clearance necessary to withdraw the shaft from the front of the cylinder block whilst the engine is in the vehicle, remove the radiator and cooling fan as described in Chapter 3 and unbolt the grille panel as described in Chapter 11.
2 Remove the braking system vacuum pump as described in Chapter 9.
3 Remove the intermediate shaft sprocket as described in Section 10.

2C•20 Diesel engine

4 Carefully withdraw the oil pump/vacuum pump drivegear, taking great care not to dislodge the oil pump driveshaft **(see illustrations)**. **Note:** *Great care must be taken to ensure that the driveshaft is not dislodged from the oil pump. If the driveshaft is dislodged it will drop down into the bottom sump. If this happens, the sump will have to be removed in order to recover the driveshaft.*

5 Unscrew the retaining bolt and washer then slide out the thrust plate and withdraw the intermediate shaft **(see illustrations)**.

Inspection

6 Examine the intermediate shaft and oil pump driveshaft for pitting, scoring or wear ridges on the bearing journals, and for chipping or wear of the gear teeth. Renew as necessary. Check the shaft bearings in the cylinder block for wear and, if worn, have these renewed by your Renault dealer or suitably-equipped engineering works. Wipe them clean if they are still serviceable.

7 Temporarily fit the thrust plate to its position on the intermediate shaft, and use a feeler blade to check that the endfloat is as given in the *Specifications*. If it is greater than the upper tolerance, a new thrust plate should be obtained, but first check the thrust surfaces on the shaft to ascertain if wear has occurred here **(see illustration)**.

Refitting

8 Clean off all traces of the old gasket or sealant from the intermediate shaft cover. Lever the old oil seal out of the cover and position a new seal in the cover and tap it into position using a suitable tubular drift, such as a socket, which bears only on the hard outer edge of the seal. Ensure that the seal lip is facing inwards.

9 Liberally lubricate the intermediate shaft, and slide it into its bearings.

10 Slide the thrust plate in position and securely tighten its retaining bolts.

11 Place a new gasket in position on the cylinder block/crankcase.

12 Carefully ease the cover over the end of the shaft and slide it into position.

13 To prevent the possibility of oil leakage smear the shafts of the cover retaining bolts with a suitable sealant (Renault recommend the use of Loctite Frenetanch).

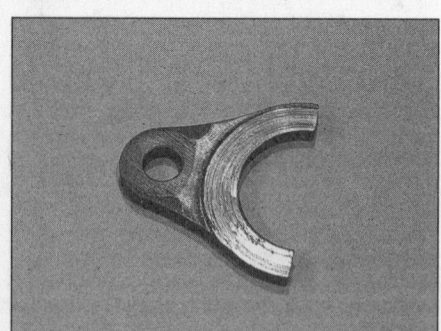

22.7 Auxiliary shaft thrust plate showing signs of excessive wear

22.4a Remove the oil pump/vacuum pump drivegear . . .

22.4b . . . and withdraw the oil pump driveshaft, taking great care not to drop it into the sump

22.5a Remove the thrust plate . . .

22.5b . . . and slide out the auxiliary shaft

14 Ensure that the gasket is correctly positioned then refit the cover retaining bolts, tightening them securely.

15 Ensure that the driveshaft is correctly engaged with the oil pump then refit the oil pump/vacuum pump drivegear, engaging it with driveshaft.

16 Refit the braking system vacuum pump as described in Chapter 9.

17 Refit the intermediate shaft sprocket as described in Section 10.

18 Refit the cooling fan and radiator as described in Chapter 3.

23 Crankshaft - removal, inspection and refitting

Refer to Chapter 2A, Section 20 **(see illustration)**.

24 Piston/connecting rod and cylinder liner assemblies - removal, inspection & refitting

Removal

1 Refer to Chapter 2A, Section 21 noting that most models are fitted with a one-piece sump but some later (J8S 708) turbo engines have the two-piece arrangement (fitted with a crankcase stiffener casting). Also note that the liners have an additional seal fitted to each of them as well as the sealing ring.

Inspection

2 Refer to Chapter 2A, Section 21 noting that the gudgeon pins are a sliding fit in the pistons and are retained by circlips. If necessary, the pistons and connecting rods can be separated as follows.

3 Using a small flat-bladed screwdriver or a pair of circlip pliers (as applicable), prise out the circlips and push out the gudgeon pin **(see illustration)**. Hand pressure should be sufficient to remove the pin. Identify the piston and rod to ensure correct reassembly and discard the circlips, new ones **must** be used on refitting.

4 Examine the gudgeon pin and connecting rod small-end bearing for signs of wear or damage. Wear will necessitate the renewal or the pin and/or connecting rod. Note that the gudgeon pins are not available separately; they are only supplied as part of the piston /liner assembly.

5 The connecting rods themselves should not be in need of renewal unless seizure or some other major mechanical failure has occurred. Check the alignment of the connecting rods visually, and if the rods are not straight, take them to an engine overhaul specialist for a more detailed check.

6 Examine all components and obtain the required parts from your Renault dealer. If new pistons are purchased they will be supplied complete with circlips. Circlips can also be purchased individually.

7 On normally-aspirated engines, position the piston so that its turbulence chamber is on the

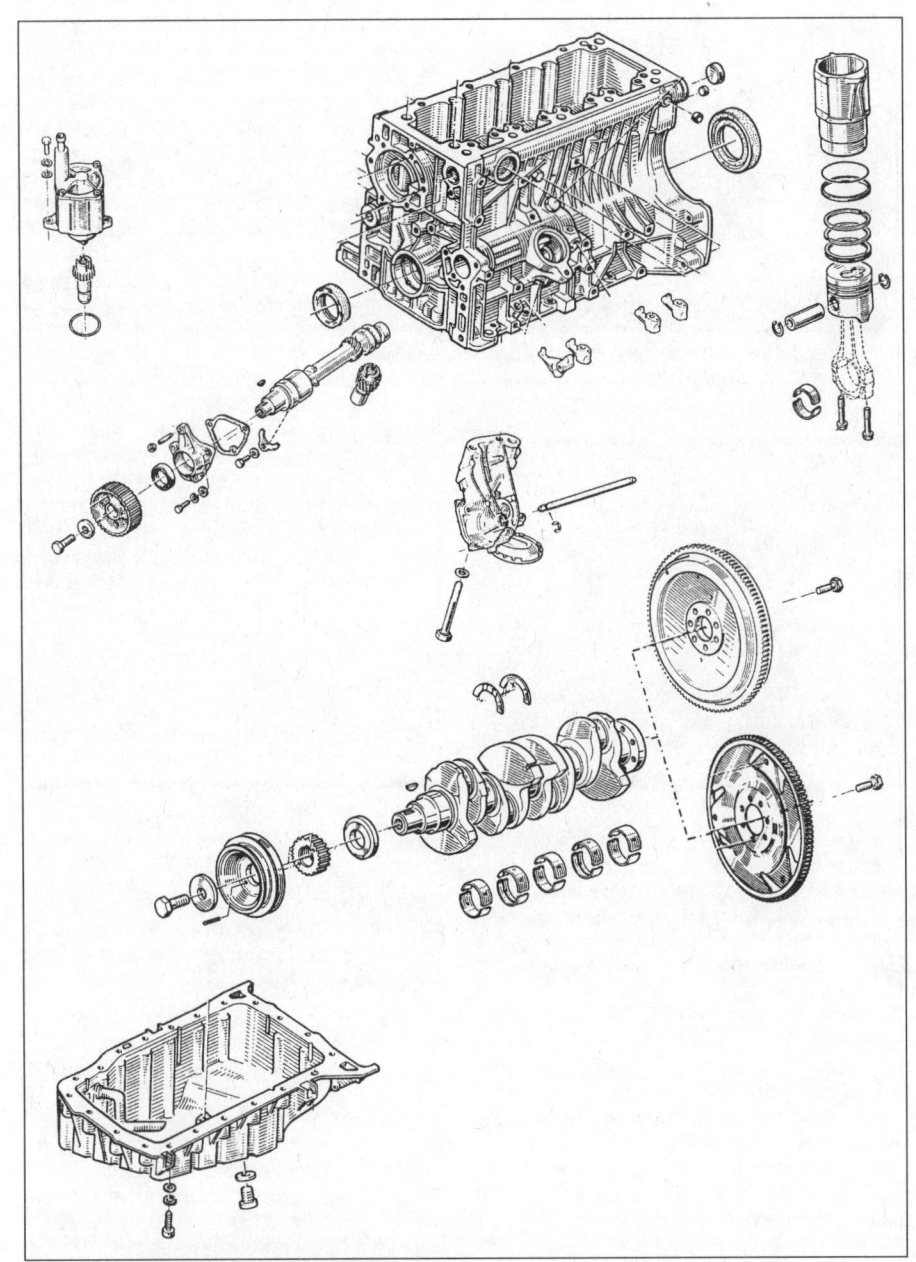

Diesel engine 2C•21

opposite side to the connecting rod big-end oil hole (see illustration).

8 On turbo engines, position the piston so that its turbulence chamber is on the opposite side to the connecting rod big-end bearing shell cutout (see illustration).

9 On all models, with the piston and connecting rod correctly mated, apply a smear of clean engine oil to the gudgeon pin and slide it into the piston and through the connecting rod small-end. Check that the piston pivots freely on the rod then secure the gudgeon pin in position with two new circlips, ensuring each circlip is correctly located in its groove in the piston (see illustrations).

Refitting

10 Refer to Chapter 2A, Section 21 ensuring that both the new sealing ring and seal are correctly fitted to each liner (see illustrations).

25 Piston rings - removal and refitting

Refer to Chapter 2A, Section 22, making sure the piston rings are correctly fitted as shown (see illustration).

26 Cylinder block/crankcase - cleaning and inspection

Refer to Chapter 2A, Section 23 noting that the piston oil jet nozzles are fitted to turbo models only.

Some early turbo models have a different oil jet nozzle arrangement to that described. On these models, all the jets are connected to a large diameter oil gallery which is bolted to the cylinder block, the gallery is then connected to the cylinder block oil way by an oil pipe with threaded end fittings. On these models undo the bolts and remove nozzles

23.1 Exploded view of the cylinder block and associated components

24.3 Using a pair of circlip pliers to remove a piston circlip

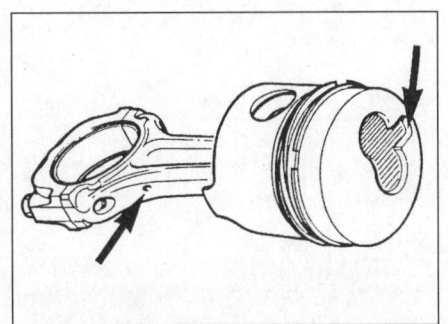

24.7 On normally aspirated models position each piston so that its turbulence chamber is opposite the connecting rod big-end oil hole

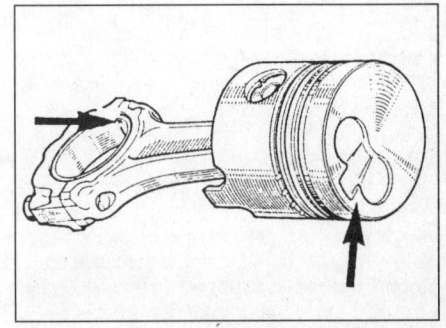

24.8 On turbo models position each piston so that its turbulence chamber is opposite the connecting rod big-end bearing shell cutouts

2C•22 Diesel engine

24.9a Position the piston/connecting rod as described in text and slide in the gudgeon pin

24.9b Fit new circlips ensuring they are correctly located in the piston groove

24.10a Fit a new O-ring to the liner seating flange ...

24.10b ... and a new sealing ring to the base of the liner

and oil gallery as an assembly **(see illustration)**. On refitting apply a few drops of locking compound (Renault recommend the use of Loctite Frenatanch) to the nozzle retaining bolt threads, do not apply too much or there is a risk of the oilways becoming blocked.

27 Engine and transmission - removal and refitting

Note: *Great care must be taken not to allow dirt to enter the fuel system during the following procedure.*

Removal

1 Park the vehicle on firm, level ground then remove the bonnet as described in Chapter 11.
2 Disconnect and remove the battery as described in Chapter 5A.
3 Drain the cooling system as described in Chapter 1.
4 If the engine is to be dismantled, drain the oil and remove the oil filter then clean and refit the drain plug, tightening it securely.
5 Firmly apply the handbrake then jack up the front of the car and support it securely on axle stands. Remove both front roadwheels.
6 Work around the plastic undershield, removing its retaining screws and prising out its retaining clips, and remove the undershield from the underneath the vehicle.

7 Remove the radiator and cooling fan as described in Chapter 3.
8 On Turbo models, remove the air cleaner housing as described in Chapter 4C and remove the inlet ducts connecting the intercooler to the turbocharger and inlet manifold.
9 On models equipped with power steering, referring to Chapter 10, unbolt and manoeuvre the power steering pump and bracket assembly clear of the engine. Tie the pump to the vehicle body to prevent any undue strain being placed on the hydraulic hoses and work back along the hoses, releasing them from any relevant retaining clips or ties securing them to the engine.
10 Disconnect the accelerator cable from the injection pump as described in Chapter 4C.

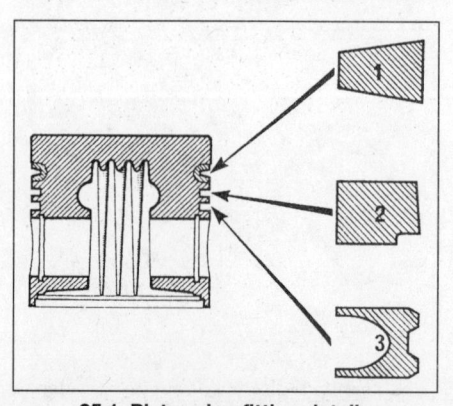

25.1 Piston ring fitting details

11 Undo the nuts/bolts and disconnect the earth leads from the engine/transmission.
12 On models with clamp-type injectors, wipe clean the injector unions and the injection pump. Undo the union bolt securing the fuel return hose to the top of each injector and the injection pump and recover the sealing washers from each hose union. With all five union bolts removed position the return hose clear of the cylinder head. Screw the union bolts back into position for safe-keeping and cover the hose and injector unions to prevent the ingress of dirt into the fuel system.
13 On models with screw-type injectors, wipe clean the injection pump fuel unions. Slacken and remove the fuel return hose union bolt from the pump and recover the sealing washer from each side of the hose union. Disconnect the return pipe from the end injector and position the hose clear of the pump. Screw the union bolt back into position on the pump for safe-keeping and cover both the hose end and union bolt to prevent the ingress of dirt into the fuel system.
14 On all models, remove the injection pump feed hose union bolt and recover the sealing washers. Remove the filter (Lucas/CAV pumps only) then detach the hose from the pump and position the hose clear of the engine. Screw the union bolt back into position on the pump for safe-keeping and cover both the hose end and union bolt to prevent the ingress of dirt into the fuel system.

Note: *The injection pump feed and return hose union bolts are not interchangeable. Great care must be taken to ensure that the bolts are not swapped.*

15 Slacken the retaining clips and disconnect the cooling system heater hoses from the engine and the two coolant hoses from the fuel filter heater housing. Also disconnect the hose connecting the expansion tank to the engine/oil cooler, from the tank so that the hose will be removed with the engine.
16 Disconnect the vacuum hose from the braking system vacuum pump. Release the hose from any relevant retaining clips or ties and position it clear of the engine.

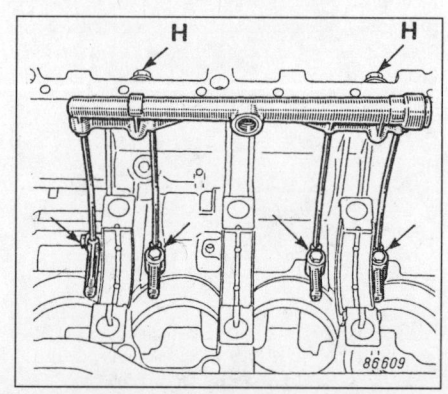

26.2 Piston oil jet nozzle arrangement - early models

17 Disconnect the wiring connectors from the various engine/transmission sensors, switches, the starter motor and the alternator. Note and label the wiring runs and connectors to avoid confusion on refitting.
18 Remove the exhaust system downpipe as described in Chapter 4C.
19 Disconnect the clutch cable from the transmission as described in Chapter 6.
20 On models equipped with air conditioning, unbolt the compressor and disengage it from the drivebelt. Position the compressor clear of the engine and support the weight of the compressor by tying it to the vehicle body to prevent any excess strain being placed on the compressor lines whilst the engine is removed. **Do not** disconnect the refrigerant lines from the compressor (see the warnings given in Chapter 3). The drivebelt can be left in position on the engine.
21 Disconnect the gearchange linkage components from the transmission as described in Chapter 7A. Withdraw the retaining clip and disconnect the speedometer cable from the transmission.
22 On models fitted with an oil cooler, unbolt the cooler assembly from the right-hand wing valance and tie it to the side of the engine.
23 Undo the retaining bolts and remove the movement damper from the engine.
24 Remove the driveshafts as described in Chapter 8.
25 Attach suitable lifting brackets to each end of the cylinder head cover and take the weight of the engine/transmission on the engine hoist. The dimensions of the brackets are shown **(see illustration)**. To prevent the engine tilting as it is removed, it is recommended that a third lifting bracket and chain should be attached to the exhaust manifold.
26 Undo the nuts securing the front engine/transmission mountings to their mounting brackets.
27 With the engine supported, unbolt the mounting brackets and remove them from the engine/transmission.
28 Make a final check that all components have been removed or disconnected that will prevent the removal of the engine/transmission from the car and ensure that components such as the driveshafts are secured so that they cannot be damaged on removal.
29 Lift the engine/transmission slightly to disengage the front mountings from the subframe and carefully lift the engine/transmission, ensuring that nothing is trapped or damaged. Tilt the engine/transmission as it is lifted and once the engine is high enough, lift it out over the front of the body and lower the unit to the ground.
30 To separate the engine and transmission first remove the starter motor, referring to Chapter 5A for further information.
31 Unscrew the remaining bolts securing the transmission to the engine, noting the correct location of the various brackets and retaining clips.
32 Gently prise the transmission off its locating dowels and move the transmission squarely away from the engine, ensuring that the clutch components are not damaged.
33 If the engine is to be overhauled, remove the clutch as described in Chapter 6.

Refitting

34 If the engine and transmission have been separated, perform the operations described below in paragraphs 35 to 39. If not, proceed as described from paragraph 83 onwards.
35 Apply a smear of high-melting-point grease to the splines of the transmission input shaft (Renault recommend the use of Molykote BR2). Do not apply too much, otherwise there is a possibility of the grease contaminating the clutch friction plate.
36 Ensure that the locating dowels are correctly positioned in the engine or transmission.
37 Carefully offer the transmission to the engine, until the locating dowels are engaged, ensuring that the weight of the transmission is not allowed to hang on the input shaft as it is engaged with the clutch friction disc.
38 If all is well, refit the transmission housing to engine bolts, ensuring that all the necessary brackets and clips are correctly positioned, and tighten them to the specified torque setting.
39 Refit the starter motor and securely tighten its retaining bolts.
40 With the aid of an assistant, reconnect the hoist and lifting tackle to the engine lifting brackets, and lift the assembly over the engine compartment.
41 Ensuring that the assembly is tilted as necessary to clear surrounding components, as during removal, and lower the assembly into position in the engine compartment, manipulating the hoist and lifting tackle as necessary.
42 Refit the mounting brackets and rubbers and securely tighten the retaining nuts and bolts.
43 Detach the hoist and lifting gear and rock the engine to settle the engine mountings in position. With the engine resting on its mountings tighten the nuts securing the mountings to the subframe to the specified torque.
44 The remainder of the refitting procedure is a direct reversal of the removal sequence, noting the following points.
 a) Tighten all nuts and bolts to their specified torque setting (where given).
 b) Refit the driveshafts as described in Chapter 8.
 c) Ensure that all coolant hoses and wiring is correctly routed and securely reconnected.
 d) Adjust the clutch cable as described in Chapter 6.
 e) Reconnect the feed and return hoses to the injection pump using new sealing washers (see Chapter 4C).
 f) Refit the power steering pump to the engine and refit the drivebelt as described in Chapters 10 and 1.
 g) Where necessary, refit the air conditioning compressor to the engine and adjust the drivebelt as described in Chapter 1.
 h) Refill the cooling system as described in Chapter 1.
 i) Refill the transmission with oil and refill the engine as described in Chapter 1.
 j) On completion, adjust the accelerator cable and bleed the fuel system as described in Chapter 4C then start the engine and check for leaks.

28 Engine - removal and refitting leaving the transmission in the vehicle

Removal

1 Carry out the operations described in paragraphs 1 to 20 of Section 27.
2 On models fitted with an oil cooler, unbolt the cooler assembly from the right-hand wing valance and tie it to the side of the engine.
3 Undo the retaining bolts and remove the movement damper from the front of the engine.
4 Remove the starter motor as described in Chapter 5A.
5 Attach suitable lifting brackets to each end of the cylinder head cover and take the weight of the engine on the hoist (see Section 27). To prevent the engine tilting as it is removed, it is recommended that a third lifting bracket and chain should be attached to the exhaust manifold.
6 Work around the clutch housing and unscrew the remaining bolts securing the transmission to the engine, noting the correct location of the various brackets and retaining clips.

27.25 Engine lifting bracket dimensions

7 From underneath the vehicle, undo the nuts securing the front engine mounting rubbers to the brackets then undo the bolts securing the mounting brackets to the engine cylinder block/crankcase. Remove both the left- and right-hand front mountings from the engine.

8 Make a final check that all components have been removed or disconnected that will prevent the removal of the engine from the car and ensure that components such as the power steering pump are secured so that they cannot be damaged on removal.

9 With the aid of an assistant, move the engine forwards to disengage it from the locating dowels then move it squarely away from the transmission, ensuring that the clutch components are not damaged.

10 Once the engine is clear of the input shaft, raise the engine ensuring that nothing is trapped or damaged. Once the engine is high enough, lift it out over the front of the body and lower the unit to the ground. If the locating dowels are a loose fit, remove them and store them with the engine for safe-keeping.

Refitting

11 Apply a smear of high-melting-point grease to the splines of the transmission input shaft (Renault recommend the use of Molykote BR2). Do not apply too much, otherwise there is a possibility of the grease contaminating the clutch friction plate. Ensure that the locating dowels are correctly positioned in the engine or transmission.

12 With the aid of an assistant, reconnect the hoist and lifting tackle to the engine lifting brackets, and lift the assembly over the engine compartment.

13 Ensuring that the assembly is tilted as necessary to clear surrounding components, as during removal, and lower the assembly into position in the engine compartment, manipulating the hoist and lifting tackle as necessary.

14 Align the engine with the transmission input shaft, and carefully move the engine rearwards until the locating dowels are engaged. Ensure that the weight of the engine is not allowed to hang on the input shaft as it is engaged with the clutch friction disc.

15 If all is well, refit the transmission housing to engine bolts, ensuring that all the necessary brackets and clips are correctly positioned, and tighten them to the specified torque setting.

16 Refit the front engine mounting assemblies to the cylinder block/crankcase. Tighten the mounting retaining bolts securely then lower the engine, aligning the mounting studs with their holes in the subframe.

17 Remove the hoist and rock the engine to settle the disturbed mountings in position. Refit the lower mounting nuts and tighten them securely.

18 The remainder of the refitting procedure is a reversal of the removal sequence, noting the following points.
a) Tighten all nuts and bolts to their specified torque setting (where given).
b) Ensure that all coolant hoses and wiring is correctly routed and securely reconnected.
c) Adjust the clutch cable as described in Chapter 6.
d) Reconnect the feed and return hoses to the injection pump using new sealing washers (see Chapter 4C).
e) Refit the power steering pump to the engine and refit the drivebelt as described in Chapters 10 and 1.
f) Where necessary, refit the air conditioning compressor to the engine and adjust the drivebelt as described in Chapter 1.
g) Refill the cooling system as described in Chapter 1.
h) Refill the engine with oil as described in Chapter 1.
i) On completion, adjust the accelerator cable and bleed the fuel system as described in Chapter 4C then start the engine and check for leaks.

29 Initial start-up after overhaul

1 With the engine refitted in the car, double-check the engine oil and coolant levels. Make a final check that everything has been reconnected and that there are no tools or rags left in the engine compartment.

2 With the injection pump stop solenoid wire disconnected, turn the engine over on the starter until the oil pressure warning lamp goes out.

3 Reconnect the solenoid wire, tightening its retaining nut securely.

4 Start the engine, noting that this may take a little longer than usual due to the fuel system components being empty.

5 While the engine is idling, check for fuel, water and oil leaks. Don't be alarmed if there are some odd smells and smoke from parts getting hot and burning off oil deposits.

6 Keep the engine idling until hot water is felt circulating through the top hose, check the idle speed (see Chapter 1), then switch it off.

7 After a few minutes, recheck the oil and coolant levels and top-up as necessary.

8 If new pistons, rings or crankshaft bearings have been fitted, the engine must be run-in for the first 500 miles (800 km). Do not operate the engine at full throttle or allow it to labour in any gear during this period. It is recommended that the oil and filter be changed at the end of this period.

Chapter 3
Cooling, heating and air conditioning systems

Contents

Air conditioning system - general information and precautions 10	Cooling system - flushing ....See Chapter 1
Air conditioning system components - removal and refitting 11	Cooling system electrical switches - testing, removal and refitting . 6
Air conditioning system refrigerant check ....See Chapter 1	Cooling system hoses - disconnection and renewal 2
Auxiliary (air conditioning compressor) drivebelt -	Electric cooling fan - testing, removal and refitting 5
check, adjustment and renewal ....See Chapter 1	General information and precautions 1
Antifreeze mixture ....See Chapter 1	Heating and ventilation system - general information 8
Coolant level check ....See "Weekly Checks"	Heating and ventilation system components - removal and refitting 9
Coolant pump - removal and refitting 7	Hose and fluid leak check ....See Chapter 1
Cooling system - draining ....See Chapter 1	Radiator - removal, inspection and refitting 3
Cooling system - filling ....See Chapter 1	Thermostat - removal, testing and refitting 4

Degrees of difficulty

Easy, suitable for novice with little experience	**Fairly easy,** suitable for beginner with some experience	**Fairly difficult,** suitable for competent DIY mechanic	**Difficult,** suitable for experienced DIY mechanic	**Very difficult,** suitable for expert DIY or professional

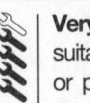

Specifications

Thermostat

Opening temperatures:
  Starts to open:
    4-cylinder engines:
      Petrol engines - early models .... 89°C
      Petrol engines - later models .... 88°C
      Diesel engines .... 81°C
    6-cylinder engines .... 86°C
  Fully-open:
    4-cylinder engines:
      Petrol engines - early models .... 101°C
      Petrol engines - later models .... 100°C
      Diesel engines .... 93°C
    6-cylinder engines .... 98°C

3•2 Cooling, heating and air conditioning systems

2.5 Slackening a coolant hose clip

1 General information and precautions

General information

The cooling system is of pressurised type, comprising a coolant pump driven by an auxiliary drivebelt, an aluminium crossflow radiator, expansion tank, electric cooling fan(s), a thermostat, heater matrix, and all associated hoses and switches.

The system functions as follows. Cold coolant in the bottom of the radiator passes through the bottom hose to the coolant pump, where it is pumped around the cylinder block and head passages, and through the oil cooler(s) (where fitted). After cooling the cylinder bores, combustion surfaces and valve seats, the coolant reaches the underside of the thermostat, which is initially closed. The coolant passes through the heater, and is returned via the cylinder block to the coolant pump.

When the engine is cold, the coolant circulates only through the cylinder block, cylinder head, and heater. When the coolant reaches a predetermined temperature, the thermostat opens, and the coolant passes through the top hose to the radiator. As the coolant circulates through the radiator, it is cooled by the inrush of air when the car is in forward motion. The airflow is supplemented by the action of the electric cooling fan(s) when necessary. Upon reaching the bottom of the radiator, the coolant has now cooled, and the cycle is repeated.

When the engine is at normal operating temperature, the coolant expands, and some of it is displaced into the expansion tank. Coolant collects in the tank, and is returned to the radiator when the system cools.

On models fitted with an engine oil cooler, the coolant is also passed through the oil cooler. Similarly, coolant also passes through the automatic transmission fluid cooler, where applicable.

The electric cooling fan(s) mounted in front of the radiator are controlled by a thermostatic switch. At a predetermined coolant temperature, the switch/sensor actuates the fan.

Precautions

Warning: *Do not attempt to remove the expansion tank filler cap, or to disturb any part of the cooling system, while the engine is hot, as there is a high risk of scalding. If the expansion tank filler cap must be removed before the engine and radiator have fully cooled (even though this is not recommended), the pressure in the cooling system must first be relieved. Cover the cap with a thick layer of cloth, to avoid scalding, and slowly unscrew the filler cap until a hissing sound is heard. When the hissing has stopped, indicating that the pressure has reduced, slowly unscrew the filler cap until it can be removed; if more hissing sounds are heard, wait until they have stopped before unscrewing the cap completely. At all times, keep well away from the filler cap opening, and protect your hands.*

Warning: *Do not allow antifreeze to come into contact with your skin, or with the painted surfaces of the vehicle. Rinse off spills immediately, with plenty of water. Never leave antifreeze lying around in an open container, or in a puddle in the driveway or on the garage floor. Children and pets are attracted by its sweet smell, but antifreeze can be fatal if ingested.*

Warning: *If the engine is hot, the electric cooling fan may start rotating even if the engine is not running. Be careful to keep your hands, hair, and any loose clothing well clear when working in the engine compartment.*

Warning: *Refer to Section 10 for precautions to be observed when working on models equipped with air conditioning.*

2 Cooling system hoses - disconnection and renewal

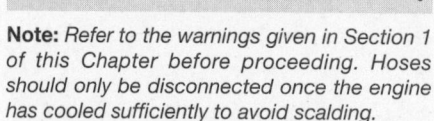

Note: *Refer to the warnings given in Section 1 of this Chapter before proceeding. Hoses should only be disconnected once the engine has cooled sufficiently to avoid scalding.*

1 If the checks described in Chapter 1 reveal a faulty hose, it must be renewed as follows.
2 First drain the cooling system (Chapter 1). If the coolant is not due for renewal, it may be re-used, providing it is collected in a clean container.
3 To disconnect a hose, proceed as follows.
4 The clips used to secure the hoses in position may be either standard worm-drive clips or disposable crimped types. The crimped type of clip is not designed to be re-used and should be replaced with a worm drive type on reassembly.
5 To disconnect a hose, use a screwdriver to slacken or release the clips, then move them along the hose, clear of the relevant inlet/outlet **(see illustration)**. Carefully work the hose free. The hoses can be removed with relative ease when new - on an older car, they may have stuck.
6 If a hose proves to be difficult to remove, try to release it by rotating its ends before attempting to free it. Gently prise the end of the hose with a blunt instrument (such as a flat-bladed screwdriver), but do not apply too much force, and take care not to damage the pipe stubs or hoses. Note in particular that the radiator inlet stub is fragile; do not use excessive force when attempting to remove the hose. If all else fails, cut the hose with a sharp knife, then slit it so that it can be peeled off in two pieces. Although this may prove expensive if the hose is otherwise undamaged, it is preferable to buying a new radiator. Check first, however, that a new hose is readily available.
7 When fitting a hose, first slide the clips onto the hose, then work the hose into position. If crimped-type clips were originally fitted, use standard worm-drive clips when refitting the hose. If the hose is stiff, use a little soapy water as a lubricant, or soften the hose by soaking it in hot water. Do not use oil or grease, which may attack the rubber.
8 Work the hose into position, checking that it is correctly routed, then slide each clip back along the hose until it passes over the flared end of the relevant inlet/outlet, before tightening the clip securely.
9 Refill the cooling system with reference to Chapter 1.
10 Check thoroughly for leaks as soon as possible after disturbing any part of the cooling system.

3 Radiator - removal, inspection and refitting

Removal

1 Disconnect the battery negative lead.
2 Remove the radiator grille panel as described in Chapter 11.
3 Where applicable, remove the headlight wiper units.
4 Where applicable, unclip the radiator blanking plate.
5 Where applicable, disconnect the wiring plugs from the cooling fan and the cooling fan thermostatic switch, mounted in the radiator.
6 Drain the cooling system as described in Chapter 1.
7 Disconnect the coolant hoses from the radiator with reference to Section 2.
8 On models with air conditioning, release the condenser mountings, but **do not** disconnect the refrigerant lines (see Section 10).
9 Unbolt the radiator upper mounting brackets, and withdraw them from the front body panel **(see illustrations)**. Note that on some models with a large-capacity radiator, it

Cooling, heating and air conditioning systems 3•3

3.9a Unscrew the securing bolts . . .

3.9b . . . and withdraw the radiator upper mounting brackets

3.10 Lifting out the radiator - six-cylinder model shown

may be necessary to unscrew the securing bolts and lift off the front body upper crossmember.

10 Lift the radiator (complete with air deflectors, where applicable) from the engine compartment, taking care not to damage the radiator fins on surrounding components **(see illustration)**.

Inspection

11 If the radiator has been removed due to suspected blockage, reverse-flush it as described in Chapter 1. Clean dirt and debris from the radiator fins, using an air line (in which case, wear eye protection) or a soft brush. Be careful, as the fins are sharp, and easily damaged.

12 If necessary, a radiator specialist can perform a "flow test" on the radiator, to establish whether an internal blockage exists.

13 A leaking radiator must be referred to a specialist for permanent repair. Do not attempt to weld or solder a leaking radiator, as damage to the plastic components may result.

14 In an emergency, minor leaks from the radiator can be cured by using a suitable radiator sealant, in accordance with its manufacturer's instructions, with the radiator in situ.

15 If the radiator is to be sent for repair or renewed, remove all hoses, and the cooling fan switch (where fitted).

16 Inspect the condition of the radiator mounting rubbers, and renew them if necessary.

Refitting

17 Refitting is a reversal of removal, bearing in mind the following points.
 a) Ensure that the lower lugs on the radiator are correctly engaged with the mounting rubbers in the body panels.
 b) On completion, refill the cooling system as described in Chapter 1.

4 Thermostat - removal, testing and refitting

4-cylinder engines

Note: *A new sealing ring may be required on refitting.*

Removal

1 The thermostat is located in a housing at the front of the cylinder head. On petrol engines, the housing is bolted to the right-hand side of the cylinder head. On diesel engine models, the housing is bolted to the front of the cylinder head **(see illustration)**.

2 Disconnect the battery negative lead.

3 Partially drain the cooling system to below the level of the thermostat housing, as described in Chapter 1.

4 Where necessary, release any relevant wiring and hoses from the retaining clips, and position clear of the thermostat housing to improve access.

5 Unscrew the securing bolts, and carefully withdraw the thermostat housing cover to expose the thermostat. Take care not to strain the coolant hose connected to the cover.

6 Lift the thermostat from the housing, and recover the sealing ring **(see illustration)**.

Testing

7 A rough test of the thermostat may be made by suspending it with a piece of string in a container full of water. Heat the water to bring it to the boil - the thermostat must open by the time the water boils. If not, renew it.

8 If a thermometer is available, the precise opening temperature of the thermostat may be determined; compare with the figures given in the *Specifications*. The opening temperature is also marked on the thermostat.

9 A thermostat which fails to close as the water cools must also be renewed.

Refitting

10 Refitting is a reversal of removal, bearing in mind the following points:
 a) Examine the sealing ring for signs of damage or deterioration, and if necessary, renew.
 b) Ensure that the thermostat is fitted the correct way round, with the spring(s) facing into the housing.
 c) Where applicable, use a new gasket when fitting the housing cover (thoroughly clean the mating faces of the cover and housing first).
 d) On completion, refill the cooling system as described in Chapter 1.

4.1 Thermostat housing (arrowed) - four-cylinder petrol engine

4.6 Removing the thermostat - four-cylinder petrol engine

3•4 Cooling, heating and air conditioning systems

6-cylinder engines

Note: *A new thermostat sealing ring may be required, and new gaskets and O-rings will be required when refitting the U-shaped section of the inlet manifold.*

Removal

11 The removal procedure is as described previously for 4-cylinder engines, but the thermostat housing is located on the top of the coolant pump at the front of the engine **(see illustration)**. If necessary, to improve access to the thermostat housing, carry out the following.
  a) Remove the air cleaner.
  b) Unbolt and remove the U-shaped section of the inlet manifold from the front of the engine. Recover the gaskets and O-rings.
  c) Where applicable, release the securing clip(s) and disconnect the throttle operating rod(s) from the throttle linkage, then unbolt the throttle linkage, and move it to one side, leaving the throttle cable connected.

4.11 Thermostat housing (arrowed) - six-cylinder engine

Testing

12 Proceed as described in paragraphs 7 to 9.

Refitting

13 Proceed as described in paragraph 10, noting the following additional points if the throttle linkage and U-shaped section of the inlet manifold have been removed.
  a) After refitting and reconnecting the throttle linkage, check the throttle cable adjustment with reference to Chapter 4.
  b) Use new gaskets and O-rings when refitting the U-shaped section of the inlet manifold.

5 Electric cooling fan - testing, removal and refitting

General

1 Petrol engine models are fitted with an electric cooling fan, controlled by the cooling fan thermostatic switch mounted in the radiator.

2 Non-turbo diesel engine models are fitted with a cooling fan bolted to the coolant pump pulley hub, which runs permanently when the engine is running.

3 Turbo diesel engine models are fitted with a viscous cooling fan, bolted to the cooling pump, which varies the fan speed with engine temperature. At low temperatures, the coupling provides very little resistance between the coolant pump pulley and the fan, so only a slight amount of drive is transmitted to the fan. As the temperature of the coupling increases, so does the internal resistance, therefore increasing drive to the cooling fan.

Electric cooling fan (petrol engines)

Testing

4 Current supply to the cooling fan(s) is via the ignition switch (see Chapter 10) and a fuse (see Chapter 12). The circuit is completed by the cooling fan thermostatic switch, which is mounted in the radiator.

5 If a fan does not appear to work, run the engine until normal operating temperature is reached, then allow it to idle. The fan should cut in within a few minutes (before the temperature gauge needle enters the red section, or before the coolant temperature warning light comes on). If not, switch off the ignition and disconnect the wiring plug from the cooling fan switch. Bridge the two contacts in the wiring plug using a length of spare wire, and switch on the ignition. If the fan now operates, the switch is probably faulty, and should be renewed.

6 If the fan still fails to operate, check that battery voltage is available at the feed wire to the switch; if not, then there is a fault in the feed wire (possibly due to a fault in the fan motor, or a blown fuse). If there is no problem with the feed, check that there is continuity between the switch earth terminal and a good earth point on the body; if not, then the earth connection is faulty, and must be re-made.

7 If the switch and the wiring are in good condition, the fault must lie in the motor itself. The motor can be checked by disconnecting it from the wiring loom, and connecting a 12-volt supply directly to it.

Removal

8 The cooling fan/shroud assembly may be riveted or bolted to the radiator.
9 If the assembly is riveted to the radiator, proceed as follows.
  a) Remove the radiator as described in Section 3.
  b) Support the radiator, and drill out the rivets securing the cooling fan shroud to the radiator. When drilling out the rivets, take great care not to damage the radiator.
10 If the assembly is bolted to the radiator, proceed as follows.
  a) Disconnect the fan motor wiring plug.
  b) Unscrew the securing bolts/nuts and withdraw the assembly from the radiator.

11 The motor may be riveted or bolted to the shroud.
12 If desired, the fan blades can be removed from the motor shaft after prising off the retaining clip.

Refitting

13 Refitting is a reversal of removal, but where applicable use new pop-rivets, and where applicable refit the radiator with reference to Section 3.

Cooling fan - non-turbo diesel engines

Removal

14 The fan blades can be removed simply by unscrewing the bolts securing the fan blades to the coolant pump pulley.

Refitting

15 Refitting is a reversal of removal, ensuring that the securing bolts are securely tightened.

Viscous cooling fan - turbo-diesel engines

16 At the time of writing, no information was available for removal and refitting of the viscous cooling fan. A special slim open-ended spanner is required to remove the fan and coupling assembly, and the coupling may have a left-hand thread. Consult a Renault dealer for further information.

6 Cooling system electrical switches - testing, removal and refitting

Electric cooling fan thermostatic switch

Testing

1 Testing of the switch is described in Section 5, as part of the electric cooling fan test procedure.

Removal

Note: *A new sealing ring or sealing compound may be required on refitting.*

2 The switch is located in the right-hand side of the radiator **(see illustration)**.

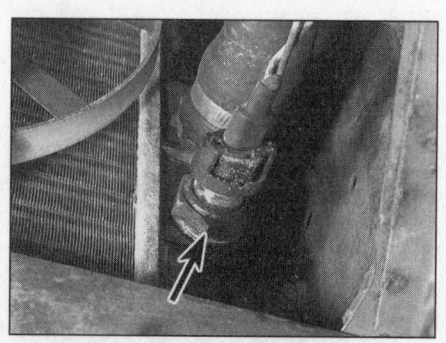

6.2 Electric cooling fan thermostatic switch (arrowed) viewed from underneath vehicle

Cooling, heating and air conditioning systems 3•5

3 Disconnect the battery negative lead.
4 Partially drain the cooling system to just below the level of the switch (see Chapter 1). Alternatively, have ready a suitable bung to plug the switch aperture in the radiator when the switch is removed. If this method is used, take great care not to damage the radiator, and do not use anything which will allow foreign matter to enter the radiator.
5 Disconnect the wiring plug from the switch.
6 Carefully unscrew the switch from the housing, and recover the sealing ring, where applicable. If the system has not been drained, plug the switch aperture to prevent further coolant loss.

Refitting

7 If the switch was originally fitted using sealing compound, clean the switch threads thoroughly, and coat them with fresh sealing compound
8 If the switch was originally fitted using a sealing ring, use a new sealing ring on refitting.
9 Refitting is a reversal of removal. Tighten the switch, and refill (or top-up) the cooling system as described in Chapter 1 or "Weekly Checks".
10 On completion, start the engine and run it until it reaches normal operating temperature. Continue to run the engine, and check that the cooling fan cuts in and out correctly.

Coolant temperature gauge/warning light sender

Testing

11 The switch is located in the thermostat housing (see Section 4), or in the right-hand side of the cylinder head, next to the thermostat housing on some 4-cylinder engines **(see illustration)**. A combined temperature gauge and warning light switch may be fitted, or separate temperature gauge and warning light switches may be fitted, depending on model. If separate senders are fitted, they are mounted together in the thermostat housing and/or in the right-hand side of the cylinder head.
12 The temperature gauge is fed with a stabilised voltage from the instrument panel feed (via the ignition switch and a fuse). The gauge earth is controlled by the sender. The sender contains a thermistor - an electronic component whose electrical resistance decreases at a predetermined rate as its temperature rises. When the coolant is cold, the sender resistance is high, current flow through the gauge is reduced, and the gauge needle points towards the blue (cold) end of the scale. As the coolant temperature rises and the sender resistance falls, current flow increases, and the gauge needle moves towards the upper end of the scale. If the sender is faulty, it must be renewed.
13 On models with a temperature warning light, the light is fed with a voltage from the instrument panel. The light earth is controlled by the sender. The sender is effectively a switch, which operates at a predetermined temperature to earth the light and complete the circuit. If the light is fitted in addition to a gauge, the senders for the gauge and light are incorporated in a single unit, with two wires, one each for the light and gauge earths.
14 If the gauge develops a fault, first check the other instruments; if they do not work at all, check the instrument panel electrical feed. If the readings are erratic, there may be a fault in the voltage stabiliser, which will necessitate renewal of the stabiliser (the stabiliser is integral with the instrument panel printed circuit board - see Chapter 12). If the fault lies in the temperature gauge alone, check it as follows.
15 If the gauge needle remains at the "cold" end of the scale when the engine is hot, disconnect the sender wiring plug, and earth the relevant wire to the cylinder head. If the needle then deflects when the ignition is switched on, the sender unit is proved faulty, and should be renewed. If the needle still does not move, remove the instrument panel (Chapter 12) and check the continuity of the wire between the sender unit and the gauge, and the feed to the gauge unit. If continuity is shown, and the fault still exists, then the gauge is faulty, and the gauge unit should be renewed.
16 If the gauge needle remains at the "hot" end of the scale when the engine is cold, disconnect the sender wire. If the needle then returns to the "cold" end of the scale when the ignition is switched on, the sender unit is proved faulty, and should be renewed. If the needle still does not move, check the remainder of the circuit as described previously.
17 The same basic principles apply to testing the warning light. The light should illuminate when the relevant sender wire is earthed.

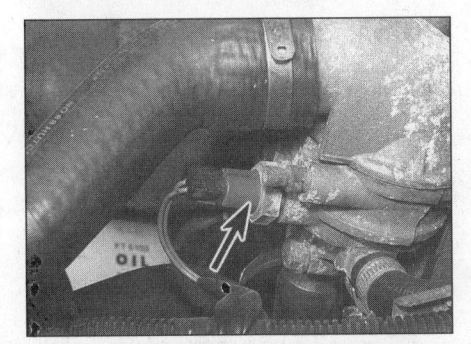

Removal and refitting

18 The procedure is similar to that described previously in this Section for the electric cooling fan thermostatic switch. On some models, access to the switch(es) is poor, and surrounding components may need to be moved to one side before the sender unit can be reached.

Coolant temperature sensors - fuel injection models

19 Refer to Chapter 4B.

7 Coolant pump - removal and refitting

Coolant pump - 4-cylinder engines

Removal

Note: *A new coolant pump gasket will be required on refitting.*

1 Disconnect the battery negative lead.
2 Drain the cooling system as described in Chapter 1.
3 To improve access, remove the radiator as described in Section 3.
4 Where applicable, disconnect the wiring from the temperature gauge switch.
5 Loosen the hose clips and disconnect the hoses from the coolant pump.
6 Loosen the coolant pump pulley bolts, then remove the auxiliary drivebelt as described in Chapter 1.
7 Remove the securing bolts, and withdraw the coolant pump pulley **(see illustration)**.
8 Unscrew the coolant pump securing bolts/nuts **(see illustration)**.
9 Where applicable, compress the timing belt tensioner plunger, using a large flat-bladed screwdriver or similar tool, and withdraw the

6.11 Coolant temperature gauge/ warning light sender (arrowed) - four-cylinder engine

7.7 Unscrewing a coolant pump pulley securing bolt - four-cylinder engine

7.8 Unscrewing a coolant pump securing bolt - four-cylinder engine

3•6 Cooling, heating and air conditioning systems

pump from the cylinder block (see illustration). Take care not to allow the tensioner plunger and spring to fly out of the housing in the pump as the pump is withdrawn. Note: *On some engines, it may be necessary to remove the timing belt tensioner, as described in Chapter 2, to allow the coolant pump to be withdrawn.*
10 Recover the gasket.

Refitting
11 Before commencing refitting, thoroughly clean the mating faces of the pump and the cylinder block.
12 Locate a new gasket on the cylinder block, using a little grease to hold it in position if necessary.
13 Where applicable, fit the timing belt tensioner plunger and spring to the housing in the pump, then refit the pump using a new gasket. Hold the timing belt tensioner plunger compressed as the pump is refitted.
14 Further refitting is a reversal of removal, bearing in mind the following points.
   a) *Where applicable, refit the timing belt tensioner as described in Chapter 2.*
   b) *Refit and tension the auxiliary drivebelt as described in Chapter 1.*
   c) *Refit the radiator with reference to Section 3.*
   d) *Refill the cooling system as described in Chapter 1.*

Coolant pump - 6-cylinder engines

Removal
Note: *New gaskets and O-rings will be required when refitting the U-shaped section of the inlet manifold. If a new pump is fitted, a new cover gasket and thermostat sealing ring will be required.*
15 Proceed as described in paragraphs 1 to 3.
16 Remove the air cleaner.
17 Unbolt and remove the U-shaped section of the inlet manifold from the front of the engine. Recover the gaskets and O-rings.
18 Where applicable, release the securing clip(s) and disconnect the throttle operating rod(s) from the throttle linkage, then unbolt the throttle linkage, and move it to one side, leaving the throttle cable connected.
19 Remove the coolant pump (auxiliary) drivebelt as described in Chapter 1.
20 Loosen the hose clips and disconnect the hoses from the coolant pump (it may not be possible to disconnect the rear hose until the pump is partially withdrawn).
21 Where applicable, disconnect the wiring plugs from the coolant temperature switch(es) mounted in the coolant pump body.
22 Unscrew the coolant pump securing bolts. The bolts can be unscrewed using a ring-spanner inserted down behind the pump pulley (see illustration).
23 The pulley is a press-fit on the pump shaft, and a suitable puller or press will be required to remove it.

Refitting
24 If a new pump is being fitted, it will be necessary to transfer the rear cover, thermostat housing and cover, thermostat, pulley, and temperature switch(es) from the old pump. Where applicable, use a new gasket when refitting the pump cover, and fit the thermostat with reference to Section 4.
25 Refitting is a reversal of removal, bearing in mind the following points.
   a) *Refit and tension the auxiliary drivebelt as described in Chapter 1.*
   b) *After refitting and reconnecting the throttle linkage, check the throttle cable adjustment with reference to Chapter 4.*
   c) *Use new gaskets and O-rings when refitting the U-shaped section of the inlet manifold.*
   d) *Refit the radiator with reference to Section 3.*
   e) *Refill the cooling system as described in Chapter 1.*

Auxiliary electric coolant pump - diesel models

General
26 On later diesel engine models, an auxiliary electric coolant pump is fitted to aid coolant circulation through the heater circuit when the engine temperature is below normal operating temperature. The pump is controlled by a thermostatic sensor mounted in the heater inlet hose. The pump is located on the steering gear crossmember in the engine compartment.

Removal
27 Access to the pump may be difficult, and it may be necessary to move surrounding components to one side before the pump can be reached.
28 Disconnect the battery negative lead.
29 Drain the cooling system as described in Chapter 1.
30 Disconnect the pump wiring plug.
31 Loosen the hose clips, and disconnect the coolant hoses from the pump. Note the hose locations to ensure correct refitting.
32 Unscrew the clamp bolt, and withdraw the pump from the crossmember.

Refitting
33 Refitting is a reversal of removal, but ensure that the hoses are correctly reconnected, and on completion refill the cooling system as described in Chapter 1.

8 Heating and ventilation system - general information

The heating/ventilation system consists of a four-speed blower motor (housed in the engine compartment scuttle, below the windscreen), a control unit mounted behind the facia, face level vents in the centre and at each end of the facia, and air ducts to the front footwells (and rear footwells on some models).

The facia-mounted controls operate flap valves to deflect and mix the air flowing through the various parts of the heating/ventilation system. The flap valves are contained in the air distribution housing, which acts as a central distribution unit, passing air to the various ducts and vents. The air closure flap (which prevents air from entering the vehicle - control in "Stop" position) is operated via a vacuum actuator, fed by vacuum from the brake servo.

Cold air enters the system through the grille at the top of the engine compartment scuttle. If required, the airflow is boosted by the blower motor, and then flows through the various ducts, according to the settings of the controls. Stale air is expelled via the vents in the rear doors. If warm air is required, the cold air is passed over the heater matrix, which is heated by the engine coolant.

On models fitted with air conditioning, a recirculation switch enables the outside air supply to be closed off, while the air inside the vehicle is recirculated. This can be useful to prevent unpleasant odours entering from outside the vehicle, but should only be used briefly, as the recirculated air inside the vehicle will soon become stale.

7.9 Withdrawing the coolant pump - four-cylinder engine

7.22 Unscrewing a coolant pump securing bolt - six-cylinder engine

Cooling, heating and air conditioning systems 3•7

9 Heating and ventilation system components - removal and refitting

Heater/ventilation control unit

Removal

1 Remove the instrument panel as described in Chapter 11.
2 Unscrew the two heater control panel securing screws, then slide the unit towards the passenger's side of the facia, and manipulate it from its housing (see illustrations).
3 Working at the rear of the assembly, unscrew the two securing screws, and remove the cover, then disconnect the wiring plugs and the heater control cables, noting their locations to aid refitting (see illustrations).
4 Withdraw the unit from the facia.

Refitting

5 Refitting is a reversal of removal, but ensure that the wiring plugs and control cables are correctly reconnected as noted before removal.

Heater/ventilation control cables

Removal

6 Remove the heater/ventilation control unit, as described previously in this Section.
7 Reach down through the aperture in the facia, and disconnect the cables from the operating levers on the heater assembly. Note that on some models, it may be necessary to remove the complete facia assembly (see Chapter 11) for access to the heater ends of the cables.
8 Release the cable sheaths from their securing clips, and withdraw the cables.

Refitting

9 Refitting is a reversal of removal, but adjust the position of the cable sheaths in the securing clips so that the positions of the operating levers on the heater unit are as shown with the control levers in the appropriate positions (see illustrations).

Heater matrix

Removal

10 Drain the cooling system (Chapter 1).
11 Remove the complete facia assembly as described in Chapter 11.
12 Working in the engine compartment, at the bulkhead, loosen the hose clips and disconnect the coolant hoses from the heater matrix pipes.
13 Working inside the vehicle, disconnect/release any wiring, cables and vacuum hoses from the heater unit.
14 Place rags or newspapers on the floor of the vehicle to minimise the effects of any coolant spillage.

9.2a Unscrew the securing screws...

9.2b ...and withdraw the heater control panel

9.3a Remove the heater control panel rear cover...

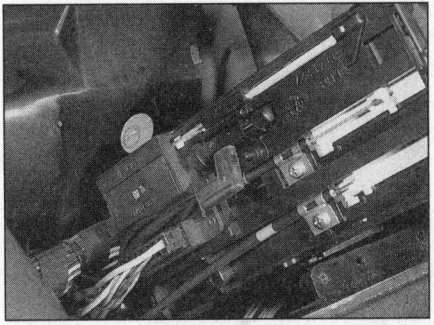

9.3b ...for access to the control cable and wiring plugs

15 Working inside the vehicle, unscrew the four securing nuts, and withdraw the heater assembly from the vehicle.
16 With the heater assembly removed, remove the securing screws, and withdraw the matrix from the heater casing.

Refitting

17 Refitting is a reversal of removal, bearing in mind the following points.
a) Ensure the foam pads are in place before sliding the matrix into the heater casing.
b) Refit the facia assembly with reference to Chapter 11.
c) Refill the cooling system as described in Chapter 1.

Heater blower motor

Removal

18 Remove the wiper arms as described in Chapter 12.
19 Working at the rear of the engine compartment, remove the securing screws,

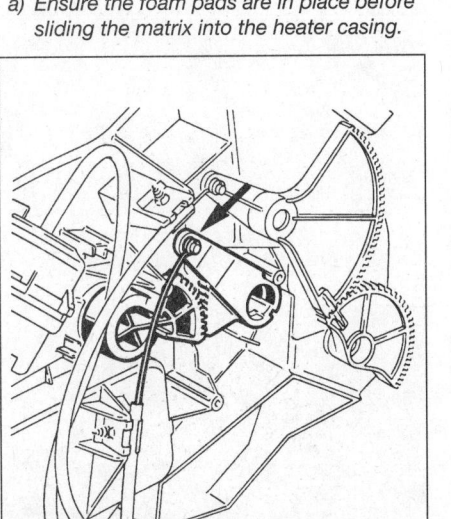

9.9a Air temperature control cable setting - control lever in cold position

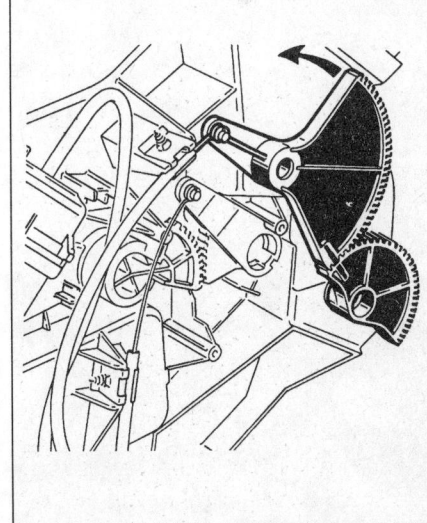

9.9b Air distribution control cable setting - control lever in windscreen demist position

3•8 Cooling, heating and air conditioning systems

and withdraw scuttle cover panels **(see illustrations)**.

20 Remove the two screws securing the heater blower motor cover to motor housing, then prise the weatherstrip from the edge of the scuttle, and lift off the motor cover, complete with weatherstrip **(see illustrations)**.

21 Disconnect the motor wiring plug, and move the wiring harness to one side **(see illustration)**.

22 Unscrew the two securing bolts, and lift the blower motor assembly from the scuttle **(see illustrations)**.

Refitting

23 Refitting is a reversal of removal, but refit the wiper arms with reference to Chapter 11.

Heater blower motor resistor
Removal

24 Remove the heater blower motor assembly as described previously in this Section.

25 Disconnect the wiring plugs from the resistor assembly.

26 Remove the securing clip, and remove the resistor.

Refitting

27 Refitting is a reversal of removal.

Heater air vent control vacuum reservoir
Removal

28 Remove the driver's side wiper arm as described in Chapter 12.

29 Working at the rear of the engine compartment, remove the securing screws, and withdraw the driver's side scuttle cover panel.

30 Remove the securing nuts, then disconnect the vacuum hoses, and withdraw the reservoir **(see illustration)**.

Refitting

31 Refitting is a reversal of removal, but refit the wiper arm with reference to Chapter 12.

10 Air conditioning system - general information and precautions

General information

1 An air conditioning system is available on certain models. It enables the temperature of incoming air to be lowered, and also

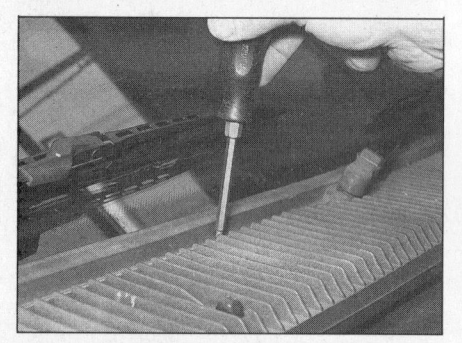

9.19a Remove the top . . .

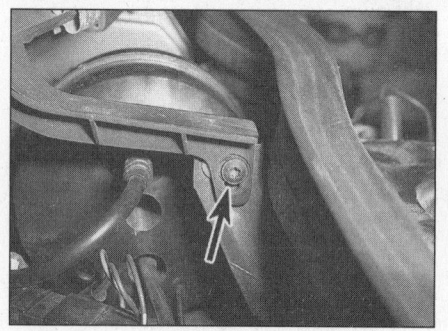

9.19b . . . and side securing screws . . .

9.19c . . . then withdraw the scuttle cover panels

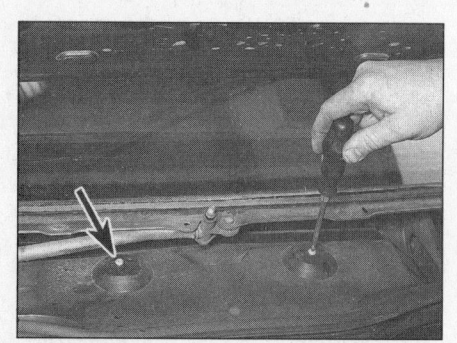

9.20a Remove the two screws . . .

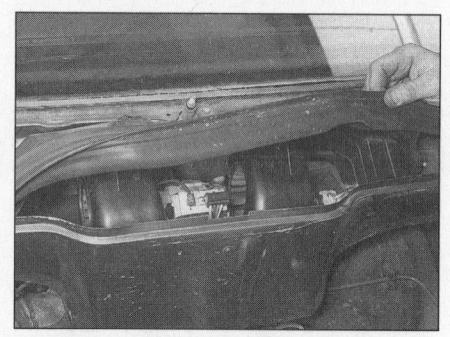

9.20b . . . then lift off the motor cover, complete with weatherstrip

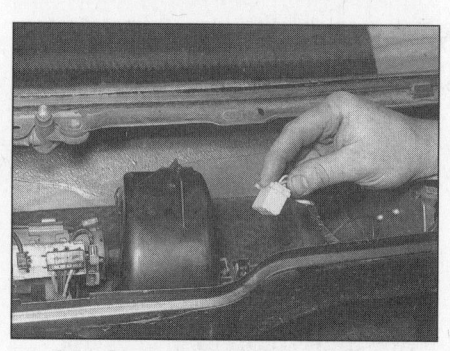

9.21 Disconnect the blower motor wiring plug

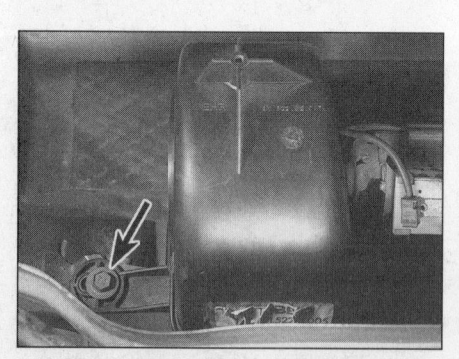

9.22a Unscrew the securing bolts . . .

9.22b . . . then lift the blower motor assembly from the scuttle

9.30 Removing the heater air vent control vacuum reservoir

Cooling, heating and air conditioning systems

dehumidifies the air, which makes for rapid demisting and increased comfort.

2 The cooling side of the system works in the same way as a domestic refrigerator. Refrigerant gas is drawn into a belt-driven compressor, and passes into a condenser mounted on the front of the radiator, where it loses heat and becomes liquid. The liquid passes through an expansion valve to an evaporator, where it changes from liquid under high pressure to gas under low pressure. This change is accompanied by a drop in temperature, which cools the evaporator. The refrigerant returns to the compressor, and the cycle begins again.

3 Air blown through the evaporator passes to the air distribution unit, where it is mixed with hot air blown through the heater matrix to achieve the desired temperature in the passenger compartment.

4 The heating side of the system works in the same way as on models without air conditioning (see Section 8).

5 Any problems with the system should be referred to a Renault dealer.

Precautions

6 When an air conditioning system is fitted, it is necessary to observe special precautions whenever dealing with any part of the system, or its associated components. If for any reason the system must be disconnected, entrust this task to your Renault dealer or a refrigeration engineer.

⚠ *Warning: The refrigeration circuit contains a liquid refrigerant (Freon), and it is therefore dangerous to disconnect any part of the system without specialised knowledge and equipment.*

7 The refrigerant is potentially dangerous, and should only be handled by qualified persons. If it is splashed onto the skin, it can cause frostbite. It is not itself poisonous, but in the presence of a naked flame (including a cigarette) it forms a poisonous gas. Uncontrolled discharging of the refrigerant is dangerous, and potentially damaging to the environment.

8 Do not operate the air conditioning system if it is known to be short of refrigerant, as this may damage the compressor.

11 Air conditioning system components - removal and refitting

⚠ *Warning: Do not attempt to open the refrigerent circuit. Refer to the precautions given in Section 10.*

The only operation which can be carried out easily without discharging the system is the renewal of the compressor drivebelt, which is described in Chapter 1. All other operations must be referred to a Renault dealer or an air conditioning specialist.

If necessary, the compressor can be unbolted and moved aside, without disconnecting the refrigerent lines, after removing the drivebelt.

Chapter 4 Part A:
Fuel and exhaust systems - carburettor engines

Contents

Accelerator cable - removal, refitting and adjustment	6
Accelerator pedal - removal and refitting	7
Air cleaner assembly - removal and refitting	2
Air cleaner filter element renewal	See Chapter 1
Carburettor - general information	9
Carburettor - removal and refitting	10
Carburettor - fault diagnosis, overhaul and adjustments	11
Exhaust manifold - removal and refitting	13
Exhaust system - general information, removal and refitting	14
Exhaust system check	See Chapter 1
Fuel filter - renewal	See Chapter 1
Fuel gauge sender unit - removal and refitting	4
Fuel pump - testing, removal and refitting	3
Fuel tank - removal and refitting	5
General fuel system checks	See Chapter 1
General information and precautions	1
Idle speed and mixture adjustment	See Chapter 1
Inlet manifold - removal and refitting	12
Unleaded petrol - general information and usage	8

Degrees of difficulty

Easy, suitable for novice with little experience	Fairly easy, suitable for beginner with some experience	Fairly difficult, suitable for competent DIY mechanic	Difficult, suitable for experienced DIY mechanic	Very difficult, suitable for expert DIY or professional

Specifications

Fuel pump
Type .... Mechanical, driven by eccentric on intermediate shaft

Carburettor
Type .... Weber DARA
Designation .... DARA 28/36 or DARA 32
Choke type .... Automatic, coolant-heated

Carburettor data	**Primary**	**Secondary**
DARA 28/36 - 0 and 0C carburettors:		
Venturi	22	29
Diffuser	3.5	4
Main jet	112	155
Idle jet	42	42
Idle air jet	150	150
Air correction jet:		
0 carburettor	200	100
0C carburettor	180	100
Emulsion tube:		
0 carburettor	F99	F56
0C carburettor	F99	F7
Accelerator pump:		
0 carburettor	50	
0C carburettor	40 double	
Needle valve	225	
Float:		
Height setting	7 mm	
Stroke	8 mm	
Initial throttle opening	0.8 mm	
Vacuum part-open:		
Compensator pushed-in	3 mm	
Compensator out	7.5 mm	
Deflooding mechanism	5.4 mm	

4A•2 Fuel and exhaust systems - carburettor engines

Carburettor data (continued)

	Primary	Secondary
Idle speed	800 ± 50 rpm	
Idle mixture CO content	1.5 ± 0.5 %	

DARA 28/36 - 1, 1C, 4 and 4C carburettors:

	Primary	Secondary
Venturi	22	29
Diffuser	3.5	4
Main jet	112	155
Idle jet	40	50
Air correction jet	200	100
Emulsion tube	F99	F56
Accelerator pump	50	
Needle valve	225	
Float:		
Height setting	7 mm	
Stroke	8 mm	
Initial throttle opening	0.9 mm	
Vacuum part-open:		
Compensator pushed-in	3 mm	
Compensator out	7.5 mm	
Deflooding mechanism	6.5 mm	
Idle speed	800 ± 50 rpm	
Idle mixture CO content	1.5 ± 0.5 %	

DARA 28/36 - 8 and 8C carburettors:

	Primary	Secondary
Venturi	22	29
Diffuser	3.5	4
Main jet	112	155
Idle jet	42	42
Air correction jet	180	100
Emulsion tube	F99	F7
Accelerator pump	40 double	
Needle valve	225	
Cold start enrichener	80	
Float:		
Height setting	7 mm	
Stroke	8 mm	
Initial throttle opening	0.8 mm	
Vacuum part-open:		
Compensator pushed-in	3.5 mm	
Compensator out	N\A mm	
Deflooding mechanism	5.4 mm	
Idle speed	800 ± 50 rpm	
Idle mixture CO content	1.5 ± 0.5 %	

DARA 32 - 48 and 48C carburettors:

	Primary	Secondary
Venturi	26	26
Diffuser	3.5	4
Main jet	135	130
Idle jet	52	45
Air correction jet	155	155
Emulsion tube	F58	F6
Accelerator pump	60	
Needle valve	225	
Float:		
Height setting	7 mm	
Stroke	8 mm	
Initial throttle opening	0.95 mm	
Vacuum part-open:		
Compensator pushed-in	5.5 mm	
Compensator out	10.0 mm	
Deflooding mechanism	9.0 mm	
Idle speed	800 ± 50 rpm	
Idle mixture CO content	1.5 ± 0.5 %	

DARA 32 - 49 and 49C carburettors:

	Primary	Secondary
Venturi	26	26
Diffuser	3.5	4
Main jet	130	135
Idle jet	57	42

Fuel and exhaust systems - carburettor engines 4A•3

Carburettor data (continued)

	Primary	Secondary
Air correction jet	155	140
Emulsion tube	F58	F6
Accelerator pump	60	
Needle valve	225	
Float:		
Height setting	7 mm	
Stroke	8 mm	
Initial throttle opening	1 mm	
Vacuum part-open:		
Compensator pushed-in	5.5 mm	
Compensator out	10.0 mm	
Deflooding mechanism	9.0 mm	
Idle speed	900 ± 50 rpm (selector lever at N)	
Idle mixture CO content	1.5 ± 0.5 %	

DARA 32 - 60 carburettor:

	Primary	Secondary
Venturi	26	26
Diffuser	3.5	4
Main jet	130	135
Idle jet	60	42-55
Air correction jet	155	140
Emulsion tube	F58	F6
Accelerator pump	60	
Needle valve	225	
Float:		
Height setting	7 mm	
Stroke	8 mm	
Initial throttle opening	1 mm	
Vacuum part-open:		
Compensator pushed-in	5.5 mm	
Compensator out	10.0 mm	
Deflooding mechanism	6.0 mm	
Idle speed	800 ± 50 rpm (selector lever at N)	
Idle mixture CO content	1.5 ± 0.5 %	

Recommended fuel

Minimum octane rating*:
- Models produced prior to January 1989 .... 98 RON unleaded (UK "super unleaded") or 98 RON leaded (UK "4-star")
- Models produced since January 1989 .... 95 RON unleaded (UK "unleaded premium") or 98 RON leaded (UK "4-star")

*Refer to your Renault dealer for further information

Torque wrench setting

	Nm	lbf ft
Carburettor nuts	24	18

1 General information and precautions

1 The fuel system consists of a fuel tank mounted under the rear of the car, a mechanical fuel pump, and a carburettor. The fuel pump is operated by an eccentric on the intermediate shaft, and is mounted on the left-hand side of the cylinder block. The air cleaner contains a disposable paper filter element, and incorporates a flap valve air temperature control system; this allows cold air from the outside of the car, and warm air from the exhaust manifold, to enter the air cleaner in the correct proportions.

2 The fuel pump lifts fuel from the fuel tank and supplies it to the carburettor via a filter which is clipped onto the inlet manifold **(see illustration)**. Excess fuel is returned from the pump to the fuel tank.

3 The carburettor is a Weber DARA twin-choke carburettor (see Section 9).

4 The exhaust system consists of four sections; the front pipe, the front silencer box, the intermediate pipe and the tailpipe and main silencer box. The system is suspended throughout its entire length by rubber mountings.

Warning: Many of the procedures in this Chapter require the removal of fuel lines and connections, which may result in some fuel spillage. Before carrying out any operation on the fuel system, refer to the precautions given in "Safety first!" at the beginning of this manual, and follow them implicitly. Petrol is a highly-dangerous and volatile liquid, and the precautions necessary when handling it cannot be overstressed.

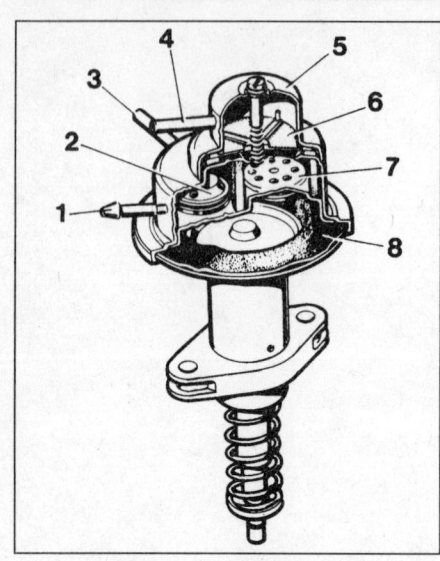

4A•4 Fuel and exhaust systems - carburettor engines

1.2 Cutaway view of fuel pump

1 Fuel return union
2 Outlet valve
3 Outlet union
4 Inlet union
5 Filter cover
6 Filter
7 Inlet valve
8 Diaphragm

3.5a Remove the filter cover from the pump and remove the seal ...

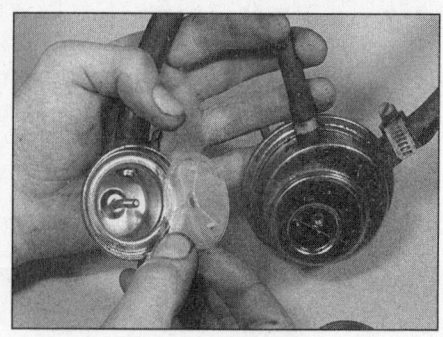

3.5b ... and filter

2 Air cleaner assembly - removal and refitting

Removal

1 Slacken the retaining clips (where fitted), and disconnect the hot and cold air hoses from the base of the housing and the rubber inlet duct from the top of the housing.
2 Unhook the rubber retaining strap and remove the air cleaner housing assembly from the engine compartment.
3 If necessary, undo the retaining screws then remove the inlet duct assembly from the top of the carburettor and recover the gasket. The air cleaner housing cold and hot air inlet ducts can also be removed if necessary.

Refitting

4 Refitting is a reversal of the removal procedure, making sure the duct/hoses are securely reconnected.

3 Fuel pump - testing, removal and refitting

Note: *Refer to the warning note in Section 1 before proceeding.*

Testing

1 To test the fuel pump on the engine, disconnect the outlet pipe which leads to the carburettor. Hold a wad of rag by the pump outlet while an assistant spins the engine on the starter. *Keep your hands away from the electric cooling fan.* Regular spurts of fuel should be ejected as the engine turns. Be careful not to spill fuel onto hot engine components.
2 The pump can also be tested by removing it. With the pump outlet pipe disconnected but the inlet and return pipes still connected, hold the wad of rag by the outlet. Operate the pump lever by hand, moving it in and out; if the pump is in a satisfactory condition, the lever should move and return smoothly, and a strong jet of fuel should be ejected.

Removal

3 Identify the pump hoses, and slacken their retaining clips. Place wads of rag beneath the hose unions to catch any spilled fuel, then disconnect the hoses from the pump; plug the hose ends to minimise fuel loss.
4 Slacken and remove the nuts and washers securing the pump to the cylinder block. Remove the pump and recover the insulating spacer and gaskets. Discard the gaskets, new ones must be used on refitting.
5 Whilst the pump is removed, undo the retaining screw and remove the filter cover from the top of the pump. Remove the seal and withdraw the filter **(see illustrations)**. Clean the filter in fresh fuel then refit the filter and seal to the cover. Fit the cover to the pump and securely tighten the retaining screw.

Refitting

6 Ensure that the pump, cylinder block and insulating spacer mating surfaces are clean and dry. Position a new gasket on each side of the spacer and fit the spacer and gasket assembly to the block.
7 Refit the pump to the cylinder block then fit the washers and securely tighten the nuts.
8 Reconnect the inlet, outlet and return hoses to the relevant pump unions, and securely tighten their retaining clips.

4 Fuel gauge sender unit - removal and refitting

Note: *Refer to the warning note in Section 1 before proceeding.*

Removal

1 Disconnect the battery negative lead.
2 Remove the rear cushion as described in Chapter 11.
3 Unclip the carpet and fold it back to gain access to the fuel gauge sender unit access cover. Remove the cover to reveal the sender unit **(see illustrations)**.
4 Disconnect the wiring connector(s) from the sender unit, and tape the connector to the vehicle body to prevent it disappearing behind the tank.
5 Where necessary, mark the hoses for identification purposes, then slacken the feed and return hose retaining clips. Where the crimped-type hose clips are fitted, cut the clips and discard them; use standard worm-drive hose clips on refitting. Disconnect both hoses from the top of the sender unit, and plug the hose ends.

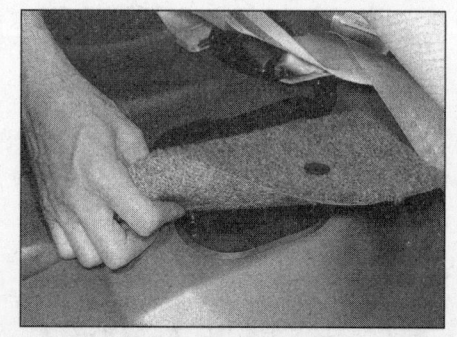

4.3a Unclip the carpet from the floor and fold it back ...

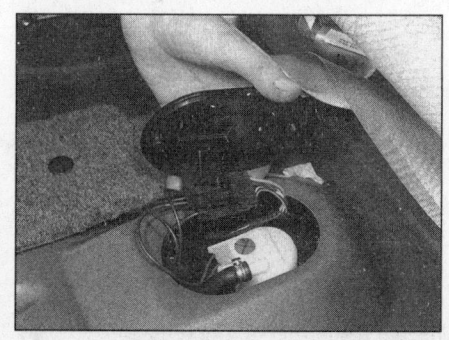

4.3b ... and remove the fuel gauge sender unit access cover

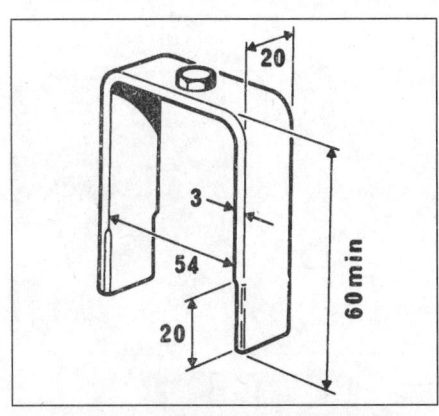

Fuel and exhaust systems - carburettor engines 4A•5

4.6 Fuel gauge sender unit retaining ring removal tool dimensions (in mm)

6 Unscrew the sender unit retaining ring from the tank. In order to loosen the ring, it may be necessary to fabricate a tool similar to that shown in **illustration 4.6**.
7 Carefully lift the sender unit from the top of the fuel tank, taking great care not to bend the sender unit float arm (where fitted), or to spill fuel onto the interior of the vehicle. Recover the rubber sealing ring and discard it - a new one must be used on refitting.

Refitting

8 Refitting is a reversal of the removal procedure, noting the following points:
  a) *Prior to refitting, fit a new rubber sealing ring to the tank.*
  b) *Refit the sender unit to the tank, taking care not to bend the float arm (where fitted). Ensure that the retaining ring is securely tightened..*
  c) *Ensure that the feed and return hoses are correctly reconnected, and are securely retained by their clips (where necessary).*

5 Fuel tank - removal and refitting

Note: *Refer to the warning note in Section 1 before proceeding.*

Removal

1 Before removing the fuel tank, all fuel must be drained from the tank. Since a fuel tank drain plug is not provided, it is therefore preferable to carry out the removal operation when the tank is nearly empty. Before proceeding, disconnect the battery negative lead, and syphon or hand-pump the remaining fuel from the tank **(see illustration)**.
2 Chock the front wheels then jack up the rear of the vehicle and support it on axle stands (see "*Jacking and vehicle support*").
3 Remove the rear bumper as described in Chapter 11.
4 Undo the mounting bolts and remove the left and right-hand support brackets from underneath the fuel tank **(see illustration)**.
5 Referring to Chapter 9, disconnect the handbrake cable from the lever and position it clear of the fuel tank.
6 Disconnect the wiring and hoses from the fuel gauge sender unit as described in Section 4.
7 Slacken the retaining clip and disconnect the main filler neck hose from the fuel tank **(see illustration)**.
8 Place a trolley jack with an interposed block of wood beneath the tank, then raise the jack until it is supporting the weight of the tank.
9 Slacken and remove the retaining bolts and slowly lower the fuel tank out of position, disconnecting any other relevant vent pipes as they become accessible (where necessary), and remove the tank from underneath the vehicle **(see illustration)**.
10 If the tank is contaminated with sediment or water, remove the sender unit (Section 4), and swill the tank out with clean fuel. On most models the tank is injection-moulded from a synthetic material (some early models have a metal tank) and if seriously damaged, it should be renewed. However, in certain cases, it may be possible to have small leaks or minor damage repaired. Seek the advice of a specialist before attempting to repair the fuel tank.

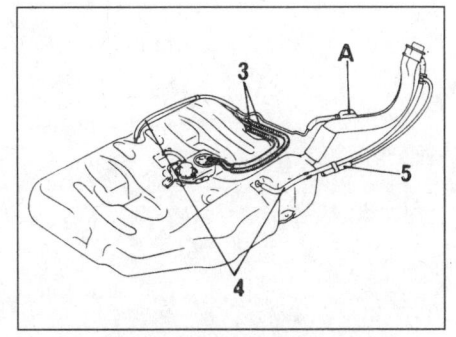

Refitting

11 Refitting is the reverse of the removal procedure, noting the following points:
  a) *When lifting the tank back into position, take care to ensure that none of the hoses become trapped between the tank and vehicle body.*
  b) *Ensure that all pipes and hoses are correctly routed, and securely held in position with their retaining clips.*
  c) *Reconnect the handbrake cable and adjust the handbrake as described in Chapter 9.*
  d) *On completion, refill the tank with a small amount of fuel, and check for signs of leakage prior to taking the vehicle out on the road.*

5.1 Typical fuel tank filler neck and breather hose arrangement

A Vent houg
3 Fuel supply and return hoses
4 Breather hoses
5 Vent hose

5.4 Fuel tank support bracket retaining bolts

5.7 Fuel tank filler neck and breather pipe connections

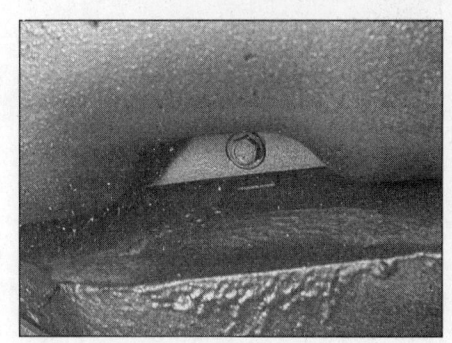

5.9 Fuel tank mounting bolt

6 Accelerator cable - removal, refitting and adjustment

Removal

1 Working in the engine compartment, free the accelerator inner cable from the throttle cam, then pull the outer cable out from its mounting bracket rubber grommet. Remove the spring clip from the cable end and the rubber grommet from the cable bracket.
2 Working back along the length of the cable, free it from any retaining clips or ties, noting its correct routing.
3 From inside the vehicle, reach up behind the facia panel, and disconnect the inner cable from the top of the accelerator pedal. If necessary, to improve access remove the

4A•6 Fuel and exhaust systems - carburettor engines

facia undercover panel (see Chapter 11). Tie a length of string to the inner cable then return to the engine compartment.
4 Release the outer cable from the bulkhead and withdraw the cable. When the end of the cable appears, untie the string and leave it in position - it can then be used to draw the cable back into position on refitting. Remove the rubber sealing grommet

Refitting

5 Fit the rubber sealing grommet to the cable. Tie the string to the end of the cable, then use the string to draw the cable into position through the bulkhead. Once the cable end is visible, untie the string, and clip the inner cable into position in the pedal.
6 Check that the cable is securely retained, then refit the facia panel (where removed).
7 From within the engine compartment, ensure that the outer cable is correctly seated in the bulkhead then work along the cable, securing it in position with the retaining clips and ties, and ensuring that the cable is correctly routed.
8 Make sure the rubber grommet is in position in the mounting bracket and fit the spring clip to the cable end.
9 Pass the outer cable through the mounting bracket grommet, and reconnect the inner cable to the throttle cam. Adjust the cable as described below.

Adjustment

10 Remove the spring clip from the accelerator outer cable. Ensuring that the throttle cam is fully against its stop, gently pull the cable out of its grommet until all free play is removed from the inner cable.
11 With the cable held in this position, refit the spring clip to the last exposed outer cable groove in front of the rubber grommet. When the clip is refitted and the outer cable is released, there should be only a small amount of free play in the inner cable.
12 Have an assistant depress the accelerator pedal, and check that the throttle cam opens fully and returns smoothly to its stop.

7 Accelerator pedal - removal and refitting

Removal

1 Disconnect the accelerator cable from the pedal as described in paragraph 3 of Section 6.
2 On right-hand drive models, unscrew the pedal pivot retaining nuts and remove the pedal from the vehicle.
3 On left-hand drive models, unscrew the pedal link arm pivot retaining nuts then unclip the link arm and remove it from the pedal. Carefully drill out the pedal retaining rivets and remove the pedal from the vehicle.

Refitting

4 Refitting is a reversal of the removal procedure, applying a little multi-purpose grease to the pivot point. On completion, adjust the accelerator cable as described in Section 6.

8 Unleaded petrol - general information and usage

Note: *The information given in this Chapter is correct at the time of writing. If updated information is thought to be required, check with a Renault dealer. If travelling abroad, consult one of the motoring organisations (or a similar authority) for advice on the fuel available.*

1 The fuel recommended by Renault is given in the Specifications Section of this Chapter, followed by the equivalent petrol currently on sale in the UK.
2 All models produced prior to January 1989 are designed to run on 98 octane petrol. Both leaded (UK "4-star") and unleaded (UK "super unleaded") petrol can be used.
3 Models produced after January 1989 are designed to run on 95 octane petrol. Both leaded (UK "4-star") and unleaded (UK "unleaded premium") petrol can be used. 98 octane unleaded (UK "super unleaded") petrol can also be used if wished, though there is no advantage in doing so.

9 Carburettor - general information

1 The carburettor is of dual barrel, downdraught type with a coolant-heated automatic choke. The throttle linkages are arranged so that the secondary throttle valve will not start to open until the primary valve is about two-thirds open, but at full throttle both valves are fully open. Choke operation is automatic by a coolant-heated arrangement, on certain versions an additional electric choke heater is also fitted.
2 The throttle valve plate block is coolant-heated. This warms the carburettor body quickly on cold starts, improving atomisation of the fuel/air mixture and preventing carburettor icing during warm-up.
3 During slow running and at idle, fuel from the float chamber passes into the idle channel through a metered idle jet. Here it is mixed with a small amount of air from a calibrated air bleed. The resulting mixture is drawn through a channel, to be discharged from the idle orifice under the primary throttle plate. A tapered mixture screw is used to vary the outlet, and this ensures fine control of the idle mixture. A constant CO idling supplementary circuit is also incorporated, the circuit is controlled by a solenoid cut-off valve **(see illustration)**.

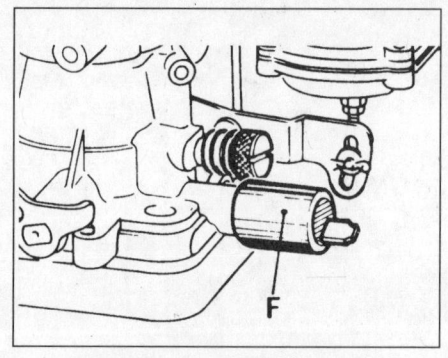

9.3 Idling supplementary circuit solenoid valve (F)

4 An idle cut-off valve is used to prevent run-on when the engine is switched off **(see illustration)**. The valve uses a solenoid plunger to block the idle jet when the ignition is switched off.
5 Under normal operating conditions, the amount of fuel discharged into the airstream is controlled by a calibrated main jet. Fuel is drawn through the main jet. The fuel is then mixed with air, drawn in through the air correction jet and through the holes in the emulsion tube. The resulting mixture is discharged from the main orifice through an auxiliary vent. A secondary barrel lock-out system is also incorporated, this uses a diaphragm to prevent the secondary throttle valve from opening when the diaphragm is subject to vacuum.
6 The carburettor also has an accelerator pump to provide an initial spurt of extra fuel during sudden acceleration. During acceleration, fuel is pumped through a ball valve located in the pump injector, and is discharged into both the primary and secondary venturis.
7 The idle speed is set by an adjustable screw. The adjustable mixture screw is sealed

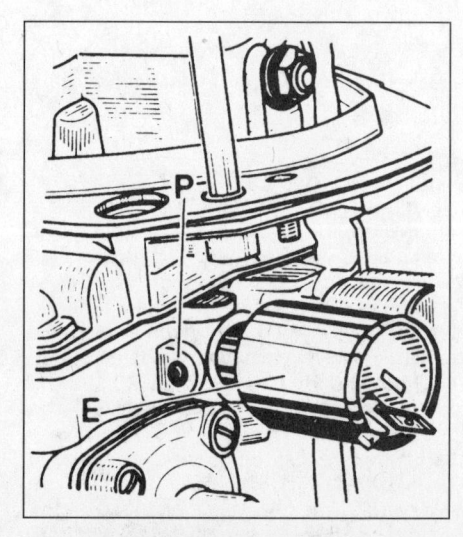

9.4 Fuel (idle) cut-off solenoid valve (E) and locking screw (P)

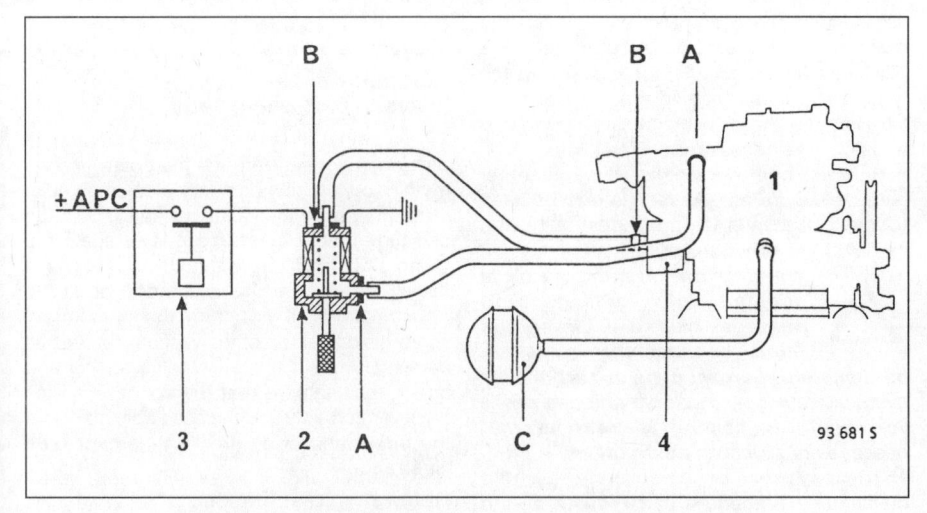

Fuel and exhaust systems - carburettor engines 4A•7

9.8 Cold enrichment system - 1990 onwards models fitted with DARA 28/36 8 and 8C carburettors

1 Carburettor
2 Solenoid valve
3 Oil temperature switch
4 Pneumatic enrichment circuit

A Green identification ring
B Brown identification ring
C Vacuum reservoir

during production with a tamperproof plug, to prevent unnecessary or inexpert adjustment.
8 On some later 1990 onwards models (fitted with DARA 28/36 8 and 8C carburettors), a pneumatic cold enrichment system is fitted to the carburettor to improve the running from cold **(see illustration)**. The system is controlled by an oil temperature switch via an electrically-operated solenoid valve. when the engine is cold (oil temperature below 15°C) the solenoid valve is open and allows vacuum to act on the carburettor diaphragm opening the auxiliary enrichment circuit. When the engine warms up (oil temperature above 40°C) the solenoid valve and diaphragm close, disabling the circuit.

10 Carburettor - removal and refitting

Note: *Refer to the warning note in Section 1 before proceeding. Where original crimped-type Renault hose clips are still fitted, the clips should be cut and discarded; obtain some standard worm-drive hose clips for refitting.*

Removal

1 Disconnect the battery negative terminal.
2 Remove the air cleaner inlet duct as described in Section 2.
3 Free the accelerator inner cable from the throttle cam, then pull the outer cable out from its mounting bracket rubber grommet, along with its spring clip.
4 Slacken the retaining clips and disconnect the coolant hoses from the carburettor. Plug the hose ends to minimise coolant loss and rinse off any spilt coolant.
5 Slacken the retaining clip, and disconnect the fuel feed hose from the carburettor. Place wads of rag around the union to catch any spilled fuel, and plug the hose as soon as it is disconnected, to minimise fuel loss.
6 Disconnect the wiring connectors from the carburettor.
7 Make a note of the correct fitted positions of all the relevant vacuum pipes and breather hoses, to ensure that they are correctly positioned on refitting, then release the retaining clips (where fitted) and disconnect them from the carburettor.
8 Unscrew the four nuts and washers securing the carburettor to the inlet manifold. Remove the carburettor assembly from the car. Remove the insulating spacer and/or gasket(s). Discard the gasket(s); new ones must be used on refitting. Plug the inlet manifold port with a wad of clean cloth, to prevent the possible entry of foreign matter.

Refitting

9 Refitting is the reverse of the removal procedure, noting the following points:
 a) Ensure that the carburettor and inlet manifold sealing faces are clean and flat. Fit a new insulating spacer and/or gasket(s), and securely tighten the carburettor retaining nuts.
 b) Use the notes made on dismantling to ensure that all hoses are refitted to their original positions and, where necessary, are securely held by their retaining clips.
 c) Refit and adjust the accelerator cable as described in Section 6.
 d) Top-up the cooling system as described in "Weekly checks".
 e) If necessary, check the idle speed and mixture settings as described in Chapter 1.

11 Carburettor - fault diagnosis, overhaul and adjustments

Fault diagnosis

1 If a carburettor fault is suspected, always check first that the ignition timing is correctly set, that the spark plugs are in good condition and correctly gapped, that the accelerator cable is correctly adjusted, and that the air cleaner filter element is clean; refer to the relevant Sections of Chapter 1, Chapter 5 or this Chapter. If the engine is running very roughly, first check the valve clearances, then check the compression pressures as described in Chapter 2.
2 If careful checking of all the above produces no improvement, the carburettor must be removed for cleaning and overhaul.
3 Prior to overhaul, check the availability of component parts before starting work; note that most sealing washers, screws and gaskets are available in kits, as are some of the major sub-assemblies. In most cases, it will be sufficient to dismantle the carburettor and to clean the jets and passages.

Overhaul

Note: *Refer to the warning note in Section 1 before proceeding.*

4 The operations described in paragraphs 5 to 10 can be carried out without the need to remove the carburettor from the engine. The adjustments described later in this Section are vital to the success of the overhaul.
5 Remove the air cleaner (Section 2), and disconnect the fuel hose from the carburettor.
6 Unscrew the plug just above the fuel inlet nozzle, remove and clean the filter gauze. Refit the gauze and the plug.
7 Extract the screws and take off the top cover. This will be as far as most overhaul work will need to go, as the jets can be removed and blown through with air from a tyre pump - never probe them with wire or their calibration will be ruined. In extreme cases of clogging a nylon bristle may be used to clear a jet **(see illustration)**.
8 Mop out dirt and sediment from the fuel bowl.
9 The tightness of the fuel inlet needle valve may be checked, but this will mean driving out the float pivot pin in order to locate a ring spanner on the valve. Fuel seeping out through the needle valve seating washer will cause too high a fuel level and consequent flooding.
10 Before fitting the carburettor cover with a new gasket, check the float height setting. Hold the cover vertically with the float hanging down so that the float tang contacts the needle valve ball. Check the float height is as given in the *Specifications*. If necessary carefully bend the float needle valve tang to adjust the height. With the float height correctly set, check the float stroke is as given

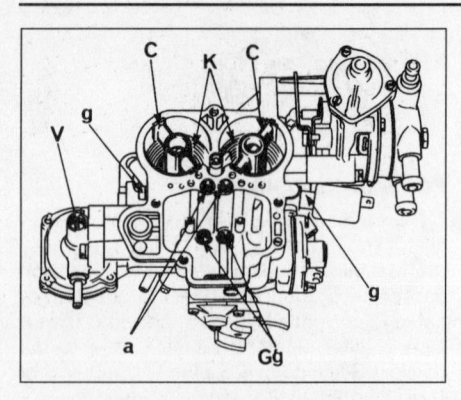

4A•8 Fuel and exhaust systems - carburettor engines

11.7 Location of carburettor jets

- a Air correction jet
- C Diffuser
- g Idle jet
- Gg Main jet
- K Venturi

in the *Specifications*. If necessary, adjust by carefully bending the tab **(see illustration)**.

11 Where it was decided to strip the carburettor completely, remove the unit, as described in Section 10, and clean away external dirt.

12 Remove the top cover and jets as previously described. A carburettor overhaul kit should now be obtained which will contain all the necessary gaskets, diaphragms and seals which will require renewal.

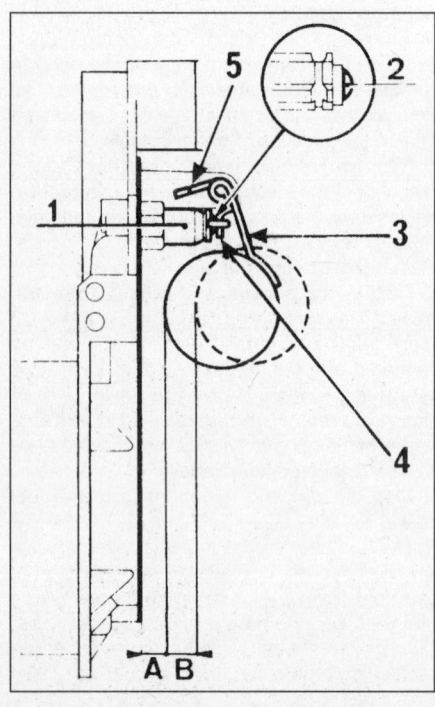

13 Remove the diaphragm units, invert the carburettor and extract the throttle valve block screws. Do not attempt to dismantle the throttle flap plates or spindles.

14 Remove the choke housing cover and housing if necessary. Clean the jets, carburettor body assemblies, float chamber and internal drillings. An air line may be used to clear the internal passages once the carburettor is fully dismantled.

⚠️ **Warning:** *If high pressure air is directed into drillings and passages were a diaphragm is fitted, the diaphragm is likely to be damaged. Aerosol cans of carburettor cleaner are widely available and can prove very useful in helping to clear internal passages of stubborn obstructions.*

15 Reassembly is a reversal of the dismantling procedure carrying out all the adjustments described below. Ensure that all jets are securely locked in position, but take great care not to overtighten them. Ensure that all mating surfaces are clean and dry, and that all body sections are correctly assembled with their fuel and air passages correctly aligned.

Adjustments

Initial throttle opening

16 To adjust the initial throttle opening, fully close the choke flaps with the fingers and set the adjusting screw on the specified (medium cold) cam step **(see illustrations)**.

17 Using a twist drill or similar gauge, check that the gap between the primary throttle flap and barrel is as given in the *Specifications*. If not, turn the screw as necessary.

Automatic choke (vacuum part-open) setting

18 To adjust the automatic choke (vacuum part-open setting) remove the cover and bi-metallic spring then manually close the choke flaps. Fully raise the vacuum capsule pushrod and turn the choke operating lever against it. Using a twist drill check that the gap between the choke flap and barrel is as given in the *Specifications*. If not, turn the adjustment screw inside the top of the vacuum capsule as necessary, Refit the bi-metallic spring and cover, making sure that the spring engages the lever correctly and the assembly marks on the cover and body are in alignment **(see illustrations)**.

Deflooding mechanism

19 To adjust the deflooding mechanism, fully close the choke flaps manually then fully open

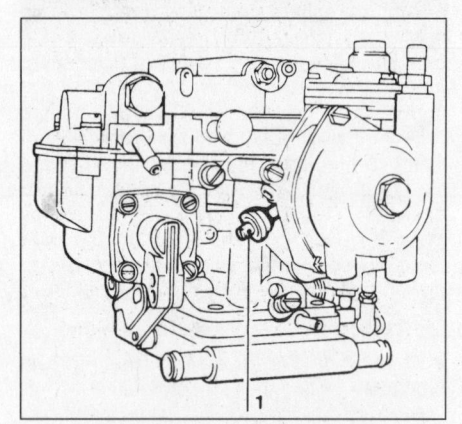

11.10 Float height setting details

1. Needle valve
2. Needle valve ball
3. Float arm
4. Needle valve tang
5. Float stroke adjusting tab
A Float height measurement
B Float stroke measurement

11.16a Initial throttle opening adjusting screw (1) - DARA 32 carburettor shown

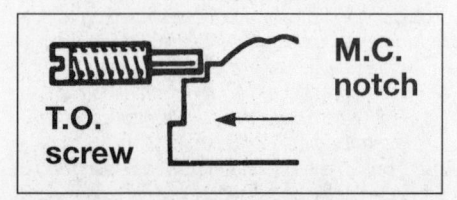

11.16b Prior to adjusting the initial throttle opening, set the adjusting screw on the "medium cold" step of the cam as shown

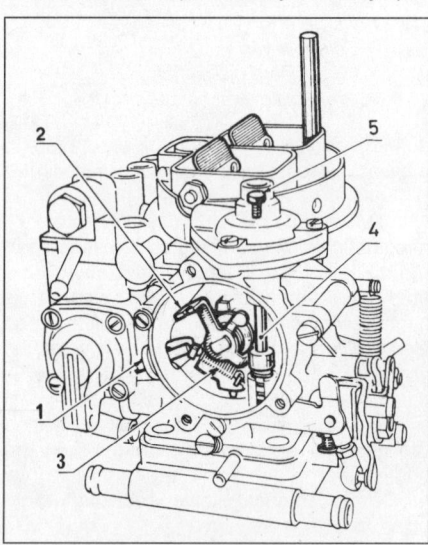

11.18a Automatic choke setting adjustment details

1. Initial throttle opening screw
2. Choke operating lever
3. Cam
4. Pushrod
5. Adjustment screw

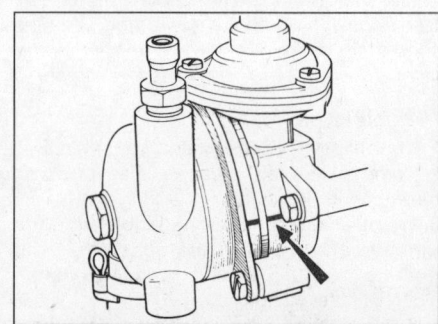

11.18b On refitting ensure the choke housing mark is aligned with the carburettor mark (arrowed)

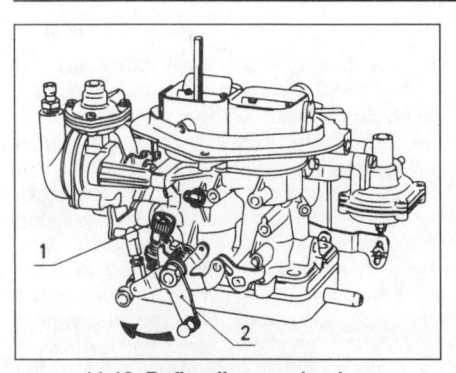

Fuel and exhaust systems - carburettor engines 4A•9

11.19 Deflooding mechanism adjustment points

1 Adjusting screw   2 Throttle lever

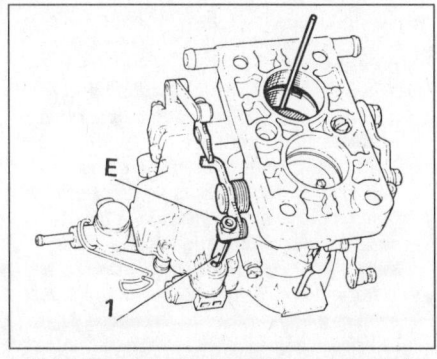

11.22 Defuming valve adjustment points

1 Valve lever   E Nut

3 Slacken the retaining clips, then disconnect the vacuum servo unit hose, breather and coolant hose(s) from the manifold.
4 Unclip the fuel filter and position it clear of the manifold.
5 Make a final check that all the necessary vacuum/breather hoses have been disconnected from the manifold.
6 Unscrew the manifold retaining nuts and washers and manoeuvre the manifold away from the head and out of the engine compartment. Recover the manifold gaskets from the studs and discard them **(see illustration)**.

Refitting

7 Refitting is the reverse of the removal procedure, noting the following points:
  a) Ensure that the manifold and cylinder head mating surfaces are clean and dry, and fit the new gaskets to the stud. Install the manifold, and securely tighten its retaining nuts.
  b) Ensure that all relevant hoses are reconnected to their original positions, and are securely held (where necessary) by their retaining clips.
  c) Refit the carburettor as described in Section 12.
  d) On completion, refill the cooling system as described in Chapter 1.

the throttle lever. Using a twist drill check that the gap between the choke flaps and barrel is as given in the *Specifications*. If not, turn the adjusting screw as necessary **(see illustration)**. After making an adjustment check the initial throttle opening, as described in paragraphs 16 and 17.
20 To adjust the float level proceed as described in paragraph 10.

Defuming valve

21 Some carburettors are fitted with a defuming valve which vents the float chamber to atmosphere when idling.
22 To adjust, hold the choke flap open and depress the defuming valve lever. Using a twist drill, measure the throttle flap opening. If it is not as specified (see *Specifications*) turn the nut **(see illustration)**.

12 Inlet manifold - removal and refitting

Note: *Refer to the warning note in Section 1 before proceeding.*

Removal

1 Remove the air cleaner housing and carburettor as described in Sections 2 and 10.
2 Drain the cooling system as described in Chapter 1.

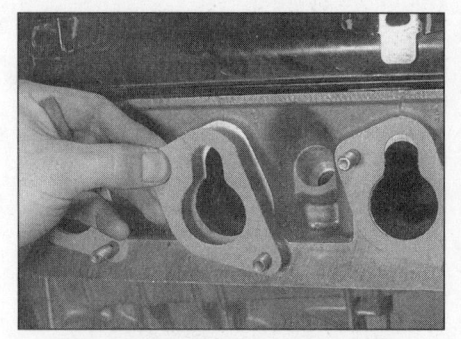

12.6 Remove the gaskets from the inlet manifold studs

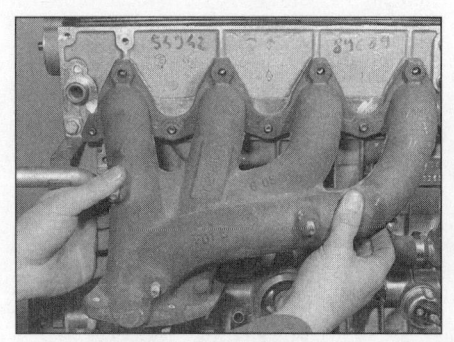

13.3b ... then remove the exhaust manifold ...

13 Exhaust manifold - removal and refitting

Removal

1 Disconnect the hot-air inlet hose from the manifold shroud, and remove it from the vehicle.
2 To improve access, firmly apply the handbrake, then jack up the front of the vehicle and support it on axle stands (see *"Jacking and vehicle support"*).
3 Undo the nuts securing the exhaust front pipe to the manifold, then remove the nut/bolt securing the front pipe to its mounting bracket. Disconnect the front pipe from the manifold, and recover the gasket **(see illustrations)**.
4 Undo the manifold retaining nuts and washers and remove the manifold from the cylinder head. Recover the gasket and discard it.

Refitting

5 Refitting is the reverse of the removal procedure, noting the following points:
  a) Examine all the exhaust manifold studs for signs of damage and corrosion; remove all traces of corrosion, and repair or renew any damaged studs.
  b) Ensure that the manifold and cylinder head sealing faces are clean and flat, and fit the new manifold gasket. Fit the manifold tightening it nuts securely.

13.3a Unscrew the retaining nuts ...

13.3c ... and recover the manifold gasket

4A

4A•10 Fuel and exhaust systems - carburettor engines

c) Reconnect the front pipe to the manifold using the information given in Section 14.

14 Exhaust system - general information, removal and refitting

General information

1 The exhaust system consists of four sections; the front pipe, the front silencer box, the intermediate pipe, and the tailpipe and main silencer box **(see illustration)**. All exhaust sections are joined by flanged joints. The front pipe joints are secured by nuts and bolts, the front silencer joint being of the spring-loaded ball type, to allow for movement in the exhaust system, and the other joints are secured by clamping rings.
2 The system is suspended throughout its entire length by rubber mountings.

Removal

3 Each exhaust section can be removed individually, or alternatively, the complete system can be removed as a unit. Even if only one part of the system needs attention, it is often easier to remove the whole system and separate the sections on the bench.
4 To remove the system or part of the system, first jack up the front or rear of the car, and support it on axle stands (see "*Jacking and vehicle support*"). Alternatively, position the car over an inspection pit, or on car ramps.

Front pipe

5 Undo the nuts securing the front pipe flange joint to the manifold, and the nut/bolt securing the front pipe to its mounting. Separate the flange joint, and collect the gasket.

6 Slacken and remove the two nuts securing the front pipe flange joint to the intermediate pipe, and recover the washers and springs. Remove the bolts, then recover the spacers and withdraw the front pipe from underneath the vehicle. Remove the gasket from the end of the front pipe.

Front silencer

7 Undo the two nuts securing the front pipe flange joint to the intermediate pipe. Recover the washers and springs then withdraw the bolts noting the spacers fitted between the flanges.
8 Slacken the intermediate pipe clamping ring nut/bolt, and disengage the clamp from the flange joint.
9 Unhook the silencer from its mounting rubbers then free it from the intermediate pipe and withdraw it from underneath the vehicle. Recover the gasket from the front pipe joint.

Intermediate pipe

10 Slacken the clamping ring nuts/bolts and disengage the clamps from the front and rear flange joints.
11 Disengage the pipe from the front silencer and tailpipe and remove it from underneath the vehicle.

Tailpipe

12 Slacken the tailpipe clamping ring nuts/bolts, and disengage the clamp from the flange joint.
13 Unscrew the tailpipe mounting nuts/bolts and remove it from underneath the vehicle.

Complete system

14 Undo the nuts securing the front pipe flange joint to the manifold, and the single bolt securing the front pipe to its mounting bracket. Separate the flange joint, and collect the gasket. Free the system from all its mounting rubbers, and lower it from under the vehicle.

Heat shield(s)

15 The heat shields are secured to the underside of the body by various nuts and bolts. Each shield can be removed once the relevant exhaust section has been removed. If a shield is being removed to gain access to a component located behind it, it may prove sufficient in some cases to remove the retaining nuts and/or bolts, and simply lower the shield, without disturbing the exhaust system.

Refitting

16 Each section is refitted by reversing the removal sequence, noting the following points:
a) Ensure that all traces of corrosion have been removed from the flanges, and renew all necessary gaskets.
b) Inspect the rubber mountings for signs of damage or deterioration, and renew as necessary.
c) When reassembling the front pipe to front silencer joint, ensure that the spacers are correctly fitted between the flanges. Refit the bolts, springs and washers and securely tighten the nuts so that both flanges are in firm contact with the spacers **(see illustration)**.
d) Where joints are secured together by a clamping ring, apply a smear of exhaust system jointing paste to the flange joint, to ensure a gas-tight seal. Where two-piece clamping rings are fitted, tighten the clamping ring nuts/bolts evenly and progressively so that the clearance between the clamp halves remains equal on either side.
e) Prior to tightening the exhaust system fasteners, ensure that all rubber mountings are correctly located, and that there is adequate clearance between the exhaust system and vehicle underbody.

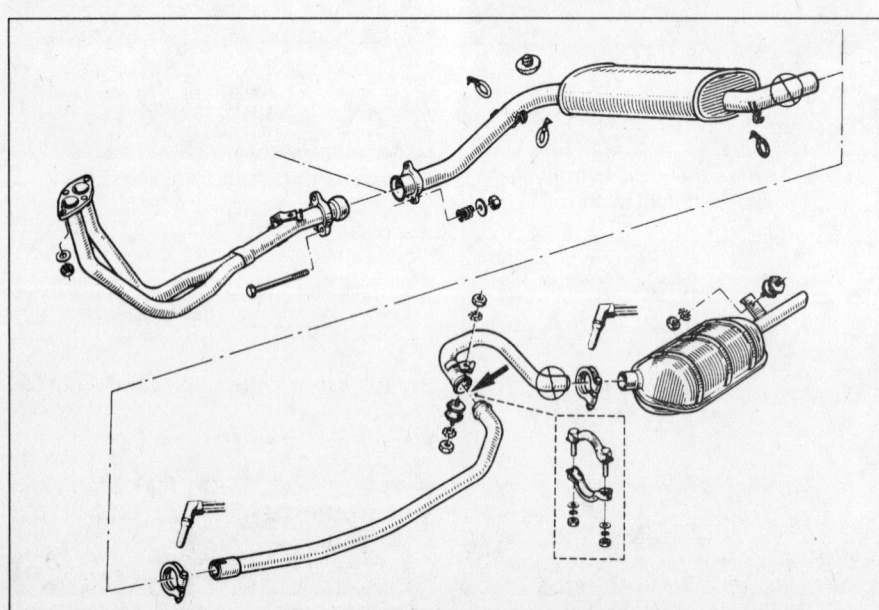

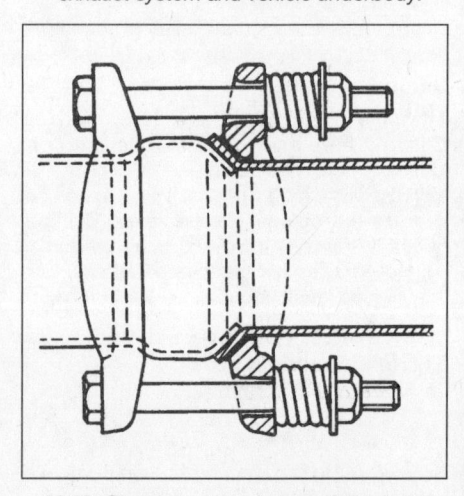

14.16 Cross-sectional view of the front pipe spring-loaded joint

14.1 Exhaust system components

Chapter 4 Part B:
Fuel and exhaust systems - fuel injection models

Contents

Accelerator cable - removal, refitting and adjustment 3	Fuel injection system - depressurisation 7
Accelerator pedal - removal and refitting 4	Fuel injection system - testing and adjustment 11
Air cleaner assembly and inlet ducts - removal and refitting 2	Fuel injection systems - general information 6
Air cleaner filter element renewal . . . . See Chapter 1	Fuel pump - removal and refitting 8
Bosch K-Jetronic injection system components -	Fuel tank - removal and refitting 10
removal and refitting 12	General fuel system checks . . . . See Chapter 1
Catalytic converter - general information and precautions 21	General information and precautions 1
Emission control systems - general information 19	Idle speed and mixture adjustment . . . . See Chapter 1
Emission control systems - testing and component renewal 20	Inlet manifold - removal and refitting 14
Exhaust manifold - removal and refitting 15	Intercooler (2458 cc models) - removal and refitting 18
Exhaust system - general information, removal and refitting 16	Renix fuel injection system components - removal and refitting 13
Exhaust system check . . . . See Chapter 1	Turbocharger (2458 cc models) -
Fuel filter - renewal . . . . See Chapter 1	general information, removal and refitting 17
Fuel gauge sender unit - removal and refitting 9	Unleaded petrol - general information and usage 5

Degrees of difficulty

Easy, suitable for novice with little experience	Fairly easy, suitable for beginner with some experience	Fairly difficult, suitable for competent DIY mechanic	Difficult, suitable for experienced DIY mechanic	Very difficult, suitable for expert DIY or professional

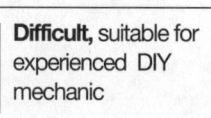

Specifications

System type
2664 cc models . . . . Bosch K-Jetronic
All other models . . . . Renix

Fuel pump
Type . . . . Electric
Regulated constant pressure (at specified idle speed):
   Bosch system . . . . 3.6 to 4.0 bar
   Renix system . . . . 2.5 to 0.2 bar

Idle speed and mixture
1995 cc and 2165 cc models . . . . 800 ± 50 rpm
2458 cc turbo models . . . . 750 ± 50 rpm
2664 cc models . . . . 900 ± 50 rpm
2849 cc models . . . . 700 ± 50 rpm
Idle mixture CO content:
   Models fitted with a catalytic converter . . . . Less than 0.5 %
   Models not fitted with a catalytic converter . . . . 1.5 ± 0.5 %

Recommended fuel
Minimum octane rating:
   Models fitted with a catalytic converter . . . . 95 RON unleaded (UK unleaded premium). Leaded fuel must **not** be used
   Models not fitted with a catalytic converter:
      Models produced prior to January 1989* . . . . 98 RON unleaded (UK "super unleaded")
      or 98 RON leaded (UK "4-star")
      Models produced since January 1989 . . . . 95 RON unleaded (UK "unleaded premium")
      or 98 RON leaded (UK "4-star")
*Some models produced before January 1989 can be run on 95 RON petrol. Refer to your Renault dealer for further information

Torque wrench setting   Nm   lbf ft
Exhaust manifold clamps - 2458 cc turbo models 25   18

4B•2 Fuel and exhaust systems - fuel injection models

1 General information and precautions

The fuel system consists of a fuel tank mounted under the rear of the car with an electric fuel pump immersed in it, a fuel filter, and the fuel feed and return lines. The fuel pump supplies fuel to the fuel distributor (2664 cc models) or fuel rail (all other models), and it is then fed to the injectors which inject the fuel into the inlet tracts near the inlet valves. A fuel filter is incorporated in the feed line from the pump to the engine compartment, to ensure that the fuel supplied to the injectors is clean.

Refer to Section 6 for further information on the operation of the relevant fuel injection system, and to Section 22 for information on the exhaust system.

Warning: *Many of the procedures in this Chapter require the removal of fuel lines and connections, which may result in some fuel spillage. Before carrying out any operation on the fuel system, refer to the precautions given in 'Safety first!' at the beginning of this manual, and follow them implicitly. Petrol is a highly-dangerous and volatile liquid, and the precautions necessary when handling it cannot be overstressed.*

Note: *Residual pressure will remain in the fuel lines long after the vehicle was last used. When disconnecting any fuel line, first depressurise the fuel system as described in Section 7.*

2 Air cleaner assembly and inlet ducts - removal and refitting

Removal

1995 cc 8-valve models

1 Release the rubber retaining strap and disconnect the duct from the base of the air cleaner housing. If necessary, unclip the duct from its inlet scoop and remove it from the engine compartment.
2 Disconnect the inlet duct from the side of the housing then unhook the air cleaner housing assembly from its mounting bracket and remove it from the engine compartment.
3 To remove the inlet duct, disconnect the wiring connector from the inlet air temperature sensor and the idle valve hose from the duct. Undo the screws securing the duct to the throttle housing then remove the duct assembly and recover the sealing ring.

1995 cc 12-valve models

4 Disconnect the ducts from the top and bottom of the air cleaner housing. Also disconnect the idle valve hose from the side of the housing. If necessary, unclip the lower duct from its inlet and remove it from the engine compartment.
5 Unhook the air cleaner housing from its mounting bracket.
6 To remove the inlet duct, disconnect the breather hose then undo the screws securing the duct to the throttle housing. Remove the duct and recover the sealing ring.

2165 cc models

7 On models where the throttle housing mounted onto the top of the inlet manifold refer to paragraphs 1 to 3.
8 On models where the throttle housing is mounted on the underside of the manifold, disconnect the ducts from the housing then release the rubber retaining strap and remove the housing from the engine compartment. If necessary, disconnect the inlet air temperature wiring connector and free the idle valve hose from the duct then release the retaining clip and remove the duct.

2458 cc turbo models

9 Slacken the retaining clip and remove the air inlet duct from the air cleaner housing. Where necessary also disconnect the evaporative emission system hose from the housing.
10 Unclip the air cleaner to turbocharger duct from the housing then unhook the rubber retaining straps and remove the air cleaner housing from the engine compartment.
11 The various ducts and hoses can be removed individually. If the throttle housing inlet duct is removed, recover the sealing ring.

2664 cc models

12 On early (pre 1986) models the air cleaner is mounted directly onto the top of the airflow meter. To remove it, slacken and remove the retaining bolts and lift the air cleaner housing assembly out of the engine compartment. Recover the sealing ring from the top of the airflow meter.
13 On later (1986 on) models the air cleaner is mounted on the left-hand side of the engine compartment. To remove the housing first unscrew the retaining bolts and remove the inlet duct and sealing ring from the top of the airflow meter. Slacken the retaining clip (where fitted) and disconnect the duct from the front of the air cleaner housing then unhook and remove the housing from the engine compartment.

2849 cc models

14 Slacken the retaining clip (where fitted) and disconnect the ducts from the housing. Unhook the retaining straps and remove the housing from the engine compartment. To remove the inlet duct disconnect the idle valve hose then undo the retaining bolts and remove the duct and sealing ring from the throttle housing.

Refitting

15 Refitting is a reversal of the removal procedure, making sure the duct/hoses are securely reconnected. Do not forget the sealing ring which is fitted to the throttle housing/airflow meter (as appropriate).

3 Accelerator cable - removal, refitting and adjustment

Removal and refitting

1 Refer to Chapter 4A, Section 6.

Adjustment

2664 cc models

2 Have an assistant depress the accelerator pedal, and check that the throttle cam opens fully and returns smoothly to its stop. With the accelerator pedal fully depressed, the throttle valve should be fully open and the compensator spring should be compressed by approximately 2 mm. Remove the spring clip from the accelerator outer cable and adjust its position as necessary.
3 If necessary the throttle valve linkage can be adjusted as follows although this should only be necessary if the linkage has been disturbed. **Note:** *Renault tool Mot. 843-08 will be needed for this operation.*
4 To adjust the linkage, insert the special tool (Mot. 843-08) between the throttle stop and the safety stop. Adjust the length of the link rod to obtain a clearance of 0.1 mm between the throttle stop screw and the stop. **Note:** *Do not disturb the throttle stop screw.* Tighten the link rod locknut on completion and recheck the accelerator cable adjustment **(see illustrations)**.

All other models

5 Remove the spring clip from the accelerator outer cable **(see illustration)**. Ensuring that the throttle cam is fully against its stop, gently pull the cable out of its grommet until all free play is removed from the inner cable.
6 With the cable held in this position, refit the spring clip to the last exposed outer cable groove in front of the rubber grommet. When the clip is refitted and the outer cable is released, there should be only a small amount of free play in the inner cable.
7 Have an assistant depress the accelerator pedal, and check that the throttle cam opens fully and returns smoothly to its stop.

4 Accelerator pedal - removal and refitting

Refer to Chapter 4A, Section 7.

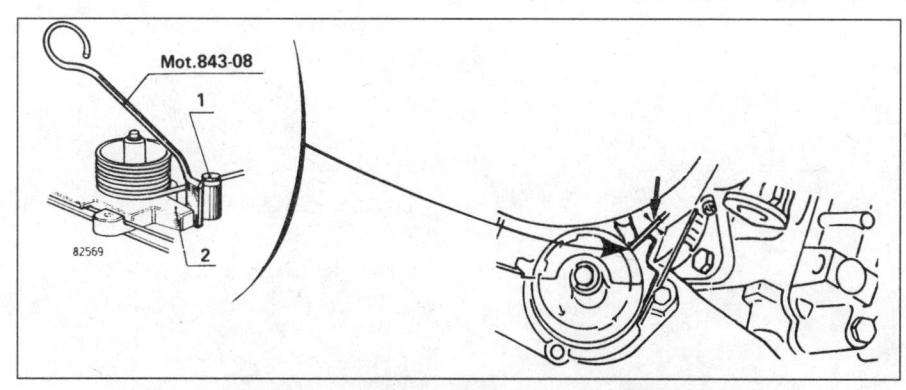

Fuel and exhaust systems - fuel injection models 4B•3

3.4a On 2664 cc models, insert the special tool (Mot.843-08) in-between the safety stop (1) and throttle stop (2) ...

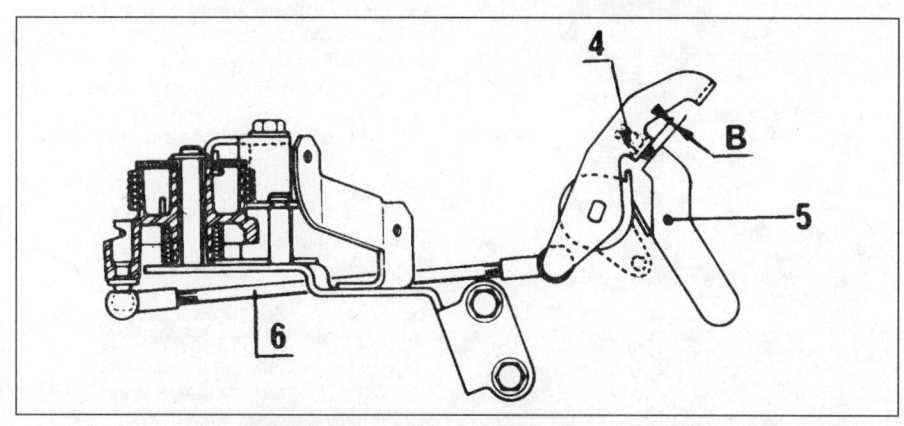

3.4b ... then adjust the length of the link rod (6) to obtain a clearance (B) of 0.1 mm between the throttle stop screw (4) and the stop (5)

5 Unleaded petrol - general information and usage

Note: *The information given in this Chapter is correct at the time of writing. If updated information is thought to be required, check with a Renault dealer. If travelling abroad, consult one of the motoring organisations (or a similar authority) for advice on the fuel available.*

1 The fuel recommended by Renault is given in the *Specifications* Section of this Chapter, followed by the equivalent petrol currently on sale in the UK.

3.5 Accelerator cable spring clip (2165 cc model shown)

2 Most models produced prior to January 1989 are designed to run on 98 octane petrol and all models produced after January 1989 are designed to run on 95 octane petrol. Some pre-1989 models can also be run on 95 octane petrol, refer to your Renault dealer for further information.

3 On models with a catalytic converter unleaded fuel **must** be used; under no circumstances should leaded (UK "4-star") fuel be used as this will damage the catalytic converter. On models not equipped with a catalytic converter, both leaded and unleaded fuels may be used.

6 Fuel injection systems - general information

Bosch K-Jetronic system - 2664 cc models

1 The Bosch K-Jetronic fuel injection system is fitted to all 2664 cc models. It is often referred to as a 'mechanical' injection system as it does not make use of an electronic control unit, although simple electrics are used.

2 The components of the system are as follows **(see illustrations)**:

a) Fuel pumps - both a lift pump and main fuel pump are fitted
b) Fuel accumulator.
c) Fuel filter.
d) Fuel metering/distributor unit - incorporates the fuel pressure regulator.
e) Airflow meter.
f) Cold start injector.
g) Thermostatic time switch.
h) Control pressure regulator.
i) Injectors.
j) Engine speed relay.
k) Auxiliary air valve - early (pre 1986) models.
l) Idle speed valve and electronic control unit - later (1986-on) models.

3 Fuel is supplied from the fuel tank by a low-pressure lift pump to the main fuel pump which is situated underneath the vehicle. The main fuel pump supplies fuel to the pressure accumulator.

4 The fuel accumulators purpose is to maintain pressure in the fuel lines when the engine is switched off. It includes a spring tensioned diaphragm which is displaced when the fuel pump is working. When the engine is switched off, the injectors close and a non-return valve in the fuel pump prevents loss of pressure into the tank. It also prevents fuel vaporising in the fuel lines and delays system pressure rise at cold starting which would result in too much fuel being injected into the engine cylinders.

5 The fuel flows through the fuel filter, which is located in the engine compartment, and onto the fuel metering/distributor housing.

6 The fuel enters the metering/distributor housing via the fuel pressure regulator. The pressure regulator keeps the supply of fuel to the control piston at a constant pressure, returning excess fuel to the tank. the fuel control piston regulates the amount of fuel supplied to each of the injectors via the differential valves (one per cylinder). The movement of the control piston is governed by the airflow meter which rises or falls according to the quantity of air entering the engine. The airflow meter is linked to the control piston by a lever ensuring that the correct amount of fuel is supplied at all times.

7 Additional control of the metering/distributor housing is by means of the control pressure regulator. The control pressure regulator contains a bi-metallic strip and vacuum diaphragm and richens the fuel/air mixture when the engine is cold or operating at full load; the bi-metallic strip controls the "cold" operation of the valve, and the vacuum diaphragm the "full load" operation.

8 For cold starting, additional enrichment is provided by an independent cold start injector. The cold start injector is controlled by the thermostatic time switch which is screwed into the cylinder head.

9 On early (pre-1986) models an auxiliary air valve is fitted to increase the air supply to the engine before it reaches normal operating temperature.

4B

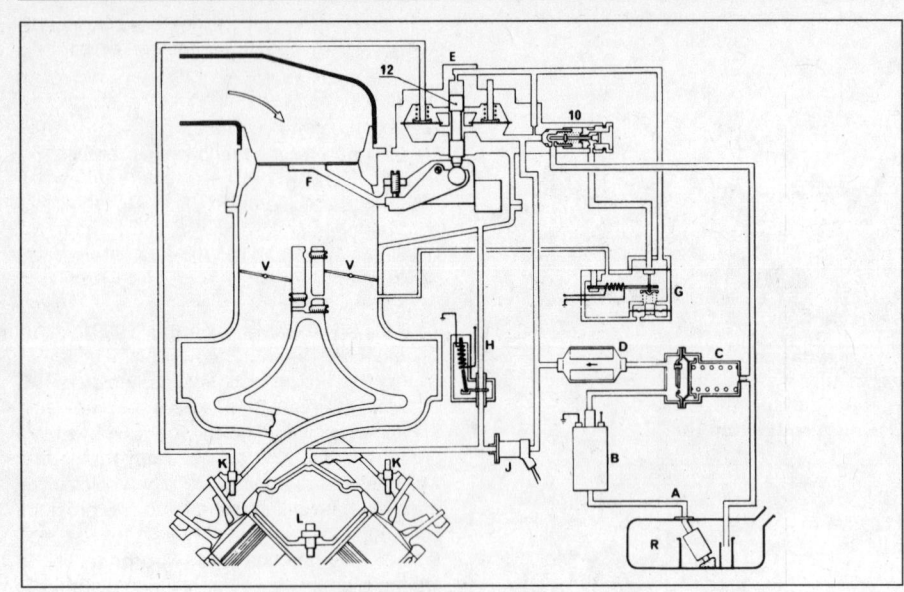

4B•4 Fuel and exhaust systems - fuel injection models

6.2a Bosch K-Jetronic fuel injection system component layout - 2664 cc models (early model shown)

A Fuel lift pump
B Main fuel pump
C Accumulator
D Fuel filter
E Fuel metering/distributor unit
F Airflow meter
G Control pressure regulator
H Auxiliary air valve
J Cold start injector
K Fuel injectors
L Thermostatic time switch
R Fuel tank
V Throttle valves
10 Fuel pressure regulator
12 Fuel metering piston

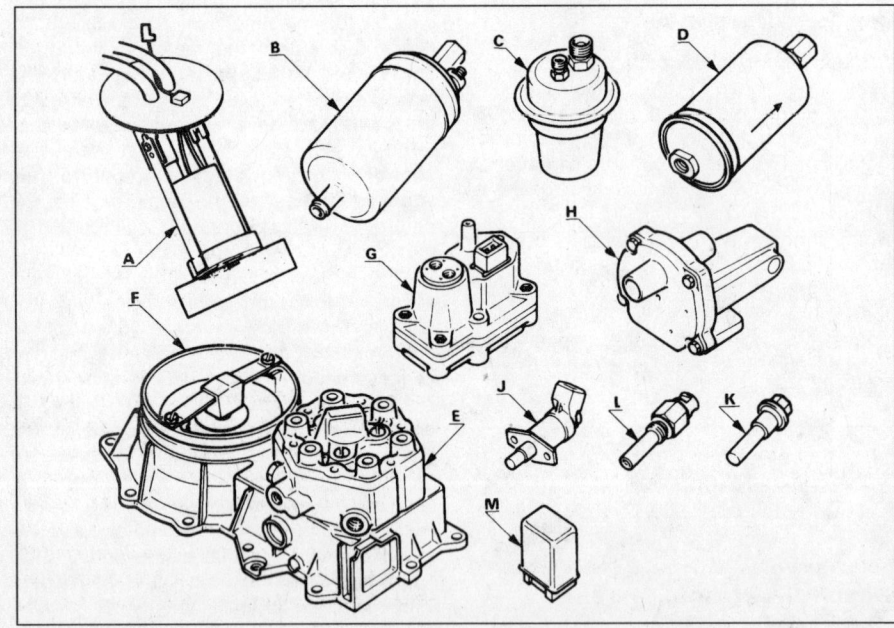

6.2b Main components of the Bosch K-Jetronic fuel injection system - 2664 cc models

A Fuel lift pump
B Main fuel pump
C Accumulator
D Fuel filter
E Fuel metering/distributor unit
F Airflow meter
G Control pressure regulator
H Auxiliary air valve - early (pre-1986) models
J Cold start injector
K Fuel injectors
L Thermostatic time switch
M Engine speed relay

10 On later (1986-on) models an idle speed control valve and electronic control unit are fitted. The electronic control unit receives signals from an additional coolant temperature sensor and a microswitch which is fitted to the throttle linkage. The control unit uses the idle speed control valve to select the appropriate idle speed for the relevant engine temperature.

Renix injection system - all other models

11 The Renix fuel injection system is fitted to all models except the 2664 cc engine model **(see illustration)**.
12 The fuel pump supplies fuel from the tank to the fuel rail, via a filter. Fuel supply pressure is controlled by the pressure regulator in the fuel rail. When the optimum operating pressure of the fuel system is exceeded, the regulator allows excess fuel to return to the tank.
13 The electrical control system consists of the ECU, along with the following sensors:
a) *Throttle switch/potentiometer (depending on model) - informs the ECU of the throttle position and, (models with a potentiometer) the rate of throttle opening/closing.*
b) *Coolant temperature sensor - informs the ECU of engine temperature.*
c) *Inlet air temperature sensor - informs the ECU of the temperature of the air passing through the throttle housing.*
d) *Crankshaft sensor - informs the ECU of the crankshaft position and speed of rotation.*
e) *Manifold Absolute Pressure (MAP) sensor - informs the ECU of the load on the engine (expressed in terms of inlet manifold vacuum).*

14 All the above signals are analysed by the ECU, and it selects the fuelling response appropriate to those values. The ECU controls the fuel injectors (varying the pulse width - the length of time the injectors are held open - to provide a richer or weaker mixture, as appropriate). The mixture is constantly varied by the ECU, to provide the best setting for cranking, starting (with either a hot or cold engine), warm-up, idle, cruising, and acceleration.
15 On all models except early 2165 cc models, the ECU also has full control over the engine idle speed, via an idle speed control valve which bypasses the throttle valve. When the throttle valve is closed, the ECU controls the opening of the valve, which in turn regulates the amount of air entering the manifold, and so controls the idle speed.
16 On models with a catalytic converter, the ECU also controls the exhaust and evaporative emission control systems, which are described in detail later in this Chapter.
17 If there is an abnormality in any of the readings obtained the ECU enters its back-up mode. In this event, it ignores the abnormal sensor signal, and assumes a pre-

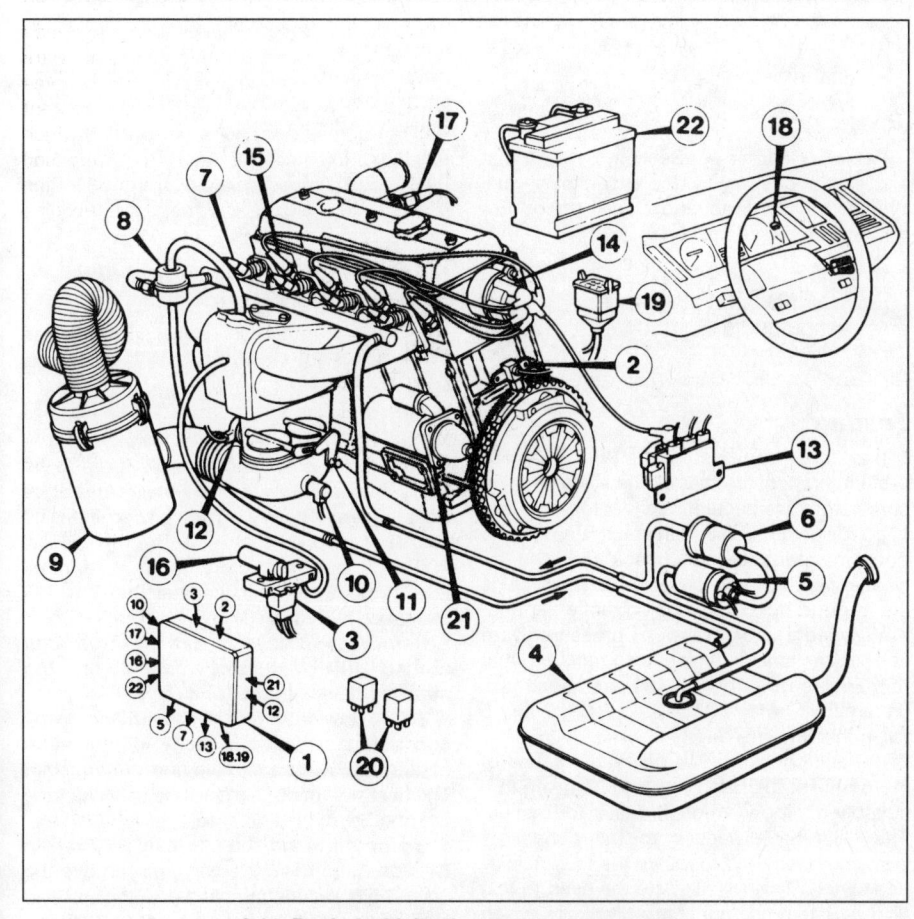

Fuel and exhaust systems - fuel injection models 4B•5

6.11 Renix fuel injection system components (early 2165 cc models shown - others similar)

1 Electronic control unit (ECU)
2 Crankshaft sensor
3 Manifold absolute pressure sensor
4 Fuel tank
5 Fuel pump
6 Fuel filter
7 Fuel injectors
8 Fuel pressure regulator
9 Air cleaner housing
10 Intake air temperature sensor
11 Throttle housing
12 Throttle switch
13 Ignition control unit and HT coil
14 Distributor
15 Spark plugs
16 Idle mixture (CO) potentiometer
17 Coolant temperature sensor
18 Instrument panel warning light
19 Diagnostic wiring connector
20 Relay
21 Starter motor
22 Battery

programmed value which will allow the engine to continue running (albeit at reduced efficiency). If the ECU enters this back-up mode, the warning light on the instrument panel will come on, and the relevant fault code will be stored in the ECU memory.

18 If the warning light comes on, the vehicle should be taken to a Renault dealer at the earliest opportunity. A complete test of the engine management system can then be carried out, using a special electronic diagnostic test unit which is simply plugged into the system's diagnostic connector.

7 Fuel system - depressurisation

Note: Refer to the warning note in Section 1 before proceeding.

⚠ **Warning: The following procedure will merely relieve the pressure in the fuel system - remember that fuel will still be present in the system components and take precautions before disconnecting any of them.**

1 The fuel system referred to in this Section is defined as the fuel pump, the fuel filter, the fuel injectors, and the accumulator and metering/distributor housing (2664 cc models), the fuel rail and the pressure regulator (all other models), and the metal pipes and flexible hoses of the fuel lines between these components. All these contain fuel which will be under pressure while the engine is running, and/or while the ignition is switched on. The pressure will remain for some time after the ignition has been switched off, and it must be relieved in a controlled fashion when any of these components are disturbed for servicing work.

2 Disconnect the battery negative terminal.

3 Place a container beneath the connection/union to be disconnected, and have a large rag ready to soak up any escaping fuel not being caught by the container.

4 Slowly loosen the connection or union nut to avoid a sudden release of pressure, and position the rag around the connection, to catch any fuel spray which may be expelled. Once the pressure is released, disconnect the fuel line. Plug the pipe ends, to minimise fuel loss and prevent the entry of dirt into the fuel system.

8 Fuel pump - removal and refitting

Note: Refer to the warning note in Section 1 before proceeding.

Removal

1 Chock the front wheels then jack up the rear of the vehicle and support it on axle stands (see "Jacking and vehicle support"). The fuel pump is mounted onto a plate on the right-hand side of the fuel tank. To minimise fuel loss, clamp the hoses on each side of the pump.

2 Bearing in mind the information given in Section 7, carefully slacken the retaining clip and disconnect the supply hose from the pump **(see illustration)**.

3 Release the feed hose union from the pump. Where the hose is secured in position with a union nut and washers unscrew the nut then disconnect the hose and recover the sealing washers; new sealing washers should be used on refitting. Where the hose end is threaded slacken the hose union; the filter can be unscrewed from the hose once its mounting clamp is released **(see illustration)**.

4 Remove the rubber insulating covers from the fuel pump terminals then unscrew the retaining nuts and disconnect the pump wiring.

5 Unscrew the fuel pump mounting clamp bolt then remove the pump from underneath the vehicle and recover the rubber mounting. On certain models it will be necessary to unbolt the pump mounting plate from the vehicle body to gain access to the mounting clamp bolt(s) which come through from the

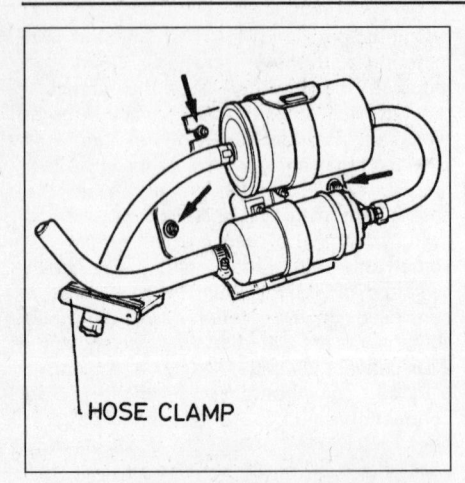

4B•6 Fuel and exhaust systems - fuel injection models

8.2 Fuel pump and filter arrangement (2165 cc engine shown) - mounting plate fasteners arrowed

8.3 Fuel pump location - 2664 cc models

rear of the plate.

Refitting

6 Refitting is the reverse of removal, noting the following points.
  a) Make sure the pump rubber mounting is correctly positioned and securely tighten the mounting clamp bolt(s).
  b) Ensure that the wiring retaining nuts are securely tightened and slide the rubber insulating covers into position.
  c) Where the feed hose is secured in position with a union nut, position a new sealing washer on each side of the hose union and securely tighten the union nut.
  d) Where the feed hose end fitting is threaded make sure it is securely tightened.
  e) On completion, start the engine and check for fuel leaks before using the vehicle on the road.

9 Fuel gauge sender unit - removal and refitting

Refer to Chapter 4A, Section 4.

10 Fuel tank - removal and refitting

Refer to Chapter 4A, Section 5, noting that it will be necessary to depressurise the fuel system as the feed and return hoses are disconnected (see Section 7).

11 Fuel injection system - testing and adjustment

Testing

1 If a fault appears in the fuel injection system, first ensure that all the system wiring connectors are securely connected and free of corrosion. Ensure that the fault is not due to poor maintenance; ie, check that the air cleaner filter element is clean, the spark plugs are in good condition and correctly gapped, the cylinder compression pressures are correct, the ignition timing is correct, and that the engine breather hoses are clear and undamaged, referring to Chapters 1, 2 and 5 for further information.

2 If these checks fail to reveal the cause of the problem, the vehicle should be taken to a suitably-equipped Renault dealer for testing. They will have access to the necessary equipment required to locate the fault quickly and simply. This will alleviate the need to test all the system components individually, which is a time-consuming operation.

Adjustment

2664 cc models

3 On 2664 cc models with Bosch K-Jetronic injection system, the mixture adjustment (exhaust gas CO level) is adjustable on all models.

4 On early (pre 1986) models the idle speed is also adjustable. Refer to Chapter 1 for details.

5 On later (1986 on) models the idle speed is automatically controlled by the idle speed control valve and control unit and cannot be adjusted. If idle speed is incorrect, there must be a fault present.

All other models

6 On all models except early 2165 cc models, the idle speed is automatically controlled by the idle speed control valve and control unit and cannot be adjusted. If idle speed is incorrect, there must be a fault present. On early 2165 cc models the idle speed can be adjusted as described in Chapter 1.

7 On models not equipped with a catalytic converter, the mixture adjustment (exhaust gas CO level) can also be adjusted as described in Chapter 1.

8 On models with a catalytic converter, the mixture adjustment is controlled by the ECU using the information supplied by the lambda sensor. No manual adjustment of the mixture (exhaust gas CO level) being possible. On these models if the exhaust gas CO level is incorrect then a fault is present in the injection system and the vehicle should be taken to a Renault dealer for testing.

12 Bosch K-Jetronic injection system components - removal and refitting

Removal

Fuel metering/distributor unit

Note: *Refer to the warning note in Section 1 before proceeding.*
Note: *Do not attempt to dismantle the metering/distributor unit; if it is thought to be faulty seek the advice of a Renault dealer or Bosch agent who will advise you on the best course of action.*

1 Disconnect the battery negative terminal and remove the air cleaner housing as described in Section 2.
2 Remove all traces of dirt from the area around the fuel pipe unions on the metering/distributor housing.
3 To ensure that the hoses are correctly connected on refitting, make identification marks between each hose and the housing. The best way to do this is to use masking tape wrapped around each hose.
4 Bearing in mind the information given in Section 7, carefully slacken and remove the union bolts securing the hoses to the unit and recover the sealing washers **(see illustration)**. Discard all washers, they should be renewed.
5 Slacken and remove the retaining screws and lift the metering/distributor unit away from the airflow meter, ensuring that the fuel control piston does not fall out (it should be retained by the tab) **(see illustration)**. Recover the sealing ring and discard; it should be renewed.
6 Refitting is the reverse of removal, noting the following.
  a) Fit a new sealing ring to the base of the unit and fit it to the airflow meter whilst ensuring that the piston remains in position. Securely tighten the retaining screws.

12.4 Slacken the union bolts and disconnect the fuel hoses from the metering/distributor unit

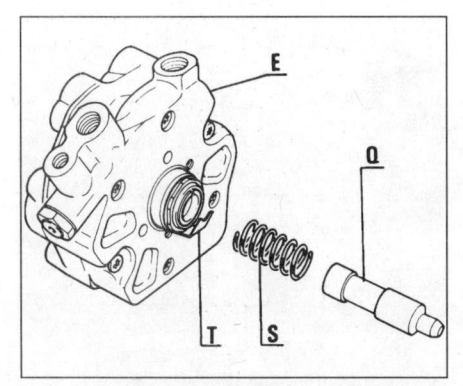

Fuel and exhaust systems - fuel injection models 4B•7

12.5 When the metering/distributor unit (E) is removed, take care not to lose the metering piston (Q) and spring (S). They should be securely held in position by the tab (T)

b) Position a new sealing washer on each side of every hose union and securely tighten the union bolts.
c) On completion, start the engine and check for signs of fuel leakage before using the vehicle on the road.

Airflow meter

Note: *Refer to the warning note in Section 1 before proceeding.*

7 Remove the fuel metering/distributor unit as described in paragraphs 1 to 5.
8 Slacken and remove the screws securing the airflow meter to the top of the throttle housing and remove the assembly from the engine. Recover the gasket.
9 Overhaul of the airflow meter is possible but should be entrusted to a Renault dealer or other suitable specialist.
10 Refitting is the reverse of removal, noting the following.
a) Ensure that the meter and throttle housing mating surfaces are clean and dry and fit a new gasket.
b) Refit the fuel metering/distributor unit (see paragraph 6).
c) On completion check and, if necessary, adjust the idle speed and mixture settings as described in Chapter 1.

Throttle housing

Note: *Refer to the warning note in Section 1 before proceeding.*

11 To remove the throttle housing complete with airflow meter and fuel metering/distributor unit, carry out the operations described in paragraphs 1 to 4. If it is wished to remove the housing separately, remove the airflow meter as described earlier.
12 Slacken the retaining clip and disconnect the hose which connects the throttle housing to the auxiliary air valve/idle speed control valve (as applicable).
13 Disconnect the throttle housing link rod from the throttle cam. To do this, pivot the retaining clip away from the link rod balljoint and slide it out of position. Where necessary, disconnect the wiring connector from the throttle valve switch.
14 Slacken and remove the bolts securing the double U-shaped section of the manifold to the front of the throttle housing and the two bolts securing the housing mounting brackets to the manifold **(see illustration)**.
15 Free the throttle housing from the manifold and remove it from the engine, disconnecting its lower vacuum hose(s) as they become accessible **(see illustration)**. Recover the locating dowels and sealing rings which are fitted between the housing and manifold. Discard the sealing rings, they should be renewed.
16 Refitting is the reverse of removal, noting the following.
a) Prior to fitting, fit the new sealing rings to the throttle housing and ensure that the locating dowels are correctly positioned. Apply a smear of engine oil to the rings to ease installation.
b) Refit the housing, not forgetting to connect the vacuum hose(s), and securely tighten the housing bolts.
c) Secure the link rod balljoint in position with the retaining clip, making sure it is clipped securely in position.
d) Refit the airflow meter and fuel metering/distributor unit referring to paragraphs 6 and 10.

Control pressure regulator

Note: *Refer to the warning note in Section 1 before proceeding.*

17 The control pressure regulator is located on the right-hand side of the engine **(see illustration)**.
18 Bearing in mind the information in Section 7, carefully slacken the union bolts and disconnect the fuel pipe unions from the regulator. Discard the sealing washers new ones should be used on refitting.

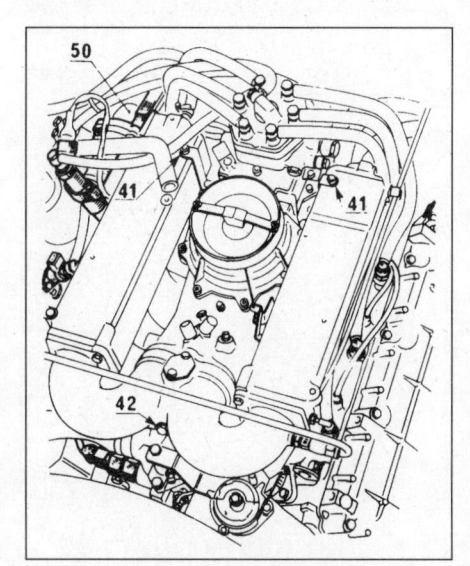

12.14 Throttle housing retaining bolts (41, 42 and 50)

12.15 Remove the throttle housing, disconnecting the vacuum hoses as they become accessible

12.17 Control pressure regulator

19 Disconnect the wiring connector then slacken the retaining clip and disconnect the vacuum hose.
20 Slacken and remove the regulator mounting bolts and remove it from the engine.
21 Refitting is the reverse of removal, using new sealing washers.

Injectors

Note: *Refer to the warning note in Section 1 before proceeding.*

22 Referring to Section 7, carefully unscrew the union bolt securing the injector pipe union to the fuel metering/distributor unit. Discard the sealing washers, they should be renewed.
23 Unclip the injector retaining clip then unscrew the fuel pipe union nut and remove the injector. Retain the injector with an open-ended spanner whilst the pipe union nut is unscrewed to prevent rotation **(see illustration)**.
24 Refitting is the reverse of removal using new sealing washers.

Fuel accumulator

Note: *Refer to the warning note in Section 1 before proceeding.*

25 The fuel accumulator is mounted underneath the rear of the vehicle, above the fuel pump.
26 Chock the front wheels then jack up the rear of the vehicle and support it on axle stands (see *"Jacking and vehicle support"*).
27 Bearing in mind the information given in Section 7, slacken the union nuts and disconnect the main fuel hoses from the accumulator. Slacken the retaining clip and

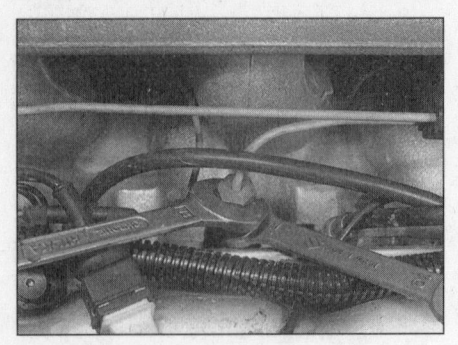

4B•8 Fuel and exhaust systems - fuel injection models

12.23 Retain the injector with an open-ended spanner whilst slackening the fuel pipe union nut

12.31 Cold start injector (early model shown)

12.39 Auxiliary air valve - early (pre-1986) models

disconnect the return hose from the accumulator.
28 Unscrew the mounting clamp bolt then remove the fuel accumulator from underneath the vehicle and recover the rubber mounting. Note that it may be necessary to unbolt the mounting plate from the vehicle body to gain access to the mounting clamp bolt(s) which come through from the rear of the plate.
29 Refitting is the reverse of removal. On completion start the engine and check for signs of fuel leakage before using the vehicle on the road.

Cold start injector
Note: *Refer to the warning note in Section 1 before proceeding.*
30 The cold start injector is situated on the right-hand side of the cylinder head.
31 Disconnect the wiring connector from the injector **(see illustration)**.
32 Bearing in mind the information in Section 7, slacken the union bolt securing the fuel pipe to the injector. Recover the sealing washers and discard them; new ones should be used on refitting.
33 Slacken and remove the injector retaining bolts and remove the injector along with its gasket/sealing ring.
34 Refitting is the reverse of removal using a new injector gasket/sealing ring (as applicable) and new fuel pipe union sealing rings.

Thermostatic time switch
35 The switch is screwed into the front of the cylinder head.
36 Disconnect the battery negative lead then either drain the cooling system to below the level of the switch (as described in Chapter 1), or have ready a suitable plug which can be used to plug the switch aperture whilst it is removed. If a plug is used, take great care not to damage the threads, and do not use anything which will allow foreign matter to enter the radiator.
37 Disconnect the wiring plug from the switch then carefully unscrew the switch and recover the sealing washer.
38 Refitting is a reversal of removal using a new sealing washer. On completion, refill the cooling system as described in Chapter 1 or top-up as described in "*Weekly checks*".

Auxiliary air valve - early (pre-1986) models
39 The auxiliary air valve is mounted on the right-hand side of the cylinder head **(see illustration)**.
40 Disconnect the wiring connector from the air valve.
41 Unscrew the upper retaining bolts which secure the vacuum piece to the top of the air valve bracket.
42 Slacken the retaining clips and disconnect the hoses valve then slide the valve out of position.
43 Refitting is the reverse of removal.

Idle speed control valve - later (1986 on) models
44 The idle speed control valve is mounted on the right-hand side of the cylinder head **(see illustration)**.
45 Disconnect the wiring connector from the idle valve then slacken the retaining clips and disconnect the hoses from the valve.
46 Slacken the retaining clamp bolt and slide the valve out of position.
47 Refitting is the reverse of removal.

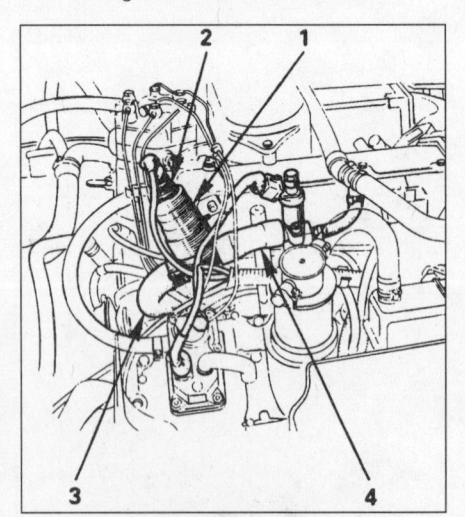

12.44 Idle speed control valve - later (1986 on) models

1 Valve
2 Wiring connector
3 Hose
4 Hose

Idle speed electronic control unit - later (1986 on) models
48 The control unit is located behind the protective cover on the left-hand side of the engine compartment **(see illustration)**.
49 Disconnect the battery negative terminal then release the retaining clip then undo the retaining screw and remove the protective cover from the control unit.
50 Undo the retaining screws then disconnect the wiring connectors and remove the control unit from the engine compartment.
51 Refitting is the reverse of removal.

Idle speed control valve temperature sensor - later (1986 on) models
52 The temperature sensor is screwed into the coolant elbow which is bolted to the top of the cylinder block.
53 To gain access to the switch remove the inlet manifold as described in Section 14.
54 Either drain the cooling system to below the level of the switch (as described in Chapter 1), or have ready a suitable plug

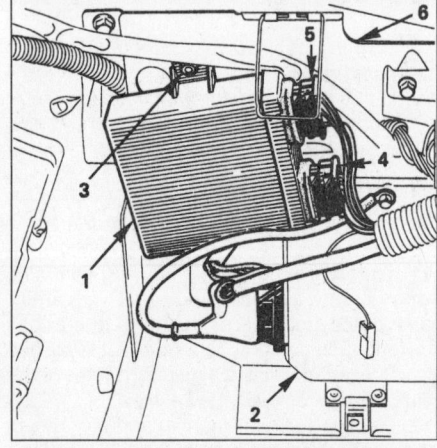

12.48 Idle speed electronic control unit - later (1986 on) models

1 Idle speed control unit
2 Ignition control unit
3 Retaining screw
4 Wiring connector
5 Wiring connector
6 Mounting plate

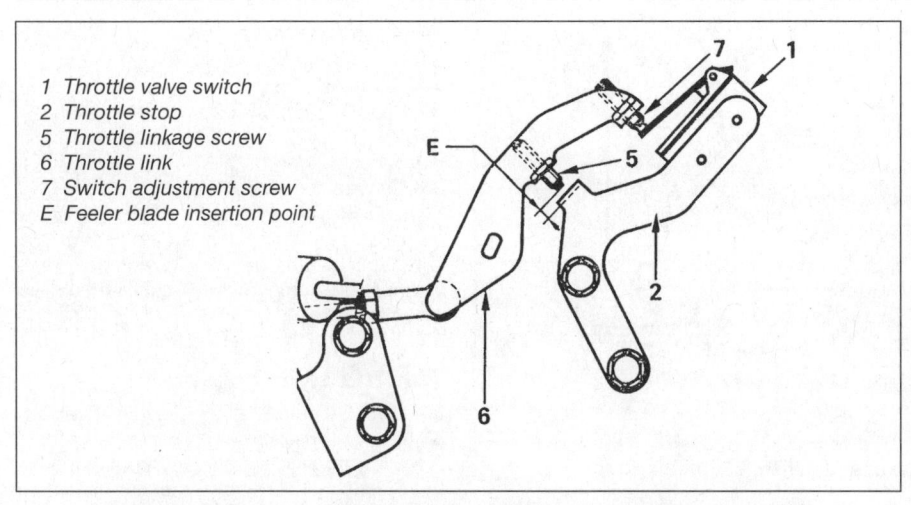

Fuel and exhaust systems - fuel injection models 4B•9

1 Throttle valve switch
2 Throttle stop
5 Throttle linkage screw
6 Throttle link
7 Switch adjustment screw
E Feeler blade insertion point

12.56 Throttle valve switch adjustment details - later (1986 on) models

which can be used to plug the switch aperture whilst it is removed. If a plug is used, take great care not to damage the threads, and do not use anything which will allow foreign matter to enter the radiator.

55 Disconnect the wiring connector then unscrew the sensor and remove it along with its sealing washer.

56 Refitting is a reversal of removal using a new sealing washer. On completion, refill the cooling system as described in Chapter 1 or top-up as described in "Weekly checks".

Throttle valve switch - later (1986 on) models

57 Disconnect the wiring connector from the throttle valve switch.
58 Undo the two retaining screws and remove the switch from the throttle housing.
59 On refitting, adjust the switch as follows (see illustration). Connect a multimeter set to the resistance scale across terminals 1 and 4 of the switch connector. Slide a 0.3 mm feeler blade into position between the linkage screw and switch mounting bracket and check that continuity exists between the terminal. Replace the feeler blade with one 0.6 mm in thickness and check that an open circuit is present. If adjustment is necessary, slacken the lock nut and adjust the switch screw (not the linkage screw) as necessary. Check the switch operation then securely tighten the adjustment screw locknut.

13 Renix fuel injection system components - removal and refitting

Throttle housing

1995 cc models

1 Disconnect the battery negative terminal.
2 Undo the retaining screws then disconnect the inlet duct from the throttle housing and recover the sealing ring.

3 On 12-valve models disconnect the accelerator cable as described in Section 3. On 8-valve models, unclip the throttle link rod from its balljoint on the throttle housing.
4 Disconnect the wiring connector from the throttle switch/potentiometer. Where necessary, also disconnect the wiring from the air temperature sensor.
5 Unscrew the retaining bolts securing the housing to the manifold and remove the housing. Recover the gasket and discard it; a new one should be used on refitting.
6 Refitting is the reverse of removal using a new gasket.

2165 cc models

7 There are two possible types of manifold and throttle housing arrangement the first is identical to that fitted to 1995 cc models and the second is very similar but is mounted on the underside of the manifold and secured in position with nuts instead of bolts. Refer to paragraphs 1 to 6 for removal and refitting details.

2458cc turbo and 2849 cc models

8 Refer to paragraphs 1 to 6.

Fuel rail and injectors

Note: Refer to the warning note in Section 1 before proceeding.

1995cc and 2165 cc models

9 Disconnect the battery negative terminal. Where necessary, to improve access remove the inlet duct as described in Section 2.
10 Bearing in mind the information given in Section 7, slacken the retaining clips and disconnect the fuel feed and return hoses from the fuel rail. Where the original crimped-type Renault hose clips are still fitted, cut them and discard; replace them with standard worm-type hose clips on refitting.
11 Disconnect the wiring connectors from the fuel injectors. On 1995 cc 12-valve models, also disconnect the vacuum pipe from the fuel pressure regulator.
12 Slacken and remove the fuel rail retaining bolts and nuts then carefully ease the fuel rail and injector assembly out from the inlet manifold and remove it from the vehicle. Remove the caps (where fitted) and sealing rings from the end of each injector and discard them; they must be renewed whenever they are disturbed (see illustrations).
13 Slide out the retaining clip(s) and remove the relevant injector(s) from the fuel rail. Remove the upper sealing ring from each disturbed injector and discard; all disturbed sealing rings must be renewed.

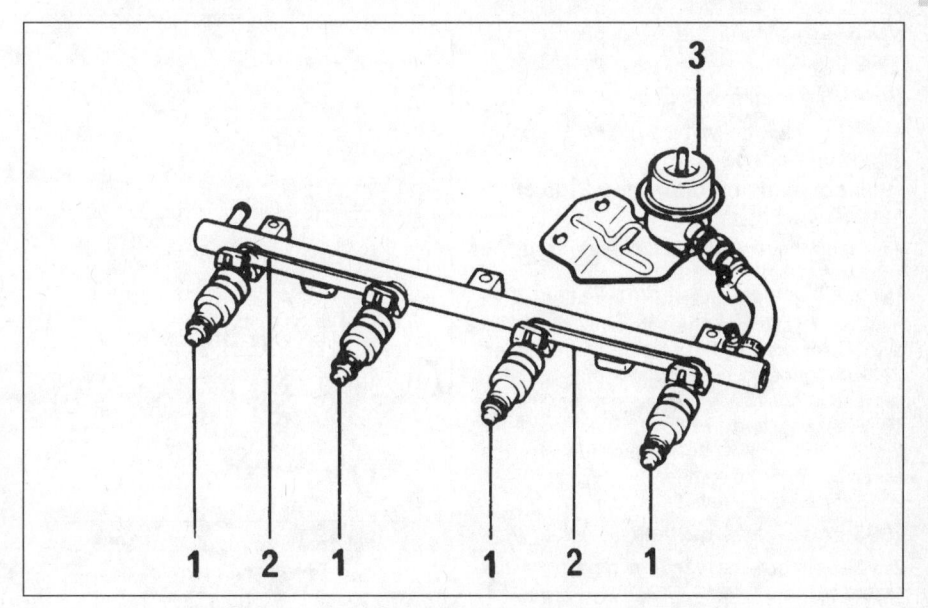

13.12a Fuel rail and injector assembly - 2165 cc model

1 Injector   2 Retaining clips   3 Fuel pressure regulator

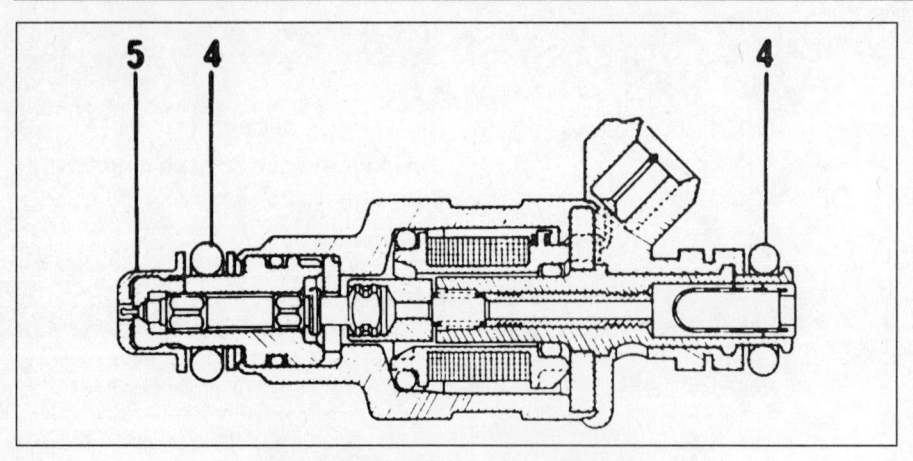

4B•10 Fuel and exhaust systems - fuel injection models

13.12b Sectional view of a fuel injector

4 Sealing ring   *5 Cap*

14 Refitting is a reversal of the removal procedure, noting the following points.
 a) Fit new sealing rings to all disturbed injector unions. Where necessary, also fit new caps to the ends of the injectors.
 b) Apply a smear of engine oil to the sealing rings to aid installation then ease the injectors and fuel rail into position ensuring that none of the sealing rings are displaced.
 c) On completion start the engine and check for fuel leaks.

2458 cc turbo and 2849 cc models

15 Disconnect the battery negative terminal and remove the relevant inlet duct(s) and hose(s) using the information in Section 2.
16 Remove and refit the injectors using the information given in paragraphs 9 to 14, noting that on 2849 cc on the left-hand fuel rail the return hose has a quick-release fitting which is disconnected by depressing the fitting tangs.

Fuel pressure regulator

Note: Refer to the warning note in Section 1 before proceeding.

1995 cc 8-valve models and 2165 cc models

17 Disconnect the vacuum pipe from the regulator.
18 Bearing in mind the information in Section 7, slacken the retaining clips and disconnect the fuel hoses from the regulator. On some models the regulator outlet pipe is secured in position with a union nut.
19 Slacken and remove the regulator mounting bracket bolts and remove the assembly from the engine.
20 Refitting is the reverse of removal.

1995 cc 12-valve models

21 Disconnect the vacuum pipe from the regulator.
22 Depressurise the fuel rail (see section 7) by slackening the retaining clip and disconnect the feed hose from the fuel rail.
23 Unscrew the retaining bolts and remove the pressure regulator retaining plate from the fuel rail **(see illustration)**.
24 Lift out the regulator along with its sealing rings.
25 Refitting is the reverse of removal using new sealing rings.

2458 cc turbo models

26 To improve access, remove the inlet duct (see Section 2).
27 Disconnect the vacuum hose from the pressure regulator.
28 Bearing the information in Section 7, slacken the retaining clips and disconnect the fuel hoses from the regulator.
29 Slacken and remove the retaining nut and washer and remove the regulator from its mounting bracket.
30 Refitting is the reverse of removal.

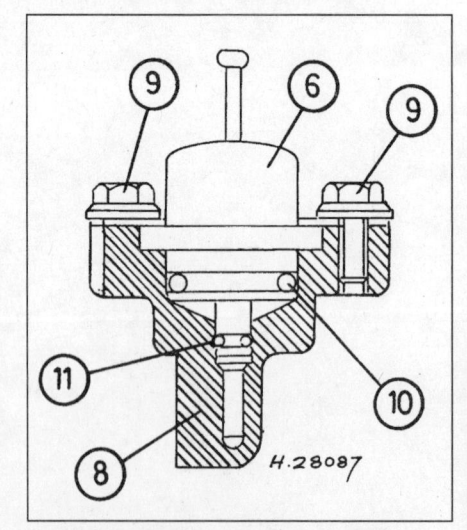

13.23 Fuel pressure regulator - 1995 cc 12-valve models

6 Fuel pressure regulator	9 Retaining bolt
8 Fuel rail	10 Sealing ring
	11 Sealing ring

2849 cc models

31 Disconnect the vacuum pipe from the regulator.
32 Bearing in mind the information in Section 7, slacken the union nut and disconnect the fuel rail hose from the regulator.
33 Depress the tabs and disconnect the feed hose quick-release fitting from the regulator.
34 Unscrew the hose fitting from the regulator then unscrew the retaining nut and washer and remove the regulator from its mounting bracket.
35 Refitting is the reverse of removal.

Throttle switch

36 Disconnect the battery negative terminal.
37 Depress the retaining clip and disconnect the wiring connector from the throttle switch.
38 Slacken and remove the two retaining screws then disengage the potentiometer from the throttle valve spindle and remove it from the vehicle.
39 Refitting is a reverse of the removal. Using a multimeter set to the resistance scale check the operation of the switch is as follows **(see illustration)**.

Throttle valve position	Resistance: terminals 2 & 18	Resistance: terminals 3 & 18
Closed	Continuity	Open circuit
Partially open (1 to 70° rotation)	Open circuit	Open circuit
Fully open (greater than 70° rotation)	Open circuit	Continuity

If necessary, slackening the retaining screws and adjust the position of the switch as required. When the switch is operating correctly, securely tighten its retaining screws and reconnect the wiring connector.

Throttle potentiometer

Note: On models fitted with a throttle potentiometer, adjustment of the potentiometer requires the use of special electronic diagnostic equipment. It is recommended that removal and refitting is therefore entrusted to a Renault dealer.

40 Disconnect the battery negative terminal.

13.39 Throttle valve switch terminal identification details

Fuel and exhaust systems - fuel injection models 4B•11

13.47 Undo the retaining screw then release the retaining clip . . .

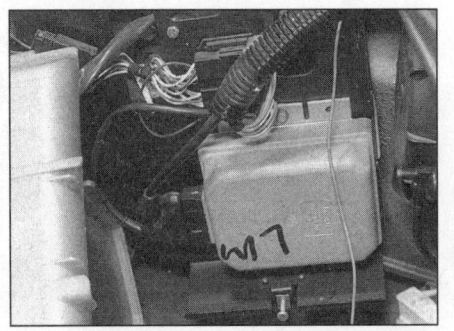

13.48 . . . and remove the protective cover to gain access to the ECU

13.54 Disconnect the idle speed control valve wiring connector (2165 cc model shown)

41 Depress the retaining clip and disconnect the wiring connector from the throttle potentiometer.
42 Using paint or a suitable marker pen, make alignment marks between the throttle housing and potentiometer.
43 Slacken and remove the two retaining screws then disengage the potentiometer from the throttle valve spindle and remove it from the vehicle.
44 Refitting is a reverse of the removal procedure ensuring that the potentiometer is correctly engaged with the throttle valve spindle. Align the marks made prior to removal and securely tighten the retaining screws. It is recommended that the operation of the potentiometer should be checked by a Renault dealer at the earliest possible opportunity.

Electronic Control Unit (ECU)

45 The ECU is located behind the protective cover on the left-hand side of the engine compartment.
46 To remove the ECU first disconnect the battery. To improve access, remove the air cleaner/inlet duct (see Section 2).
47 Release the retaining clip then undo the retaining screw and remove the protective cover **(see illustration)**.
48 Disconnect the wiring connectors from the ECU and relays and remove the ECU from the engine compartment **(see illustration)**.
49 Refitting is a reverse of the removal procedure ensuring that the wiring connector is securely reconnected.

Idle speed control valve

1995cc 12-valve models

50 The idle speed valve is mounted onto the front of the inlet manifold.
51 Disconnect the battery negative terminal then disconnect the wiring connector from the valve.
52 Unscrew the retaining bolts and remove the valve and gasket from the manifold.
53 Refitting is the reverse of removal using a new gasket.

All other models

54 The idle speed valve is located on the front left-hand side of the cylinder on 4-cylinder models, and towards the rear of the cylinder head on 6-cylinder models **(see illustration)**.
55 To remove it first disconnect the battery negative terminal.
56 Depress the retaining clip and disconnect the wiring connector from the air valve.
57 Slacken the retaining clips and disconnect both vacuum hoses from the end of the auxiliary air valve.
58 Slacken the clamp bolt then slide the valve out and remove it from the engine compartment.
59 Refitting is a reversal of the removal procedure.

Manifold absolute pressure (MAP) sensor

60 The MAP sensor is situated on the left-hand side of the engine compartment where it is mounted onto the suspension turret **(see illustration)**. To remove it, first disconnect the battery negative terminal.
61 Undo the retaining nut and bolt free the MAP sensor from the bracket.
62 Disconnect the wiring connector and vacuum hose and remove the MAP sensor from the engine compartment.
63 Refitting is the reverse of the removal procedure.

Coolant temperature sensor

1995 cc and 2165 cc models

64 The coolant temperature sensor is screwed into the thermostat housing on the front, right-hand side of the cylinder head **(see illustration)**.
65 Either partially drain the cooling system to just below the level of the sensor (as described in Chapter 1), or have ready a suitable plug which can be used to plug the sensor aperture whilst it is removed. If a plug is used, take great care not to damage the sensor unit aperture, and do not use anything which will allow foreign matter to enter the cooling system.
66 Disconnect the battery negative lead then disconnect the wiring from the sensor.
67 Unscrew the sensor unit and recover its sealing washer.
68 Refitting is the reverse of removal using a new sealing washer. Refill the cooling system as described in Chapter 1 or top-up as described in *"Weekly checks"*.

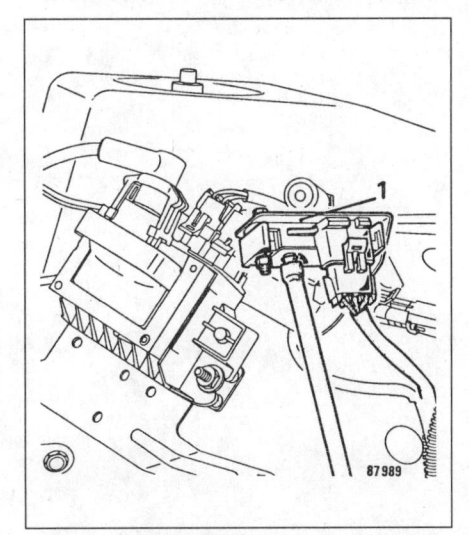

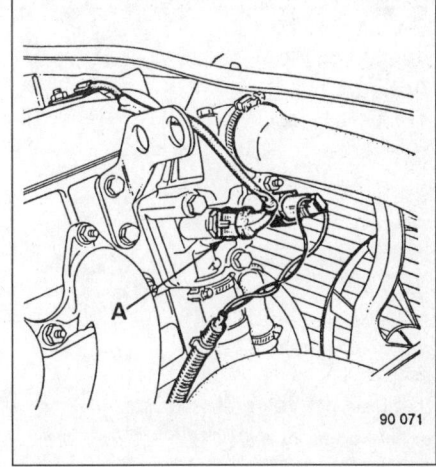

13.60 Manifold absolute pressure (MAP) sensor (1) is mounted on the left-hand suspension turret

13.64 On 1995 cc and 2165 cc models the coolant temperature sensor (A) is screwed into the thermostat housing

4B•12 Fuel and exhaust systems - fuel injection models

2458 cc turbo models

69 The coolant temperature sensor is screwed into the thermostat housing which is mounted on the from the cylinder block **(see illustration)**. Remove and refit as described in paragraphs 65 to 68.

2849 cc models

70 The coolant temperature is screwed into the coolant elbow which is bolted to the top of the cylinder block **(see illustration)**. To gain access to the switch remove the inlet manifold as described in Section 14. The switch can then be removed and refitted as described in paragraphs 65 to 68.

Inlet air temperature sensor

1995 cc and 2165 cc models

71 On 1995 cc 12-valve models the inlet air temperature sensor is screwed into the top of the inlet manifold and on 8-valve and 2165 cc models it is screwed into the inlet duct linking the air cleaner to the throttle housing. To remove the sensor first disconnect the battery negative terminal.
72 Disconnect the wiring connector then unscrew the sensor and remove it from the vehicle.
73 Refitting is the reverse of removal.

2458 cc turbo and 2849 cc models

74 The inlet air temperature sensor is screwed into the side of the throttle housing. To improve access to the sensor, undo the retaining screws and position the inlet duct clear of the housing.
75 Disconnect the wiring connector then unscrew the sensor and remove it from the housing.
76 Refitting is the reverse of removal, ensuring that the inlet duct sealing ring is correctly fitted to the throttle housing.

Crankshaft sensor

77 On 2849 cc models access to the crankshaft sensor can be gained from above. On all other models, firmly apply the handbrake then jack up the front of the vehicle and support on axle stands so access to the sensor can be gained from below.
78 To remove the sensor first disconnect the battery negative terminal.
79 Trace the wiring back from the sensor to the wiring connector and disconnect it from the main harness.
80 Slacken and remove the sensor retaining bolts and remove it from the engine **(see illustration)**. On some models a protective cover is fitted over the sensor.
81 Refitting is a reverse of the removal procedure tighten the retaining bolts securely. No adjustment of the sensor is necessary.

Knock sensor

2458 cc turbo models

82 On 2458 cc turbo models, the knock sensor is screwed into the top of the cylinder head. To gain access to the sensor, undo the regulator mounting bolts and position the regulator clear of the sensor. If necessary, to further improve access remove the fuel pressure regulator as described earlier in this Section.
83 Disconnect the sensor wiring connector then unscrew it and remove it from the engine.
84 Refitting is the reverse of removal ensuring that the sensor is securely tightened.

All other models

85 The knock sensor is screwed into the left-hand side of the cylinder block. To gain access to the sensor, firmly apply the handbrake then jack up the front of the vehicle and support it on axle stands (see "Jacking and vehicle support").
86 Trace the wiring back from the sensor and disconnect it at the wiring connector **(see illustration)**.
87 Unscrew the sensor and remove it from the engine.
88 Refitting is the reverse of removal ensuring that the sensor is securely tightened.

14 Inlet manifold - removal and refitting

1 Disconnect the battery negative terminal and proceed as described under the relevant sub-heading.

1995 cc and 2165 cc models

Removal

2 Remove the air cleaner housing and inlet duct as described in Section 2.
3 On 1995 cc 12-valve models and 2165 cc models where the throttle housing is on the underside of the manifold, disconnect the accelerator cable as described in Section 3.
4 On all other models where the throttle linkage is mounted onto the cylinder head cover, unclip the throttle link rod from the throttle housing.
5 Bearing mind the information given in Section 7, disconnect the fuel hoses from the fuel rail.
6 Disconnect the wiring connectors from the throttle valve switch/potentiometer, the injectors and (where necessary) the inlet air temperature sensor. Position the wiring clear of the manifold.
7 Release the retaining clips (where fitted) and disconnect all the relevant vacuum and

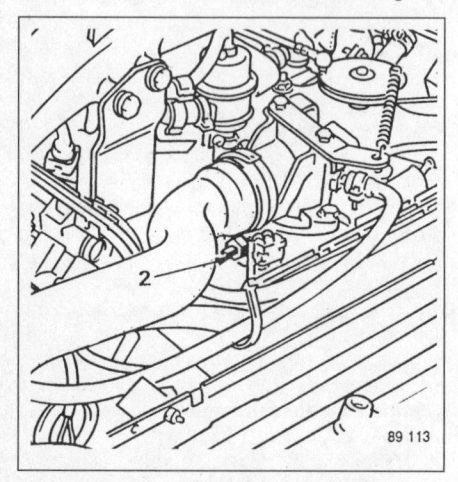

13.69 On 2458 cc turbo models the coolant temperature sensor (2) is screwed into the thermostat housing

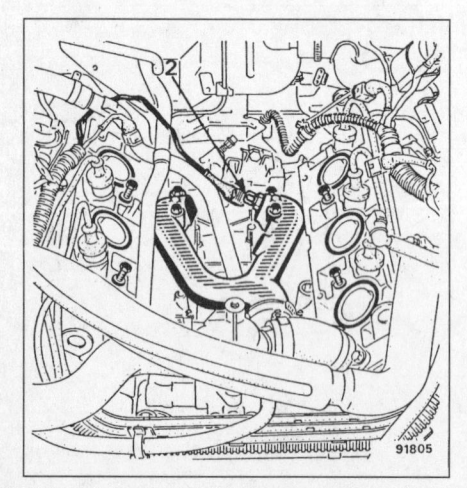

13.70 Coolant temperature sensor location (shown with inlet manifold removed) - 2849 cc models

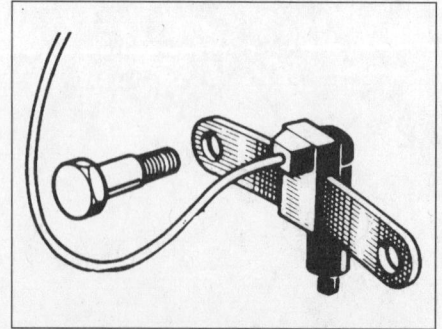

13.80 Crankshaft sensor and shouldered retaining bolt

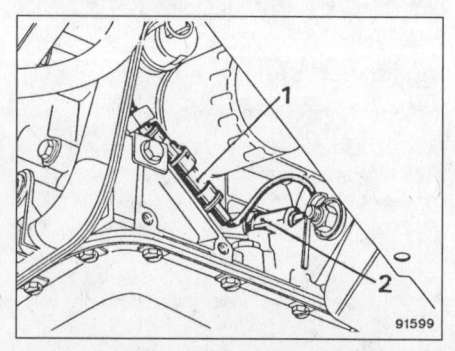

13.86 Knock sensor (2) and wiring connector (1) - 2849 cc model

breather hoses from the manifold. Make identification marks on the hoses to ensure that they are connected correctly on refitting.
8 Undo the retaining bolts and remove the support bracket from the underside of the manifold.
9 Undo the manifold retaining nuts and withdraw the manifold from the engine compartment. Recover the manifold gasket(s) and discard; new gasket(s) must be used on refitting.

Refitting

10 Refitting is the reverse of removal, noting the following points.
a) Ensure that the manifold and cylinder head mating surfaces are clean and dry, then fit the new gaskets(s) to the cylinder head studs. Refit the manifold and securely tighten its retaining nuts.
b) Ensure that all relevant hoses are reconnected to their original positions and are securely held (where necessary) by the retaining clips.
c) Adjust the accelerator cable as described in Section 3.

2458 cc turbo models

Removal

11 Unbolt and remove the inlet ducts from above the engine (see Section 2).
12 Unclip the throttle link rod from the housing then unbolt the throttle cam assembly and position it clear of the manifold.
13 Remove both fuel rail and injector assemblies as described in Section 13.
14 Disconnect the wiring connectors from the throttle valve switch/potentiometer and inlet air temperature sensor and position the wiring clear of the manifold.
15 Release the retaining clips (where fitted) and disconnect all the relevant vacuum and breather hoses from the manifold. Make identification marks on the hoses to ensure that they are connected correctly on refitting.
16 Slacken and remove the manifold retaining bolts, noting the correct fitted positions of the brackets retained by the them. Lift off the manifold and remove the sealing rings from the cylinder head. Discard the sealing rings, new ones should be used on refitting.

Refitting

17 Refitting is the reverse of removal, noting the following.
a) Fit the new rings in the cylinder head recesses and fit the manifold, tightening its retaining bolts securely.
b) Ensure that all relevant hoses are reconnected to their original positions and are securely held (where necessary) by the retaining clips.

2664 cc models

Removal

18 Remove the throttle housing as described in Section 12.
19 Release the retaining clips (where fitted) and disconnect all the relevant vacuum and breather hoses from the manifold. Make identification marks on the hoses to ensure that they are connected correctly on refitting.
20 Slacken and remove the four retaining bolts and lift the manifold carefully away from the engine. Remove the sealing rings from the cylinder head ports and discard; new ones should be used on refitting.

Refitting

21 Refitting is the reverse of removal, noting the following.
a) Fit the new rings in the cylinder head recesses and fit the manifold, ensuring that the throttle link rod is correctly positioned. Securely tighten the manifold retaining bolts.
b) Ensure that all relevant hoses are reconnected to their original positions and are securely held (where necessary) by the retaining clips.
c) Refit the throttle housing as described in Section 12.

2849 cc models

22 Refer to the information given in paragraphs 11 to 17, noting that it will also be necessary to unbolt the fuel pressure regulator mounting bracket from the manifold.

15 Exhaust manifold - removal and refitting

1995 cc and 2165 cc models

Removal

1 To improve access, firmly apply the handbrake, then jack up the front of the vehicle and support it on axle stands (see "Jacking and vehicle support").
2 Undo the nuts securing the exhaust front pipe to the manifold, then remove the nut/bolt securing the front pipe to its mounting bracket. Disconnect the front pipe from the manifold, and recover the gasket. On 1995 cc 12-valve models unbolt the manifold support bracket.
3 Undo the manifold retaining nuts and washers and remove the manifold from the cylinder head. Recover the gasket and discard it.

Refitting

4 Refitting is the reverse of the removal procedure, noting the following points:
a) Examine all the exhaust manifold studs for signs of damage and corrosion; remove all traces of corrosion, and repair or renew any damaged studs.
b) Ensure that the manifold and cylinder head sealing faces are clean and flat, and fit the new manifold gasket. Fit the manifold and securely tighten the retaining nuts.
c) Reconnect the front pipe to the manifold using a new gasket.

2458 cc turbo models

Note: *All disturbed manifold clamp nuts must be renewed.*

Removal

5 In this Section, the manifold is defined as all the sections linking the cylinder heads to the base of the turbocharger. Each side can be removed individually.
6 Firmly apply the handbrake then jack up the front of the vehicle and support it on axle stands so access can be gained both from above and below. To improve access from above, release the retaining clips and position the wiring/hoses clear of the manifold(s).
7 Slacken the clamp nuts and remove both connecting pieces and gaskets linking the manifold to the turbocharger.
8 Unscrew the retaining nuts and remove the heatshield from the manifold.
9 Undo the manifold retaining nuts and washers and remove the manifold and gaskets from the cylinder head. Discard all gaskets, new ones should be used on refitting.
10 If necessary, repeat paragraphs 7 to 9 on the opposite cylinder head.

Refitting

11 Ensure that all joints are clean and dry and examine all the exhaust manifold studs for signs of damage and corrosion; remove all traces of corrosion, and repair or renew any damaged studs.
12 Fit the new manifold gaskets to the cylinder head studs and refit the manifold. Tighten the retaining nuts lightly only at this stage.
13 Fit the new nuts to the connecting piece clamps and manoeuvre the connecting pieces, clamps and new gaskets into position. Position the clamps as shown in **illustration 15.13** and lightly tighten the clamp nuts.
14 With all manifold sections in position, securely tighten the manifold retaining nuts, then tighten the clamp nuts to the specified torque.
15 Start the engine and run it for approximately 15 minutes, stop the engine and check all the manifold clamp nuts are still tightened to the specified torque. **Do not** loosen the nuts.
16 With all the nuts retightened, refit the manifold heatshield(s) and securely tighten the retaining nuts.

2664 cc and 2849 cc models

Removal

17 Depending on model, there are two possible types of manifold arrangement. On 2664 cc and some early 2849 cc models, the manifolds each have an individual front pipe. On all other models, the manifolds are linked together and a single front pipe is fitted.
18 Firmly apply the handbrake then jack up

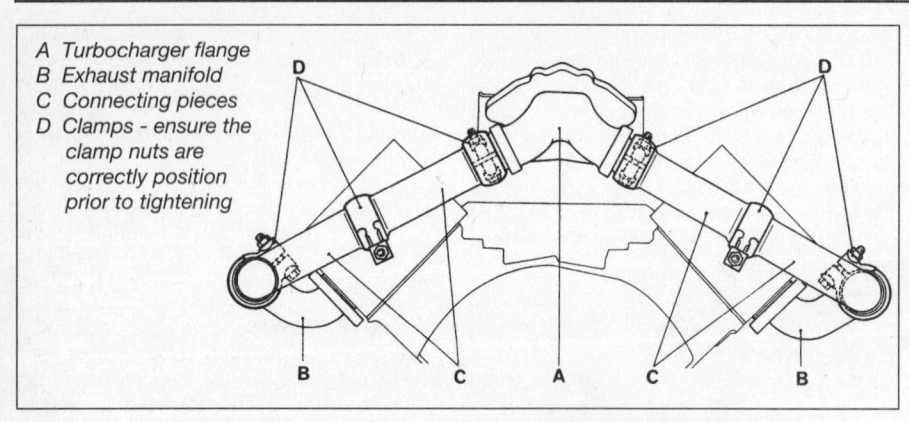

4B•14 Fuel and exhaust systems - fuel injection models

A Turbocharger flange
B Exhaust manifold
C Connecting pieces
D Clamps - ensure the clamp nuts are correctly position prior to tightening

15.13 Exhaust manifold components - 2458 cc turbo models

the front of the vehicle and support it on axle stands (see "*Jacking and vehicle support*").
19 Slacken the retaining nuts and disconnect the front pipe(s) from the manifold(s) and recover the gasket(s).
20 On models where the manifolds are linked by a joining manifold, unscrew the retaining nuts and remove the joining manifold and gaskets.
21 Unscrew the retaining nuts and remove the heatshield (where fitted) from the manifold.
22 Undo the manifold retaining nuts and washers and remove the manifold and gaskets from the cylinder head. Discard all disturbed gaskets, new ones should be used on refitting.

Refitting
23 Refitting is the reverse of the removal procedure, noting the following points:
a) Examine all the exhaust manifold studs for signs of damage and corrosion; remove all traces of corrosion, and repair or renew any damaged studs.
b) Ensure that the manifold and cylinder head sealing faces are clean and flat, and fit the new manifold gasket. Refit the manifold and securely tighten the retaining nuts.
c) Refit the jointing manifold (where fitted) an connect the front pipe(s) new gaskets.

16 Exhaust system - general information, removal and refitting

1995 cc and 2165 cc models not fitted with a catalytic converter
1 The exhaust system consists of four sections; the front pipe, the front silencer box, the intermediate pipe, and the tailpipe and main silencer. All exhaust sections are joined by flanged joints. The front pipe joints are secured by nuts and bolts, the front silencer joint being of the spring-loaded ball type, to allow for movement in the exhaust system; the other joints are secured by clamping rings.

2 Refer to Chapter 4A, Section 14 for removal and refitting details.

1995 cc and 2165 cc models with a catalytic converter
3 On these models the exhaust system consists of five sections; the front pipe, the catalytic converter, the front silencer box, the intermediate pipe, and the tailpipe and main silencer box. All exhaust sections are joined by flanged joints. The front pipe joints are secured by nuts and bolts, the catalytic converter joint being of the spring-loaded ball type, to allow for movement in the exhaust system, and the other joints are secured by clamping rings.
4 Refer to Chapter 4A, Section 14 for removal and refitting details, bearing in mind the above differences. Note that it will be necessary to disconnect the lambda sensor wiring connector when removing the catalytic converter/complete system.

2458 cc turbo models
5 On models not fitted with a catalytic converter, the exhaust system consists of four sections; the front pipe, the front silencer box, the intermediate pipe, and the tailpipe and main silencer box (see illustration). On models with a catalytic converter the exhaust system consists of five sections, the catalytic converter being fitted in-between the front pipe and (shorter) front silencer section. All exhaust sections are joined by flanged joints. The front pipe joints are secured by nuts and bolts, the front silencer/catalytic converter joint being of the spring-loaded ball type, to allow for movement in the exhaust system, and the other joints are secured by clamping rings.
6 Refer to Chapter 4A, Section 14 for removal and refitting details, bearing in mind the above changes. Also note the following points.

a) The front pipe to turbocharger nuts are secured in position with a locking plate which must be renewed whenever it is disturbed. Bend down the locking tabs prior to unscrewing the nuts then cut the plate and remove it from the front pipe. On refitting cut the new plate to allow it to the fitted to the front pipe. Tighten the retaining nuts securely and lock them in position by bending the locking tabs against the flats of the nuts and the front pipe flange.
b) On models with a catalytic converter it will be necessary to disconnect the lambda sensor wiring connector when removing the front pipe/complete system.

2664 cc models
7 On 2664 cc models the exhaust system consists of five sections; two front pipes (one for each manifold), the front silencer, the intermediate pipe, and the tailpipe and main silencer (see illustration). All exhaust sections are joined by flanged joints. The front pipes being secured to the manifold secured by nuts and the other joints are secured by clamping rings.

1 Locking plate (cut the plate at the indicated point to facilitate renewal - see text)
2 Front pipe to turbocharger flange
3 Front pipe to front silencer flange
4 Gasket
5 Gasket

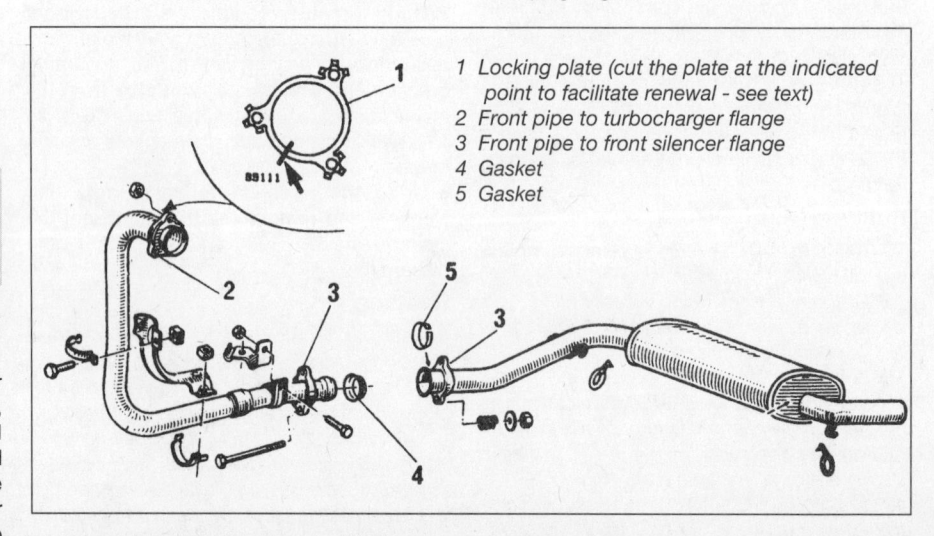

16.5 Exhaust system front section details - 2458 cc turbo models (non-catalyst models shown)

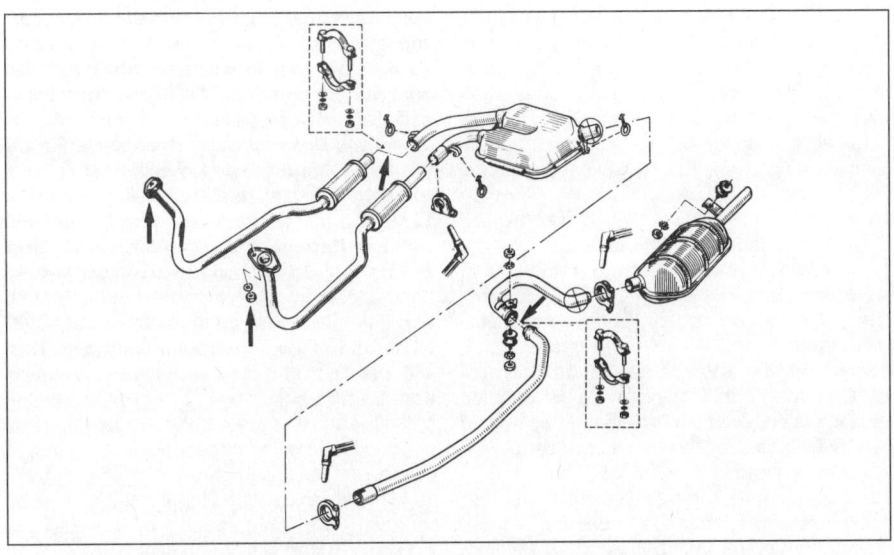

Fuel and exhaust systems - fuel injection models 4B•15

2 Energy for the operation of the turbocharger comes from the exhaust gas. The gas flows through a specially-shaped housing (the turbine housing) and in so doing, spins the turbine wheel. The turbine wheel is attached to a shaft, at the end of which is another vaned wheel known as the compressor wheel. The compressor wheel spins in its own housing and compresses the inducted air on the way to the inlet manifold.

3 Between the turbocharger and the throttle housing, the compressed air passes through an intercooler. This is an air-to-air heat exchanger, mounted on the right-hand side of the engine compartment, and supplied with cooling air ducted through the front spoiler. The purpose of the intercooler is to remove from the inducted air some of the heat gained in being compressed. Because cooler air is denser, removal of this heat further increases engine efficiency.

4 Boost pressure (the pressure in the inlet manifold) is limited by a wastegate, which

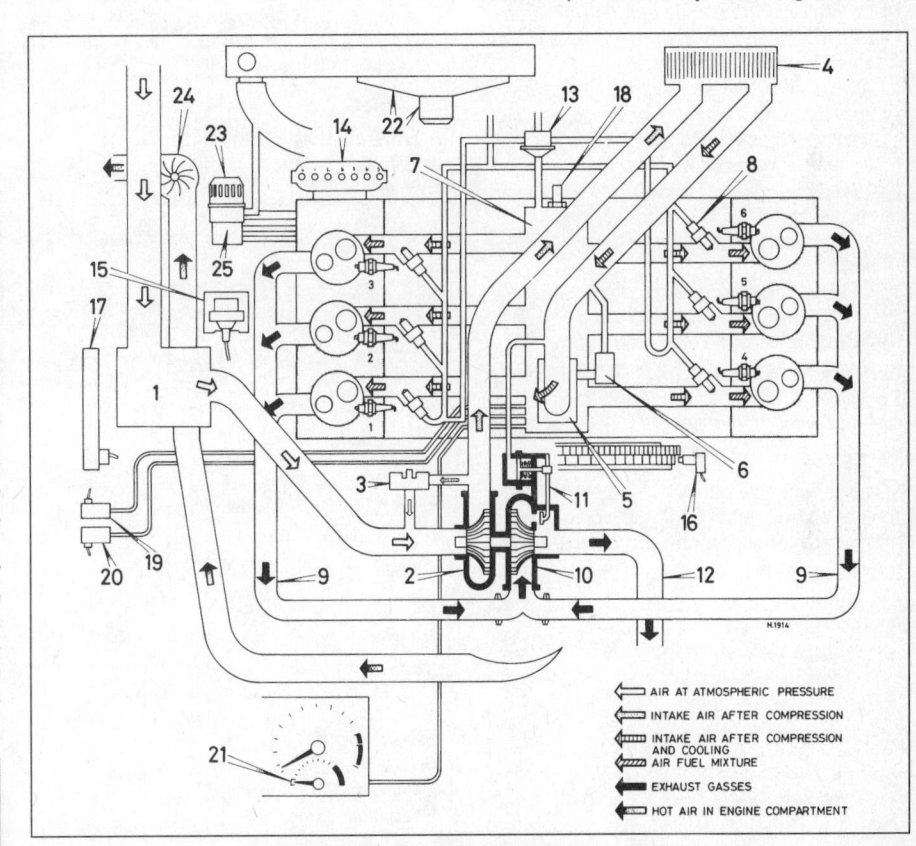

16.7 Exploded view of exhaust system components - 2664 cc models

8 Refer to Chapter 4A, Section 14 for removal and refitting details, bearing in mind the above information, and ignore the information about the spring-loaded front pipe joint.

2849 cc models

9 On some manual transmission models, a similar exhaust system to that fitted to the 2664 cc models is used (see paragraph 7). On all other models the exhaust system consists of the following five sections; the front pipe, the front intermediate pipe or catalytic converter (as applicable), the centre silencer, the rear intermediate pipe, and the tailpipe and main silencer. All exhaust sections are joined by flanged joints. The front pipe joints are secured by nuts and bolts, the front intermediate pipe/catalytic converter joint being of the spring-loaded ball type, to allow for movement in the exhaust system, and the other joints are secured by clamping rings.

10 Refer to Chapter 4A Section 14 for removal and refitting details, noting the above information. On models with a catalytic converter, note that it will be necessary to disconnect the lambda sensor wiring connector when removing the catalytic converter/complete system.

17 Turbocharger (2458 cc models) - general information, removal and refitting

General information

1 A turbocharger is fitted to all 2458 cc models. It increases engine efficiency by raising the pressure in the inlet manifold above atmospheric pressure. Instead of the air simply being sucked into the cylinders, it is forced in **(see illustration)**. Additional fuel is supplied by the injectors in proportion to the increased air inlet.

17.1 Turbocharging system - 2458 cc models

1 Air cleaner
2 Compressor
3 Compressor spill valve
4 Intercooler
5 Plenum chamber
6 Idle speed control valve
7 Inlet manifold
8 Injectors
9 Exhaust manifolds
10 Turbine
11 Turbocharger control (aneroid and wastegate)
12 Exhaust front pipe
13 Fuel pressure regulator
14 Ignition distributor
15 Ignition control unit and HT coil
16 Crankshaft sensor
17 Electronic control unit (ECU)
18 Knock sensor
19 Manifold absolute pressure sensor
20 Safety pressure sensor
21 Boost pressure sensor
22 Radiator and cooling fan
23 Oil filter
24 Hot air extractor fan
25 Oil cooler

- AIR AT ATMOSPHERIC PRESSURE
- INTAKE AIR AFTER COMPRESSION
- INTAKE AIR AFTER COMPRESSION AND COOLING
- AIR FUEL MIXTURE
- EXHAUST GASSES
- HOT AIR IN ENGINE COMPARTMENT

4B

4B•16 Fuel and exhaust systems - fuel injection models

diverts the exhaust gas away from the turbine wheel in response to a pressure-sensitive actuator.

5 The turbo shaft is pressure-lubricated by an oil feed pipe from the main oil gallery. The shaft 'floats' on a cushion of oil. A drain pipe returns the oil to the sump.

6 On later models the turbocharger is water-cooled and is linked into the engine cooling system. The turbocharger cooling circuit is supplied by an auxiliary electric pump which circulates the coolant around the turbocharger for some time after the engine is stopped to prevent overheating.

7 The turbocharger operates at extremely high speeds and temperatures. Certain precautions must be observed to avoid premature failure of the turbo or injury to the operator.

a) Do not operate the turbo with any parts exposed. Foreign objects falling onto the rotating vanes could cause excessive damage and (if ejected) personal injury.
b) Do not race the engine immediately after start-up, especially if it is cold. Give the oil a few seconds to circulate.
c) Always allow the engine to return to idle speed before switching it off - do not blip the throttle and switch off, as this will leave the turbo spinning without lubrication.
d) Allow the engine to idle for several minutes before switching off after a high-speed run.
e) Observe the recommended intervals for oil and filter changing, and use a reputable oil of the specified quality. Neglect of oil changing, or use of inferior oil, can cause carbon formation on the turbo shaft and subsequent failure.

Removal

8 Disconnect the battery negative terminal then unbolt and remove the turbocharger heat shield, the inlet scoop and the cooling air fan duct (see illustration).

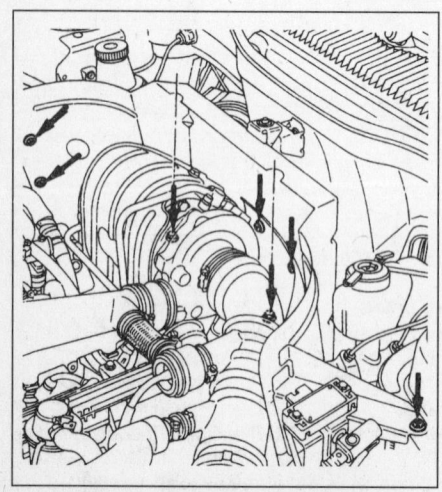

9 Referring to Section 16, release the locking tabs then undo the retaining nuts and disconnect the front pipe from the turbocharger. Note that a new locking plate will be required on refitting. On models with a catalytic converter disconnect the lambda sensor wiring connector to prevent any strain being placed on the wiring.

10 Remove the heatshield and disconnect the inlet duct (see illustration).

11 Slacken the oil feed and return pipe mounting bracket bolts then undo the retaining bolts and disconnect the pipe unions (see illustration). Remove the gaskets and discard, new ones will be needed on refitting.

12 On later models, slacken the union bolts securing the coolant pipes to the turbocharger. Remove the sealing washers and discard, new ones will be needed on refitting.

13 Slacken and remove the turbocharger fixing nuts and remove the turbocharger from the engine compartment. In order to slacken the nuts it may be necessary to modify a 17 mm ring spanner as shown (see illustration). Discard the nuts, new ones must be used on refitting.

Refitting

14 Ensure that the turbocharger and manifold flange faces are clean and dry and fit the turbocharger. Fit the new self-locking nuts and tighten securely.

15 Fit new sealing washers to the coolant pipe unions and securely tighten the union bolts (where necessary).

16 Fit the lower oil pipe union connection to the turbocharger, using a new gasket, and securely tighten its retaining bolts.

17 Prime the turbocharger unit by filling it with oil through the oil supply hole (see illustration). Disconnect the three-way wiring connector from the ignition control unit then turn the engine over on the starter motor until oil starts to flow from the turbocharger. This will ensure that the oil supply circuit is free from the air locks.

18 Wipe up all spilt oil, then reconnect the upper oil pipe to the turbocharger using a new gasket and securely tighten its retaining bolts.

19 The remainder of refitting, is the reverse of removal. On completion, reconnect the ignition unit and start the engine, running it at idle speed only for a short while until normal oil supply is restored to the turbocharger. Where necessary, top-up the cooling system as described in "Weekly checks".

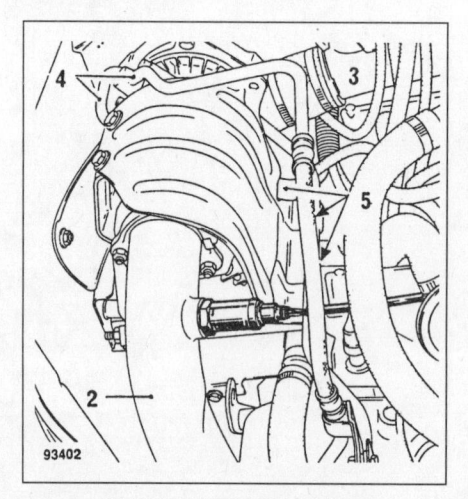

17.8 Turbocharger heatshield, intake scoop and air fan ducting screws

17.10 Turbocharger fixings

2 Exhaust front pipe
3 Pressurised outlet
4 Oil and coolant supply pipes
5 Heatshield fixings

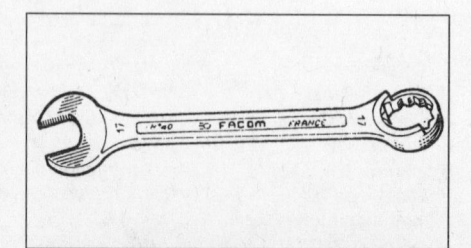

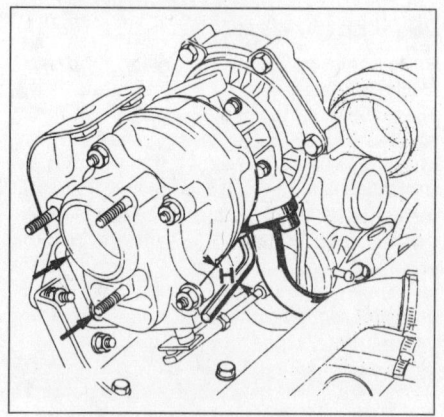

17.11 An Allen key no longer than 30 mm (H) in length will be required to unscrew the oil pipe bolts

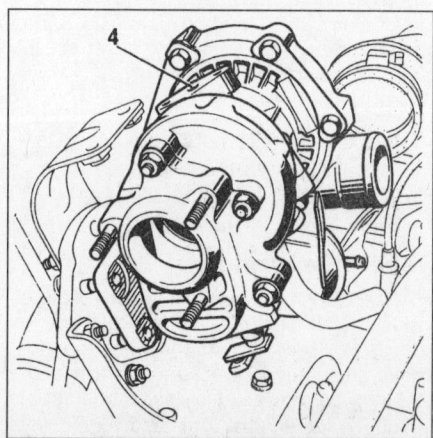

17.13 Modified ring spanner for removing the turbocharger fixing nuts

17.17 On refitting fill the turbocharger with clean engine oil through oil hole (4)

18 Intercooler (2458 cc models) - removal and refitting

Removal

1 Disconnect and remove the battery.
2 Remove the ABS control unit as described in Chapter 9.
3 Unbolt and remove the battery support.
4 Disconnect the intercooler ducts, remove the mounting bolts and withdraw the intercooler (see illustration).

Refitting

5 Refitting is a reversal of removal.

19 Emission control systems - general information

1 To help minimise harmful emissions, all models are equipped the crankcase emission-control system described below. Certain models are also equipped with a catalytic converter and an evaporative emission control system.
2 The emission control systems function as follows.

Crankcase emission control

3 To reduce the emission of unburned hydrocarbons from the crankcase into the atmosphere, the engine is sealed and the blow-by gases and oil vapour are drawn from inside the crankcase, through a wire mesh oil separator, into the inlet tract to be burned by the engine during normal combustion.
4 Under conditions of high manifold depression (idling, deceleration) the gases will be sucked positively out of the crankcase. Under conditions of low manifold depression (acceleration, full-throttle running) the gases are forced out of the crankcase by the (relatively) higher crankcase pressure; if the engine is worn, the raised crankcase pressure (due to increased blow-by) will cause some of the flow to return under all manifold conditions.

Exhaust emission control

5 To minimise the amount of pollutants which escape into the atmosphere, some models are fitted with a catalytic converter in the exhaust system. On all models where a catalytic converter is fitted, the system is of the closed-loop type, in which a lambda sensor in the exhaust system provides the fuel-injection system ECU with constant feedback, enabling the ECU to adjust the mixture to provide the best possible conditions for the converter to operate.
6 The lambda sensor has a heating element built-in that is controlled by the ECU through the lambda sensor relay to quickly bring the sensor's tip to an efficient operating temperature. The sensor's tip is sensitive to oxygen and sends the ECU a varying voltage depending on the amount of oxygen in the exhaust gases; if the inlet air/fuel mixture is too rich, the exhaust gases are low in oxygen so the sensor sends a low-voltage signal, the voltage rising as the mixture weakens and the amount of oxygen rises in the exhaust gases. Peak conversion efficiency of all major pollutants occurs if the inlet air/fuel mixture is maintained at the chemically-correct ratio for the complete combustion of petrol of 14.7 parts (by weight) of air to 1 part of fuel (the 'stoichiometric' ratio). The sensor output voltage alters in a large step at this point, the ECU using the signal change as a reference point and correcting the inlet air/fuel mixture accordingly by altering the fuel injector pulse width.

Evaporative emission control

7 To minimise the escape into the atmosphere of unburned hydrocarbons, an evaporative emissions control system is fitted to models equipped with a catalytic converter. The fuel tank filler cap is sealed and a charcoal canister is situated in the engine compartment to collect the petrol vapours generated in the tank when the car is parked. It stores them until they can be cleared from the canister into the inlet tract to be burned by the engine during normal combustion.
8 To ensure that the engine runs correctly when it is cold and/or idling and to protect the catalytic converter from the effects of an over-rich mixture, the vapours are only allowed to pass into the inlet tract under certain conditions.
9 On most models an electrical solenoid valve is used to control the evaporative emission system; the is not opened by the ECU until the engine has warmed up, and the engine is under load; the valve solenoid is then modulated on and off to allow the stored vapour to pass into the inlet tract.
10 On other models the flow of the stored vapour is controlled by a vacuum diaphragm unit built into the canister itself.

20 Emission control systems - testing and component renewal

Crankcase emission control

1 The components of this system require no attention other than to check that the hose(s) are clear and undamaged at regular intervals.

Evaporative emission control system

Testing

2 If the system is thought to be faulty, disconnect the hoses from the charcoal canister and purge control valve and check that they are clear by blowing through them. If the purge control valve (where fitted) or charcoal canister are thought to be faulty, they must be renewed.

Charcoal canister - renewal

3 On 2458 cc turbo models, to gain access to the canister, remove the air cleaner housing as described in Section 2.
4 On all other models, firmly apply the handbrake then jack up the front of the vehicle and support it on axle stands to gain access to the canister which is behind the right-hand front wing.
5 Unhook and remove the canister retaining strap.
6 Mark the hoses for identification purposes then disconnect them and remove the canister from the vehicle. Where the crimped-type hose clips are fitted, cut the clips and discard them, replace them with standard worm-drive hose clips on refitting.
7 Refitting is a reverse of the removal procedure ensuring that the hoses are correctly reconnected.

Purge valve - renewal

8 To renew the purge valve, disconnect the battery negative terminal then depress the retaining clip and disconnect the wiring connector from the valve.
9 Disconnect the hoses from either end of the valve then undo the retaining nuts/bolts and remove it from the engine compartment.
10 Refitting is a reversal of the removal procedure ensuring that the valve is fitted the correct way around and the hoses are securely connected.

Exhaust emission control

Testing

11 The performance of the catalytic converter can be checked only by measuring the exhaust gases using a good-quality, carefully-calibrated exhaust gas analyser as described in Chapter 1.
12 If the CO level is too high, the vehicle should be taken to a Renault dealer so that the complete fuel-injection and ignition systems, including the lambda sensor, can be

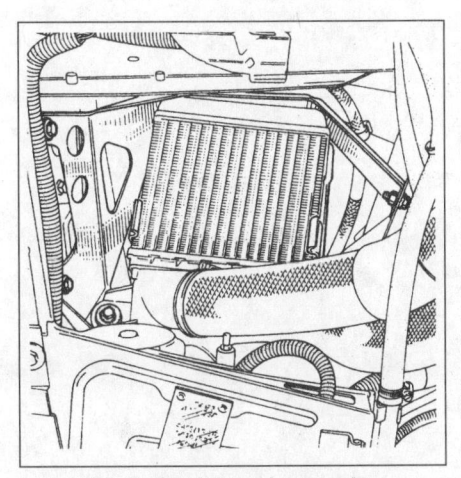

18.4 Turbocharger intercooler - 2458 cc models

thoroughly checked using the special diagnostic equipment. Once these have been checked and are known to be free from faults, the fault must be in the catalytic converter.

Catalytic converter - renewal

13 Refer to Section 16.

Lambda sensor - renewal

Note: *The lambda sensor is delicate and will not work if it is dropped or knocked, if its power supply is disrupted, or if any cleaning materials are used on it.*

14 On 2458 cc turbo models the sensor is screwed into the top of the exhaust system front pipe and can be reached from within the engine compartment. On all other models, it will be necessary to raise the vehicle and support it on axle stands to gain access to the lambda sensor.

15 On all models, trace the wiring back from the lambda sensor and disconnect it at the connector **(see illustration)**.

16 Unscrew the sensor and remove it from the front pipe/catalytic converter.

17 Refitting is a reverse of the removal procedure using a new sealing washer. Prior to installing the sensor apply a smear of high temperature grease to the sensor threads. Ensure that the sensor is securely tightened and that the wiring is correctly routed and in no danger of contacting either the exhaust system or engine.

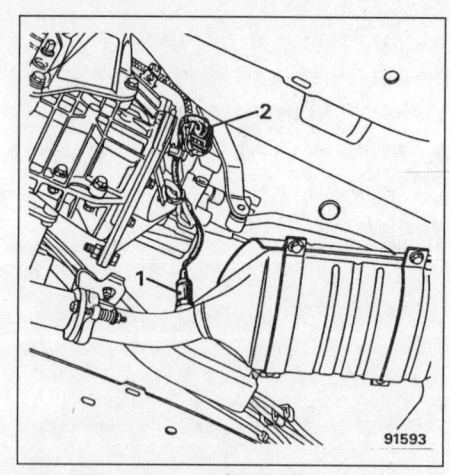

20.15 Exhaust system lambda sensor (1) and wiring connector (2) (2165 cc model shown)

21 Catalytic converter - general information and precautions

1 The catalytic converter is a reliable and simple device which needs no maintenance in itself, but there are some facts of which an owner should be aware if the converter is to function properly for its full service life.
  a) DO NOT use leaded petrol in a car equipped with a catalytic converter - the lead will coat the precious metals, reducing their converting efficiency and will eventually destroy the converter.
  b) Always keep the ignition and fuel systems well-maintained in accordance with the manufacturer's schedule.
  c) If the engine develops a misfire, do not drive the car at all (or at least as little as possible) until the fault is cured.
  d) DO NOT push- or tow-start the car - this will soak the catalytic converter in unburned fuel, causing it to overheat when the engine does start.
  e) DO NOT switch off the ignition at high engine speeds.
  f) DO NOT use fuel or engine oil additives - these may contain substances harmful to the catalytic converter.
  g) DO NOT continue to use the car if the engine burns oil to the extent of leaving a visible trail of blue smoke.
  h) Remember that the catalytic converter operates at very high temperatures. DO NOT, therefore, park the car in dry undergrowth, over long grass or piles of dead leaves after a long run.
  i) Remember that the catalytic converter is FRAGILE - do not strike it with tools during servicing work.
  J) In some cases a sulphurous smell (like that of rotten eggs) may be noticed from the exhaust. This is common to many catalytic converter-equipped cars and once the car has covered a few thousand miles the problem should disappear.
  k) The catalytic converter, used on a well-maintained and well-driven car, should last for between 50 000 and 100 000 miles - if the converter is no longer effective it must be renewed.

Chapter 4 Part C:
Fuel and exhaust systems - diesel models

Contents

Accelerator cable - removal, refitting and adjustment	11	Fuel system - priming and bleeding	2
Accelerator pedal - removal and refitting	12	Fuel tank - removal and refitting	14
Air cleaner housing - removal and refitting	18	General information and precautions	1
Air filter renewal	See Chapter 1	Idle speed and anti-stall speed checking and adjustment See Chapter	1
Boost pressure fuel delivery (LDA) corrector (Turbo models) - general information	20	Injection timing (Bosch fuel injection pump) - checking and adjustment	9
Emission control systems - general information	21	Injection timing (Lucas fuel injection pump) - checking and adjustment	8
Emission control systems - testing and component renewal	22	Injection timing - checking methods and adjustment	7
Exhaust system - general information and component renewal	19	Intercooler - removal and refitting	17
Fast idle thermostatic sensor - removal, refitting, testing and adjustment	4	Manifolds - removal and refitting	15
Fuel filter renewal	See Chapter 1	Maximum speed - checking and adjustment	3
Fuel filter water draining	See Chapter 1	Stop solenoid - description, removal and refitting	5
Fuel injection pump - removal and refitting	6	Turbocharger - general information, removal and refitting	16
Fuel injectors - testing, removal and refitting	10		
Fuel level sender - removal and refitting	13		

Degrees of difficulty

Easy, suitable for novice with little experience	Fairly easy, suitable for beginner with some experience	Fairly difficult, suitable for competent DIY mechanic	Difficult, suitable for experienced DIY mechanic	Very difficult, suitable for expert DIY or professional

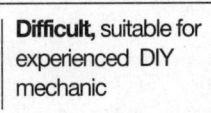

Specifications

General
System type .... Rear-mounted fuel tank, distributor fuel injection pump with integral transfer pump, indirect injection. Turbocharger and intercooler on some engines
Firing order .... 1-3-4-2 (No 1 at flywheel end)

Maximum speed
Bosch injection pump:
  Turbo models .... 4700 to 4800 rpm
  Normally aspirated models .... 4900 ± 100 rpm
Lucas injection pump .... 4750 to 4900 rpm

Lucas injection pump
Direction of rotation .... Clockwise viewed from sprocket end
Static timing:
  Engine position .... Number 1 cylinder at TDC (see Text)
  Pump position:
    Early type pump .... 1.80 mm
    Later type pump .... Value shown on pump (see Text)
Dynamic timing (at specified idle speed) .... 9.5° BTDC

Injection pump (Bosch)
Direction of rotation .... Clockwise viewed from sprocket end
Static timing*:
  Engine position .... Number 1 cylinder at TDC (see Text)
  Pump position:
    J8S 736 engine .... 0.75 ± 0.02 mm
    All other engines .... 0.70 ± 0.02 mm
Dynamic timing (at specified idle speed)*:
  J8S 736 engine .... 14° BTDC
  All other engines .... 13.5° BTDC

*Engine code is stamped on the engine number plate - see "Vehicle identification numbers" for details

4C•2 Fuel and exhaust systems - diesel models

Injectors
Type*:
J8S 736 engine	Screw-type pintle injector
All other engines	Clamp-type pintle injector

Opening pressure:
Bosch injectors	125 to 138 bar
Lucas injectors	113 to 125 bar
Maximum difference between any two injectors	8 bar

*Engine code is stamped on the engine number plate - see "Vehicle identification numbers" for details

Torque wrench settings

	Nm	lbf ft
Injectors:		
Clamp-type:		
Retaining nuts	17	12
Return hose union bolts	10	7
Screw-type injectors	70	52
Injector pipe union nuts	25	18
Injection pump:		
Feed and return hose union bolts	25	18
Mounting nuts and bolts	25	18
Stop solenoid	20	15

1 General information and precautions

General information

1 The fuel system consists of a rear-mounted fuel tank, a fuel filter with integral water separator, a fuel injection pump, injectors and associated components. A turbocharger and intercooler are fitted to some models **(see illustration)**.

2 Fuel is drawn from the fuel tank to the fuel injection pump by a vane-type transfer pump incorporated in the fuel injection pump. Before reaching the pump, the fuel passes through a fuel filter, where foreign matter and water are removed. Excess fuel lubricates the moving components of the pump, and is then returned to the tank.

3 The fuel injection pump is driven at half-crankshaft speed by the timing belt. The high pressure required to inject the fuel into the compressed air in the swirl chambers is achieved by a cam plate acting on a single piston on the Bosch pump, or by two opposed pistons forced together by rollers running in a cam ring on the Lucas pump. The fuel passes through a central rotor with a single outlet drilling which aligns with ports leading to the injector pipes.

4 Fuel metering is controlled by a centrifugal governor, which reacts to accelerator pedal position and engine speed. The governor is linked to a metering valve, which increases or decreases the amount of fuel delivered at each pumping stroke. On turbocharged models, a separate device also increases fuel delivery with increasing boost pressure.

5 Basic injection timing is determined when the pump is fitted. When the engine is running, it is varied automatically to suit the prevailing engine speed by a mechanism which turns the cam plate or ring.

6 The four fuel injectors produce a homogeneous spray of fuel into the swirl chambers located in the cylinder head. The injectors are calibrated to open and close at critical pressures to provide efficient and even combustion. Each injector needle is lubricated by fuel, which accumulates in the spring chamber and is channelled to the injection pump return hose by leak-off pipes.

7 Bosch or Lucas fuel system components may be fitted, depending on model. Components from the latter manufacturer are marked either "CAV", "Roto-Diesel" or "Con-Diesel", depending on their date and place of manufacture. With the exception of the fuel filter assembly, replacement components must be of the same make as those originally fitted.

8 Cold starting is assisted by pre-heater or "glow" plugs fitted to each swirl chamber. A thermostatic sensor in the cooling system operates a fast idle lever on the injection pump to increase the idling speed when the engine is cold.

9 A stop solenoid cuts the fuel supply to the injection pump rotor when the ignition is switched off, and there is also a hand-operated stop lever for use in an emergency.

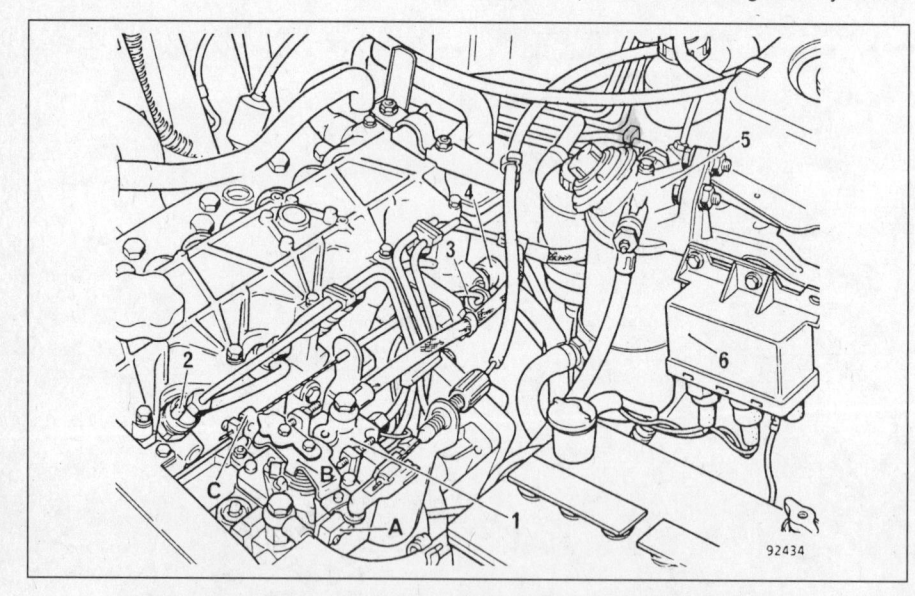

1.1 Fuel system components - J8S 736 engine shown (others similar)

1 Injection pump
  A No-load switch
  B Accelerator lever
  C Fast idle lever
2 Injectors
3 Fast idle thermostatic valve
4 Post-heating cut-off temperature switch
5 Fuel filter
6 Preheating system control unit

Fuel and exhaust systems - diesel models 4C•3

10 Provided that the specified maintenance is carried out, the fuel injection equipment will give long and trouble-free service. The injection pump itself may well outlast the engine. The main potential cause of damage to the injection pump and injectors is dirt or water in the fuel.

11 Servicing of the injection pump and injectors is very limited for the home mechanic, and any dismantling or adjustment other than that described in this Chapter must be entrusted to a Renault dealer or fuel injection specialist.

Precautions

Warning: *It is necessary to take certain precautions when working on the fuel system components, particularly the fuel injectors. Before carrying out any operations on the fuel system, refer to the precautions given in "Safety first!" at the beginning of this manual, and to any additional warning notes at the start of the relevant Sections.*

2 Fuel system - priming and bleeding

1 After disconnecting part of the fuel supply system or running out of fuel, it is necessary to prime the system and bleed off any air which may have entered the system components.

2 All models are fitted with a hand-operated priming pump, consisting of a plunger-type pump on the fuel filter housing, which is located on the left-hand side of the engine compartment **(see illustration)**.

3 To prime the system, loosen the bleed screw, located either on the filter outlet union bolt (Lucas filter arrangement) or the top of the fuel filter head (Bosch filter arrangement). If no bleed screw is fitted, loosen the outlet union itself (note that on certain models, a bleed screw is also provided on the fuel inlet union at the injection pump).

4 Pump the priming plunger until fuel free from air bubbles emerges from the outlet union or bleed screw (as applicable) then tighten it securely.

5 Switch on the ignition (to activate the stop solenoid) and continue pumping the priming plunger until firm resistance is felt, then pump a few more times.

6 If a large amount of air has entered the pump, place a wad of rag around the fuel return union on the pump (to absorb spilt fuel), then slacken the union. Operate the priming plunger (with the ignition switched on to activate the stop solenoid), or crank the engine on the starter motor in 10 second bursts, until fuel free from air bubbles emerges from the fuel union. Tighten the union and mop up split fuel. **Note:** *Be prepared to stop the engine if it should fire, to avoid excessive fuel spray and spillage.*

7 If air has entered the injector pipes, place wads of rag around the injector pipe unions at the injectors (to absorb spilt fuel), then slacken the unions. Crank the engine on the starter motor until fuel emerges from the unions, then stop cranking the engine and retighten the unions. Mop up spilt fuel (see note in previous paragraph).

8 Start the engine with the accelerator pedal fully depressed. Additional cranking may be necessary to finally bleed the system before the engine starts.

3 Maximum speed - checking and adjustment

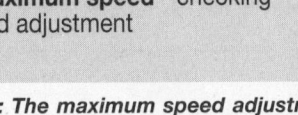

Caution: *The maximum speed adjustment screw is sealed by the manufacturers at the factory, using paint or a locking wire and a lead seal. There is no reason why it should require adjustment. Do not disturb the screw if the vehicle is still within the warranty period, otherwise the warranty will be invalidated. This adjustment requires the use of a tachometer - refer to Chapter 1 for alternative methods.*

1 Run the engine to normal operating temperature.

2 Have an assistant fully depress the accelerator pedal, and check that the maximum engine speed is as given in the *Specifications*. Do not keep the engine at maximum speed for more than two or three seconds.

3 If adjustment is necessary, stop the engine, then loosen the locknut, turn the maximum speed adjustment screw as necessary, and retighten the locknut.

4 Repeat the procedure in paragraph 2 to check the adjustment.

5 Stop the engine and disconnect the tachometer.

4 Fast idle thermostatic valve - removal, refitting, testing and adjustment

Removal

Models with valve screwed into the cylinder head

1 Disconnect the battery negative terminal.

2 Partially drain the cooling system as described in Chapter 1.

3 Loosen the clamp screw or nut (as applicable) and slide the fast idle cable end fitting off from the injection pump end of the inner cable.

4 Free the fast idle cable from the bracket on the fuel injection pump. Remove the cable fitting (where fitted) from the pump lever and store it with the valve for safe-keeping.

5 Using a suitable open-ended spanner, unscrew the thermostatic valve from the cylinder head, and remove the valve and cable assembly. Discard the sealing ring, a new one should be used on refitting.

Models with valve mounted on the side of then injection pump

6 Disconnect the battery negative terminal.

7 Using a hose clamp or similar, clamp both the fast idle valve coolant hoses to minimise coolant loss during the subsequent operation.

8 Slacken the retaining clips and disconnect both coolant hoses from the valve whilst being prepared for some coolant spillage. Wash off any spilt coolant immediately with cold water.

9 Loosen the clamp screw and nut and slide the fast idle cable end fitting arrangement off from the end of the cable.

10 Undo the retaining bolts and nuts and remove the mounting bracket from the rear of the injection pump.

11 Slacken and remove the bolts securing the fast idle valve mounting bracket to the rear of the pump. Remove the bracket and valve assembly from the pump, freeing the cable from the fast idle lever. Remove the cable fitting (where fitted) from the pump lever and store it with the valve for safe-keeping. If necessary, the valve assembly can be dismantled as follows.

12 Slacken the two screws securing the two halves of the valve housing together. As the screws near the ends of their threads, compress the two halves of the valve together to relieve spring pressure on the bolts. With the bolts removed, carefully separate the valve, gradually relieving spring pressure, removing both halves from the bracket.

13 Remove the rubber gaiter and cable guide from the front half of the valve housing then withdraw the cable and springs.

14 Using a suitable peg spanner, unscrew the thermostatic capsule ring nut from the rear half of the valve. Lift out the capsule and recover the sealing ring.

2.2 Priming the fuel system, note bleed screw (arrowed) - Lucas filter shown

Refitting

Models with valve screwed into the cylinder head

15 Fit a new sealing ring to the valve and screw the valve into position in the cylinder head, tightening it securely.
16 Refit the cable fitting to the pump fast idle lever and insert the cable through the pump bracket and pass the inner cable through the fitting. Slide the end fitting onto the inner and lightly tighten its clamp screw or nut (as applicable).
17 Refill the cooling system as described in Chapter 1.
18 Adjust the cable as described under the following sub-heading.

Models with valve mounted on the side of then injection pump

19 If the valve has been dismantled proceed as follows, if not proceed straight to paragraph 23.
20 Fit a new sealing ring to the thermostatic capsule and fit the capsule to the rear half of the valve. Refit the ring nut and tighten it securely.
21 Fit the springs to the cable and insert the cable through the front half of the valve. Locate the cable guide in the rubber gaiter then slide the gaiter into position ensuring it is correctly seated in the groove on the front of the valve.
22 Position the valve halves on either side of the mounting bracket. With the aid of an assistant, compress the two halves and refit the retaining bolts, tightening them securely.
23 Refit the cable fitting to the pump lever and manoeuvre the fast idle valve and bracket assembly into position, passing the cable through its fitting.
24 Refit the injection pump rear mounting bracket and securely tighten both the fast idle valve and mounting bracket retaining nuts and bolts.
25 Connect the coolant hoses to the fast idle valve and securely tighten their retaining clips. Remove the hose clamps and top-up the cooling system as described in "*Weekly checks*".
26 Slide the cable end fitting arrangement onto the cable and lightly tighten its clamp screw and nut.
27 Adjust the cable as described under the following sub-heading.

Testing and adjustment

Lucas injection pump

28 With the engine cold, push the lever fully to the end of its travel (towards the rear of the pump). Hold it in this position and slide the cable end fitting along the cable until its abuts the fast idle lever and securely tighten its clamp nut or screw (as applicable).
29 Start the engine, warm it up to its normal operating temperature. As the engine warms up the fast idle cable should extend so that the fast idle lever returns to is stop.
30 Once the cooling fan has cut in, measure the clearance between the fast idle lever and the cable end fitting. There should be a gap of approximately 2 to 3 mm. If not, slacken the clamp screw or nut (as applicable), move the end fitting to the correct position and securely retighten the screw or nut.

Bosch injection pump - models with fast idle valve screwed into cylinder head

31 Warm the engine up to normal operating temperature and adjust the idle speed, anti-stall speed and fast idle speed as described in Chapter 1.
32 Pull the inner cable tight to remove any slack, then measure the clearance between the cable end fitting and the lever. There should be a gap of approximately 6 mm. If not, slacken the clamp screw or nut (as applicable), move the end fitting to the correct position and securely retighten the screw or nut.
33 Switch off the engine and allow it to cool. As the engine cools the fast idle valve cable should retract and eventually pull the lever back towards the rear of the pump.

Bosch injection pump - models with fast idle valve mounted on the injection pump

34 Loosen the injection pump fast idle lever stop retaining screw and move the stop away from the lever.
35 With the cable end fitting positioned clear of the fast idle lever, move the lever towards the rear of the pump until the position is reached where resistance is felt; this is the point where the lever is starting to act on the fast idle mechanism in the pump. Hold the fast idle lever in this position and set the lever stop so that there is a gap of 0.5 mm between the stop and the lever **(see illustration)**. With the stop correctly positioned securely tighten its retaining screw.
36 With the engine cold, accurately measure the temperature of the fast idle valve thermostatic capsule located in the rear half of the valve. Referring to the following table, obtain the relevant dimensions (in mm) 'A' and 'B' which correspond to the temperature of capsule.

Capsule temperature	Dim A	Dim B
Less than 18°C	6.5	4.5
22°C	5.9	3.5
25°C	5.5	2.7
30°C	4.75	1.5
35°C	4	2
40°C	3.25	0

37 Insert shims equal in thickness to dimension 'A' in between the pump fast idle lever and its stop **(see illustration)**. Remove all slack from the cable then slide the end fitting along the cable until it abuts the fast idle lever and securely tighten its clamp screw and nut. Withdraw the shims and check the clearance between the fast idle lever and stop is equal to dimension 'A'. If not repeat the adjustment procedure.
38 With dimension 'A' correctly set, slacken the fast idle lever balljoint nut and slide the balljoint away from the accelerator lever.
39 Insert shims equal in thickness to

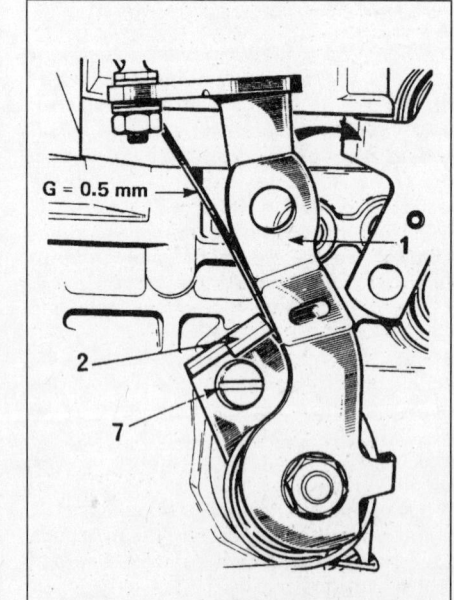

4.35 Position the fast idle lever (1) as described in the Text and adjust the gap 'G' by slackening the screw (7) and repositioning the lever stop (2) - Bosch injection pump with fast idle valve mounted on pump

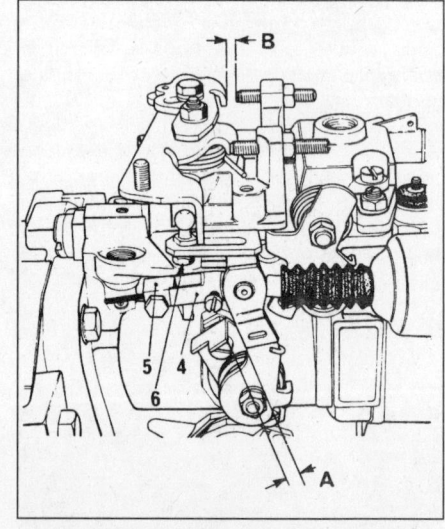

4.37 Fast idle vale adjustment details - Bosch injection pump with fast idle valve mounted on pump

4 Cable end fitting front section
5 Cable end fitting rear section
6 Balljoint
A Fast idle lever-to-stop clearance
B Idle speed screw-to-accelerator lever clearance

Fuel and exhaust systems - diesel models 4C•5

dimension 'B' in between the idle speed adjusting screw and the accelerator lever. Slide the balljoint along its slot until it abuts the accelerator lever and securely tighten its retaining nut. Withdraw the shims and check the clearance between the accelerator lever and the idle speed adjusting screw is equal to dimension 'B'. If not repeat the adjustment procedure.

40 Start the engine and warm it up to its normal operating temperature. As the engine warms up the cable end fitting should slowly extend until the fast idle lever returns to its stop and the cable end fitting is clear of the lever. At the same time the accelerator lever should be back against the idle speed adjusting screw.

5 Stop solenoid - description, removal and refitting

Caution: Be careful not to allow dirt into the injection pump during this procedure.

Description

1 The stop solenoid is located on the end of the fuel injection pump. Its purpose is to cut the fuel supply when the ignition is switched off. If an open-circuit occurs in the solenoid or supply wiring, it will be impossible to start the engine, as the fuel will not reach the injectors. The same applies if the solenoid plunger jams in the "stop" position. If the solenoid jams in the "run" position, the engine will not stop when the ignition is switched off.

2 If the solenoid has failed and the engine will not run, a temporary repair may be made by removing the solenoid as described in the following paragraphs. Refit the solenoid body without the plunger and spring. Tape up the wire so that it cannot touch earth. The engine can now be started as usual, but it will be necessary to use the manual stop lever (see Chapter 1) on the fuel injection pump (or to stall the engine in gear) to stop it.

Removal

3 Disconnect the battery negative lead.
4 Withdraw the rubber boot (where applicable), then unscrew the terminal nut and disconnect the wire from the top of the solenoid.
5 Carefully clean around the solenoid, then unscrew and withdraw the solenoid, and recover the sealing washer or O-ring (as applicable). Recover the solenoid plunger and spring if they remain in the pump. Operate the hand-priming pump as the solenoid is removed, to flush away any dirt.

Refitting

6 Refitting is a reversal of removal, using a new sealing washer or O-ring and tightening the solenoid to the specified torque setting.

6 Fuel injection pump - removal and refitting

Caution: Be careful not to allow dirt into the injection pump or injector pipes during this procedure. New sealing rings should be used on the fuel pipe banjo unions when refitting.

Removal

1 Disconnect the battery negative terminal and remove the radiator and cooling fan as described in Chapter 3. **Note:** *If necessary, the radiator can be left in position with the coolant hoses connected and simply moved as far forwards as possible; this removes the need to drain and refill the cooling system but does not give as good access.*
2 Referring to Chapter 2C, rotate the crankshaft until No.1 cylinder is positioned on TDC at the end of its compression stroke.
3 Remove the timing belt cover as described in Chapter 2C.
4 Check that the sprocket timing marks are positioned as described in Chapter 2C, Section 9, paragraph 5.
5 On models fitted with a Bosch injection pump, rotate the crankshaft backwards slightly so that the injection pump sprocket timing mark moves back by three teeth, ie. so that the tooth three in front of the timing mark is aligned with the pointer in the timing cover window.
6 On models fitted with a Lucas injection pump, rotate the crankshaft backwards slightly so that the injection pump sprocket timing mark moves back by one tooth, ie. so that the tooth one in front of the timing mark is aligned with the pointer in the timing cover window.
7 Remove the injection pump sprocket as described in Section 9 of Chapter 2C. Note that the crankshaft and camshaft must not be rotated whilst the sprocket is removed.
8 Cover the alternator with a clean cloth or plastic bag to prevent the possibility of fuel being spilt onto it during the following operations.
9 On models where the fast idle valve is screwed into the cylinder head, loosen the clamp screw or nut (as applicable) and slide the end fitting off the end of the fast idle inner cable. Where necessary, remove the cable fitting from the pump lever and store it with the valve for safe-keeping.
10 On models where the fast idle valve is mounted onto the injection pump, using a hose clamp or similar, clamp both the fast idle valve coolant hoses to minimise coolant loss. Slacken the retaining clips and disconnect both coolant hoses from the valve whilst being prepared for some coolant spillage. Wash off any spilt coolant immediately with cold water.
11 On all models, unclip the accelerator cable end fitting from the lever balljoint and free the cable from the injection pump bracket.
12 Wipe clean the fuel feed and return unions on the injection pump.
13 Slacken and remove the fuel feed hose union bolt from the pump and recover the sealing washer from each side of the hose union. Also recover the filter (where fitted) from the hose union. Position the hose clear of the pump and screw the union bolt back into position on the pump for safe-keeping. Cover both the hose end and union bolt to prevent the ingress of dirt into the fuel system.
14 Detach the fuel return hose from the pump as described in the previous paragraph. **Note:** *The injection pump feed and return hose union bolts are not interchangeable. Great care must be taken to ensure that the bolts are not swapped.*
15 Wipe clean the pipe unions then slacken the union nut securing the injector pipes to each injector and the four union nuts securing the pipes to the rear of the injection pump; as each pump union nut is slackened, retain the adapter with a suitable open-ended spanner to prevent it being unscrewed from the pump. With all the union nuts undone remove the injector pipes from the engine.
16 Undo the retaining nut and disconnect the wiring from the injection pump stop solenoid. Where necessary, trace the wiring back from the pump microswitch(es) and disconnect it at the connector(s) (as applicable). Free all wiring from any relevant retaining clips.
17 Using a scriber or suitable marker pen, make alignment marks between the injection pump front flange and the front mounting bracket (see illustration). These mark can then be used to ensure that the pump is correctly positioned on refitting.
18 Undo the retaining nuts/bolts securing the injection pump rear mounting bracket in position (see illustration).
19 Slacken and remove the three nuts securing the pump to its front mounting bracket and manoeuvre the pump away from the bracket and out of the engine compartment (see illustrations). Do not rotate the crankshaft or camshaft whilst the pump is removed.

Refitting

20 If a new pump is being installed transfer the alignment mark from the original pump onto the mounting flange of the new pump.
21 Manoeuvre the pump into position and refit its three front retaining nuts. Align the marks made prior to removal then securely tighten the retaining nuts.
22 Refit the rear bracket to the injection pump and securely tighten its retaining nuts/bolts.
23 Refit the injection pump sprocket as described in Section 9 of Chapter 2C. When aligning the camshaft and injection pump sprocket marks with their TDC marks, note that it will also be necessary to rotate the crankshaft slightly and insert the locking pin

4C•6 Fuel and exhaust systems - diesel models

6.17 Mark the injection pump in relation to the mounting bracket (arrowed)

6.18 Unscrew the injection pump rear mounting nuts ...

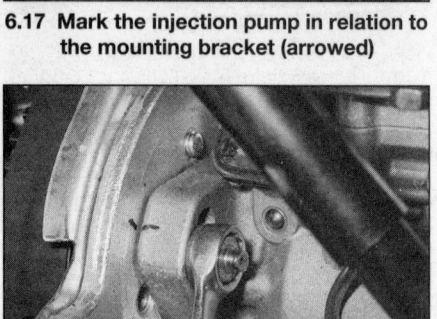

6.19a ... and front mounting nuts ...

6.19b ... and remove the pump from the engine unit

(see Section 3 of Chapter 2C) to position the crankshaft at TDC.

24 With the timing belt correctly fitted, set up the injection pump timing as described in Sections 7 to 9 (as applicable).

25 Reconnect all the relevant wiring to the pump.

26 Reconnect the fuel feed and return hose unions to the pump, not forgetting to fit the filter to the feed hose union (where necessary). Position a new sealing washer on each side of both unions and tighten the union bolts to the specified torque setting.

27 Refit the injector pipes and tighten their union nuts to the specified torque setting.

28 On models where the fast idle valve is mounted onto the injection pump, reconnect the coolant hoses to the fast idle valve and securely tighten their retaining clips. Remove the hose clamps and top-up the cooling system as described in "Weekly Checks".

29 Mop up any spilt fuel/coolant then remove the cover from the alternator.

30 Reconnect the accelerator cable and adjust as described in Section 11.

31 Reconnect the fast idle valve cable and adjust as described in Section 4.

32 Reconnect the battery negative lead and bleed the fuel system using the information in Section 2.

33 On completion, start the engine, and check the fuel injection pump adjustments as described in Chapter 1.

7 Injection timing - checking methods and adjustment

1 Checking the injection timing is not a routine operation. It is only necessary after the injection pump has been disturbed.

2 Dynamic timing equipment does exist, but it is unlikely to be available to the home mechanic. The equipment works by converting pressure pulses in an injector pipe into electrical signals. If such equipment is available, use it in accordance with its maker's instructions.

3 Static timing as described in this Chapter gives good results if carried out carefully. A dial test indicator will be needed, with probes and adapters appropriate to the type of injection pump. Read through the procedures before starting work, to find out what is involved.

8 Injection timing (Lucas fuel injection pump) - checking and adjustment

Caution: The maximum engine speed and transfer pressure settings, together with timing access plugs, are sealed by the manufacturers at the factory using locking wire and lead seals. Do not disturb the wire if the vehicle is still within the warranty period otherwise the warranty will be invalidated. Also do not attempt the timing procedure unless accurate instrumentation is available. Suitable special tools for carrying out pump timing are available from motor factors, and a dial test indicator will be required regardless of the method used. Refer to the precautions given in Section 1 of this Chapter before proceeding.

1 There are two different methods of checking the injection timing on the Lucas pump. The relevant method depends on the type of pump fitted.

2 On models fitted with the earlier type of pump, the timing is checked via the lower of the two covers on the side of the injection pump, whereas on models fitted with the later type of pump the injection timing is checked via the guide on the top of the pump body. The pump type can be easily identified using the suffix letter of the pump identification number which is stamped on a label attached to the side of the pump body.

3 All early pump numbers carry the suffix letter `A' (eg. DPCR 8443 A400A), whereas the later pump numbers carry the suffix letters `B' and onwards (eg. DPCR 8443 B403C). The later type pump is also easily identified by the fact that the lower of the two covers on the side of the pump is replaced with a blanking plug; details of the injection timing are stamped on this plug (see paragraph 29).

Checking and adjustment

Early type pump

4 If the injection timing is being checked with the pump in position on the engine, rather than as part of the pump refitting procedure, disconnect the battery negative lead and cover the alternator with a clean cloth or plastic bag to prevent the possibility of fuel being spilt onto it. Remove the injector pipes as described in paragraph 15 of Section 6.

5 Referring to Chapter 2C, rotate the crankshaft until No.1 cylinder is positioned on TDC at the end of its compression stroke.

6 With No.1 cylinder positioned on TDC at the end of its compression stroke, rotate the crankshaft backwards slightly so that the injection pump sprocket timing mark moves back by one tooth, ie. so that the tooth in front of the timing mark will be aligned with the TDC mark/cover pointer.

7 Slacken and remove the lower of the two large covers on the side of the injection pump. Note: Do not under any circumstances unscrew the upper cover. As the lower cover is removed, position a suitable container beneath the plug to catch any escaping fuel. Mop up any split fuel with a clean cloth.

8 Inside the plug aperture there is a guide which is located in the end of a circlip. The position of this circlip, which represents the point where the injection commences, is determined at the factory and great care must be taken to ensure that the circlip is not moved.

9 Insert the timing probe through the guide

until it contacts the injection pump rotor. If access to the special timing probe can not be gained, (Renault tool number Mot. 877) then a suitable timing probe must be fabricated. The probe should be a snug fit in the guide and have a pointed tip to ensure it fully engages with the groove in the pump rotor **(see illustration)**.

10 With the probe in position, securely mount the dial gauge onto the injection pump so that its tip is in contact with the end of the probe. Ensure that the gauge is directly in line with the probe and is positioned so that its plunger is at the mid-point of its travel.

11 Loosen the pump mounting nuts and bolts and slowly rotate the pump body backwards and forwards whilst observing the dial gauge, to determine the centre point of the groove in the pump rotor. When the centre point of the groove is found zero the dial gauge.

12 Carefully withdraw the timing probe slightly so that it no longer contacts the rotor, whilst taking great care not to disturb the dial gauge. *Note: If the crankshaft is rotated whilst the probe is located in the rotor groove, there is a risk of the probe getting jammed resulting in either the probe breaking or the pump timing circlip being moved.*

13 Rotate the crankshaft through two rotations in the normal direction of rotation (clockwise) until number 1 piston is back at TDC. Referring to Section 3 of Chapter 2C, unscrew the plug from the cylinder block/crankcase and lock the crankshaft in position by inserting a suitable locking pin.

14 Slide the timing probe back into position so that it is contacting the pump rotor. The reading obtained on the dial gauge should be equal to the specified pump timing measurement given in the *Specifications* at the start of this Chapter. If adjustment is necessary, slacken the front and rear pump mounting nuts and bolts and slowly rotate the pump body until the point is found where the specified reading is obtained. When the pump is correctly positioned, tighten both its front and rear mounting nuts and bolts to the specified torque.

15 Withdraw the timing probe slightly, so that it is positioned clear of the pump rotor, and remove the crankshaft locking pin. Rotate the crankshaft through two rotations in the normal direction of rotation, stopping just before No.1 piston reaches TDC so that the injection pump sprocket timing mark is approximately 2 teeth away from its TDC mark.

16 Slide the timing probe back into position so that it is contacting the pump rotor then slowly rotate the crankshaft until the centre point of the groove is found. Zero the gauge, then slightly withdraw the timing probe and continue rotating the crankshaft until No.1 piston reaches TDC and the crankshaft locking pin can be reinserted. Recheck the timing measurement.

17 If adjustment is necessary, slacken the pump mounting nuts and bolts and repeat the operations in paragraphs 14 to 16.

18 When the pump timing is correctly set remove the dial gauge and withdraw the timing probe.

19 Refit the cover and sealing washer to the side of the injection pump and tighten it securely.

20 If the procedure is being carried out as part of the pump refitting sequence, proceed as described in Section 6.

21 If the procedure is being carried out with the pump fitted to the engine, refit the injector pipes tightening their union nuts to the specified torque setting. Reconnect the battery then bleed the fuel system as described in Section 2. Start the engine and adjust the idle speed and anti-stall speeds as described in Chapter 1.

Later type pump

Note: To check the injection pump timing a special timing probe (the probe is part of Renault tool kit Mot. 1079) is required. Without access to this probe, injection pump timing checking will have to be entrusted to a Renault dealer.

22 If the injection timing is being checked with the pump in position on the engine, rather than as part of the pump refitting procedure, disconnect the battery negative lead and cover the alternator with a clean cloth or plastic bag to prevent the possibility of fuel being spilt onto it. Remove the injector pipes as described in paragraph 15 of Section 6.

23 Referring to Section 3 of Chapter 2C, rotate the crankshaft until No.1 cylinder is positioned on TDC at the end of its compression stroke and lock the crankshaft in position with the locking pin.

24 Unscrew the access plug from the guide on the top of the pump body and recover the sealing washer **(see illustration)**.

25 Insert the timing probe (part of Renault tool kit Mot. 1079).

26 With the probe in position, securely mount the dial gauge (dial test indicator) onto the injection pump so that its tip is in contact with the end of the probe. Ensure that the gauge is directly in line with the probe and is positioned so that its plunger is at the mid-point of its travel.

27 Remove the crankshaft locking tool, then turn the crankshaft backwards (anti-clockwise) approximately a quarter-turn. Check that the timing probe is correctly seated against the guide sealing washer surface, then zero the dial gauge. *Note: The timing probe must be seated against the guide sealing washer surface and not the upper lip of the guide for the measurement to be accurate.*

28 Rotate the crankshaft slowly in the correct direction of rotation until the crankshaft locking tool can be re-inserted (bringing the engine back to TDC) **(see illustration)**.

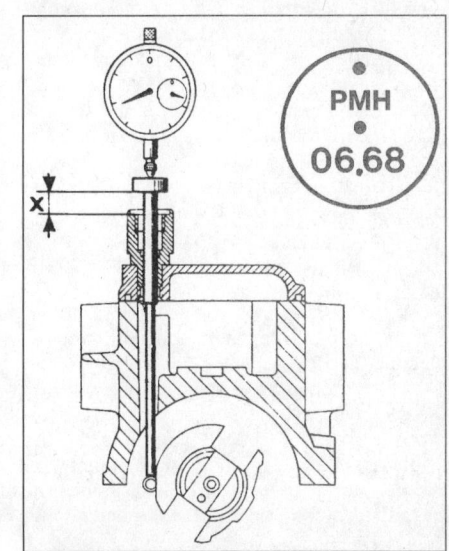

8.24 Removing the injection pump timing access guide plug - later Lucas injection pump

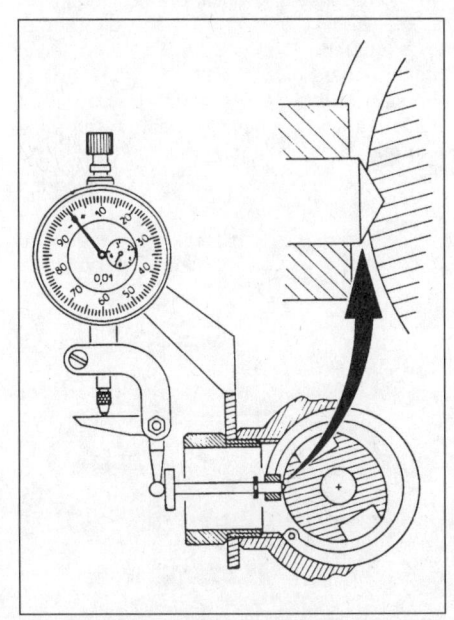

8.9 Checking the injection timing on early type Lucas injection pump

8.28 Checking the injection timing on later type Lucas injection pump. Dimension 'X' must equal the timing value stamped on the plastic disc (inset)

4C•8 Fuel and exhaust systems - diesel models

8.29a Checking the injection pump timing - later Lucas pump

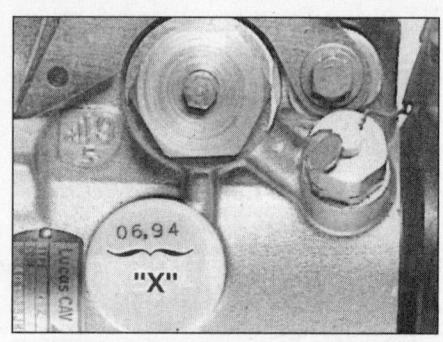

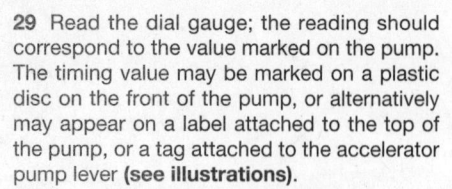

8.29b On later Lucas pumps the timing value `X' is marked on plastic disc on the front of pump . . .

8.29c . . . and/or on a label on the top of the pump and on the accelerator pump lever

29 Read the dial gauge; the reading should correspond to the value marked on the pump. The timing value may be marked on a plastic disc on the front of the pump, or alternatively may appear on a label attached to the top of the pump, or a tag attached to the accelerator pump lever **(see illustrations)**.

30 If adjustment is necessary, slacken the front and rear pump mounting nuts and bolts, then slowly rotate the pump body until the point is found where the specified reading is obtained on the dial gauge. When the pump is correctly positioned, tighten both its front and rear mounting nuts and bolts to the specified torque setting.

31 Withdraw the timing probe slightly, so that it is positioned clear of the pump rotor dowel, and remove the crankshaft locking pin. Rotate the crankshaft through one and three quarter rotations in the normal direction of rotation.

32 Slide the timing probe back into position ensuring that it is correctly seated against the guide sealing washer surface, not the upper lip, then zero the dial gauge.

33 Rotate the crankshaft slowly in the correct direction of rotation until the crankshaft locking tool can be re-inserted (bringing the engine back to TDC). Recheck the timing measurement.

34 If adjustment is necessary, slacken the pump mounting nuts and bolts and repeat the operations in paragraphs 30 to 33.

35 When the pump timing is correctly set remove the dial gauge and withdraw the timing probe.

36 Refit the screw and sealing washer to the guide and tighten it securely.

37 If the procedure is being carried out as part of the pump refitting sequence, proceed as described in Section 6.

38 If the procedure is being carried out with the pump fitted to the engine, refit the injector pipes tightening their union nuts to the specified torque setting. Reconnect the battery then bleed the fuel system as described in Section 2. Start the engine and adjust the idle speed and anti-stall speeds as described in Chapter 1.

9 Injection timing (Bosch fuel injection pump) - checking and adjustment

Caution: *Some of the injection pump settings and access plugs may be sealed by the manufacturers at the factory, using paint or locking wire and lead seals. Do not disturb the seals if the vehicle is still within the warranty period, otherwise the warranty will be invalidated. Also do not attempt the timing procedure unless accurate instrumentation is available.*

1 If the injection timing is being checked with the pump in position on the engine, rather than as part of the pump refitting procedure, disconnect the battery negative lead and cover the alternator with a clean cloth or plastic bag to prevent the possibility of fuel being spilt onto it. Remove the injector pipes as described in paragraph 15 of Section 6 **(see illustration)**.

2 If not already having done so, slacken the clamp screw and/or nut (as applicable) and slide the fast idle cable end fitting arrangement along the cable so that its no longer in contact with the pump fast idle lever (ie. so the fast idle lever returns to its stop).

3 Referring to Section 3 of Chapter 2C, rotate the crankshaft until No.1 cylinder is positioned on TDC at the end of its compression stroke, then turn the crankshaft backwards (anti-clockwise) approximately a quarter-turn.

4 Unscrew the access screw, situated in the centre of the four injector pipe unions, from the rear of the injection pump. As the screw is removed, position a suitable container beneath the pump to catch any escaping fuel. Mop up any split fuel with a clean cloth.

5 Screw the adapter into the rear of the pump and mount the dial gauge in the adapter. If access to the special adapter cannot be gained (Part of Renault tool kit Mot. 1079), they can be purchased from most good motor factors. Position the dial gauge so that its plunger is at the mid-point of its travel and securely tighten the adapter locknut **(see illustration)**.

6 Slowly rotate the crankshaft back and forth whilst observing the dial gauge, to determine when the injection pump piston is at the bottom of its travel (BDC). When the piston is correctly positioned, zero the dial gauge.

7 Rotate the crankshaft slowly in the correct direction of rotation to bring No.1 piston to TDC. Referring to Section 3 of Chapter 1, unscrew the plug from the cylinder lock/crankcase and lock the crankshaft in position by inserting a suitable locking pin.

8 The reading obtained on the dial gauge should be equal to the specified pump timing measurement given in the *Specifications* at the start of this Chapter **(see illustration)**. If adjustment is necessary, slacken the front and rear pump mounting nuts and bolts and

9.1 Unscrewing an injector pipe-to-fuel pump union nut

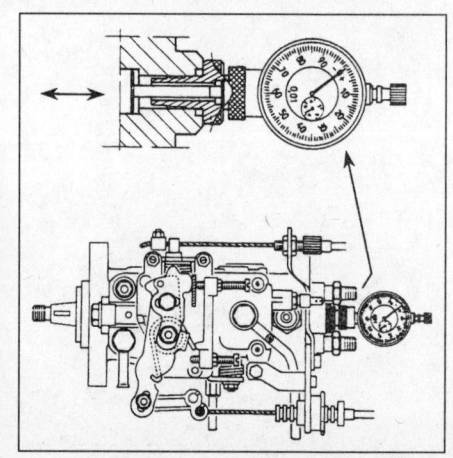

9.5 Checking the injection timing on the Bosch fuel injection pump

Fuel and exhaust systems - diesel models 4C•9

9.8 Checking the injection pump timing - Bosch pump

slowly rotate the pump body until the point is found where the specified reading is obtained. When the pump is correctly positioned, tighten both its front and rear mounting nuts and bolts securely.
9 Rotate the crankshaft through one and three quarter rotations in the normal direction of rotation. Find the injection pump piston BDC as described in paragraph 6 and zero the dial gauge.
10 Rotate the crankshaft slowly in the correct direction of rotation until the crankshaft locking tool can be re-inserted (bringing the engine back to TDC). Recheck the timing measurement.
11 If adjustment is necessary, slacken the pump mounting nuts and bolts and repeat the operations in paragraphs 8 to 10.
12 When the pump timing is correctly set unscrew the adapter and remove the dial gauge.
13 Refit the screw and sealing washer to the pump and tighten it securely.
14 If the procedure is being carried out as part of the pump refitting sequence, proceed as described in Section 6.
15 If the procedure is being carried out with the pump fitted to the engine, refit the injector pipes tightening their union nuts to the specified torque setting. Reconnect the battery then bleed the fuel system as described in Section 2. Start the engine and adjust the idle speed and anti-stall speeds as described in Chapter 1.

10 Fuel injectors - testing, removal and refitting

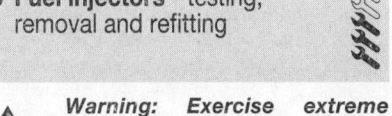

Warning: *Exercise extreme caution when working on the fuel injectors. Never expose the hands or any part of the body to injector spray, as the high working pressure can cause the fuel to penetrate the skin, with possibly fatal results. You are strongly advised to have any work which involves testing the injectors under pressure carried out by a dealer or fuel injection specialist.*

Testing

1 Injectors do deteriorate with prolonged use, and it is reasonable to expect them to need reconditioning or renewal after 60 000 miles (100 000 km) or so. Accurate testing, overhaul and calibration of the injectors must be left to a specialist. A defective injector which is causing knocking or smoking can be located without dismantling as follows.
2 Run the engine at a fast idle. Slacken each injector union in turn, placing rag around the union to catch spilt fuel, and being careful not to expose the skin to any spray. When the union on the defective injector is slackened, the knocking or smoking will stop.

Removal

Note: *Take great care not to allow dirt into the injectors or fuel pipes during this procedure. Do not drop the injectors or allow the needles at their tips to become damaged. The injectors are precision-made to fine limits and must not be handled roughly. In particular, do not mount them in a bench vice.*

Clamp-type injector

3 Disconnect the battery negative lead and cover the alternator with a clean cloth or plastic bag to prevent the possibility of fuel being spilt onto it.
4 Wipe clean the injector then unscrew the union bolt and disconnect the return hose from the top of the injector **(see illustration)**. Recover the sealing washer positioned on each side of the hose union and cover the hose and injector union to prevent the entry of dirt into the system.
5 Slacken the union nut and free the injector pipe from the side of the injector. Cover the hose and injector union to prevent the entry of dirt into the system.
6 Slacken and remove the two retaining nuts and washers and lift off the injector retaining clamp.
7 Lift out the injector and recover the sealing and flame shield washers. Also remove the injector sleeve if it is a loose fit in the head.

Screw-type injector

8 Disconnect the battery negative lead and cover the alternator with a clean cloth or

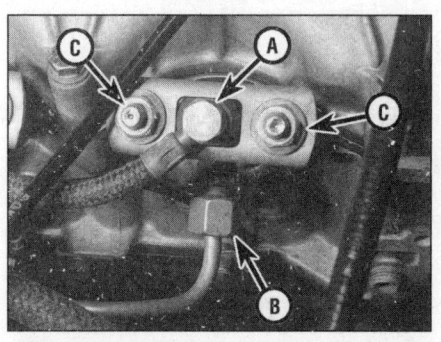

10.4 View of a typical clamp-type injector

A *Return hose union bolt*
B *Injector pipe union nut*
C *Retaining nuts*

plastic bag to prevent the possibility of fuel being spilt onto it.
9 Carefully clean around the injectors and pipe union nuts and disconnect the return pipe(s) from the injector **(see illustration)**.
10 Wipe clean the pipe unions then slacken the union nut securing the injector pipes to each injector and the four union nuts securing the pipes to the rear of the injection pump; as each pump union nut is slackened, retain the adapter with a suitable open-ended spanner to prevent it being unscrewed from the pump **(see illustration)**. With all the union nuts undone remove the injector pipes from the engine. Cover the injector and pipe unions to prevent the entry of dirt into the system.
11 Unscrew the injector, using a deep socket or box spanner, and remove it from the cylinder head **(see illustrations)**.
12 Recover the sealing and flame shield washers. Also remove the injector sleeve if it is a loose fit in the cylinder head **(see illustrations)**.

Refitting

Clamp-type injector

13 Obtain a new sealing washer and flame shield washer. Where removed, also renew the injector sleeve if it is damaged.
14 Where necessary, refit the injector sleeve to the cylinder head.

10.9 On screw-type injectors, disconnect the return pipe(s) ...

10.10 ... and unscrew the injector pipe union nut

4C•10 Fuel and exhaust systems - diesel models

10.11a Unscrew the injector...

10.11b ...and remove it from the cylinder head

10.12a Recover the sealing washer...

10.12b ...and flame shield washer from the injector aperture

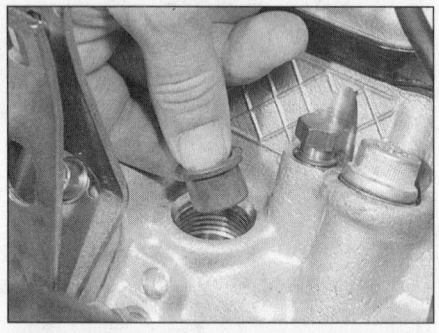

10.12c Also remove the injector sleeve if it is a loose fit in the cylinder head

15 Fit the new flame shield washer to the sleeve, noting that it should be fitted with its convex side downwards (facing the cylinder head).
16 Fit the new sealing washer to the top of the sleeve.
17 Slide the injector into position ensuring that it enters the sleeve squarely.
18 Reconnect the injector pipe and tighten its union nut by hand only at this stage.
19 Install the injector clamp and refit the washers and retaining nuts. Tighten the retaining nuts evenly and progressively to the specified torque setting.
20 Position a new sealing washer on each side of the return hose union and refit the union bolt to the top of the injector. Tighten both the union bolt and the injector pipe union nut to their specified torque settings.
21 Start the engine. If difficulty is experienced, bleed the fuel system as described in Section 2.

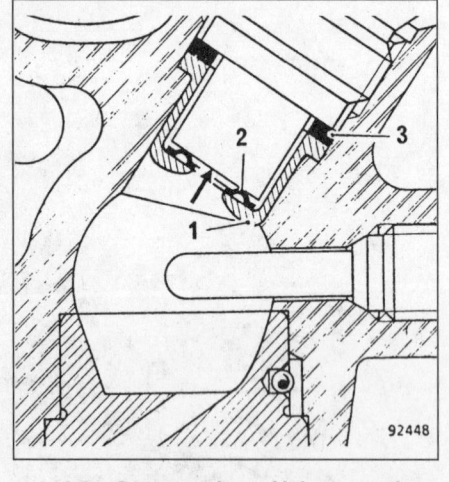

Screw-type injector

22 Obtain a new sealing washer and flame shield washer. Where removed, also renew the injector sleeve if it is damaged.
23 Where necessary, refit the injector sleeve to the cylinder head.
24 Fit the new flame shield washer to the sleeve making sure its should be fitted with the convex side upwards (facing the injector) **(see illustration)**.
25 Fit the new sealing washer to the top of the sleeve.
26 Screw the injector into position and tighten it to the specified torque.
27 Refit the injector pipes and tighten the union nuts to the specified torque setting. Position any clips attached to the pipes as noted before removal.
28 Reconnect the return pipe securely to the injector.
29 Start the engine. If difficulty is experienced, bleed the fuel system as described in Section 2.

10.24 Cutaway view of injector and associated components - screw-type injector

1 Injector sleeve
2 Flame shield washer
3 Sealing washer

11 Accelerator cable - removal, refitting and adjustment

Removal

1 Working in the engine compartment, free the accelerator inner cable from the injection pump lever, then pull the outer cable out from its mounting bracket rubber grommet. Remove the spring clip from the cable end and the rubber grommet from the cable bracket.
2 Working back along the length of the cable, free it from any retaining clips or ties, noting its correct routing.
3 From inside the vehicle, reach up behind the facia panel, and disconnect the inner cable from the top of the accelerator pedal. If necessary, to improve access remove the facia undercover panel (see Chapter 11). Tie a length of string to the inner cable then return to the engine compartment.
4 Release the outer cable from the bulkhead and withdraw the cable. When the end of cable appears, untie the string and leave it in position - it can then be used to draw the cable back into position on refitting. Remove the rubber sealing grommet

Refitting

5 Fit the rubber sealing grommet to the cable. Tie the string to the end of the cable, then use the string to draw the cable into position through the bulkhead. Once the cable end is visible, untie the string, and clip the inner cable into position in the pedal.
6 Check that the cable is securely retained, then refit the facia panel (where removed).
7 From within the engine compartment, ensure that the outer cable is correctly seated in the bulkhead then work along the cable, securing it in position with the retaining clips and ties, and ensuring that the cable is correctly routed.
8 Make sure the rubber grommet is in position in the mounting bracket and fit the spring clip to the cable end.
9 Pass the outer cable through the mounting bracket grommet, and reconnect the inner

cable to the throttle cam. Adjust the cable as described below.

Adjustment

10 Remove the spring clip from the accelerator outer cable. Ensuring that the injection pump lever is fully against its stop, gently pull the cable out of its grommet until all free play is removed from the inner cable.
11 With the cable held in this position, refit the spring clip to the last exposed outer cable groove in front of the rubber grommet. When the clip is refitted and the outer cable is released, there should be only a small amount of free play in the inner cable.
12 Have an assistant depress the accelerator pedal, and check that the injection pump lever opens fully and returns smoothly to its stop.

12 Accelerator pedal - removal and refitting

Refer to Chapter 4A, Section 7.

13 Fuel level sender - removal and refitting

Refer to Chapter 4A, Section 4.

14 Fuel tank - removal and refitting

Refer to Chapter 4A, Section 5.

15 Manifolds - removal and refitting

Removal

1 Although the manifolds are separate, they are retained by the same nuts, since the stud holes are split between the manifold flanges. Since they share the same gasket both manifolds should be removed to allow the gasket to be replaced.
2 Note the location of any wiring or hose brackets/clips attached to the manifolds, and remove them. Where necessary also unbolt the dipstick tube from the manifold.
3 Release the retaining clip, and disconnect the inlet duct from the inlet manifold.
4 On early normally-aspirated models where the air filter is housed in the manifold, remove the filter and housing as described in Chapter 1.
5 On Turbo models remove the turbocharger as described in Section 16.
6 On normally-aspirated models disconnect the exhaust system from the manifold using the information given in Section 19.
7 Where necessary, undo the two nuts and remove the heatshield from the inlet manifold.
8 On models equipped with an EGR system, undo the two bolts/nuts (as applicable) and disconnect the EGR pipe from the exhaust manifold. Recover the gasket and discard it, a new one should be used on refitting.
9 Disconnect the relevant breather/vacuum hoses from the inlet manifold, noting their correct fitted location.
10 Progressively unscrew the nuts/bolts (as applicable) securing the inlet and exhaust manifolds. Where necessary, free the PCV pipes from the manifold studs and position them clear of the manifolds.
11 Withdraw the inlet and exhaust manifolds from the cylinder head and recover the manifold gasket.

Refitting

12 Refitting is a reversal of removal, bearing in mind the following points.
  a) Ensure that the cylinder head and manifold mating surfaces are clean and use a new gasket.
  b) On Turbo models refit the turbocharger as described in Section 16.
  c) On normally-aspirated models reconnect the exhaust system to the manifold with reference to Section 19.
  d) Ensure that any wiring or hose brackets/clips are positioned as noted before removal.

16 Turbocharger - general information, removal and refitting

General information

1 Refer to Chapter 4B, Section 17.

Removal

2 Disconnect the battery negative terminal then undo the retaining nuts and remove the heatshield from the turbocharger.
3 Slacken the retaining clips and disconnect both inlet ducts from the turbocharger.
4 Undo the retaining bolt securing the support bracket to the turbocharger.
5 Undo the two nuts and free the exhaust system front pipe from the base of the turbocharger.
6 Whilst being prepared from some oil spillage, disconnect the oil supply pipe from the top of the top of the turbocharger. Where the pipe is secured in position with a union nut, simply slacken the nut and disconnect. Where a union bolt is used, slacken and remove the bolt and discard the sealing washers; new ones should be used on refitting. Where the pipe union is secured in position by two bolts, slacken and remove the bolts and discard the gasket fitted between the union and turbocharger; a new one must be used on refitting.
7 Be prepared for oil spillage and disconnect the oil return pipe/hose from the base of the turbocharger. Where a hose is fitted, slacken the retaining clip and detach the hose. Where a pipe is fitted unscrew the retaining bolts and discard the gasket fitted between the pipe union and turbocharger; a new gasket will be needed on refitting
8 Plug the hose and pipe ends to minimise oil loss and prevent the entry of dirt into the system.
9 Slacken and remove the four nuts retaining nuts and remove the turbocharger from the exhaust manifold. Discard the four self-locking retaining nuts; new ones must be used on refitting.

Refitting

10 Ensure the manifold and turbocharger mating surfaces are clean and dry and manoeuvre the turbocharger into position.
11 Fit the new turbocharger retaining nuts and tighten them securely.
12 Reconnect the exhaust system front pipe, ensuring its mating surface is clean, and securely tighten its retaining nuts.
13 Refit the support bracket retaining bolt and tighten it securely.
14 Where an oil return hose is fitted, reconnect the hose and securely tighten its retaining clip. On models where a pipe is fitted, fit a new gasket between the pipe union and turbocharger and securely tighten the union retaining bolts.
15 Using a suitable oil can, inject some clean engine oil into the turbocharger oil supply pipe hole. Once the pipe is full of oil, reconnect the feed pipe. On models with a union nut, securely tighten the nut. On models where a union bolt is used, position a new sealing washer on each side of the hose union and securely tighten the union bolt. On other models fit a new gasket and securely tighten the union retaining bolts.
16 Reconnect the inlet ducts and securely tighten the retaining clips.
17 Refit the turbocharger heatshield.
18 Disconnect the wiring from the injection pump stop solenoid and reconnect the battery.
19 Crank the engine on the starter motor until the instrument panel oil pressure warning lamp goes out (this may take several seconds).
20 Reconnect the wiring to the stop solenoid, then start the engine using the normal procedure. Run the engine at idle speed, and check the turbocharger oil unions for leakage. Rectify any problems without delay.

4C•12 Fuel and exhaust systems – diesel models

17 Intercooler – removal and refitting

Refer to Chapter 4B, Section 18.

18 Air cleaner housing – removal and refitting

Removal

Models with the air cleaner mounted in the inlet manifold

1 Remove the air cleaner filter and housing as described in Chapter 1. If necessary, the ducts can be unclipped and removed, the resonator housing is retained by a rubber strap.

All other models

2 Release the retaining clips and disconnect the inlet ducts from the base of the air cleaner housing. Release the rubber retaining strap and remove the air cleaner housing from the engine compartment. The various ducts can be removed once their retaining clips have been released

Refitting

3 Refitting is a reversal of the relevant removal procedure ensuring all inlet ducts are securely reconnected.

19 Exhaust system – general information and component renewal

The exhaust system consists of the four sections; the front pipe, the front intermediate pipe, the rear intermediate pipe, and the tailpipe and main silencer. All exhaust sections are joined by flanged joints. The front pipe joints are secured by nuts and bolts, the front intermediate pipe joint being of the spring-loaded ball type, to allow for movement in the exhaust system, and the other joints are secured by clamping rings.

Refer to Chapter 4A, Section 14 for removal and refitting details, noting the above information.

20 Boost pressure fuel delivery (LDA) corrector (Turbo models) – general information

1 This device is mounted on the top of the injection pump and is connected to the inlet manifold by a vacuum pipe. Its purpose is to adjust the injection pump fuel metering in relation to the turbocharger boost pressure. Effectively, the quantity of fuel-injected is increased as the boost pressure increases.

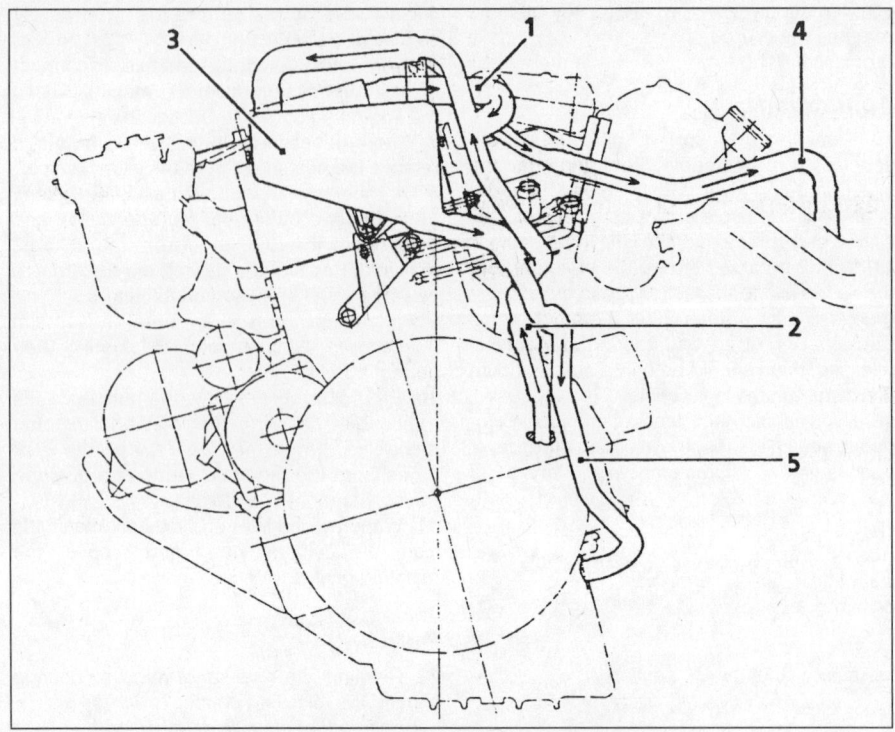

2 An adjustment screw is provided, but this is sealed at factory, and no attempt should be made to carry out adjustments without the use of specialist test equipment.
3 If a fault with the device is suspected, consult a Renault dealer or a suitably qualified specialist.

21 Emission control systems – general information

1 Certain engines are equipped with systems designed to reduce the emission of harmful by-products of the combustion process into the atmosphere.
2 The following systems may be fitted according to model.

Crankcase emission control system

3 A crankcase ventilation system is fitted to all models.
4 Oil fumes and piston blow-by gases (combustion gases which have passed by the piston rings) are drawn from the crankcase through an auxiliary oil separator, through the main oil separator, into the air inlet tract **(see illustration)**. The gases are then drawn into the engine together with fresh air/fuel mixture. Condensed oil vapour is returned from the main oil separator to the engine sump.

21.4 Crankcase emission control system

1 Cylinder head cover hose
2 Sump hose
3 Oil separator
4 Inlet manifold hose (Turbo models only)
5 Return pipe to the sump

Exhaust gas recirculation (EGR) system

5 This system is fitted to later turbo models (fitted with J8S 738 engine).
6 The system is designed to recirculate small quantities of exhaust gas into the inlet tract, and therefore into the combustion process. This process reduces the level of oxides of nitrogen present in the final exhaust gas which is released into the atmosphere, and also lowers the combustion temperature.
7 The volume of exhaust gas recirculated is controlled by vacuum supplied from the brake servo vacuum pump, via a solenoid valve controlled by a micro-switch mounted on the injection pump. A temperature valve fitted in the vacuum supply line cuts off the vacuum supply until the engine reaches normal operating temperature.
8 A vacuum-operated recirculation valve is fitted to regulate the quantity of exhaust gas recirculated. The valve is operated by the vacuum supplied by the brake vacuum pump via the solenoid valve and the temperature valve.
9 Between idling speed and a pre-determined engine load, the micro-switch on the injection pump closes to supply power to the solenoid valve, which allows the recirculation valve to open. Under full-load conditions, the micro-switch opens to cut-off exhaust gas recirculation. Additional control is provided by the temperature valve, which cuts off the

vacuum supply until the coolant temperature reaches a predetermined temperature (approximately 40°C), preventing the recirculation valve from opening during the engine warm-up phase.

22 Emission control systems - testing and component renewal

Crankcase emission control system components

Testing

1 If the system is thought to be faulty, firstly, check that the hoses are unobstructed. On high mileage vehicles, particularly when regularly used for short journeys, a jelly-like deposit may be evident inside the system hoses and oil separators. If excessive deposits are present, the relevant component(s) should be removed and cleaned.
2 Periodically inspect the system components for security and damage, and renew them as necessary. Note that damaged or loose hoses can cause various engine running problems (erratic idle speed, stalling, etc) which can be difficult to trace.

Component renewal

3 Renewal procedures for the hoses and oil separator are self-evident.

Exhaust gas recirculation system

Testing

4 Start the engine, and run it until it reaches normal operating temperature (the cooling fan should have cut in and out at least once).
5 With the engine idling, disconnect the vacuum hose from the recirculation valve. As the hose is disconnected, it should be possible to hear the valve click shut. If no click is heard, proceed as follows.
6 Check that vacuum is present at the recirculation valve end of the vacuum hose. If a vacuum gauge is available, check that the vacuum is at least 500 mbars. If vacuum is present, it is likely that the recirculation valve is faulty (jammed or pierced diaphragm). If no vacuum is present, carry out the following checks.
7 Check the security of all vacuum hose connections.
8 Check the electrical feed to the solenoid valve.
9 Check the operation of the temperature valve. This can be done by checking that vacuum will pass through the valve with the engine at normal operating temperature. Stop the engine and disconnect the temperature valve vacuum hoses at the recirculation valve and the solenoid valve, and check that it is possible to blow through the hoses. If not, it is likely that the temperature valve is faulty, or the hoses are obstructed.
10 Check the operation of the micro-switch on the injection pump, using a continuity tester or an ohmmeter. With the pump accelerator lever in the idle position, the switch should be open (no continuity/infinite resistance). As the accelerator lever is moved from the idle position, the switch should close (continuity/zero resistance), until the lever reaches a predetermined position towards the end of its travel, when the switch should open again. Note that adjustment of the switch should not be attempted. If it is suspected that adjustment is required, or if a new switch is fitted, consult a Renault dealer who will have access to the specialist calibration equipment required to carry out accurate adjustment.

Recirculation valve - renewal

11 The valve is mounted onto the inlet duct just above the intercooler and is connected to the exhaust manifold via a braided hose.
12 Disconnect the vacuum hose from the valve.
13 Undo the retaining bolts and release the valve from the inlet duct. Undo the two bolts securing the valve hose to the exhaust manifold and remove the valve and hose assembly from the engine compartment. Recover the gaskets and discard them; new ones must be used on refitting.
14 If a new valve is to be fitted, unscrew the retaining bolts and transfer the pipe to the new valve, using a new gasket.
15 Refitting is a reversal of removal, using new gaskets and ensuring that all retaining bolts are securely tightened.

Solenoid valve - renewal

16 The valve is mounted on a bracket in the engine compartment. To locate the valve, trace the vacuum hose back from the recirculation valve.
17 Disconnect the wiring plug and the vacuum hoses from the valve.
18 Unscrew the securing nut(s), and withdraw the valve complete with its bracket.
19 Refitting is a reversal of removal, ensuring that the vacuum hoses are securely reconnected.

Temperature valve

20 The temperature valve is screwed into the cooling system hose at the rear of the cylinder head.
21 To remove the valve, simply disconnect the vacuum hoses and unscrew the valve from the hose. Be prepared for coolant spillage and recover the sealing ring. Plug the hose aperture to minimise coolant loss.
22 Refitting is reversal of removal using a new sealing ring. Check for coolant leaks on completion, and if necessary, top-up the coolant level as described in *Weekly checks*.

Chapter 5 Part A:
Starting and charging systems

Contents

Alternator - removal and refitting	7	Electrical system check	See "Weekly Checks"
Alternator testing and overhaul	8	General information and precautions	1
Alternator drivebelt - removal, refitting and tensioning	6	Ignition switch - removal and refitting	12
Battery - removal and refitting	4	Oil level sensor - removal and refitting	14
Battery - testing and charging	3	Oil pressure warning light switch - removal and refitting	13
Battery check	See "Weekly Checks"	Starter motor - removal and refitting	10
Charging system - testing	5	Starter motor - testing and overhaul	11
Electrical fault finding - general information	2	Starting system - testing	9

Degrees of difficulty

Easy, suitable for novice with little experience	**Fairly easy,** suitable for beginner with some experience	**Fairly difficult,** suitable for competent DIY mechanic	**Difficult,** suitable for experienced DIY mechanic	**Very difficult,** suitable for expert DIY or professional

Specifications

System type .... 12-volt, negative earth

Battery
Charge condition:
  Poor .... 12.5 volts
  Normal .... 12.6 volts
  Good .... 12.7 volts

Alternator
Type .... Paris-Rhone

Starter motor
Type .... Valeo, Bosch or Paris-Rhone (depending on model)

1 General information and precautions

General information

1 The engine electrical system consists mainly of the charging and starting systems. Because of their engine-related functions, these components are covered separately from the body electrical devices such as the lights, instruments, etc (which are covered in Chapter 12). On petrol engine models refer to Part B for information on the ignition system, and on diesel models refer to Part C for information on the preheating system.

2 The electrical system is of the 12-volt negative earth type.

3 The battery is of the low maintenance or "maintenance-free" (sealed for life) type and is charged by the alternator, which is belt-driven from the crankshaft pulley.

4 The starter motor is of the pre-engaged type incorporating an integral solenoid. On starting, the solenoid moves the drive pinion into engagement with the flywheel ring gear before the starter motor is energised. Once the engine has started, a one-way clutch prevents the motor armature being driven by the engine until the pinion disengages from the flywheel.

Precautions

5 Further details of the various systems are given in the relevant Sections of this Chapter. While some repair procedures are given, the usual course of action is to renew the component concerned. The owner whose interest extends beyond mere component renewal should obtain a copy of the *"Automobile Electrical & Electronic Systems Manual"*, available from the publishers of this manual.

6 It is necessary to take extra care when working on the electrical system to avoid damage to semi-conductor devices (diodes and transistors), and to avoid the risk of personal injury. In addition to the precautions given in *"Safety first!"* at the beginning of this manual, observe the following when working on the system:

7 *Always remove rings, watches, etc before working on the electrical system.* Even with the battery disconnected, capacitive discharge could occur if a component's live

5A•2 Starting and charging systems

terminal is earthed through a metal object. This could cause a shock or nasty burn.

8 *Do not reverse the battery connections.* Components such as the alternator, electronic control units, or any other components having semi-conductor circuitry could be irreparably damaged.

9 If the engine is being started using jump leads and a slave battery, connect the batteries *positive-to-positive* and *negative-to-negative* (see *"Booster battery (jump) starting"*). This also applies when connecting a battery charger.

10 Never disconnect the battery terminals, the alternator, any electrical wiring or any test instruments when the engine is running.

11 Do not allow the engine to turn the alternator when the alternator is not connected.

12 Never "test" for alternator output by 'flashing' the output lead to earth.

13 Never use an ohmmeter of the type incorporating a hand-cranked generator for circuit or continuity testing.

14 Always ensure that the battery negative lead is disconnected when working on the electrical system.

15 Before using electric-arc welding equipment on the car, disconnect the battery, alternator and components such as the fuel injection/ignition electronic control unit to protect them from the risk of damage.

16 The radio/cassette unit fitted as standard equipment by Renault is equipped with a built-in security code to deter thieves. If the power source to the unit is cut, the anti-theft system will activate. Even if the power source is immediately reconnected, the radio/cassette unit will not function until the correct security code has been entered. Therefore, if you do not know the correct security code for the radio/cassette unit **do not** disconnect the battery negative terminal of the battery or remove the radio/cassette unit from the vehicle. Refer to *"Radio/cassette unit anti-theft system precaution"* in the *Reference* Section at the end of this manual for further information.

2 Electrical fault finding - general information

Refer to Chapter 12.

3 Battery - testing and charging

Standard and low maintenance battery - testing

1 If the vehicle covers a small annual mileage it is worthwhile checking the specific gravity of the electrolyte every three months to determine the state of charge of the battery. Use a hydrometer to make the check and compare the results with the following table. Note that the specific gravity readings assume an electrolyte temperature of 15°C (60°F); for every 10°C (48°F) below 15°C (60°F) subtract 0.007. For every 10°C (48°F) above 15°C (60°F) add 0.007.

	Above 25°C(77°F)	Below 25°C(77°F)
Fully-charged	1.210 to 1.230	1.270 to 1.290
70% charged	1.170 to 1.190	1.230 to 1.250
Discharged	1.050 to 1.070	1.110 to 1.130

2 If the battery condition is suspect, first check the specific gravity of electrolyte in each cell. A variation of 0.040 or more between any cells indicates loss of electrolyte or deterioration of the internal plates.

3 If the specific gravity variation is 0.040 or more, the battery should be renewed. If the cell variation is satisfactory but the battery is discharged, it should be charged as described later in this Section.

Maintenance-free battery - testing

4 In cases where a "sealed for life" maintenance-free battery is fitted, topping-up and testing of the electrolyte in each cell is not possible. The condition of the battery can therefore only be tested using a battery condition indicator or a voltmeter.

5 On some models a "Delco" type maintenance-free battery, with a built-in charge condition indicator may be fitted. The indicator is located in the top of the battery casing, and indicates the condition of the battery from its colour. If the indicator shows green, then the battery is in a good state of charge. If the indicator turns darker, eventually to black, then the battery requires charging, as described later in this Section. If the indicator shows clear/yellow, then the electrolyte level in the battery is too low to allow further use, and the battery should be renewed. **Do not** attempt to charge, load or jump start a battery when the indicator shows clear/yellow.

6 If testing the battery using a voltmeter, connect the voltmeter across the battery and compare the result with those given in the *Specifications* under "charge condition". The test is only accurate if the battery has not been subjected to any kind of charge for the previous six hours. If this is not the case, switch on the headlights for 30 seconds, then wait four to five minutes before testing the battery after switching off the headlights. All other electrical circuits must be switched off, so check that the doors and tailgate are fully shut when making the test.

7 If the voltage reading is less than 12.2 volts, then the battery is discharged, whilst a reading of 12.2 to 12.4 volts indicates a partially discharged condition.

8 If the battery is to be charged, remove it from the vehicle (Section 4) and charge it as described later in this Section.

Standard and low maintenance battery - charging

Note: *The following is intended as a guide only. Always refer to the manufacturer's recommendations (often printed on a label attached to the battery) before charging a battery.*

9 Charge the battery at a rate of 3.5 to 4 amps and continue to charge the battery at this rate until no further rise in specific gravity is noted over a four hour period.

10 Alternatively, a trickle charger charging at the rate of 1.5 amps can safely be used overnight.

11 Specially rapid 'boost' charges which are claimed to restore the power of the battery in 1 to 2 hours are not recommended, as they can cause serious damage to the battery plates through overheating.

12 While charging the battery, note that the temperature of the electrolyte should never exceed 37.8°C (100°F).

Maintenance-free battery - charging

Note: *The following is intended as a guide only. Always refer to the manufacturer's recommendations (often printed on a label attached to the battery) before charging a battery.*

13 This battery type takes considerably longer to fully recharge than the standard type, the time taken being dependent on the extent of discharge, but it can take anything up to three days.

14 A constant voltage type charger is required, to be set, when connected, to 13.9 to 14.9 volts with a charger current below 25 amps. Using this method, the battery should be usable within three hours, giving a voltage reading of 12.5 volts, but this is for a partially discharged battery and, as mentioned, full charging can take considerably longer.

15 If the battery is to be charged from a fully discharged state (condition reading less than 12.2 volts), have it recharged by your Renault dealer or local automotive electrician, as the charge rate is higher and constant supervision during charging is necessary.

4 Battery - removal and refitting

Removal

1 The battery is located on the right-hand side of the engine compartment on petrol models, and the left-hand side on diesel models.

2 Unscrew the fastener from the top of the battery negative (earth) terminal and disconnect the lead(s).

Starting and charging systems 5A•3

3 Remove the insulation cover (where fitted) and disconnect the positive terminal lead(s) in the same way.
4 Unscrew the nut/bolts (as applicable) and remove battery retaining clamp.
5 Lift the battery out of the engine compartment. If necessary, undo the retaining bolts then release all the relevant clips securing the wiring to the tray and remove the battery tray from the engine compartment.
6 If necessary, with the tray removed, undo the retaining bolts and remove the mounting plate from the top of the left-hand engine/transmission mounting.

Refitting

7 Refitting is a reversal of removal, but smear petroleum jelly on the terminals when reconnecting the leads, and always reconnect the positive lead first, and the negative lead last.

5 Charging system - testing

Note: *Refer to the warnings given in "Safety first!" and in Section 1 of this Chapter before starting work.*
1 If the ignition warning light fails to illuminate when the ignition is switched on, first check the alternator wiring connections for security. If satisfactory, check that the warning light bulb has not blown, and that the bulbholder is secure in its location in the instrument panel. If the light still fails to illuminate, check the continuity of the warning light feed wire from the alternator to the bulbholder. If all is satisfactory, the alternator is at fault and should be renewed or taken to an auto-electrician for testing and repair.
2 If the ignition warning light illuminates when the engine is running, stop the engine and check that the drivebelt is correctly tensioned (see Chapter 1) and that the alternator connections are secure. If all is so far satisfactory, have the alternator checked by an auto-electrician for testing and repair.
3 If the alternator output is suspect even though the warning light functions correctly, the regulated voltage may be checked as follows.
4 Connect a voltmeter across the battery terminals and start the engine.
5 Increase the engine speed until the voltmeter reading remains steady; the reading should be approximately 12 to 13 volts, and no more than 14 volts.
6 Switch on as many electrical accessories (eg, the headlights, heated rear window and heater blower) as possible, and check that the alternator maintains the regulated voltage at around 13 to 14 volts.
7 If the regulated voltage is not as stated, the fault may be due to worn brushes, weak brush springs, a faulty voltage regulator, a faulty diode, a severed phase winding or worn or damaged slip rings. The alternator should be renewed or taken to an auto-electrician for testing and repair

6 Alternator drivebelt - removal, refitting and tensioning

Refer to the procedure given for the auxiliary drivebelt(s) in Chapter 1.

7 Alternator - removal and refitting

Removal
1 Disconnect the battery negative lead. Where necessary, to improve access to the alternator remove the air cleaner housing (see Chapter 4).
2 Slacken the auxiliary drivebelt as described in Chapter 1 and disengage it from the alternator pulley.
3 Remove the rubber covers (where fitted) from the alternator terminals, then unscrew the retaining nut(s) and disconnect the wiring from the rear of the alternator **(see illustration)**.
4 Undo the retaining screws and remove the cover plate from the alternator (where fitted).
5 Slacken and remove the nut and bolt securing the base of the alternator to the adjuster bracket then unscrew the nut and withdraw the alternator pivot bolt **(see illustration)**.
6 Manoeuvre the alternator away from its mounting brackets and out from the engine compartment.

Refitting
7 Refitting is a reversal of removal, tensioning the auxiliary drivebelt as described in Chapter 1 and ensuring that the alternator mountings are securely tightened.

7.3 Alternator wiring connections

8 Alternator - testing and overhaul

If the alternator is thought to be suspect, it should be removed from the vehicle and taken to an auto-electrician for testing. Most auto-electricians will be able to supply and fit brushes at a reasonable cost. However, check on the cost of repairs before proceeding as it may prove more economical to obtain a new or exchange alternator.

9 Starting system - testing

Note: *Refer to the precautions given in "Safety first!" and in Section 1 of this Chapter before starting work.*
1 If the starter motor fails to operate when the ignition key is turned to the appropriate position, the following possible causes may be to blame.
 a) *The battery is faulty.*
 b) *The electrical connections between the switch, solenoid, battery and starter motor are somewhere failing to pass the necessary current from the battery through the starter to earth.*
 c) *The solenoid is faulty.*
 d) *The starter motor is mechanically or electrically defective.*
2 To check the battery, switch on the headlights. If they dim after a few seconds, this indicates that the battery is discharged - recharge (see Section 3) or renew the battery. If the headlights glow brightly, operate the ignition switch and observe the lights. If they dim, then this indicates that current is reaching the starter motor, therefore the fault must lie in the starter motor. If the lights continue to glow brightly (and no clicking sound can be heard from the starter motor solenoid), this indicates that there is a fault in the circuit or solenoid - see following paragraphs. If the starter motor turns slowly when operated, but the battery is in good condition, then this indicates that either the

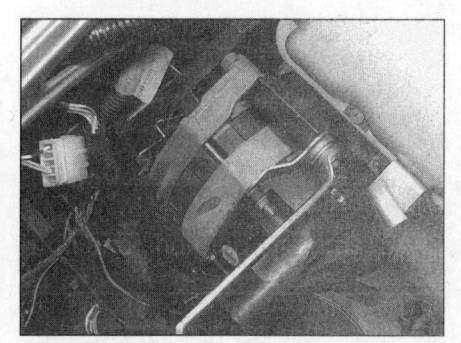

7.5 Unscrewing the alternator pivot bolt

5A•4 Starting and charging systems

starter motor is faulty, or there is considerable resistance somewhere in the circuit.

3 If a fault in the circuit is suspected, disconnect the battery leads (including the earth connection to the body), the starter/solenoid wiring and the engine/transmission earth strap. Thoroughly clean the connections, and reconnect the leads and wiring, then use a voltmeter or test lamp to check that full battery voltage is available at the battery positive lead connection to the solenoid, and that the earth is sound. Smear petroleum jelly around the battery terminals to prevent corrosion - corroded connections are amongst the most frequent causes of electrical system faults.

4 If the battery and all connections are in good condition, check the circuit by disconnecting the wire from the solenoid blade terminal. Connect a voltmeter or test lamp between the wire end and a good earth (such as the battery negative terminal), and check that the wire is live when the ignition switch is turned to the `start' position. If it is, then the circuit is sound - if not the circuit wiring can be checked as described in Chapter 12.

5 The solenoid contacts can be checked by connecting a voltmeter or test lamp between the battery positive feed connection on the starter side of the solenoid, and earth. When the ignition switch is turned to the `start' position, there should be a reading or lighted bulb, as applicable. If there is no reading or lighted bulb, the solenoid is faulty and should be renewed.

6 If the circuit and solenoid are proved sound, the fault must lie in the starter motor. In this event, it may be possible to have the starter motor overhauled by a specialist, but check on the cost of spares before proceeding, as it may prove more economical to obtain a new or exchange motor.

10 Starter motor - removal and refitting

Removal

1 Disconnect the battery negative lead.
2 So that access to the motor can be gained both from above and below, firmly apply the handbrake then jack up the front of the vehicle and support it on axle stands (see "Jacking and vehicle support"). Proceed as described under the relevant sub-heading.

4-cylinder models

3 Slacken and remove the retaining nuts and disconnect the wiring from the starter motor solenoid (see illustration).
4 Unscrew the mounting nuts and bolts and remove the mounting brackets from the rear of the starter motor (see illustration).
5 Undo the three mounting bolts, supporting the motor as the bolts are withdrawn.

10.3 Starter motor solenoid wiring connections and upper rear mounting bracket

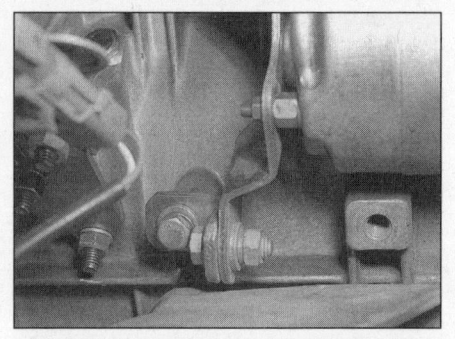

10.4 Starter motor rear lower mounting bracket

6 Manoeuvre the starter motor out from the engine compartment and recover the locating dowel(s) from the motor/transmission (as applicable).

6-cylinder models

7 On turbo models, slacken and remove the bolt securing the oil cooler hose support bracket in position. Wipe clean the area around the hoses then unscrew the union nuts and disconnect both hoses from the cylinder block. Be prepared for some oil spillage and plug the hoses ends and block unions to minimise oil loss and prevent dirt entry. Discard the hose sealing rings, new ones must be used on refitting (see illustrations).
8 On non-turbo models, position a suitable container beneath the oil filter. Unscrew the filter using an oil filter removal tool if necessary, and drain the oil into the container. If the oil filter is damaged or distorted during removal, it must be renewed. Given the low cost of a new oil filter relative to the cost of repairing the damage which could result if a re-used filter springs a leak, it is probably a good idea to renew the filter in any case.
9 On all models, slacken and remove the retaining nuts and disconnect the wiring from the starter motor solenoid.

10 Support the starter motor then unscrew the three motor mounting bolts. Remove the bolts noting the correct fitted position of each one.
11 Manoeuvre the starter motor out of position and recover the protective plate and locating dowel from the motor/transmission. **Note:** *It may be necessary to unbolt the left-hand mounting assembly and raise the engine slightly to gain the necessary clearance required to remove the motor.*

Refitting

4-cylinder models

12 Ensure that the locating dowel is in position and fit the starter motor to the transmission. Fit the three bolts securing the motor to the transmission and tighten them securely.
13 Refit the rear mounting brackets to the motor. First tighten the bolts securing the brackets to the cylinder block and then tighten the bolts securing the brackets to the motor.
14 Reconnect the wiring and securely tighten the retaining nuts then lower the vehicle to the ground and reconnect the battery.

6-cylinder models

15 Fit the locating dowel and protective plate to the transmission and manoeuvre the starter motor into position. Fit the motor retaining bolts, noting that the uppermost bolt is the shortest, and tighten them securely.
16 Reconnect the wiring and securely tighten the retaining nuts

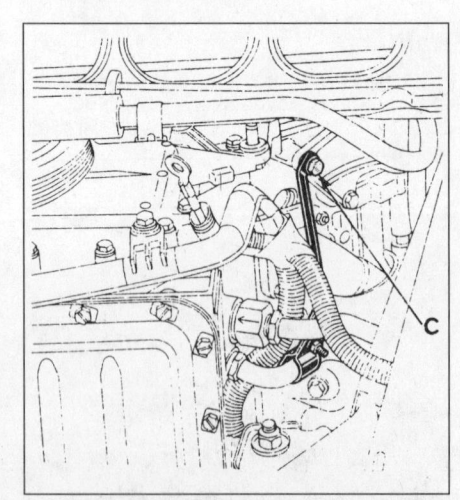

10.7a Oil cooler hose support bracket retaining bolt (C) - turbo models

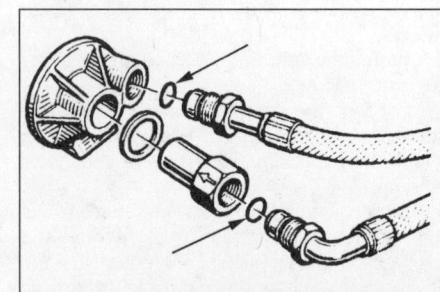

10.7b Oil cooler hose union connections at cylinder block - turbo models (sealing rings arrowed)

11 Starter motor - testing and overhaul

If the starter motor is thought to be suspect, it should be removed from the vehicle and taken to an auto-electrician for testing. Most auto-electricians will be able to supply and fit brushes at a reasonable cost. However, check on the cost of repairs before proceeding as it may prove more economical to obtain a new or exchange motor.

12 Ignition switch - removal and refitting

The ignition switch is integral with the steering column lock, and can be removed as described in Chapter 10.

17 On non-turbo models, fit a new oil filter as described in Chapter 1.
18 On turbo models, fit new sealing rings to the oil cooler hoses and connect the hoses to the cylinder block. Refit the support bracket retaining bolt, tighten it securely, then tighten both the union nuts securely.
19 On all models, lower the vehicle to the ground and reconnect the battery. Check the engine oil level as described in *"Weekly Checks"*.

13 Oil pressure warning light switch/pressure sensor - removal and refitting

Removal

4-cylinder models

1 On models not fitted with an engine oil cooler, the switch is located on the side of the cylinder block, above the oil filter mounting. On models equipped with an oil cooler, the switch is screwed into the top of the oil cooler assembly which is mounted on the right-hand wing valance. On some models an oil pressure gauge sensor may also be fitted in addition to the switch; the pressure sensor can be easily identified as it is cylindrical in appearance.
2 Remove the protective sleeve from the wiring plug (where applicable), then disconnect the wiring from the switch/sensor.
3 Unscrew the switch/sensor and recover the sealing ring, where fitted. Be prepared for oil spillage, and if the switch is to be left removed from the engine for any length of time, plug the hole to minimise oil loss.

6-cylinder models

5 On 6-cylinder models, the switch is screwed into the right-hand side of the cylinder block, towards the front of the block. On models equipped with an oil pressure sensor, the sensor is screwed into the left-hand side of the cylinder, towards the front of the block.
6 Remove the switch/sensor as described in paragraphs 2 and 3.

Refitting

7 Refitting is the reverse of removal. Where no sealing washer is fitted, clean the switch/sensor threads thoroughly, then coat them with fresh sealing compound prior to refitting. Where a sealing ring is fitted, use a new one.

14 Oil level sensor - removal and refitting

Where fitted, the sensor is mounted on the right-hand side of the cylinder block on 4-cylinder engines, and on the left-hand side of the block on 6-cylinder engines. Removal and refitting is as described in Section 13, noting that on some engines the sensor is held in position by a retaining plate and bolt.

Notes

Chapter 5 Part B:
Ignition system (petrol models)

Contents

Control unit - removal and refitting 5
Distributor - removal and refitting 4
Ignition HT coil - removal, testing and refitting 3
Ignition system - general information 1
Ignition system - testing 2
Ignition timing - checking and adjustment 6
Spark plug renewal .... See Chapter 1

Degrees of difficulty

| **Easy,** suitable for novice with little experience | **Fairly easy,** suitable for beginner with some experience | **Fairly difficult,** suitable for competent DIY mechanic | **Difficult,** suitable for experienced DIY mechanic | **Very difficult,** suitable for expert DIY or professional |

Specifications

System type
2664 cc models .... Transistorised ignition system
All other models .... Integral electronic (Renix) ignition system

Firing order
4-cylinder models .... 1-3-4-2 (No 1 cylinder at transmission end)
6-cylinder models .... 1-6-3-5-2-4 (No 1 cylinder at transmission end of left-hand cylinder head)

Ignition HT coil resistances
Primary windings* .... 0.4 to 0.8 ohms
Secondary windings* .... 2 to 12 K ohms
*The above results are approximate values only - see text

Ignition timing
2664 cc models .... 10° BTDC @ idle speed
All other models .... 10° BTDC @ idle speed*

*On these models the ignition timing is controlled by the electronic control unit and no adjustment is possible - see text

1 Ignition system - general information

1 The type of ignition system varies according to vehicle model; see *Specifications*.
2 Both ignition systems require the minimum of maintenance; even ignition timing is carried out automatically according to engine load and speed requirements with the integral electronic ignition system.

Transistorised ignition system - 2664 cc models

3 The main components of the system are shown in **illustration 1.3**.
4 An impulse goes from the distributor to the control unit. This makes and breaks the current to the ignition coil with the help of impulses from the impulse sender. The dwell angle is regulated electronically. From the ignition coil, the high tension impulse goes as usual to the spark plugs via the distributor rotor arm.

5 The ignition advance is controlled by means of centrifugal weights and a vacuum diaphragm unit.
6 The impulse sender opens and closes a magnetic, as opposed to an electrical, circuit. This induces impulses in the magnetic pick-up or coil. The impulse sender is made up of four main parts: the stator, the magnetic pick-up (or coil), the rotor and the magnet. Whilst the stator, pick-up and magnet are connected to the distributor housing, the rotor is connected to, and rotated with, the distributor shaft.

5B•2 Ignition system (petrol models)

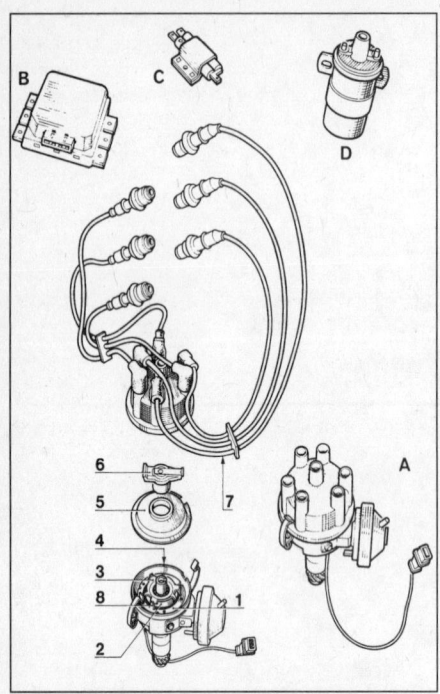

1.3 Transistorised ignition system components - 2664 cc models

1. Permanent magnet
2. Pick-up coil
3. Rotor
4. No1 cylinder alignment mark
5. Shield
6. Rotor arm
7. HT leads and distributor cap assembly
8. Impulse sender segment
A. Complete distributor assembly
B. Control unit
C. Resistor
D. Ignition HT coil

10 The control unit determines the correct ignition timing and dwell by processing signals from the crankshaft sensor, the vacuum capsule and knock sensor (where fitted).

2 Ignition system - testing

> **Warning:** Voltages produced by an electronic ignition system are considerably higher than those produced by conventional ignition systems. Extreme care must be taken when working on the system with the ignition switched on. Persons with surgically-implanted cardiac pacemaker devices should keep well clear of the ignition circuits, components and test equipment

7 The control unit is of a solid state design employing transistors. It converts and amplifies the impulses from the impulse sender and sends them on to the coil. It also performs a second function, control of the dwell angle. The rotor has six teeth and the stator has three teeth. The magnet creates a magnetic field which passes through the stator. When the pole teeth are opposite each other, the magnetic circuit is closed; when the teeth are apart, the circuit is open. In this way the rotor opens and closes the magnetic field as it rotates; this generates the current pulses in the magnetic pick-up.

8 A wiring plug is incorporated into the ignition system for use with special electronic diagnostic equipment available to Renault dealers. The plug is of little use to the home mechanic.

Integral electronic (Renix) ignition system - all other models

9 The system is fully computerised and the main ignition functions take place within the control unit **(see illustration)**. The distributor is considerably reduced in size as it only incorporates the rotor arm.

Note: *Refer to the warning given in Section 1 of Part A of this Chapter before starting work. Always switch off the ignition before disconnecting or connecting any component and when using a multi-meter to check resistances.*

General

1 The components of electronic ignition systems are normally very reliable; most faults are far more likely to be due to loose or dirty connections or to "tracking" of HT voltage due to dirt, dampness or damaged insulation than to the failure of any of the system's components. **Always** check all wiring thoroughly before condemning an electrical component and work methodically to eliminate all other possibilities before deciding that a particular component is faulty.

2 The old practice of checking for a spark by holding the live end of an HT lead a short distance away from the engine is not

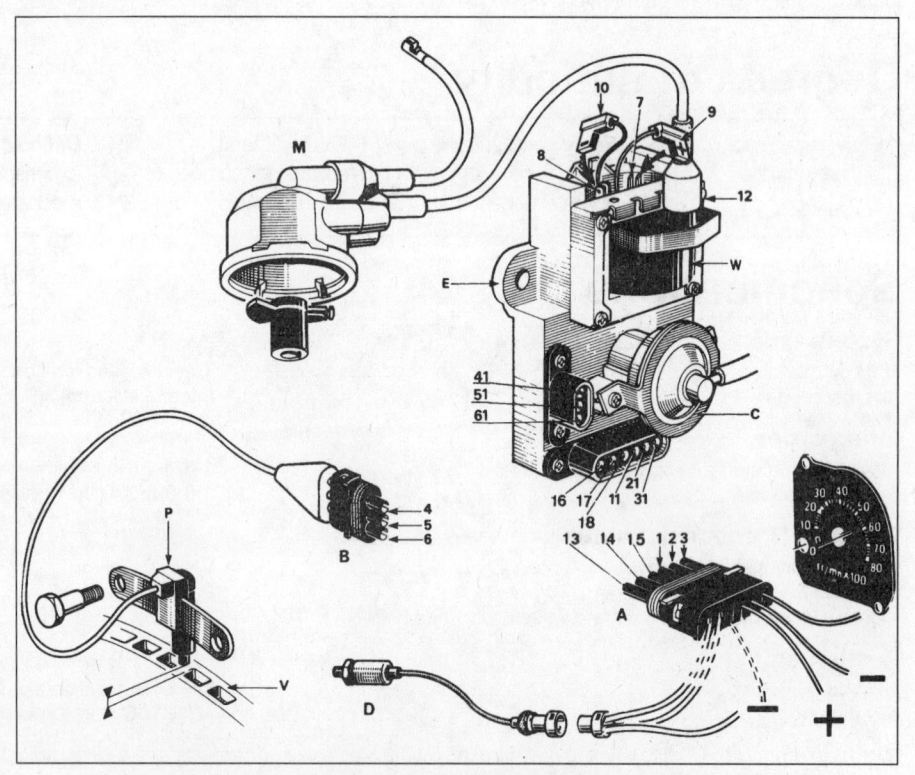

1.9 Integral electronic ignition system components and wiring connectors - early 4-cylinder model shown (others similar)

1. + Feed
2. Earth
3. Tachometer
4. TDC sensor
5. TDC sensor
6. Screening
7. Ignition coil + terminal
8. Ignition coil - terminal
9. Ignition coil + lead
10. Ignition coil - lead
11. Control unit + entry
12. Ignition coil HT terminal
13. Knock sensor or earth
14. Knock sensor or earth
15. knock sensor or earth
16. Terminal for +2° offset on flywheel
17. Terminal for -8° offset on flywheel
18. Terminal for +4° offset on flywheel
21. Control unit earth
31. Tachometer terminal
41. TDC sensor terminal
51. TDC sensor terminal
61. Screening

A. Control unit wiring connector
B. TDC sensor wiring connector
C. Vacuum unit
D. Knock sensor (where fitted)
E. Control unit
M. Distributor cap
P. TDC sensor
V. Flywheel
W. Ignition HT coil

Ignition system (petrol models) 5B•3

recommended; not only is there a high risk of a powerful electric shock, but the HT coil or amplifier unit will be damaged. Similarly, **never** try to "diagnose" misfires by pulling off one HT lead at a time.

Engine will not start

3 If the engine either will not turn over at all, or only turns very slowly, check the battery and starter motor. Connect a voltmeter across the battery terminals (meter positive probe to battery positive terminal), disconnect the ignition coil HT lead from the distributor cap and earth it, then note the voltage reading obtained while turning over the engine on the starter for (no more than) ten seconds. If the reading obtained is less than approximately 9.5 volts, first check the battery, starter motor and charging system as described in the relevant Sections of this Chapter.

4 If the engine turns over at normal speed but will not start, check the HT circuit by connecting a timing light (following the manufacturer's instructions) and turning the engine over on the starter motor; if the light flashes, voltage is reaching the spark plugs, so these should be checked first. If the light does not flash, check the HT leads themselves followed by the distributor cap, carbon brush and rotor arm using the information given in Chapter 1.

5 If there is a spark, check the fuel system for faults referring to the relevant part of Chapter 4 for further information.

6 If there is still no spark, check the voltage at the ignition HT coil "+" terminal; it should be the same as the battery voltage (ie, at least 11.7 volts). If the voltage at the coil is more than 1 volt less than that at the battery, check the feed back through the fusebox and ignition switch to the battery and its earth until the fault is found.

7 If the feed to the HT coil is sound, check the coil's primary and secondary winding resistance as described later in this Section; renew the coil if faulty, but be careful to check carefully the condition of the LT connections themselves before doing so, to ensure that the fault is not due to dirty or poorly-fastened connectors.

8 If the HT coil is in good condition, the fault is probably within the control unit or distributor stator assembly (transistorised ignition system) or in the control unit or crankshaft sensor (integral electronic ignition system). Testing of these components should be entrusted to a Renault dealer.

Engine misfires

9 An irregular misfire suggests either a loose connection or intermittent fault on the primary circuit, or an HT fault on the coil side of the rotor arm.

10 With the ignition switched off, check carefully through the system ensuring that all connections are clean and securely fastened. If the equipment is available, check the LT circuit as described above.

11 Check that the HT coil, the distributor cap and the HT leads are clean and dry. Check the leads themselves and the spark plugs (by substitution, if necessary), then check the distributor cap, carbon brush and rotor arm as described in Chapter 1.

12 Regular misfiring is almost certainly due to a fault in the distributor cap, HT leads or spark plugs. Use a timing light (paragraph 4 above) to check whether HT voltage is present at all leads.

13 If HT voltage is not present on any particular lead, the fault will be in that lead or in the distributor cap. If HT is present on all leads, the fault will be in the spark plugs; check and renew them if there is any doubt about their condition.

14 If no HT is present, check the HT coil; its secondary windings may be breaking down under load.

3 Ignition HT coil - removal, testing and refitting

Removal

1 Disconnect the battery negative terminal.
2 Disconnect the HT lead from the coil.

2664 cc models

3 Unscrew the retaining nuts and disconnect the LT wiring from the coil **(see illustration)**.
4 Unscrew the retaining nut/bolt and remove the coil assembly from the engine compartment.

All other models

5 Undo the retaining screws and remove the coil from the control unit assembly. On early models, as the coil is removed disconnect the LT wiring, noting each wires correct fitted location.

Testing

6 Testing of the coil consists of using a multimeter set to its resistance function, to check the primary (LT `+' to `-' terminals) and secondary (LT `+' to HT lead terminal) windings for continuity, bearing in mind that on the four output, static type HT coil there are two sets of each windings.. Compare the results obtained to those given in the *Specifications* at the start of this Chapter. Note the resistance of the coil windings will vary slightly according to the coil temperature, the results in the *Specifications* are approximate values for when the coil is at 20°C.

7 Check that there is no continuity between the HT lead terminal and the coil body/mounting bracket.

8 If the coil is thought to be faulty, have your findings confirmed by a Renault dealer before renewing the coil.

Refitting

9 Refitting is a reversal of the relevant removal procedure ensuring that the wiring connectors (where necessary) and HT lead are correctly and securely reconnected.

4 Distributor - removal and refitting

1995 cc and 2165 cc engines

Removal

1 Undo the retaining screws then lift the distributor cap assembly away and position it clear of the distributor **(see illustration)**.
2 Remove the rotor arm and the shield **(see illustrations)**.
3 Slacken and remove the distributor retaining bolts and remove the distributor body from the end of the cylinder head. Recover the sealing ring from the distributor groove **(see illustration)**.

Refitting

4 Refitting is the reverse of removal using a new sealing ring.

2458 cc turbo and 2849 cc engines

Removal

5 Release the retaining clips and remove the cover from the top of the distributor which is fitted to the front end of the left-hand cylinder head **(see illustration)**.
6 Undo the retaining screws then remove the distributor cap and position it clear from the distributor.
7 Undo the three screws and remove the rotor arm.
8 Remove the insulator from the cylinder head along with its sealing ring.

Refitting

9 Refitting is the reverse of removal using a new sealing ring.

2664 cc models

Removal

10 Remove the throttle housing assembly as described in Chapter 4B.
11 Position number 1 cylinder at TDC on its

3.3 Ignition HT coil - 2664 cc models

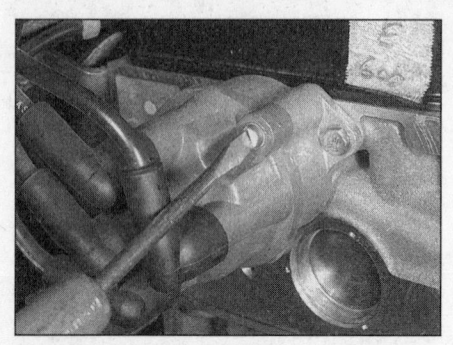

5B•4 Ignition system (petrol models)

4.1 On 1995 cc and 2165 cc models, undo the retaining screws and remove the cap assembly from the distributor

4.2a Pull off rotor arm . . .

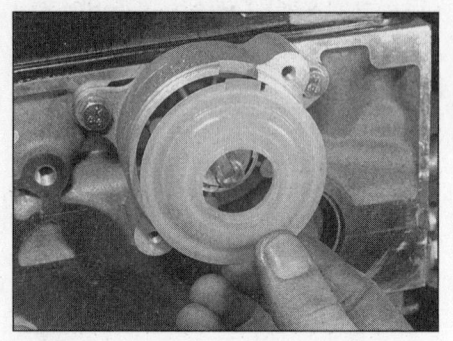

4.2b . . . and remove the shield from the distributor

4.3 Remove the distributor from the cylinder head and recover the sealing ring (arrowed)

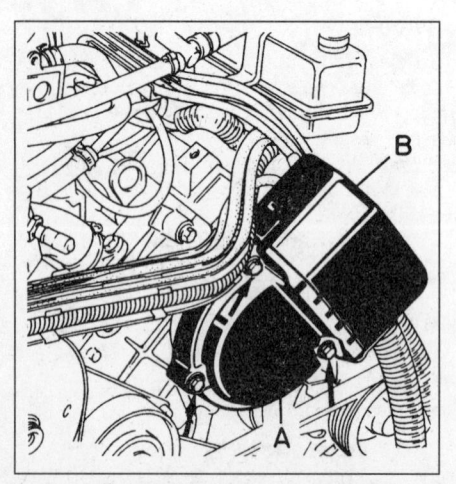

4.5 Distributor assembly cover (B) and cap retaining screws (A) - 2458 cc turbo models

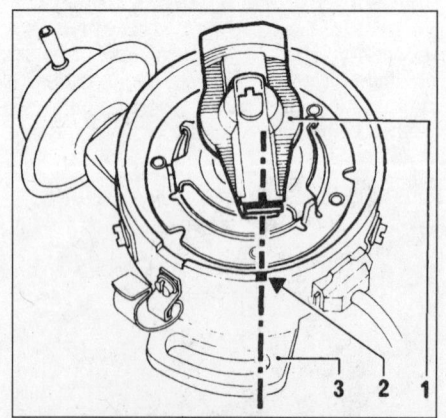

4.12 On 2664 cc and 2849 cc (Z7W 702 engine) models, with No1 cylinder at TDC the rotor arm (1) should align with the mark (2) on the distributor body (3)

compression stroke as described in Chapter 2.

12 Release the retaining clips and position the distributor cap clear of the distributor. Check that the rotor arm is pointing at the No1 cylinder spark plug contact so that it is aligned with the mark on the distributor body **(see illustration)**. If necessary, make a suitable alignment mark using paint or a marker pen.

13 Using paint or a suitable marker pen, make alignment marks between the distributor body and the cylinder head.

14 Disconnect the distributor wiring connector.

15 Slacken and remove the retaining nuts and washers and withdraw the distributor from the cylinder head. Recover the sealing ring from the distributor end. Do not rotate the crankshaft whilst the distributor is removed.

Refitting

16 Fit a new sealing ring to the end of the distributor and lubricate it with a smear of engine oil to aid installation.

17 Check that No1 cylinder is still position at TDC on its compression stroke and align the rotor arm with the mark on the distributor body.

18 Bearing in mind that the rotor arm will twist slightly as its drive gear engages, move the arm a few degrees to the side of the mark and fit the distributor to the head.

19 With the distributor fully in position check the rotor arm is correctly aligned then refit the washer and distributor retaining nuts. Align the marks made prior to removal then tighten the distributor nuts securely.

20 Refit the distributor cap making sure it is clipped securely in position.

21 Refit the inlet manifold as described in Chapter 4.

22 On completion check the ignition timing as described in Section 6.

5 Control unit - removal and refitting

Removal

Transistorised ignition system - 2664 cc models

1 The ignition system control unit is located behind the plastic cover on the left-hand side of the engine compartment.

2 Disconnect the battery negative terminal then release the fasteners and remove protective cover.

3 Disconnect the wiring connector from the control then undo the retaining screw(s) and remove it from the engine compartment.

4 Refitting is the reverse of removal.

Integral electronic ignition system - all other models

5 Disconnect the battery negative terminal.

6 Disconnect the HT lead, wiring connectors and vacuum hose from the control unit assembly.

7 Undo the retaining bolts and remove the control unit from the engine compartment. **Do not** attempt to remove the vacuum unit from the control unit. If necessary the ignition HT coil can be removed as described in Section 3.

Refitting

8 Refitting is the reverse of removal.

6 Ignition timing - checking and adjustment

Transistorised ignition system - 2664 cc models

1 To check the ignition timing, a stroboscopic timing light will be required. It is also recommended that the flywheel/driveplate timing mark is highlighted as follows.

2 Using a socket and suitable extension bar on the crankshaft pulley bolt, slowly turn the engine over until the timing mark (a straight line or arrow) scribed on the edge of the flywheel/driveplate appears in the transmission housing aperture. Highlight the line with quick-drying white paint - typist's correction fluid is ideal.

3 Start the engine, allow it to warm up to normal operating temperature, and then stop it.

4 Disconnect the vacuum hose from the distributor diaphragm, and plug the hose end.

5 Connect the timing light to No 1 cylinder spark plug lead as described in the timing light manufacturer's instructions.

6 Start the engine, allowing it to idle at the specified speed, and point the timing light at the transmission housing aperture. The flywheel/driveplate timing mark should be aligned with the appropriate notch on the housing (refer to the *Specifications* for the correct timing setting). The numbers on the plate indicate degrees Before Top Dead Centre (BTDC).

7 If adjustment is necessary, slacken the distributor retaining nuts, then slowly rotate the distributor body as required until the flywheel/driveplate and relevant housing mark are brought into alignment. Once the marks are correctly aligned, hold the distributor stationary and securely tighten its mounting nuts. Recheck that the timing marks are still correctly aligned and, if necessary, repeat the adjustment procedure.

8 When the timing is correctly set, increase the engine speed, and check that the pulley mark advances to beyond the beginning of the housing reference marks, returning to the specified mark when the engine is allowed to idle. This shows that the centrifugal advance mechanism is functioning; if a detailed check is thought necessary, this must be left to Renault dealer having the necessary equipment. Reconnect the vacuum hose to the distributor, and repeat the check. The rate of advance should significantly increase if the vacuum diaphragm is functioning correctly, but again a detailed check must be left to a Renault dealer.

9 When the ignition timing is correct, stop the engine and disconnect the timing light.

Integral electronic ignition system - all other models

10 On models with an integral electronic ignition system, no adjustment of the ignition timing is possible. The timing is constantly being monitored and adjusted by the control unit and the value given in the *Specifications* should be treated as a nominal value only. The timing can be checked as described above but should it be incorrect, then a fault must be present in the ignition system.

Notes

Chapter 5 Part C:
Preheating system (diesel models)

Contents

Coolant temperature switch (post-heating cut-off) - testing,
   removal and refitting 5
Fast idle speed system (models equipped with air conditioning)
   - general information and testing 6
Glow plugs - removal, inspection and refitting 2
No-load switch - testing, adjustment, removal and refitting 4
Preheating system - description and testing 1
Preheating system control unit - removal and refitting 3

Degrees of difficulty

Easy, suitable for novice with little experience	Fairly easy, suitable for beginner with some experience	Fairly difficult, suitable for competent DIY mechanic	Difficult, suitable for experienced DIY mechanic	Very difficult, suitable for expert DIY or professional

Specifications

Torque wrench setting   Nm   lbf ft
Glow plugs 25   18

1 Preheating system - description and testing

Description

1 Each swirl chamber has a heater plug (commonly called a glow plug) screwed into it. The plugs are electrically operated before, during, and a short time after start-up when the engine is cold. Electrical feed to the glow plugs is controlled by the preheating control unit.

2 The glow plugs also provide a 'post-heating' function, whereby the glow plugs remain switched on for a period after the engine has started. Once the starter has been switched off, the glow plugs are supplied with full current for 10 seconds, then the plugs are supplied with half the 'preheating' current, alternately, in pairs, for up to 3 minutes. The exact 'post-heating' time is controlled by the preheating control unit, and is dependant on the prevailing engine operating conditions. The supply to the plugs will be interrupted by:
   a) *Opening of the 'no-load' switch. The supply is cut off 3 seconds after the switch opens (ie 3 seconds after the accelerator pedal is depressed). The current supply is restored as soon as the switch closes.*
   b) *A coolant temperature of more than 60°C.*

3 A warning light in the instrument panel tells the driver that preheating is taking place. When the light goes out, the engine is ready to be started. The voltage supply to the glow plugs continues for several seconds after the light goes out. If no attempt is made to start, the timer then cuts off the supply in order to avoid draining the battery and overheating of the glow plugs.

Testing

4 If the system malfunctions, testing is ultimately by substitution of known good units, but some preliminary checks may be made as follows.

5 Connect a voltmeter or 12-volt test lamp between the glow plug supply cable and earth (engine or vehicle metal). Make sure that the live connection is kept clear of the engine and bodywork.

6 Have an assistant switch on the ignition, and check that voltage is applied to the glow plugs. Note the time for which the warning light is lit, and the total time for which voltage is applied before the system cuts out. Switch off the ignition.

7 At an under-bonnet temperature of 20°C, typical times noted should be 5 or 6 seconds for warning light operation, followed by a further 10 seconds supply after the light goes out. Warning light time will increase with lower

5C•2 Preheating system (diesel models)

temperatures and decrease with higher temperatures.

8 If there is no supply at all, the control unit or associated wiring is at fault.

9 To locate a defective glow plug, disconnect the main supply cable and the interconnecting wire or strap from the top of the glow plugs. Be careful not to drop the nuts and washers.

10 Use a continuity tester, or a 12-volt test lamp connected to the battery positive terminal, to check for continuity between each glow plug terminal and earth. The resistance of a glow plug in good condition is very low (less than 1 ohm), so if the test lamp does not light or the continuity tester shows a high resistance, the glow plug is certainly defective.

11 If an ammeter is available, the current draw of each glow plug can be checked. After an initial surge of 15 to 20 amps, each plug should draw 12 amps. Any plug which draws much more or less than this is probably defective.

12 As a final check, the glow plugs can be removed and inspected as described in the following Section.

2 Glow plugs - removal, inspection and refitting

Removal

Caution: *If the preheating system has just been energised, or if the engine has been running, the glow plugs will be very hot.*

1 Disconnect the battery negative lead.
2 Unscrew the nuts from the glow plug terminals, and recover the washers. Note that the main supply cable is connected to one of the glow plugs. Remove the interconnecting wire from the top of the glow plugs.
3 Where applicable, carefully move any obstructing pipes or wires to one side to enable access to the glow plugs.
4 Unscrew the glow plugs and remove them from the cylinder head.

Inspection

5 Inspect the glow plugs for physical damage. Burnt or eroded glow plug tips can be caused by a bad injector spray pattern. Have the injectors checked if this sort of damage is found.
6 If the glow plugs are in good physical condition, check them electrically using a 12 volt test lamp or continuity tester as described in the previous Section.
7 The glow plugs can be energised by applying 12 volts to them to verify that they heat up evenly and in the required time. Observe the following precautions.
  a) Support the glow plug by clamping it carefully in a vice or self-locking pliers. Remember it will become red-hot.
  b) Make sure that the power supply or test lead incorporates a fuse or overload trip

to protect against damage from a short-circuit.
  c) After testing, allow the glow plug to cool for several minutes before attempting to handle it.

8 A glow plug in good condition will start to glow red at the tip after drawing current for 5 seconds or so. Any plug which takes much longer to start glowing, or which starts glowing in the middle instead of at the tip, is defective.

Refitting

9 Refit by reversing the removal operations. Apply a smear of copper-based anti-seize compound to the plug threads and tighten the glow plugs to the specified torque. Do not overtighten, as this can damage the glow plug element.

3 Preheating system control unit - removal and refitting

Removal

1 The unit is located on the left-hand side of the engine compartment where it is mounted behind the battery.
2 Disconnect the battery negative lead, then disconnect the wiring from the base of the unit, noting the connector locations.
3 Unscrew the retaining bolts and withdraw the unit from the vehicle.

Refitting

4 Refitting is a reversal of removal, ensuring that the wiring connectors are correctly connected.

4 No-load switch - testing, adjustment, removal and refitting

Testing and adjustment

1 Ensure that the idle speed and anti-stall settings are correctly adjusted as described in Chapter 1.
2 To test and adjust the switch, first disconnect the switch wiring connector.
3 Connect a continuity tester or an ohmmeter across wiring connector terminals.
4 Insert feeler blades of different thicknesses between the injection pump accelerator lever and the anti-stall adjustment screw, and note the readings on the continuity tester or ohmmeter, as applicable **(see illustration)**. The readings obtained should be as follows.

Spacer thickness	Test reading
Up to 11 mm	Continuity/zero resistance
More than 12 mm	No continuity/infinite resistance

5 To adjust the switch, either slacken the switch retaining screws and repositioning the switch, or slacken the retaining nut and repositioning the operating cam (as applicable).
6 If the switch is permanently open or closed, renew it.

Removal

7 Disconnect the switch wiring connector, then remove the two securing screws, and withdraw the switch from its bracket on the injection pump.

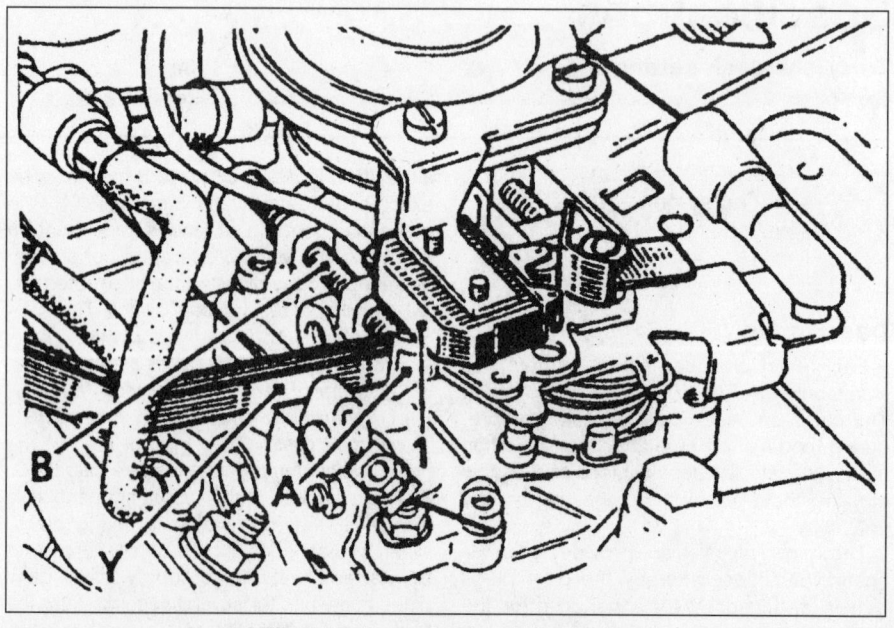

4.4 Checking the no-load switch adjustment

1 No-load switch
A Accelerator lever
B Anti-stall adjustment screw
Y Shim of required thickness

Refitting

8 Refitting is a reversal of removal, but where applicable, before tightening the securing screws adjust the switch as described previously.

5 Coolant temperature switch (post-heating cut-off) - testing, removal and refitting

Testing

1 Remove the switch as described in the following paragraph.
2 Connect a continuity tester or an ohmmeter across the switch terminals. There should be continuity/zero resistance between the terminals.
3 Now suspend the switch in a container of water (using wire or string), along with a thermometer.
4 Gently heat the water, keeping note of the rise in temperature.
5 Periodically remove the switch from the water, ensure that the terminals are dry, and again check for continuity/resistance. The results obtained should be as follows.

Water temperature below 55°C ± 2°C - Continuity/zero resistance
Water temperature above 65°C ± 2°C - No continuity/infinite resistance

6 If the results obtained are not as specified, the switch is proved faulty and should be renewed.

Removal

7 The switch is either screwed into the one of the coolant hoses at the rear of the cylinder head or into the cylinder head itself.
8 Drain the cooling system as described in Chapter 1. Alternatively have the new switch or a suitable bung to hand.
9 Disconnect the switch wiring plug, then unscrew and remove the switch. Recover the sealing ring, where applicable.

Refitting

10 Refitting is a reversal of removal, using a new sealing ring, where applicable.
11 On completion, check the coolant level, and top-up or refill and bleed the cooling system, as necessary, as described in Chapter 1.

6 Fast idle speed system (with air conditioning) - general information and testing

General information

1 On models equipped with air conditioning, the fast idle speed is controlled by the preheating system control unit via and electrically-operated solenoid valve and a vacuum diaphragm unit. The vacuum to operated the diaphragm unit is supplied by the braking system vacuum pump (see Chapter 9).
2 The preheating control unit supplies the solenoid valve with voltage at the same time as the glow plugs (see Section 1). The solenoid valve then opens and allows the vacuum supply to act on the diaphragm unit which in turn operates the injection pump fast idle lever through a cable. When fast idling/post-heating is no longer need, the control unit switches the solenoid valve off which in turn cuts off the vacuum supply to the diaphragm unit. The diaphragm unit then releases the cable and returns the fast idle lever to its stop. In addition to this the control unit also switches the solenoid valve on every time the air conditioning compressor is in use.

Testing

3 Check the vacuum hoses for signs of cracking and splitting. If the hoses are in good condition disconnect the hose from the vacuum diaphragm unit. Apply a vacuum to the diaphragm unit and check that the cable moves. If the diaphragm unit is faulty, renew it.
4 If the vacuum hoses and diaphragm unit appear to be in good condition, check the voltage supply to the solenoid valve as described in Section 1. In addition also check that the solenoid valve is supplied with voltage every time the air conditioning system is switched on. If the voltage supply is incorrect, it is likely that the preheating control unit is faulty.
5 If the voltage supply is correct, check the operation of the solenoid valve using a 12 volt battery. Apply voltage to the terminals and check the solenoid valve switches correctly. If the valve is faulty renew it.

Chapter 6
Clutch

Contents

Clutch - removal, inspection and refitting	7	Clutch release mechanism - removal, inspection and refitting	8
Clutch cable - removal, refitting and adjustment	2	Clutch slave cylinder - removal, overhaul and refitting	4
Clutch hydraulic system - bleeding	5	Fluid level check	See *Weekly Checks*
Clutch master cylinder - removal, overhaul and refitting	3	General information	1
Clutch pedal - removal and refitting	6		

Degrees of difficulty

Easy, suitable for novice with little experience | **Fairly easy,** suitable for beginner with some experience | **Fairly difficult,** suitable for competent DIY mechanic | **Difficult,** suitable for experienced DIY mechanic | **Very difficult,** suitable for expert DIY or professional

Specifications

General
Type .... Single dry plate, with diaphragm spring and sealed ball release bearing in constant contact with diaphragm spring fingers. Cable operated on NG type transmission and hydraulically operated on UN type transmission (see Chapter 7A)

Friction plate diameter
Petrol engine models:
- 1995 cc and 2165 cc models .... 215 mm
- 2458 cc models .... 235 mm
- 2664 cc and 2849 cc models .... 228.6 mm

Diesel engine models:
- Turbo models .... 215 mm
- Non-turbo models .... 200 mm

Torque wrench settings

	Nm	lbf ft
Master cylinder mounting nuts	15	11
Pressure plate cover to flywheel:		
7 mm bolts	25	18
8 mm bolts	30	22
Slave cylinder mounting bolts	15	11

6•2 Clutch

1 General information

1 The clutch consists of a friction plate, a pressure plate assembly, a release bearing and the release mechanism; all of these components are contained in the large cast-aluminium alloy bellhousing, sandwiched between the engine and the transmission. The release mechanism is mechanical, being operated by a cable on models with an NG type transmission or hydraulically on models with a UN type transmission (see Chapter 7A).

2 The friction plate is fitted between the engine flywheel and the clutch pressure plate, and is allowed to slide on the transmission input shaft splines.

3 The pressure plate assembly is bolted to the engine flywheel. When the engine is running, drive is transmitted from the crankshaft, via the flywheel, to the friction plate (these components being clamped securely together by the pressure plate assembly) and from the friction plate to the transmission input shaft.

4 On models with a cable-operated clutch, depressing the clutch pedal pulls the control cable inner wire, and this rotates the release fork by acting on the lever at the fork's upper end **(see illustration)**. The release fork then presses the release bearing against the pressure plate spring fingers. This causes the springs to deform and releases the clamping

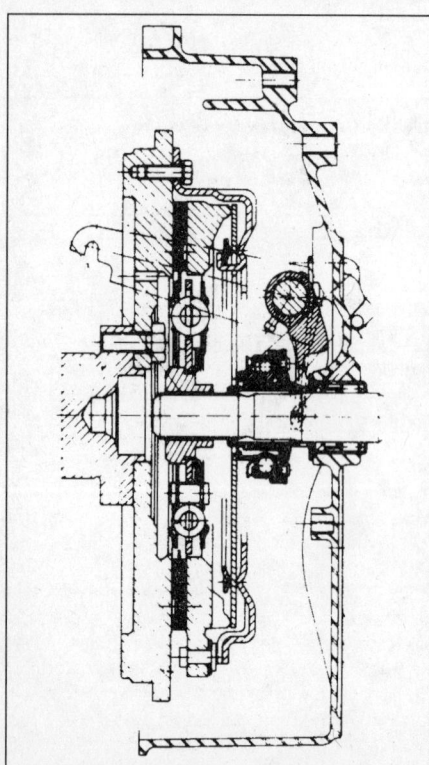

1.4 Sectional view of a cable-operated clutch - NG type transmission

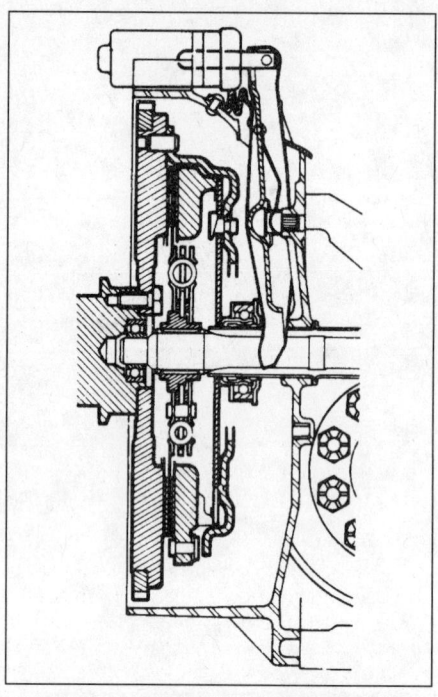

1.5 Sectional view of a hydraulically-operated clutch - UN type transmission

force on the pressure plate. To ensure correct operation, the clutch must be regularly adjusted.

5 On models with a hydraulically-operated clutch, depressing the pedal pushes on the master cylinder pushrod. This hydraulically forces the slave cylinder piston is connected to the end of the clutch release fork **(see illustration)**. The release fork acts on its pivot and presses the release bearing against the pressure plate spring fingers. This causes the springs to deform and releases the clamping force on the pressure plate. The hydraulic clutch is self-adjusting and requires no manual adjustment.

2 Clutch cable - removal, refitting and adjustment

Removal

1 Unscrew the locknuts on the cable end fitting at the clutch release lever to obtain maximum freeplay in the cable.

2 Detach the inner cable from the clutch release lever then free the outer cable from the mounting bracket and recover the rubber spacer.

3 Work back along the length of the cable, noting its correct routing, and free it from any relevant clips or ties.

4 From inside the vehicle, reach up behind the facia panel, and disconnect the inner cable from the clutch pedal arm link. If necessary, to improve access remove the facia undercover panel (see Chapter 11).

5 Return to the engine compartment, then release the cable guide from the bulkhead and withdraw the cable forwards and remove it from the vehicle.

6 Examine the cable, looking for worn end fittings or a damaged outer casing, and for signs of fraying of the inner wire. Check the cable's operation; the inner wire should move smoothly and easily through the outer casing. Remember that a cable that appears serviceable when tested off the car may well be much heavier in operation when in its working position. Renew the cable if it shows signs of excessive wear or any damage.

Refitting

7 Apply a thin smear of multi-purpose grease to the cable end fittings, then pass the cable through the engine compartment bulkhead.

8 From inside the vehicle, hook the inner cable onto the clutch pedal, and check that it is correctly located.

9 Ensure that the cable is correctly routed and retained by all the relevant retaining clips and guides.

10 Fit the rubber spacer to the lower end of the cable then pass the cable through the mounting bracket and engage the inner cable with the release lever. Adjust the cable as follows.

Adjustment

11 Support the clutch pedal in the fully raised position.

12 Adjust the locknuts on the cable end fitting until there is a freeplay (G) at the end of the rubber spacer of between 3.0 and 4.0 mm **(see illustration)**.

13 Apply grease to the cable end fitting cut-out in the release lever.

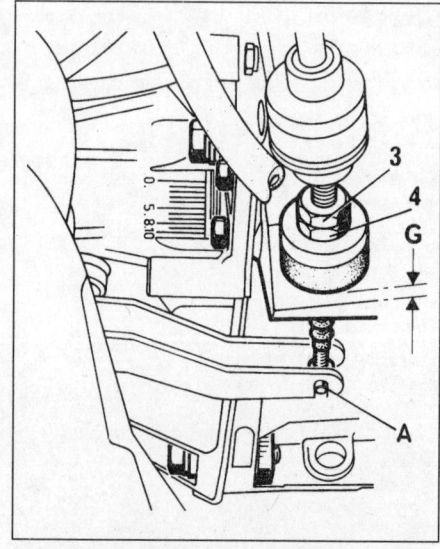

2.12 Transmission end of clutch cable

- A Inner cable
- G Clutch cable freeplay measuring point
- 3 Upper locknut
- 4 Lower locknut

Clutch 6•3

3 Clutch master cylinder - removal, overhaul and refitting

Removal

1 On left-hand drive cars, to improve access to the master cylinder, firmly apply the handbrake then jack up the front of the vehicle and support it on axle stands (see "Jacking and vehicle support"). Unclip the plastic cover from under the wheelarch.

2 To minimise fluid loss, clamp the fluid supply hose connecting the master cylinder to the fluid reservoir.

3 Remove all traces of dirt from the master cylinder then disconnect the fluid supply hose from the master cylinder. Unscrew the union nut and disconnect the pipeline which serves the slave cylinder (see illustration). Catch any fluid in a suitable container and plug the hose ends and master cylinder ports to minimise fluid loss and prevent the entry of dirt into the system.

4 From inside the vehicle, reach up behind the facia to gain access to the inner end of the master cylinder. If necessary, to improve access remove the facia undercover panel (see Chapter 11). On models were the pushrod is connected to the pedal by a clevis pin, remove the split pin and withdraw the clevis pin. Where the pushrod is secured in position with a clip, carefully remove the clip and washer then detach the pushrod from the pedal and recover the end fitting bushes.

5 Slacken and remove the master cylinder retaining nuts then return to the engine compartment and remove the master cylinder from the vehicle.

Overhaul

6 Remove all traces of dirt from the outside of the cylinder and carefully clamp the cylinder in a vice equipped with soft jaws.

7 Release the retaining clip and remove the fluid inlet union and sealing ring from the top of the cylinder.

8 Depress the pushrod and extract the piston retaining pin from the cylinder port.

9 Using circlip pliers, remove the circlip from the end of the cylinder and withdraw the pushrod, piston assembly and spring.

10 Examine the surfaces of the piston and cylinder. If they are scored or corroded, renew the master cylinder complete.

11 If the cylinder is in good condition, obtain a piston repair kit which will contain all the necessary renewable items.

12 Ensure that all components are clean and dry and lubricate the piston assembly with fresh hydraulic fluid.

13 Fit the spring to the piston then carefully enter the assembly into the cylinder. Ease the piston into position with a twisting motion, taking great care not to trap the seal lips.

14 Depress the piston and secure it in position with the retaining pin. Fit the pushrod to the end of the piston and install the circlip, making sure it is correctly located in the cylinder groove.

15 Refit the fluid inlet union and sealing ring and secure it in position with the retaining clip.

Refitting

16 Prior to refitting, check the length "L" of the pushrod as shown in illustration 3.16. If necessary, adjust the length by slackening the locknut and repositioning the end fitting as necessary. Once the pushrod is the correct length, securely tighten the locknut.

17 Refit the master cylinder to the engine compartment bulkhead.

18 Where the pushrod is secured to the pedal by a clevis pin, apply a smear of multi-purpose grease to the pin. Ensure that the pushrod is correctly engaged with the pedal then fit the pin and secure it in position with a new split pin.

19 Where the pushrod is retained by a clip, apply a smear of grease to the pivot and bushes and fit the bushes to the pushrod end. Locate the end fitting on the pedal pivot then refit the washer and secure it in position with the retaining clip.

20 On all models, refit the master cylinder retaining nuts and tighten them to the specified torque.

21 Reconnect the slave cylinder pipe and securely tighten its union nut. Connect the fluid supply hose to the cylinder and remove the clamp.

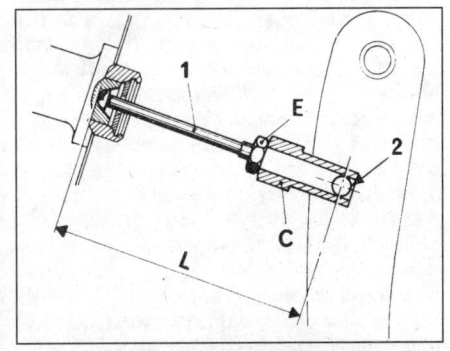

3.16 Clutch master cylinder pushrod setting diagram

1 Pushrod
2 End fitting hole
C End fitting
E Locknut
L Pushrod length (see text)

3.3 Clutch master cylinder (arrowed)

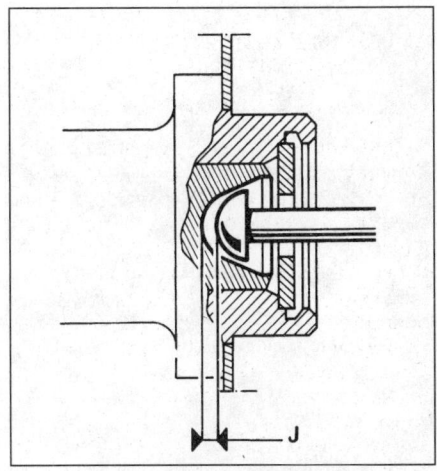

3.22 Clutch master cylinder pushrod freeplay

J = 0.2 to 0.5 mm

22 With the master cylinder correctly installed, it is necessary to check the pushrod freeplay. With the pedal in the at-rest position the clearance between the end of the pushrod and the master cylinder piston should be as shown in illustration 3.22. This clearance amounts to approximately 1 to 2.5 mm of freeplay. If necessary adjust the pushrod freeplay by slackening the locknut and screwing the pushrod into/out of (as applicable) the end fitting. Once the correct clearance is obtained securely tighten the locknut.

23 On completion bleed the hydraulic system as described in Section 5.

4 Clutch slave cylinder - removal, overhaul and refitting

Removal

1 On V6 engine models, to improve access to the slave cylinder, unbolt the power steering pump and move it to one side of the engine compartment (see Chapter 10).

2 On Turbo models, to further improve access, unbolt and remove the heatshield and inlet scoop fitted between the turbocharger and bulkhead (see relevant part of Chapter 4).

3 Where necessary, undo the retaining bolts and remove the heatshield from around the slave cylinder (see illustration).

4 On all models, minimise fluid loss by first removing the master cylinder reservoir cap, and then tightening it down onto a piece of polythene, to obtain an airtight seal.

5 Wipe away all traces of dirt around the hydraulic pipe union on the slave cylinder and unscrew the union nut. Carefully ease the pipe out of the wheel cylinder, and plug or tape over its end to prevent dirt entry (see illustrations). Wipe off any spills immediately. Where necessary, also unscrew the union nut

6•4 Clutch

4.3 Clutch slave cylinder heatshield

4.5a Clutch slave cylinder (fluid inlet union arrowed) - 2165 cc model

4.5b Clutch slave cylinder (arrowed) - 2664 cc model

and disconnect the bleed pipe from the cylinder.
6 Unscrew the two cylinder retaining bolts and carefully draw the cylinder off the pushrod which will remain attached to the clutch release fork.

Overhaul

7 Remove all traces of dirt from the outside of the cylinder.
8 Remove the dust cover from the cylinder and withdraw the piston assembly and spring.
9 Examine the surfaces of the piston and cylinder. If they are scored or corroded, renew the slave cylinder complete.
10 If the cylinder is in good condition, obtain a piston repair kit which will contain all the necessary renewable items.
11 Ensure that all components are clean and dry and lubricate the piston assembly with fresh hydraulic fluid.
12 Fit the spring to the piston then carefully enter the assembly into the cylinder. Ease the piston into position with a twisting motion, taking great care not to trap the seal lips.
13 Depress the piston and fit the dust cover

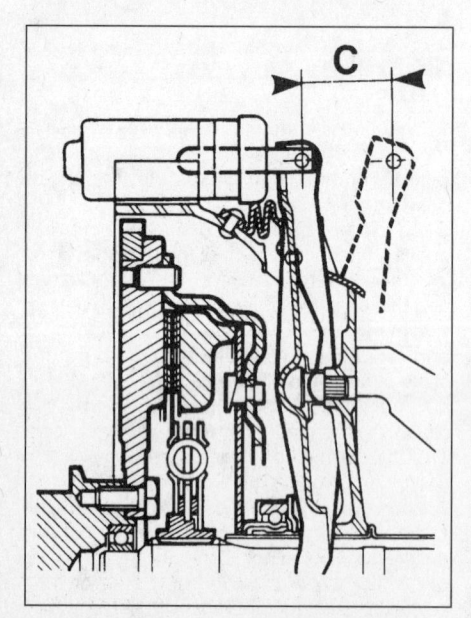

Refitting

14 Apply a smear of grease to the release fork pushrod and ease the slave cylinder onto the rod. Fit the cylinder retaining bolts and tighten them to the specified torque.
15 Reconnect the pipe(s) to the slave cylinder and securely tighten the union nut(s) (as applicable). Remove the polythene from the master cylinder.
16 Bleed the hydraulic system as described in Section 5.
17 With the hydraulic system bled, have an assistant operate the clutch pedal whilst you check the release lever pushrod travel. The pushrod should travel at least 11 mm **(see illustration)**.
18 If the cylinder is operating correctly, refit the heatshield and any remaining components removed for access.

5 Clutch hydraulic system - bleeding

⚠️ **Warning: Hydraulic fluid is poisonous; wash off immediately and thoroughly in the case of skin contact, and seek immediate medical advice if any fluid is swallowed or gets into the eyes. Certain types of hydraulic fluid are flammable, and may ignite when allowed into contact with hot components; when servicing any hydraulic system, it is safest to assume that the fluid is flammable, and to take precautions against the risk of fire as though it is petrol that is being handled. Hydraulic fluid is also an effective paint stripper, and will attack plastics; if any is spilt, it should be washed off immediately, using copious quantities of fresh water. Finally, it is hygroscopic (it absorbs moisture from the air) - old fluid may be contaminated and unfit for further use. When topping-up or renewing the fluid, always use the recommended type, and ensure that it comes from a freshly opened sealed container.**

1 The correct operation of any hydraulic system is only possible after removing all air from the components and circuit; this is achieved by bleeding the system.

2 During the bleeding procedure, add only clean, unused hydraulic fluid of the recommended type; never re-use fluid that has already been bled from the system. Ensure that sufficient fluid is available before starting work.
3 If there is any possibility of incorrect fluid being already in the system, the hydraulic circuit must be flushed completely with uncontaminated, correct fluid.
4 If hydraulic fluid has been lost from the system, or air has entered because of a leak, ensure that the fault is cured before continuing further.
5 On all models except the Turbo, the bleed screw is screwed directly into the slave cylinder body. On Turbo models the bleed screw is located on the end of an extension pipe which is attached to the base of the transmission; to gain access firmly apply the handbrake then jack up the front of the vehicle and support it on axle stands (see "Jacking and vehicle support") **(see illustration)**.
6 Check that all pipes and hoses are secure, unions tight and the bleed screw is closed. Clean any dirt from around the bleed screw.
7 Unscrew the master cylinder fluid reservoir

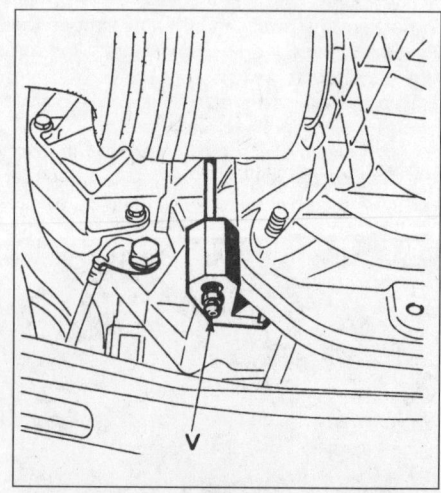

4.17 Clutch slave cylinder pushrod travel (C) should be at least 11 mm

5.5 On Turbo models the clutch bleed nipple (V) is located on the end of an extension tube and is fixed to the base of the transmission unit

cap, and top the master cylinder reservoir up to the "MAX" level line; refit the cap loosely, and remember to maintain the fluid level at least above the "DANGER" level line throughout the procedure, or there is a risk of further air entering the system.

8 There are a number of one-man, do-it-yourself brake bleeding kits currently available from motor accessory shops. It is recommended that one of these kits is used whenever possible, as they greatly simplify the bleeding operation, and reduce the risk of expelled air and fluid being drawn back into the system. If such a kit is not available, the basic (two-man) method must be used, which is described in detail below.

9 If a kit is to be used, prepare the vehicle as described previously, and follow the kit manufacturer's instructions, as the procedure may vary slightly according to the type being used; generally, they are as outlined below in the relevant sub-section.

Bleeding - basic (two-man) method

10 Collect a clean glass jar, a suitable length of plastic or rubber tubing which is a tight fit over the bleed screw, and a ring spanner to fit the screw. The help of an assistant will also be required.

11 Remove the dust cap from the slave cylinder bleed screw. Fit the spanner and tube to the screw, place the other end of the tube in the jar, and pour in sufficient fluid to cover the end of the tube.

12 Ensure that the fluid level is maintained at least above the "DANGER" level line in the reservoir throughout the procedure.

13 Have the assistant fully depress the clutch pedal several times to build up pressure, then maintain it on the final downstroke.

14 While pedal pressure is maintained, unscrew the bleed screw (approximately one turn) and allow the compressed fluid and air to flow into the jar. The assistant should maintain pedal pressure and should not release it until instructed to do so. When the flow stops, tighten the bleed screw again, have the assistant release the pedal slowly, and recheck the reservoir fluid level.

15 Repeat the steps given in paragraphs 13 and 14 until the fluid emerging from the bleed screw is free from air bubbles. If the master cylinder has been drained and refilled allow approximately five seconds between cycles for the master cylinder passages to refill.

16 When no more air bubbles appear, tighten the bleed screw to the specified torque, remove the tube and spanner, and refit the dust cap. Do not overtighten the bleed screw.

Bleeding - using a one-way valve kit

17 As their name implies, these kits consist of a length of tubing with a one-way valve fitted, to prevent expelled air and fluid being drawn back into the system; some kits include a translucent container, which can be positioned so that the air bubbles can be more easily seen flowing from the end of the tube.

18 The kit is connected to the bleed screw, which is then opened. The user returns to the driver's seat, depresses the clutch pedal with a smooth, steady stroke, and slowly releases it; this is repeated until the expelled fluid is clear of air bubbles.

19 Note that these kits simplify work so much that it is easy to forget the clutch fluid reservoir level; ensure that this is maintained at least above the "DANGER" level line at all times.

Bleeding - using a pressure-bleeding kit

20 These kits are usually operated by the reservoir of pressurised air contained in the spare tyre. However, note that it will probably be necessary to reduce the pressure to a lower level than normal; refer to the instructions supplied with the kit.

21 By connecting a pressurised, fluid-filled container to the clutch fluid reservoir, bleeding can be carried out simply by opening the bleed screw and allowing the fluid to flow out until no more air bubbles can be seen in the expelled fluid.

22 This method has the advantage that the large reservoir of fluid provides an additional safeguard against air being drawn into the system during bleeding.

23 Pressure-bleeding is particularly effective when bleeding "difficult" systems, or when bleeding the complete system at the time of routine fluid renewal.

All methods

24 When bleeding is complete, and correct pedal feel is restored, wash off any spilt fluid, tighten the bleed screw securely and refit the dust cap. Where necessary, lower the vehicle to the ground.

25 Check the hydraulic fluid level in the master cylinder reservoir, and top-up if necessary (refer to "Weekly Checks" and Chapter 1).

26 Discard any hydraulic fluid that has been bled from the system; it will not be fit for re-use.

27 Check the feel of the clutch pedal. If the clutch is still not operating correctly, air must still be present in the system, and further bleeding is required. Failure to bleed satisfactorily after a reasonable repetition of the bleeding procedure may be due to worn master cylinder/slave cylinder seals.

6 Clutch pedal - removal and refitting

Removal

1 Remove the facia undercover panel (see Chapter 11).

2 On models with a cable-operated clutch, unscrew the locknuts on the cable end fitting at the clutch release lever to obtain maximum freeplay in the cable. From inside the vehicle, reach up behind the facia panel, and disconnect the inner cable from the clutch pedal arm link.

3 On models with a hydraulically-operated clutch, detach the master cylinder pushrod from the pedal as described in paragraph 4 of Section 3.

Left-hand drive models

4 On later (1989 on) Turbo models were an assist spring assembly is fitted to the clutch pedal, it is necessary to clamp the spring assembly in its compressed position before removing the pedal. In the absence of the special Renault tool (Emb.1082) it will be necessary to fabricate a suitable alternative (see illustration). Alternately, hold the spring in its compressed using a strong cable-tie. With the spring securely compressed, remove the retaining clip and withdraw the pin securing the spring to the pedal.

5 On all models, using pliers, carefully unhook the spring from the pedal.

6 Remove the retaining clips from each end of the pedal pivot shaft.

7 Carefully slide the pivot shaft out towards the right until it is disengaged from the clutch pedal. Remove the clutch pedal and recover the spring and spacer and slide the pivot shaft back in.

8 Examine all components for signs of wear or damage and renew as necessary.

Right-hand drive models

9 Using a pair of pliers, carefully unhook the assist spring and remove it from the pedal.

10 Remove the pedal as described in paragraphs 6 to 8.

Refitting

11 Refitting is a reverse of the removal procedure, noting the following.

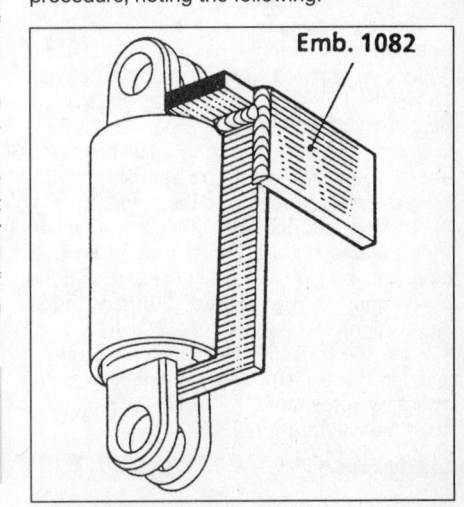

6.4 Renault clutch pedal assist spring holding tool fitted to the spring assembly - later (1989 on) LHD Turbo models

6•6 Clutch

a) Prior to refitting, apply multi-purpose grease to the pedal pivot bushes and the pivot shaft.
b) Ensure that the pedal pivot shaft retaining clips making are correctly engaged with the shaft grooves and mounting brackets.
c) On later left-hand drive Turbo models, attach the assist spring assembly to the pedal, securing its pivot pin in position with the retaining clip, and remove the compressor tool/cable-tie (as applicable).
d) On models with a cable operated clutch, reconnect the cable to the pedal and adjust as described in Section 2.
e) On models with a hydraulically-operated clutch, reconnect the master cylinder pushrod to the pedal and check the pushrod freeplay as described in Section 3.

7 Clutch - removal, inspection and refitting

Warning: *Dust created by clutch wear and deposited on the clutch components may contain asbestos, which is a health hazard. DO NOT blow it out with compressed air, or inhale any of it. DO NOT use petrol or petroleum-based solvents to clean off the dust. Brake system cleaner or methylated spirit should be used to flush the dust into a suitable receptacle. After the clutch components are wiped clean with rags, dispose of the contaminated rags and cleaner in a sealed, marked container.*

Note: *Although some friction materials may no longer contain asbestos, it is safest to assume that they do, and to take precautions accordingly.*

Removal

1 Unless the complete engine/transmission is to be removed from the car and separated for major overhaul (see Chapter 2), the clutch can be reached by removing the transmission as described in Chapter 7A.
2 Before disturbing the clutch, use paint or a marker pen to mark the relationship of the pressure plate assembly to the flywheel.
3 Working in a diagonal sequence, slacken the pressure plate bolts by half a turn at a time, until spring pressure is released and the bolts can be unscrewed by hand **(see illustration)**.
4 Prise the pressure plate assembly off its locating dowels, and collect the friction plate, noting which way round the friction plate is fitted **(see illustrations)**.

Inspection

Note: *Due to the amount of work necessary to remove and refit clutch components, it is usually considered good practice to renew the clutch friction plate, pressure plate assembly* and release bearing as a matched set, even if only one of these is actually worn enough to require renewal. It is also worth considering the renewal of the clutch components on a preventive basis if the engine and/or transmission have been removed for some other reason.

5 Remove the clutch assembly.
6 When cleaning clutch components, read first the warning at the beginning of this Section; remove dust using a clean, dry cloth, and working in a well-ventilated atmosphere.
7 Check the friction plate facings for signs of wear, damage or oil contamination. If the friction material is cracked, burnt, scored or damaged, or if it is contaminated with oil or grease (shown by shiny black patches), the friction plate must be renewed.
8 If the friction material is still serviceable, check that the centre boss splines are unworn, that the torsion springs are in good condition and securely fastened, and that all the rivets are tight. If any wear or damage is found, the friction plate must be renewed.
9 If the friction material is fouled with oil, this must be due to an oil leak from the crankshaft oil seal, from the sump-to-cylinder block joint, or from the transmission input shaft. Renew the seal or repair the joint, as appropriate, as described in Chapter 2 or 7, before installing the new friction plate.
10 Check the pressure plate assembly for obvious signs of wear or damage; shake it to check for loose rivets or worn or damaged fulcrum rings, and check that the drive straps securing the pressure plate to the cover do not show signs (such as a deep yellow or blue discoloration) of overheating. If the diaphragm spring is worn or damaged, or if its pressure is in any way suspect, the pressure plate assembly should be renewed.
11 Examine the machined bearing surfaces of the pressure plate and of the flywheel; they should be clean, completely flat, and free from scratches or scoring. If either is discoloured from excessive heat, or shows signs of cracks, it should be renewed - although minor damage of this nature can sometimes be polished away using emery paper.
12 Check that the release bearing contact surface rotates smoothly and easily, with no sign of noise or roughness. Also check that the surface itself is smooth and unworn, with no signs of cracks, pitting or scoring. If there is any doubt about its condition, the bearing must be renewed.

Refitting

13 On reassembly, ensure that the bearing surfaces of the flywheel and pressure plate are completely clean, smooth, and free from oil or grease. Use solvent to remove any protective grease from new components.
14 Fit the friction plate so that its spring hub assembly faces away from the flywheel.
15 Refit the pressure plate assembly, aligning the marks made on dismantling (if the original pressure plate is re-used), and locating the pressure plate on its locating dowels. Fit the pressure plate bolts, but tighten them only finger-tight, so that the friction plate can still be moved.
16 The friction plate must now be centralised, so that when the transmission is

7.3 Evenly and progressively slacken and remove the clutch retaining bolts . . .

7.4a . . . then remove the pressure plate assembly . . .

7.4b . . . and friction disc from the flywheel

7.17 Centralising the friction plate with the special tool

Clutch 6•7

refitted, its input shaft will pass through the splines at the centre of the friction plate.
17 Centralisation can be achieved by passing a screwdriver or other long bar through the friction plate and into the hole in the crankshaft; the friction plate can then be moved around until it is centred on the crankshaft hole. Alternatively, a clutch-aligning tool can be used to eliminate the guesswork; these can be obtained from most accessory shops **(see illustration)**. A home-made aligning tool can be fabricated from a length of metal rod or wooden dowel which fits closely inside the crankshaft hole, and has insulating tape wound around it to match the diameter of the friction plate splined hole.
18 When the friction plate is centralised, tighten the pressure plate bolts evenly and in a diagonal sequence to the specified torque setting.
19 Refit the transmission as described in Chapter 7A.

8.3a Clutch release bearing - hydraulically-operated clutch (retaining clip arrowed)

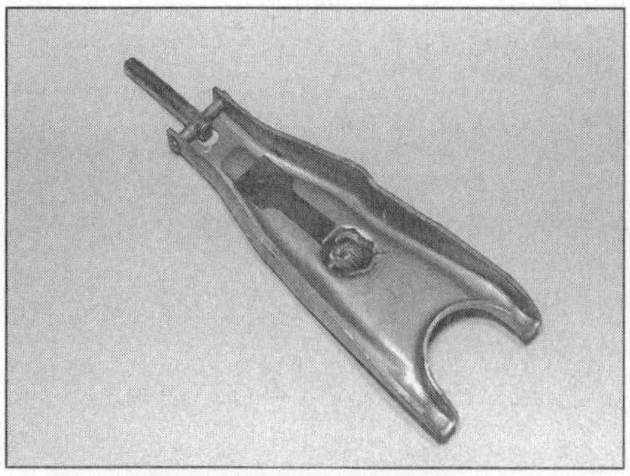

8.3b Clutch release fork - hydraulically-operated clutch

8 Clutch release mechanism - removal, inspection and refitting

Removal

1 With the transmission removed in order to provide access to the clutch, attention can be given to the release bearing located in the clutch housing.
2 On models with a cable-operated clutch (NG type transmission), prise the retaining spring legs apart and withdraw the release bearing from its guide tube. To remove the fork, extract the roll pins, then withdraw the shaft and remove the fork and spring from the transmission.
3 On models with a hydraulically-operated clutch (UN type transmission), tilt the release fork, then unclip the bearing and slide it from its guide tube. The release fork can then be simply slid from its ball-stud pivot **(see illustrations)**.

Inspection

4 Check the release mechanism, renewing any component which is worn or damaged. Carefully check all bearing surfaces and points of contact.
5 When checking the release bearing itself, note that it is often considered worthwhile to renew it as a matter of course. Check that the contact surface rotates smoothly and easily, with no sign of noise or roughness, and that the surface itself is smooth and unworn, with no signs of cracks, pitting or scoring. If there is any doubt about its condition, the bearing must be renewed.
6 Where fitted, check the condition of the bearing in the centre of the crankshaft rear mounting flange for the flywheel. If it is worn or rattles, renew it. Draw the bearing out of position with a puller and tap the new one into position with a socket which bears only on the bearing outer race. If the bearing is a loose fit in the crankshaft/flywheel, apply thread locking compound to its outer circumference before fitting.

Refitting

7 Refitting is the reverse of removal, noting the following points.
a) Apply a little high-melting-point grease (Renault recommend the use of Molykote BR2 grease) to the contact surfaces of the release bearing, release fork and guide sleeve. Also apply grease to the release fork shaft/ball-stud pivot points (as applicable).
b) On models with a cable-operated clutch, ensure that the release fork roll pins are correctly installed **(see illustration)**.
c) On models with a hydraulically-operated clutch ensure that the release fork spring goes behind the head of the ball-stud and holds the lever firmly in position **(see illustration)**.
d) Make sure the release bearing is correctly clipped onto the release fork before refitting the transmission to the engine **(see illustration)**.

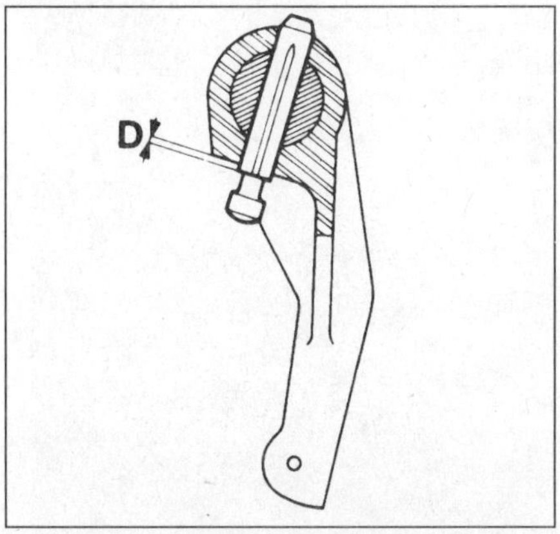

8.7a On models with a cable-operated clutch, ensure the release fork roll pins are pressed in so that they project (measurement D) 1 mm from the fork

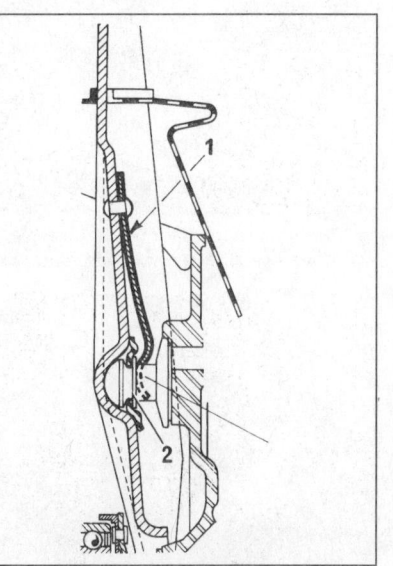

8.7b On models with a hydraulically-operated clutch, ensure the release fork spring (1) is correctly located behind the ball-stud (2) as shown

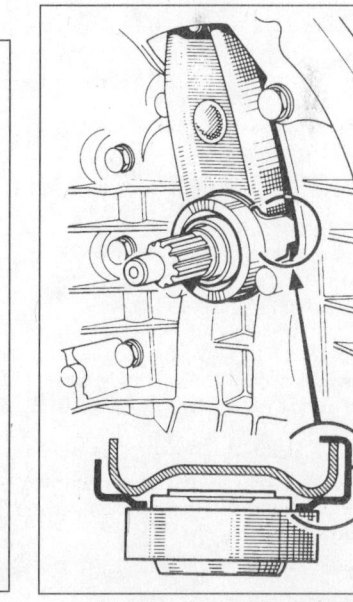

8.7c Ensure the release bearing is correctly engaged with the release fork so that it is securely retained by its clips (hydraulically-operated clutch shown)

6

Chapter 7 Part A:
Manual transmission

Contents

Differential shaft oil seal - renewal 4
Gearchange linkage - removal and refitting 2
General information 1
Reverse gear interlock cable - removal and refitting 3
Reversing light switch - testing, removal and refitting 5
Transmission - removal and refitting 6
Transmission oil level check .... See Chapter 1
Transmission oil renewal .... See Chapter 1
Transmission overhaul - general information 7

Degrees of difficulty

Easy, suitable for novice with little experience	Fairly easy, suitable for beginner with some experience	Fairly difficult, suitable for competent DIY mechanic	Difficult, suitable for experienced DIY mechanic	Very difficult, suitable for expert DIY or professional

Specifications

General

Type .... Five forward speeds (all synchromesh) and reverse
Designation:
  Petrol models:
    1995 cc and 2165 cc (except 1988 on B29E) models* .... NG3 (with cable-operated clutch)
    1988 on (B29E) 2165 cc models* .... UN1 (with hydraulically-operated clutch)
    2458 cc, 2664 cc and 2849 cc models .... UN1 (with hydraulically-operated clutch)
  Diesel models:
    Turbo models .... NG3 (with cable-operated clutch)
    Non-turbo models .... NG1 (with cable-operated clutch)

Vehicle code is stamped on the vehicle identification plate - see "Vehicle identification numbers" for details

Torque wrench settings	Nm	lbf ft
Gearchange linkage remote control rod:		
Rod to transmission lever nut	45	33
Rod to gearchange lever nut	20	15
Rod mounting bracket screws	15	11
Reverse gear interlock cable to transmission	20	15
Transmission to engine mounting bolts	50	37

1 General information

1 The transmission is contained in a cast-aluminium alloy casing bolted to the rear of the engine, and consists of the gearbox and final drive differential.

2 Drive is transmitted from the crankshaft via the clutch to the input shaft, which has a splined extension to accept the clutch friction plate, and rotates in sealed ball-bearings. From the input shaft, drive is transmitted to the output shaft, which rotates in a roller bearing at its right-hand end, and a sealed ball-bearing at its left-hand end. From the output shaft, the drive is transmitted to the differential crownwheel, which rotates with the differential case and planetary gears, thus driving the sun gears and driveshafts. The rotation of the planetary gears on their shaft allows the inner roadwheel to rotate at a slower speed than the outer roadwheel when the car is cornering.

3 The input and output shafts are arranged side by side, parallel to the crankshaft and driveshafts, so that their gear pinion teeth are in constant mesh. In the neutral position, the output shaft gear pinions rotate freely, so that drive cannot be transmitted to the crownwheel.

4 Gear selection is via a floor-mounted lever and selector rod mechanism. The selector rod causes the appropriate selector fork to move its respective synchro-sleeve along the shaft, to lock the gear pinion to the synchro-hub. Since the synchro-hubs are splined to the output shaft, this locks the pinion to the shaft, so that drive can be transmitted. To ensure that gear-changing can be made quickly and quietly, a synchro-mesh system is fitted to all forward gears, consisting of baulk rings and spring-loaded fingers, as well as the gear pinions and synchro-hubs. The synchro-mesh cones are formed on the mating faces of the baulk rings and gear pinions.

5 Two different manual transmissions are used on the models covered in this manual. All V6 engine models are fitted with the UN-type transmission which has a hydraulically-operated clutch. Nearly all four cylinder engine models (both petrol and diesel) are fitted with the NG-type transmission which has a cable-operated clutch; the exception to this being later (1989 on) vehicles, with the model code B29E (see "Buying spare parts and vehicle identification"), which are fitted with a UN-type transmission and hydraulically-operated clutch.

7A•2 Manual transmission

2 Gearchange linkage - removal and refitting

Removal

1 Firmly apply the handbrake then jack up the front of the vehicle and support it on axle stands (see "*Jacking and vehicle support*").
2 Unscrew the nut and washer securing the gearchange linkage remote control rod to the transmission selector arm **(see illustration)**. Slacken and remove the nut and washer securing the rod to the gearchange lever then detach the rod and manoeuvre it out from underneath the vehicle. Recover the pivot bushes from the gearchange lever and, if loose, the front and rear mounting bushes.
3 Using a large, flat-bladed screwdriver, carefully unclip the selector rod from its balljoints and remove it from underneath the vehicle.
4 Slacken and remove the four screws and remove the remote control rod mounting bracket.
5 To remove the gearchange lever, extract the roll pin securing the reverse gear interlock cable to the lever.
6 Working inside the car, remove the centre console as described in Chapter 11.
7 Unscrew and remove the gear lever knob then slide the rubber gaiter, reverse gear interlock knob, spring and cable stop off from the lever. Withdraw the gear lever.
8 Examine the gearchange linkage components for signs of wear or damage and renew as necessary.

Refitting

9 Refitting is the reverse of removal, applying a smear of multi-purpose grease (Renault recommend the use of Molykote 33 Medium) to the remote control rod and selector rod pivots. Tighten the remote control rod and mounting bracket fixings to their specified torque settings.

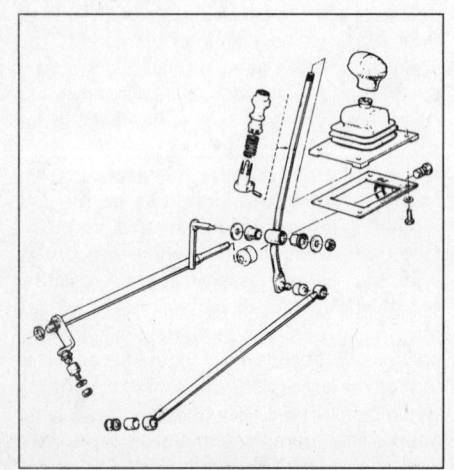

2.2 Gearchange linkage components

3 Reverse gear interlock cable - removal and refitting

Removal

1 Firmly apply the handbrake then jack up the front of the vehicle and support it on axle stands (see "*Jacking and vehicle support*").
2 Unscrew the cable end fitting from the transmission. Plug the cable aperture to prevent oil loss/dirt entry whilst the cable is removed.
3 Release the outer cable from the gearchange lever stop then unhook the inner cable from the from the gearchange lever knob and remove the cable from underneath the vehicle. To improve access to the cable, unclip the gaiter from the centre console and fold it back over the gearchange lever.

Refitting

4 Refitting is the reverse of removal. Apply a smear of sealant (Renault recommend the use of CAF 4/60 Thixo) to the threads of the cable end fitting prior to refitting it to the transmission and tighten it to the specified torque.

4 Differential shaft oil seal - renewal

Models with UN type transmission (hydraulically-operated clutch)

1 Detach the driveshaft from the transmission as described in Chapter 8. Be prepared for some oil loss as the seal is removed.
2 Note the correct fitted position of the seal in the ring nut/housing then carefully prise it out of position, using a flat-bladed screwdriver. Take great care not to damage the transmission housing when levering the seal out.
3 Remove all traces of dirt from the area around the oil seal aperture and wrap insulation tape around the shaft splines to protect the new seal.
4 Apply a smear of grease to the oil seal lip and, making sure it is the correct way around, carefully ease it over the end of the shaft. Using a tubular spacer which bears only on the seal's hard outer edge, press the seal squarely into position until it is correctly positioned in relation to the ring nut/housing (as applicable).
5 Remove the tape from the shaft and refit the driveshaft as described in Chapter 8.
6 On completion check and, if necessary, top-up the transmission oil level as described in Chapter 1.

Models with NG type transmission (cable-operated clutch)

7 Detach the driveshaft from the transmission as described in Chapter 8. Be prepared for some oil loss or drain the transmission as described in Chapter 1.
8 Using paint or a suitable marker pen, mark the differential seal ring nut in relation to the transmission casing.
9 Undo the retaining bolt/nut and remove the ring nut locking plate.
10 Using a suitable peg spanner, unscrew the ring nut and remove it from the transmission, noting the exact number of turns necessary to do so. Recover the sealing ring from the outside of the ring nut.
11 With the ring nut removed, carefully press/lever the seal out of the rear of the nut, noting which way around it is fitted.
12 Making sure the seal is the correct way around, press the new oil seal in from the rear of the ring nut, using a socket which bears only on the hard outer edge of the seal. The seal should be pressed in until its face is flush with the ring nut inner face **(see illustration)**.
13 Fit a new sealing ring to the outside of the ring nut and lubricate it with a smear of engine oil to ease installation.
14 Wrap insulation tape around the end of the shaft splines to protect the oil seal lips.
15 Apply a smear of grease to the oil seal lip and ease the ring nut assembly over the end of the shaft and into position.
16 Using the peg spanner, screw the ring nut into the transmission by the exact amount of turns counted on removal. Align the marks made prior to removal then refit the locking plate and securely tighten its retaining nut/bolt.
17 Remove the tape from the shaft and refit the driveshaft as described in Chapter 8.
18 On completion check and, if necessary, top-up the transmission oil level as described in Chapter 1.

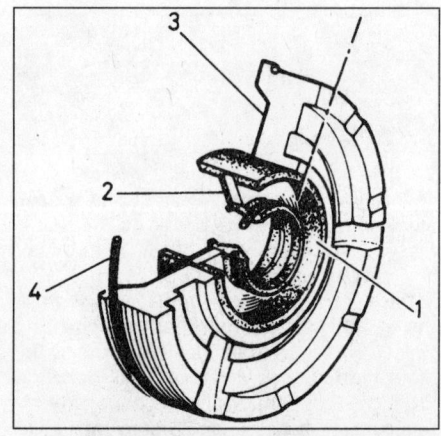

4.12 On NG type transmission, position the new seal (1) so that its face (2) is flush with the ring nut inner face (3). Prior to installation fit a new sealing ring (4) to the ring nut

Manual transmission 7A•3

5 Reversing light switch - testing, removal and refitting

Testing

1 The reversing light circuit is controlled by a plunger-type switch that is screwed into the side of the transmission casing. If a fault develops in the circuit, first ensure that the circuit fuse has not blown.
2 To test the switch, disconnect the wiring connector, and use a multimeter (set to the resistance function) or a battery-and-bulb test circuit to check that there is continuity between the switch terminals only when reverse gear is selected. If this is not the case, and there are no obvious breaks or other damage to the wires, the switch is faulty, and must be renewed.

Removal

3 Firmly apply the handbrake then jack up the front of the vehicle and support it on axle stands (see "*Jacking and vehicle support*").
4 Disconnect the wiring connector, then unscrew the switch and remove it from the transmission casing **(see illustration)**.

Refitting

5 Clean the threads of the switch and apply a smear of fresh sealing compound to its threads. Fit the switch to the transmission and tighten securely.
6 Reconnect the wiring connector, and test the operation of the circuit. Lower the vehicle to the ground.

6 Transmission - removal and refitting

Removal

1 Disconnect the battery negative terminal.
2 Drain in the transmission oil as described in Chapter 1.
3 Firmly apply the handbrake then chock the rear wheels and raise the front of the vehicle. The front end must be sufficiently high for the clutch bellhousing to pass under the car when the transmission is withdrawn.
4 Remove both driveshafts as described in Chapter 8.
5 Remove the starter motor as described in Chapter 5A.
6 Disconnect the leads from the reversing lamp.
7 Remove the exhaust front pipe as described in the relevant Part of Chapter 4. Where necessary, also remove the catalytic converter.
8 Unbolt the crankshaft sensor and tie it out of the way.
9 Disconnect the gearchange linkage remote control rod, selector rod and reverse gear interlock cable from the transmission as described in Sections 2 and 3.
10 Unbolt the cover plate from the lower front face of the flywheel housing.

NG type transmission (cable-operated clutch)

11 Referring to Chapter 6, disconnect the clutch cable and position it clear of the transmission.
12 On models with an electronic speedometer drive, release the speedometer sender and withdraw it from the transmission. On models with a speedometer cable, withdraw the retaining pin and disconnect the cable from the transmission.
13 Place a jack with a block of wood beneath the engine, to take the weight of the engine. Alternatively, fit a hoist or support bar to take the engine weight.
14 Place a jack and block of wood beneath the transmission, and raise the jack to take the weight of the transmission.
15 Disconnect the transmission flexible mountings and brackets.
16 Unscrew and remove the bolts which connect the transmission to the engine.
17 Support the engine using a jack or hoist and place a second jack, preferably of trolley type, under the transmission.
18 Raise the transmission slightly to clear the crossmember during withdrawal to the rear.
19 Remove the transmission from under the vehicle, noting the correct fitted location of the locating dowels.

UN type transmission (hydraulically-operated clutch)

20 Referring to Chapter 6, unbolt the slave cylinder and position it clear of the transmission. Note that there is no need to disconnect the hydraulic circuit, but do not depress the clutch pedal until the cylinder is refitted.
21 On V6 engine models, where necessary, unbolt the engine movement damper and mounting bracket from the left-hand side of the transmission **(see illustration)**.
22 Remove the engine as described in paragraph 13 to 19, noting that the bolt directly below the clutch slave cylinder cannot be withdrawn

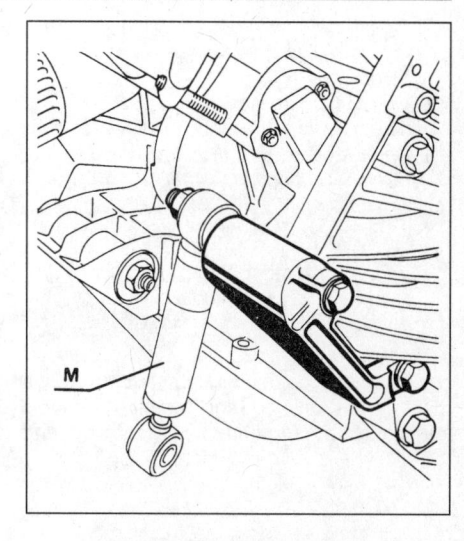

6.21 On V6 engine models, unbolt the mounting bracket and movement damper (M) from the left-hand side of the transmission

Refitting

23 The transmission is refitted by a reversal of the removal procedure, bearing in mind the following points:
a) Apply a little high-melting-point grease (Renault recommend the use of Molykote BR2) to the splines of the transmission input shaft. Do not apply too much, otherwise there is a possibility of the grease contaminating the clutch friction plate.
b) Ensure the locating dowels are in position before lifting the transmission into position.
c) On UN type transmission fit the transmission to engine bolt which is positioned beneath the slave cylinder, to the transmission prior to refitting the transmission (see illustration).

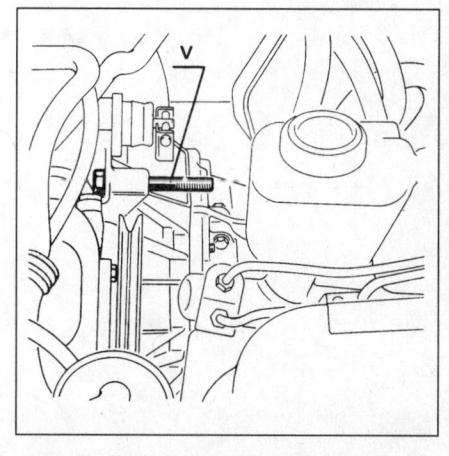

6.23 On UN type transmission, be sure to fit the bolt (V) to the transmission prior to installation as it cannot be installed once the transmission is in position

5.4 Reversing light switch is screwed into the side of the transmission

d) Tighten all nuts and bolts to the specified torque and renew those of self-locking type.
e) On NG type transmission, adjust the clutch cable as described in Chapter 6.
f) On all models, refill the transmission with oil as described in Chapter 1.

7 Transmission overhaul - general information

1 Overhauling a manual transmission is a difficult and involved job for the DIY home mechanic. In addition to dismantling and reassembling many small parts, clearances must be precisely measured and, if necessary, changed by selecting shims and spacers. Internal transmission components are also often difficult to obtain, and in many instances, extremely expensive. Because of this, if the transmission develops a fault or becomes noisy, the best course of action is to have the unit overhauled by a specialist repairer, or to obtain an exchange reconditioned unit.

2 Nevertheless, it is not impossible for the more experienced mechanic to overhaul the transmission, provided the special tools are available, and the job is done in a deliberate step-by-step manner, so that nothing is overlooked.

3 The tools necessary for an overhaul include internal and external circlip pliers, bearing pullers, a slide hammer, a set of pin punches, a dial test indicator, and possibly a hydraulic press. In addition, a large, sturdy workbench and a vice will be required.

4 During dismantling of the transmission, make careful notes of how each component is fitted, to make reassembly easier and more accurate.

5 Before dismantling the transmission, it will help if you have some idea what area is malfunctioning. Certain problems can be closely related to specific areas in the transmission, which can make component examination and replacement easier. Refer to the fault diagnosis in the *Reference* Section for more information.

Chapter 7 Part B:
Automatic transmission

Contents

Differential oil seal - renewal 8	Selector cable - removal, refitting and adjustment 6
Fluid cooler - removal and refitting 9	Selector lever - removal and refitting 7
Fluid level check . . . . See Chapter 1	Transmission - removal and refitting 10
General information 1	Transmission - draining and refilling 2
Governor cable (4141 transmission) - adjustment 5	Transmission fluid filter - renewal 3
Kickdown switch - testing and adjustment 4	Transmission overhaul - general information 11

Degrees of difficulty

Easy, suitable for novice with little experience / **Fairly easy,** suitable for beginner with some experience / **Fairly difficult,** suitable for competent DIY mechanic / **Difficult,** suitable for experienced DIY mechanic / **Very difficult,** suitable for expert DIY or professional

Specifications

General
Type . . . . Three- or four-speed automatic, computer-controlled
Designation:
  1995 cc models*:
    Models with vehicle code B297 . . . . MJ3 transmission
    Models with vehicle code B29H . . . . AR4 transmission
  2165 cc models*:
    Models with vehicle code B29B:
      Early (pre 1989) models . . . . MJ3 transmission
      Later (1989 on) models . . . . AR4 transmission
    Models with vehicle code B29E:
      Early (pre 1988) models . . . . MJ3 transmission
      Later (1988 on) models . . . . AR4 transmission
  2664 cc models . . . . 4141 transmission
  2849 cc models . . . . AR4 transmission

*Vehicle code is stamped on the vehicle identification plate - see "Vehicle identification numbers" for details

Transmission fluid
Fluid type:
  4141 transmission:
    Transmission unit . . . . Dexron type ATF (Duckhams Uni-Matic)
    Final drive unit . . . . Tranself TRX 80W140 **only** (available from your Renault dealer)
  MJ3 transmission . . . . Dexron type ATF (Duckhams Uni-Matic)
  AR4 transmission:
    Transmission unit . . . . Dexron type ATF (Duckhams Uni-Matic)
    Final drive unit . . . . Tranself TRX 80W140 **only** (available from your Renault dealer)
Capacity (approximate):
  4141 transmission:
    Transmission unit:
      From dry . . . . 6.0 litres
      At fluid change . . . . 2.5 litres
    Final drive unit . . . . 1.6 litres
  MJ3 transmission:
    From dry . . . . 6.0 litres
    At fluid change . . . . 2.5 litres
  AR4 transmission:
    Transmission unit:
      From dry . . . . 6.0 litres
      At fluid change . . . . 4.0 litre
    Final drive unit . . . . 0.85 litres

7B•2 Automatic transmission

Torque wrench settings

	Nm	lbf ft
Driveplate to torque converter bolts:		
4141 transmission	30	22
MJ3 transmission	30	22
AR4 transmission	35	26
Fluid cooler hose union nut/bolt	20	15
Fluid filter bolts:		
4141 transmission	5	4
MJ3 transmission	9	6
AR4 transmission	5	4
Sump fixing bolts:		
4141 transmission	13	10
MJ3 transmission	6	4
AR4 transmission	10	7
Transmission to engine mounting bolts	60	44

1 General information

Note: *The automatic transmission is a relatively complex unit and therefore, should problems occur, it is recommended that the fault be discussed with your Renault dealer, who should be able to advise you on the best course of action to be taken. Items that can be attempted by the home mechanic are given in the following Sections in this Chapter. To obtain trouble-free operation and maximum life expectancy from your automatic transmission, it must be serviced as described and not be subjected to abuse.*

1 One of the three types of automatic transmission may be fitted. The transmission type can be determined from the identification plate which is attached to the top of the unit **(see illustration)**.

2 All types of transmission are similar in operating characteristics, but vary slightly in design detail and maintenance requirements. Due to the complexity of the automatic transmission, any repair or overhaul work must be left to a Renault dealer with the necessary special equipment for fault diagnosis and repair. The contents of this Chapter are therefore confined to supplying general information, and any service information and instructions that can be used by the owner.

4141 transmission

3 Fully automatic gearchanging is provided without the use of a clutch, but override selection is still available to the driver.

4 The automatic transmission consists of three main assemblies: the torque converter, the final drive and the gearbox. For a cutaway view of the transmission, with the three main assemblies sub-divided **see illustration 1.4**.

5 The torque converter takes the place of the conventional clutch and transmits the drive automatically from the engine to the gearbox, providing increased torque when starting off.

6 The converter receives its lubricant from the pump mounted on the rear of the gearcase, and is driven directly by the engine. This pump also distributes fluid to the respective gears, clutch and brake assemblies within the gearbox.

7 The gearbox comprises an epicyclic geartrain giving three forward and one reverse gear; selection of which is dependent on the hydraulic pressure supplied to the respective clutches and brakes. The hydraulic pressure is regulated by the hydraulic distributor, and gear selection is determined by two solenoid valves. These are actuated by the electrically-operated governor/computer. The exact hydraulic pressure is regulated by a vacuum capsule and pilot valve operating according to engine loading.

8 The clutches (E1 and E2) and brakes (F1 and F2) are multi-disc oil bath type and, according to the hydraulic loading, engage or release the epicyclic geartrain components.

9 The governor is, in effect, a low output alternator which provides variable current to the computer. It is driven by a worm gear on the final drive pinion and its output depends on the vehicle speed and engine loading.

10 The computer acts upon the variation of current from the governor combined with the selected lever position to open or close the solenoid valves accordingly. In addition it acts as a safety device to prevent the 1st gear

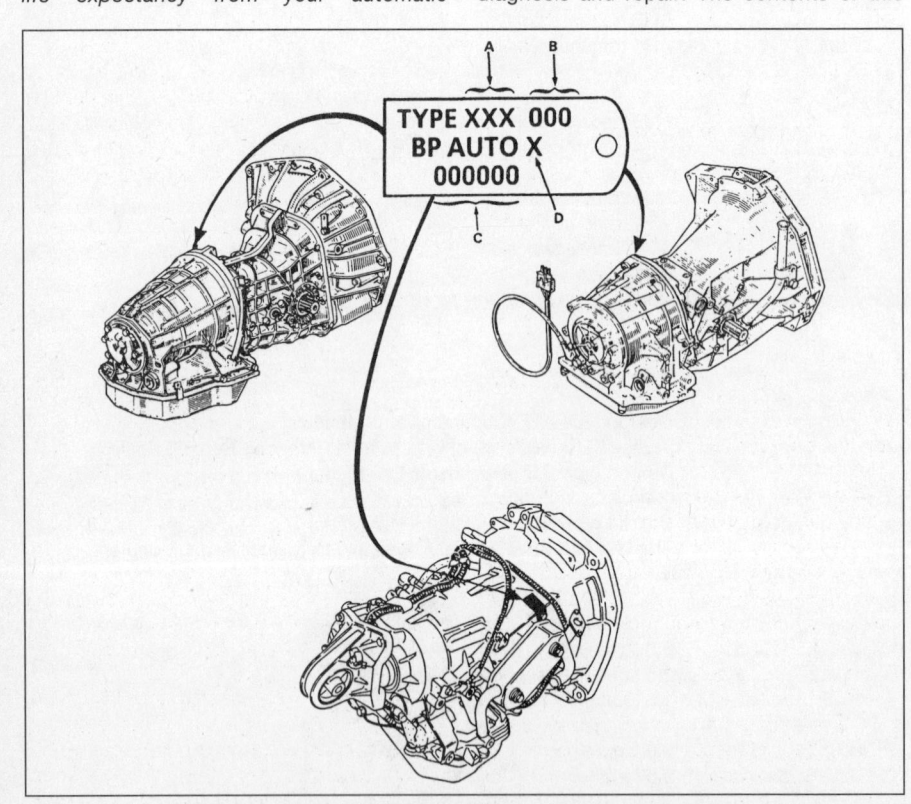

1.1 Automatic transmission identification plate location

  A Transmission type   C Fabrication number
  B Suffix   D Identification number

Automatic transmission 7B•3

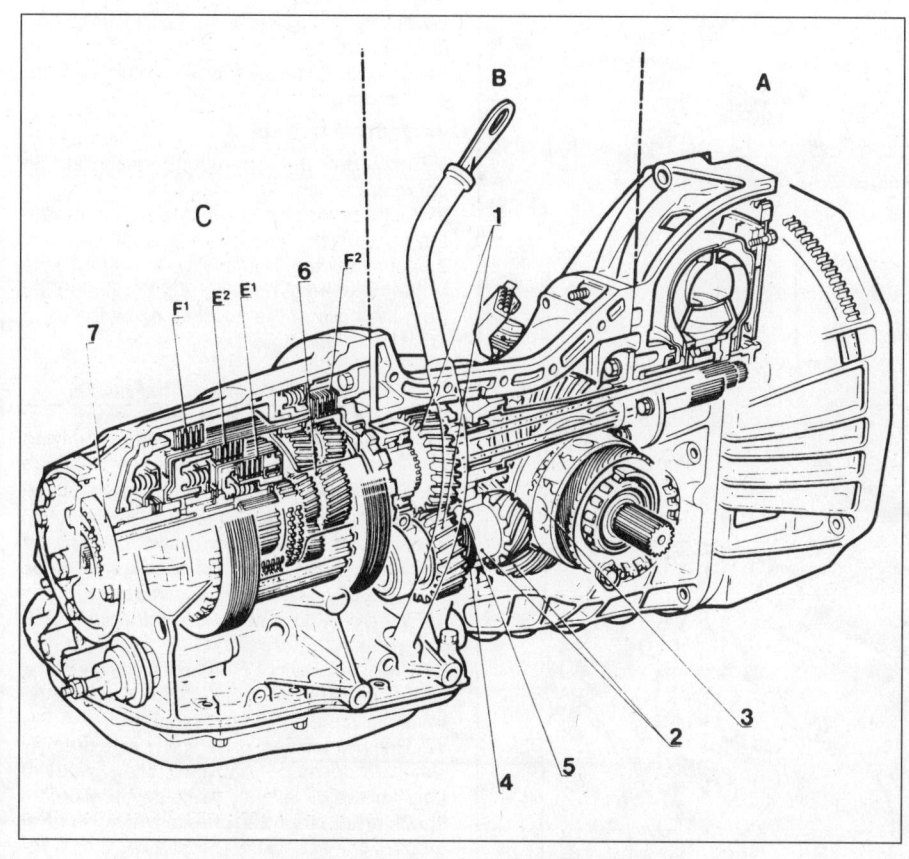

1.4 Cutaway view of a 4141 transmission unit

1 Reduction gears
2 Crownwheel and pinion
3 Differential
4 Worm gear
5 Governor
6 Epicyclic geartrain
7 Oil pump

A Torque converter housing
B Final drive
C Gearbox
E1 Multi-disc clutch
E2 Multi-disc clutch
F1 Brake discs
F2 Brake discs

"hold" position being selected at a speed in excess of 22 mph (35 kph) at light throttle.

11 The system also incorporates a kickdown switch, operated by pressing the throttle pedal to its fully open position, at which point under certain engine loading and speeds, the computer will be activated and a lower gear automatically selected.

12 The drive selected in the gearbox is transferred to the differential unit via stepdown gears, which compensate for the difference in levels between the main gear assemblies in the gearbox and the level of the crownwheel and pinion in the differential housing.

13 The selector lever is centrally situated within the car and has six positional alternatives, as follows:

P (Park): With the lever in this position, the transmission is neutralised and the drive wheels are locked

R (Reverse): Reverse gear position, which when selected also actuates the reversing light switch

N (Neutral): The transmission is in neutral
1 (1st gear): 1st gear hold position
2 (2nd gear): Automatic operation between 1st and 2nd gears
A (Automatic) or D (Drive): Gears engage automatically according to engine loading and car speed.

14 In addition to the above, the kickdown switch causes a lower gear to be selected at a higher speed than normal when the throttle pedal is suddenly pressed fully down. This device is designed to give sudden acceleration when required, such as for overtaking.

15 Because of the obvious hazards of starting the car when in gear, a starter inhibitor switch is fitted and only allows the starter to be operated when the selector is in Park or Neutral position. The inhibitor switch is fitted below the transmission governor/computer units. Its removal necessitates withdrawal of the oil sump plate.

MJ3 transmission

16 The mechanical function of the Type MJ3 is similar to that of the Type 4141 just described, but the controls are fully computerised (see illustration).

17 The computer or module is continually supplied with signals from the speed sensor, load potentiometer, multi-purpose switch, kickdown switch and its own control unit, and from this information it transmits signals to the pilot solenoid valves to select the correct gear range.

18 The kickdown switch is located at the bottom of the accelerator pedal travel and a lower gear is selected when the pedal is fully depressed.

19 The multi-purpose switch is located on the rear of the transmission and is operated by the range selector lever. It controls the engine starting circuit, the reversing light circuit and the pilot solenoid valve circuit.

20 The load potentiometer is operated by the throttle valve and provides a variable voltage to the computer, dependent on the throttle position.

21 The speed sensor is mounted on the left-hand side of the transmission and it provides an output signal dependent on the speed of the parking pawl wheel, which is proportional to the speed of the car.

22 A vacuum capsule is connected directly to the inlet manifold to regulate transmission fluid pressure according to engine loading.

AR4 transmission

23 The four-speed AR4 transmission is computer-controlled, and is similar in operation to the three-speed MJ3 transmission described above. The hydraulic pressure in the AR4 transmission is controlled by the computer, via one of the many electrical solenoid valves. Note that there is no vacuum unit on the AR4 transmission.

2 Transmission - draining and refilling

4141 transmission

Note: On the 4141 transmission, the final drive is lubricated by its own separate oil bath. Final drive unit oil draining and refilling is described as a separate operation to avoid confusion.

Transmission fluid

1 Take the vehicle on a short run, to warm the transmission up to normal operating temperature.

2 Park the car on level ground, then switch off the ignition and apply the handbrake firmly. For improved access, jack up the front of the car and support it securely on axle stands (see "Jacking and vehicle support"). Note that, when refilling and checking the fluid level, the car must be lowered to the ground, and level, to ensure accuracy.

3 Remove the dipstick, then position a suitable container under the transmission.

7B•4 Automatic transmission

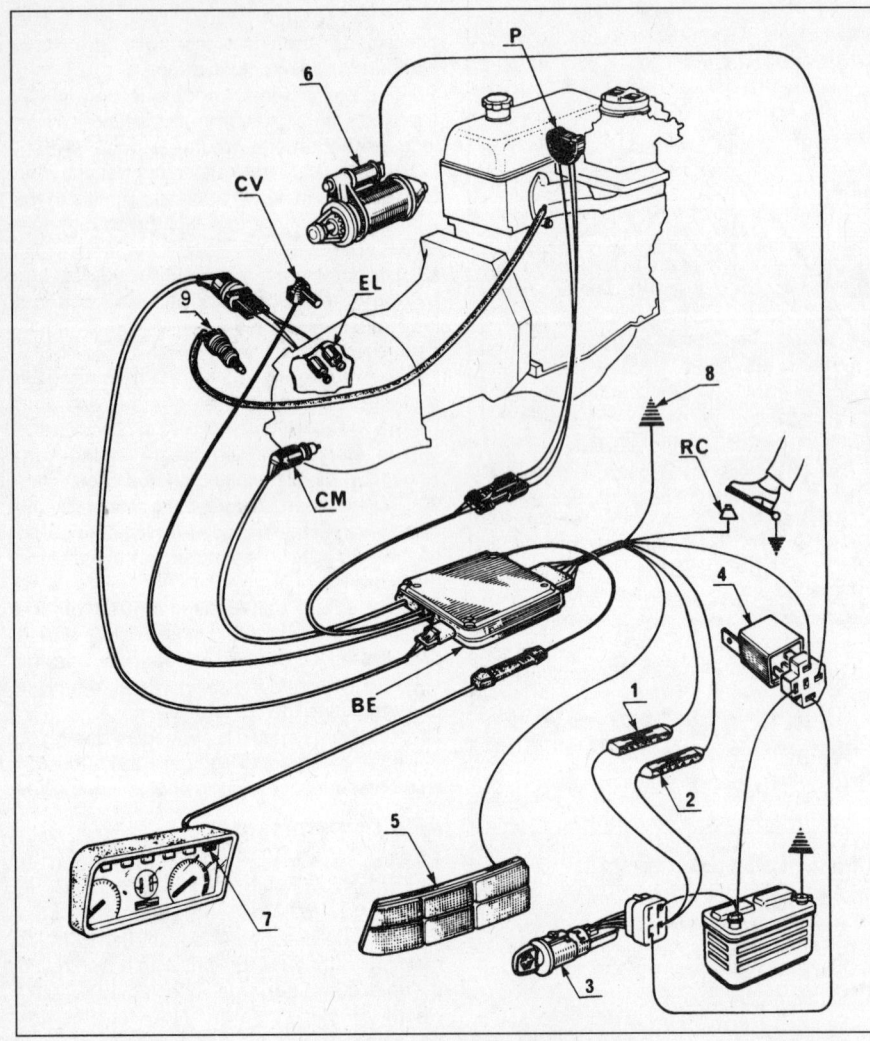

1.16 MJ3 transmission control units

1. Reversing light fuse
2. Supply fuse
3. Ignition switch
4. Starter relay
5. Reversing lights
6. Starter motor
7. Instrument panel warning light
8. Earth lead
9. Vacuum capsule
BE Computer
CM Multi-purpose switch
CV Speed sensor
EL Pilot solenoid valves
RC Kickdown switch

Unscrew the transmission drain plug from the sump and allow the fluid to drain completely into the container **(see illustration)**.

> **Warning:** *If the fluid is hot, take precautions against scalding.*

4 Clean the drain plug, being especially careful to wipe any metallic particles off the magnetic insert. Discard the original sealing washer; this should be renewed whenever it is disturbed.

5 When the fluid has finished draining, clean the drain plug threads and those of the transmission casing. Fit a new sealing washer to the drain plug, and refit the plug to the transmission, tightening it securely. If the car was raised for the draining operation, now lower it to the ground. Make sure that the car is level (front-to-rear and side-to-side).

6 Refilling the transmission is an awkward operation, adding the specified type of fluid to the transmission a little at a time via the dipstick tube. Use a funnel with a fine mesh gauze, to avoid spillage, and to ensure that no foreign matter enters the transmission. Allow plenty of time for the fluid level to settle properly.

7 Once the level is up to the "MAX" mark on the dipstick, refit the dipstick. Start the engine, and allow it to idle for a few minutes. Switch the engine off, then recheck the level, as described in Chapter 1, topping-up if necessary. Take the car on a short run to fully distribute the new fluid around the transmission, then recheck the fluid level on your return.

Final drive unit oil

8 Carry out the operations described in paragraphs 1 and 2.

9 Position a suitable container underneath the transmission and, referring to **illustration 2.3**, unscrew the filler/level plug from the side of the differential housing and the drain plug from the base of the housing. Allow the oil to drain completely into the container.

> **Warning:** *If the fluid is hot, take precautions against scalding.*

10 Clean the filler/level and drain plugs, being especially careful to wipe any metallic particles off the magnetic insert. Discard the original sealing washers; these should be renewed whenever they are disturbed.

11 When the fluid has finished draining, clean the drain plug threads and those of the transmission casing. Fit a new sealing washer to the drain plug, and refit the plug to the transmission, tightening it securely.

12 Refilling the final drive unit is an extremely awkward operation. Above all, allow plenty of time for the oil level to settle properly before checking it. Note that the car must be level ground when checking the oil level.

13 Refill with the exact amount of the specified type of oil via the filler/level plug hole and check that the fluid level is up to the lower edge. **Note:** *If the correct amount was poured into the final drive and a large amount flows out, refit the filler/level plug and take the car on a short journey so that the new oil is distributed fully around the transmission*

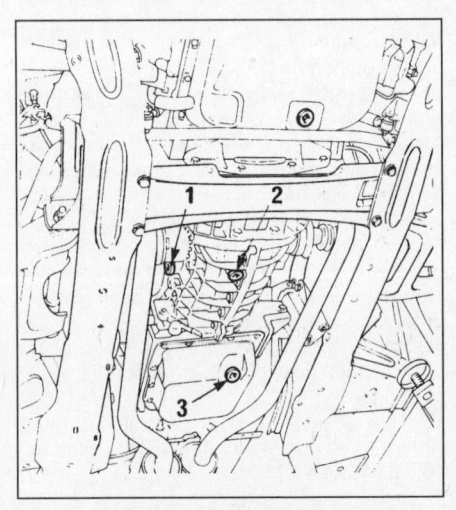

2.3 Drain and filler/level plugs - 4141 transmission

1. Final drive filler/level plug
2. Final drive drain plug
3. Transmission fluid drain plug

Automatic transmission 7B•5

components, then check the level again on your return.

14 Allow excess oil to drain from the final drive then fit a new sealing washer to the filler/level plug and refit it to the transmission tightening it securely. Lower the vehicle to the ground.

15 To ensure that the oil level is correct, take the vehicle on a short drive to distribute the oil around the final drive components. With the vehicle parked on level ground, stop the engine and allow the oil level to settle for a few minutes. Unscrew the filler/level plug from the side of the differential housing and check the oil level is up to the lower edge of the hole. If necessary, top-up with the specified type of oil until the oil trickles out of the hole. The level will be correct when the flow of oil ceases. Once the level is correct, refit the filler/level plug and sealing washer and tighten securely.

MJ3 transmission

16 On the MJ3 transmission both the transmission and final drive unit share the same fluid. The transmission fluid can be renewed as described in paragraphs 1 to 7 (see illustration).

AR4 transmission

Note: *On the AR4 transmission, the final drive is lubricated by its own separate oil bath. Final drive unit oil draining and refilling is described as a separate operation to avoid confusion.*

Transmission fluid

Note: *Special Renault test equipment is necessary to check the fluid level, and the home mechanic would be well-advised to take the vehicle to a Renault dealer to have the work carried out (see Chapter 1). If the task is to be undertaken, take great care not to introduce dirt into the transmission.*

17 Apply the handbrake, jack up the front of the vehicle, and support it on axle stands (see "Jacking and vehicle support").

18 Place a container beneath the sump, and unscrew the fluid level plug from the transmission (see illustration). Note that there is no transmission drain plug, and only a small amount of fluid will drain out of the level hole. When the flow of fluid stops, clean the threads of the level plug and refit it to the transmission, tightening it securely.

19 With the container still positioned underneath the transmission, slacken the sump retaining bolts slightly, and allow the transmission fluid to drain into the container. When most of the fluid has drained, remove the retaining bolts completely, and lower the sump away from the transmission, tipping the contents of the sump into the container. Recover the sump seal from the base of the transmission.

20 Remove the magnet from inside the sump, and clean all traces of metal filings from it. The filings (if any) should be very fine; any sizeable chips of metal indicate a worn component in the transmission. Examine the sump seal for signs of damage or deterioration, and renew if necessary.

21 Ensure the seal is correctly located in the sump groove. Refit the sump to the transmission, tightening its retaining bolts to the specified torque.

22 Lower the vehicle to the ground, then fill the transmission with the specified type and amount of fluid via the filler tube (see illustration). Use a funnel with a fine mesh gauze, to avoid spillage, and to ensure that no foreign matter enters the transmission. Allow plenty of time for the fluid level to settle properly.

23 Take the vehicle on a short journey, then check the transmission fluid level as described in Chapter 1.

Final drive unit oil

24 The final drive unit oil level is checked and topped-up via the filler/level plug, and drained

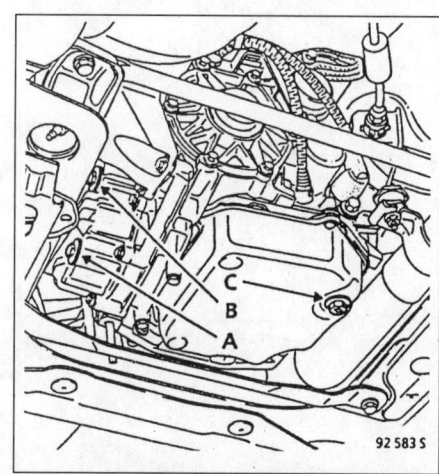

2.18 Transmission drain and level plugs - AR4 transmission

A Final drive drain plug
B Final drive filler/level plug
C Transmission fluid level plug

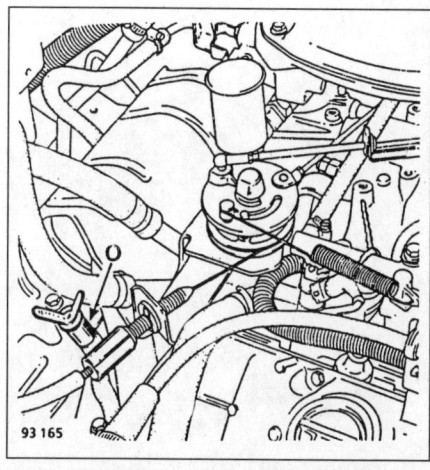

2.22 On the AR4 transmission unit, refill the transmission through the filler tube (O) on the right-hand side of the transmission

via the drain plug shown in **illustration 2.18**. The oil level can be checked as described in paragraph 15 or renewed as described in paragraphs 8 to 15.

3 Transmission fluid filter - renewal

4141 and MJ3 transmissions

1 Drain the transmission fluid as described in Section 2.

2 With the container still positioned underneath the transmission, slacken the sump retaining bolts slightly, and allow the transmission fluid to drain into the container. When most of the fluid has drained, remove the retaining bolts completely, and lower the sump away from the transmission, tipping the contents of the sump into the container. Recover the sump seal/gasket.

3 Unbolt the filter from the base of the transmission and recover the filter seal/gasket (as applicable) (see illustration).

4 Remove the magnet(s) from inside the sump, noting their correct fitted positions, and clean all traces of metal filings. The filings (if any) should be very fine; any sizeable chips of metal indicate a worn component in the transmission. Refit the magnet(s) in the correct position(s).

5 Fit a new gasket/seal on top of the filter element, and offer the filter to the transmission. Refit the retaining bolts, ensuring that the gasket/seal is still correctly positioned, and tighten them to the specified torque.

6 Ensure the seal/gasket is correctly located on the sump and refit the sump to the transmission, tightening its retaining bolts to the specified torque.

7 Lower the vehicle to the ground, then fill the transmission with the specified type and amount of fluid as described in Section 2.

2.16 Transmission drain plug (3) - MJ3 transmission

7B•6 Automatic transmission

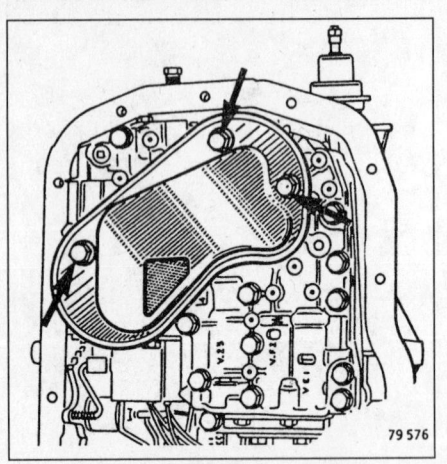

3.3 Transmission fluid filter retaining bolts (4141 transmission shown)

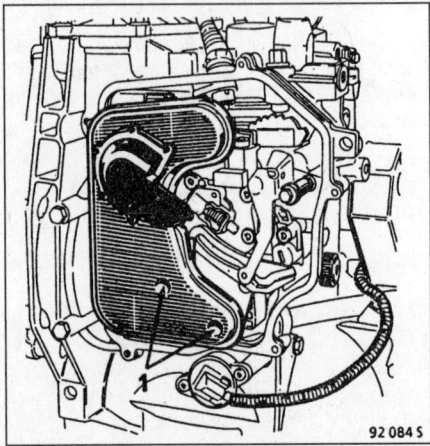

3.9 Transmission fluid filter retaining bolts - AR4 transmission

AR4 transmission

8 Drain the transmission fluid as described in Section 2.
9 Slacken and remove the filter retaining bolts, and remove the filter element from the base of the transmission, along with its sealing gasket **(see illustration)**. Discard the gasket; a new one must be used on refitting.
10 Position a new gasket on the top of the filter element, and offer the filter to the transmission. Refit the retaining bolts, ensuring that the gasket is still correctly positioned, and tighten them to the specified torque.
11 Refit the sump and refill the transmission as described in Section 2.

4 Kickdown switch - testing and adjustment

4141 transmission

1 Disconnect the lead from the kickdown switch and then connect a test lamp between the switch and the battery (+) terminal.
2 Switch on the ignition and have an assistant depress the accelerator pedal fully. The test bulb should illuminate.
3 If it does not, adjust in the following way. Mark the kickdown cable sleeve with a line 3.0 mm from the switch cover **(see illustration)**.
4 With the accelerator fully depressed, the line should be flush with the switch cover. If it is not, reposition the clip (F).
5 Recheck the operation using the test lamp.

MJ3 transmission

6 The switch is incorporated in the throttle potentiometer. The correct function of the switch can only be checked using special Renault equipment.

AR4 transmission

7 On most models, the kickdown switch is incorporated in the throttle potentiometer, as on MJ3 transmission. However, some models have a kickdown switch incorporated in the accelerator cable. The switch is situated at the pedal end of the outer cable, where it can be found behind the facia. The correct function of the switch can only be checked using special Renault equipment.

5 Governor cable (4141 transmission) - adjustment

1 The throttle linkage and kickdown switch must be correctly adjusted. Firmly apply the handbrake then jack up the front of the vehicle and support it on axle stands (see "Jacking and vehicle support").
2 Working at the transmission end of the cable, slacken the locknut and screw the cable adjuster in as far as it will go **(see illustration)**.
3 Working at the upper end of the cable, have an assistant depress the accelerator pedal fully and then remove all freeplay from the cable by slackening the locknut and rotating the throttle housing end adjuster **(see illustration)**. Once the adjuster is correctly positioned, securely tighten the locknut.
4 Return to the transmission end of the cable and position the cable adjuster to obtain a clearance of between 0.3 and 0.5 mm at the quadrant stop pin **(see illustration)**. With the clearance correctly set, securely tighten the locknut and lower the vehicle to the ground.

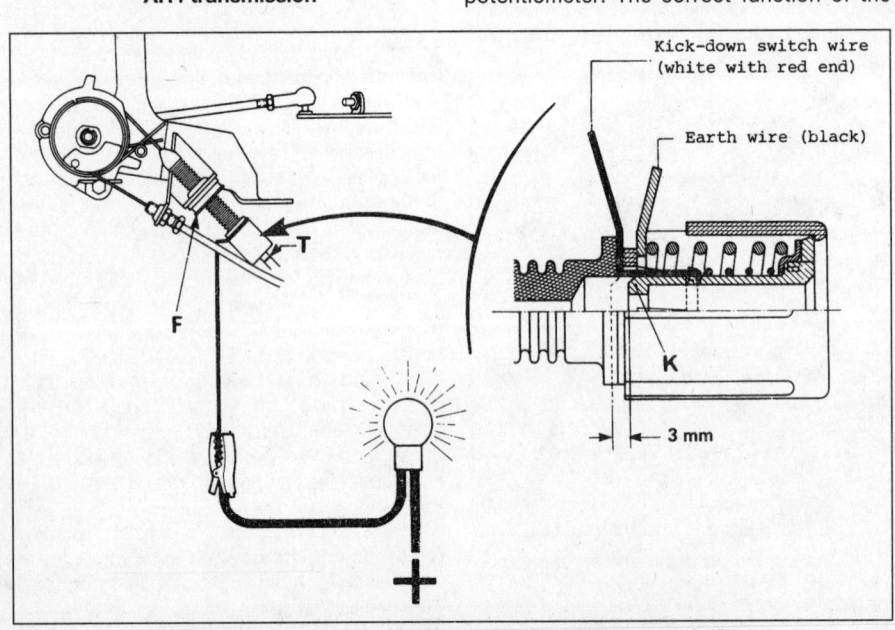

4.3 Kickdown switch adjustment details - 4141 transmission

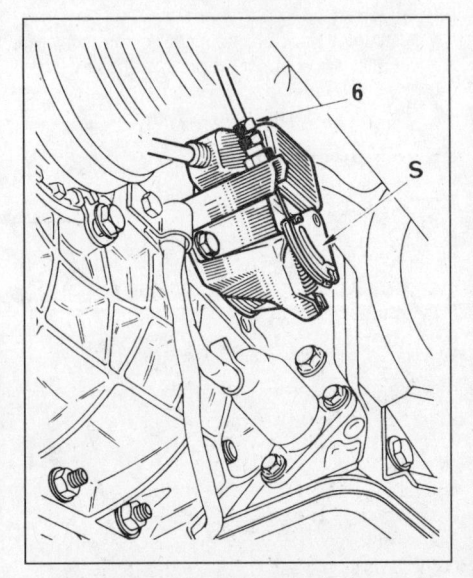

F Clip   K Stop   T Alignment mark

5.2 Governor cable transmission end adjuster (6) and quadrant (s) - 4141 transmission

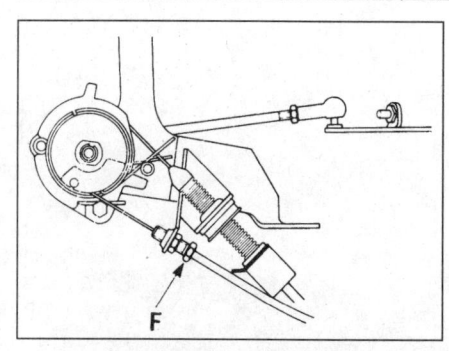

Automatic transmission 7B•7

5.3 Governor cable throttle housing end adjuster - 4141 transmission (F)

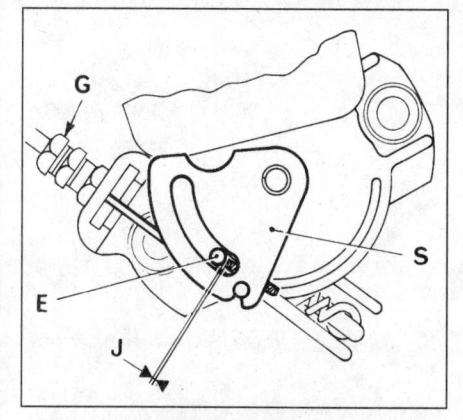

5.4 Governor cable adjustment details - 4141 transmission

E Stop pin
G Adjuster
J Clearance measuring point (0.3 to 0.5 mm)
S Quadrant

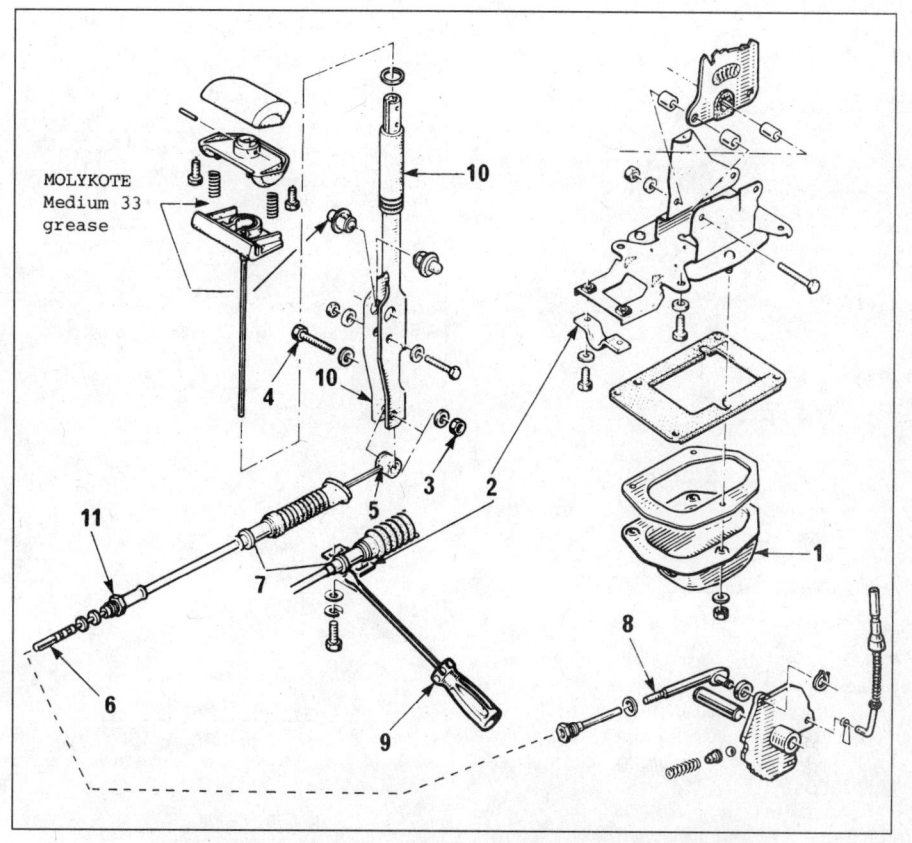

6.2 Selector lever and cable components - 4141 transmission

1 Cover
2 Cable retaining clamp
3 Nut
4 Pivot bolt
5 Cable end fitting
6 Cable threaded end fitting
7 Outer cable stop
8 Quadrant end fitting (situated inside the transmission unit)
9 Screwdriver showing cable adjustment point
10 Selector lever
11 Outer cable threaded end fitting

6 Selector cable - removal, refitting and adjustment

4141 transmission

Removal

1 Firmly apply the handbrake then jack up the front of the vehicle and support it on axle stands (see "Jacking and vehicle support"). Position the selector lever in position "N".

2 Working under the car at the selector lever end, undo the retaining nuts and washers and remove the cover and seal(s) from the base of the selector lever (see illustration).

3 Unscrew the two retaining bolts and remove the cable retaining clamp.

4 Unscrew the retaining nut then withdraw the pivot bolt and detach the selector cable from the lever. Withdraw the spacer from the cable end fitting.

5 Working at the transmission end of the cable, unscrew the outer cable end fitting from the transmission.

6 Unscrew the inner cable from its transmission end fitting then withdraw the cable from the transmission, noting its is likely to be a tight fit due to the sealing washers at its lower end.

7 If the transmission end fitting is broken, it will be necessary to remove the sump (see Section 3) and renew it. The end fitting is secured in position with a retaining clip.

Refitting and adjustment

8 With the automatic transmission still positioned in neutral "N", move the selector lever to position "P".

9 Screw the outer cable end fitting into the transmission and tighten securely.

10 Screw the inner cable into the transmission end fitting until the cable end is flush with the sleeve (see illustration).

11 Pull on the inner cable and move the transmission into position "P". This will ensure that the inner cable is correctly joined to its end fitting.

12 Ensure the cable is correctly routed, then connect the inner cable to the selector lever and securely tightening the pivot bolt nut.

13 Refit the outer cable retaining clamp, tightening its retaining bolts lightly only.

Remove all freeplay from the cable by levering between the cable fitting and retaining clamp then securely tighten the clamp retaining bolts (see illustration).

14 Check the operation of the selector mechanism then refit the cover and seal(s), tightening the retaining nuts securely. Lower the vehicle to the ground

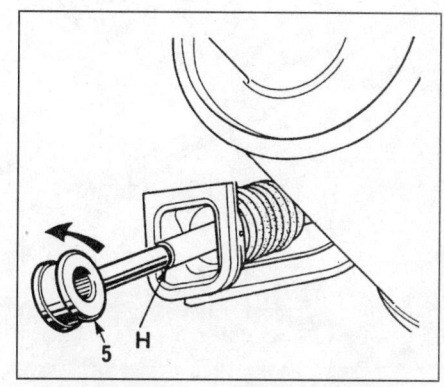

6.10 On 4141 transmission, screw the inner cable (5) into position until the cable end is flush with the sleeve (H)

7B•8 Automatic transmission

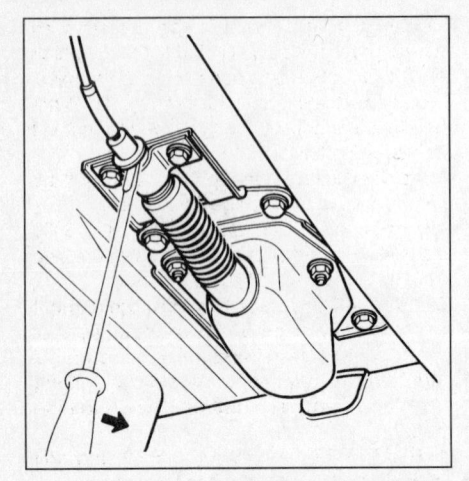

6.13 Remove all freeplay from the cable by levering in-between the outer cable stop and retaining clamp as shown

MJ3 transmission

Removal

15 Carry out the operations described in paragraphs 1 to 4.
16 Working at the transmission end of the cable, undo the retaining bolt(s) and remove the outer cable retaining clamp **(see illustration)**.
17 Unclip the inner cable end fitting from the selector lever and remove the cable from underneath the vehicle.

Refitting and adjustment

18 With the automatic transmission still positioned in neutral "N", move the selector lever to position "P".
19 Connect the cable end fitting to the transmission selector lever making sure its retaining clips are facing towards the lever **(see illustration)**.
20 Refit the outer cable retaining clamp and securely tighten its retaining bolt(s).

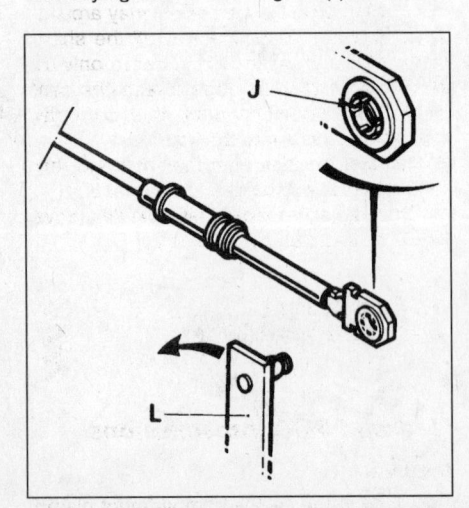

6.19 On MJ3 transmission ensuring the selector cable clips (J) are facing towards the transmission selector lever (L)

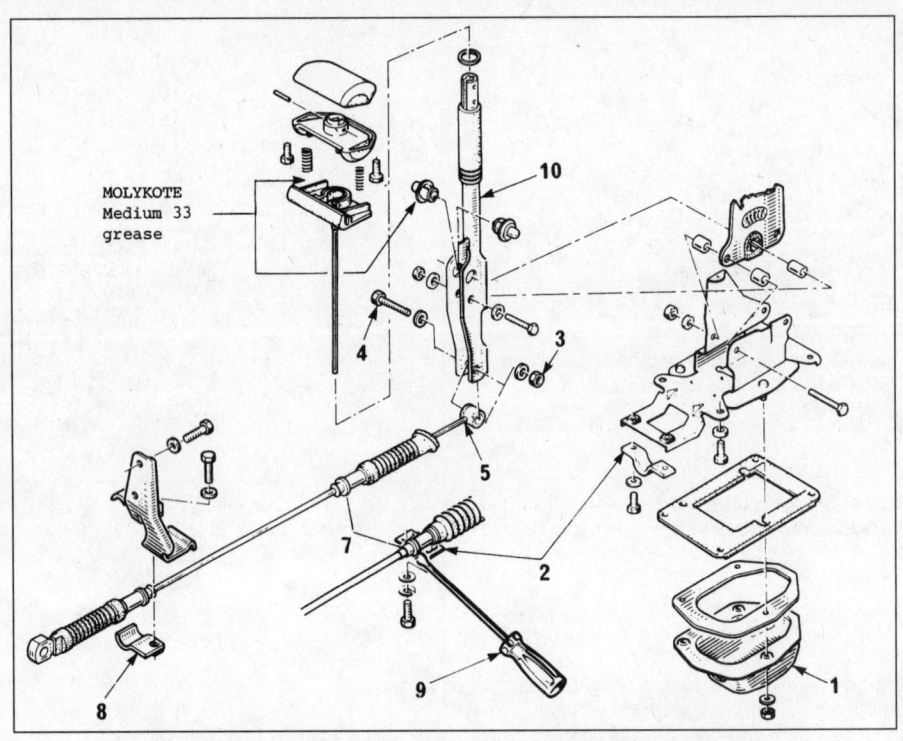

6.16 Selector lever and cable components - MJ3 transmission

1. Cover
2. Cable retaining clamp
3. Nut
4. Pivot bolt
5. Cable end fitting
6.
7. Outer cable stop
8. Cable retaining clamp
9. Screwdriver showing cable adjustment point
10. Selector lever

21 Carry out the operations described in paragraphs 11 to 14.

AR4 transmission

Removal

22 Position the selector lever in position "N" then remove the handle from the selector lever by lifting it upwards.
23 Carefully unclip and remove the selector lever trim panel from the top of the centre console.
24 Firmly apply the handbrake then jack up the front of the vehicle and support it on axle stands (see *"Jacking and vehicle support"*).
25 Working under the car, slacken and remove the retaining bolts and lower the selector lever out of position, disconnecting the wiring as it becomes accessible. Recover the lever seal.
26 Working at the transmission end of the cable, unclip the cable balljoint from the selector lever.
27 Unbolt the transmission support bracket and remove the selector lever and cable assembly from underneath the vehicle. Release the retaining clip and remove the support bracket from the cable end.
28 Unclip the lower cover from the selector lever base and unclip the cable balljoint from the lever. Slide out the outer cable retaining clip, noting which way around it is fitted, then depress the retaining clips and separate the cable and lever **(see illustration)**.

Refitting and adjustment

29 Clip the cable into the selector lever housing and secure it in position with the retaining clip. Make sure the clip is fitted the correct way around and clip the inner cable balljoint onto the selector lever. Check the cable is securely retained and refit the lower cover.

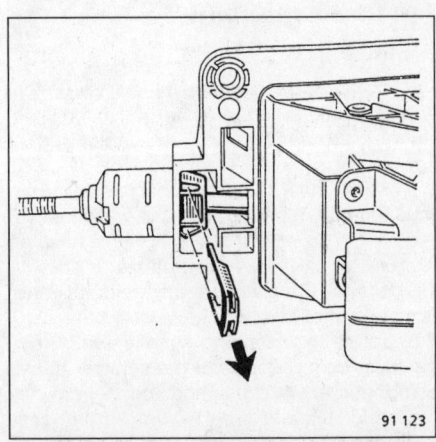

6.28 On AR4 transmission, slide out the retaining clip and detach the selector cable from the lever

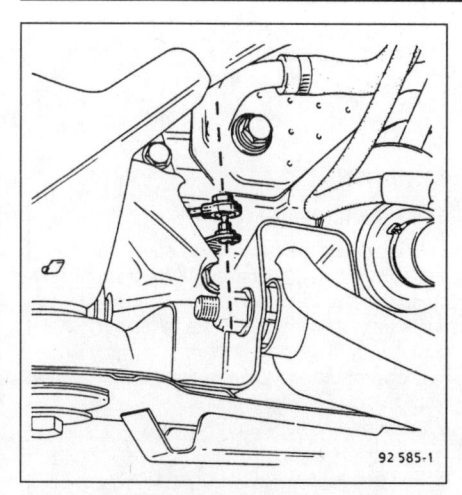

Automatic transmission 7B•9

6.35a On AR4 transmission, align the inner cable end fitting with the transmission selector lever balljoint . . .

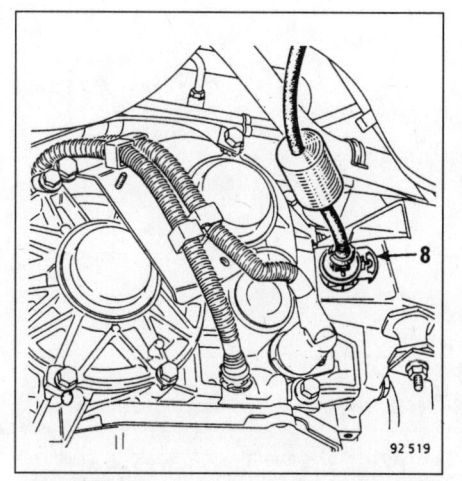

6.35b . . . then secure the outer cable in position by sliding in the adjuster clip (8)

30 Clip the transmission support bracket onto the cable and release the cable adjuster by pulling out its clip.
31 Manoeuvre the lever and cable assembly into position and fit the support bracket to the transmission, tightening its retaining bolts and nuts securely.
32 Ensure the seal is in position and reconnect the wiring to the selector lever. Lift the selector lever upwards into position and securely tighten the retaining bolts.
33 Refit the selector lever trim panel to the console and refit the handle to the lever.
34 Move the selector lever to the 1st gear "1" position and manually move the transmission selector lever to the 1st gear position (selector lever fully to the rear).
35 Align the cable balljoint with the transmission lever then secure the outer cable in position by sliding in the cable adjuster clip **(see illustrations)**.
36 Clip the cable balljoint onto the transmission selector lever then check the operation of the selector mechanism and lower the vehicle to the ground.

7 Selector lever - removal and refitting

4141 and MJ3 transmission

Removal

1 Detach the cable from the selector lever as described in paragraphs 1 to 4 of Section 6.
2 Undo the two retaining screws and unclip the cover from the selector lever handle.
3 Tap out the roll pin and lift the handle assembly upwards and away from the selector lever. Take care not to lose the handle springs.
4 Remove the centre console as described in Chapter 11.

5 Slacken and remove the retaining bolts and remove the lever assembly from the vehicle and recover the seal.

Refitting

6 Refit the lever assembly and seal and securely tighten its retaining bolts.
7 Refit the centre console as described in Chapter 11.
8 Insert the handle assembly into the selector lever, making sure the springs are correctly fitted, and secure it in position with the roll pin. Refit the handle cover and screws.
9 Reconnect the selector cable and adjust as described in paragraphs 12 to 14 of Section 6.

AR4 transmission

10 The selector lever is removed and refitted complete with the cable as described in Section 6.

8 Differential shaft oil seal - renewal

4141 transmission

1 Detach the driveshaft from the transmission as described in Chapter 8. Be prepared for some oil loss or drain the final drive unit as described in Section 2.
2 Using paint or a suitable marker pen, mark the differential seal ring nut in relation to the transmission casing.
3 Undo the retaining bolt/nut and remove the ring nut locking plate.
4 Using a suitable peg spanner, unscrew the ring nut and remove it from the transmission, noting the exact number of turns necessary to do so. Recover the sealing ring from the outside of the ring nut.
5 With the ring nut removed, note the correct fitted position of the seal in the ring nut then carefully press/lever the seal out of position, noting which way around it is fitted.

6 Making sure the seal is the correct way around, press the seal in using a socket which bears only on the hard outer edge of the seal.
7 Fit a new sealing ring to the outside of the ring nut and lubricate it with a smear of engine oil to ease installation.
8 Wrap insulation tape around the end of the shaft splines to protect the oil seal lips.
9 Apply a smear of grease to the oil seal lip and ease the ring nut assembly over the end of the shaft and into position.
10 Using the peg spanner, screw the ring nut into the transmission by the exact amount of turns counted on removal. Align the marks made prior to removal then refit the locking plate and securely tighten its retaining nut/bolt.
11 Remove the tape from the shaft and refit the driveshaft as described in Chapter 8.
12 On completion, top-up/refill the final drive oil level as described in Section 2.

MJ3 transmission

13 Renewal of the differential oil seals is a complex procedure which involves removing the cover plate from the transmission and moving the differential assembly. Seal renewal should therefore be entrusted to a Renault dealer.

AR4 transmission

14 Detach the driveshaft from the transmission as described in Chapter 8. Be prepared for some oil loss or drain the final drive unit as described in Section 2.
15 Note the correct fitted position of the seal in the housing then carefully prise it out of position, using a flat-bladed screwdriver. Take great care not to damage the transmission housing when levering the seal out.
16 Remove all traces of dirt from the area around the oil seal aperture and wrap insulation tape around the shaft splines to protect the new seal.
17 Apply a smear of grease to the oil seal lip and, making sure it is the correct way around, carefully ease it over the end of the shaft. Using a tubular spacer which bears only on the seal's hard outer edge, press the seal squarely into position until it is correctly positioned in relation to the housing.
18 Remove the tape from the shaft and refit the driveshaft as described in Chapter 8.
19 On completion, top-up/refill the final drive oil level as described in Section 2.

9 Fluid cooler - removal and refitting

4141 and MJ3 transmissions

Removal

1 Working in the engine compartment clamp the coolant hoses as close to the cooler as possible to minimise coolant loss **(see illustrations)**.

7B•10 Automatic transmission

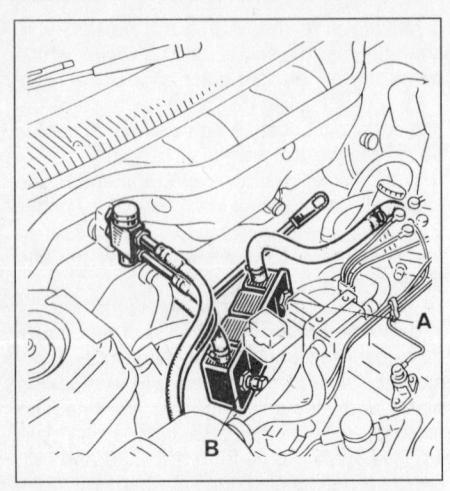

9.1a Transmission fluid cooler assembly - 4141 transmission (fluid hoses A and B)

2 Slacken the retaining clips and disconnect the coolant hoses from the cooler.
3 Wipe clean the area around the transmission fluid hoses then unscrew the hose union nuts/nut and bolt (as applicable). Disconnect the hoses and plug the hose ends and cooler ports to minimise fluid loss and prevent entry of dirt into the hydraulic system. Discard the sealing washers/rings, new ones should be used on refitting.
4 Undo the retaining bolts and remove the cooler from the engine compartment.

Refitting

5 Refitting is the reverse of removal using new fluid hose sealing washers/rings. On completion check the coolant and transmission fluid levels as described in Chapter 1 and *Weekly checks*.

AR4 transmission

6 On the AR4 transmission there are two possible types of fluid cooler. Some models are fitted with a coolant-cooled cooler which is bolted onto the transmission itself, where as other models are fitted with an air-cooled cooler which is situated behind the right-hand wing, where it is below the battery tray.

Coolant-cooled cooler - removal and refitting

7 Working in the engine compartment clamp the coolant hoses as close to the cooler as possible to minimise coolant loss. Slacken the retaining clips and disconnect the coolant hoses from the cooler.
8 Wipe clean the area around the fluid cooler then unscrew the two cooler mounting bolts. Remove the cooler from the transmission and recover the sealing rings fitted between the cooler and transmission. Also remove the sealing washers which are fitted to the mounting bolts. Both washers and sealing rings should be renewed.
9 On refitting, locate the new sealing rings in the recesses in the cooler and fit the new sealing washers to the mounting bolts. Fit the

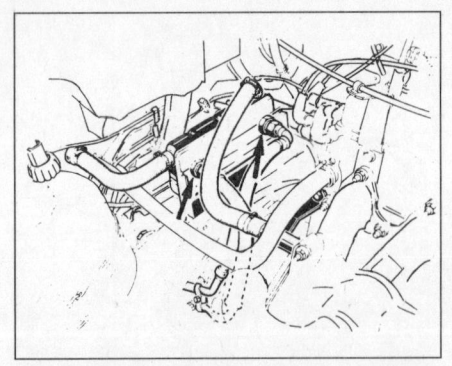

9.1b Transmission fluid cooler assembly - MJ3 transmission (fluid hoses arrowed)

cooler to the transmission, making sure the sealing rings remain correctly positioned, then fit the mounting bolts and tighten them to the specified torque. Reconnect the coolant hoses and check the transmission fluid and coolant levels as described in Chapter 1 and *Weekly checks*.

Air cooled cooler - removal and refitting

10 Firmly apply the handbrake then jack up the front of the vehicle and support it on axle stands (see "*Jacking and vehicle support*"). To improve access remove the battery as described in Chapter 5.
11 Disconnect the cooler fan wiring connector.
12 Wipe clean the area around the cooler hose unions then unscrew the hose union nuts. Disconnect the hoses and plug the hose ends and cooler ports to minimise fluid loss and prevent entry of dirt into the hydraulic system. Discard the sealing rings, new ones should be used on refitting.
13 Unbolt the cooler and remove the cooler and fan assembly from the engine compartment. If necessary the fan and cooler can then be separated.
14 Refitting is the reverse of removal, using new sealing rings. On completion, check the transmission fluid level as described in Chapter 1.

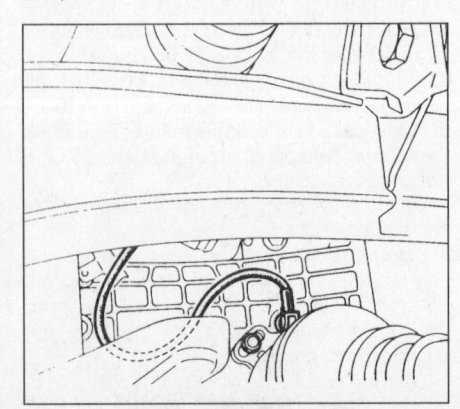

10.11 Speedometer sensor and retaining plate

10 Transmission - removal and refitting

Removal

4141 transmission

1 Disconnect the battery negative terminal.
2 Drain the transmission fluid and final drive oil as described in Section 2.
3 Release the computer and speedometer wiring from the body clips.
4 Disconnect the transmission fluid hoses from the cooler (see Section 9).
5 Disconnect the vacuum pipe from the inlet manifold.
6 Disconnect the scuttle drain pipes.
7 Unbolt the power steering pump and tie it to the side of the engine compartment (see Chapter 10).
8 Disconnect the exhaust front pipe from the manifold (see Chapter 4).
9 Remove the transmission dipstick guide tube and disconnect the reversing light switch wiring (where fitted).
10 Unscrew and remove the torque converter housing-to-engine upper connecting bolts.
11 Remove the speedometer sensor from the transmission **(see illustration)**.
12 Remove the starter motor (see Chapter 5).
13 Disconnect the driveshafts from the transmission, as described in Chapter 8.
14 Remove the crankshaft sensor and bracket.
15 Disconnect the selector control cable from the transmission as described in Section 6.
16 On six-cylinder engines, unbolt and remove the engine movement damper and its mounting bracket from the engine **(see illustration)**.
17 Disconnect the governor cable (see Section 5).

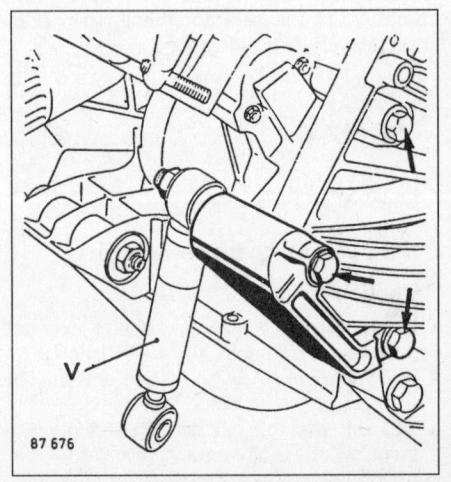

10.16 On six-cylinder engines, unbolt and remove the engine movement damper and mounting bracket

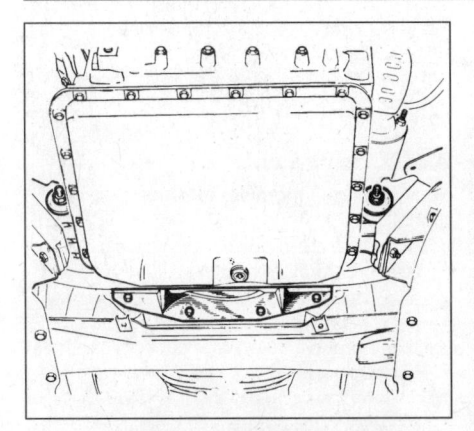

Automatic transmission 7B•11

10.19 Lower driveplate cover - 4141 transmission

18 Disconnect the earth strap from the left-hand transmission mounting.
19 Undo the retaining bolts and remove the lower driveplate cover plate from the transmission, to gain access to the torque converter retaining bolts (see illustration). Slacken and remove the visible bolt then, using a socket and extension bar to rotate the crankshaft pulley, undo the remaining bolts securing the torque converter to the driveplate as they become accessible. There are three in total.
20 Unless the car is over an inspection pit, or up on a car lift, the rear end must now be raised and securely supported so that enough clearance will be provided for the transmission to be withdrawn rearwards and removed from under the car.
21 Place a jack with a block of wood beneath the engine, to take the weight of the engine. Alternatively, attach a hoist or support bar to take the weight of the engine. Support the transmission on a trolley jack and remove both mountings from the transmission.
22 Unscrew and remove the remaining lower engine/transmission connecting bolts.
23 Lower the trolley jack and withdraw the transmission. Guide the computer out with the transmission. The engine should rest safely on the front crossmember.

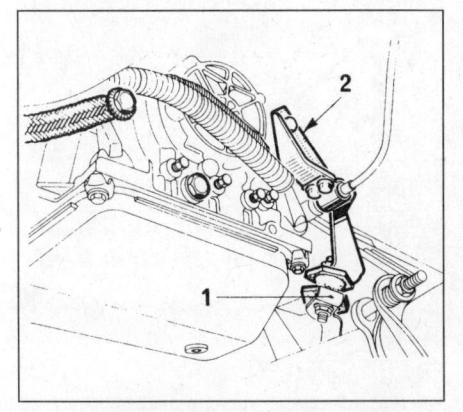

24 As the transmission is removed take great care to ensure that the torque converter does not fall out. It is recommended that the converter ie secured in position using a length of metal strip bolted to one of the transmission bolt holes.

MJ3 transmission

25 Remove the air cleaner (see Chapter 4).
26 Carry out the operations described in paragraphs 1 to 15.
27 Working from the top, unscrew and remove the engine/transmission connecting bolts.
28 Unbolt and remove the exhaust and selector cable mounting brackets (see illustration).
29 Disconnect the transmission earth strap.
30 On 1995 cc engined models, unbolt the cover plate from the lower front face of the torque converter housing (see illustration).
31 Working through the starter motor aperture, disconnect the driveplate from the torque converter using the information in paragraph 19.
32 Unless the car is over an inspection pit, or up on a car lift, the rear end must now be raised and securely supported so that enough clearance will be provided for the transmission to be withdrawn rearwards and removed from under the car.
33 Place a jack with a block of wood beneath the engine, to take the weight of the engine. Alternatively, attach a hoist or support bar to take the weight of the engine. Support the transmission on a trolley jack and remove both mountings from the transmission.
34 Unbolt and remove the left-hand transmission mounting, complete with brackets.
35 Disconnect the vacuum pipe from the transmission capsule.
36 Remove the right-hand mounting, complete with brackets and remove the transmission described in paragraphs 22 to 24.

AR4 transmission

37 Disconnect the battery negative terminal.
38 Drain the transmission fluid and final drive oil as described in Section 2.

39 Disconnect the driveshafts from the transmission as described in Chapter 8.
40 Remove the starter motor as described in Chapter 5.
41 Remove the exhaust front pipe as described in Chapter 4.
42 Referring to Section 9, on models with a coolant-cooled transmission cooler, clamp the hoses and disconnect the coolant hoses from the cooler. On models with an air-cooled transmission cooler, unscrew the union nuts and disconnect the fluid cooler pipes from the union on the front of the transmission (see illustration). Recover the sealing rings (these must be renewed) and plug the hose/pipe ends to prevent dirt entry into the system.
43 Unbolt the driveplate coverplates from the transmission.
44 On six cylinder engines, unbolt and remove the engine movement damper and its mounting bracket from the engine.
45 Disconnect the selector control cable from the transmission as described in Section 6.
46 Disconnect all the necessary wiring connectors and breather hoses from the transmission noting the correct routing of the wiring.
47 Remove the transmission as described in paragraphs 20 to 24, ignoring the remark about the computer.

Refitting

4141 transmission

48 Refitting is a reversal of removal, but observe the following points.
a) Make sure that the transmission locating dowels are in position on the engine mating face and that the power steering pump mounting bolt is in place in the transmission before the transmission is offered to the engine, as it cannot be fitted later (see illustration).

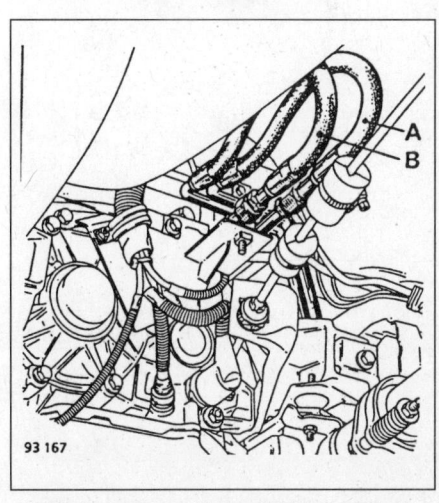

10.28 Exhaust and selector cable mounting brackets - MJ3 transmission

10.30 Driveplate cover bolts - 1995 cc models

10.42 On model with an AR4 transmission unit with an air-cooled fluid cooler, unscrew the union nuts and disconnect the hoses (A and B) from transmission

7B•12 Automatic transmission

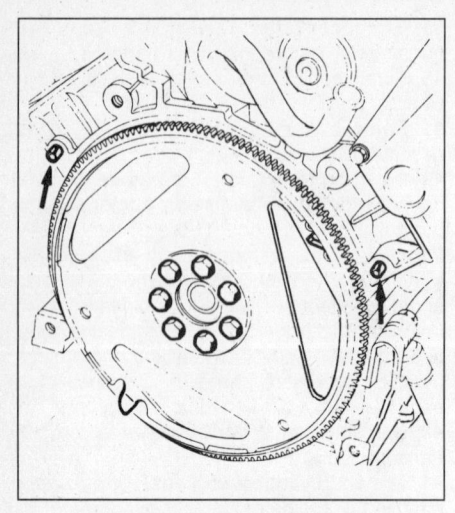

10.48 Transmission locating dowel locations - 4141 transmission

b) Ensure the torque converter is correctly located on the transmission and apply a little Molykote BR2 grease to the torque converter locating hub. Do not apply too much, otherwise there is a possibility of the grease contaminating the torque converter.
c) When connecting the driveplate to the torque converter, no particular alignment is required but tighten the bolts to the specified torque.
d) Note that the longest socket-headed screw for the right-hand transmission mounting goes at the top.
e) Adjust the governor cable and selector control, as described in Sections 5 and 6.
f) Refill the final drive and the transmission as described in Section 2.

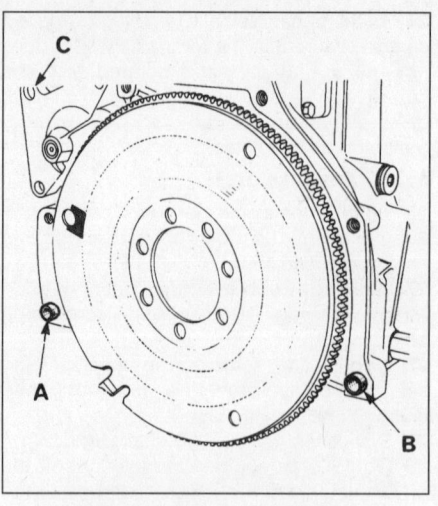

10.49a Transmission locating dowel locations - MJ3 transmission

A Engine locating dowel
B Engine locating dowel
C Starter motor locating dowel

MJ3 transmission

49 Refitting is a reversal of removal, but observe the following points.
a) Check that the positioning dowels are in place before offering the transmission to the engine, also grease the recess in the crankshaft which accepts the torque converter with Molykote BR2 grease **(see illustration)**.
b) Align the converter in relation to the driveplate, as shown **(see illustration)**. Tighten the driveplate bolts to the specified torque.
c) Adjust the selector cable as described in Section 6.
d) Refill the transmission as described in Section 2.

AR4 transmission

50 Refitting is the reverse of removal noting the following.
a) Check that the positioning dowels are in place before offering the transmission to the engine, also grease the recess in the crankshaft which accepts the torque converter with Molykote BR2 grease **(see illustration)**.
b) Tighten all fixings to the specified torque settings (where given).
c) Adjust the selector cable as described in Section 6.
d) Refill the transmission as described in Section 2.

11 Transmission overhaul - general information

In the event of a fault occurring with the transmission, it is first necessary to determine whether it is of an electrical, mechanical or hydraulic nature, and to do this, special test equipment is required. It is therefore essential to have the work carried out by a Renault dealer if a transmission fault is suspected.

Do not remove the transmission from the car for possible repair before professional fault diagnosis has been carried out, since most tests require the transmission to be in the vehicle.

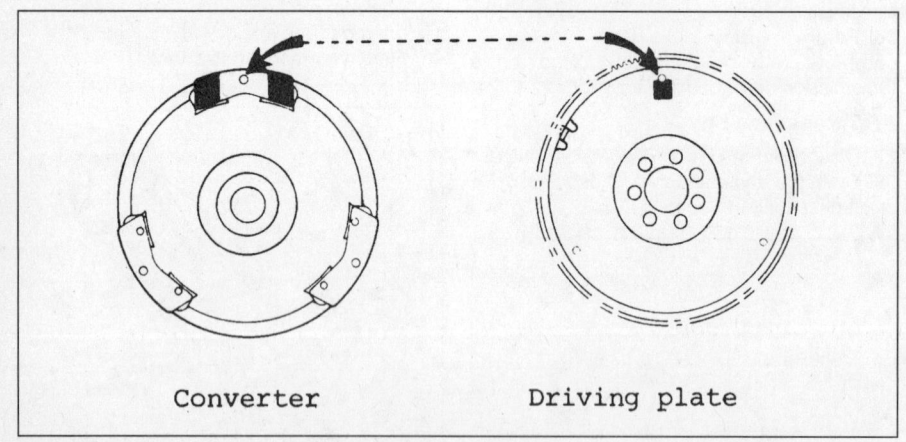

10.49b On refitting the MJ3 transmission, ensure the torque converter is correctly engaged with the driveplate lug

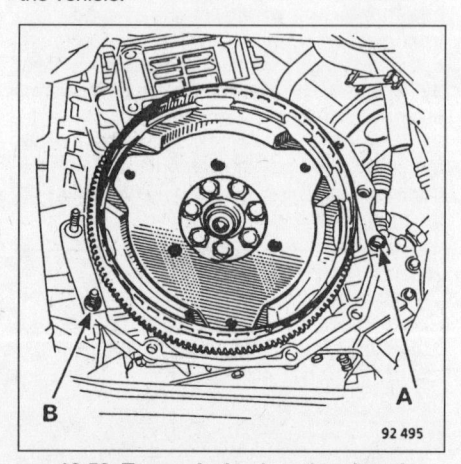

10.50 Transmission locating dowel locations (A and B) - AR4 transmission

# Chapter 8
Driveshafts

Contents

Driveshafts - removal and refitting 2
Driveshaft check .... See Chapter 1
Driveshaft overhaul - general information 4
Driveshaft rubber gaiters - renewal 3
General information 1

Degrees of difficulty

| **Easy,** suitable for novice with little experience | **Fairly easy,** suitable for beginner with some experience | **Fairly difficult,** suitable for competent DIY mechanic | **Difficult,** suitable for experienced DIY mechanic | **Very difficult,** suitable for expert DIY or professional |

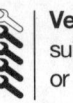

Specifications

General
Type .... Tubular steel with constant velocity (CV) joint at each end
Lubrication .... Special grease supplied in sachets with gaiter kits - joints are otherwise pre-packed with grease, and sealed

Torque wrench setting   Nm   lbf ft
Driveshaft nut 250   185

1 General information

Drive is transmitted from the differential to the front wheels by means of two tubular steel driveshafts of equal length.

Both driveshafts are splined at their outer ends, to accept the wheel hubs, and are threaded so that each hub can be fastened by a large nut. The inner end of each driveshaft is splined, to accept the differential sun gear.

Constant velocity (CV) joints are fitted to each end of the driveshafts, to ensure that the smooth and efficient transmission of power at all suspension and steering angles. The inner end of each driveshaft is secured to the sun gear with a roll-pin.

The inboard driveshaft joint is of the tripod type, while the outboard joint may be of tripod or ball-and-cage type, depending on model.

2 Driveshafts - removal and refitting

Removal
Note: *A new driveshaft nut, a new hub carrier upper balljoint nut, and a new track-rod end balljoint nut will be required on refitting, and new roll-pins will be required to secure the inner end of the driveshaft on refitting. Sealant may be required to coat the outer end of the driveshaft. A balljoint separator tool will be required for this operation.*

1 Apply the handbrake, then jack up the front of the vehicle and support securely on axle stands (see *"Jacking and vehicle support"*). Remove the relevant front roadwheel.
2 On models with ABS, it is advisable to remove the ABS wheel sensor as described in Chapter 9, to avoid any possibility of damage during the removal procedure.

> **HAYNES HiNT** *On models where access to the driveshaft nut can be obtained by removing the wheel trims, before jacking up the vehicle, loosen the driveshaft nut as follows.*
> *a) Chock the front wheels, and remove the wheel trim.*
> *b) Have an assistant firmly apply the footbrake.*
> *c) Loosen the driveshaft nut using a socket and extension..*

3 If the driveshaft nut has been loosened, proceed to paragraph 5, otherwise proceed as follows.
4 Refit at least two roadwheel bolts to the front hub, and tighten them securely. Have an assistant firmly depress the brake pedal to prevent the front hub from rotating, then using a socket and a long extension bar, slacken and remove the driveshaft retaining nut. Alternatively, a tool can be fabricated from two lengths of steel strip (one long, one short) and a nut and bolt; the nut and bolt forming the pivot of a forked tool. Bolt the tool to the hub using two wheel bolts, and hold the tool to prevent the hub from rotating as the driveshaft retaining nut is undone **(see Tool Tip)**. This nut is very tight; make sure that there is no risk of pulling the car off the axle stands. (If the roadwheel trim allows access to the driveshaft nut, the initial slackening can be done with the wheels chocked and on the ground.)

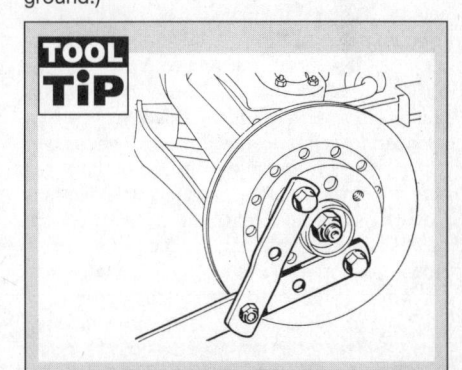

Using a fabricated tool to hold the front hub stationary whilst the driveshaft nut is slackened

5 Unbolt the brake caliper from the hub carrier as described in Chapter 9. Note that there is no need to disconnect the fluid hose - suspend the caliper from the suspension strut using wire or string, ensuring that the hose is not strained.

8•2 Driveshafts

2.7 Release the track-rod balljoint using a balljoint separator tool

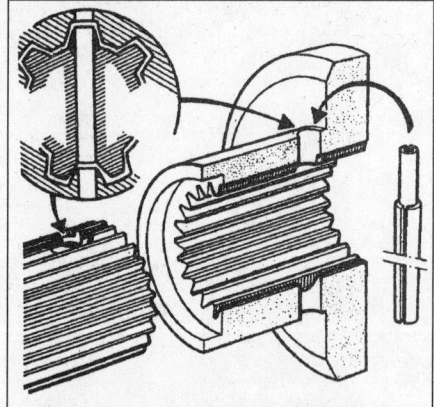

2.8 Drive out the double roll-pins securing the driveshaft to the sun gear shaft

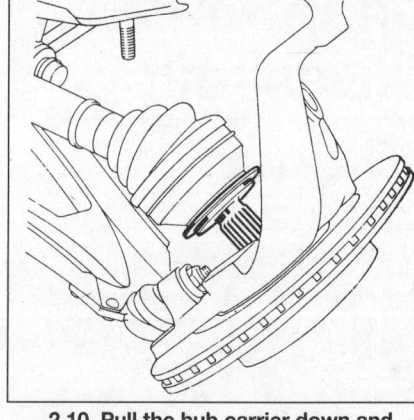

2.10 Pull the hub carrier down and disconnect the driveshaft from the hub

6 Slacken and partially unscrew the hub carrier upper balljoint nut (unscrew the nut as far as the end of the threads on the balljoint to prevent damage to the threads as the joint is released), then release the balljoint using a balljoint separator tool. Remove the nut, and discard it - a new nut must be used on refitting.
7 Similarly, release the balljoint and disconnect the track-rod from the steering arm on the hub carrier **(see illustration)**.
8 Working at the transmission end of the driveshaft, where applicable remove the sealant from the ends of the roll-pins securing the inner end of the driveshaft to the sun gear shaft, then drive out the double roll-pins, using a pin-punch **(see illustration)**.
Caution: Where applicable, when driving out the roll-pins from the right-hand driveshaft, turn the driveshaft so that there is no risk of damage to the speedometer sensor as the roll-pins are driven out.
9 Unscrew the driveshaft nut from the hub carrier end of the driveshaft. Recover the washer. Discard the nut, a new one must be used on refitting.
10 Pull the hub carrier from the upper balljoint, and tilt the hub carrier downwards until the splined end of the driveshaft can be released from it **(see illustration)**. If necessary tap the end of the driveshaft using a hammer and a soft metal drift to release it from the hub - **do not** strike the end of the driveshaft hard, as this may cause damage to the joints.
Note: *On some models the driveshaft ends are fitted to the hub carriers using locking compound. Renault use a special extractor tool to release the driveshaft ends on these models, but if the driveshaft cannot be released by hand, it should be possible to use a three-legged puller as follows.*
a) Remove the brake disc with reference to Chapter 9.
b) Temporarily refit the driveshaft nut to protect the threads on the end of the driveshaft.
c) Fit the puller, with the arms bearing on the hub, and the centre screw bearing on the end of the driveshaft.

d) Use the puller to release the hub from the end of the driveshaft. Note that it is possible that the hub will be pulled from the hub bearing assembly (the bearing front half inner race will remain in position on the hub) - this is not a problem provided that the bearing is kept clean, and repacked with grease on refitting.
11 Place a container beneath the transmission end of the driveshaft to catch escaping oil/fluid which may be released as the end of the driveshaft is withdrawn.
12 Pull the driveshaft from the transmission, then withdraw the assembly from under the vehicle **(see illustration)**. Where applicable, recover the O-ring from the end of the sun gear shaft.

Refitting

Note: *If a new driveshaft is being fitted, it may be supplied with a protective cardboard cover fitted over the outer driveshaft gaiter. In this case, do not remove the cover until the completion of the refitting procedure.*
13 On models where the driveshaft end was fitted to the hub carrier using locking compound, thoroughly clean all traces of old locking compound from the end of the driveshaft.
14 Before installing the driveshaft, examine the driveshaft oil seal in the transmission for signs of damage or deterioration and, if necessary, renew it, referring to the appropriate Part of Chapter 7 for further information. (Having got this far it is worth renewing the seal as a matter of course.)
15 Thoroughly clean the driveshaft splines, and the apertures in the transmission and hub assembly. Apply a thin film of grease to the oil seal lips, and to the driveshaft splines and shoulders. Check that all gaiter clips are securely fastened.
16 Where applicable, fit a new O-ring to the sun gear shaft.
17 Engage the inboard end of the driveshaft with the sun gear, ensuring that the roll-pin holes in the driveshaft and gear shaft are aligned.
18 Where applicable, coat the hub end of the driveshaft with locking compound (Renault recommend the use of "Loctite Scelbloc"), then engage the end of the driveshaft with the hub. **Note:** *If the hub has been pulled from the bearing during the removal procedure, ensure that the bearing is clean and packed with the appropriate grease, then refit the hub as described in the "Front wheel bearing - renewal" procedure in Chapter 10.*
19 Reconnect the hub carrier upper balljoint, and tighten a new nut to the specified torque.
20 Secure the inner end of the driveshaft to the sun gear using new roll-pins. Coat the ends of the roll-pins with sealant **(see illustration)**.
21 Screw the **new** driveshaft nut onto the end of the driveshaft as far as possible by

2.12 Pull the driveshaft from the transmission

2.20 Driving in new driveshaft roll-pins

Driveshafts 8•3

hand, ensuring that the washer is in place, then tighten the nut until the end of the driveshaft is fully engaged with the hub. Do not fully tighten the nut at this stage.

22 Reconnect the track-rod to the hub carrier, and tighten a new balljoint nut to the specified torque.

23 Where applicable, refit the brake disc, then refit the brake caliper, with reference to Chapter 9.

24 Use the method employed on removal to prevent the hub from rotating, and tighten the driveshaft retaining nut to the specified torque. Check that the hub rotates freely.

25 Where applicable, refit the ABS wheel sensor, with reference to Chapter 9.

26 Where applicable, tear off the protective cover from the outer driveshaft joint gaiter. Do not use a sharp tool which may damage the gaiter.

HAYNES HiNT *On models where access to the driveshaft nut can be obtained by removing the wheel trim, the driveshaft nut can be tightened with the footbrake firmly applied, and the vehicle resting on its wheels.*

27 Refit the roadwheel, then lower the vehicle to the ground and tighten the roadwheel bolts to the specified torque.

28 On completion, check the transmission oil/fluid level using the information given in Chapter 1.

3 Driveshaft rubber gaiters - renewal

Outer joint

Tripod joint

1 With the driveshaft removed as described in Section 2, proceed as follows.

2 Remove the inner driveshaft joint, as described later in this Section.

3 Remove the larger gaiter securing clip. (see Tool Tip).

4 Pull back the gaiter, and wipe out as much grease as possible.

5 Using a screwdriver, prise up the locking plate arms one by one, and separate the joint from the driveshaft (see illustration).

6 Recover the thrust button, shim and spring from the centre of the joint spider (see illustration).

7 Remove the smaller gaiter securing clip, then slide the gaiter from the inner end of the driveshaft. Where applicable, also slide the plastic bush from the driveshaft (see illustration).

8 Thoroughly clean the joint using paraffin, or a suitable solvent, and dry it thoroughly. Check the tripod joint bearings and joint outer member for signs of wear, pitting or scuffing on their bearing surfaces. Check that the bearing rollers rotate smoothly and easily around the tripod joint, with no traces of roughness.

9 If on inspection, the tripod joint or outer member reveal signs of wear or damage, it will be necessary to renew the complete driveshaft assembly, since the joint is not available separately. If the joint is in satisfactory condition, obtain a repair kit consisting of a new gaiter, retaining clips, and the correct type and quantity of grease.

Where a wire-type clip is fitted, the clip can be released using two lengths of drilled rod or stout metal tube to lever the ends of the clip

Although not strictly necessary, it is also recommended that the outer constant velocity joint gaiter is renewed, regardless of its apparent condition.

10 Where applicable, slide the plastic bush onto the driveshaft.

11 Tape over the splines on the inner end of the driveshaft (to protect the gaiter), then carefully slide the outer gaiter onto the driveshaft.

12 Fit the spring and thrust button to the centre of the joint spider.

13 Align the locking plate so that the arms are centrally positioned between the spider rollers (see illustration).

14 Slide the joint onto the driveshaft, then tilt the shaft to engage one of the locking plate arms in its slot, then use a screwdriver to engage the remaining two.

15 Slide the thrust button shim into position under the head of the thrust button (see illustration).

16 Pack the joint with the grease supplied in the repair kit. Work the grease well into the joint, and fill the gaiter with any excess.

17 Locate the gaiter ends in their grooves on the driveshaft and joint, or on the plastic bush, as applicable, then lift the edges of the gaiter to equalise air pressure in the gaiter.

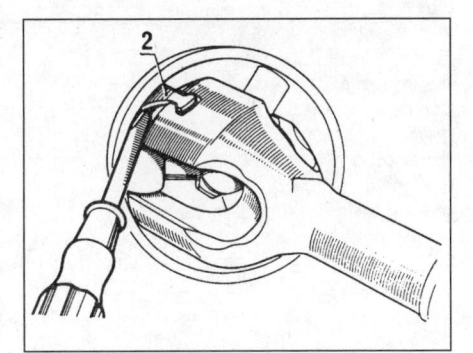

3.5 Prising up a locking plate arm (2) - tripod-type outer joint

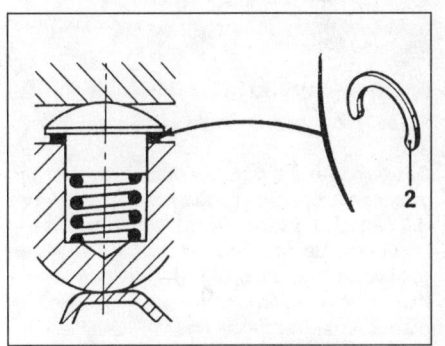

3.6 Recover the thrust button, shim (2) and spring from the joint spider - tripod-type outer joint

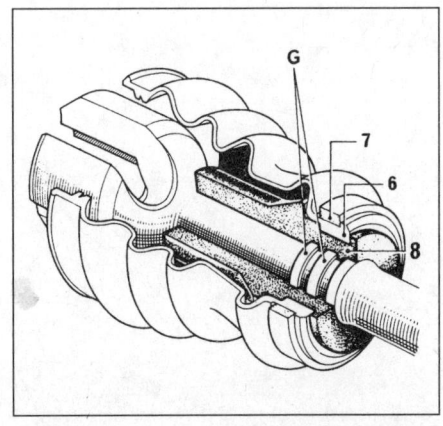

3.7 Tripod-type outer driveshaft joint plastic bush location

G Shaft ribs
6 Gaiter
7 Gaiter securing clip
8 Plastic bush

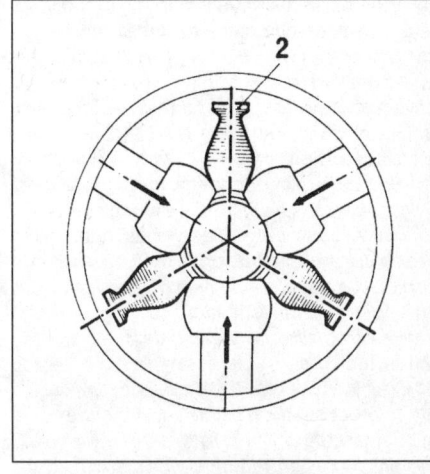

3.13 Locking plate (2) correctly positioned - tripod-type outer joint

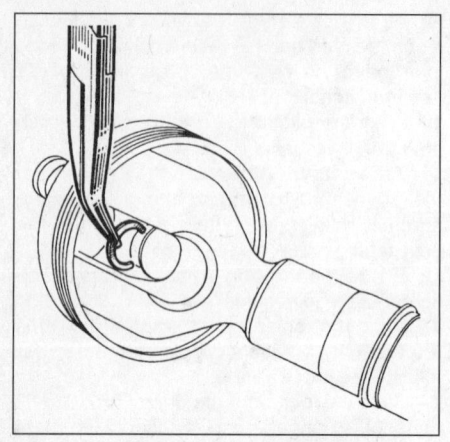

8•4 Driveshafts

3.15 Fitting the thrust button shim - tripod-type outer joint

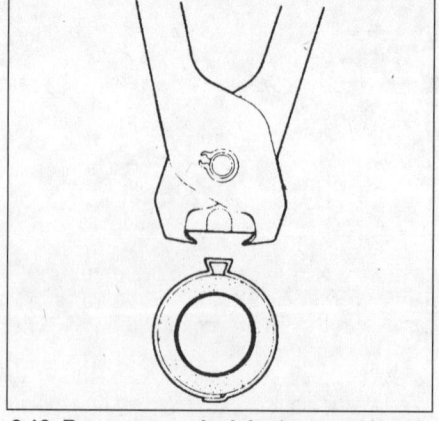

3.18 Remove any slack in the metal band-type clip using a pair of side cutters

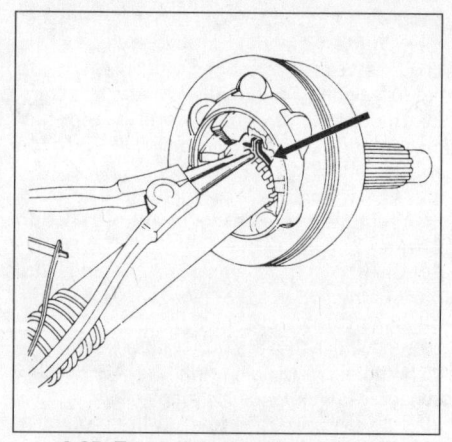

3.25 Expand the circlip (arrowed) and tap the joint from the driveshaft - ball-and-cage-type outer joint

18 Fit both the inner and outer retaining clips, and locate them in the grooves, then secure them. To secure the metal band-type clip, lock the ends of the clip together, then remove any slack in the clip by carefully compressing the raised section of the clip using a pair of side cutters **(see illustration)**.
19 Refit the inner driveshaft joint as described later in this Section.
20 Refit the driveshaft as described in Section 2.

Ball-and-cage joint - removable joint

Note: *Two types of ball-and-cage joint have been used in production. The first type is removable, and is retained by a circlip which is visible once the gaiter has been slid back from the joint (see paragraph 25). The second type of joint is retained by a hidden circlip, and is bonded to the end of the driveshaft - this joint cannot be removed, and the inner driveshaft joint must be removed in order to renew the outer gaiter.*

21 Remove the driveshaft (see Section 2).
22 Remove the gaiter securing clips.
23 Secure the driveshaft in a vice.
24 Pull back the gaiter from the joint and wipe out as much grease as possible.
25 Using a pair of circlip pliers, expand the circlip securing the joint to the end of the driveshaft, and at the same time, tap the joint from the driveshaft using a soft-faced mallet **(see illustration)**.
26 Slide the gaiter and the rubber collar from the end of the driveshaft **(see illustration)**.
27 Thoroughly clean the joint using paraffin, or a suitable solvent, and dry it thoroughly. Carry out a visual inspection of the joint.
28 Move the inner splined driving member from side to side, to expose each ball in turn at the top of its track. Examine the balls for cracks, flat spots, or signs of surface pitting.
29 Inspect the ball tracks on the inner and outer members. If the tracks have widened, the balls will no longer be a tight fit. At the same time, check the ball cage windows for wear or cracking between the windows.

30 If on inspection, the constant velocity joint components are found to be worn or damaged, it will be necessary to renew the complete driveshaft assembly, since the joint is not available separately. If the joint is in satisfactory condition, obtain a repair kit consisting of a new gaiter, retaining clips, and the correct type and quantity of grease **(see illustration)**. Although not strictly necessary, it is also recommended that the outer constant velocity joint gaiter is renewed, regardless of its apparent condition.
31 Tape over the splines on the end of the driveshaft, then slide the rubber collar and gaiter onto the shaft. Locate the inner end of the gaiter on the driveshaft, and secure it in position with the rubber collar.
32 Remove the tape, then slide the constant velocity joint coupling onto the driveshaft until the internal circlip locates in the driveshaft groove.
33 Check that the circlip holds the joint securely on the driveshaft, then pack the joint with the grease supplied. Work the grease well into the ball tracks, and fill the gaiter with any excess.
34 Locate the outer lip of the gaiter in the groove on the joint outer member. With the coupling aligned with the driveshaft, lift the lip of the gaiter to equalise the air pressure. Secure the gaiter in position with the large retaining clip. Where a wire-type retaining clip is fitted, the clip can be secured using two lengths of drilled rod or stout metal tube to lever the ends of the clip.
35 Check that the constant velocity joint moves freely in all directions, then refit the driveshaft to the vehicle as described in Section 2.

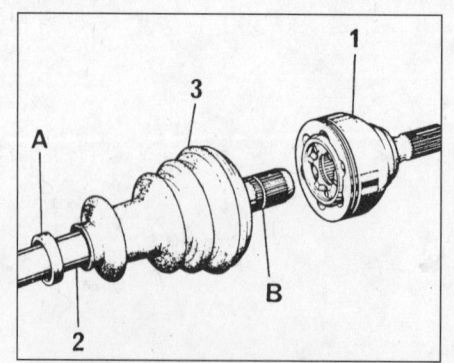

Ball-and-cage joint - bonded joint

Note: *Refer to the note at the beginning of paragraph 21.*

36 With the driveshaft removed as described in Section 2, remove the inner driveshaft joint and the joint spider, as described later in this Section.
37 Remove the outer gaiter securing clips.
38 Slide the gaiter from the inner end of the driveshaft, or cut the gaiter off and discard it.
39 Proceed as described in paragraphs 27 to 30.
40 Apply a little grease to the smaller hole in the new gaiter, to help it to slide up the driveshaft, then slide the gaiter and the securing clips onto the driveshaft.

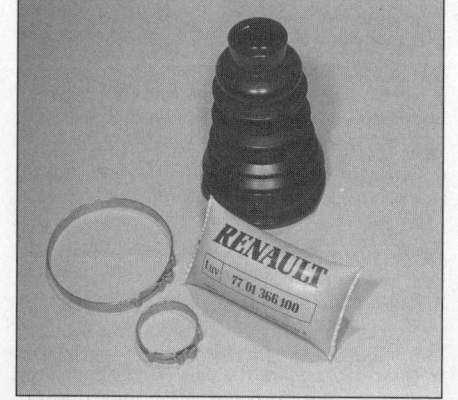

3.26 Ball-and-cage-type outer driveshaft joint components

1 Outer joint
2 Driveshaft
3 Gaiter
A Gaiter rubber collar
B Circlip groove

3.30 Renault driveshaft gaiter repair kit

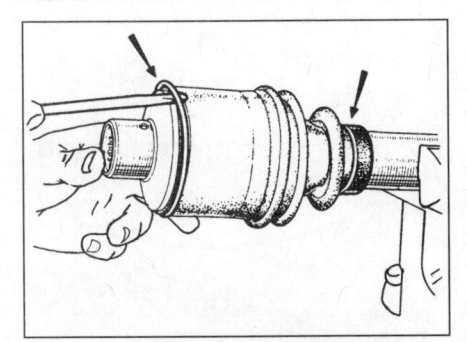

3.47 Removing the driveshaft gaiter securing clips (arrowed) - GI 76-type inner joint

41 Pack the joint with the grease supplied. Work the grease well into the ball tracks, and fill the gaiter with any excess.

42 Locate the outer lip of the gaiter in the groove on the joint outer member. Secure the gaiter in position with the large retaining clip. Where a wire-type retaining clip is fitted, the clip can be secured using two lengths of drilled rod or stout metal tube to lever the ends of the clip.

43 With the coupling aligned with the driveshaft, lift the lip of the gaiter to equalise the air pressure, then locate the inner lip of the gaiter in the groove in the driveshaft, and secure the gaiter with the smaller retaining clip.

44 Refit the inner joint spider and the inner joint as described later in this Section.

45 Check that the constant velocity joint moves freely in all directions, then refit the driveshaft to the vehicle as described in Section 2.

Inner joint

Note: *Three different types of inner joint may be fitted, and the joints can be identified as follows.*

a) *GI 76-type joint - has a smooth circular outer member, and the outer member can be slid from the joint spider once the gaiter has been removed.*

b) *GI 82-type joint - has a smooth circular outer member, and the outer member is secured to the joint spider by an anti-separation plate (visible with the gaiter removed).*

c) *RC 490-type joint - has a recessed outer member, which appears clover leaf-shaped when viewed end-on.*

GI 76-type joint

46 Remove the driveshaft, as described in Section 2.

47 Grip the driveshaft in the jaws of a vice and remove the gaiter securing clips **(see illustration)**.

48 Cut the gaiter along its length, and remove and discard it.

49 Wipe out as much grease as possible.

50 Slide the outer member off the joint spider **(see illustration)**. Be prepared to hold the rollers in place, otherwise they may fall off the

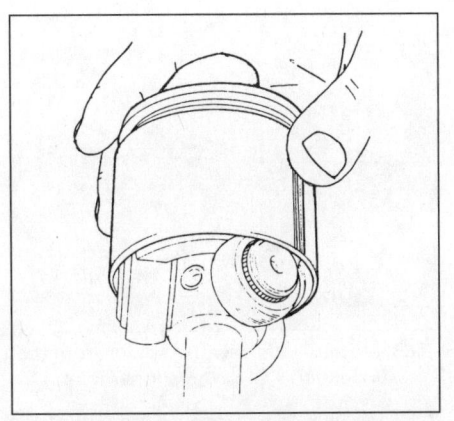

3.50 Sliding the outer member from the joint spider - GI 76-type inner joint

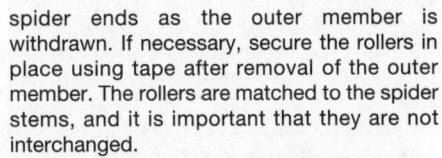

spider ends as the outer member is withdrawn. If necessary, secure the rollers in place using tape after removal of the outer member. The rollers are matched to the spider stems, and it is important that they are not interchanged.

51 Use a press or three-legged extractor to remove the spider **(see illustration)**. Do not tap the shaft out of the spider as the rollers will be marked and the needles damaged.

52 Wipe clean the joint components, taking care not to remove the alignment marks made on dismantling. **Do not** use paraffin or other solvents to clean this type of joint.

53 Examine the joint spider, rollers and outer member for any signs of scoring or wear. Check that the rollers move smoothly on the spider stems. If wear is evident, the spider and roller assembly can be renewed, but it is not possible to obtain a replacement outer member. Obtain a new gaiter, clips, and a quantity of the special lubricating grease. These parts are available in the form of a repair kit from your Renault dealer.

54 Smear the driveshaft with oil and slide the new gaiter small securing clip and gaiter onto the shaft.

55 Using a piece of tubing, drive the spider onto the driveshaft. Peen the spider splines to

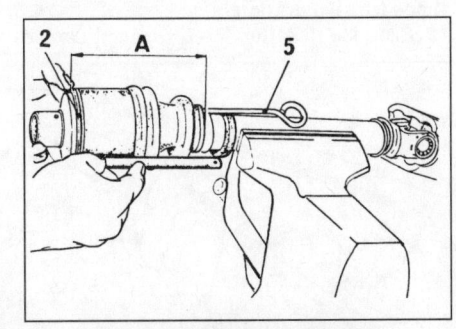

3.60 Driveshaft gaiter setting - GI 76-type inner joint

2 Gaiter
5 Rod inserted to release air
A = 156.0 mm

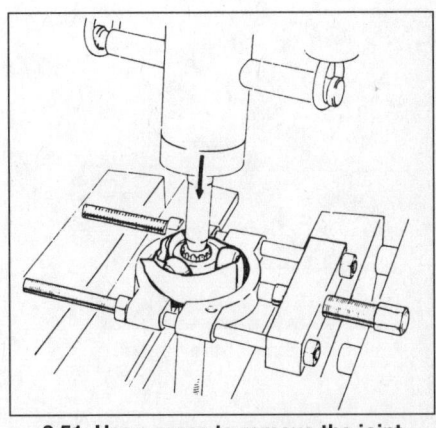

3.51 Use a press to remove the joint spider - GI 76-type inner joint

the shaft at three equidistant points using a centre punch.

56 Distribute the grease pack supplied equally between the gaiter and the joint outer member.

57 Remove the temporary adhesive tape and fit the outer member to the spider.

58 Push the gaiter into position so that the lip of the gaiter engages in the outer member groove.

59 Fit the large diameter gaiter securing clip.

60 Insert a rod under the narrower neck of the gaiter to expel any trapped air and then set the overall length of the gaiter as shown **(see illustration)**.

61 Remove the rod and fit the small gaiter securing clip.

62 Refit the driveshaft as described in Section 2.

GI 82-type joint

63 Remove the driveshaft as described in Section 2.

64 Release the large retaining spring and the inner retaining collar, then slide the gaiter down the shaft to expose the joint.

65 Wipe out as much grease as possible.

66 Using pliers, carefully bend up the anti-separation plate tangs at their corners **(see illustration)**. Slide the outer member off the joint spider. Be prepared to hold the rollers in place, otherwise they may fall off the spider ends as the outer member is

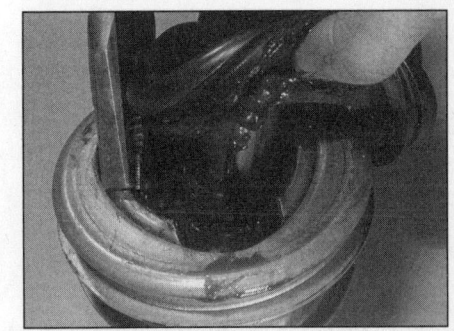

3.66 Bending up an anti-separation plate tang - GI 82-type inner joint

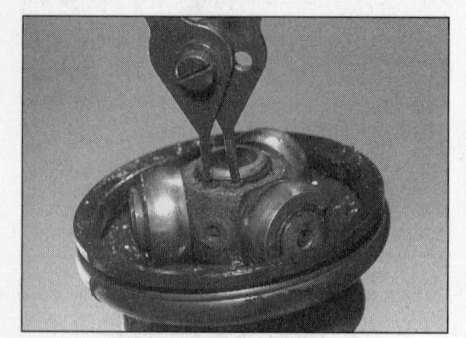

8•6 Driveshafts

3.67 Extract the circlip...

3.68 ...and withdraw the spider from the driveshaft - Gl 82-type inner joint

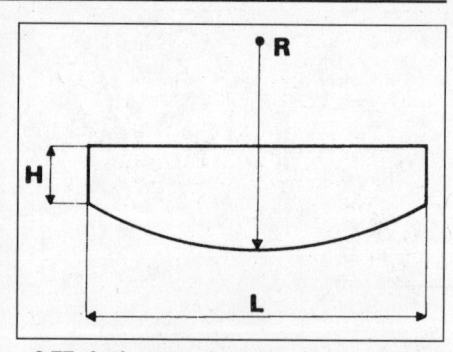

3.77 Anti-separation plate tang support plate - Gl 82-type inner joint

$H = 6.0$ mm
$L = 40.0$ mm
$R = 45.0$ mm

withdrawn. If necessary, secure the rollers in place using tape after removal of the outer member. The rollers are matched to the spider stems, and it is important that they are not interchanged.

67 Using circlip pliers, extract the circlip securing the joint spider to the driveshaft **(see illustration)**. Note that on some models, the joint may be staked in position; if so, relieve the staking using a file. Mark the position of the tripod in relation to the driveshaft, using a dab of paint or a punch.

68 Withdraw the spider from the end of the driveshaft **(see illustration)**. If necessary, use a press or three-legged extractor to remove the spider. Do not tap the shaft out of the spider as the rollers will be marked and the needles damaged.

69 With the spider removed, slide the gaiter and inner retaining collar off the end of the driveshaft.

70 Wipe clean the joint components, taking care not to remove the alignment marks made on dismantling. **Do not** use paraffin or other solvents to clean this type of joint.

71 Examine the joint spider, rollers and outer member for any signs of scoring or wear. Check that the rollers move smoothly on the spider stems. If wear is evident, the spider and roller assembly can be renewed, but it is not possible to obtain a replacement outer member. Obtain a new gaiter, clips, and a quantity of the special lubricating grease. These parts are available in the form of a repair kit from your Renault dealer.

72 Tape over the splines on the end of the driveshaft, then carefully slide the inner retaining collar and gaiter onto the shaft.

73 Remove the tape, then, aligning the marks made on dismantling, engage the joint spider with the driveshaft splines. Use a hammer and soft metal drift to tap the spider onto the shaft, taking great care not to damage the driveshaft splines or joint rollers. Alternatively, support the driveshaft, and press the spider into position using a hydraulic press and suitable tubular spacer which bears only on the centre of the spider.

74 Secure the spider in position with the circlip, ensuring that it is correctly located in the driveshaft groove. Where no circlip is fitted, secure the spider in position by staking the end of the driveshaft in three places, at intervals of 120°, using a hammer and punch.

75 Evenly distribute the grease contained in the repair kit around the spider and inside the outer member. Pack the gaiter with the remainder of the grease.

76 Slide the outer member into position over the spider.

77 Using a piece of 2.5 mm thick steel or similar material, make up a support plate to the dimensions shown **(see illustration)**.

78 Position the support plate under each anti-separation plate tang in the outer member in turn, and tap the tang down onto the support plate. Remove the plate when all the tangs have been returned to their original shape **(see illustration)**.

79 Slide the gaiter up the driveshaft. Locate the gaiter in the grooves on the driveshaft and outer member.

80 Slide the inner retaining collar into place over the inner end of the gaiter.

81 Using a blunt rod, carefully lift the outer lip of the gaiter to equalise the air pressure. With the rod in position, compress the joint until the dimension from the inner end of the gaiter to the flat end face of the outer member is as shown **(see illustration)**. Hold the outer member in this position and withdraw the rod.

82 Slip the new retaining spring into place to secure the outer lip of the gaiter to the outer member. Take care to ensure that the retaining spring is not overstretched during the fitting process.

83 Check that the constant velocity joint moves freely in all directions, then refit the driveshaft as described in Section 2.

RC 490-type joint

84 Remove the driveshaft as described in Section 2.

85 Using a pair of grips, bend up the metal joint cover at the points where it has been staked into the outer member recesses.

86 Using a pair of snips, cut the gaiter inner retaining clip.

87 Using a soft metal drift, tap the metal joint cover off the outer member. Slide the outer member off the end of the joint spider **(see illustrations)**. Be prepared to hold the rollers in place, otherwise they may fall off the spider ends as the outer member is withdrawn. If

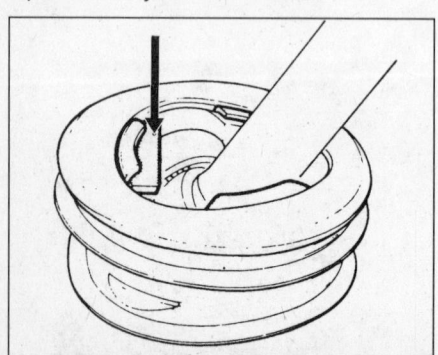

3.78 Support plate (arrowed) positioned under anti-separation plate tang - Gl 82-type inner joint

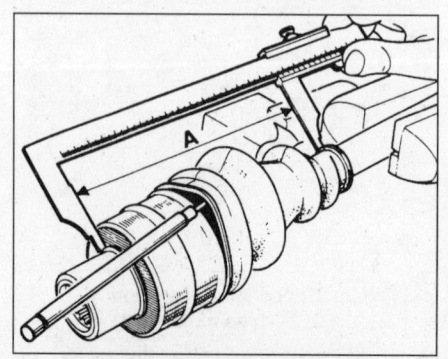

3.81 Driveshaft gaiter setting - Gl 82-type inner joint
$A = 162.0 \pm 1.0$ mm

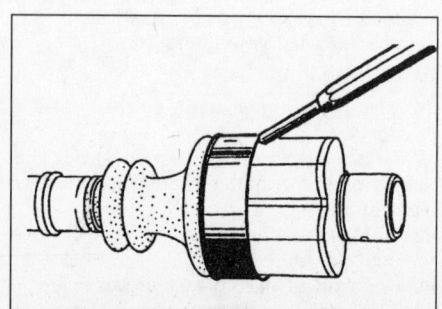

3.87a Tap the joint cover from the outer member...

3.87b ... and slide the outer member off the joint spider - RC 490-type inner joint

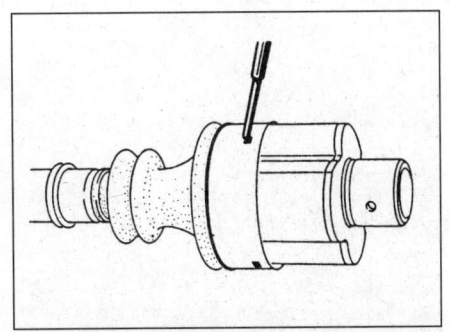

3.95 Staking the joint cover into the outer member recess - RC 490-type inner joint

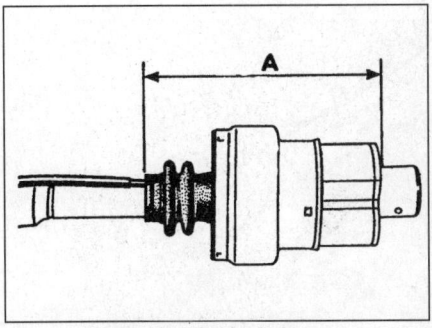

3.96 Driveshaft gaiter setting - RC 490-type inner joint

A = 156.0 ± 1.0 mm

necessary, secure the rollers in place using tape after removal of the outer member. The rollers are matched to the spider stems, and it is important that they are not interchanged.
88 Wipe out as much grease as possible.
89 Remove the spider and gaiter assembly, and examine the joint components for wear, using the information given in paragraphs 68 to 71 of this Section. Make alignment marks between the spider and the shaft for use when refitting. Obtain a repair kit consisting of a gaiter, retaining clip, metal insert and joint cover, and the correct type and amount of special grease.
90 Fit the metal insert into the inside of the gaiter, then locate the gaiter assembly inside the metal joint cover.
91 Tape over the driveshaft splines, and slide the gaiter and joint cover assembly onto the driveshaft.
92 Refit the spider as described in paragraphs 73 and 74.
93 Evenly distribute the special grease contained in the repair kit around the spider and inside the outer member. Pack the gaiter with the remainder of the grease.
94 Slide the outer member into position over the spider.
95 Slide the metal joint cover onto the outer member until it is flush with the outer member

guide panel. Secure the joint cover in position by staking it into the recesses in the outer member, using a hammer and a round-ended punch **(see illustration)**.
96 Using a blunt rod, carefully lift the inner lip of the gaiter to equalise the air pressure. With the rod in position, compress the joint until the dimension from the inner end of the gaiter to the flat end face of the outer member is as shown **(see illustration)**. Hold the outer member in this position and withdraw the rod.
97 Fit the small retaining clip to the inner end of the gaiter. To secure the metal band-type clip, lock the ends of the clip together, then remove any slack in the clip by carefully compressing the raised section of the clip using a pair of side cutters.
98 Check that the constant velocity joint moves freely in all directions, then refit the driveshaft as described in Section 2.

4 Driveshaft overhaul - general information

If any of the checks described in Chapter 1 reveal wear in a driveshaft joint, first remove the roadwheel trim or centre cap (as appropriate) and check that the driveshaft

retaining nut is still correctly tightened; if in doubt, use a torque wrench to check it. Refit the centre cap or trim, and repeat the check on the other driveshaft.

Road test the vehicle, and listen for a metallic clicking from the front as the vehicle is driven slowly in a circle on full-lock. If a clicking noise is heard, this indicates wear in the outer constant velocity joint.

If vibration, consistent with road speed, is felt through the vehicle when accelerating, there is a possibility of wear in the inner constant velocity joints.

Constant velocity joints can be dismantled and inspected for wear as described in Section 3.

Wear in the outer constant velocity joint can only be rectified by renewing the driveshaft. This is necessary since no outer joint components are available separately. For the inner joint, the tripod joint and roller assembly is available separately, but wear in any of the other components will also necessitate driveshaft renewal.

On models with ABS, the reluctor ring should be removed from the old driveshaft and fitted to the new one. See Chapter 9.

Chapter 9
Braking system

Contents

Anti-lock braking system (ABS) - general information	22
Anti-lock braking system (ABS) components - removal and refitting	23
Brake fluid level check	See "Weekly Checks"
Brake fluid renewal	See Chapter 1
Brake pedal - removal and refitting	17
Brake vacuum pump - removal and refitting	24
Brake vacuum pump - testing and overhaul	25
Front brake caliper - removal, overhaul and refitting	10
Front brake disc - inspection, removal and refitting	7
Front brake pad check	See Chapter 1
Front brake pads - renewal	4
General information	1
Handbrake cables - removal and refitting	20
Handbrake check and adjustment	See Chapter 1
Handbrake lever - removal and refitting	19
Hydraulic pipes and hoses - renewal	3
Hydraulic system - bleeding	2
Master cylinder - removal, overhaul and refitting	13
Rear brake caliper - removal, overhaul and refitting	11
Rear brake disc - inspection, removal and refitting	8
Rear brake drum - removal, inspection and refitting	9
Rear brake pad check	See Chapter 1
Rear brake pads - renewal	5
Rear brake pressure regulating valve - testing, adjustment, removal and refitting	18
Rear brake shoe check	See Chapter 1
Rear brake shoes - inspection and renewal	6
Rear wheel cylinder - removal, overhaul and refitting	12
Stop-light switch - adjustment, removal and refitting	21
Vacuum servo unit - testing, removal and refitting	14
Vacuum servo unit air filter - renewal	16
Vacuum servo unit non-return valve - removal, testing and refitting	15

Degrees of difficulty

Easy, suitable for novice with little experience	**Fairly easy,** suitable for beginner with some experience	**Fairly difficult,** suitable for competent DIY mechanic	**Difficult,** suitable for experienced DIY mechanic	**Very difficult,** suitable for expert DIY or professional

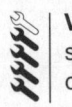

Specifications

General
System type .... Dual hydraulic circuit, split diagonally with servo assistance. Anti-lock braking system (ABS) available as an option. Front disc brakes on all models. Rear disc brakes fitted as standard on some models, and on all models with ABS. Rear drum brakes on certain models. On diesel models, vacuum provided by engine-driven pump. Cable-operated handbrake acting on rear brakes

Front brakes
Disc thickness:
  New:
    All except 6-cylinder engine models .... 20.0 mm
    6-cylinder engine models .... 22.0 mm
  Minimum thickness:
    All except 6-cylinder engine models .... 18.0 mm
    6-cylinder engine models .... 20.0 mm
Maximum disc run-out .... 0.07 mm
Pad thickness (including backing):
  New .... 18.0 mm
  Minimum thickness .... 6.0 mm

9•2 Braking system

Rear disc brakes
Disc thickness:
  New:
    All except 6-cylinder engine models .... 10.5 mm
    6-cylinder engine models .... 12.0 mm
  Minimum thickness:
    All except 6-cylinder engine models .... 9.5 mm
    6-cylinder engine models .... 11.0 mm
Maximum disc run-out .... 0.07 mm
Pad thickness (including backing):
  New .... 14.0 mm
  Minimum thickness .... 6.0 mm

Rear drum brakes
Drum internal diameter:
  New .... 28.6 mm
  Maximum diameter .... 229.6 mm
Shoe thickness (including backing):
  New .... 7.0 mm
  Minimum thickness .... 2.5 mm

Torque wrench settings

	Nm	lbf ft
Fluid bleed screws	7	5
Master cylinder-to-vacuum servo nuts	13	10
Vacuum servo bolts	20	15
Brake fluid hose and pipe unions	13	10
ABS wheel sensor bolts	10	7
Front ABS wheel sensor mounting bracket bolts	25	18
Front brake caliper guide pin bolts	35	26
Front brake caliper mounting bracket-to-hub carrier bolts	100	74
Rear caliper mounting bracket-to-hub carrier bolts	65	48

1 General information

The braking system is of the servo-assisted, dual-circuit hydraulic type. The arrangement of the hydraulic system is such that each circuit operates one front and one rear brake from a tandem master cylinder. Under normal circumstances, both circuits operate in unison. However, in the event of hydraulic failure in one circuit, full braking force will still be available at two wheels.

Some large-capacity engine models have disc brakes all round as standard; other models are fitted with front disc brakes and rear drum brakes. ABS is fitted as standard to certain models, and is offered as an option on most other models (refer to Section 22 for further information on ABS operation).

The front disc brakes are actuated by single-piston sliding type calipers, which ensure that equal pressure is applied to each disc pad.

On models with rear drum brakes, the rear brakes incorporate leading and trailing shoes, which are actuated by twin-piston wheel cylinders. A self-adjust mechanism is incorporated, to automatically compensate for brake shoe wear. As the brake shoe linings wear, the footbrake operation automatically operates the adjuster mechanism, which effectively lengthens the shoe strut and repositions the brake shoes, to remove the lining-to-drum clearance.

On models with rear disc brakes, the brakes are actuated by single-piston sliding calipers which incorporate mechanical handbrake mechanisms.

A load-sensitive pressure-regulating valve is fitted to regulate the hydraulic pressure applied to the rear brakes. The regulating valve helps to prevent rear wheel lock-up during emergency braking.

On all models, the handbrake provides an independent mechanical means of rear brake application.

On diesel engines, there is insufficient vacuum in the inlet manifold to operate the braking system servo effectively at all times. To overcome this problem, a vacuum pump is fitted to the engine, to provide sufficient vacuum to operate the servo unit. The vacuum pump is driven from the engine auxiliary shaft.

Note: *When servicing any part of the system, work carefully and methodically; also observe scrupulous cleanliness when overhauling any part of the hydraulic system. Always renew components (in axle sets, where applicable) if in doubt about their condition, and use only genuine Renault replacement parts, or at least those of known good quality. Note the warnings given in "Safety first" and at relevant points in this Chapter concerning the dangers of asbestos dust and hydraulic fluid.*

2 Hydraulic system - bleeding

Warning: *Hydraulic fluid is poisonous; wash off immediately and thoroughly in the case of skin contact, and seek immediate medical advice if any fluid is swallowed or gets into the eyes. Certain types of hydraulic fluid are inflammable, and may ignite when allowed into contact with hot components; when servicing any hydraulic system, it is safest to assume that the fluid IS inflammable, and to take precautions against the risk of fire as though it is petrol that is being handled. Hydraulic fluid is also an effective paint stripper, and will attack plastics; if any is spilt, it should be washed off immediately, using copious quantities of clean water. Finally, it is hygroscopic (it absorbs moisture from the air). The more moisture is absorbed by the fluid, the lower its boiling point becomes, leading to a dangerous loss of braking under hard use. Old fluid may be contaminated and unfit for further use. When topping-up or renewing the fluid, always use the recommended type, and ensure that it comes from a freshly-opened sealed container.*

General

1 The correct functioning of the brake hydraulic system is only possible after removing all air from the components and circuit; this is achieved by bleeding the system.
2 During the bleeding procedure, add only clean, fresh hydraulic fluid of the specified type; never re-use fluid that has already been bled from the system. Ensure that sufficient fluid is available before starting work.
3 If there is any possibility of incorrect fluid being used in the system, the brake lines and components must be completely flushed with uncontaminated fluid and new seals fitted to the components.
4 If brake fluid has been lost from the master cylinder due to a leak in the system, ensure that the cause is traced and rectified before proceeding further.
5 Park the vehicle on level ground, switch off the ignition and select first gear. Chock the wheels and release the handbrake.
6 Check that all pipes and hoses are secure, unions tight, and bleed screws closed. Remove the dust caps and clean any dirt from around the bleed screws.
7 Unscrew the brake fluid reservoir cap, and top-up the reservoir to the "MAX" level line. Refit the cap loosely, and remember to maintain the fluid level at least above the "MIN" level line throughout the procedure, otherwise there is a risk of further air entering the system.
8 There are a number of one-man, do-it-yourself, brake bleeding kits currently available from motor accessory shops. It is recommended that one of these kits is used wherever possible, as they greatly simplify the bleeding operation, and also reduce the risk of expelled air and fluid being drawn back into the system. If such a kit is not available, the basic (two-man) method must be used, which is described in detail below.
9 If a kit is to be used, prepare the vehicle as described previously, and follow the kit manufacturer's instructions, as the procedure may vary slightly according to the type being used; generally, they are as outlined below in the relevant sub-section.
10 Whichever method is used, the correct sequence must be followed to ensure that the removal of all air from the system.

Bleeding sequence

11 If the hydraulic system has only been partially disconnected and suitable precautions were taken to minimise fluid loss, it should only be necessary to bleed that part of the system (ie the primary or secondary circuit).
12 If the complete system is to be bled, then it should be done in the following sequence:
 a) Right-hand rear brake.
 b) Left-hand front brake.
 c) Left-hand rear brake.
 d) Right-hand front brake.

Bleeding - basic (two-man) method

13 Collect a clean glass jar and a suitable length of plastic or rubber tubing, which is a tight fit over the bleed screw, and a ring spanner to fit the screws. The help of an assistant will also be required.
14 If not already done, remove the dust cap from the bleed screw of the first wheel to be bled and fit the bleed tube to the screw **(see illustrations)**.
15 Immerse the other end of the bleed tube in the jar, which should contain enough fluid to cover the end of the tube.
16 Ensure that the reservoir fluid level is maintained at least above the "MIN" level line throughout the procedure.
17 Open the bleed screw approximately half a turn, and have your assistant depress the brake pedal with a smooth steady stroke down to the floor, and then hold it there. When the flow of fluid through the tube stops, tighten the bleed screw and have your assistant release the pedal slowly.
18 Repeat this operation (paragraph 17) until clean brake fluid, free from air bubbles, can be seen flowing from the end of the tube.
19 When no more air bubbles appear, tighten the bleed screw, remove the bleed tube and refit the dust cap. Repeat these procedures on the remaining calipers in sequence until all air is removed from the system and the brake pedal feels firm again.

Bleeding - using a one-way valve kit

20 As their name implies, these kits consist of a length of tubing with a one-way valve fitted, to prevent expelled air and fluid being drawn back into the system; some kits incorporate a translucent container, which can be positioned so that the air bubbles can be more easily seen flowing from the end of the tube.
21 The kit is connected to the bleed screw, which is then opened. The user returns to the driver's seat, depresses the brake pedal with a smooth steady stroke, and slowly releases it; this is repeated until the expelled fluid is clear of air bubbles.
22 Note that these kits simplify work so much that it is easy to forget the reservoir fluid level; ensure that this is maintained at least above the "MIN" level line at all times.

Bleeding - using a pressure-bleeding kit

23 These kits are usually operated by the reserve of pressurised air contained in the spare tyre. However, note that it will probably be necessary to reduce the pressure to a lower level than normal; refer to the instructions supplied with the kit.
24 By connecting a pressurised, fluid-filled container to the fluid reservoir, bleeding is then carried out by simply opening each bleed screw in turn (in the specified sequence) and allowing the fluid to run out, rather like turning on a tap, until no air bubbles can be seen in the expelled fluid.
25 This method has the advantage that the large reservoir of fluid provides an additional safeguard against air being drawn into the system during bleeding.
26 Pressure bleeding is particularly effective when bleeding "difficult" systems, or when bleeding the complete system at the time of routine fluid renewal. It is also the method recommended by Renault if the hydraulic system has been drained either wholly or partially.

All methods

27 When bleeding is completed, check and top-up the fluid level in the reservoir.
28 Check the feel of the brake pedal. If it feels at all spongy, air must still be present in the system, and further bleeding is indicated. Failure to bleed satisfactorily after a reasonable repetition of the bleeding operations may be due to worn master cylinder seals.
29 Discard brake fluid which has been bled from the system; it will not be fit for re-use.

3 Hydraulic pipes and hoses - renewal

Note: *Before starting work, refer to the warning at the beginning of Section 2 concerning the dangers of hydraulic fluid.*

2.14a Rear wheel cylinder bleed screw (arrowed)

2.14b Bleed tube and container connected to front caliper

9•4 Braking system

Inspection

1 The hydraulic pipes, hoses, hose connections and pipe unions should be regularly examined.
2 First check for signs of leakage at the pipe unions, then examine the flexible hoses for signs of cracking, chafing and fraying.
3 The brake pipes should be examined carefully for signs of dents, corrosion or other damage. Corrosion should be scraped off, and if the depth of pitting is significant, the pipes renewed. This is particularly likely in those areas underneath the vehicle body where the pipes are exposed and unprotected.

Removal

4 If any pipe or hose is to be renewed, minimise fluid loss by removing the fluid reservoir cap and then tightening it down onto a piece of polythene (taking care not to damage the level sender unit) to obtain an airtight seal. Alternatively, flexible hoses can be sealed, if required, using a proprietary brake hose clamp; metal brake pipe unions can be plugged (if care is taken not to allow dirt into the system) or capped immediately they are disconnected. Place a wad of rag under any union that is to be disconnected, to catch any spilt fluid. If a section of pipe is to be removed from the master cylinder, the reservoir should be emptied by siphoning out the fluid or drawing out the fluid with a pipette.
5 If a flexible hose is to be disconnected, unscrew the brake pipe union nut before removing the spring clip which secures the hose to its mounting bracket (see illustration).
6 To unscrew the union nuts, it is preferable to obtain a brake pipe spanner of the correct size (11 mm/13 mm split ring); these are available from motor accessory shops (see illustration). Failing this, a close-fitting open-ended spanner will be required, though if the nuts are tight or corroded, their flats may be rounded off if the spanner slips. In such a case, a self-locking wrench is often the only way to unscrew a stubborn union, but it follows that the pipe and the damaged nuts must be renewed on reassembly. Always clean a union and surrounding area before disconnecting it. If disconnecting a component with more than one union, make a careful note of the connections before disturbing any of them.
7 If a brake pipe is to be renewed, it can be obtained, cut to length and with the union nuts and end flares in place, from Renault dealers. All that is then necessary is to bend it to shape, following the line of the original, before fitting it to the vehicle. Alternatively, most motor accessory shops can make up brake pipes from kits, but this requires very careful measurement of the original to ensure that the replacement is of the correct length. The safest answer is usually to take the original to the shop as a pattern.

Refitting

8 On refitting, do not overtighten the union nuts. The specified torque wrench settings (where given) are not high, and it is not necessary to exercise brute force to obtain a sound joint.
9 Ensure that the pipes and hoses are correctly routed with no kinks, and that they are secured in the clips or brackets provided. In the case of flexible hoses, make sure that they cannot contact other components during movement of the steering and/or suspension assemblies.
10 After fitting, remove the polythene from the reservoir (or remove the plugs or clamps, as applicable), and bleed the hydraulic system as described in Section 2. Wash off any spilt fluid, and check carefully for fluid leaks.

3.5 Brake pipe spring clip (arrowed)

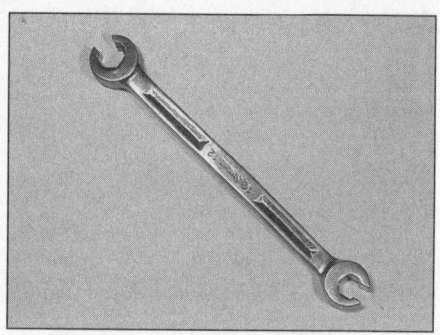

3.6 Brake pipe union spanner

4 Front brake pads - renewal

> **Warning:** *Disc brake pads must be renewed on both front wheels at the same time - never renew the pads on only one wheel, as uneven braking may result. Also, the dust created by wear of the pads may contain asbestos, which is a health hazard. Never blow it out with compressed air and don't inhale any of it. An approved filtering mask should be worn when working on the brakes. DO NOT use petroleum based solvents to clean brake parts. Use brake cleaner or methylated spirit only.*

Note: *Thread-locking compound will be required to coat the threads of the caliper guide pin bolts on refitting.*

1 Apply the handbrake then jack up the front of the vehicle and support it securely on axle stands (see "Jacking and vehicle support"). Remove the front roadwheels.
2 Pull the caliper body outwards, away from the centre of the car. This will push the piston back into its bore to facilitate removal and refitting of the pads.
3 Disconnect the brake pad wear warning sensor wiring at the connector (see illustration).
4 Unscrew the upper and lower guide pin bolts using a suitable spanner, while holding the guide pins with a second spanner (see illustrations).

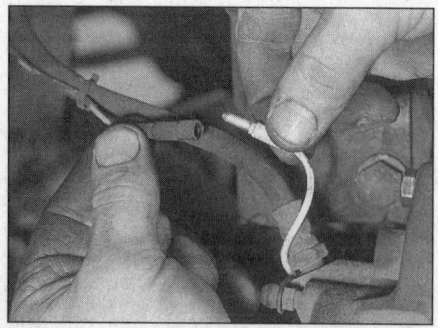

4.3 Disconnecting the brake pad wear warning sensor wiring

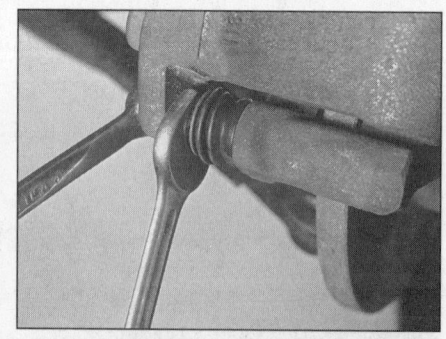

4.4a Slacken the guide pin bolts whilst holding the guide pins with an open-ended spanner . . .

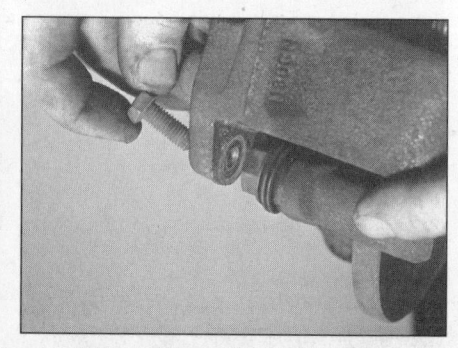

4.4b . . . then withdraw the bolts and lift off the brake caliper

Braking system 9•5

4.6 Removing the outer brake pad from the carrier bracket

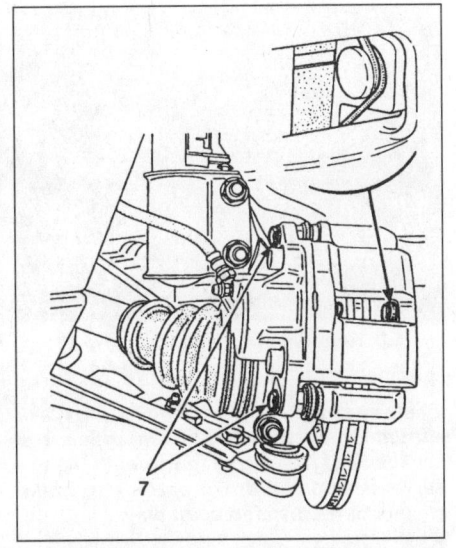

4.8 Front caliper guide pin bolts (7) and correct fitted position of anti-rattle spring

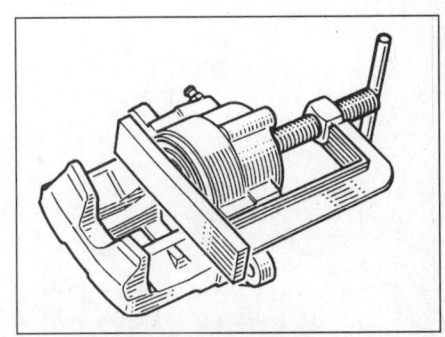

4.9 Using a clamp to retract the caliper piston

5 With the guide pins removed, lift the caliper off the brake pads and carrier bracket, and tie it up in a convenient place under the wheelarch. Do not allow the caliper to hang unsupported on the flexible brake hose.

6 Withdraw the two brake pads from the carrier bracket **(see illustration)**. If required, the thickness of the pads can be checked at this stage using a steel rule.

7 Before refitting the pads, check that the guide pins are free to slide in the carrier bracket and check that the rubber dust excluders around the guide pins are undamaged. Brush the dust and dirt from the caliper and piston but do not inhale it as it is injurious to health. Inspect the dust excluder around the piston for damage and inspect the piston for evidence of fluid leaks, corrosion or damage. If attention to any of these components is necessary, refer to Section 10.

8 To refit the pads, place them in position on the carrier bracket, noting that the pad with the warning sensor wire must be nearest to the centre of the car. The anti-rattle springs must be located as shown **(see illustration)**.

9 Make sure that the caliper piston is fully retracted in its bore. If not, carefully push it in, preferably using a G-clamp or, alternatively, using a flat bar or screwdriver as a lever **(see illustration)**. As the piston is retracted, the fluid level in the reservoir will rise - if necessary, syphon out some fluid to allow for this.

10 Position the caliper over the pads, then fit the lower guide pin bolt, having first coated its threads with locking fluid. Apply locking fluid to the upper guide pin bolt, press the caliper into position, then fit the bolt. Tighten the bolts to the specified torque, starting with the lower bolt.

11 Reconnect the brake pad wear warning sensor wiring, then refit the roadwheel and repeat the renewal procedure on the remaining front brake.

12 On completion, check the hydraulic fluid level in the reservoir, then depress the brake pedal two or three times to bring the pads into contact with the disc. Lower the vehicle to the ground.

5 Rear brake pads - renewal

Warning: Disc brake pads must be renewed on both rear wheels at the same time - never renew the pads on only one wheel, as uneven braking may result. Also, the dust created by wear of the pads may contain asbestos, which is a health hazard. Never blow it out with compressed air and don't inhale any of it. An approved filtering mask should be worn when working on the brakes. DO NOT use petroleum based solvents to clean brake parts. Use brake cleaner or methylated spirit only.

Note: *New caliper sliding key locking clips will be required on refitting.*

1 Chock the front wheels, engage reverse gear (or "P" on models with automatic transmission) and release the handbrake. Jack up the rear of the vehicle and support it securely on axle stands (see *"Jacking and vehicle support"*). Remove the relevant roadwheel.

2 Disconnect the handbrake cable from the lever on the caliper, with reference to Section 20.

3 Extract the two locking clips and then tap out the two sliding keys **(see illustrations)**.

4 Lift the caliper off the brake pads and carrier bracket, and tie it up in a convenient place under the wheelarch **(see illustration)**. Do not allow the caliper to hang unsupported on the flexible brake hose.

5 Withdraw the disc pads and springs. If required, the thickness of the pads can be checked at this stage using a steel rule.

6 Before refitting the pads, brush the dust and dirt from the caliper and piston but do not inhale it as it is injurious to health. Inspect the dust excluder around the piston for damage and inspect the piston for evidence of fluid leaks, corrosion or damage. If attention to any of these components is necessary, refer to Section 11.

7 The caliper piston must now be fully retracted into the cylinder. Do this by turning the piston with the square section shaft of a screwdriver, or a square bar, until the piston

5.3a Extract the locking clips (arrowed) ...

5.3b ... then tap out the sliding keys (arrowed)

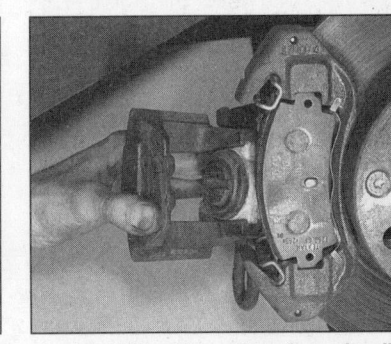

5.4 Lifting the caliper from the disc

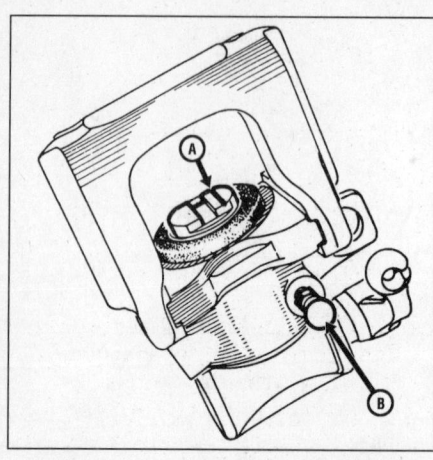

9•6 Braking system

5.7 The rear caliper piston must be aligned as shown before refitting

Groove (A) must be in line with the caliper bleed screw (B)

continues to turn but will not go in any further. As the piston is retracted, the fluid level in the reservoir will rise - if necessary, syphon out some fluid to allow for this. Turn the piston so that the narrow groove in the piston face is uppermost, in line with the caliper bleed screw **(see illustration)**.
8 Locate the springs under the pads, then fit the pads **(see illustration)**.
9 Fit the caliper by engaging one end of the caliper between the spring clip and the keyway on the bracket. Compress the springs and engage the opposite end of the caliper.
10 Insert the first key, then insert a screwdriver in the second key slot and use it as a lever until the second key can be inserted.
11 Fit new key locking clips.
12 Reconnect the handbrake cable to the lever on the caliper with reference to Section 20.
13 Refit the roadwheel, then repeat the renewal procedure on the remaining rear brake.
14 Apply the footbrake several times to position the pads against the discs.
15 On completion, check the hydraulic fluid level in the reservoir, then depress the brake pedal two or three times to bring the pads into contact with the disc. Lower the vehicle to the ground.

6 Rear brake shoes - inspection and renewal

⚠ **Warning:** *Brake shoes must be renewed on both rear wheels at the same time - never renew the shoes on only one wheel, as uneven braking may result. Also, the dust created by wear of the shoes may contain asbestos, which is a health hazard. Never blow it out with compressed air and don't*

5.8 Rear disc pad spring (arrowed) correctly fitted to caliper

inhale any of it. An approved filtering mask should be worn when working on the brakes. DO NOT use petroleum based solvents to clean brake parts. Use brake cleaner or methylated spirit only.

Inspection
1 Remove the brake drum as described in Section 9.
2 Carefully remove all traces of brake dust from the brake drum, backplate and shoes.
3 Measure the thickness of each brake shoe (friction material and shoe) at several points. If either shoe is worn at any point to the specified minimum thickness or less, all four shoes must be renewed as a set. The shoes should also be renewed if they are fouled with oil or grease - there is no satisfactory way of degreasing friction material once contaminated.
4 If any of the brake shoes are worn unevenly, or fouled with oil or grease, trace and rectify the cause before reassembly.

Renewal
Note: *A new rear hub nut will be required on refitting.*

Bendix brake shoes
5 With the brake drum removed as described in Section 9, proceed as follows.
6 Using a screwdriver, carefully lever out the dust cap from the centre of the hub.
7 Using a socket and extension bar, unscrew and remove the rear hub nut.
8 Withdraw the hub/bearing assembly from the stub axle.
9 Note carefully the fitted positions of the leading and trailing shoes and the spring and strut components.
10 Using a pair of pliers, disconnect the upper return spring.
11 Disconnect the handbrake cable from the lever on the trailing shoe.
12 Remove the two shoe hold-down springs. Do this by inserting a screwdriver into the centre of the spring and rotating it **(see illustration)**.
13 Move the automatic adjuster lever (attached to the leading shoe) as far as possible in the direction of the stub axle **(see illustration)**.
14 Pull both shoes away from the backplate and then disconnect the link strut from the leading shoe **(see illustration)**.
15 Move the adjuster lever back to its original position. Rotate the adjuster toothed segment to do this.
16 Twist the leading shoe at right-angles to the backplate and disconnect the lower shoe return spring, using a suitable pair of pliers.

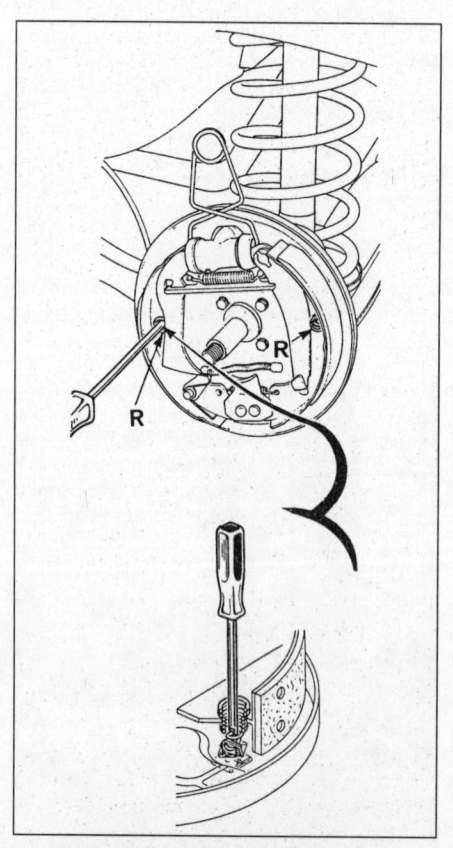

6.12 Remove the two hold-down springs (R) using a screwdriver - Bendix brakes

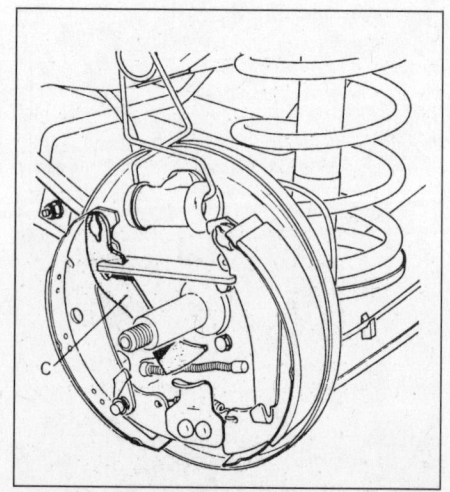

6.13 Move the automatic adjuster lever (C) as far as possible towards the stub axle - Bendix brakes

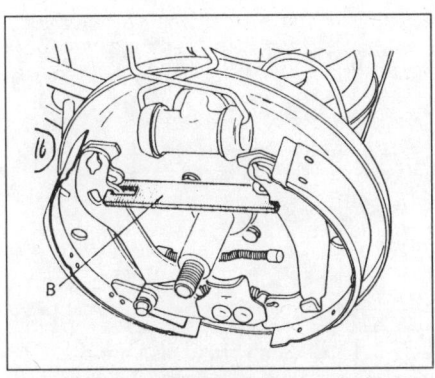

Braking system 9•7

6.14 Disconnect the link strut (B) from the leading shoe

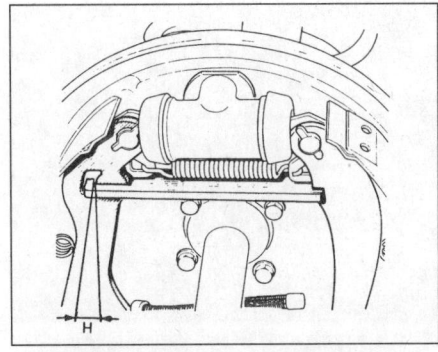

6.20 Bendix brake shoe self-adjuster link strut-to-adjuster lever slot gap
H = 1.0 mm

17 Remove both shoes. While the shoes are removed, do not touch the brake pedal, or the wheel cylinder pistons will be ejected.

> **HAYNES HiNT** Wrap an elastic band or a cable-tie around the wheel cylinder to retain the pistons.

18 Place the old shoes on the bench and transfer the handbrake lever and the automatic adjuster lever and toothed segment to the new shoes. The levers are retained by spring clips, which should also be renewed.
19 Release the retaining spring, and transfer the link strut and spring to the new trailing shoe.
20 Fit the new shoes by reversing the removal operations, bearing in mind the following points.
  a) Apply a smear of high melting-point grease to the shoe contact spots on the brake backplate, to the shoe lower anchor block recesses and to the end faces of the wheel cylinder pistons.
  b) Ensure that the shoes and return springs are positioned as noted before removal.
  c) Move the automatic adjuster lever fully towards the stub axle.

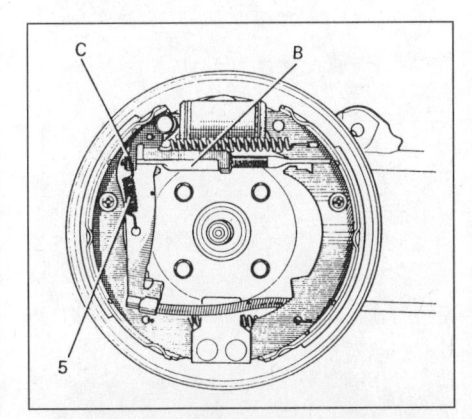

6.24 Girling rear drum brake components

  B Threaded adjuster rod
  C Adjuster lever
  5 Adjuster lever spring

  d) Before refitting the drum, the self-adjuster mechanism must be set. To do this, move the handbrake lever hard up against the trailing shoe, then measure the gap between the inner end of the slot in the link strut, and the inner edge of the strut hole in the adjuster lever **(see illustration)**. If the gap is not as specified, renew the link strut spring and the upper and lower shoe return springs.
  e) Fit a new hub nut, and tighten to the specified torque.
  f) Refit the brake drum with reference to Section 9.

21 Repeat the procedure on the remaining rear brake, then adjust the handbrake cables as described in Section 20.
22 Refit the roadwheels, then lower the vehicle to the ground.

Girling brake shoes

23 Proceed as described for Bendix brake shoes in paragraphs 5 to 11.
24 Disconnect the adjuster lever spring, and withdraw the adjuster lever **(see illustration)**.
25 Grip one of the shoe hold-down spring caps with a pair of pliers, depress the cup and turn it through 90° so that it will pass over the tee shaped head of the hold-down pin. Remove the cup, spring and pin. Remove the other hold-down components in a similar way.
26 Remove the threaded adjuster rod.
27 Pull the shoes away from the backplate and cross the upper ends of the shoes over each other so that the lower shoe return spring can be released from behind the anchor block.
28 Remove the shoes. While the shoes are removed, do not touch the brake pedal, or the wheel cylinder pistons will be ejected.

> **HAYNES HiNT** Wrap an elastic band or a cable-tie around the wheel cylinder to retain the pistons.

29 If the new shoes are not supplied complete with a new handbrake lever, transfer the old one to the new trailing shoe.
30 Fit the new shoes by reversing the removal operations, bearing in mind the following points.
  a) Clean the threads of the threaded adjuster rod and lightly oil them.
  b) Apply a smear of high melting-point grease to the shoe contact spots on the brake backplate, to the shoe lower anchor block recesses and to the end faces of the wheel cylinder pistons.
  c) Make sure that the shoe lower return spring locates behind the anchor block tab.
  d) Before connecting the upper return spring, adjust the adjuster rod by turning the star wheel until the diameter of the brake shoes (held in position on the backplate) is approximately 228.0 mm.
  e) Check that the ends of the shoe return springs are engaged with the holes in the shoes as shown **(see illustration)**.
  f) Fit a new hub nut, and tighten to the specified torque.
  g) Refit the brake drum with reference to Section 9.

7 Front brake disc - inspection, removal and refitting

Note: Before starting work, refer to the warning at the beginning of Section 4 concerning the dangers of asbestos dust. If either disc requires renewal, both should be renewed at the same time, to ensure even and consistent braking. In principle, new pads should be fitted also.

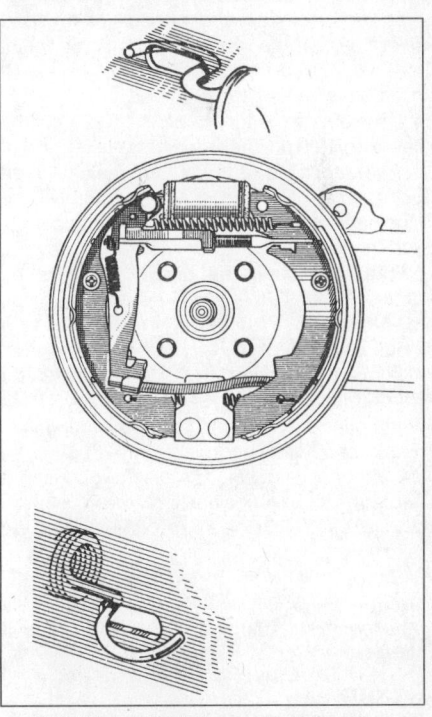

6.30 Check that the ends of the shoe return springs are engaged with the shoes as shown - Girling brakes

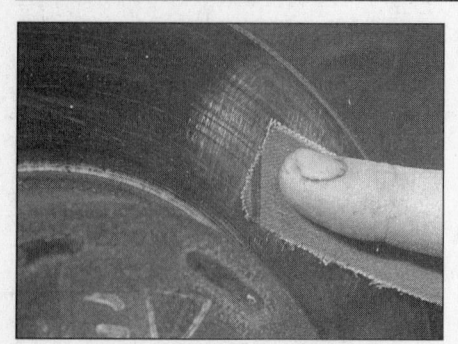

9•8 Braking system

7.2 Using emery tape to remove light scoring from the disc

7.3 Measuring brake disc thickness with a micrometer

7.4 Checking brake disc run-out with a dial gauge

Inspection

1 Apply the handbrake, then jack up the front of the vehicle and support it securely on axle stands (see *"Jacking and vehicle support"*). Remove the appropriate front roadwheel.
2 Slowly rotate the brake disc so that the full area of both sides can be checked; remove the brake pads, as described in Section 4, if better access is required to the inboard surface. Light scoring is normal in the area swept by the brake pads, and can be removed using emery tape **(see illustration)**. If heavy scoring is found, the disc must be renewed.
3 It is normal to find a lip of rust and brake dust around the disc's perimeter; this can be scraped off if required. If, however, a lip has formed due to wear of the brake pad swept area, the disc thickness must be measured using a micrometer **(see illustration)**. Take measurements at several places around the disc at the inside and outside of the pad swept area; if the disc has worn at any point to the specified minimum thickness or less, it must be renewed.
4 If the disc is thought to be warped, it can be checked for run-out, ideally by using a dial gauge mounted on any convenient fixed point, while the disc is slowly rotated **(see illustration)**. In the absence of a dial gauge, use feeler blades to measure (at several points all around the disc) the clearance between the disc and a fixed point such as the caliper mounting bracket. If the measurements obtained are at the specified maximum or beyond, the disc is excessively warped, and must be renewed; however, it is worth checking first that the hub bearing is in good condition (Chapters 1 and 10). Also try the effect of removing the disc and turning it through 180° to reposition it on the hub; if run-out is still excessive, the disc must be renewed.
5 Check the disc for cracks (especially around the wheel bolt holes), and for any other wear or damage. Renew the disc if necessary.

Removal

Note: *Suitable thread-locking compound will be required to coat the threads of the brake caliper mounting bracket bolts on refitting.*

6 Unscrew the two bolts securing the brake caliper mounting bracket to the hub carrier, and slide the caliper assembly, complete with pads, off the disc. Using a piece of wire or string, tie the caliper to the front suspension coil spring, to avoid placing any strain on the hydraulic brake hose.
7 If the same disc is to be refitted, use chalk or paint to mark the relationship of the disc to the hub. Remove the screw(s) securing the brake disc to the hub, and remove the disc. If it is tight, lightly tap its rear face with a hide or plastic mallet.

Refitting

8 Refitting is the reverse of the removal procedure, noting the following points:
  a) Ensure that the mating surfaces of the disc and hub are clean and flat.
  b) If applicable, align the marks made on removal.
  c) Securely tighten the disc retaining screws.
  d) If a new disc has been fitted, use a suitable solvent to wipe any preservative coating from the disc before refitting the caliper.
  e) Apply locking fluid to the threads of the brake caliper mounting bracket bolts, and tighten them to the specified torque.
  f) Refit the roadwheel, then lower the vehicle to the ground and tighten the roadwheel bolts to the specified torque.
  g) On completion, depress the brake pedal several times to bring the brake pads into contact with the disc.

8 Rear brake disc - inspection, removal and refitting

Note: *Before starting work, refer to the warning at the beginning of Section 4 concerning the dangers of asbestos dust. If either disc requires renewal, both should be renewed at the same time, to ensure even and consistent braking. In principle, new pads should be fitted also.*

Inspection

1 Chock the front wheels, engage reverse gear (or "P" on models with automatic transmission) and release the handbrake. Jack up the rear of the vehicle and support it securely on axle stands (see *"Jacking and vehicle support"*). Remove the relevant roadwheel.
2 Proceed as described for the inspection of the front brake discs in Section 7.

Removal

Note: *Suitable thread-locking compound will be required to coat the threads of the brake caliper mounting bracket bolts on refitting.*

3 Remove the rear brake pads as described in Section 5.
4 Unscrew the two bolts securing the caliper mounting bracket to the hub carrier, and withdraw the caliper mounting bracket **(see illustration)**.
5 If the same disc is to be refitted, use chalk or paint to mark the relationship of the disc to the hub.
6 Remove the securing screw(s) and lift off the brake disc **(see illustrations)**. If it is tight, lightly tap its rear face with a hide or plastic mallet.

Refitting

7 Refitting is the reverse of the removal procedure, noting the following points:
  a) Ensure that the mating surfaces of the disc and hub are clean and flat.
  b) If applicable, align the marks made on removal.
  c) Securely tighten the disc retaining screws.

8.4 Withdrawing the rear caliper mounting bracket

Braking system 9•9

8.6a Remove the securing screws . . .

8.6b . . . and withdraw the brake disc

d) *If a new disc has been fitted, use a suitable solvent to wipe any preservative coating from the disc before refitting the caliper.*
e) *Apply locking fluid to the threads of the brake caliper mounting bracket bolts, and tighten them to the specified torque.*
f) *Refit the rear brake pads as described in Section 5.*
g) *Refit the roadwheel, then lower the vehicle to the ground and tighten the roadwheel bolts to the specified torque.*
h) *On completion, depress the brake pedal several times to bring the brake pads into contact with the disc.*

9 Rear brake drum - removal, inspection and refitting

Note: *Before starting work, refer to the warning at the beginning of Section 6 concerning the dangers of asbestos dust. If either drum requires renewal, both should be renewed at the same time, to ensure even and consistent braking. In principle, new shoes should be fitted also.*

Removal

1 Chock the front wheels, engage reverse gear (or "P" on models with automatic transmission) and release the handbrake. Jack up the rear of the vehicle and support it securely on axle stands (see *"Jacking and vehicle support"*). Remove the relevant roadwheel.
2 Unscrew the drum securing screw(s), then pull the drum from the hub. If the drum is tight, try gently tapping with a plastic or copper-faced mallet, or levering using two larger screwdrivers.
3 If the drum is still difficult to remove, proceed as follows.
 a) *Prise out the blanking plug from the brake backplate.*
 b) *Insert a screwdriver through the hole in the backplate, and push on the trailing shoe handbrake lever until the lever stop-lug passes over the edge of the shoe.*
 c) *Once the stop-lug has been released, push the lever back to free the shoes.*

Additionally, it may be possible to release the adjuster mechanism by inserting a thin screwdriver through one of the roadwheel bolt holes in the drum.

Inspection

4 Working carefully, remove all traces of brake dust from the drum, but *avoid inhaling the dust, as it is injurious to health.*
5 Scrub clean the outside of the drum, and check it for obvious signs of wear or damage such as cracks around the roadwheel bolt holes; renew the drum if necessary.
6 Carefully examine the inside of the drum. Light scoring of the friction surface is normal, but if heavy scoring is found, the drum must be renewed. It is usual to find a lip on the drum's inboard edge which consists of a mixture of rust and brake dust; this should be scraped away to leave a smooth surface which can be polished with fine (120 to 150 grade) emery paper. If the lip is due to the friction surface being recessed by wear, then the drum must be refinished (within the specified limits) or renewed.
7 If the drum is thought to be excessively worn or oval, its internal diameter must be measured at several points using an internal micrometer. Take measurements in pairs, the second at right-angles to the first, and compare the two to check for signs of ovality. Minor ovality can be corrected by machining; otherwise, renew the drum.

Refitting

8 If a new brake drum is to be installed, use a suitable solvent to remove any preservative coating that may have been applied to its interior.
9 Ensure that the handbrake lever stop-peg is correctly repositioned against the edge of the brake shoe web.
10 Slide the brake drum over the brake shoes, lining up the roadwheel bolt holes in the drum with those in the hub. If necessary, back off the shoes adjuster mechanism to allow the drum to pass over the shoes.
11 Refit and tighten the drum securing screw(s).
12 Depress the footbrake several times to operate the self-adjusting mechanism.
13 Check the adjustment of the handbrake as described in Chapter 1.

14 On completion, refit the roadwheel(s), lower the vehicle to the ground and tighten the wheel bolts to the specified torque.

10 Front brake caliper - removal, overhaul and refitting

⚠️ **Warning:** *Before starting work, refer to the warnings at the beginning of Sections 2 and 4 concerning the dangers of hydraulic fluid and asbestos dust.*

Removal

1 Apply the handbrake, then jack up the front of the vehicle and support it securely on axle stands (see *"Jacking and vehicle support"*). Remove the appropriate roadwheel.
2 Minimise fluid loss by using a brake hose clamp, a G-clamp, or a similar tool with protected jaws, to clamp the flexible hose leading to the caliper **(see illustration)**.
3 Clean around the hose union on the caliper, then loosen the brake hose union nut.
4 Slacken and remove the upper and lower caliper guide pin bolts, using a slim open-ended spanner to prevent the guide pin itself from rotating.
5 With the guide pin bolts removed, lift the caliper away from the brake disc, then unscrew the caliper from the end of the brake hose. Note that the brake pads need not be disturbed, and can be left in position in the caliper mounting bracket.

Overhaul

Note: *Ensure that an appropriate overhaul kit can be obtained before dismantling the caliper.*
6 With the caliper on the bench, wipe away all traces of dust and dirt, but *avoid inhaling the dust, as it is injurious to health.*
7 Using a small flat-bladed screwdriver, carefully prise the dust seal retaining clip out of the caliper bore.
8 Withdraw the partially-ejected piston from the caliper body and remove the dust seal. The piston can be withdrawn by hand, or if necessary forced out by applying compressed air to the union bolt hole.
Caution: *The piston may be ejected with*

10.2 Clamp the brake hose

9•10 Braking system

some force. Only low pressure should be required, such as is generated by a foot pump.

9 Extract the piston hydraulic seal using a blunt instrument such as a knitting needle or a feeler blade, taking care not to damage the caliper bore **(see illustration)**.

10 Withdraw the guide pins from the caliper mounting bracket and remove the rubber gaiters.

11 Thoroughly clean all components using only methylated spirit, isopropyl alcohol or clean hydraulic fluid as a cleaning medium. Never use mineral-based solvents, such as petrol or paraffin, which will attack the hydraulic system rubber components. Dry the components immediately, using compressed air or a clean, lint-free cloth. Use compressed air to blow clear the fluid passages.

12 Check all components and renew any that are worn or damaged. Check particularly the cylinder bore and piston; if they are scratched, worn or corroded in any way, they must be renewed (note that this means the renewal of the complete body assembly). Similarly, check the condition of the guide pins and their bores; they should be undamaged and (when cleaned) a reasonably tight sliding fit in the caliper mounting bracket bores. If there is any doubt about the condition of a component, renew it.

13 If the assembly is fit for further use, obtain the appropriate repair kit; the components are available from Renault dealers, in various combinations.

14 Renew all rubber seals, dust covers and caps disturbed on dismantling as a matter of course; these should never be re-used.

15 Before commencing reassembly, ensure that all components are absolutely clean and dry.

16 Dip the piston and the new piston (fluid) seal in clean hydraulic fluid. Smear clean fluid on the cylinder bore surface.

17 Fit the new piston (fluid) seal, using only the fingers to manipulate it into the cylinder bore groove. Fit the new dust seal to the piston. Refit the piston to the cylinder bore using a twisting motion, ensuring that the piston enters squarely into the bore. Press the piston fully into the bore, then press the dust seal into the caliper body.

18 Install the dust seal retaining clip, ensuring that it is correctly seated in the caliper groove.

19 Apply the grease supplied in the repair kit (or a good quality high-temperature brake grease or anti-seize compound) to the guide pins. Fit the pins to the caliper mounting bracket. Fit the new rubber gaiters, ensuring that they are correctly located in the grooves on both the pin, and mounting bracket.

Refitting

Note: Suitable thread-locking compound will be required to coat the threads of the brake caliper guide pin bolts on refitting.

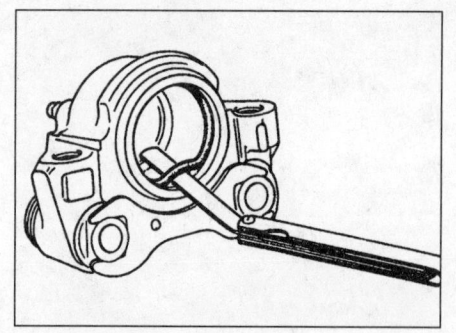

10.9 Using a feeler blade to remove the piston seal from the caliper bore

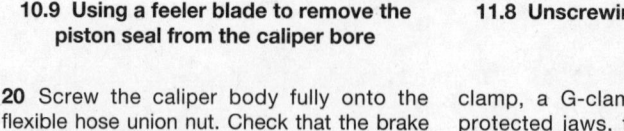

20 Screw the caliper body fully onto the flexible hose union nut. Check that the brake pads are still correctly fitted in the caliper mounting bracket.

21 Position the caliper over the pads. Clean the threads of the lower guide pin bolt, then coat the threads with locking fluid, and fit the bolt. Similarly, apply locking fluid to the upper guide pin bolt, press the caliper into position, and fit the bolt. Tighten the guide pin bolts to the specified torque, starting with the lower bolt.

22 Tighten the brake hose union nut to the specified torque.

23 Remove clamp from the caliper fluid hose.

24 Apply the footbrake several times to position the pads against the discs.

25 Bleed the hydraulic system as described in Section 2. Providing the precautions described were taken to minimise brake fluid loss, it should only be necessary to bleed the relevant front brake.

26 Refit the roadwheel, then lower the vehicle to the ground and tighten the roadwheel bolts to the specified torque.

11 Rear brake caliper - removal, overhaul and refitting

⚠ *Warning: Before starting work, refer to the warnings at the beginning of Sections 2 and 4 concerning the dangers of hydraulic fluid and asbestos dust.*

Removal

Note: New caliper sliding key locking clips will be required on refitting.

1 Chock the front wheels, engage reverse gear (or "P" on models with automatic transmission) and release the handbrake. Jack up the rear of the vehicle and support it securely on axle stands (see *"Jacking and vehicle support"*). Remove the relevant roadwheel.

2 Disconnect the handbrake cable from the lever on the caliper, with reference to Section 20.

3 Minimise fluid loss by using a brake hose

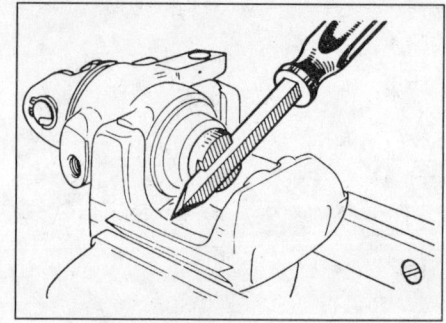

11.8 Unscrewing the rear caliper piston

clamp, a G-clamp, or a similar tool with protected jaws, to clamp the flexible hose leading to the caliper.

4 Clean the area around the fluid hose union on the caliper, then loosen the fluid hose union nut.

5 Extract the two locking clips and then tap out the two sliding keys.

6 Lift the caliper off the brake pads and carrier bracket, then unscrew the caliper from the end of the brake hose. Note that the brake pads need not be disturbed, and can be left in position in the caliper mounting bracket.

Overhaul

7 Clean away external dirt from the caliper and grip it in the jaws of a vice fitted with jaw protectors.

8 Remove the dust excluder from around the piston and then unscrew the piston using the square section shaft of a screwdriver **(see illustration)**. Once the piston turns freely but does not come out any further, apply low air pressure to the fluid inlet hole and eject the piston. Only low air pressure is required such as is generated by a foot-operated tyre pump.

9 Inspect the surfaces of the piston and cylinder bore. If there is evidence of scoring or corrosion, renew the caliper cylinder as follows.
   a) To remove the cylinder a wedge will have to be made in accordance with the dimensions shown **(see illustration)**.
   b) Drive the wedge in to slightly separate the cylinder support bracket arms and so slide out the caliper cylinder, which can be done once the spring-loaded locating pin has been depressed (using a pin-punch, drill or similar tool) **(see illustration)**.
   c) Depress the locating pin, **(see illustration)** then slide the new cylinder into position until the cylinder locating pin engages with the hole in the support bracket.
   d) Tap the wedge out from the caliper support bracket arms.

10 If the piston and cylinder are in good condition, then the cylinder will not have to be removed from the support bracket, but can be overhauled in the following way.

11 Extract the piston hydraulic seal using a

Braking system 9•11

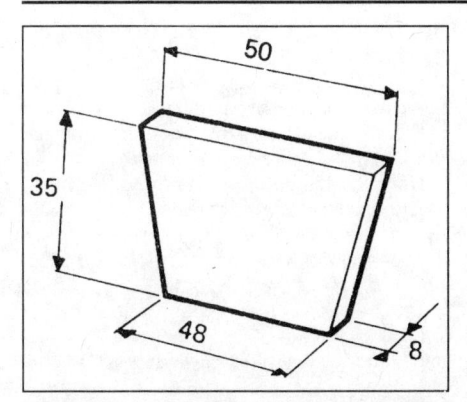

11.9a Rear caliper cylinder removal wedge

All dimensions in mm

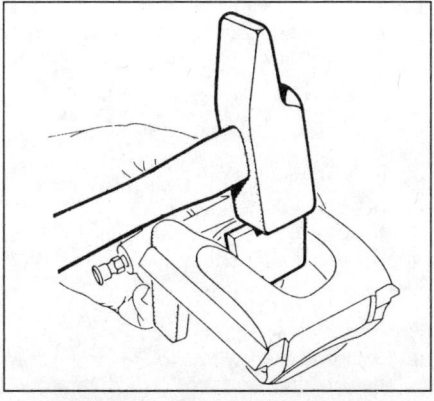

11.9b Driving in the wedge to separate the cylinder support bracket arms

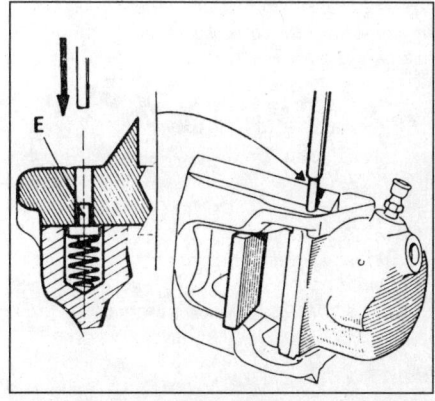

11.9c Depressing the rear caliper cylinder locating pin (E)

blunt instrument such as a knitting needle or a feeler blade, taking care not to damage the cylinder bore.

12 Thoroughly clean all components using only methylated spirit, isopropyl alcohol or clean hydraulic fluid as a cleaning medium. Never use mineral-based solvents, such as petrol or paraffin, which will attack the hydraulic system rubber components. Dry the components immediately, using compressed air or a clean, lint-free cloth. Use compressed air to blow clear the fluid passages.

13 If the assembly is fit for further use, obtain the appropriate repair kit from a Renault dealer.

14 If the handbrake operating mechanism is thought to be worn or faulty, dismantle it in the following way before renewing the piston seal, otherwise proceed to paragraph 23.

15 Grip the caliper in a vice fitted with jaw protectors.

16 Remove the dust cover from the rear of the handbrake mechanism, then extract the circlip from the end of the handbrake lever shaft **(see illustration)**.

17 Compress the spring washers on the adjusting screw, and pull out the handbrake lever shaft.

18 Remove the handbrake lever shaft cam, spring, adjusting screw and plain washer.

19 Slide the spring washers from the adjusting screw.

20 Drive out the adjusting screw sleeve, using a pin-punch. Take care not to damage the cylinder bore. Remove the sealing ring from the base of the sleeve.

21 Clean all components and renew any that are worn.

22 Reassembly is a reversal of removal but observe the following points.
 a) *Fit a new sealing ring, then drive in the adjusting screw sleeve until it is flush with the rear face of the cylinder.*
 b) *Make sure that the spring washers are fitted convex face-to-convex face. Note that the if the spring washers are weak,*
  the handbrake operating lever may not occupy its correct "rest" position.
 c) *It will be necessary to compress the spring washers on the adjusting screw (as during dismantling) to enable the handbrake operating lever shaft to be fitted.*

23 Fit the new piston seal, using only the fingers to manipulate it into the cylinder bore groove.

24 Dip the piston in clean hydraulic fluid and slide the piston into the cylinder. Ensure that the piston enters squarely into the bore.

25 Using the square section shaft of a screwdriver, turn the piston until it turns, but will not go into the cylinder any further. Turn the piston so that the narrow groove in the piston face is uppermost, in line with the caliper bleed screw.

26 Fit the new dust excluder to the end of the cylinder bore.

Refitting

27 Ensure that the caliper piston is aligned as described in paragraph 25.

28 Screw the caliper onto the fluid hose, but do not fully tighten the union nut at this stage.

29 Fit the caliper by engaging one end of the caliper between the spring clip and the keyway on the bracket. Compress the springs and engage the opposite end of the caliper.

30 Insert the first key, then insert a screwdriver in the second key slot and use it as a lever until the second key can be inserted.

31 Fit new key locking clips.

32 Tighten the fluid hose union nut to the specified torque, then remove the clamp from the hose.

33 Apply the footbrake several times to position the pads against the discs.

34 Reconnect the handbrake cable to the lever on the caliper with reference to Section 20.

35 Bleed the hydraulic system as described in Section 2. Providing the precautions described were taken to minimise brake fluid loss, it should only be necessary to bleed the relevant front brake.

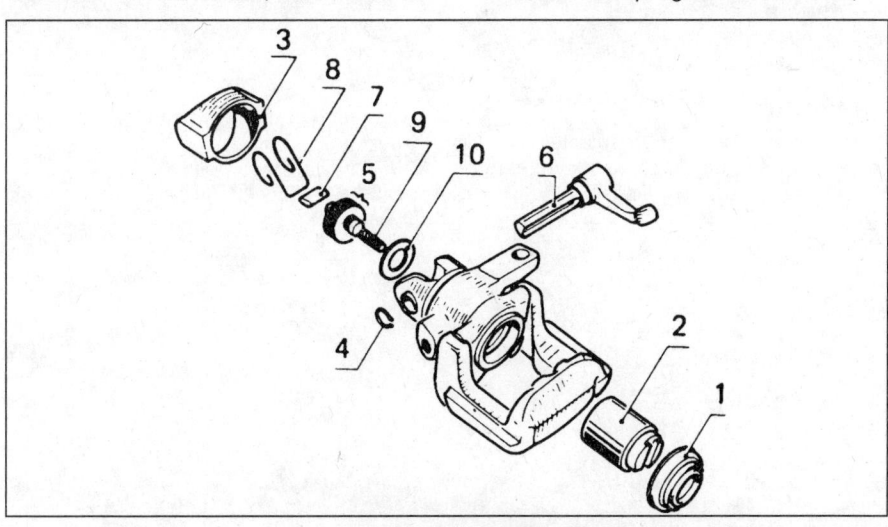

11.16 Exploded view of rear caliper

1 Dust excluder
2 Piston
3 Rear dust cover
4 Circlip
5 Spring washers
6 Handbrake lever shaft
7 Lever shaft cam
8 Spring
9 Adjusting screw
10 Plain washer

9•12 Braking system

36 Refit the roadwheel, then lower the vehicle to the ground and tighten the roadwheel bolts to the specified torque.

12 Rear wheel cylinder - removal, overhaul and refitting

Note: *Before starting work, refer to the warnings at the beginning of Section 2 concerning the dangers of hydraulic fluid, and at the beginning of Section 6 concerning the dangers of asbestos dust.*

Removal

1 Remove the brake drum as described in Section 9.
2 Using pliers, carefully unhook the brake shoe upper return spring and remove it from the brake shoes. Pull the upper ends of the shoes away from the wheel cylinder to disengage them from the pistons.
3 Minimise fluid loss, by using a brake hose clamp, a G-clamp or a similar tool with protected jaws to clamp the flexible hose at the nearest convenient point to the wheel cylinder.
4 Wipe away all traces of dirt around the brake pipe union at the rear of the wheel cylinder, and unscrew the union nut. Carefully ease the pipe out of the wheel cylinder, and plug or tape over its end to prevent dirt entry. Wipe off any spilt fluid immediately.
5 Where applicable, extract the spring clip securing the fluid hose to the support bracket.
6 Unscrew the two wheel cylinder retaining bolts from the rear of the backplate **(see illustration)**. Remove the cylinder, taking care not to allow hydraulic fluid to contaminate the brake shoe linings.

Overhaul

7 Clean away external dirt, then remove the dust excluders and shake out the pistons, seals and spring **(see illustration)**.
8 Examine the condition of the piston and cylinder bore surfaces. If they are scored or corroded, renew the complete cylinder assembly.
9 If the pistons and cylinder are in good condition then discard the seals and wash the components in methylated spirit.
10 Obtain a repair kit which will contain all the necessary renewable items.
11 Reassemble the wheel cylinder after dipping each component in clean hydraulic fluid.

Refitting

12 Ensure that the backplate and wheel cylinder mating surfaces are clean, then spread the brake shoes and manoeuvre the wheel cylinder into position.
13 Engage the brake pipe, and screw in the union nut two or three turns to ensure that the thread has started.
14 Insert the two wheel cylinder retaining bolts, and tighten them securely. Now fully tighten the brake pipe union nut.
15 Where applicable, refit the clip securing the fluid hose to the support bracket.
16 Remove the clamp from the brake hose.
17 Ensure that the brake shoes are correctly located in the cylinder pistons. Carefully refit the brake shoe upper return spring, using a screwdriver or long-nosed pliers to stretch the spring into position.
18 Refit the brake drum as described in Section 9.
19 Bleed the brake hydraulic system as described in Section 2. Providing suitable precautions were taken to minimise loss of fluid, it should only be necessary to bleed the relevant rear brake.

13 Master cylinder - removal, overhaul and refitting

Removal

1 Syphon the fluid from the master cylinder reservoir. Use a syringe, a clean battery hydrometer or a poultry baster to do this, **never** use the mouth to suck the fluid out through a tube.
2 Disconnect the leads from the low fluid level warning light switch and pull the reservoir upwards out of the sealing grommets.
3 Note the locations of the hydraulic pipes and then disconnect them from the master cylinder by unscrewing the unions. Place a wad of rag under the master cylinder to catch any fluid which may drain out.
4 Unscrew the nuts securing the master cylinder to the front face of the vacuum servo, and withdraw the master cylinder.

Overhaul

5 A faulty master cylinder cannot be overhauled, as no spare parts are available. If the master cylinder is faulty or worn, the complete assembly must be renewed.

Refitting

6 Before fitting the master cylinder, check that the servo operating rod protrusion is as specified **(see illustration 14.14)**. If necessary adjust by turning the operating rod adjusting nut.
7 Place the master cylinder in position on the servo, then refit and tighten the securing nuts.
8 Reconnect the brake fluid pipes and tighten the union nuts.
9 Push the reservoir firmly into its grommets.
10 Fill the reservoir with clean fluid and bleed the complete hydraulic system as described in Section 2.
11 On completion, check that the length of the servo pushrod is as specified - see Section 14.

14 Vacuum servo unit - testing, removal and refitting

Testing

1 Operation of the servo can be checked in the following way.
2 With the engine stopped, depress the brake pedal several times. The pedal travel should

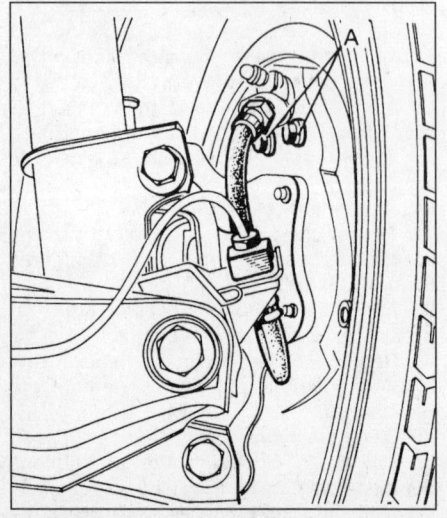

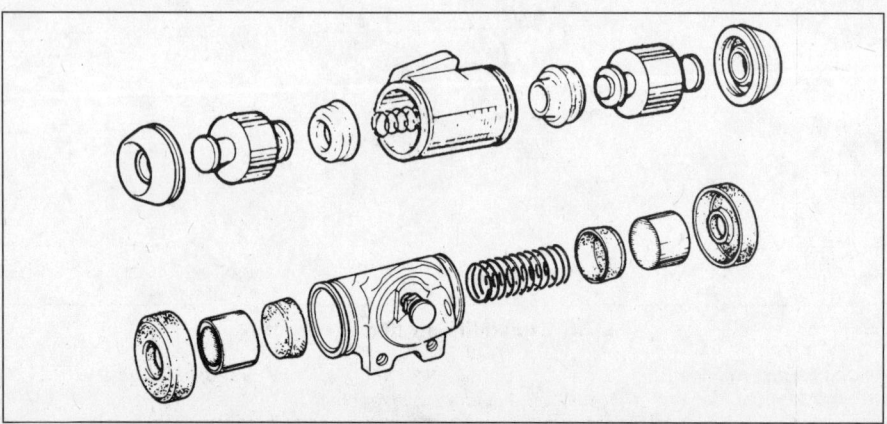

12.6 Rear wheel cylinder securing bolts (A)

12.7 Exploded views of alternative types of rear wheel cylinder

Braking system 9•13

remain the same each time the pedal is depressed.

3 Depress the brake pedal fully and hold it down, then start the engine. It should be possible to feel the pedal move down slightly.

4 Hold the pedal depressed with the engine running, then switch off the engine, whilst still holding the pedal depressed. The pedal should not rise nor fall.

5 Start the engine and run it for at least a minute. Stop the engine, then depress the brake pedal several times. The pedal travel should decrease with each application, and it should be possible to detect a "hissing" sound from the servo as the pedal is depressed. After about four or five depressions of the pedal, no further hissing should be heard, and the pedal should feel considerably firmer.

6 If the foregoing tests do not prove satisfactory, check the servo vacuum hose and non-return valve for security and leakage at the valve grommet (see illustration).

7 If the brake servo operates properly in the test, but still gives less effective service on the road, the air filter through which air flows into the servo should be inspected. A dirty filter will reduce the effectiveness of the servo.

8 The servo unit itself cannot be repaired and therefore renewal is necessary if the unit proves to be faulty.

Removal

9 Remove the master cylinder, as described in Section 13. Note that on some models, it may be possible to move the master cylinder to one side, without disconnecting the brake fluid pipes.

10 Disconnect the vacuum hose from the servo unit.

11 Working in the driver's footwell, remove the driver's side lower facia trim panel, then disconnect the servo pushrod from the brake pedal by extracting the split pin or spring clip (as applicable) and the clevis pin.

12 Again working in the footwell, unscrew the brake servo mounting nuts (see illustration).

13 Withdraw the servo from the engine compartment.

Refitting

14 Before refitting the servo, check that the protrusion of the servo operating rod is as specified (see illustration). The protrusion is measured from the front face of the servo to the end of the operating rod. If necessary, adjust the operating rod protrusion by turning the adjuster nut on the end of the rod.

15 Similarly, check the length of the servo pushrod. The length is measured from the rear face of the servo to the centre of the hole in the pushrod clevis. If necessary, adjust the length of the pushrod by loosening the locknut and turning the clevis. Tighten the locknut on completion.

16 Refitting is a reversal of removal, but refit the master cylinder with reference to Section 13.

14.6 Brake servo non-return valve (arrowed)

15 Vacuum servo unit non-return valve - removal, testing and refitting

Removal

1 Slacken the clip and disconnect the vacuum hose from the non-return valve on the front face of the servo unit.

2 Withdraw the valve from its rubber sealing grommet by pulling and twisting. Pull the sealing grommet from the servo.

Testing

3 Examine the non-return valve and sealing grommet for damage and signs of deterioration, and renew if necessary. The valve can be tested by blowing through it in both directions - it should only be possible to blow from the servo end to the manifold end.

Refitting

4 Fit a new grommet to the servo, then push the valve into position in the grommet. A smear of rubber grease will aid fitting. Do not push too hard, as it is possible to push the sealing grommet into the servo.

5 Reconnect the vacuum hose.

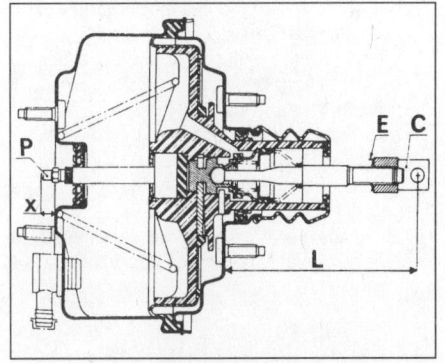

14.14 Servo operating rod protrusion and pushrod length

C Pushrod clevis
E Locknut
L = 128.5 mm
P Operating rod adjuster nut
X = 9.0 mm

14.12 Brake servo securing nuts (arrowed)

16 Vacuum servo unit air filter - renewal

1 Working in the driver's footwell, remove the lower facia trim panel, then pull the dust excluder from the rear of the servo, and slide it up the pushrod.

2 Using a scriber or similar pointed tool, prise the filter from its housing, and cut it to allow it to pass over the pushrod (see illustration).

3 Cut the new filter, and push it into position, ensuring that it is correctly seated.

4 Push the dust excluder into position.

17 Brake pedal - removal and refitting

Removal

1 Working in the driver's footwell, remove the lower facia trim panel, then disconnect the servo pushrod from the brake pedal arm by extracting the spring clip and pushing out the clevis pin.

2 Release the return spring from the pedal, noting its location, then remove the spring clip from the end of the pedal pivot shaft.

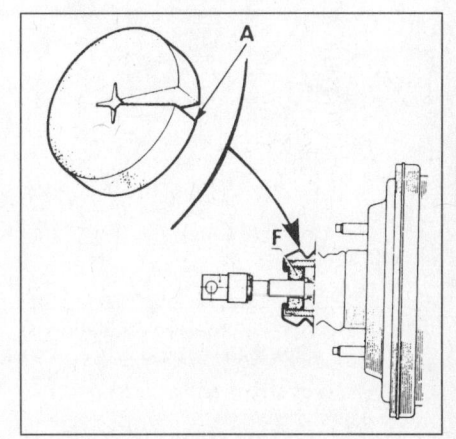

16.2 Vacuum servo unit air filter

A Cut   F Filter location

9•14 Braking system

3 Slide the pivot shaft out of the pedal bracket until the pedal can be removed.

Refitting

4 Refitting is a reversal of removal. Ensure that the return spring is correctly engaged with the pedal.

18 Rear brake pressure regulating valve - testing, adjusting, removal and refitting

Testing and adjustment

1 Testing and adjustment of the pressure regulating valve requires the use of special pressure gauges and adapters, and should be entrusted to a Renault dealer.

Removal

2 Remove the clip securing the suspension arm link rod to the lugs on the suspension arms and the lug on the end of the valve operating arm (see illustration).
3 Unclip the lug from the end of the valve operating arm.
4 Place a container beneath the valve, then unscrew the union nuts and disconnect the brake fluid pipes from the valve. Plug or cover the open ends of the pipes and valve to reduce fluid loss and prevent dirt ingress.
5 Unscrew the securing bolts, and withdraw the valve from its mounting bracket, sliding the valve operating arm from the suspension arm link rod as the valve is withdrawn.

Refitting

6 If a new valve is being fitted, it will be supplied with two clips which are used to preset the valve (see illustration). Do not remove the clips before the valve is fitted.

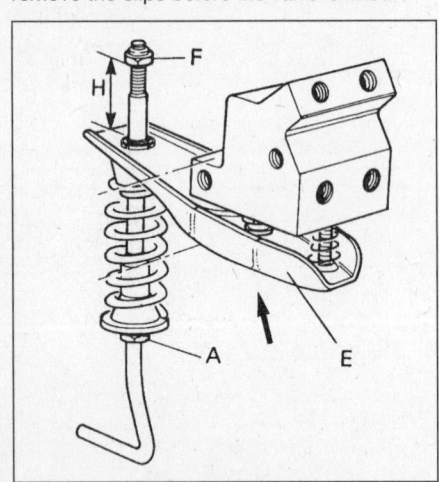

18.13 Rear brake pressure regulating valve initial setting

A Valve operating arm adjuster nut
E Valve operating arm
F Nut - **DO NOT MOVE**
H = 32.0 mm
Press arm in direction of arrow

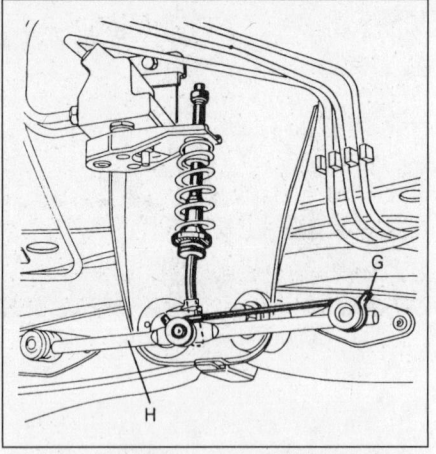

18.2 Rear brake pressure regulating valve

G Clip
H Suspension arm link rod

7 Before refitting the valve, slacken the valve operating arm adjuster nut.
8 Locate the end of the valve operating arm in the hole in the suspension arm link rod, then clip the lug into position on the end of the valve operating arm. Refit the clip to secure the suspension arm link rod to the lugs on the suspension rods and valve operating arm.
9 Manoeuvre the valve into position on its mounting bracket, then refit and tighten the securing bolts.
10 Reconnect the brake pipes to the valve and tighten the union nuts.
11 Tighten the valve operating arm adjuster nut.
12 If a new valve has been fitted, the clips used to preset the valve can now be removed.
13 The initial setting of the valve should now be checked as follows.
  a) Push the valve operating lever upwards towards the valve.
  b) Measure the distance from the top of the valve operating arm to the bottom of the nut at the top of the valve operating arm (see illustration).
  c) If the distance is not as specified, adjust by slackening the valve operating arm adjuster nut and turning the sleeve. **Do not** disturb the nut at the top of the valve operating arm.
14 Bleed the brake hydraulic system as described in Section 2.
15 Have the valve adjustment checked by a Renault dealer at the earliest opportunity (see paragraph 1).

19 Handbrake lever - removal and refitting

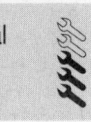

Removal

1 Jack up the vehicle and support on axle stands (see "Jacking and vehicle support").
2 Working under the vehicle, where applicable, remove the cover from the vehicle floor to reveal the handbrake adjuster sleeve.
3 Slacken the adjuster locknuts and unscrew the sleeve from the handbrake rod.
4 Working inside the vehicle, remove the centre console as described in Chapter 11.
5 Unscrew the two bolts securing the handbrake lever assembly to the floor, then withdraw the handbrake lever/rod assembly (see illustration).

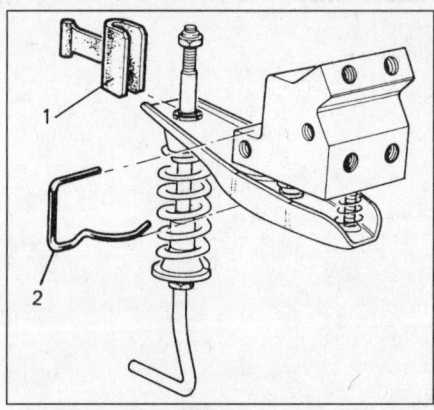

Refitting

6 Refitting is a reversal of removal, but on completion, check the handbrake cable adjustment as described in Chapter 1.

20 Handbrake cables - removal and refitting

Removal

1 There are two handbrake cables, one cable running from each rear brake assembly to the cable equaliser under the vehicle floor.
2 Chock the front wheels, engage reverse gear (or "P" on models with automatic transmission) and release the handbrake. Jack up the rear of the vehicle and support it securely on axle stands (see "Jacking and vehicle support").

18.6 New rear brake pressure regulating valve clips (1 and 2)

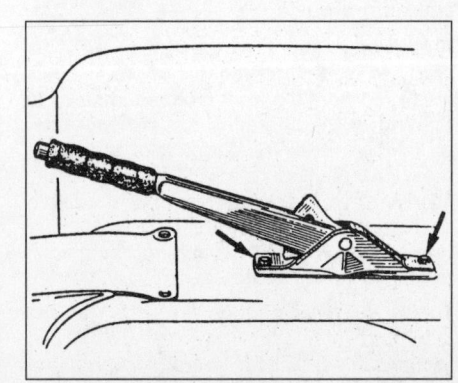

19.5 Handbrake lever securing bolts (arrowed)

Braking system 9•15

3 On models with rear brakes, remove the relevant brake drums as described in Section 9, then disconnect the handbrake cable from the lever on the trailing shoe. If necessary, back-off the handbrake cable adjustment, with reference to Chapter 1.
4 On models with rear disc brakes, disconnect the handbrake cable from the lever on the caliper. Again, if necessary back-off the handbrake cable adjustment with reference to Chapter 1.
5 Working at the cable equaliser, remove the securing clip, then withdraw the clevis pin and disconnect the equaliser from the rod.
6 Slide the cable from the equaliser, then release the cable from the clips on the body, and withdraw the cable.

Refitting
7 Refitting is a reversal of removal, but lightly grease the end of the cable at the equaliser, and on completion, check the cable adjustment as described in Chapter 1.

21 Stop-light switch - adjustment, removal and refitting

Adjustment
1 The switch is located on the pedal mounting bracket (see illustration).
2 To improve access, remove the driver's side lower facia panel.
3 The switch should be adjusted so that the stop-lights come on when the switch plunger move out approximately 1.0 mm as the brake pedal is depressed.
4 If adjustment is required, where applicable slacken the locknuts, then screw the switch in or out of its mounting bracket as necessary.

Removal
5 Disconnect the battery negative lead.
6 To improve access, remove the driver's side lower facia panel.
7 Disconnect the wiring from the switch.
8 Where applicable, unscrew the locknuts, then unscrew or withdraw the switch from the bracket.

Refitting
9 Refitting is a reversal of removal, but adjust the position of the switch as described previously in this Section.

22 Anti-lock braking system (ABS) - general information

1 ABS is available as an option on certain models covered by this manual, and is fitted as standard equipment on some models. The purpose of the system is to prevent the wheel(s) locking during heavy braking. This is achieved by automatic release of the brake on

21.1 Stop-light switch location (arrowed)

the relevant wheel, followed by re-application of the brake. The system comprises an electronic control module, a hydraulic modulator block, hydraulic solenoid valves and accumulators, an electrically-driven return pump, and four roadwheel sensors.
2 Models with ABS are fitted with front and rear disc brakes.
3 The solenoids (which control the fluid pressure to the calipers) are controlled by the electronic control unit, which itself receives signals from the wheel sensors. The wheel sensors monitor the speed of rotation of each wheel. By comparing these speed signals from the four wheels, the control unit can determine when a wheel is decelerating at an abnormal rate, compared to the speed of the vehicle. Using this information, the control unit can predict when a wheel is about to lock, and is able to reduce the fluid pressure to the brake on the relevant wheel to prevent it from locking.
4 During normal operation, the system functions in the same way as a conventional non-ABS braking system.

23 Anti-lock braking system (ABS) components - removal and refitting

Front wheel sensor
Removal
1 Disconnect the battery negative lead.
2 Apply the handbrake, then jack up the front of the vehicle and support securely on axle stands (see "Jacking and vehicle support"). To improve access, remove the relevant roadwheel.
3 Locate the sensor wiring connector under the wheel arch, then separate the two halves of the connector. Release the wiring from any clips and support brackets.
4 Unscrew the securing bolt, then withdraw the sensor from the hub carrier (see illustration).

Refitting
5 Refitting is a reversal of removal, bearing in mind the following points.
a) Smear the hub carrier contact faces of the sensor with a little grease before fitting.

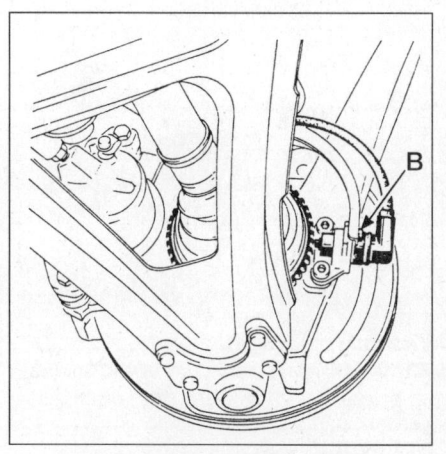

23.4 ABS front wheel sensor securing bolt

b) Tighten the securing bolt to the specified torque - do not overtighten the bolt.
c) Ensure that the wiring connector is securely reconnected.

Rear wheel sensor
6 The procedure is as described previously for the front wheel sensor, but note that the wiring connector is located under the rear of the vehicle (see illustration).

Electronic control unit
Removal
7 The electronic control unit is located at the front right-hand corner of the engine compartment.
8 Remove the battery, with reference to Chapter 5.
9 Unclip the plastic cover from the control unit.
10 Unscrew the two securing bolts, then lift the control unit from the engine compartment, and disconnect the wiring plug (see illustration).

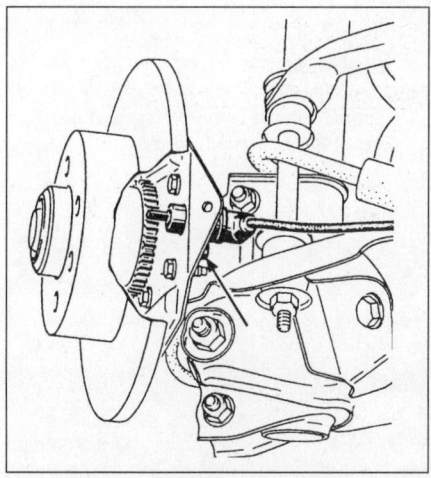

23.6 ABS rear wheel sensor securing bolt (arrowed)

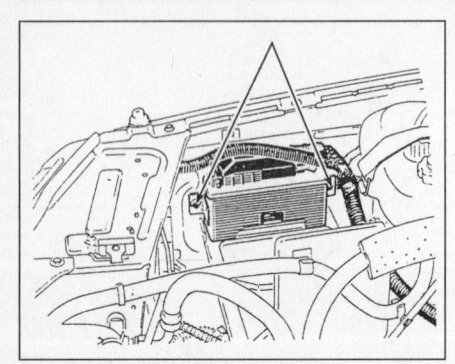

9•16 Braking system

23.10 ABS electronic control unit securing bolts (arrowed)

Refitting

11 Refitting is a reversal of removal, but ensure that the wiring connector is securely reconnected.

Hydraulic unit

Removal

12 Disconnect the battery negative lead.
13 Mark the fluid pipes to identify them for location, then unscrew the union nuts, and disconnect the fluid pipes from the hydraulic unit. Be prepared for fluid spillage, and plug or cover the open ends of the pipes to minimise fluid loss, and to prevent dirt ingress.
14 Withdraw the cover from hydraulic unit wiring connector (see illustration).
15 Loosen the clamp screws, and release the wiring from the clamp, then disconnect the wiring connector from the hydraulic unit.
16 Unscrew the three nuts securing the hydraulic unit to the mounting bracket, then withdraw the unit from the engine compartment.

Refitting

17 Refitting is a reversal of removal, bearing in mind the following points.
 a) Ensure that the fluid pipes are correctly reconnected as noted before removal - note that the top of the unit is engraved with letters which correspond to the various hydraulic circuits (see illustration).
 b) Do not switch on the ignition until the complete brake hydraulic system has been bled as described in Section 2.

24 Brake vacuum pump - removal and refitting

Note: A new O-ring will be required on refitting.

Removal

1 Disconnect the battery negative lead.
2 Slacken the retaining clip and disconnect the vacuum hose from the top of the pump which is situated on the left-hand side of the cylinder block (see illustrations).

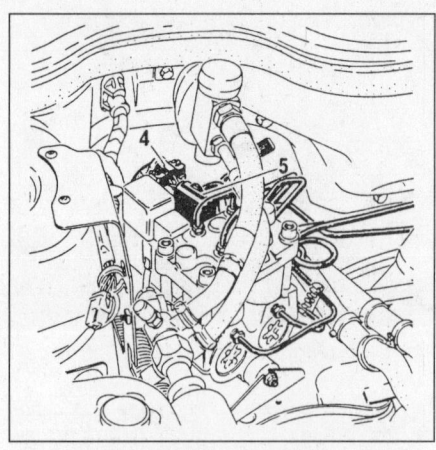

23.14 ABS hydraulic unit wiring clamp (4) and connector (5)

3 Undo the two retaining bolts and carefully lift the pump away from the engine. Remove the O-ring from the pump sealing groove and discard it; a new one should be used on refitting.
4 If necessary, carefully withdraw the oil pump/vacuum pump drivegear, taking great care not to dislodge the oil pump driveshaft.
Note: Great care must be taken to ensure that the driveshaft is not dislodged from the oil pump. If the driveshaft is dislodged it will drop down into the bottom sump. If this happens, the sump will have to be removed in order to recover the driveshaft.

Refitting

5 Where necessary, ensure that the driveshaft is correctly engaged with the oil pump then refit the oil pump/vacuum pump drivegear, engaging it with driveshaft.
6 Fit a new O-ring to the vacuum pump groove and apply a smear of engine oil to it to aid installation.
7 Slide the vacuum pump into position aligning its drive dog with the slot in the drivegear.
8 Refit the pump retaining bolts and tighten them securely.
9 Reconnect the vacuum hose to the pump.

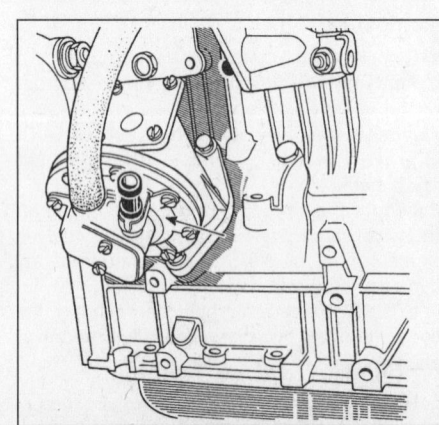

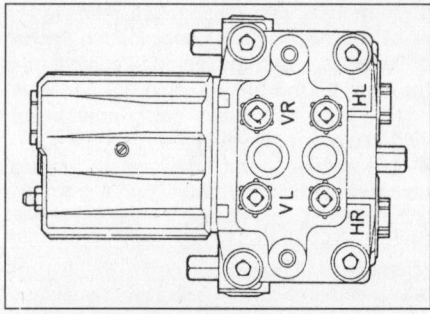

23.17 ABS hydraulic unit fluid pipe identification marks

VL Front left   HL Rear left
VR Front right   HR Rear right

10 Lower the vehicle to the ground then reconnect the battery. Check the braking system before taking the vehicle on the road.

25 Brake vacuum pump - testing and overhaul

Testing

Note: A vacuum gauge will be required for this test.

1 The operation of the vacuum pump can be checked using a vacuum gauge.
2 Disconnect the vacuum hose from the pump, and connect the gauge to the pump union, using a suitable length of hose.
3 Start the engine and allow it to idle, then increase the engine speed to at least 2000 rpm. As a guide, at 2000 rpm, the minimum vacuum should be at least 770 mbar (570 mm Hg). If the vacuum registered is significantly less than this, it is likely that the pump is faulty. However, seek the advice of a Renault dealer before condemning the pump.

Overhaul

4 Overhaul of the pump is not possible, since no spare parts are available. If faulty, the complete assembly must be renewed.

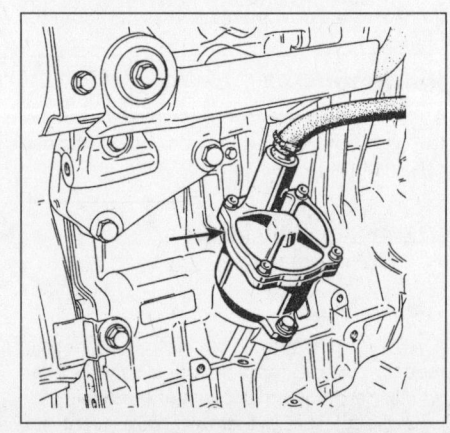

24.2a Brake vacuum pump location (arrowed)

24.2b Alternative type of brake vacuum pump (arrowed)

Chapter 10
Suspension and steering

Contents

Auxiliary (power steering pump) drivebelt check, adjustment and renewal . . . . See Chapter 1	
Electronic variable-rate suspension - general information 16	
Electronic variable rate suspension components - removal and refitting 17	
Front anti-roll bar components 5	
Front hub bearing - renewal 3	
Front hub carrier - removal and refitting 2	
Front shock absorber and coil spring - removal, inspection and refitting 4	
Front suspension lower arm and balljoint - removal, inspection and refitting 7	
Front suspension radius rod - removal, refitting and adjustment . . . 8	
Front suspension upper arm and balljoint - removal, inspection and refitting 6	
General information 1	
Ignition switch/steering column lock - removal and refitting 19	
Power steering fluid level check . . . . See Chapter 1	
Power steering pump - removal and refitting 23	
Power steering system - bleeding 24	
Rear anti-roll bar - removal and refitting 13	
Rear coil spring - removal, inspection and refitting 12	
Rear hub bearings - renewal 10	
Rear hub carrier - removal, inspection and refitting 9	
Rear shock absorber - removal, inspection and refitting 11	
Rear suspension lower arm - removal, inspection and refitting 14	
Rear suspension radius rod - removal, inspection and refitting 15	
Steering column and intermediate shaft - removal and refitting 20	
Steering gear - removal, overhaul and refitting 22	
Steering gear gaiters - renewal 21	
Steering wheel - removal and refitting 18	
Suspension and steering check . . . . See Chapter 1	
Track-rod and inner balljoint - removal, inspection and refitting 26	
Track-rod end balljoint - removal and refitting 25	
Wheel and tyre maintenance and tyre pressure checks . . . . See "Weekly Checks"	
Wheel alignment and steering angles - general information, checking and adjustment 27	

Degrees of difficulty

Easy, suitable for novice with little experience	**Fairly easy,** suitable for beginner with some experience	**Fairly difficult,** suitable for competent DIY mechanic	**Difficult,** suitable for experienced DIY mechanic	**Very difficult,** suitable for expert DIY or professional

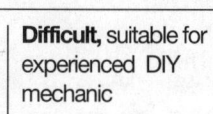

Specifications

Front suspension
Type . . . . Independent, by upper and lower arms, with coil springs mounted over shock absorbers

Rear suspension
Type . . . . Independent, with lower arms, coil springs and shock absorbers

Steering
Type . . . . Rack-and-pinion with collapsible safety column. Power steering available on some models

Wheel alignment and steering angles
Front wheel castor angle:
  Manual steering:
    Suspension setting (H5 - H2 - see Section 27):
      10.0 mm . . . . 2°
      35.0 mm . . . . 1°30'
      60.0 mm . . . . 1°
      85.0 mm . . . . 0°30'
      110.0 mm . . . . 0°
  Power steering:
    Suspension setting (H5 - H2 - see Section 27):
      10.0 mm . . . . 4°
      35.0 mm . . . . 3°30'
      60.0 mm . . . . 3°
      85.0 mm . . . . 2°30'
      110.0 mm . . . . 2°

10•2 Steering and suspension

Wheel alignment and steering angles (continued)

Maximum difference in castor angle from side-to-side	1°
Front wheel camber angle	0° ±30'
Maximum difference in camber angle from side-to-side	1°
Steering axis inclination	12°30' ± 30'
Maximum difference in steering axis inclination from side-to-side	1°
Front wheel toe setting	0°30' ± 10' toe-out (3.0 ± 1.0 mm toe-out)
Rear wheel camber angle	1°15' ±30' negative
Rear wheel toe setting	0°10' ± 10' toe-in (1.0 ± 1.0 mm toe-in)

Torque wrench settings*

	Nm	lbf ft
Front suspension		
Driveshaft nut	250	185
Upper arm (-to-hub carrier) balljoint nut	65	48
Lower arm (-to-hub carrier) balljoint nut	65	48
Front hub bearing carrier bolts	20	15
Front shock absorber upper mounting nut:		
Models up to 1988	15	11
Models from 1989	20	
Front shock absorber lower mounting through-bolt and nut	80	59
Front shock absorber lower mounting locknut	60	44
Anti-roll bar clamp nuts	15	11
Anti-roll bar link-to-upper arm/shock absorber nut	80	59
Upper arm inner pivot bolt and nut	95	70
Upper arm balljoint-to-upper arm nuts and bolts	25	18
Lower arm pivot pin nuts - "Phase I" models	130	96
Lower arm front pivot bolt and nut - "Phase II" models	135	100
Lower arm rear pivot-to-body nuts and bolts - "Phase II" models	45	33
Radius rod-to-upper arm nuts	70	52
Rear suspension		
Rear hub nut	160	118
Rear hub carrier-to-lower arm nut and bolt	100	74
Rear shock absorber upper mounting nut	30	22
Rear shock absorber lower mounting nuts and bolts	110	81
Anti-roll bar-to-lower arm through-bolt and nut	17	13
Anti-roll bar clamp nuts	15	11
Lower arm inboard pivot bolt and nut	80	59
Radius rod-to-lower arm nut and bolt	34	25
Radius rod-to-body nut and bolt	45	33
Steering		
Track-rod end balljoint nut	40	30
Track-rod end locknut	35	26
Track-rod inner balljoint-to-steering gear	50	37
Steering wheel hub screws	15	11
Roadwheels		
Roadwheel bolts:		
4-bolt fixing	90	66
5-bolt fixing	100	74

*Renew all "Nyloc"-type self-locking nuts on refitting

1 General information

The independent front suspension comprises upper and lower arms, with coils springs mounted over shock absorbers. The hub carriers, which carry the wheel bearings, brake calipers, and the hub/disc assemblies are located by the upper and lower arms via balljoints. The upper and lower arms have rubber inner mounting bushes, and the upper arms are located at their outer ends by radius rods which are adjustable to allow adjustment of the castor angle. A front anti-roll bar is rubber-mounted onto the body, and connects the upper arms via drop-links.

The independent rear suspension comprises lower arms, coil springs and shock absorbers. The hub carriers, which carry the wheel bearings, brake calipers, and hub/drum or disc assemblies are bolted to the lower arms at their lower ends, and are located by the shock absorbers. The lower arms have rubber inner mounting bushes, and are located at their outer ends by radius rods. A rear anti-roll bar is rubber-mounted onto the body, and connects to the lower arms via drop links.

Certain high-specification models are fitted with an electronically-controlled variable-rate suspension - refer to Section 16 for further details.

The steering column is connected by a universal joint to an intermediate shaft, which is in turn connected to the steering gear via a second universal joint. The steering gear is mounted on a body crossmember, and is connected via the track-rods to the steering arms projecting from the rear of the hub carriers. The track-rod ends are adjustable to

Steering and suspension 10•3

enable adjustment of wheel alignment. Power steering is fitted to most models, with hydraulic pressure provided by a belt-driven pump which is driven from the crankshaft pulley.

2 Front hub carrier - removal and refitting

Removal

Note: *A new driveshaft nut will be required on refitting, and all "Nyloc"-type self-locking nuts should be renewed. A balljoint separator tool will be required for this operation. Sealant may be required to coat the outer end of the driveshaft.*

1 Apply the handbrake, then jack up the front of the vehicle and support securely on axle stands (see *"Jacking and vehicle support"*).
2 On models with ABS, it is advisable to remove the ABS wheel sensor as described in Chapter 9, to avoid any possibility of damage during the removal procedure.

HAYNES HINT: On models where access to the driveshaft nut can be obtained by removing the wheel trims, before jacking up the vehicle, loosen the driveshaft nut as follows.
a) Chock the front wheels, and remove the wheel trim.
b) Have an assistant firmly apply the footbrake.
c) Loosen the driveshaft nut using a socket and extension.

3 If the driveshaft nut has been loosened, proceed to paragraph 5, otherwise proceed as follows.
4 Refit at least two roadwheel bolts to the front hub, and tighten them securely. Have an assistant firmly depress the brake pedal to prevent the front hub from rotating, then using a socket and a long extension bar, slacken and remove the driveshaft retaining nut. Alternatively, a tool can be fabricated from two lengths of steel strip (one long, one short)

TOOL TIP

Using a fabricated tool to hold the front hub stationary whilst the driveshaft retaining nut is slackened

and a nut and bolt; the nut and bolt forming the pivot of a forked tool. Bolt the tool to the hub using two wheel bolts, and hold the tool to prevent the hub from rotating as the driveshaft retaining nut is undone **(see Tool Tip)**. This nut is very tight; make sure that there is no risk of pulling the car off the axle stands. (If the roadwheel trim allows access to the driveshaft nut, the initial slackening can be done with the wheels chocked and on the ground.)
5 Unbolt the brake caliper from the hub carrier as described in Chapter 9. Note that there is no need to disconnect the fluid hose - suspend the caliper from the suspension strut using wire or string, ensuring that the hose is not strained.
6 Remove the securing screw(s) and withdraw the brake disc.
7 Slacken and partially unscrew the hub carrier upper balljoint nut (unscrew the nut as far as the end of the threads on the balljoint to prevent damage to the threads as the joint is released), then release the balljoint using a balljoint separator tool **(see illustration)**. Remove the nut and discard it - a new nut must be used on refitting.
8 Similarly, release the balljoint and disconnect the track-rod from the steering arm on the hub carrier.
9 Unscrew the driveshaft nut from the end of the driveshaft. Recover the washer. Discard the nut - a new one must be used on refitting.
10 Pull the hub carrier from the upper balljoint, and tilt the hub carrier downwards until the splined end of the driveshaft can be released from it **(see illustration)**. If necessary tap the end of the driveshaft using a hammer and a soft metal drift to release it from the hub - **do not** strike the end of the driveshaft hard, as this may cause damage to the joints.

Note: *On some models the driveshaft ends are fitted to the hub carriers using locking compound. Renault use a special extractor tool to release the driveshaft ends on these models, but if the driveshaft cannot be released by hand, it should be possible to use a three-legged puller as follows.*
a) Temporarily refit the driveshaft nut to protect the threads on the end of the driveshaft.
b) Fit the puller, with the arms bearing on the hub, and the centre screw bearing on the end of the driveshaft.
c) Use the puller to release the hub from the end of the driveshaft. Note that it is possible that the hub will be pulled from the hub bearing assembly (the bearing front half inner race will remain in position on the hub) - this is not a problem provided that the bearing is kept clean, and repacked with grease on refitting.

11 Support the end of the driveshaft by suspending it using wire or string - do not allow the end of the driveshaft to hang down under its own weight, as this may damage the CV joints.
12 Support the hub carrier, then disconnect the hub carrier lower balljoint as described previously in paragraph 7 for the upper balljoint.
13 Withdraw the hub carrier.
14 If desired, the hub/bearing assembly can be unbolted from the hub carrier. The bearing carrier bolts can be reached through the holes in the hub flange **(see illustration)**. Note that the bearing carrier bolts also secure the brake disc shield.

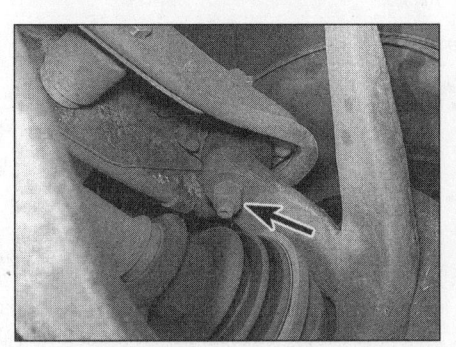

2.7 Hub carrier upper balljoint nut (arrowed)

2.10 Pull the hub carrier from the upper balljoint

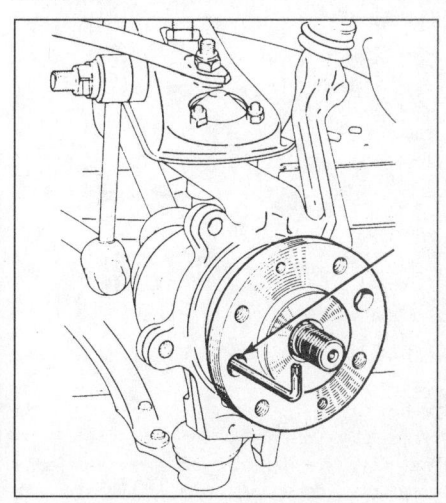

2.14 Using an Allen key (arrowed) through the hole in the hub flange to unscrew a bearing carrier bolt

10•4 Steering and suspension

Refitting

15 Where applicable, fit the hub/bearing assembly to the hub carrier, ensuring that the brake disc shield is in place, and tighten the bearing carrier securing bolts to the specified torque.
16 Fit the hub carrier to the lower balljoint, and tighten the new securing nut as far as possible by hand.
17 On models where the driveshaft end was fitted to the hub carrier using locking compound, thoroughly clean all traces of old locking compound from the end of the driveshaft. Thoroughly clean the driveshaft aperture in the hub.
18 Where applicable, coat the end of the driveshaft with locking compound (Renault recommend the use of "Loctite Scelbloc"), then engage the end of the driveshaft with the hub. **Note:** *If the hub has been pulled from the bearing during the removal procedure, ensure that the bearing is clean and packed with the appropriate grease, then refit the hub as described during the front wheel bearing renewal procedure in Section 3.*

HAYNES HiNT *On models where access to the driveshaft nut can be obtained by removing the wheel trim, the driveshaft nut can be tightened with the footbrake firmly applied, and the vehicle resting on its wheels.*

19 Reconnect the hub carrier upper balljoint, and tighten the new nut to the specified torque.
20 Tighten the new hub carrier lower balljoint nut to the specified torque.
21 Screw the **new** driveshaft nut onto the end of the driveshaft as far as possible by hand, ensuring that the washer is in place, then tighten the nut until the end of the driveshaft is fully engaged with the hub **(see illustration)**. Do not fully tighten the nut at this stage.

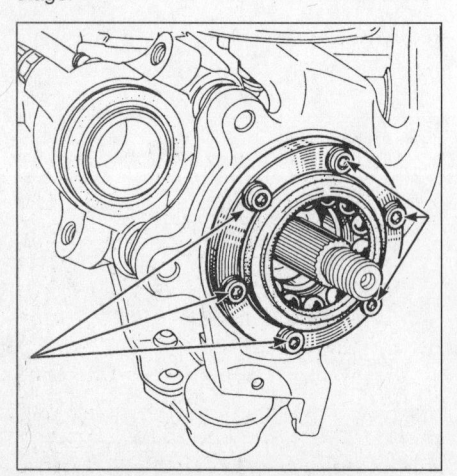

2.21 Fitting a new driveshaft nut

22 Reconnect the track-rod to the hub carrier, and tighten the new balljoint nut to the specified torque.
23 Refit the brake disc, then refit the brake caliper, with reference to Chapter 9.
24 Use the method employed on removal to prevent the hub from rotating, and tighten the driveshaft retaining nut to the specified torque. Check that the hub rotates freely.
25 Where applicable, refit the ABS wheel sensor, with reference to Chapter 9.
26 Refit the roadwheel, then lower the vehicle to the ground and tighten the roadwheel bolts to the specified torque.

3 Front hub bearing - renewal

Note: *A new driveshaft nut will be required on refitting - this should be supplied in the new bearing kit. All "Nyloc"-type self-locking nuts should be renewed. A balljoint separator tool will be required for this operation. Sealant may be required to coat the outer end of the driveshaft.*

1 Proceed as described in Section 2, paragraphs 1 to 6.
2 Unscrew the driveshaft nut from the end of the driveshaft. Recover the washer.
3 The end of the driveshaft must now be released from the hub. If necessary tap the end of the driveshaft using a hammer and a soft metal drift to release it from the hub - **do not** strike the end of the driveshaft hard, as this may cause damage to the joints.

Note: *On some models the driveshaft ends are fitted to the hub carriers using locking compound. Renault use a special extractor tool to release the driveshaft ends on these models, but if the driveshaft cannot be released by hand, it should be possible to use a three-legged puller as follows.*
 a) *Temporarily refit the driveshaft nut to protect the threads on the end of the driveshaft.*
 b) *Fit the puller, with the arms bearing on the hub, and the centre screw bearing on the end of the driveshaft.*
 c) *Use the puller to release the hub from the end of the driveshaft. Note that the hub may be pulled from the bearing carrier as it is released from the driveshaft.*

4 Unscrew the securing bolts, and withdraw the bearing carrier from the hub carrier. If the hub is still in place in the bearing carrier, the bearing carrier bolts can be reached through the holes in the hub flange **(see illustration)**.
5 If the hub is still in place in the bearing carrier, use two roadwheel bolts and two steel packing pieces to force the hub from the bearing carrier. Screw the bolts in evenly and gradually so that the ends of the bolts bear on the packing pieces to force out the hub. Note that the bearing front half inner race will remain in place on the hub **(see illustrations)**.
6 The bearing half inner race must now be removed from the hub using a suitable bearing puller. Alternatively, support the bearing race on thin steel bars, and press or drive the hub from the bearing race. Recover the bearing plastic cap if it is loose.
7 The bearing outer race and ball assembly is integral with the bearing carrier, and must be renewed as an assembly.
8 Obtain a bearing kit, which will consist of a bearing assembly, grease and a new driveshaft nut.
9 Thoroughly clean the bearing contact face of the hub, the splines in the hub, and the driveshaft splines.
10 Ensure that the bearing plastic cap is in position on the hub, then fit the new bearing front half inner race to the hub. Use a metal tube (with an internal diameter of

3.4 Front hub bearing carrier-to-hub carrier bolts (arrowed)

3.5a Using two roadwheel bolts and packing pieces . . .

3.5b . . . to force the hub from the bearing carrier

Steering and suspension 10•5

approximately 41.0 mm) to press or drive the race into position until it contacts the flange on the hub.

11 Pack the new bearing assembly with the grease supplied in the repair kit, then fit the rear half inner race to the bearing.

12 Offer the bearing carrier/bearing assembly into position on the hub carrier, then refit the securing bolts and tighten to the specified torque.

13 Where applicable, coat the end of the driveshaft with locking compound (Renault recommend the use of "Loctite Scelbloc"), then engage the hub with the end of the driveshaft.

14 Slide the hub onto the driveshaft (if necessary, gently tap it into position, using a soft-faced mallet) until the **new** driveshaft nut can be screwed on by a few threads. Ensure that the washer is in place under the driveshaft nut.

15 Tighten the driveshaft nut until the hub is fully engaged with the end of the driveshaft. Do not fully tighten the nut at this stage.

16 Proceed as described in Section 2, paragraphs 23 to 26.

4 Front shock absorber and coil spring - removal, inspection and refitting

The coil spring is mounted over the shock absorber, and is located between the lower spring seat on the shock absorber body, and the upper spring seat on the body suspension turret. Although the spring and shock absorber are not integral, the two components must be removed and refitted as an assembly. The spring must be compressed using suitable tools before attempting to remove the assembly.

Due to the layout of the suspension components and the bodywork, it is not possible to use conventional spring compressor tools to compress the spring. Renault special tools are available for this purpose, and a Renault dealer will have access to these tools.

Due to the requirement for special tools, it is considered that the removal and refitting of the front shock absorber and coil spring is beyond the scope of the DIY mechanic, and the job **must** be entrusted to a Renault dealer.

⚠ **Warning:** Do not attempt to remove the shock absorber and coil spring using conventional spring compressor tools, or improvised tools. Any attempt to remove the coil spring and shock absorber without Renault special tools and the knowledge to use them safely could result in damage and/or serious personal injury.

5 Front anti-roll bar components - removal and refitting

Anti-roll bar

Note: *All "Nyloc"-type self-locking nuts should be renewed on refitting.*

Removal

1 Apply the handbrake, then jack up the front of the vehicle and support it securely on axle stands (see *"Jacking and vehicle support"*). To improve access, remove the roadwheels.

2 Working on one side of the vehicle, counterhold the shock absorber lower mounting through-bolt, noting that the bolt also secures the anti-roll bar drop-link, and unscrew the nut from the end of the bolt. Slide the through-bolt back far enough to release the anti-roll bar drop-link **(see illustration)**.

3 Repeat the procedure given in paragraph 2 to disconnect the remaining drop link.

4 Unscrew the nuts securing the anti-roll bar mounting to the body.

5 Carefully manipulate the anti-roll bar out from under the vehicle.

Refitting

6 Check the condition of the rubber mounting bushes, and renew them if there is any sign of damage or deterioration.

7 Refitting is a reversal of removal, but do not fully tighten the fixings until the weight of the vehicle is resting on its wheels, and tighten the fixings to the specified torque.

Drop links

Removal

8 Proceed as described in paragraphs 1 and 2.

9 Using circlip pliers, extract the circlip securing the lower end of the drop-link to the anti-roll bar. Recover the spacer.

10 Slide the drop-link from the anti-roll bar.

Refitting

11 Examine the condition of the drop-link rubber bushes, and renew the drop-link if there is any sign of damage or deterioration.

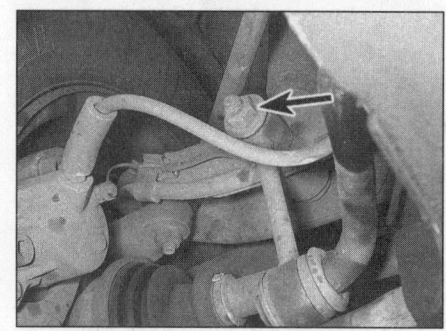

5.2 Shock absorber/front anti-roll bar drop link securing nut (arrowed)

12 Refitting is a reversal of removal, but do not fully tighten the fixings until the weight of the vehicle is resting on its wheels, and tighten the fixings to the specified torque.

6 Front suspension upper arm and balljoint - removal, inspection and refitting

Upper arm

Note: *and all "Nyloc"-type self-locking nuts should be renewed. A balljoint separator tool will be required for this operation.*

Removal

1 Apply the handbrake, then jack up the front of the vehicle and support it securely on axle stands (see *"Jacking and vehicle support"*). Remove the relevant roadwheel.

2 Position a jack under the suspension lower arm, and raise the arm until it is horizontal.

3 Loosen the shock absorber lower mounting locknut.

4 Unscrew the securing nuts and bolts, and disconnect the radius rod from the upper arm **(see illustration)**.

5 Slacken and partially unscrew the hub carrier upper balljoint nut (unscrew the nut as far as the end of the threads on the balljoint to prevent damage to the threads as the joint is released), then release the balljoint using a balljoint separator tool. Remove the nut.

6 Counterhold the shock absorber lower mounting through-bolt, noting that the bolt also secures the anti-roll bar drop-link, and unscrew the nut from the end of the bolt. Slide the through-bolt from the upper arm, noting the locations of the washers.

7 Unscrew the bottom of the shock absorber from the lower mounting.

8 Counterhold the upper arm inner pivot bolt, and unscrew the nut **(see illustration)**. Withdraw the pivot bolt, noting the locations of any washers, then withdraw the upper arm from under the wheel arch.

Inspection

9 Examine the inner pivot bush for wear or deterioration, and if necessary renew the bush as follows.

6.4 Radius rod-to-upper arm securing nuts (arrowed)

10•6 Steering and suspension

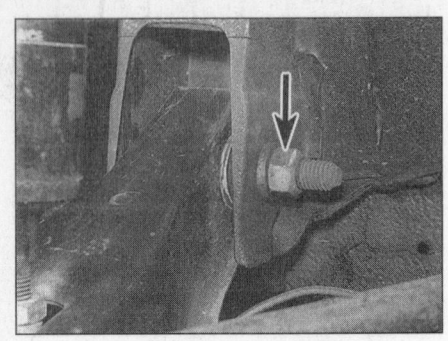

6.8 Front suspension upper arm pivot bolt nut (arrowed)

10 Use a length of tubing with a bolt, nut and washers to draw out the old bush and fit the new one. Make sure that the new bush is centralised as shown **(see illustration)**.
11 If desired, the balljoint can be renewed as described in the following sub-Section.

Refitting

12 Refitting is a reversal of removal, bearing in mind the following points.
  a) Do not fully tighten the upper arm inner pivot bolt or the shock absorber lower mounting through-bolt until the vehicle is resting on its wheels.
  b) Ensure that all washers are positioned as noted before removal.
  c) Renew all "Nyloc"-type self-locking nuts.
  d) Tighten all fixings to the specified torque.

Upper arm balljoint

Note: *A balljoint separator tool will be required for this operation.*

Removal

13 Apply the handbrake, then jack up the front of the vehicle and support it securely on axle stands (see *"Jacking and vehicle support"*). Remove the relevant roadwheel.
14 Slacken and partially unscrew the hub carrier upper balljoint nut (unscrew the nut as far as the end of the threads on the balljoint to prevent damage to the threads as the joint is

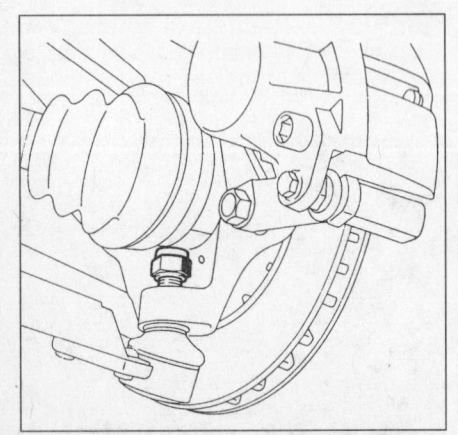

7.2 Force the lower arm balljoint nut against the driveshaft casing to release the balljoint

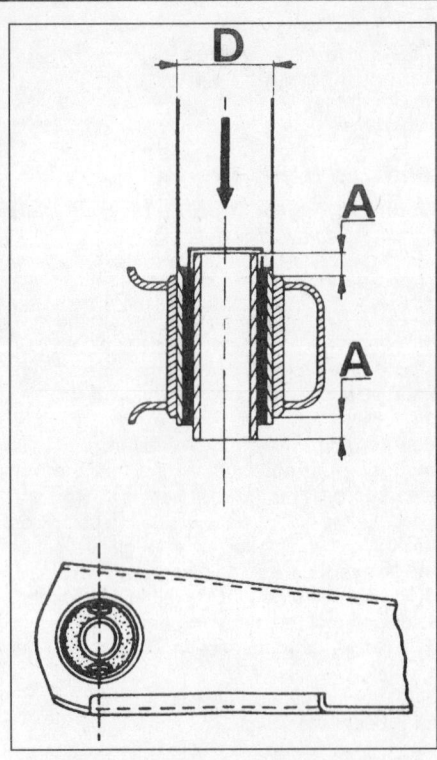

6.10 Front suspension upper arm bush fitting dimensions

A = 7.5 mm
D Bush removal tube
(34.5 mm outside diameter)

released), then release the balljoint using a balljoint separator tool. Remove the nut.
15 Unscrew the nuts securing the radius rod to the upper arm, noting that one of the nuts also secures the upper arm balljoint.
16 Unscrew the remaining nuts and bolts securing the balljoint to the upper arm, then withdraw the balljoint.

Refitting

17 Refitting is a reversal of removal, but all "Nyloc"-type self-locking nuts should be renewed, and tighten all fixings to the specified torque. On completion, check the front wheel alignment at the earliest opportunity (see Section 27).

7.3 Lower arm pivot pin securing nut (A) and castor adjusting shim (B) – "Phase I" models

7 Front suspension lower arm and balljoint - removal, inspection and refitting

Lower arm - "Phase I" models

Note: *A new balljoint nut will be required on refitting, and all "Nyloc"-type self-locking nuts should be renewed. A balljoint separator tool will be required for this operation.*

Removal

1 Apply the handbrake, then jack up the front of the vehicle and support it securely on axle stands (see *"Jacking and vehicle support"*). Remove the relevant roadwheel.
2 Loosen the lower arm balljoint nut until it contacts the driveshaft joint casing. Continue to unscrew the nut, forcing the nut against the driveshaft to release the balljoint **(see illustration)**. Once the balljoint it free, unscrew the nut and discard it - a new nut must be used on refitting.
3 Unscrew the lower arm pivot pin rear securing nut, and withdraw the pivot pin towards the front of the vehicle, noting the position and orientation if the castor adjusting shim on the pin (the shim is fitted to the front of the pin on models with manual steering, or to the rear of the pin on models with power steering) **(see illustration)**.
4 Withdraw the lower arm.

Inspection

5 Examine the inner pivot bushes for wear or deterioration, and if necessary renew the bush(es) as follows.
6 Use a length of tubing with a bolt, nut and washers to draw out the old bush and fit the new one. Make sure that the new bush is positioned as shown **(see illustration)**.
7 If both rubber bushes are being renewed, preventing one of them from being used as a reference point for measuring the fitting dimension, mark the position of one of the external bush tubes on the lower arm, then remove the bush, and fit the new bush in the same place. Remove out the second bush and measure the distance from the first bush.

Refitting

8 Smear the pivot pin with a little grease, then offer the lower arm into position, and refit the pivot pin (from the front of the vehicle). Ensure that the castor adjusting shim is correctly

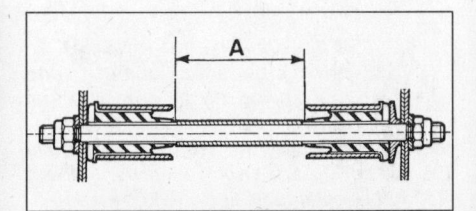

7.6 Lower arm pivot bush fitting dimension – "Phase I" models

A = 112.6 mm

Steering and suspension 10•7

orientated and positioned as noted before removal. The shim fits at the front of the pin on models with manual steering, or at the rear of the pin on models with power steering.

9 Fit a new pivot pin nut, but do not fully tighten at this stage.

10 Reconnect the lower arm balljoint, then fit a new balljoint nut and tighten to the specified torque.

11 Refit the roadwheel and lower the vehicle to the ground, then tighten the lower arm pivot pin nut to the specified torque.

Lower arm - "Phase II" models

Note: A new balljoint nut will be required on refitting, and all "Nyloc"-type self-locking nuts should be renewed. A balljoint separator tool will be required for this operation.

Removal

12 Proceed as described in paragraphs 1 and 2.
13 Counterhold the lower arm front pivot bolt, and unscrew the nut.
14 Similarly, unscrew the two nuts securing the lower arm rear pivot to the body, and withdraw the bolts **(see illustration)**.
15 Withdraw the front pivot bolt, noting the position and orientation of the castor adjusting shim on the pin (the shim is fitted to the front of the pin on models with manual steering, or to the rear of the pin on models with power steering), and the locations of any washers.
16 Withdraw the lower arm.

Inspection

17 Examine the inner pivot bushes for wear or deterioration, and if necessary renew the bush(es) as follows.
18 To renew the front pivot bushes, saw off the flange of one of the bushes, then use a suitable tube to press both bushes from the lower arm. Press the new bushes into position until the flanges are resting on the surface of the lower arm.

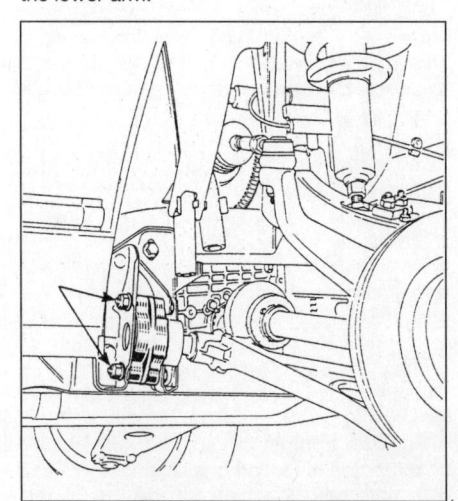

19 To renew the rear pivot bush, pull or press the bush off the lower arm pin, then press the new bush into position to give the dimension shown, between the outer faces of the front and rear bushes. Note that the dimension varies depending on whether manual or power steering is fitted. Also ensure that the angle of the centreline passing through the pivot bush bolt holes is as shown in relation to a vertical line at right-angles to the lower arm horizontal axis **(see illustration)**.

Refitting

20 If a new lower arm is being fitted, note that it will be supplied with a front pivot bush fitted in the position required for models with power steering. If the vehicle being worked on has manual steering, the bush must be pressed further onto the lower arm pin to the position specified for models with manual steering - see paragraph 19.
21 Smear the front pivot bolt with a little grease, then offer the lower arm into position, and refit the pivot bolt. Ensure that the castor adjusting shim is correctly orientated and positioned as noted before removal. The shim fits at the front of the pin on models with manual steering, or at the rear of the pin on models with power steering.
22 Fit a new front pivot bolt nut, but do not fully tighten it at this stage.
23 Smear the two lower arm rear pivot bolts with a little grease, then refit the bolts (the bolts fit from the front of the vehicle) and fit new nuts. Do not fully tighten the nuts and bolts at this stage.

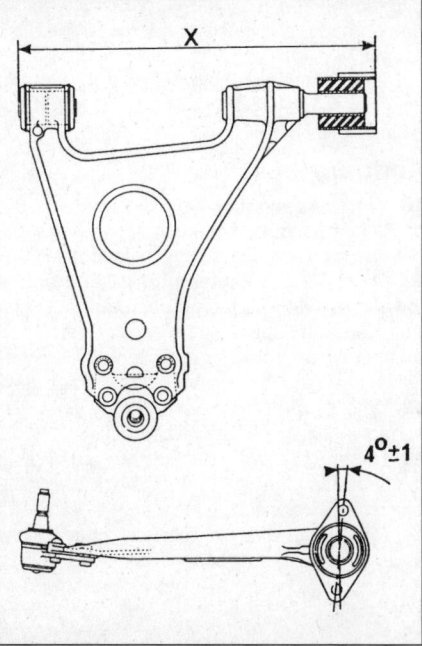

7.19 Lower arm rear pivot bush fitting position - "Phase II" models

$X = 382.4 \pm 0.5$ mm (manual steering)
$X = 393.0$ mm (power steering)

24 Reconnect the lower arm balljoint, then fit a new balljoint nut and tighten to the specified torque.
25 Refit the roadwheel and lower the vehicle to the ground, then tighten the lower arm front and rear pivot nuts and bolts to the specified torque.

Lower arm balljoint - models with balljoint secured by rivets

Removal

26 Remove the lower arm as described previously in this Section.
27 Clamp the lower arm in a vice, and drill out the rivets securing the balljoint.

Refitting

28 Fit the new balljoint using the nuts and bolts supplied, noting the following points.
a) The balljoint should be positioned at the underneath the lower arm.
b) The bolts should fit from above the lower arm.

29 Refit the lower arm as described previously in this Section.
30 On completion, check the front wheel alignment at the earliest opportunity (see Section 27).

Lower arm balljoint - models with balljoint secured by nuts and bolts

Note: A new balljoint nut will be required on refitting. A balljoint separator tool will be required for this operation.

Removal

31 Proceed as described in paragraphs 1 and 2.
32 Unscrew the securing nuts and bolts, and withdraw the balljoint.

Refitting

33 Proceed as described in paragraph 28.
34 Reconnect the lower arm balljoint, then fit a new balljoint nut and tighten to the specified torque.
35 Refit the roadwheel and lower the vehicle to the ground.
36 On completion, check the front wheel alignment at the earliest opportunity (see Section 27).

8 Front suspension radius rod - removal, refitting and adjustment

Removal

Note: Do not unscrew the locknut and alter the position of the radius rod end, as this will alter the castor angle. All "Nyloc" type self-locking nuts should be renewed.

1 Apply the handbrake, then jack up the front of the vehicle and support it securely on axle stands (see "Jacking and vehicle support"). Remove the relevant roadwheel.

7.14 Lower arm rear pivot-to-body securing nuts (arrowed) - "Phase II" models

10

10•8 Steering and suspension

2 Position a jack under the suspension lower arm, and raise the arm until it is horizontal.
3 Unscrew the securing nuts and bolts, and disconnect the radius rod from the upper arm.
4 Counterhold the pivot bolt, and unscrew the nut, then withdraw the pivot bolt, noting the locations of any washers. Withdraw the radius rod.

Refitting

5 Offer the radius rod into position, then refit the pivot bolt (from the rear of the vehicle), and refit the nut, ensuring that any washers are in place as noted before removal. Do not fully tighten the pivot bolt at this stage.
6 Reconnect the radius rod to the upper arm, then refit the nuts and bolts and tighten to the specified torque.
7 Remove the jack from under the lower arm, then refit the roadwheel and lower the vehicle to the ground.
8 Tighten the radius rod pivot bolt to the specified torque with the vehicle resting on its wheels.
9 Check the front wheel alignment at the earliest opportunity (see Section 27).

Adjustment

10 The radius rod can be adjusted to alter the front wheel castor angle. There is normally no need to do this unless the front suspension has been damaged, or new components have been fitted.
11 To adjust the castor angle, suitable alignment gauges are required, and the work should be entrusted to a Renault dealer or wheel alignment specialist.
12 For reference, the radius rod can be adjusted as follows.
 a) Proceed as described in paragraphs 1 to 3.
 b) Loosen the locknut on the radius rod end.
 c) Turn the radius rod (by half a turn at a time), then refit the nuts and bolts securing the rod to the upper arm, and tighten the rod end locknut.
 d) Refit the roadwheel and lower the vehicle to the ground.
 e) Check the castor angle and front wheel alignment (see Section 27).
 f) If necessary, repeat the procedure until the front wheel alignment is correct.
Note that one full turn of the radius rod will alter the castor angle by 15'.

9 Rear hub carrier - removal and refitting

Removal

Note: all "Nyloc"-type self-locking nuts should be renewed.
1 Remove the appropriate rear shock absorber as described in Section 11.
2 Disconnect the handbrake cable from the brake caliper or shoe lever, as applicable, as described in Chapter 9.

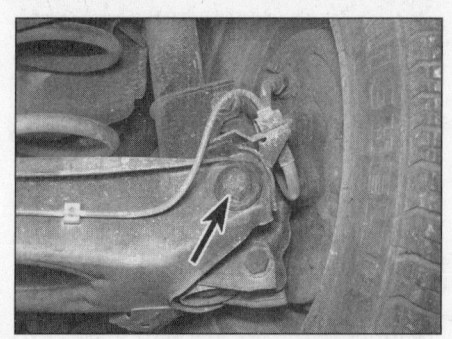

9.5 Rear hub carrier-to-lower arm securing bolt (arrowed)

3 Clamp the brake hose to the caliper or wheel cylinder, as applicable, then unscrew the union and disconnect the brake hose from the caliper or wheel cylinder. Again, refer to Chapter 9 if necessary.
4 Where applicable, remove the ABS rear wheel sensor as described in Chapter 9.
5 Counterhold the bolt and unscrew the nut securing the hub carrier to the lower arm (see illustration). Note the locations of any washers and/or spacers to ensure correct refitting.
6 Withdraw the hub carrier assembly, complete with the hub and brake components.
7 If desired, the brake components and the hub can be removed from the hub carrier with reference to Chapter 9, and Section 10 of this Chapter.

Inspection

8 Examine the mounting bush for wear or deterioration, and if necessary renew the bush as follows.
9 Use a length of tubing with a bolt, nut and washers to draw out the old bush and fit the new one.

Refitting

10 Where applicable, refit the hub and the brake components as described in Section 10 of this Chapter and Chapter 9 respectively.
11 Offer the hub carrier into position on the lower arm, and refit the bolt and nut, ensuring that any washers and/or spacers are positioned as noted before removal. Do not fully tighten the nut and bolt at this stage.
12 Where applicable refit the ABS wheel sensor as described in Chapter 9.
13 Reconnect the handbrake cable as described in Chapter 9, and check the cable adjustment as described in Chapter 1.
14 Reconnect the brake hose to the caliper or wheel cylinder (as applicable), then bleed the brake hydraulic system as described in Chapter 9.
15 Refit the shock absorber as described in Section 11.
16 On completion, refit the roadwheel and lower the vehicle to the ground, then tighten the hub carrier-to-lower arm nut and bolt to the specified torque.

10 Rear hub bearings - renewal

Note: A new rear hub nut will be required on refitting.
1 The hub bearings are integral with the rear hubs, and if the hub bearings are worn or damaged, the complete hub/bearing assembly must be renewed.
2 Chock the front wheels, engage reverse gear (or "P" on models with automatic transmission) and release the handbrake. Jack up the rear of the vehicle and support it securely on axle stands (see "Jacking and vehicle support"). Remove the relevant roadwheel.
3 Using a screwdriver, carefully lever out the dust cap from the centre of the hub.
4 Remove the brake disc or drum, as applicable, as described in Chapter 9.
5 Using a socket and extension bar, unscrew and remove the rear hub nut.
6 Withdraw the hub/bearing assembly from the stub axle.
7 Thoroughly clean the stub axle, then lubricate the stub axle with a little clean gear oil.
8 Slide the hub/bearing assembly into position, then fit a new hub nut, and tighten to the specified torque.
9 Refit the brake disc or drum, as applicable, as described in Chapter 9.
10 Refit the dust cap to the centre of the hub.
11 Refit the roadwheel and lower the vehicle to the ground.

11 Rear shock absorber - removal, inspection and refitting

Removal

Note: All "Nyloc"-type self-locking nuts should be renewed.
1 Chock the front wheels, engage reverse gear (or "P" on models with automatic transmission) and release the handbrake. Jack up the rear of the vehicle and support it securely on axle stands (see "Jacking and vehicle support"). Remove the relevant roadwheel.
2 Position a jack under the suspension lower arm, and raise the jack slightly to support the suspension.
3 Working in the luggage compartment, remove the parcel shelf side trim panel (see Chapter 11) to expose the shock absorber top mounting.
4 Pull the rubber cover from the shock absorber top mounting (see illustration).
5 Counterhold the flats on the top of the shock absorber piston, and unscrew the mounting nut. Recover the mounting plates and rubber from the top of the mounting,

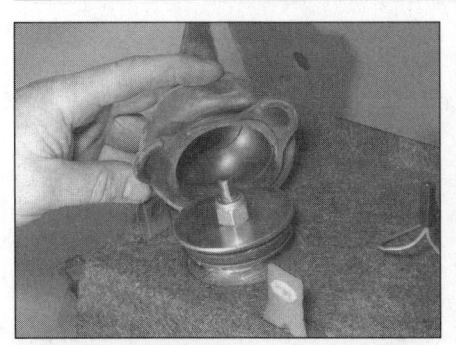

Steering and suspension 10•9

11.4 Pull off the rubber cover to expose the rear shock absorber top mounting

11.6 Rear shock absorber lower mounting bolts (arrowed)

noting their fitted order to ensure correct refitting.
6 Working under the wheel arch, counterhold the nuts and unscrew the bolts securing the lower end of the shock absorber to the hub carrier **(see illustration)**.
7 Compress the shock absorber, and withdraw it from under the vehicle. Recover the upper mounting plates and rubber, again noting their locations to ensure correct refitting.

Inspection

8 Examine the shock absorber for signs of leakage. Check the shock absorber piston for signs of pitting along its entire length, and check the body for signs of damage. Test the operation of the shock absorber, while holding it in an upright position, by moving the piston through full stroke, and then through short strokes of 50.0 to 100.0 mm. In both cases, the resistance felt should be smooth and continuous. If the resistance is jerky, or uneven, or if there is any visible sign of wear or damage to the strut, renewal is necessary.

Refitting

9 Refitting is a reversal of removal, bearing in mind the following points.
 a) Ensure that the shock absorber top mounting plates and rubbers are correctly fitted as noted before removal.
 b) Note that the lower shock absorber securing bolts fit from the rear of the vehicle.
 c) All "Nyloc"-type self-locking nuts should be renewed.
 d) Tighten all fixings to the specified torque.

12 Rear coil spring - removal, inspection and refitting

Note: *A spring compressor tool will be required for this operation - see text. All "Nyloc"-type self-locking nuts should be renewed.*

Removal

1 Remove the relevant shock absorber, as described in Section 11.

2 Using a suitable spring compressor tool (or tools), compress the spring lightly to release the load on the rear anti-roll bar. If it is not possible to fit a conventional compressor tool, a suitable alternative can be made up as follows.
 a) Obtain a length of stout threaded bar, suitable nuts and washers, and a metal plate (with a hole drilled in the centre) which will pass horizontally through the lower spring coils.
 b) Secure the upper end of the threaded rod in the shock absorber top mounting hole.
 c) Fit the metal plate through the lower spring coils, and secure it to the end of the threaded rod with a washer and nut.
 d) Tighten the nut at one end of the threaded bar to compress the coil spring until the anti-roll bar is no longer under tension.

3 Counterhold the through-bolt and unscrew the nut securing the anti-roll bar to the lower arm. Recover the sleeves, washers, spacers and rubbers, noting their orientation and locations to ensure correct refitting.
4 Disconnect the brake pressure regulating valve operating rod from the lower arm, with reference to Chapter 9.
5 If a conventional spring compressor tool has been used, lower the jack supporting the lower arm until the spring can be removed complete with the compressor. Take care not to strain the brake fluid hose.
6 If a threaded rod has been used to compress the spring, as described in paragraph 2, slowly loosen the nut on the threaded rod, at the same time lowering the jack under the lower arm, until the coil spring is free from tension. Again, take care not to strain the brake fluid hose. Unscrew the nut from the end of the threaded rod, and withdraw the coil spring.

Inspection

7 If any doubt exists about the condition of the coil spring, gradually release the spring compressor (where applicable), and check the spring for distortion and signs of cracking. Since no minimum free length is specified by Renault, the only way to check the tension of the spring is to compare it to a new

component. Renew the spring if it is damaged or distorted, or if there is any doubt as to its condition.

Refitting

8 Refitting is a reversal of removal, bearing in mind the following points.
 a) Take care not to strain the brake hose when moving the lower arm.
 b) When reconnecting the anti-roll bar to the lower arm, ensure that all sleeves, washers, spacers and rubbers are correctly orientated and located as noted before removal.
 c) Do not fully tighten the anti-roll bar through-bolt and nut until the vehicle is resting on its wheels.
 d) Reconnect the brake pressure regulating valve operating rod, with reference to Chapter 9.
 e) Refit the shock absorber as described in Section 11.
 f) All "Nyloc"-type self-locking nuts should be renewed.
 g) Tighten all fixings to the specified torque.
 h) On completion, have the adjustment of the brake pressure regulating valve checked at the earliest opportunity (see Chapter 9).

13 Rear anti-roll bar - removal and refitting

Removal

Note: *All "Nyloc"-type self-locking nuts should be renewed.*

1 It is unlikely that the anti-roll bar will ever require removal, as the clamp rubbers and the end link mounting rubbers can be renewed without removing it. If it is necessary to remove the anti-roll bar, proceed as follows.
2 Remove the fuel tank as described in Chapter 4.
3 Working on one side of the vehicle, position a jack under the lower arm and raise the lower arm slightly to relieve the tension in the anti-roll bar.
4 Counterhold the through-bolt and unscrew the nut securing the anti-roll bar to the lower arm **(see illustration)**. Recover the sleeves, washers, spacers and rubbers, noting their orientation and locations to ensure correct refitting.
5 Repeat the procedure in paragraphs 3 and 4 on the remaining side of the vehicle.
6 Unscrew the nuts securing the anti-roll bar clamps to the body. Recover the washers.
7 The anti-roll bar can now be manipulated out from the body member cut-outs, and withdrawn from under the vehicle.

Refitting

8 Refitting is a reversal of removal, bearing in mind the following points.

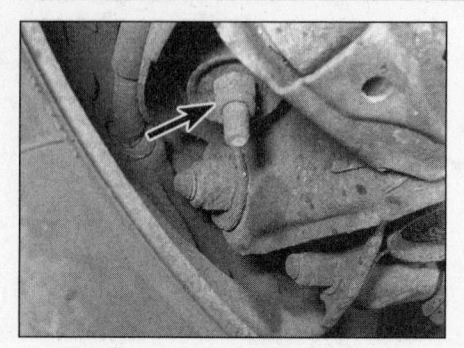

10•10 Steering and suspension

13.4 Rear anti-roll bar-to-lower arm securing nut (arrowed)

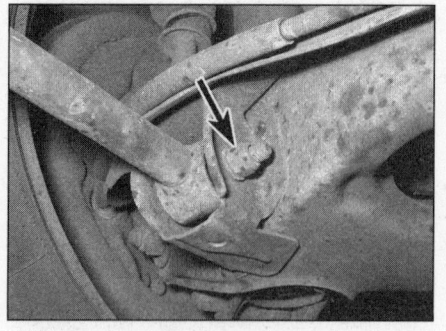

14.7 Rear suspension radius rod-to-lower arm nut (arrowed)

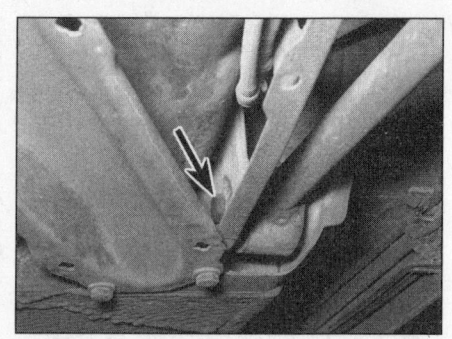

15.4 Rear suspension radius rod-to-body bracket bolt (arrowed)

a) *Ensure that the sleeves, washers, spacers and rubbers on the anti-roll bar-to-lower arm through-bolt are orientated and located as noted before removal.*
b) *Do not fully tighten the anti-roll bar fixings until the vehicle is resting on its wheels.*
c) *Refit the fuel tank as described in Chapter 4.*
d) *All "Nyloc"-type self-locking nuts should be renewed.*
e) *Tighten all fixings to the specified torque.*

14 Rear suspension lower arm - removal, inspection and refitting

Removal

Note: *All "Nyloc"-type self-locking nuts should be renewed.*

1 Remove the shock absorber and the coil spring as described in Sections 11 and 12 respectively.
2 Disconnect the handbrake cable from the brake caliper or shoe lever, as applicable, as described in Chapter 9.
3 Disconnect the brake pressure regulating valve operating rod from the lower arm, with reference to Chapter 9.
4 Clamp the brake hose to the caliper or wheel cylinder, as applicable, then unscrew the union and disconnect the brake hose from the caliper or wheel cylinder. Again, refer to Chapter 9 if necessary.
5 Where applicable, remove the rear ABS wheel sensor as described in Chapter 9.
6 Release any pipes, hoses, and/or wiring from the clips on the lower arm.
7 Counterhold the through-bolt and unscrew the nut securing the radius rod to the lower arm **(see illustration)**. Recover the washer(s) and withdraw the through-bolt.
8 Counterhold the pivot bolt, and unscrew the nut securing the inboard end of the lower arm. Recover the washers and shims, noting their locations to ensure correct refitting (the shims are used to set the wheel alignment), then withdraw the pivot bolt.
9 Withdraw the lower arm, complete with the hub carrier.
10 If desired, the hub carrier can be unbolted

from the lower arm. Again, note the location of the any washers and/or spacers.

Inspection

11 Examine the inner pivot bush for wear or deterioration, and if necessary renew the bush as follows.
12 Use a length of tubing with a bolt, nut and washers to draw out the old bush and fit the new one.

Refitting

13 Where applicable, refit the hub carrier to the lower arm, but do not fully tighten the nut and bolt at this stage.
14 Offer the lower arm into position, and refit the pivot bolt (noting that the pivot bolt fits from the rear of the vehicle), fitting the washers and shims in their original locations. If a new lower arm is being fitted, fit two shims either side of the arm. Note that the shims are used to set the rear wheel alignment, and it may be necessary to adjust their position on completion of refitting, once the wheel alignment has been checked.
15 Fit the pivot bolt nut, but do not fully tighten at this stage.
16 Reconnect the radius rod to the lower arm, then refit the through-bolt and nut (noting that the bolt fits from the roadwheel side of the radius rod), ensuring that the washers are positioned as noted before removal. Do not fully tighten the through-bolt and nut at this stage.
17 Refit the coil spring and shock absorber as described in Sections 12 and 11 respectively.
18 Reconnect the brake pressure regulating valve operating rod, as described in Chapter 9.
19 Reconnect the handbrake cable as described in Chapter 9, and check the cable adjustment as described in Chapter 1.
20 Where applicable, refit the rear ABS wheel sensor as described in Chapter 9.
21 Clip any wires, hoses and/or pipes into position in the clips on the lower arm.
22 Reconnect the brake hose to the caliper or wheel cylinder, as applicable, then remove the clamp from the hose, and bleed the brake hydraulic system as described in Chapter 9.
23 On completion, refit the roadwheel and

lower the vehicle to the ground, then tighten the lower arm pivot nut and bolt, and the radius rod through-bolt and nut to the specified torque. Where applicable, also tighten the hub carrier-to-lower arm nut and bolt.
24 Check the rear wheel alignment at the earliest opportunity, with reference to Section 27, and have the adjustment of the brake pressure regulating valve checked (see Chapter 9).

15 Rear suspension radius rod - removal, inspection and refitting

Note: *New securing nuts must be used on refitting.*

Removal

1 Chock the front wheels, engage reverse gear (or "P" on models with automatic transmission) and release the handbrake. Jack up the rear of the vehicle and support it securely on axle stands (see *"Jacking and vehicle support"*). Remove the relevant roadwheel.
2 Unbolt the relevant fuel tank support member, and withdraw it for access to the radius rod front mounting bolt and nut.
3 Counterhold the bolt, and unscrew the nut securing the rear of the radius rod to the lower arm. Withdraw the bolt and recover the washers.
4 Repeat the procedure and unbolt the front end of the radius rod from the body bracket, then manipulate the radius rod from under the vehicle **(see illustration)**.

Inspection

5 Examine the mounting bushes for wear or deterioration, and if necessary renew as follows.
6 Use a length of tubing with a bolt, nut and washers to draw out the old bush and fit the new one.

Refitting

7 Refitting is a reversal of removal, bearing in mind the following points.
a) *Lightly grease the radius rod securing bolts before fitting.*

Steering and suspension 10•11

b) Note that the rear radius rod securing bolt fits from the roadwheel side of the rod, and the front bolt fits from the fuel tank side.
c) Fit new nuts to the radius rod bolts.
d) Do not fully tighten the radius rod mounting nuts and bolts until the vehicle is resting on its wheels.

16 Electronic variable-rate suspension - general information

Certain high-specification models are fitted with an electronically-controlled variable-rate suspension.
The system comprises the following components.
a) Four special shock absorbers with integral solenoid valves.
b) Vehicle speed sensor.
c) Longitudinal, transverse and vertical accelerometers.
d) A control switch assembly, with "Sport" button.
e) An electronic control unit (ECU).
The system also receives a signal from the stop light switch.
The system provides three levels of ride quality, by varying the damping effect of the shock absorbers. This is achieved by using the solenoid valves to control the fluid flow in the shock absorbers. "Comfort", "Medium" and "Sport" settings are provided, and these settings are selected automatically by the ECU according to the signals received from the various sensors. The "Sport" setting can be selected manually using a switch mounted on the centre console.
If a fault in the system is suspected, the problem should be referred to a Renault dealer who will have access to the appropriate specialist diagnostic equipment.

17 Electronic variable rate suspension components - removal and refitting

Front shock absorbers

1 This procedure is considered to be beyond the scope of the DIY mechanic due to the requirement for Renault special tools to carry out the procedure safely. Refer to Section 4 for further information.

Warning: Do not attempt to remove the shock absorber and coil spring using conventional spring compressor tools, or improvised tools. Any attempt to remove the coil spring and shock absorber without Renault special tools and the knowledge to use them safely could result in damage and/or serious personal injury.

Rear shock absorbers

2 The procedure is identical to that described in Section 11 for conventional shock absorbers, but the solenoid wiring connector must be disconnected (disconnect the battery negative lead first), and the wiring must be released from any clips and brackets under the body. The wiring connector is located under the under the rear right-hand side of the vehicle (see illustration). Note that the left-hand shock absorber wiring connector has an orange colour code marking, and the right-hand wiring connector has a black colour code mark.

Electronic control unit (ECU)

Removal

3 The ECU is mounted along with the accelerometers on a plate behind the glovebox.
4 Disconnect the battery negative lead.
5 Remove the glovebox as described in Chapter 11 to reveal the ECU/accelerometer mounting bracket.
6 Disconnect the 9-pin multi-plug from the assembly.
7 Disconnect the accelerometer wiring connectors and free the wiring from the clips. Note the routing of all wiring to ensure correct refitting.
8 Unscrew the three securing nuts, and withdraw the mounting plate assembly from the facia. Note the locations of any spacers on the mountings to ensure correct refitting.
9 Disconnect the ECU multi-plug, then withdraw the ECU from the mounting plate.

Refitting

10 Refitting is a reversal of removal, bearing in mind the following points.
a) Ensure that the ECU multi-plug is securely reconnected.
b) Ensure that any spacers are positioned as noted before removal.
c) Ensure that the wiring is routed as noted before removal.
d) Do not overtighten the mounting nuts, which may cause the mounting plate to distort.

Accelerometers

Removal

11 The accelerometers are mounted together, along with the ECU on a plate behind the glovebox.
12 Remove the mounting plate as described previously in paragraphs 5 to 9.
13 Unbolt the relevant accelerometer, and unclip the wiring, noting its routing to ensure correct refitting.

Refitting

14 Refitting is a reversal of removal, but ensure that the wiring is correctly routed, and refit the mounting plate with reference to paragraph 11.

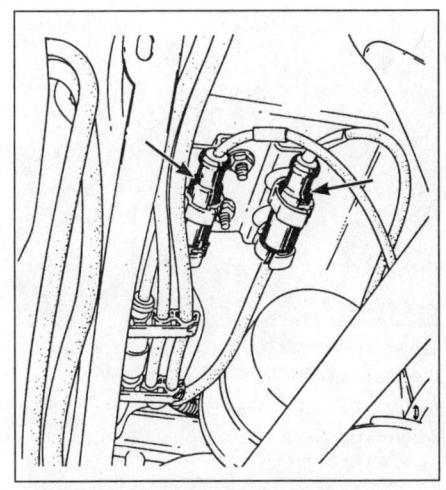

17.2 Electronic rear shock absorber wiring connectors (arrowed)

Vehicle speed sensor

15 On models with a cable-driven speedometer, the vehicle speed sensor is positioned in the speedometer cable run between the transmission and the speedometer. On models with an electronically-operated speedometer, the sensor is integral with the speedometer sensor in the transmission. No information was available for the sensor at the time of writing.

18 Steering wheel - removal and refitting

Removal

1 Set the front wheels in the straight-ahead position.
2 Pull the cover from the centre of the steering wheel (see illustration).
3 Make alignment marks between the end of the steering column shaft, the steering wheel, and the steering wheel hub.
4 Using circlip pliers, extract the circlip from the top of the steering column shaft (see illustration).
5 Remove the two screws from the steering

18.2 Removing the steering wheel cover

10•12 Steering and suspension

18.4 Extracting the circlip from the top of the steering column shaft

18.5 Steering column hub screws screwed into holes ready to force off steering wheel

18.13 Fine adjustment of the steering wheel can be made by sliding the wheel within the elongated holes (arrowed)

column hub, then screw them into the two alternative holes provided, and tighten them progressively to force the hub from the top of the column shaft **(see illustration)**.
6 Withdraw the hub and the steering wheel.
7 Where applicable, disconnect the steering wheel switch wiring plugs.

Refitting

8 Refit the steering wheel and the hub to the steering column shaft, aligning the marks made before removal.
9 Where applicable, reconnect the switch wiring plugs.
10 Hold the steering wheel in position, then refit the two screws to their original locations, and tighten them to draw the hub into position on the shaft.
11 Refit the circlip to the end of the shaft.
12 Refit the cover to the centre of the steering wheel.

13 Road test the vehicle to check the steering wheel position. If necessary, fine adjustment of the steering wheel position can be made by slackening the two screws securing the hub, and sliding the steering wheel within the elongated bolt holes **(see illustration)**.

19 Ignition switch/ steering column lock - removal and refitting

Removal

1 Disconnect the battery negative lead.
2 Remove the steering column shrouds as described in Chapter 11.
3 Remove the securing screws, and withdraw the driver's side lower facia panel **(see illustration)**.
4 To improve access, on models with an adjustable steering column, fully lower the column. If necessary, loosen the four steering column securing nuts, and lower the column.
5 Insert the ignition key, and turn the key to the arrow position (between "A" and "M").
6 Trace the wiring back from the switch and disconnect the wiring plugs **(see illustration)**. Note the routing of the wiring to aid refitting.
7 Unscrew the ignition switch securing screw from the top of the lock barrel, then depress the switch securing lug, and withdraw the assembly using the key. Feed the wiring through the switch housing as the switch is withdrawn **(see illustrations)**.
8 If desired, the switch can be separated from the lock by unscrewing the two securing screws **(see illustration)**.

19.3 Removing the driver's side lower facia panel

19.6 Ignition switch wiring connectors (arrowed)

19.7a Unscrew the ignition switch securing screw (arrowed) . . .

19.7b . . . then depress the switch securing lug (arrowed) . . .

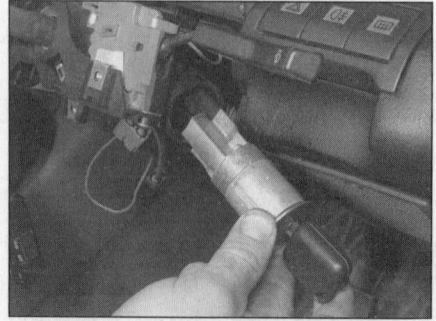

19.7c . . . and withdraw the switch assembly

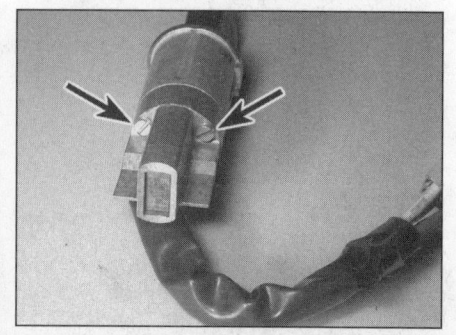

19.8 Ignition switch securing screws (arrowed)

Steering and suspension 10•13

Refitting

9 Refitting is a reversal of removal, bearing in mind the following points.
 a) If the switch has been separated from the lock, ensure that the switch wiper engages correctly with the lock on refitting.
 b) Ensure that the switch wiring is correctly routed, and clipped in position as noted before removal.

20 Steering column and intermediate shaft - removal and refitting

Note: *All "Nyloc"-type self-locking nuts should be renewed.*

Steering column

Removal

1 Disconnect the battery negative lead.
2 Remove the steering wheel as described in Section 18.
3 Remove the steering column shrouds as described in Chapter 11.
4 Remove the steering column stalk switches as described in Chapter 12.
5 Remove the driver's side lower facia panel.
6 Working under the steering column, disconnect the ignition switch wiring plugs.
7 Working at the bottom of the steering column shaft, make alignment marks between the end of the steering column shaft and the intermediate shaft.
8 Unscrew the universal joint pinch-bolt and nut, and slide the steering column shaft from the intermediate shaft universal joint **(see illustration)**.
9 Working under the facia, unscrew the four securing nuts, then lower the steering column, and withdraw it **(see illustration)**.

Refitting

10 Refitting is a reversal of removal, bearing in mind the following points.
 a) Ensure that the marks made on the column shaft and the intermediate shaft before removal are aligned.
 c) All "Nyloc"-type self-locking nuts should be renewed.
 d) Tighten the steering column nuts securely.
 e) Refit the steering wheel as described in Section 18.

Intermediate shaft - models with manual steering

Removal

11 Working in the engine compartment, prise out the plastic clips which secure the intermediate shaft rubber gaiter to the bulkhead. Push the gaiter into the vehicle interior, and remove the split plastic bush **(see illustration)**.
12 Unscrew the pinch-bolt and nut securing

20.8 Steering column shaft universal joint pinch-bolt (arrowed)

the intermediate shaft to the steering gear pinion.
13 Working inside the vehicle, remove the driver's side lower facia panel.
14 Working at the bottom of the steering column shaft, unscrew the universal joint pinch-bolt and nut, and slide the steering column shaft from the intermediate shaft universal joint.
15 Withdraw the intermediate shaft through the bulkhead into the passenger compartment.

Inspection

16 Measure the length of the intermediate shaft, and check that it is within the specified limits **(see illustrations)**. The shaft is collapsible (to prevent injury to the driver in the event of an accident), and must be renewed if the length is not as specified.

Refitting

17 Refitting is a reversal of removal, bearing in mind the following points.
 a) All "Nyloc"-type self-locking nuts should be renewed.
 b) Tighten the pinch-bolts and nuts securely.
 c) If the universal joint pivot bolts and nuts have been loosened or removed for any reason, note that they must be tightened when the joint body is aligned with the relevant shaft **(see illustration)**. Turn the shafts as necessary to achieve the correct alignment before tightening the pinch-bolts and nuts.

20.9 Steering column securing nut (arrowed)

 d) Ensure that the split plastic bush is correctly refitted to the bottom of the shaft.

Intermediate shaft - models with power steering

18 The procedure is as described previously for models with manual steering, but both intermediate shaft universal joint pinch-bolts can be reached from inside the passenger compartment.

21 Steering gear gaiters - renewal

1 Remove the track rod end balljoint as described in Section 25.
2 Mark the fitted position of the gaiter on the track rod. Release the retaining clips, and slide the gaiter off the steering gear housing and track rod end.
3 Thoroughly clean the track rod and the steering gear housing, using fine abrasive paper to polish off any corrosion, burrs or sharp edges which might damage the sealing lips of the new gaiter on installation.
4 Recover the grease from inside the old gaiter. If it is uncontaminated with dirt or grit, apply it to the track rod inner balljoint. If the old grease is contaminated, or it is suspected that some has been lost, apply some new molybdenum disulphide grease.

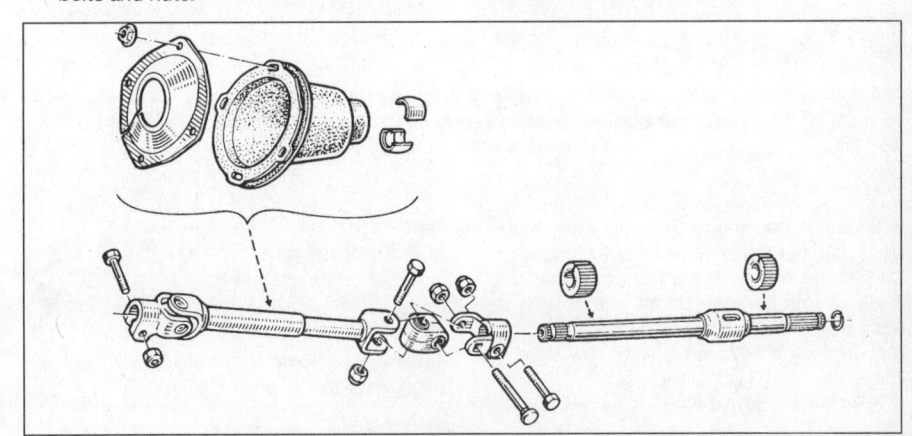

20.11 Steering intermediate shaft components

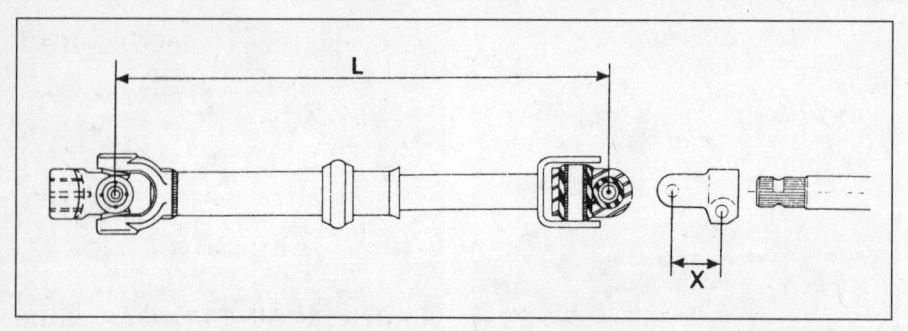

10•14 Steering and suspension

20.16a Steering intermediate shaft measurement - models with manual steering

Where X = 30.0 mm
L = 306.0 ± 1.0 mm

Where X = 33.0 mm
L = 303.0 ± 1.0 mm

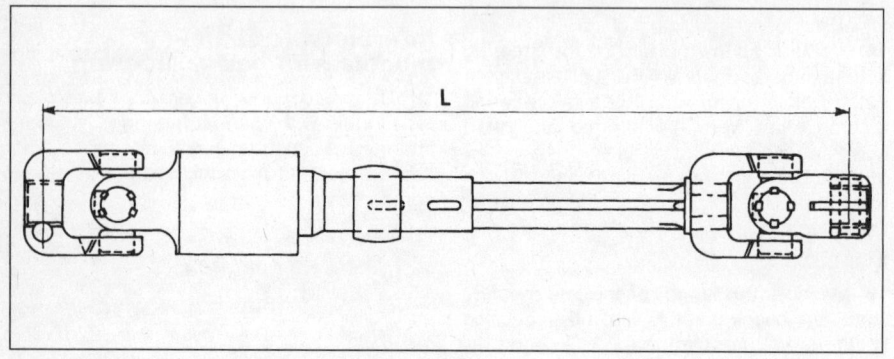

20.16b Steering intermediate shaft measurement - models with power steering

L = 341.0 ± 1.0 mm

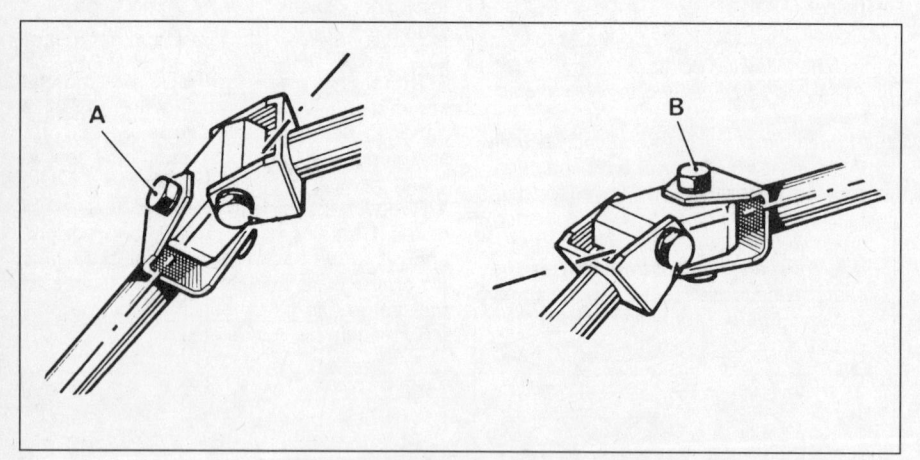

20.17 Steering intermediate shaft/steering column universal joint alignment for tightening of pivot bolts (A and B)

5 Grease the inside of the new gaiter. Carefully slide the gaiter onto the track rod, and locate it on the steering gear housing. Align the outer edge of the gaiter with the mark made on the track rod prior to removal, then secure it in position with new retaining clips.
6 Refit the track rod balljoint as described in Section 25.

22 Steering gear - removal, overhaul and refitting

Note: *A balljoint separator tool will be required for this operation.*

Removal

1 Apply the handbrake, then jack up the front of the vehicle and support securely on axle stands (see *"Jacking and vehicle support"*). Remove the roadwheels.
2 Working on each side of the vehicle in turn, slacken and partially unscrew the track-rod end balljoint nut (unscrew the nut as far as the end of the threads on the balljoint to prevent damage to the threads as the joint is released), then release the balljoint using a balljoint separator tool. Remove the nut.
3 Working in the engine compartment, on models with the V6 Turbo engine, unbolt the turbocharger heat shield, then unscrew the nuts securing the steering intermediate shaft heat shield.
4 Move the steering to full left-hand lock.
5 Make alignment marks on the steering intermediate shaft lower universal joint and the steering gear pinion.
6 Unscrew the lower universal joint clamp bolt, and disconnect the intermediate shaft from the steering gear pinion.
7 On models with power steering gear, clamp the fluid pipes leading from the fluid reservoir. Place a suitable container under the fluid pipe connections on the steering gear, then unscrew the unions, and disconnect the fluid pipes. Collect the escaping fluid.
8 Where applicable, on models with ABS, to allow sufficient clearance to remove the steering gear, unbolt the ABS hydraulic unit mounting bracket, and release the hydraulic modulator from its mountings **without** disconnecting the fluid pipes.
9 Unscrew the four bolts securing the steering gear to the body crossmember.
10 Move the steering gear fully to the right then, where applicable, remove the intermediate shaft heat shield, and withdraw the steering gear, complete with the track-rods through the right-hand wheel arch **(see illustration)**. On models with ABS, lift the hydraulic unit and the mounting bracket as necessary as the steering gear is withdrawn.

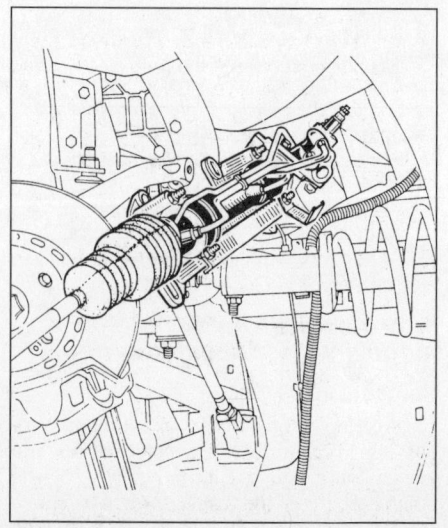

22.10 Removing the steering gear through the right-hand wheel arch

Steering and suspension 10•15

Overhaul

General

11 Renewal procedures for the gaiters, track-rod ends and track-rods are given in Sections 21, 25 and 26 respectively.

12 Examine the steering gear assembly for signs of wear or damage. Check that the rack moves freely over the full length of its travel, with no signs of roughness or excessive free-play between the steering gear pinion and rack. Internal wear or damage can only be cured by renewing the steering gear assembly, but note the points in the following paragraphs.

Thrust plunger adjustment

13 If there is excessive free-play of the rack in the steering gear housing, accompanied by a knocking noise, it may be possible to correct this by adjusting the rack thrust plunger.

14 Relieve the staking on the plunger adjusting nut (see illustration).

15 Using a 10 mm Allen key, tighten the adjusting nut (by no more than three turns) until the free-play disappears. Check that the rack still moves freely over its full travel, then secure the adjusting nut by staking it.

Refitting

16 Refitting is a reversal of removal, bearing in mind the following points.
a) Before refitting, ensure that the rack is centred in the housing.- this can be done by moving the rack to full lock in one direction, then counting the number of turns of the pinion as the rack is turned to the full opposite lock. Turn the rack back so that the pinion rotates half the number of turns counted from lock-to-lock to centre the rack.
b) Align the marks made on the intermediate shaft lower joint and the steering gear pinion before removal.
c) Tighten all fixings to the specified torque.
d) On models with ABS, ensure that the hydraulic modulator and mounting bracket are correctly secured.
e) On models with power steering, bleed the power steering hydraulic circuit as described in Section 24.
f) If the track-rods have been renewed, or if new steering gear has been fitted, check the front wheel alignment as described in Section 27.

23 Power steering pump - removal and refitting

Removal

1 The power steering pump on four-cylinder and most six-cylinder models is driven from the crankshaft pulley. On certain six-cylinder non-turbo models, the power steering pump is driven from the rear of the left-hand camshaft.

2 Remove the power steering pump drivebelt as described in Chapter 1.

3 Using brake hose clamps, clamp both the supply and return hoses near the power steering fluid reservoir. This will minimise fluid loss during subsequent operations.

4 Where applicable, release the fluid hoses/pipes from any clips or brackets to enable the hoses/pipes to be moved to one side, clear of the pump.

5 Slacken the retaining clips, and disconnect the fluid supply and feed hoses from the pump. If necessary, to disconnect the feed pipe, unscrew the union nut and recover the O-ring. Be prepared for some fluid spillage as the hoses/pipes are disconnected, and plug the hose/pipe and pump openings to minimise fluid loss and to prevent the entry of dirt into the system.

6 Unscrew the pump mounting bolts, and the adjuster bolts where applicable, and withdraw the pump from its mounting bracket (see illustrations).

7 The pump is a sealed unit, and if faulty, it must be renewed.

Refitting

8 Refitting is a reversal of removal, bearing in mind the following points.
a) Where applicable, use a new O-ring when reconnecting the feed pipe union.
b) Refit the pump drivebelt, and tension it as described in Chapter 1.
c) On completion, remove the hose clamps, and bleed the power steering hydraulic system as described in Section 24.

24 Power steering system - bleeding

1 This procedure will only be necessary when any part of the hydraulic system has been disconnected, or if air has entered because of leakage.

2 Remove the fluid reservoir filler cap, and top-up the fluid level to the maximum mark, using only the specified fluid. Refer to Chapter 1 "Lubricants and fluids" for fluid specifications, and again to Chapter 1 for details of the different types of fluid reservoir markings.

3 With the engine stopped, slowly move the steering from lock-to-lock several times to expel trapped air, then top-up the level in the fluid reservoir. Repeat this procedure until the fluid level in the reservoir does not drop any further.

4 Start the engine. Slowly move the steering from lock-to-lock several times to expel any air remaining in the system. Repeat this procedure until bubbles cease to appear in the fluid reservoir.

5 If, when turning the steering, an abnormal noise is heard from the fluid pipes, it indicates that there is still air in the system. Check this by turning the wheels to the straight-ahead position and switching off the engine. If the fluid level in the reservoir rises, air is still present in the system, and further bleeding is necessary.

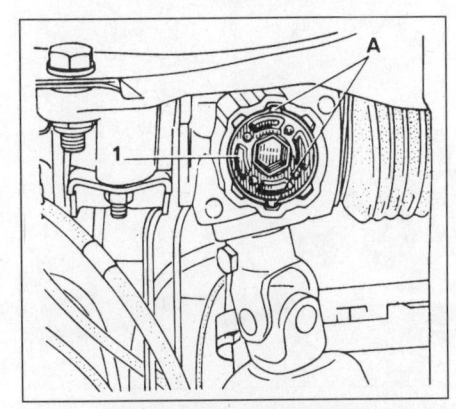

22.14 Steering gear plunger adjusting hut (1) and staking (A)

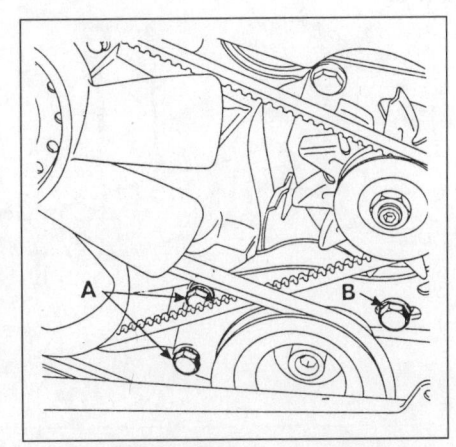

23.6a Power steering pump bracket mounting bolts (A) and adjuster bolt (B) - crankshaft pulley-driven pump

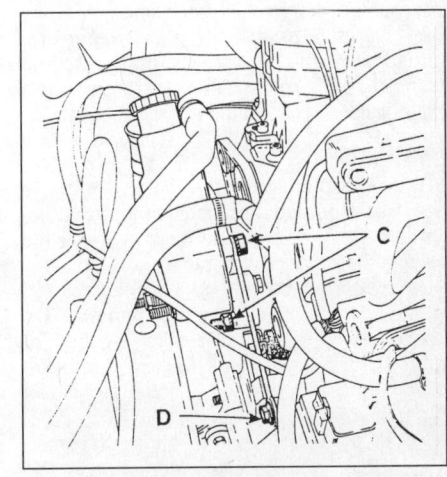

23.6b Power steering pump bracket mounting bolts (C) and adjuster bolt (D) - camshaft pulley-driven pump

10•16 Steering and suspension

6 Once all traces of air have been removed, stop the engine and allow the system to cool. Once cool, check that the fluid level is up to the maximum mark on the power steering fluid reservoir; top-up if necessary.

25 Track-rod end balljoint - removal and refitting

Note: *A balljoint separator tool will be required for this operation. A new balljoint nut will be required on refitting.*

Removal

1 Apply the handbrake, then jack up the front of the vehicle and support it securely on axle stands (see *"Jacking and vehicle support"*). Remove the appropriate front roadwheel.
2 If the track-rod end is to be re-used, use a straight-edge and a scriber, or similar, to mark its relationship to the track rod.
3 Holding the track-rod end, unscrew its locknut by one quarter of a turn **(see illustration)**. Do not move the locknut from this position, as it will serve as a reference mark on refitting.
4 Remove the nut securing the track-rod end to the hub carrier. Release the balljoint tapered shank using a universal balljoint separator. If the track-rod end is to be re-used, protect the threaded end of the shank by screwing the nut back on a few turns before using the separator **(see illustration)**.
5 Counting the **exact** number of turns necessary to do so, unscrew the track-rod end from the from the track rod.
6 Count the number of exposed threads between the end of the track-rod end and the locknut, and record this figure. If a new track-rod end is to be fitted, unscrew the locknut from the old track-rod end.
7 Carefully clean the balljoint and the threads. Renew the track-rod end if the balljoint movement is sloppy or if it is too stiff, if it is excessively worn, or if it is damaged in any way. Carefully check the shank taper and threads. If the gaiter is damaged, the complete track-rod end must be renewed; it is not possible to obtain the gaiter separately.

Refitting

8 If applicable, screw the locknut onto the new track-rod end, and position it so that the same number of exposed threads are visible as were noted prior to removal.
9 Screw the track-rod end into the track rod by the number of turns noted on removal. This should bring the locknut to within a quarter of a turn of the end of the track rod, with the alignment marks that were made on removal (if applicable) lined up.
10 Refit the balljoint shank to the hub carrier, and tighten the new retaining nut to the specified torque. If difficulty is experienced due to the balljoint shank rotating, jam it by exerting pressure on the top of the balljoint.

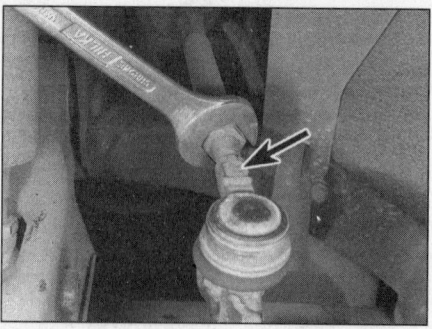

25.3 Hold the track-rod end, and unscrew the locknut (arrowed)

11 Refit the roadwheel, lower the vehicle to the ground and tighten the roadwheel bolts to the specified torque.
12 Check the front wheel alignment as described in Section 27, then tighten the track-rod end locknut.

26 Track-rod and inner balljoint - removal, inspection and refitting

Note: *A new lockwasher must be used on refitting.*

Removal

1 Remove the track-rod end balljoint as described in Section 25.
2 Cut the retaining clips, and slide the steering gear gaiter off the track-rod.
3 Using a suitable pair of grips, unscrew the track-rod inner balljoint from the steering rack end. Prevent the steering rack from turning by holding the balljoint lockwasher with a second pair of grips. Take care not to mark the surfaces of the rack and balljoint.
4 Remove the track-rod/inner balljoint assembly and discard the lockwasher; a new one must be used on refitting.

25.4 Release the track-rod end balljoint using a balljoint separator tool

Inspection

5 Examine the inner balljoint for signs of slackness or tight spots. Check that the track-rod itself is straight and free from damage. If necessary, renew the track-rod/inner balljoint; the new one will be supplied complete with a new lockwasher and a new end balljoint. It is also recommended that the steering gear gaiter is renewed.

Refitting

6 If a new track-rod is being installed, remove the outer balljoint from the track-rod end.
7 Locate the new lockwasher assembly on the end of the steering rack, ensuring that its locating tabs are correctly located with the flats on the rack end **(see illustration)**.
8 Apply a few drops of locking fluid to the inner balljoint threads. Screw the balljoint into the steering rack and tighten it securely. Again, take care not to damage or mark the balljoint or steering rack.
9 Slide the new gaiter onto the end of the track-rod, and locate it on the steering gear housing. Turn the steering from lock-to-lock to check that the gaiter is correctly positioned, then secure it with new retaining clips.
10 Refit the track-rod end balljoint as described in Section 25.

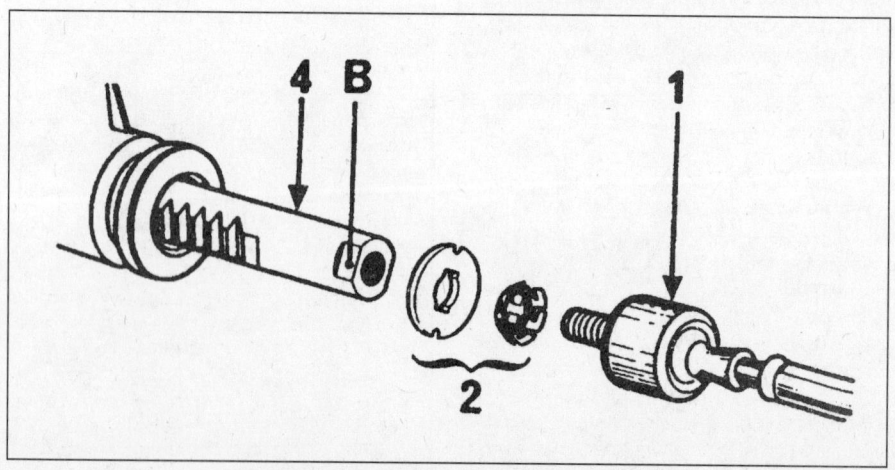

26.7 Track-rod inner balljoint components

1 Inner balljoint
2 Lockwasher assembly
4 Steering rack
B Lockwasher locating flats

27 Wheel alignment and steering angles - general information

General information

1 A vehicle's steering and suspension geometry is defined in four basic settings - all angles are expressed in degrees (toe settings are also expressed as a measurement). The steering axis is defined as an imaginary line drawn through the axis of the shock absorber, extended where necessary to contact the ground.

2 **Camber** is the angle between each roadwheel and a vertical line drawn through its centre and tyre contact patch, when viewed from the front or rear of the car. "Positive" camber is when the roadwheels are tilted outwards from the vertical at the top; "negative" camber is when they are tilted inwards.

3 Camber is not adjustable. Values are given for reference only. Checking is possible using a camber checking gauge, but if the figure obtained is significantly different from that specified, the vehicle must be taken for careful checking by a professional. Wrong camber settings can only be caused by wear or damage to the body or suspension components.

4 **Castor** is the angle between the steering axis and a vertical line drawn through each roadwheel's centre and tyre contact patch, when viewed from the side of the car. "Positive" castor is when the steering axis is tilted so that it contacts the ground ahead of the vertical; "negative" castor is when it contacts the ground behind the vertical.

5 Castor is adjustable by altering the length of the front suspension radius rods - see Section 8.

6 **Steering axis inclination/SAI** - also known as **kingpin inclination/KPI** - is the angle between the steering axis and a vertical line drawn through each roadwheel's centre and tyre contact patch, when viewed from the front or rear of the car.

7 SAI/KPI is not adjustable, and is given for reference only.

8 **Toe** is the difference, viewed from above, between lines drawn through the roadwheel centres and the car's centreline. "Toe-in" is when the roadwheels point inwards, towards each other at the front, while "toe-out" is when they splay outwards from each other at the front.

9 The front wheel toe setting is adjusted by screwing the track-rod ends in or out of their track rods to alter the effective length of the track rod assemblies.

10 Rear wheel toe setting is adjusted by adjusting the shims on the rear suspension lower arm pivot bolts - see Section 14.

Checking and adjustment
General

11 Due to the special measuring equipment necessary to check the wheel alignment, and the skill required to use it properly, the checking and adjustment of these settings is best left to a Renault dealer or similar expert. Most tyre-fitting specialists now possess sophisticated checking equipment.

12 For accurate checking, the vehicle must be at the kerb weight specified in *"Dimensions and weights"* in the *Reference Section* at the rear of this manual.

13 Before starting work, check first that the tyre sizes and types are as specified, then check tyre pressures and tread wear. Also check roadwheel run-out, the condition of the hub bearings, the steering wheel free play and the condition of the front suspension components (Chapter 1). Correct any faults found.

14 Park the vehicle on level ground, with the front roadwheels in the straight-ahead position. Rock the rear and front ends to settle the suspension. Release the handbrake and roll the vehicle backwards approximately 1 metre (3 feet), then forwards again, to relieve any stresses in the steering and suspension components.

Front wheel toe setting

15 Two methods are available to the home mechanic for checking the front wheel toe setting. One method is to use a gauge to measure the distance between the front and rear inside edges of the roadwheels. The other method is to use a scuff plate, in which each front wheel is rolled across a movable plate which records any deviation, or scuff, of the tyre from the straight-ahead position as it moves across the plate. Such gauges are available in relatively inexpensive form from accessory outlets. It is up to the owner to decide whether the expense is justified, in view of the small amount of use such equipment would normally receive.

16 Prepare the vehicle as described in paragraphs 12 to 14 above.

17 If the measurement procedure is being used, carefully measure the distance between the front edges of the roadwheel rims and the rear edges of the rims. Subtract the rear measurement from the front measurement, and check that the result is within the specified range. If not, adjust the toe setting as described in paragraph 19.

18 If scuff plates are to be used, roll the vehicle backwards, check that the roadwheels are in the straight-ahead position, then roll it across the scuff plates so that each front roadwheel passes squarely over the centre of its respective plate. Note the angle recorded by the scuff plates. To ensure accuracy, repeat the check three times, and take the average of the three readings. If the roadwheels are running parallel, there will of course be no angle recorded; if a deviation value is shown on the scuff plates, compare the reading obtained for each wheel with that specified. If the value recorded is outside the specified tolerance, the toe setting is incorrect, and must be adjusted as follows.

19 Apply the handbrake, then jack up the front of the vehicle and support it securely on axle stands (see *"Jacking and vehicle support"*). Turn the steering wheel onto full-left lock, and record the number of exposed threads on the right-hand track-rod end. Now turn the steering onto full-right lock, and record the number of threads on the left-hand side. If there are the same number of threads visible on both sides, then subsequent adjustment should be made equally on both sides. If there are more threads visible on one side than the other, it will be necessary to compensate for this during adjustment. **Note:** *It is important to ensure that, after adjustment, the same number of threads are visible on each track rod end.*

20 First clean the track-rod threads; if they are corroded, apply penetrating fluid before starting adjustment. Release the rubber gaiter outboard clips, then peel back the gaiters and apply a smear of grease, so that both gaiters are free and will not be twisted or strained as their respective track-rods are rotated.

21 Use a straight-edge and a scriber or similar to mark the relationship of each track-rod to its track-rod end. Holding each track-rod in turn, unscrew its locknut fully.

22 Alter the length of the track-rods, bearing in mind the note in paragraph 19, by screwing them into or out of the track-rod ends. Rotate the track-rod using an open-ended spanner fitted to the flats provided. Shortening the track-rods (screwing them onto their track-rod ends) will reduce toe-in and increase toe-out. Note that one turn of the track-rods will give an adjustment of 30' (3 mm).

23 When the setting is correct, hold the track-rods and securely tighten the track-rod end locknuts. Count the exposed threads on the end of each track-rod. If the number of threads exposed is not the same on both sides, then the adjustment has not been made equally, and problems will be encountered with tyre scrubbing in turns; also, the steering wheel spokes will no longer be centralised when the wheels are in the straight-ahead position.

24 When the track-rod lengths are the same, lower the vehicle to the ground and re-check the toe setting; readjust if necessary. When the setting is correct, tighten the track-rod end locknuts. Ensure that the rubber gaiters are seated correctly and are not twisted or strained, then secure them in position with new retaining clips.

Front wheel castor angle

25 To check the castor angle, a suitable wheel alignment gauge will be required - it is not possible to improvise an accurate alternative. Checking of the castor angle should therefore be entrusted to a Renault

10•18 Steering and suspension

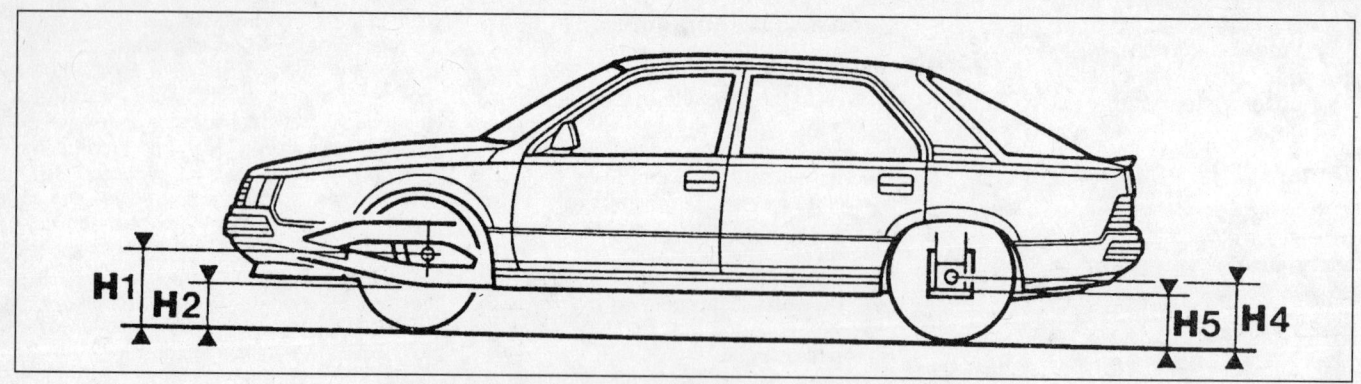

27.25 Vehicle suspension setting heights - see *Specifications*.

dealer or wheel alignment specialist, who will have access to the necessary equipment and data. Note that the castor angle measured will depend on the vehicle suspension setting (ride height) **(see illustration)**.

26 Adjustment of the front wheel castor angle is made by altering the length of the front suspension radius rods - refer to Section 8 for details.

Rear wheel toe setting

27 The procedure for checking the rear wheel toe setting is the same as that described previously for checking the front wheel toe-setting.

28 Adjustment is made by adjusting the shims on the rear suspension lower arm pivot bolts - see Section 14.

Chapter 11
Bodywork and fittings

Contents

Body exterior fittings - removal and refitting	22
Bonnet - removal, refitting and adjustment	8
Bonnet lock - removal and refitting	10
Bonnet release cable - removal and refitting	9
Bumpers - removal and refitting	6
Central locking system components - removal and refitting	17
Centre console - removal and refitting	26
Door - removal, refitting and adjustment	11
Door handles and lock components - removal and refitting	13
Door inner trim panel - removal and refitting	12
Door window glass and regulator - removal and refitting	14
Electric window components - removal and refitting	18
Exterior mirrors and associated components - removal and refitting	19
Facia panel assembly - removal and refitting	27
General information	1
Interior trim - removal and refitting	25
Maintenance - bodywork and underframe	2
Maintenance - upholstery and carpets	3
Major body damage - repair	5
Minor body damage - repair	4
Radiator grille panel - removal and refitting	7
Seats - removal and refitting	23
Seat belt components - removal and refitting	24
Sunroof - general information	21
Tailgate and support struts - removal, refitting and adjustment	15
Tailgate lock components - removal and refitting	16
Windscreen and tailgate glass - general information	20

Degrees of difficulty

Easy, suitable for novice with little experience	Fairly easy, suitable for beginner with some experience	Fairly difficult, suitable for competent DIY mechanic	Difficult, suitable for experienced DIY mechanic	Very difficult, suitable for expert DIY or professional

1 General information

The bodyshell is of five-door Hatchback configuration, and is made of pressed steel sections. Most components are welded together, but some use is made of structural adhesives. The front wings are bolted on.

The bonnet, doors and some other vulnerable panels are made of zinc-coated metal, and are further protected by being coated with an anti-chip primer prior to being sprayed.

Extensive use is made of plastic materials, mainly in the interior, but also in exterior components. The front and rear bumpers and the front grille are injection-moulded from a synthetic material which is very strong, and yet light. Plastic components such as wheel arch liners are fitted to the underside of the vehicle, to improve the body's resistance to corrosion.

2 Maintenance - bodywork and underframe

The general condition of a vehicle's bodywork is the one thing that significantly affects its value. Maintenance is easy, but needs to be regular. Neglect, particularly after minor damage, can lead quickly to further deterioration and costly repair bills. It is important also to keep watch on those parts of the vehicle not immediately visible, for instance the underside, inside all the wheel arches, and the lower part of the engine compartment.

The basic maintenance routine for the bodywork is washing - preferably with a lot of water, from a hose. This will remove all the loose solids which may have stuck to the vehicle. It is important to flush these off in such a way as to prevent grit from scratching the finish. The wheel arches and underframe need washing in the same way, to remove any accumulated mud which will retain moisture and tend to encourage rust. Oddly enough, the best time to clean the underframe and wheel arches is in wet weather, when the mud is thoroughly wet and soft. In very wet weather, the underframe is usually cleaned of large accumulations automatically, and this is a good time for inspection.

Periodically, except on vehicles with a wax-based underbody protective coating, it is a good idea to have the whole of the underframe of the vehicle steam-cleaned, engine compartment included, so that a thorough inspection can be carried out to see what minor repairs and renovations are necessary. Steam-cleaning is available at many garages, and is necessary for the removal of the accumulation of oily grime, which sometimes is allowed to become thick in certain areas. If steam-cleaning facilities are not available, there are one or two excellent grease solvents available, which can be brush-applied; the dirt can then be simply hosed off. Note that these methods should not be used on vehicles with wax-based underbody protective coating, or the coating will be removed. Such vehicles should be inspected annually, preferably just prior to Winter, when the underbody should be washed down, and any damage to the wax coating repaired. Ideally, a completely fresh coat should be applied. It would also be worth considering the use of such wax-based protection for injection into door panels, sills, box sections, etc, as an additional safeguard against rust damage, where such protection is not provided by the vehicle manufacturer.

After washing paintwork, wipe off with a chamois leather to give an unspotted clear finish. A coat of clear protective wax polish will give added protection against chemical pollutants in the air. If the paintwork sheen has dulled or oxidised, use a cleaner/polisher combination to restore the brilliance of the shine. This requires a little effort, but such dulling is usually caused because regular washing has been neglected. Care needs to be taken with metallic paintwork, as special non-abrasive cleaner/polisher is required to avoid damage to the finish. Always check that the door and ventilator opening drain holes and pipes are completely clear, so that water can be drained out. Brightwork should be treated in the same way as paintwork. Windscreens and windows can be kept clear of the smeary film which often appears, by the use of proprietary glass cleaner. Never use any form of wax or other body or chromium polish on glass.

3 Maintenance - upholstery and carpets

Mats and carpets should be brushed or vacuum-cleaned regularly, to keep them free of grit. If they are badly stained, remove them from the vehicle for scrubbing or sponging, and make quite sure they are dry before refitting. Seats and interior trim panels can be kept clean by wiping with a damp cloth. If they do become stained (which can be more apparent on light-coloured upholstery), use a little liquid detergent and a soft nail brush to scour the grime out of the grain of the material. Do not forget to keep the headlining clean in the same way as the upholstery. When using liquid cleaners inside the vehicle, do not over-wet the surfaces being cleaned. Excessive damp could get into the seams and padded interior, causing stains, offensive odours or even rot. If the inside of the vehicle gets wet accidentally, it is worthwhile taking some trouble to dry it out properly, particularly where carpets are involved. *Do not leave oil or electric heaters inside the vehicle for this purpose.*

4 Minor body damage - repair

Repairs of minor scratches in bodywork

If the scratch is very superficial, and does not penetrate to the metal of the bodywork, repair is very simple. Lightly rub the area of the scratch with a paintwork renovator, or a very fine cutting paste, to remove loose paint from the scratch, and to clear the surrounding bodywork of wax polish. Rinse the area with clean water.

Apply touch-up paint to the scratch using a fine paint brush; continue to apply fine layers of paint until the surface of the paint in the scratch is level with the surrounding paintwork. Allow the new paint at least two weeks to harden, then blend it into the surrounding paintwork by rubbing the scratch area with a paintwork renovator or a very fine cutting paste. Finally, apply wax polish.

Where the scratch has penetrated right through to the metal of the bodywork, causing the metal to rust, a different repair technique is required. Remove any loose rust from the bottom of the scratch with a penknife, then apply rust-inhibiting paint, to prevent the formation of rust in the future. Using a rubber or nylon applicator, fill the scratch with bodystopper paste. If required, this paste can be mixed with cellulose thinners, to provide a very thin paste which is ideal for filling narrow scratches. Before the stopper-paste in the scratch hardens, wrap a piece of smooth cotton rag around the top of a finger. Dip the finger in cellulose thinners, and quickly sweep it across the surface of the stopper-paste in the scratch; this will ensure that the surface of the stopper-paste is slightly hollowed. The scratch can now be painted over as described earlier in this Section.

Repairs of dents in bodywork

When deep denting of the vehicle's bodywork has taken place, the first task is to pull the dent out, until the affected bodywork almost attains its original shape. There is little point in trying to restore the original shape completely, as the metal in the damaged area will have stretched on impact, and cannot be reshaped fully to its original contour. It is better to bring the level of the dent up to a point which is about 3 mm below the level of the surrounding bodywork. In cases where the dent is very shallow anyway, it is not worth trying to pull it out at all. If the underside of the dent is accessible, it can be hammered out gently from behind, using a mallet with a wooden or plastic head. Whilst doing this, hold a suitable block of wood firmly against the outside of the panel, to absorb the impact from the hammer blows and thus prevent a large area of the bodywork from being "belled-out".

Should the dent be in a section of the bodywork which has a double skin, or some other factor making it inaccessible from behind, a different technique is called for. Drill several small holes through the metal inside the area - particularly in the deeper section. Then screw long self-tapping screws into the holes, just sufficiently for them to gain a good purchase in the metal. Now the dent can be pulled out by pulling on the protruding heads of the screws with a pair of pliers.

The next stage of the repair is the removal of the paint from the damaged area, and from an inch or so of the surrounding "sound" bodywork. This is accomplished most easily by using a wire brush or abrasive pad on a power drill, although it can be done just as effectively by hand, using sheets of abrasive paper. To complete the preparation for filling, score the surface of the bare metal with a screwdriver or the tang of a file, or alternatively, drill small holes in the affected area. This will provide a really good "key" for the filler paste.

To complete the repair, see the Section on filling and respraying.

Repairs of rust holes or gashes in bodywork

Remove all paint from the affected area, and from an inch or so of the surrounding "sound" bodywork, using an abrasive pad or a wire brush on a power drill. If these are not available, a few sheets of abrasive paper will do the job most effectively. With the paint removed, you will be able to judge the severity of the corrosion, and therefore decide whether to renew the whole panel (if this is possible) or to repair the affected area. New body panels are not as expensive as most people think, and it is often quicker and more satisfactory to fit a new panel than to attempt to repair large areas of corrosion.

Remove all fittings from the affected area, except those which will act as a guide to the original shape of the damaged bodywork (eg headlamp shells etc). Then, using tin snips or a hacksaw blade, remove all loose metal and any other metal badly affected by corrosion. Hammer the edges of the hole inwards, in order to create a slight depression for the filler paste.

Wire-brush the affected area to remove the powdery rust from the surface of the remaining metal. Paint the affected area with rust-inhibiting paint; if the back of the rusted area is accessible, treat this also.

Before filling can take place, it will be necessary to block the hole in some way. This can be achieved by the use of glass-fibre matting, aluminium or plastic mesh, or aluminium tape.

Aluminium or plastic mesh, or glass-fibre matting is probably the best material to use for a large hole. Cut a piece to the approximate size and shape of the hole to be filled, then position it in the hole so that its edges are below the level of the surrounding bodywork. It can be retained in position by several blobs of filler paste around its periphery.

Aluminium tape should be used for small or very narrow holes. Pull a piece off the roll, trim it to the approximate size and shape required, then pull off the backing paper (if used) and stick the tape over the hole; it can be overlapped if the thickness of one piece is insufficient. Burnish down the edges of the tape with the handle of a screwdriver or similar, to ensure that the tape is securely attached to the metal underneath.

Bodywork repairs - filling and respraying

Before using this Section, see the Sections on dent, deep scratch, rust holes and gash repairs.

Many types of bodyfiller are available, but generally speaking, those proprietary kits which contain a tin of filler paste and a tube of resin hardener are best for this type of repair. A wide, flexible plastic or nylon applicator will be found invaluable for imparting a smooth and well-contoured finish to the surface of the filler.

Mix up a little filler on a clean piece of card or board - measure the hardener carefully (follow the maker's instructions on the pack), otherwise the filler will set too rapidly or too slowly. Using the applicator, apply the filler paste to the prepared area; draw the applicator across the surface of the filler to achieve the correct contour and to level the surface. As soon as a contour that approximates to the correct one is achieved, stop working the paste - if you carry on too long, the paste will become sticky and begin to "pick-up" on the applicator. Continue to add thin layers of filler paste at 20-minute intervals, until the level of the filler is just proud of the surrounding bodywork.

Once the filler has hardened, the excess can be removed using a metal plane or file. From then on, progressively-finer grades of abrasive paper should be used, starting with a 40-grade production paper, and finishing with a 400-grade wet-and-dry paper. Always wrap the abrasive paper around a flat rubber, cork, or wooden block - otherwise the surface of the filler will not be completely flat. During the smoothing of the filler surface, the wet-and-dry paper should be periodically rinsed in water. This will ensure that a very smooth finish is imparted to the filler at the final stage.

At this stage, the "dent" should be surrounded by a ring of bare metal, which in turn should be encircled by the finely "feathered" edge of the good paintwork. Rinse the repair area with clean water, until all of the dust produced by the rubbing-down operation has gone.

Spray the whole area with a light coat of primer - this will show up any imperfections in the surface of the filler. Repair these imperfections with fresh filler paste or bodystopper, and once more smooth the surface with abrasive paper. If bodystopper is used, it can be mixed with cellulose thinners, to form a really thin paste which is ideal for filling small holes. Repeat this spray-and-repair procedure until you are satisfied that the surface of the filler, and the feathered edge of the paintwork, are perfect. Clean the repair area with clean water, and allow to dry fully.

The repair area is now ready for final spraying. Paint spraying must be carried out in a warm, dry, windless and dust-free atmosphere. This condition can be created artificially if you have access to a large indoor working area, but if you are forced to work in the open, you will have to pick your day very carefully. If you are working indoors, dousing the floor in the work area with water will help to settle the dust which would otherwise be in the atmosphere. If the repair area is confined to one body panel, mask off the surrounding panels; this will help to minimise the effects of a slight mis-match in paint colours. Bodywork fittings (eg chrome strips, door handles etc) will also need to be masked off. Use genuine masking tape, and several thicknesses of newspaper, for the masking operations.

Before commencing to spray, agitate the aerosol can thoroughly, then spray a test area (an old tin, or similar) until the technique is mastered. Cover the repair area with a thick coat of primer; the thickness should be built up using several thin layers of paint, rather than one thick one. Using 400 grade wet-and-dry paper, rub down the surface of the primer until it is really smooth. While doing this, the work area should be thoroughly doused with water, and the wet-and-dry paper periodically rinsed in water. Allow to dry before spraying on more paint.

Spray on the top coat, again building up the thickness by using several thin layers of paint. Start spraying in the centre of the repair area, and then, using a circular motion, work outwards until the whole repair area and about 2 inches of the surrounding original paintwork is covered. Remove all masking material 10 to 15 minutes after spraying on the final coat of paint.

Allow the new paint at least two weeks to harden, then, using a paintwork renovator or a very fine cutting paste, blend the edges of the paint into the existing paintwork. Finally, apply wax polish.

Plastic components

With the use of more and more plastic body components by the vehicle manufacturers (eg bumpers, spoilers, and in some cases major body panels), rectification of more serious damage to such items has become a matter of either entrusting repair work to a specialist in this field, or renewing complete components. Repair of such damage by the DIY owner is not really feasible, owing to the cost of the equipment and materials required for effecting such repairs. The basic technique involves making a groove along the line of the crack in the plastic, using a rotary burr in a power drill. The damaged part is then welded back together, using a hot air gun to heat up and fuse a plastic filler rod into the groove. Any excess plastic is then removed, and the area rubbed down to a smooth finish. It is important that a filler rod of the correct plastic is used, as body components can be made of a variety of different types (eg polycarbonate, ABS, polypropylene).

Damage of a less serious nature (abrasions, minor cracks etc) can be repaired by the DIY owner using a two-part epoxy filler repair. Once mixed in equal, this is used in similar fashion to the bodywork filler used on metal panels. The filler is usually cured in twenty to thirty minutes, ready for sanding and painting.

If the owner is renewing a complete component himself, or if he has repaired it with epoxy filler, he will be left with the problem of finding a suitable paint for finishing which is compatible with the type of plastic used. At one time, the use of a universal paint was not possible, owing to the complex range of plastics encountered in body component applications. Standard paints, generally speaking, will not bond to plastic or rubber satisfactorily, but suitable paints to match any plastic or rubber finish, can be obtained from dealers. However, it is now possible to obtain a plastic body parts finishing kit which consists of a pre-primer treatment, a primer and coloured top coat. Full instructions are normally supplied with a kit, but basically, the method of use is to first apply the pre-primer to the component concerned, and allow it to dry for up to 30 minutes. Then the primer is applied, and left to dry for about an hour before finally applying the special-coloured top coat. The result is a correctly-coloured component, where the paint will flex with the plastic or rubber, a property that standard paint does not normally posses.

5 Major body damage - repair

Where serious damage has occurred, or large areas need renewal due to neglect, it means that complete new panels will need welding-in, and this is best left to professionals. If the damage is due to impact, it will also be necessary to check completely the alignment of the bodyshell, and this can only be carried out accurately by a Renault dealer using special jigs. If the body is left misaligned, it is primarily dangerous, as the car will not handle properly, and secondly, uneven stresses will be imposed on the steering, suspension and possibly transmission, causing abnormal wear, or complete failure, particularly to such items as the tyres.

11•4 Bodywork and fittings

6 Bumpers - removal and refitting

Front bumper - "Phase I" models

Removal

1 An integral front bumper/spoiler assembly is fitted.
2 If desired, to improve access, apply the handbrake, then jack up the front of the vehicle and support securely on axle stands (see *"Jacking and vehicle support"*).
3 On models with lights mounted in the bumper, disconnect the battery negative lead.
4 Working at the lower rear edges of the bumper, remove the two screws on each side securing the lower edges of the bumper assembly (**see illustration**).
5 Working at the lower front edge of the bumper, prise out the cover plate on each side, and unscrew the two lower securing screws (**see illustration**).
6 The bumper side fixing bolts must now be unscrewed (**see illustration**). On most models, these bolts are accessible from underneath the vehicle, but on some models it may be necessary to remove surrounding components as follows.

   a) Where applicable, on 4-cylinder engine models, unclip the electronic control unit from the left-hand wing panel, and move it to one side (without disconnecting the wiring), then unscrew the three bolts and remove the control unit mounting bracket.
   b) Where applicable, on 6-cylinder engine models, unscrew the three bolts securing the ignition module from the left-hand wing panel, and move the assembly to one side, taking car not to strain the wiring (see Chapter 5).
   c) On Turbo-diesel models, remove the securing screw and release the intercooler from its bracket on the right-hand wing panel (see Chapter 4), and move it to one side, then unscrew the two upper mounting bracket bolts, and move the bracket to one side for access to the bumper bolts.

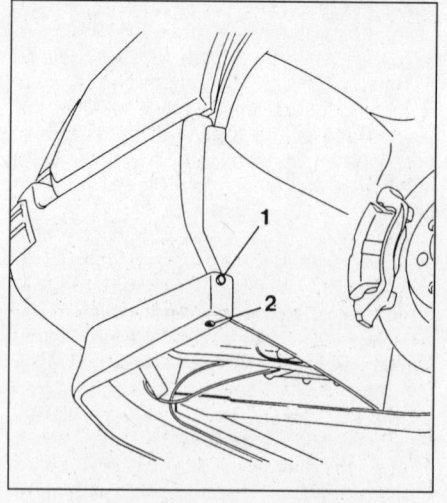

6.4 Front bumper rear securing screws (1 and 2) - "Phase I" models

7 Pull the bumper assembly forwards then, where applicable, trace the wiring back from the light units, and disconnect the light wiring connectors.
8 Withdraw the bumper assembly from the vehicle.

Refitting

9 Refitting is a reversal of removal.

Front bumper - "Phase II" models

10 The procedure is as described previously for the bumper on "Phase I" models, noting the following differences (**see illustrations**).
   a) There is only one screw on each side securing the lower rear edges of the bumper/spoiler assembly to the wheel arch liners.
   b) The two lower front bumper securing screws fit from underneath the assembly, and screw into brackets on the front body panel. There are no screw covers.

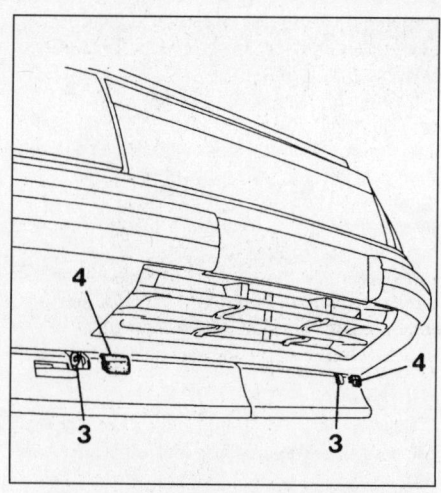

Rear bumper

Removal

11 Disconnect the battery negative lead.
12 Open the tailgate, and remove the plastic

6.10a Front bumper lower securing screw (arrowed) - "Phase II" model

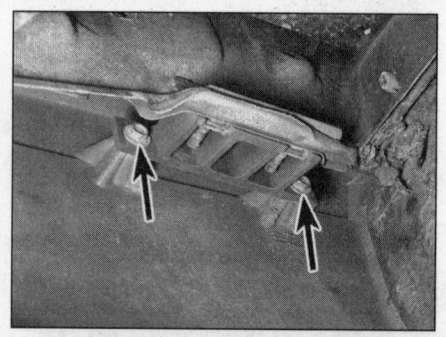

6.10b Front bumper side fixing bolts (arrowed) - "Phase II" model

6.5 Front bumper cover plates (4) and lower securing screws (3) - "Phase I" models

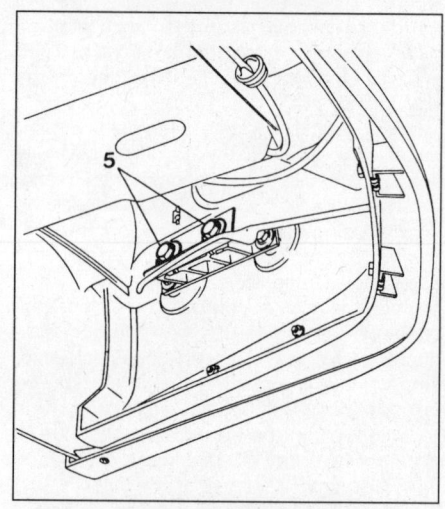

6.6 Front bumper side fixing bolts (5) - "Phase I" models

trim panels from the rear corners of the luggage compartment to expose the bumper side fixing bolts (**see illustrations**).
13 Unscrew the two bumper side fixing bolts from each side of the luggage compartment.
14 Working at the lower rear edge of the bumper, unscrew the two rear bumper securing screws (**see illustration**). The screws may be accessible from the rear of the bumper, or from underneath the bumper, depending on model.
15 Where applicable, unscrew the single screw on each side securing the front lower corners of the bumper.
16 Withdraw the bumper from the rear of the vehicle, and disconnect the number plate light wiring connectors.

Refitting

17 Refitting is a reversal of removal.

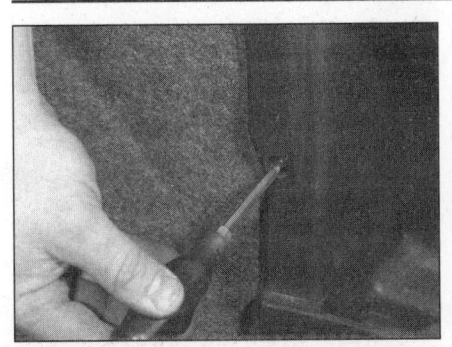

Bodywork and fittings 11•5

6.12a Remove the securing screws...

6.12b ...and withdraw the plastic trim panels from the luggage compartment...

6.12c ...to expose the bumper side fixing bolts (arrowed)

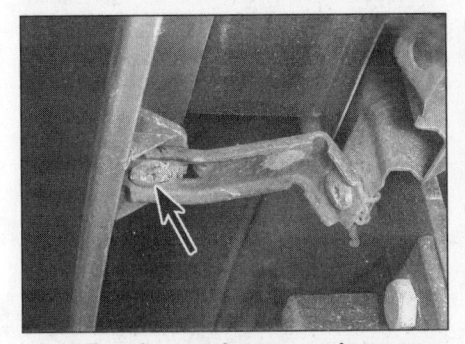

6.14 Rear bumper lower securing screw (arrowed)

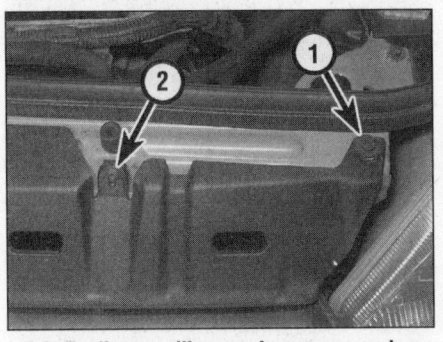

7.3 Radiator grille panel upper securing bolt (1) and screw (2)

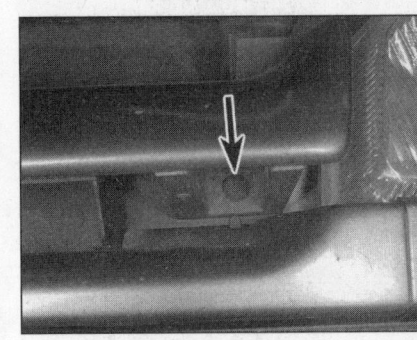

7.4 Radiator grille panel lower securing screw (arrowed)

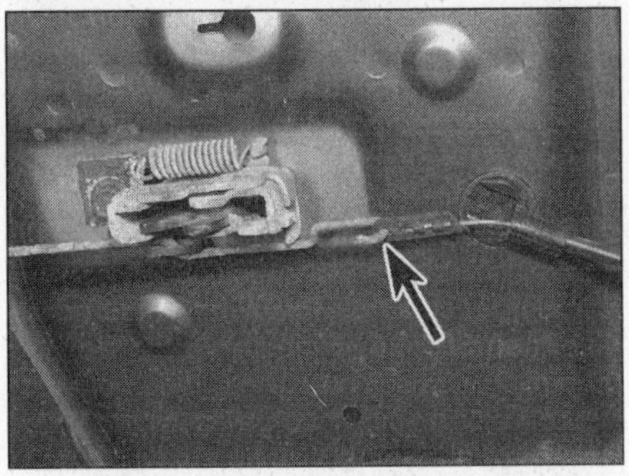

8.3 Unhook the bonnet release cable from the lock operating rod (arrowed)

7 Radiator grille panel - removal and refitting

Removal

1 Open the bonnet.
2 Unscrew the two upper securing bolts from each top corner of the panel.
3 Unscrew the two upper securing screws (see illustration).
4 Working through the holes in the lower edge of the grille panel, unscrew the two lower securing screws, then withdraw the panel (see illustration).

Refitting

5 Refitting is a reversal of removal.

8 Bonnet - removal, refitting and adjustment

Removal

1 Open the bonnet and, using a pencil or felt-tip pen, mark the outline of each bonnet hinge relative to the bonnet, to use as a guide on refitting.
2 Fold back the sound insulation on the driver's side of the bonnet for access to the bonnet release cable.
3 Unhook the end of the bonnet release cable from the lock operating rod, then feed the cable down through the openings in the bonnet (see illustration). Note the cable routing to ensure correct refitting.
4 Where applicable, disconnect the battery negative lead, then disconnect the "bonnet closed" switch wiring connector. Where applicable, feed the wiring through the bonnet. Note the routing of the wiring to aid refitting.
5 Have an assistant support the bonnet, then unscrew the four bolts securing the bonnet to the hinges (note that there is no need to disconnect the bonnet support struts), and lift the bonnet from the vehicle.

Refitting

6 Refitting is a reversal of removal, but route the bonnet release cable, and the wiring (where applicable) as noted before removal.
7 On completion, check the alignment of the bonnet and the operation of the lock mechanism, and adjust if necessary as described in the following paragraphs.

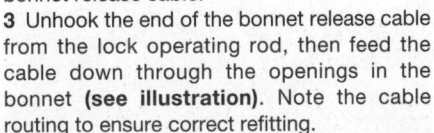

> **HAYNES HINT**: If the original bonnet is to be refitted, tie a length of string to the end of the bonnet release cable. Feed the cable through the bonnet, then untie the string. Leave the string in position to aid refitting of the cable.

Adjustment

8 Close the bonnet and check for alignment with the surrounding body panels. If necessary, slacken the hinge bolts and re-align the bonnet to suit. Once the bonnet is correctly aligned, tighten the hinge bolts.
9 The height of the edges of the bonnet can be adjusted using the rubber buffers on the body panels. To adjust the height, screw the rubbers in or out of the body panel as required.
10 Once the bonnet is correctly aligned, check that the bonnet lock fastens and releases in a satisfactory manner. If adjustment is required, slacken the bonnet lock striker plate securing nuts, and move the strikers within the elongated holes in the body panel. If necessary, the bonnet closure height can be adjusted by means of the shims located under the lock striker plates on the body (see illustration). Tighten the striker plate nuts on completion of adjustment.

11•6 Bodywork and fittings

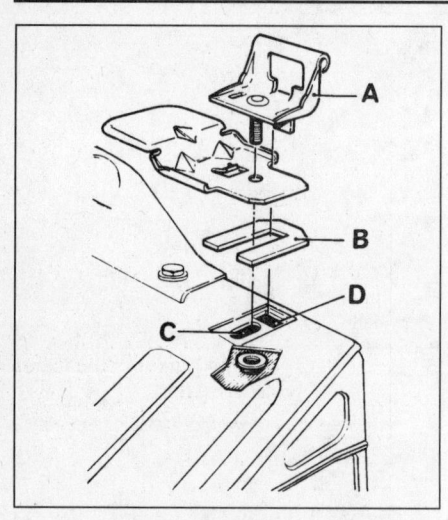

8.10 Bonnet lock striker details

A Striker plate
B Height adjustment shim
C Elongated hole
D Elongated hole

9 Bonnet release cable - removal and refitting

Removal

1 Working under the bonnet, unhook the end of the bonnet release cable from the lock operating rod, then feed the cable down through the openings in the bonnet. Note the cable routing to ensure correct refitting.

> **HAYNES HiNT** If the bonnet release cable has broken, the bonnet can be opened as follows.
> a) Apply the handbrake, then jack up the front of the vehicle and support on axle stands (see "Jacking and vehicle support").
> b) Using a long extension and a suitable socket, reach up and locate the nuts securing the lock striker plates to the front body panel. Unscrew the nuts.
> c) The bonnet can now be lifted sufficiently to release the safety catch, allowing the bonnet to be opened.

2 Working under the driver's side of the facia, unscrew the two bolts securing the bonnet release lever to the side of the footwell.
3 Pull the release lever from under the facia, and draw the cable through the grommet in the bulkhead into the passenger compartment.

Refitting

4 Refitting is a reversal of removal, but where applicable use the string to pull the cable into position, and ensure that the cable is routed as noted before removal.

> **HAYNES HiNT** Tie a length of string to the end of the bonnet release cable. Feed the cable through the bonnet, then untie the string. Leave the string in position to aid refitting of the cable.

10 Bonnet lock - removal and refitting

Removal

1 Open the bonnet, and pull back the sound insulation from the front edge of the bonnet to expose the bonnet lock operating rod (see illustration).
2 Unhook the end of the bonnet release cable from the driver's side end of the lock operating rod.
3 Unhook the spring from the opposite end of the lock operating rod.
4 Release the lock operating rod from the clip at the centre of the bonnet.
5 Unscrew the two bolts securing the relevant lock, then release the operating rod from the lock and withdraw the lock.
6 If desired, the lock strikers can be removed from the body front panel by unscrewing the securing nuts. Similarly, the bonnet safety catch can be unbolted from the bonnet.

Refitting

7 Refitting is a reversal of removal, but on completion, check the operation of the bonnet lock mechanism, and adjust the position of the lock strikers if necessary, with reference to Section 8.

11 Door - removal, refitting and adjustment

Note: *The door hinges are bolted to the body, and welded to the door. The door hinge pins should not be disturbed unless the hinge assembly is to be renewed, or the hinge pin is seriously worn or damaged. The hinge pins are a press-fit, and will be destroyed, along with their pivot bushes, if removed. Although it may be possible to drive out the hinge pins with a drift, this is extremely difficult due to the tight fit of the pins, and the limited access - it is recommended that hinge pin renewal is entrusted to a Renault dealer or an automotive engineer with access to a suitable hydraulic hinge pin removal tool.*

Front door

Removal

Note: *New pop-rivets may be required to secure the door hinge cover panel on refitting.*
1 Disconnect the battery negative lead.
2 Remove the door inner trim panel and the sealing sheet, as described in Section 12.
3 Working through the apertures in the door, disconnect the wiring connectors, and release the wiring harness(es) from the clips on the door. Note the routing of the harness(es) to aid refitting.
4 Pull the wiring grommet from the front edge of the door, then feed the wiring harness(es) through the aperture in the front edge of the door.
5 Support the lower edge of the door on blocks of wood or a jack, with pads of rag between the blocks/jack and the door to protect the paintwork.
6 Remove the two bolts securing the door check strap to the body pillar (see illustration).
7 Apply the handbrake, then jack up the front of the vehicle and support on axle stands (see "Jacking and vehicle support"). Remove the appropriate roadwheel.
8 Remove the wheel arch liner with reference to Section 22.
9 Working at the rear of the wheel arch, remove the cover panel to expose the door hinges (see illustration). On some models, the cover panel may be secured by rivets, in which case the rivets must be drilled out to remove the panel.
10 Have an assistant support the door then,

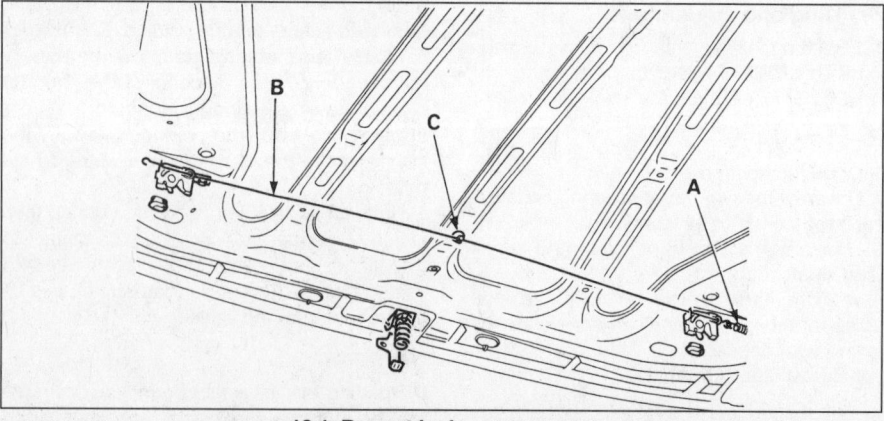

10.1 Bonnet lock components

A Spring   B Lock operating rod   C Clip

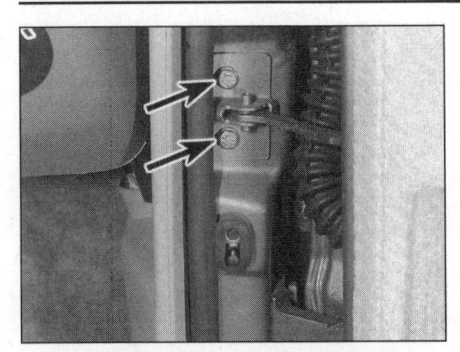

Bodywork and fittings 11•7

11.6 Door check strap-to-body pillar securing bolts (arrowed)

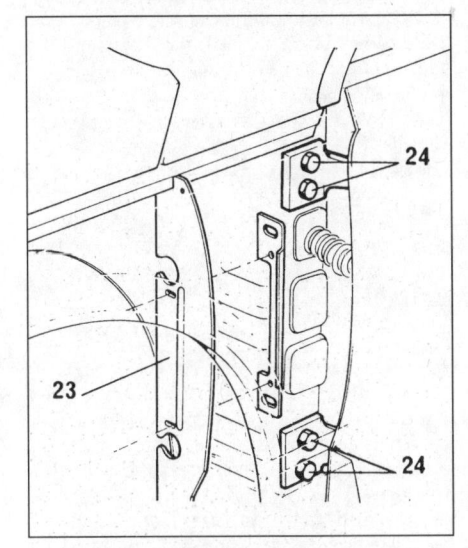

11.9 Front door hinge arrangement

23 Wheel arch cover panel
24 Door hinge bolts

using a spanner, reach up through the wheel arch and unscrew the bolts securing the hinges to the body.
11 With the aid of the assistant, lift the door from the vehicle.

Refitting and adjustment
12 Refitting is a reversal of removal, bearing in mind the following points.
a) Do not fully tighten the hinge bolts until the door alignment has been checked. Slight adjustment can be carried out by moving the door within the limits of the hinge bolt holes.
b) Where applicable, use new pop-rivets to secure the door hinge cover panel.
c) Route the wiring harness(es) as noted before removal
d) Refit the sealing sheet and the door inner trim panel with reference to Section 12.
e) On completion, check the operation of the door lock. If necessary, loosen the lock striker, and adjust its position on the body pillar to achieve satisfactory lock operation.

Rear door
Removal
13 On models with electrical components mounted in the rear door, proceed as described in paragraphs 1 to 4.
14 Support the lower edge of the door on blocks of wood or a jack, with pads of rag between the blocks/jack and the door to protect the paintwork.
15 Remove the two bolts securing the door check strap to the body pillar.
16 Open the front door to expose the hinge bolts on the body pillar.
17 Have an assistant support the door, then unscrew the hinge bolts and lift the door from the vehicle.

Refitting and adjustment
18 Proceed as described in paragraph 12.

12 Door inner trim panel - removal and refitting

Front door
Removal
1 Open the door, then fully lower the window glass.
2 Disconnect the battery negative lead.
3 Working at the top rear corner of the door, use a small screwdriver to depress the door lock button tab, then pull the knob from the end of the lock operating rod **(see illustration)**.
4 Where applicable remove the securing screw, then unclip the mirror trim panel from the front corner of the door **(see illustration)**.
5 Grip the loudspeaker cover panel, and turn it through a quarter-turn anti-clockwise to release it from the door trim panel **(see illustration)**.
6 Working at the front of the armrest assembly, unscrew the bolt securing the switch/door handle housing.
7 Pull the armrest hinged cover upwards, and tilt it against the door to expose the upper armrest fixing screws.
8 Carefully prise the switch/door handle housing from the front of the armrest **(see illustration)**.
9 On models with manually-adjustable door mirrors, proceed as follows **(see illustrations)**.
a) Working at the rear of the switch/door handle housing, loosen the mirror adjustment switch mounting screw.
b) Push the mirror adjustment switch downwards out of the housing.

10 Disconnect the wiring plugs from the rear of the switches in the housing, and remove the switch/door handle housing assembly.
11 Remove the three upper and two lower armrest securing screws, then remove the armrest securing nut, which is accessible through the switch housing at the front of the armrest **(see illustrations)**. Withdraw the armrest assembly.
12 Using a suitable forked tool, work around the edge of the door trim panel, and release the panel securing clips.
13 Pull the panel up to release it from the

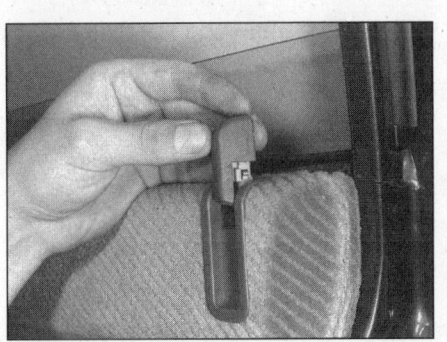

12.3 Removing the door lock button

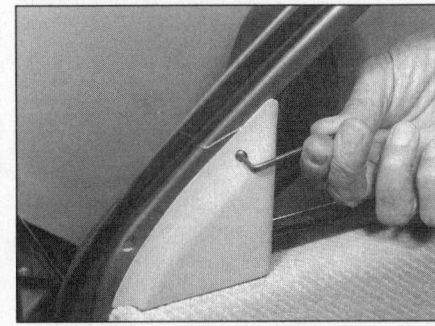

12.4 Removing the mirror trim plate securing screw

12.5 Removing the loudspeaker cover panel

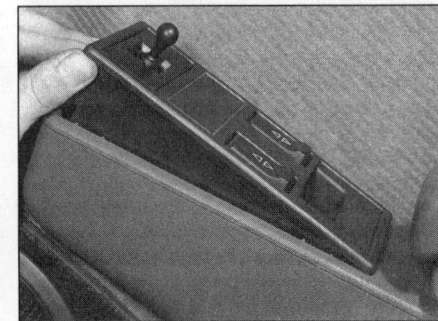

12.8 Prise the switch/door handle housing from the armrest

11•8 Bodywork and fittings

12.9a Loosen the mirror adjustment switch mounting screw . . .

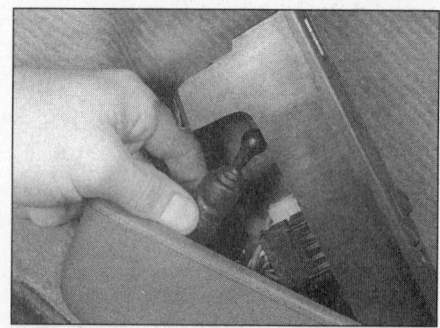

12.9b . . . and push the mirror from the housing

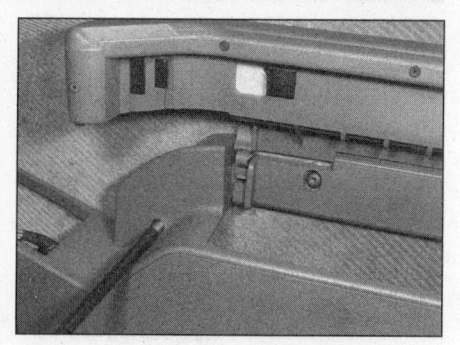

12.11a Remove the armrest securing screws . . .

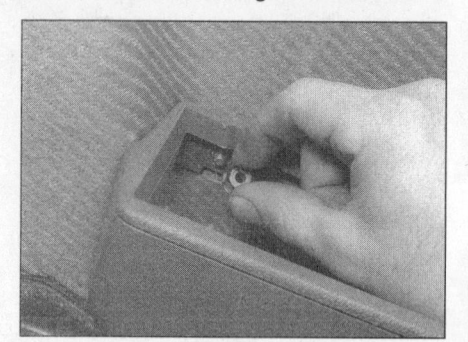

12.11b . . . and the securing nut

12.14 Peeling back the door plastic sealing sheet

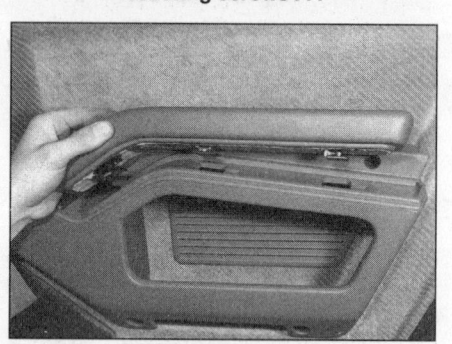

12.16 Removing the rear door armrest top trim panel

weatherstrip at the top of the door, then feed the wiring through the apertures in the panel and withdraw the panel.

14 If desired, the plastic sealing sheet can be peeled back from the door after removing the loudspeaker, and the electronic control unit and associated wiring clips, where applicable **(see illustration)**. If care is taken, the sealing sheet can be re-used.

Refitting

15 Refitting is a reversal of removal but, where applicable, bearing in mind the following points.
 a) Where applicable, make sure that the plastic sealing sheet is securely refitted. If necessary, use new sealant, and if the sheet was broken during removal, use a new sheet.
 b) Before refitting, check whether any of the trim panel securing clips were broken on removal, and renew them as necessary.

Rear door
Removal

16 Open the door, and tap the armrest top trim panel forwards to release the securing clips, and withdraw the trim panel **(see illustration)**.
17 Where applicable, prise the ashtray from its housing.
18 Remove the securing screws (located underneath the door handle and/or in the ashtray housing), and lift the door handle/ashtray housing from the armrest **(see illustration)**.
19 On models with manual rear windows, remove the window regulator handle. If necessary, use a forked tool similar to that shown, together with a piece of cloth to protect the trim panel.
20 Remove the five armrest securing screws, two at the top, two at the bottom, and one at

the top of the cut-out in the armrest **(see illustrations)**.
21 Unscrew the armrest securing nut, accessible through the door handle/ashtray aperture.
22 Working at the top front corner of the door, use a small screwdriver to depress the door lock button tab, then pull the knob from the end of the lock operating rod.
23 Using a suitable forked tool, work around the edge of the door trim panel, and release the panel securing clips.
24 Pull the panel up to release it from the weatherstrip at the top of the door, and withdraw the panel.
25 If desired, the plastic sealing sheet can be peeled back from the door. If care is taken, the sealing sheet can be re-used.

Refitting

26 Refer to paragraph 15.

12.18 Rear door handle/ashtray housing securing screw (arrowed)

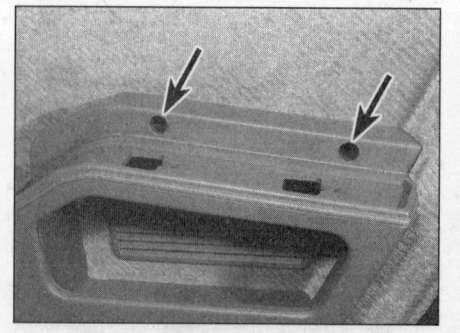

12.20a Remove the armrest upper (arrowed) . . .

12.20b . . . and lower securing screws

Bodywork and fittings 11•9

13 Door handles and lock components - removal and refitting

Front door interior handle

Removal

1 The interior door handle is integral with the switch/door handle housing in the door armrest. The handle operates a bellcrank, bolted to the door, which is connected to the lock operating rod.
2 To remove the handle, disconnect the battery negative lead, then prise the switch/door handle housing from the front of the armrest.
3 On models with manually-adjustable door mirrors, proceed as follows.
  a) Working at the rear of the switch/door handle housing, loosen the mirror adjustment switch mounting screw.
  b) Push the mirror adjustment switch downwards out of the housing.
4 Disconnect the wiring plugs from the rear of the switches in the housing, and remove the switch/door handle housing assembly.
5 If desired, the bellcrank assembly can be removed as follows.
  a) Remove the door inner trim panel and the plastic sealing sheet, as described in Section 12.
  b) Unscrew the bolt securing the bellcrank assembly to the door.
  c) Manipulate the assembly through the door aperture and disconnect the lock operating rod.

Refitting

6 Refitting is a reversal of removal, but where applicable, refit the door inner trim panel with reference to Section 12.

Rear door interior handle

Removal

7 The interior handle is integral with the door handle/ashtray housing in the door armrest. The handle operates a bellcrank, bolted to the door, which is connected to the lock operating rod.
8 Where applicable, prise the ashtray from its housing.
9 Remove the securing screws (located underneath the door handle and/or in the ashtray housing), and lift the door handle/ashtray housing from the armrest.
10 If desired, the bellcrank assembly can be removed with reference to paragraph 5.

Refitting

11 Refitting is a reversal of removal, but where applicable, refit the door inner trim panel with reference to Section 12.

Front door exterior handle

Removal

12 Remove the door inner trim panel as described in Section 12.
13 Working through the aperture in the top rear corner of the door, unscrew the exterior handle securing nut (see illustration).
14 Working from outside the door, release the handle from the door, by pulling the front edge of the handle out from the door, then sliding the handle towards the front of the door.
15 Pull the handle from the door, and disconnect the lock operating rods (see illustration). If the rods cannot be disconnected from the handle, disconnect them at the lock end, noting that it may be necessary to unscrew the lock securing screws from the rear edge of the door, and move the lock slightly to enable the rods to be disconnected.

Refitting

16 Refitting is a reversal of removal, but ensure that the lock operating rods are correctly reconnected, and check the operation of the handle and lock mechanism before refitting the door trim panel.

Rear door exterior handle

Removal

17 Remove the door inner trim panel as described in Section 12.
18 Working through the aperture in the top rear corner of the door, unscrew the exterior handle securing nut.
19 Working from outside the door, release the handle from the door, by pulling the rear edge of the handle out from the door, then sliding the handle towards the rear of the door.
20 Pull the handle from the door, and disconnect the lock operating rod.

13.13 Front door exterior handle securing nut (arrowed)

Refitting

21 Refitting is a reversal of removal, but ensure that the lock operating rod is correctly reconnected, and check the operation of the handle and lock mechanism before refitting the door trim panel.

Front door lock cylinder

Removal

22 Remove the door exterior handle as described previously in this Section.
23 Disconnect the operating rod from the lock cylinder, if not already done.
24 Prise out the securing clip, and withdraw the lock cylinder from the handle assembly (see illustration).

Refitting

25 Refitting is a reversal of removal, noting that the lock cylinder will only fit in one position. Ensure that the lock operating rods are correctly reconnected, and check the operation of the handle and lock mechanism before refitting the door trim panel.

Door lock

Removal

26 On models fitted with central locking, disconnect the battery negative lead.

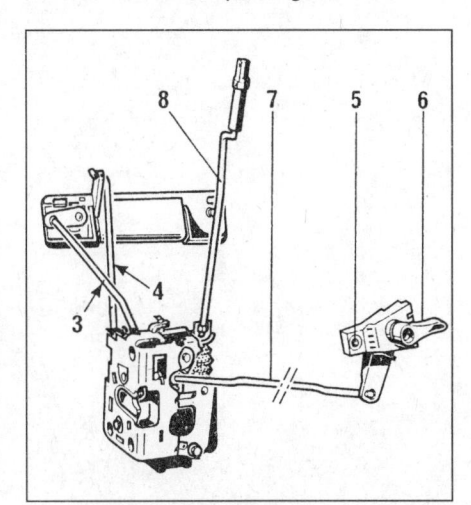

13.15 Front door lock and handle mechanism

3 Exterior handle operating rod
4 Exterior handle operating rod
5 Bellcrank securing bolt
6 Bellcrank
7 Interior handle operating rod
8 Lock button operating rod

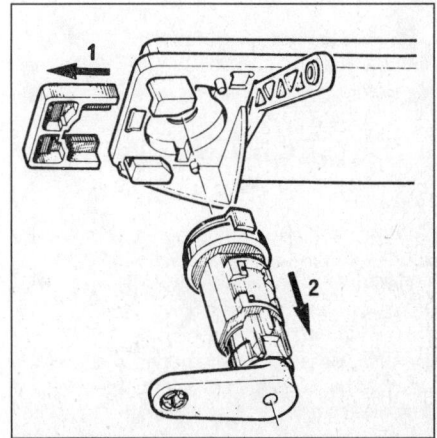

13.24 Front door lock cylinder removal

1 Prise out the securing clip
2 Withdraw the lock cylinder

11•10 Bodywork and fittings

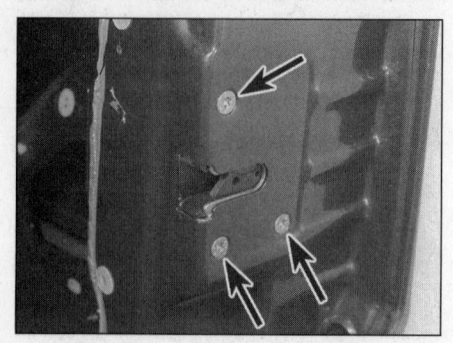

13.31 Door lock securing screws (arrowed)

27 Remove the door inner trim panel as described in Section 12.
28 Remove the door exterior handle as described previously in this Section.
29 Unscrew the bolt securing the door interior handle bellcrank assembly to the door, then disconnect the lock operating rod from the bellcrank, and withdraw the bellcrank assembly from the door.
30 If working on the rear door, unclip the lock button operating rod bellcrank from the front of the door, and release the operating rod from the clips inside the door.
31 Working at the rear edge of the door, unscrew the three lock securing screws (if not already done) **(see illustration)**.
32 Manipulate the lock assembly, complete with the operating rods, out through the door aperture. Note the routing of the lock operating rods to aid refitting. On models with central locking, disconnect the lock motor wiring plug as the lock is withdrawn, noting the routing of the wiring.

Refitting

33 Refitting is a reversal of removal, but make sure that the lock operating rods are correctly reconnected, and check the operation of the handle and lock mechanism before refitting the door trim panel.

14 Door window glass and regulator - removal and refitting

Front door window glass

Removal

1 With the window fully raised, remove the door inner trim panel (see Section 12).
2 Carefully pull the inner weatherstrip from the lower edge of the window aperture.
3 Working through the two holes at the top of the door, unscrew the two bolts securing the window glass to the regulator mechanism **(see illustration)**.
4 Hold the glass in the raised position (ideally with the aid of an assistant), and fully lower the regulator mechanism (temporarily reconnect the battery and the window switch to enable the window to be lowered).

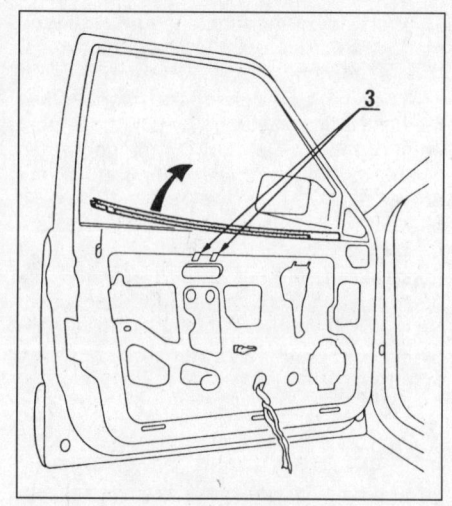

14.3 Pull the weatherstrip (arrowed) from the window aperture

3 Window glass securing bolt access holes

5 Tilt the glass to release it from the guide, then lift the glass out from the outside of the door.

Refitting

6 Refitting is a reversal of removal, noting the following points.
a) Ensure that the glass engages correctly with the guide.
b) Refit the weatherstrip with the window fully lowered.
c) Check the operation of the window mechanism before refitting the door trim panel.

Front door regulator

Removal

7 Remove the window glass, as described previously in this Section.
8 Reach inside the door, and disconnect the wiring plug from the window regulator motor.
9 Unscrew the two upper bolts and the lower nut securing the regulator guide rail to the door, and unscrew the three nuts securing the regulator motor assembly **(see illustration)**.
10 Tilt the regulator mechanism, and withdraw it through the larger door aperture.

Refitting

11 Manoeuvre the regulator mechanism into position, and refit the securing nuts and bolts. Do not fully tighten the nuts and bolts at this stage.
12 With the regulator mechanism in the fully lowered position, refit the window glass, and engage it with the guide. The guide should hold the glass in the raised position, but as a precaution, it is advisable to secure the glass to the top of the door window aperture using tape.
13 Reconnect the regulator motor wiring plug, then fully raise the regulator mechanism

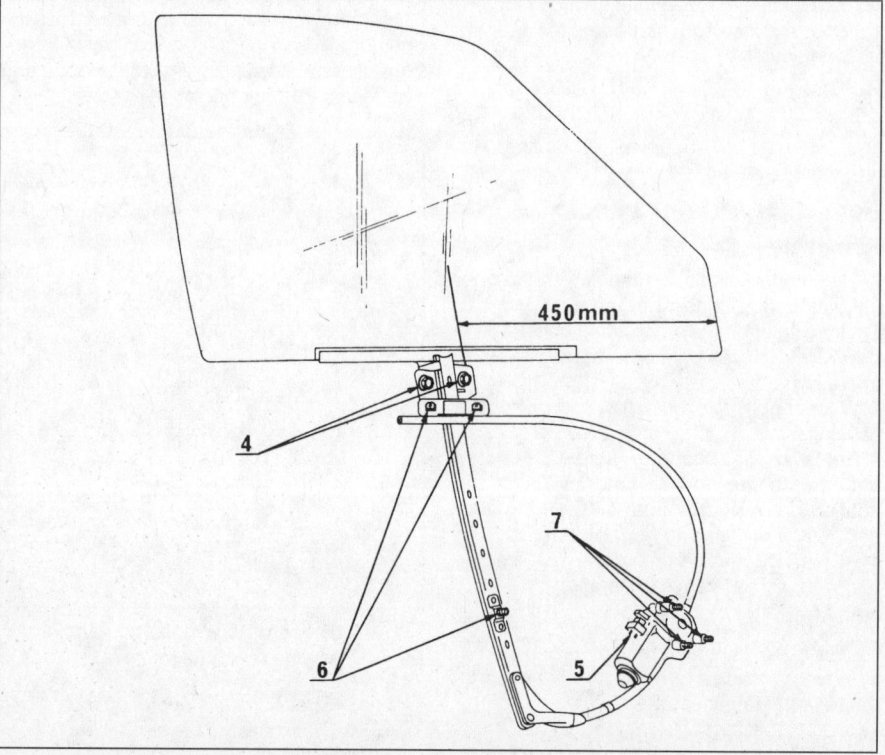

14.9 Front door glass regulator components

4 Glass securing bolts
5 Motor wiring plug
6 Regulator guide rail securing bolts and nut
7 Motor securing nuts

Bodywork and fittings 11•11

(if not already done, temporarily reconnect the battery and the window switch to enable the mechanism to be raised).
14 Align the lower edge of the glass with the regulator mechanism, then refit the two glass securing bolts, but do not fully tighten them at this stage.
15 Tighten the regulator mechanism securing nuts and bolts.
16 Tighten the glass securing bolts.
17 Check the operation of the mechanism, and check that the glass is correctly aligned when in the fully raised position. If necessary, the glass alignment can be adjusted by loosening the glass securing bolts, and sliding the glass as necessary within the elongated bolt holes.
18 Refit the weatherstrip to the window aperture with the window fully lowered.
19 Make a final check on the operation of the window mechanism before refitting the door trim panel.

Rear door sliding window glass

Removal

20 Remove the window regulator mechanism and the fixed window glass, as described later in this Section.
21 Lift the sliding window glass, and withdraw it towards the rear of the door, from the outside of the window aperture.

Refitting

22 Lower the sliding window glass into the door, then refit the fixed window glass and the regulator mechanism, as described later in this Section.

Rear door fixed window glass

Removal

23 Remove the window regulator mechanism as described later in this Section.
24 Carefully lower the sliding window glass to the bottom of the door.

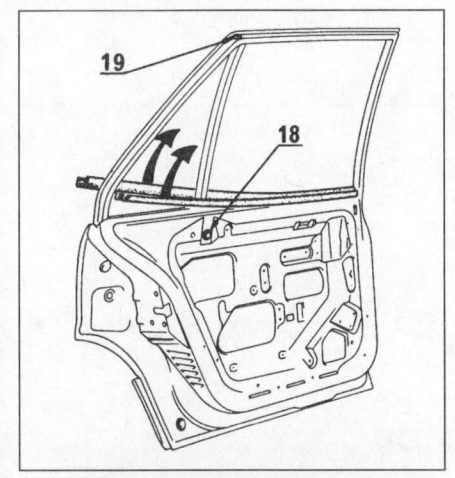

25 Pull the inner and outer weatherstrips from the lower edge of the window aperture.
26 Unscrew the upper and lower bolts securing the rear window guide rail to the door **(see illustration)**.
27 Unclip the weatherstrip from the window aperture and the guide rail, then withdraw the guide rail from the door **(see illustration)**.
28 Pull the fixed window glass from the door, complete with its seal.

Refitting

29 Manoeuvre the fixed window glass into position.
30 Refit the rear window guide rail, and clip the weatherstrip into position on the window aperture and guide rail. Do not fit the guide rail securing bolts at this stage.
31 Lift the sliding window glass to the fully raised position, and hold it in position (with the aid of an assistant, or adhesive tape).
32 Refit and tighten the rear window guide rail securing bolts.
33 Fit the inner and outer weatherstrips to the lower edge of the window aperture.
34 Refit the window regulator mechanism as described later in this Section.

Rear door regulator

Removal

35 Remove the door inner trim panel as described in Section 12.
36 Working through the two holes at the top of the door, unscrew the two bolts securing the window glass to the regulator mechanism.
37 Hold the glass in the raised position (ideally with the aid of an assistant, or using adhesive tape), and fully lower the regulator mechanism (temporarily reconnect the battery and the window switch, or refit the regulator handle, to enable the window to be lowered).
38 Unscrew the three bolts securing the regulator guide rail to the door, and unscrew the three nuts securing the regulator motor assembly **(see illustration)**.
39 Unscrew the bolt securing the door interior handle bellcrank, then disconnect the

lock operating rod from the bellcrank, and withdraw the bellcrank assembly.
40 Unclip the lock button operating rod bellcrank from the front of the door, and release the operating rod from the clips inside the door. Lower the rods into the door, to clear the rear upper aperture in the door.
41 On models with electric windows, reach inside the door and disconnect the wiring plug from the regulator motor.
42 Manipulate the regulator mechanism out through the rear upper aperture in the door.

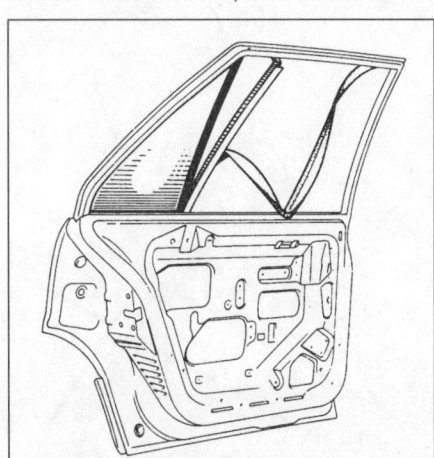

Refitting

43 Manoeuvre the regulator mechanism into position, and refit the securing nuts and bolts. Do not fully tighten the nuts and bolts at this stage.
44 Manipulate the lock button operating rods and bellcrank into position, and clip them to the door.
45 Refit the door interior handle bellcrank, then reconnect the lock operating rod, and refit the bellcrank securing bolt.
46 Reconnect the regulator motor wiring plug, then fully raise the regulator mechanism (if not already done, temporarily reconnect the battery and the window switch to enable the mechanism to be raised).
47 Align the lower edge of the glass with the regulator mechanism, then refit the two glass securing bolts, but do not fully tighten them at this stage.

14.26 Pull the weatherstrips (arrowed) from the rear door window aperture, then remove the upper (19) and lower (18) guide rail securing bolts

14.27 Unclip the weatherstrip from the rear door window aperture and guide rail

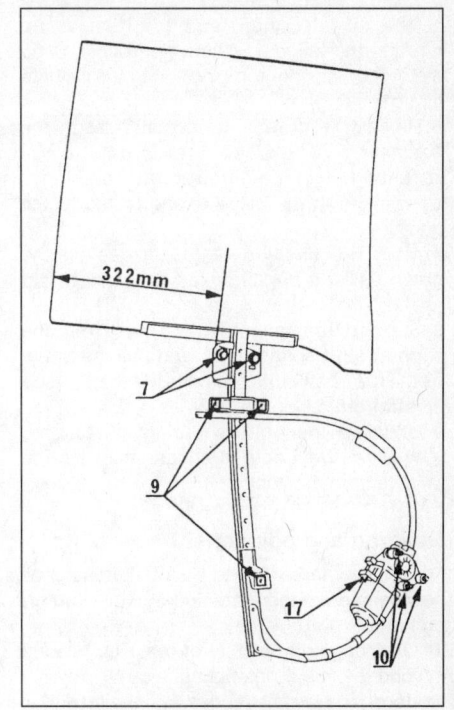

14.38 Rear door glass regulator components

7 Glass securing bolts
9 Regulator guide rail securing bolts
10 Motor securing nuts
17 Motor

11•12 Bodywork and fittings

48 Tighten the regulator mechanism securing nuts and bolts.
49 Tighten the glass securing bolts.
50 Check the operation of the mechanism, and check that the glass is correctly aligned when in the fully raised position. If necessary, the glass alignment can be adjusted by loosening the glass securing bolts, and sliding the glass as necessary within the elongated bolt holes.
51 Make a final check on the operation of the window mechanism before refitting the door trim panel.

15 Tailgate and support struts - removal, refitting and adjustment

Tailgate

Removal

1 Disconnect the battery negative lead.
2 Open the tailgate, then remove the securing screws, and remove the trim panel from the rear of the tailgate **(see illustration)**.
3 Disconnect the wiring plugs from the components in the tailgate.
4 It the original tailgate is to be refitted, tie a length of string to the end of the wiring harness.
5 Pull the wiring grommet from the top corner of the tailgate, then feed the harness up through the tailgate. Where applicable, untie the string, leaving it in position in the tailgate to aid refitting.
6 Release the rear trim panel from the headlining to expose the tailgate hinge securing bolts **(see illustration)**. The hinges are welded to the tailgate, and bolted to the body.
7 Mark the position of the hinges on the body, using a pencil or felt top pen, to aid refitting.
8 Support the tailgate, then prise out the support strut spring clips, and pull the struts from the balljoints on the tailgate **(see illustration)**.
9 Unscrew the bolts securing the tailgate hinges to the body, then carefully lift the tailgate from the vehicle.

Refitting and adjustment

10 If a new tailgate is to be fitted, transfer all serviceable components (locks, wiper motor, etc) to it.
11 Refitting is a reversal of removal, bearing in mind the followings points.
  a) Align the tailgate hinges with the marks made on the body before removal.
  b) If the original tailgate is being refitted, draw the wiring harness through the tailgate using the string.
  c) Before refitting the rear trim panel to the headlining, check the alignment of the tailgate with the surrounding body panels.

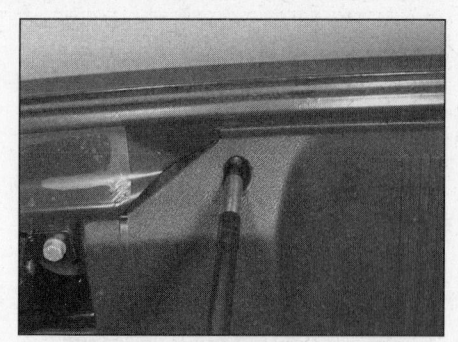

15.2 Removing a tailgate trim panel securing screw

If adjustment is necessary, slacken the tailgate hinge bolts, and move the tailgate slightly within the elongated holes.
  d) On completion, check the tailgate lock for satisfactory operation. If necessary, adjust the position of the lock striker on the rear body panel to achieve satisfactory operation. The striker can be moved once the securing bolts have been slackened.

Support struts

Removal

12 Support the tailgate in the open position, with the help of an assistant, or using a stout piece of wood.
13 Using a flat-bladed screwdriver, release the spring clip, and pull the support strut from its balljoint on the tailgate.
14 Similarly, release the strut from the balljoint on the body, and withdraw the strut from the vehicle.

Refitting

15 Refitting is a reversal of removal, but ensure that the spring clips are correctly engaged.

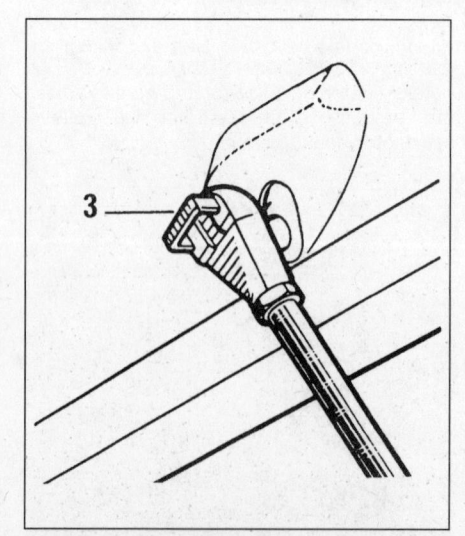

15.8 Prise out the tailgate support strut spring clips (3)

15.6 Tailgate hinge securing bolt (arrowed)

16 Tailgate lock components - removal and refitting

Tailgate lock

Removal

1 On models with central locking, disconnect the battery negative lead.
2 Open the tailgate, then remove the securing screws, and remove the trim panel from the rear of the tailgate.
3 Where applicable, reach inside the tailgate, and disconnect the wiring plug from the lock motor. Where applicable, also separate the two halves of the lock switch wiring connector.
4 Remove the two (manual lock) or three (models with central locking) lock securing screws, then withdraw the lock from the tailgate and unhook the lock operating rod **(see illustration)**.

Refitting

5 Refitting is a reversal of removal, but on completion check the operation of the lock, and if necessary adjust by moving the lock striker within its elongated bolt holes on the rear body panel.

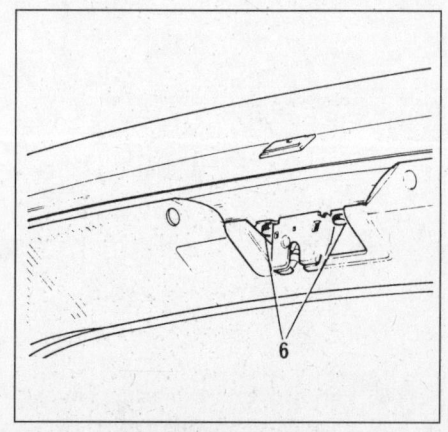

16.4 Manual tailgate lock securing screws (6)

Bodywork and fittings 11•13

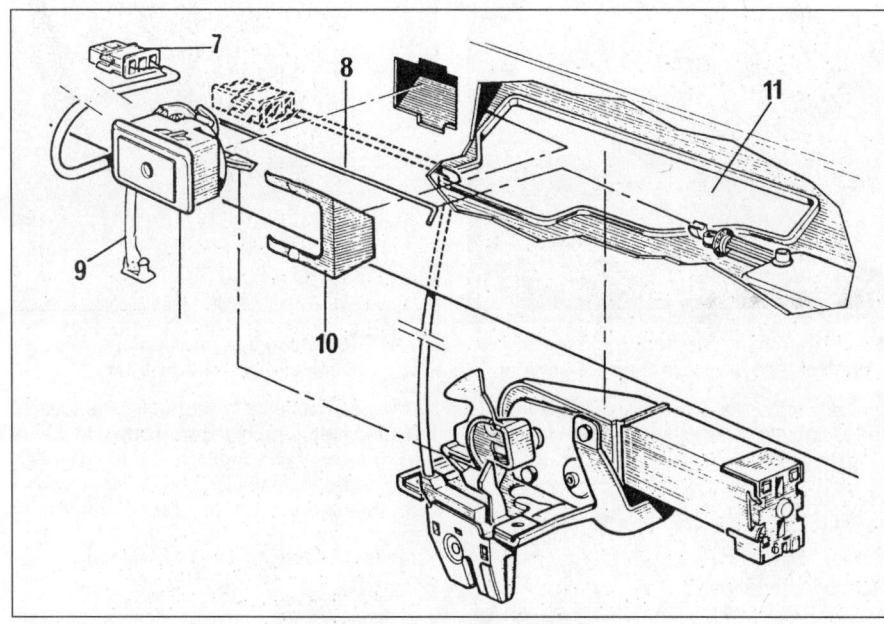

16.8 Tailgate lock components

7 Lock cylinder wiring connector
8 Lock operating rod
9 Water drain tube
10 Lock cylinder securing clip
11 Tailgate aperture

Tailgate lock cylinder

Removal

6 Proceed as described in paragraphs 1 and 2.
7 On models with central locking, remove the tailgate lock, as described previously in this Section.
8 Reach inside the tailgate, and separate the two halves of the lock cylinder wiring connector **(see illustration)**.
9 Unclip the water drain tube from the lock cylinder.
10 Working through the aperture in the tailgate, prise the securing clip from the rear of the lock cylinder.
11 Withdraw the lock cylinder from outside the tailgate, disconnecting the lock operating rod as the lock is withdrawn.

Refitting

12 Refitting is a reversal of removal, but check the operation of the lock mechanism before refitting the tailgate trim panel.

Tailgate lock striker

Removal

13 The lock striker securing screw can be reached through the access hole in the luggage compartment rear trim panel, but it may be necessary to remove the trim panel to allow clearance for the striker assembly to be removed.

Refitting

14 Refitting is a reversal of removal, but before tightening the securing screw, adjust the position of the striker to achieve satisfactory lock operation.

17 Central locking system components - removal and refitting

Door lock motor

Removal

1 The door lock motors are integral with the locks.
2 Remove the relevant door lock as described in Section 13, then remove the securing screws and separate the lock motor from the lock assembly.

> **HAYNES HiNT** *If the fuel filler flap motor fails to operate, the motor can be operated manually by pulling the lever located behind the luggage compartment side trim panel.*

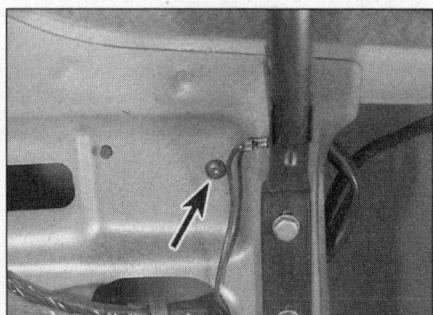

Refitting

3 Refitting is a reversal of removal.

Tailgate lock motor

4 The lock motor is integral with the tailgate lock. Removal and refitting of the lock is described in Section 16.

Fuel filler flap motor

Removal

5 Working in the luggage compartment, remove the securing screws and withdraw the side trim panel from the luggage compartment **(see illustration)**.
6 Remove the securing screw **(see illustration)**.
7 Withdraw the lock motor and disconnect the wiring plug.

Refitting

8 Refitting is a reversal of removal.

Remote control receiver unit

Removal

9 The unit is mounted at the front of the roof panel, between the sun visors.

17.5 Removing the luggage compartment side trim panel . . .

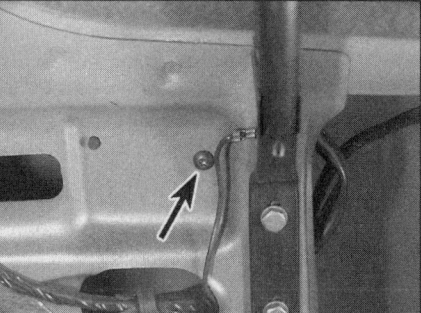

17.6 . . . for access to the fuel filler flap motor securing screw (arrowed)

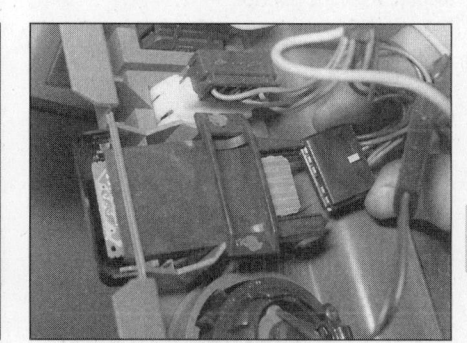

17.12 Disconnecting the wiring plug from the door locking remote control receiver unit

11•14 Bodywork and fittings

10 Disconnect the battery negative lead.
11 Fold the sunvisors away from the roof console.
12 Remove the two securing screws, then lower the roof console, and disconnect the wiring plugs **(see illustration)**.
13 Unclip the cover from the front of the remote control receiver unit.
14 Depress the securing clips, and push the receiver unit from its housing in the roof console **(see illustration)**.

Refitting
15 Refitting is a reversal of removal.

Remote control transmitter batteries - renewal
16 Remove the securing screw(s) and lift the transmitter cover.
17 Withdraw the old batteries, and fit the new ones, ensuring that they are fitted the correct way round **(see illustration)**. The polarity is marked on the transmitter casing.
18 Close the cover and refit the securing screw(s).

18 Electric window components - removal and refitting

Window switches
1 Refer to Chapter 12.

Window regulator motors
2 The regulator motors are integral with the regulator assemblies, and cannot be obtained separately.
3 Removal and refitting details for the regulator assemblies are given in Section 14.

19 Exterior mirrors and associated components - removal and refitting

Manually-adjustable mirror
Removal
1 Open the door, and carefully prise the switch/door handle housing from the front of the armrest.
2 Working at the rear of the switch/door handle housing, loosen the mirror adjustment switch mounting screw. Push the mirror adjustment switch downwards out of the housing.
3 Working at the front corner of the door, where applicable, remove the securing screw, then unclip the mirror trim panel from the front corner of the door.
4 To aid refitting of the mirror control cable, tie a length of string to the end of the adjuster switch.
5 Unscrew the three mirror securing nuts, and withdraw the mirror from the door. Feed the control cable through the door, then untie the

17.14 Removing the door locking remote control receiver from the roof console

string from the adjuster switch and leave the string in position in the door to aid refitting.
Refitting
6 Refitting is a reversal of removal, but use the string to pull the control cable into position in the door.

Electric mirror
Removal
7 Disconnect the battery negative lead.
8 Working at the front corner of the door, where applicable, remove the securing screw, then unclip the mirror trim panel from the front corner of the door.
9 Separate the two halves of the mirror wiring connector.
10 Unscrew the three mirror securing nuts, and withdraw the mirror from the door.
Refitting
11 Refitting is a reversal of removal.

20 Windscreen and tailgate glass - general information

These areas of glass are secured by the tight fit of the weatherstrip in the body aperture, and are bonded in position with a special adhesive. Renewal of such fixed glass is a difficult, messy and time-consuming task, which is considered beyond the scope of the home mechanic. It is difficult, unless one has plenty of practice, to obtain a secure, waterproof fit. Furthermore, there is a high risk of breaking the glass; this applies especially to the laminated glass windscreen. In view of this, owners are strongly advised to have this work carried out by one of the many specialist windscreen fitters.

21 Sunroof - general information

The factory-fitted sunroof is of the electric tilt/slide type.
Due to the complexity of the sunroof mechanism, considerable expertise is required to repair, replace or adjust the

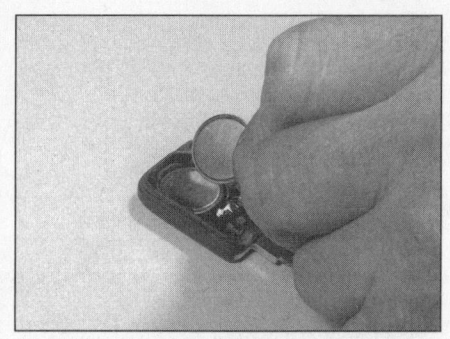

17.17 Removing a central locking remote control transmitter battery

sunroof components successfully. Removal of the roof first requires the headlining to be removed, which is a tedious operation, and no a task to be undertaken lightly. Any problems with the sunroof should be referred to a Renault dealer.
For sunroof switch removal, see Chapter 12.

22 Body exterior fittings - removal and refitting

Wheel arch liners and mud shields
1 The wheel arch liners are secured by a combination of self-tapping screws, and push-fit clips. Removal is self-evident, and normally the clips can be released by pulling the liner away from the wheel arch.
2 The mud shields are secured in a similar manner, although certain panels may be secured using pop-rivets. Where applicable, drill out the pop-rivets, and use new rivets on refitting.

Body trim strips and badges
3 The various body trim strips and badges are held in position with a special adhesive tape (or plastic and metal clips on some trim strips). Removal of a trim/badge secured by adhesive tape requires the trim/badge to be heated, to soften the adhesive, and then cut away from the surface. Due to the high risk of damage to the vehicle paintwork during this operation, it is recommended that this task should be entrusted to a Renault dealer.

23 Seats - removal and refitting

Front seats
Removal
1 On models with electrically-adjustable seats, disconnect the battery negative lead.
2 Move the seat fully forwards, and tilt the backrest forwards.

Bodywork and fittings 11•15

3 Remove the bolts (one on each side) securing the seat rails to the floor. Recover the washers, where applicable.
4 Move the seat fully rearwards.
5 Remove the bolts (one on each side) securing the front of the seat rails to the floor. Again, recover the washers where applicable.
6 On models with electrically-adjustable seats, tilt the seat and disconnect the wiring plug(s).
7 Lift the seat out through the door aperture.

Refitting

8 Refitting is a reversal of removal, but tighten the securing bolts to the specified torque.

Rear seat back

Removal

9 Release the securing catches and fold the rear seats down.
10 The rear seat back side squabs must now be removed as follows (see illustrations).
  a) Working at the bottom of the seat squab, loosen the lower seat back hinge securing nut.
  b) Pull the seat squab upwards to release the upper hook from the securing lug on the body.
  c) Unhook the lower end of the squab and withdraw it.
11 Working at each side of the seat back, remove the screws securing the seat cushion connecting rods to the seat back (see illustration).
12 Unscrew the two hinge nuts at each side (securing the seat back hinges to floor), and withdraw seat back (see illustration).

Refitting

13 Refitting is a reversal of removal, but do not fully tighten the lower seat back hinge bolts until the seat side squabs have been fitted.

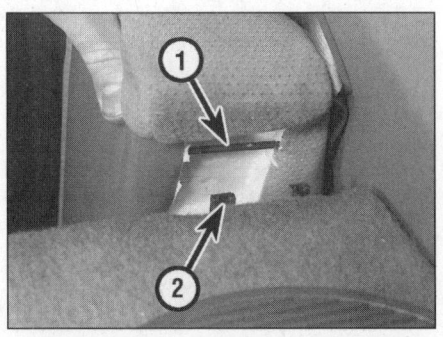

Rear seat cushion

Removal

14 Release the securing catches and fold the rear seats down.
15 Working at each side of the seat cushion, remove the screws securing the seat cushion connecting rods to the seat back.
16 Working at the lower edge of the seat cushion, squeeze the retaining lugs together, and tap out the pins securing the seat cushion to the lower hinges.
17 Withdraw seat cushion.

Refitting

18 Refitting is a reversal of removal.

24 Seat belt components - removal and refitting

Front seat belt

Removal

1 Prise off the cover plate, and unscrew the

23.10a Pull the seat back side squab upwards to release the upper hook (1) from the lug (2) - viewed from luggage compartment

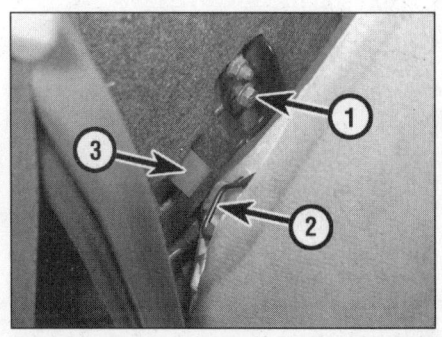

23.10b Unhook the lower end of the squab
1 Lower seat back hinge nut (must be loosened)
2 Hook
3 Lug

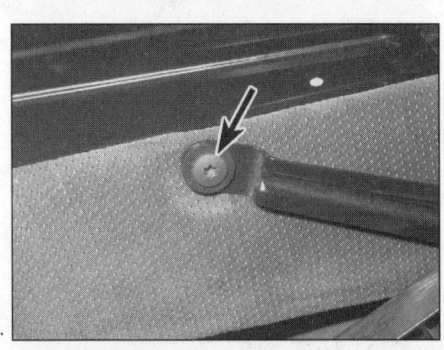

23.11 Seat cushion connecting rod-to-seat back screw (arrowed)

23.12 Unscrewing a seat back hinge nut

seat belt upper anchor bolt (see illustration). Note the location of any washers and spacers on the anchor bolt.
2 Unclip the lens, then prise out the body pillar-mounted courtesy light.
3 Remove the upper securing screw, then unclip the upper trim panel from the body pillar (see illustration).
4 Prise off the cover plate, and unscrew the seat belt lower anchor bolt. Again, note the location of any washers and spacers on the anchor bolt.
5 Unclip the seat belt surround from the upper trim panel, and feed the seat belt through the slot in the panel (see illustration).
6 Remove the two upper securing screws from the lower pillar trim panel (see illustration).

7 Remove the screws securing the lower pillar trim panel to the sill, and pull the trim panel back for access to the inertia reel (see illustrations).
8 Unscrew two of the screws securing the belt guide to the body pillar, and slide the belt from the guide (see illustration).
9 Unscrew the bolt securing the inertia reel to the body pillar, and withdraw the seat belt assembly. Again, note the location of any washers and spacers on the anchor bolt.

Refitting

10 Refitting is a reversal of removal, but ensure that any washers and spacers on the anchor bolts are fitted as noted before removal. Ensure that the seat belt anchor bolts are securely tightened.

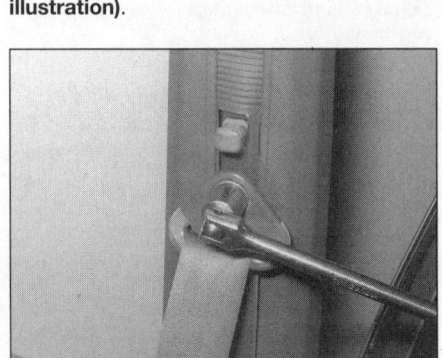

24.1 Unscrewing the front seat belt upper anchor bolt

24.3 Body centre pillar upper trim panel securing screw (arrowed)

11•16 Bodywork and fittings

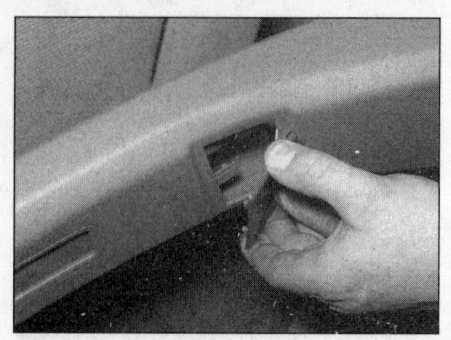

24.5 Unclip the seat belt surround and feed the seat belt through the slot in the panel

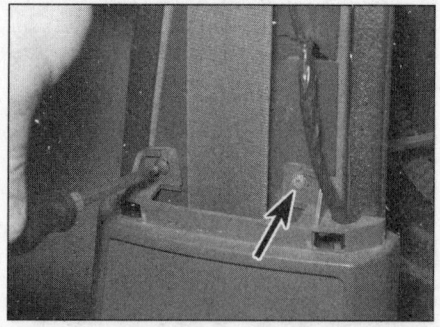

24.6 Remove the upper securing screws from the lower pillar trim panel

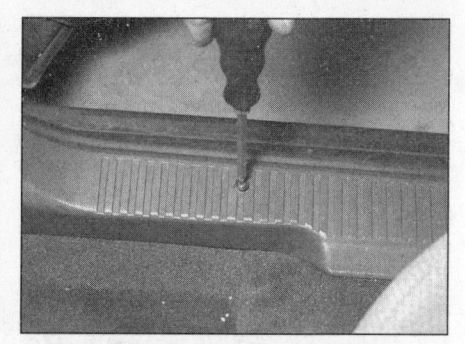

24.7a Remove the securing screws . . .

24.7b . . . and pull the trim panel back for access to the inertia reel

24.8 Loosen two of the screws (arrowed) securing the belt guide

24.18 Unclip the rear pillar trim panel for access to the inertia reel

Front seat belt stalk

Removal

11 For access to the seat belt stalk securing bolt, remove the seat (see Section 23), or remove the centre console as described in Section 26.
12 Unscrew the securing bolt and remove the stalk assembly. Note the locations of any spacers and washers on the securing bolt.

Refitting

13 Refitting is a reversal of removal, ensuring that any washers and spacers are located as noted before removal.

Rear side seat belt

Removal

14 Release the securing catches and fold the rear seats down.
15 Loosen the lower seat back hinge securing bolt.
16 Pull the seat back side squab upwards to release the upper hook from the securing lug on the body, then unhook the lower end of the squab, and withdraw the squab.
17 Remove the two front screws, and the two side screws, and withdraw the parcel shelf side support panel. Disconnect the loudspeaker wiring .
18 Unclip the rear pillar trim panel for access to the inertia reel **(see illustration)**.
19 Unscrew the lower seat belt anchor bolt, noting the location of any washers and spacers.
20 Prise the seat belt surround from the C-pillar trim panel, and feed seat belt webbing through the panel.
21 Unscrew the inertia reel securing bolt, again noting the location of any washers and spacers, and withdraw the seat belt assembly.

Refitting

22 Refitting is a reversal of removal, but ensure that any washers and spacers are positioned as noted before removal, and ensure that the seat belt anchor bolts are securely tightened.

Rear centre belt and buckles

Removal

23 The assemblies can simply be unbolted from the floor panel, after folding the rear seats forwards. Note the location of any washers and spacers to ensure correct refitting.

Refitting

24 Refitting is a reversal of removal. Ensure that any washers and spacers are positioned as noted before removal, and securely tighten the anchor bolts.

25 Interior trim - removal and refitting

General

1 The interior trim panels are secured by a combination of screws and clips, and removal is usually self-explanatory. The following paragraphs describe the removal of panels which often need to be removed for access to other components.

Door trim panels

2 Refer to Section 12.

Rear parcel shelf support panels

Removal

3 Remove the rear parcel shelf.
4 Disconnect the wiring from the loudspeaker.
5 Remove the two front and two side securing screws, and withdraw the panel **(see illustrations)**.

Refitting

6 Refitting is a reversal of removal.

Steering column shrouds

Removal

7 Working under the steering column, unscrew the three screws securing the lower shroud **(see illustration)**.
8 Withdraw the lower shroud, then turn the steering wheel as necessary for access to the two upper shroud securing screws. Remove the screws **(see illustrations)**.
9 Lift off the upper shroud and, where applicable, slacken the two clamp screws securing the radio/cassette player remote control switch to the upper shroud **(see illustration)**.

Bodywork and fittings 11•17

25.5a Remove the two front screws...

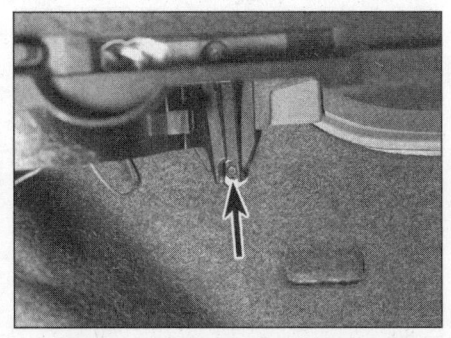

25.5b ...and the side screws...

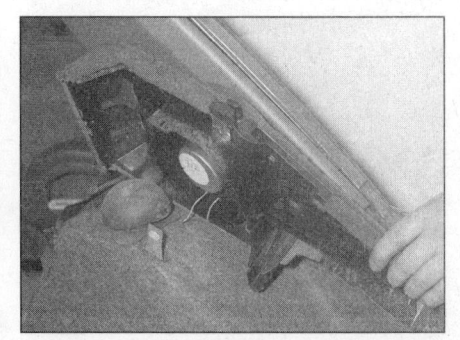

25.5c ...and withdraw the rear parcel shelf support panel

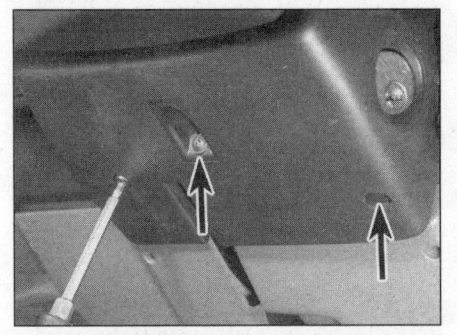

25.7 Unscrew the three lower steering column shroud securing screws

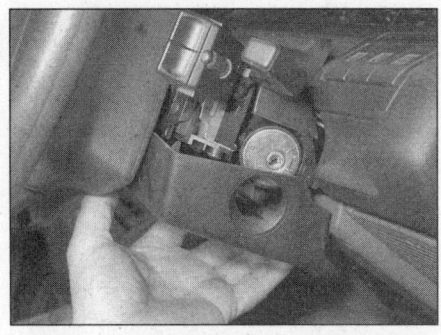

25.8a Withdraw the lower shroud...

25.8b ...then unscrew the upper shroud securing screws

10 Lift the upper shroud upwards and, where applicable, manipulate the radio/cassette player remote control switch from the shroud (see illustration).

Refitting

11 Refitting is a reversal of removal.

Carpets

12 The passenger compartment floor carpet is in several pieces, and is secured along the edges by screws or various types of clips.
13 Carpet removal and refitting is reasonably straightforward, but time-consuming, due to the fact that all adjoining trim panels must be released, and the seats and centre console must be removed.

Headlining

14 The headlining is clipped to the roof, and can be withdrawn only once all fittings such as the grab handles, sun visors, sunroof, windscreen, centre and rear pillar trim panels, and associated components have been removed. The door, tailgate and sunroof aperture weatherstrips will also have to be prised clear.
15 Note that headlining removal requires considerable skill and experience if it is to be carried out without damage, and is therefore best entrusted to an expert.

26 Centre console - removal and refitting

Removal

1 Disconnect the battery negative lead.
2 Remove the radio/cassette player as described in Chapter 12.

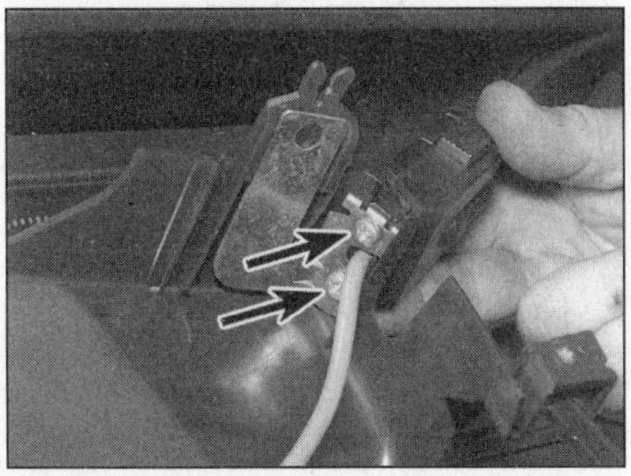

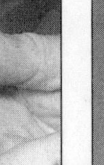

25.9 Slacken the two clamp screws (arrowed)...

25.10 ...and manipulate the remote control switch from the shroud

3 On models with a manual gearbox, unclip the gear lever gaiter from the centre console.
4 On models with automatic transmission, unclip the gear selector position indicator plate from the centre console.
5 Working at the sides of the console, unscrew the front and rear securing screws, then lift the console from the floor for access to the switch wiring plugs (see illustrations).
6 Disconnect the wiring plugs from the centre console-mounted switches, then withdraw the console from the vehicle.

Refitting

7 Refitting is a reversal of removal.

27 Facia panel assembly - removal and refitting

Glovebox assembly

Removal

1 Open the glovebox.
2 Prise the centre pins from the hinge pins at the sides of the glovebox, withdraw the pins, and lift out the glovebox (see illustrations).

Refitting

3 Refitting is a reversal of removal.

Main facia panel

Removal

4 Disconnect the battery negative lead.

11•18 Bodywork and fittings

26.5a Remove the front . . .

26.5b . . . and rear centre console securing screws (arrowed)

27.2a Prise out the centre pins . . .

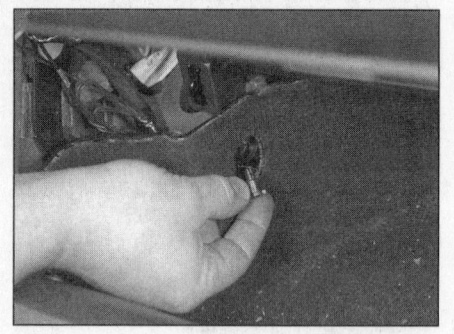

27.2b . . . then withdraw the hinge pins . . .

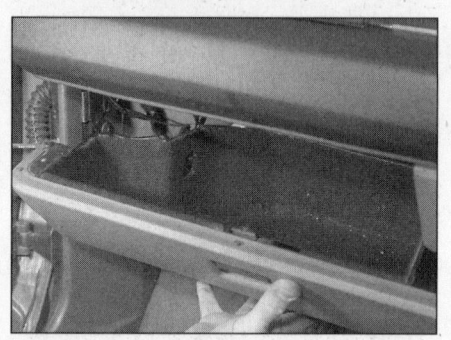

27.2c . . . and withdraw the glovebox

27.6 Removing the driver's side lower facia panel

5 Remove the instrument panel as described in Chapter 12.
6 Remove the securing screws, and withdraw the driver's side lower facia trim panel **(see illustration)**.
7 Remove the centre console as described in Section 26.
8 Remove the steering column as described in Chapter 10
9 Pull the knob from the headlight beam adjustment switch, then depress the securing clip, and remove the switch surround. Unscrew the switch securing screw(s), then depress the securing clip and push the switch from the facia (see Chapter 12).
10 Unscrew the heater control panel securing screws. Manipulate the heater control panel from the facia, and leave it in position in the vehicle.
11 Reach up under the facia, and disconnect all relevant wiring harness plugs, noting their locations and routing to aid reconnection.
12 Working at the bulkhead in the engine compartment, unscrew the two upper facia securing nuts **(see illustration)**.
13 Working at the lower corners of the facia in the passenger compartment, unscrew the lower facia securing nut and screw **(see illustrations)**.
14 With that aid of an assistant, pull the facia assembly back from the bulkhead, keeping it level to release it from the centre locating clip.
15 As the facia is pulled forwards, check that all necessary wiring has been disconnected and released from any securing clips and brackets.
16 Note the routing of all wiring to ensure correct refitting.

Refitting

17 Refitting is a reversal of removal, bearing in mind the following points.
   a) Ensure that the facia centre locating clip is correctly engaged.
   b) Ensure that all wiring is routed as noted before removal.
   c) Manipulate the heater control assembly into position before finally pushing the facia into position.
   d) Refit the steering column as described in Chapter 10.

27.12 Upper facia securing nut (arrowed) in engine compartment

27.13a Facia lower right-hand securing nut (arrowed) viewed through headlight beam adjustment switch aperture

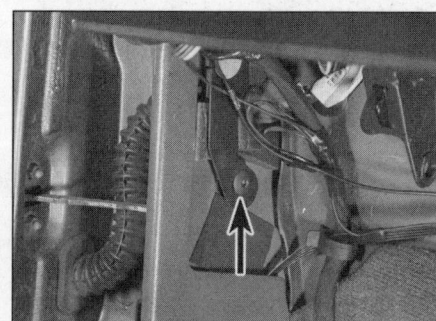

27.13b Facia lower left-hand securing screw (arrowed)

Chapter 12
Body electrical systems

Contents

Battery - removal and refitting	See Chapter 5
Battery check and maintenance	See "Weekly Checks"
Bulbs (exterior lights) - renewal	5
Bulbs (interior lights) - renewal	6
Cigarette lighter - removal and refitting	11
Clock/temperature display unit - removal and refitting	12
Cruise control system - general information	22
Cruse control system components - removal and refitting	23
Electrical fault-finding - general information	2
Electrical system check	See Chapter 1
Exterior light units - removal and refitting	7
Fuses and relays - general information	3
General information and precautions	1
Headlight beam alignment - general information	8
Horn(s) and compressor - removal and refitting	13
Instrument panel - removal and refitting	9
Instrument panel components - removal and refitting	10
Loudspeakers - removal and refitting	20
Radio aerial - removal and refitting	21
Radio/cassette player - removal and refitting	19
Reversing light switch (models with manual transmission) - removal and refitting	See Chapter 7A
Speedometer cable/sensor - removal and refitting	14
Starter inhibitor/reversing light switch (models with automatic transmission) - removal and refitting	See Chapter 7B
Stop light switch - removal and refitting	See Chapter 9
Switches - removal and refitting	4
Tailgate wiper motor - removal and refitting	17
Trip computer - general information	24
Voice synthesiser/warning systems - general information	25
Washer fluid level check	See "Weekly Checks"
Wiper arm - removal and refitting	15
Wiper blade check and renewal	See "Weekly Checks"
Windscreen wiper motor and linkage - removal and refitting	16
Windscreen/tailgate/headlight washer system components - removal and refitting	18

Degrees of difficulty

Easy, suitable for novice with little experience	Fairly easy, suitable for beginner with some experience	Fairly difficult, suitable for competent DIY mechanic	Difficult, suitable for experienced DIY mechanic	Very difficult, suitable for expert DIY or professional

Specifications

General
System type .... 12-volt negative earth

Fuses
Note: *Fuse applications are indicated on the label attached to the base of the fusebox. The following applications are typical.*

No	Rating (amps)	Circuit(s) protected
1	20	Rear screen heater
2	15	Windscreen wiper park
3	15	Cigarette lighter/interior lights
4	25	Central door locking
5	-	Not used
6	-	Not used
7	2	Automatic transmission
8	5	Heater controls illumination and air conditioning
9	-	Not used
10	-	Not used
11	30	Left-hand front seat adjuster
12	30	Right-hand front seat adjuster
13	25	Heating/air conditioning
14	10	Radio/cassette player

Fuses (continued)

Note: *Fuse applications are indicated on the label attached to the base of the fusebox. The following applications are typical.*

No	Rating (amps)	Circuit(s) protected
15	7.5	Rear foglight
16	10	Hazard/direction indicator flasher unit
17	5	Left-hand side/tail light
18	5	Right-hand side/tail light
19	10	Windscreen wipers
20	5	Reversing lights
21	3	Instrument panel
22	10	Stop lights switch/cruise control and "lights on" buzzer
23	30	Left-hand electric window/electric mirrors
24	30	Right-hand electric window/sunroof

Bulbs

Note: *The type, and rating of bulbs fitted depends on the age and model of the vehicle. The following is intended as a guide only.*

Bulb	Rating
Headlight main/dipped beam	Halogen BH1 or H4 60/55
Headlight main beam	H1 60
Direction indicator lights	21
Front sidelight	3 or 4
Tail/stop light	5/21
Tail light	5
Rear foglight	21
Reversing light	21
Front foglight	H3 55
Direction indicator repeater	5
Rear number plate	5
Instrument panel	1.4
Interior	5

1 General information and precautions

Warning: Before carrying out any work on the electrical system, read through the precautions given in "Safety first!" at the beginning of this manual, and in Chapter 5.

The electrical system is of 12-volt negative earth type. Power for the lights and all electrical accessories is supplied by a lead/acid type battery, which is charged by the alternator.

This Chapter covers repair and service procedures for the various electrical components not associated with engine. Information on the battery, alternator and starter motor can be found in Chapter 5.

It should be noted that, prior to working on any component in the electrical system, the battery negative terminal should first be disconnected, to prevent the possibility of electrical short-circuits and/or fires.

Caution: If the radio/cassette player fitted to the vehicle is one with an anti-theft security code, as the standard unit is, refer to the information given in the preliminary Sections of this manual before disconnecting the battery.

2 Electrical fault-finding - general information

Note: *Refer to the precautions given in "Safety first!" and in Section 1 of this Chapter before starting work. The following tests relate to testing of the main electrical circuits, and should not be used to test delicate electronic circuits (such as anti-lock braking systems), particularly where an electronic control module is used.*

General

1 A typical electrical circuit consists of an electrical component, any switches, relays, motors, fuses, fusible links or circuit breakers related to that component, and the wiring and connectors which link the component to both the battery and the chassis. To help to pinpoint a problem in an electrical circuit, wiring diagrams are included at the end of this manual.

2 Before attempting to diagnose an electrical fault, first study the appropriate wiring diagram, to obtain a more complete understanding of the components included in the particular circuit concerned. The possible sources of a fault can be narrowed down by noting whether other components related to the circuit are operating properly. If several components or circuits fail at one time, the problem is likely to be related to a shared fuse or earth connection.

3 Electrical problems usually stem from simple causes, such as loose or corroded connections, a faulty earth connection, a blown fuse, a melted fusible link, or a faulty relay (refer to Section 3 for details of testing relays). Visually inspect the condition of all fuses, wires and connections in a problem circuit before testing the components. Use the wiring diagrams to determine which terminal connections will need to be checked, in order to pinpoint the trouble-spot.

4 The basic tools required for electrical fault-finding include a circuit tester or voltmeter (a 12-volt bulb with a set of test leads can also be used for certain tests); a self-powered test light (sometimes known as a continuity tester); an ohmmeter (to measure resistance); a battery and set of test leads; and a jumper wire, preferably with a circuit breaker or fuse incorporated, which can be used to bypass suspect wires or electrical components. Before attempting to locate a problem with test instruments, use the wiring diagram to determine where to make the connections.

5 To find the source of an intermittent wiring fault (usually due to a poor or dirty connection, or damaged wiring insulation), a "wiggle" test can be performed on the wiring. This involves wiggling the wiring by hand, to see if the fault occurs as the wiring is moved. It should be possible to narrow down the source of the fault to a particular section of wiring. This method of testing can be used in conjunction with any of the tests described in the following sub-Sections.

Body electrical systems 12•3

6 Apart from problems due to poor connections, two basic types of fault can occur in an electrical circuit - open-circuit, or short-circuit.
7 Open-circuit faults are caused by a break somewhere in the circuit, which prevents current from flowing. An open-circuit fault will prevent a component from working, but will not cause the relevant circuit fuse to blow.
8 Short-circuit faults are caused by a "short" somewhere in the circuit, which allows the current flowing in the circuit to "escape" along an alternative route, usually to earth. Short-circuit faults are normally caused by a breakdown in wiring insulation, which allows a feed wire to touch either another wire, or an earthed component such as the bodyshell. A short-circuit fault will normally cause the relevant circuit fuse to blow.

Finding an open-circuit

9 To check for an open-circuit, connect one lead of a circuit tester or voltmeter to either the negative battery terminal or a known good earth.
10 Connect the other lead to a connector in the circuit being tested, preferably nearest to the battery or fuse.
11 Switch on the circuit, bearing in mind that some circuits are live only when the ignition switch is moved to a particular position.
12 If voltage is present (indicated either by the tester bulb lighting or a voltmeter reading, as applicable), this means that the section of the circuit between the relevant connector and the battery is problem-free.
13 Continue to check the remainder of the circuit in the same fashion.
14 When a point is reached at which no voltage is present, the problem must lie between that point and the previous test point with voltage. Most problems can be traced to a broken wire, corroded or loose connection.

Finding a short-circuit

15 To check for a short-circuit, first disconnect the load(s) from the circuit (loads are the components which draw current from a circuit, such as bulbs, motors, heating elements, etc).
16 Remove the relevant fuse from the circuit, and connect a circuit tester or voltmeter to the fuse connections.
17 Switch on the circuit, bearing in mind that some circuits are live only when the ignition switch is moved to a particular position.
18 If voltage is present (indicated either by the tester bulb lighting or a voltmeter reading, as applicable), this means that there is a short-circuit.
19 If no voltage is present, but the fuse still blows with the load(s) connected, this indicates an internal fault in the load(s).

Finding an earth fault

20 The battery negative terminal is connected to "earth" - the metal of the engine/transmission and the car body - and most systems are wired so that they only receive a positive feed, the current returning via the metal of the car body. This means that the component mounting and the body form part of that circuit. Loose or corroded mountings can therefore cause a range of electrical faults, ranging from total failure of a circuit, to a puzzling partial fault. In particular, lights may shine dimly (especially when another circuit sharing the same earth point is in operation), motors (eg wiper motors or the radiator cooling fan motor) may run slowly, and the operation of one circuit may have an apparently unrelated effect on another. Note that on many vehicles, earth straps are used between certain components, such as the engine/transmission and the body, usually where there is no metal-to-metal contact between components, due to flexible rubber mountings, etc.
21 To check whether a component is properly earthed, disconnect the battery, and connect one lead of an ohmmeter to a known good earth point. Connect the other lead to the wire or earth connection being tested. The resistance reading should be zero; if not, check the connection as follows.
22 If an earth connection is thought to be faulty, dismantle the connection, and clean back to bare metal both the bodyshell and the wire terminal or the component earth connection mating surface. Be careful to remove all traces of dirt and corrosion, then use a knife to trim away any paint, so that a clean metal-to-metal joint is made. On reassembly, tighten the joint fasteners securely; if a wire terminal is being refitted, use serrated washers between the terminal and the bodyshell, to ensure a clean and secure connection. When the connection is remade, prevent the onset of corrosion in the future by applying a coat of petroleum jelly or silicone-based grease, or by spraying on (at regular intervals) a proprietary ignition sealer.

3 Fuses and relays - general information

Fuses

1 Fuses are designed to break a circuit when a predetermined current is reached, in order to protect the components and wiring which could be damaged by excessive current flow. Any excessive current flow will be due to a fault in the circuit, usually a short-circuit (see Section 2).
2 The main fuses are located in the fusebox, below the driver's side of the facia.
3 For access to the fuses, squeeze the securing clips and lower the fusebox from the facia.
4 A blown fuse can be recognised from its melted or broken wire.
5 To remove a fuse, first ensure that the relevant circuit is switched off.
6 Using the plastic tool provided in the fusebox, pull the fuse from its location (see illustration).
7 Spare fuses are provided in the blank terminal positions in the fusebox.
8 Before renewing a blown fuse, trace and rectify the cause, and always use a fuse of the correct rating. Never substitute a fuse of a higher rating, or make temporary repairs using wire or metal foil; more serious damage, or even fire, could result.
9 Note that the fuses are colour-coded as follows. Refer to the *Specifications* and wiring diagrams for details of the fuse ratings and the circuits protected.

Colour	Rating
Orange	5A
Red	10A
Blue	15A
Yellow	20A
Clear or white	25A
Green	30A

Relays

10 A relay is an electrically-operated switch, which is used for the following reasons:
 a) *A relay can switch a heavy current remotely from the circuit in which the current is flowing, allowing the use of lighter-gauge wiring and switch contacts.*
 b) *A relay can receive more than one control input, unlike a mechanical switch.*
 c) *A relay can have a timer function - for example, the intermittent wiper relay.*
11 Most of the relays are located under the facia, above the main fusebox.
12 Access to the relays is most easily obtained by removing the securing screw and withdrawing the facia blanking plate, or the voice synthesiser loudspeaker, as applicable. The relays can then be reached through the aperture in the facia (see illustrations).
13 If a circuit or system controlled by a relay develops a fault, and the relay is suspect, operate the system. If the relay is functioning, it should be possible to hear it "click" as it is energised. If this is the case, the fault lies with the components or wiring of the system. If the relay is not being energised, then either the relay is not receiving a main supply or a switching voltage, or the relay itself is faulty.

3.6 Removing a fuse from the fusebox

12•4 Body electrical systems

Testing is by the substitution of a known good unit, but be careful - while some relays are identical in appearance and in operation, others look similar but perform different functions.

14 To remove a relay, first ensure that the relevant circuit is switched off. The relay can then simply be pulled out from the socket, and pushed back into position.

4 Switches - removal and refitting

Note: *Disconnect the battery negative lead before removing any switch, and reconnect the lead after refitting the switch.*

Ignition switch/steering column lock
1 Refer to Chapter 10.

Steering column combination switches
2 Remove the steering column shrouds as described in Chapter 11.
3 Unscrew the two securing screws, then lift the switch from the steering column and disconnect the wiring plug(s). Withdraw the switch **(see illustrations)**.
4 Refitting is a reversal of removal.

Radio/cassette player remote control switch
5 Remove the steering column shrouds as described in Chapter 11.
6 Once the switch has been released from the upper column shroud, trace the wiring back from the switch, and separate the two halves of the wiring connector.
7 Refitting is a reversal of removal.

Facia-mounted pushbutton switches
8 Remove the instrument panel as described in Section 9.
9 Working at the rear of the panel, disconnect the switch wiring plug(s), then depress the securing clips, and push the switch out from the front of the panel **(see illustration)**.
10 Refitting is a reversal of removal.

3.12a Withdraw the facia blanking plate . . .

3.12b . . . for access to the relays

4.3a Remove the securing screws . . .

4.3b . . . and withdraw the steering column combination switch

Centre console-mounted switches
11 Using a small flat-bladed screwdriver, carefully prise the switch from the centre console, and disconnect the wiring plug. Take care not to damage the surround panel.
12 Refitting is a reversal of removal.

Instrument panel illumination rheostat
13 Remove the instrument panel as described in Section 9.
14 Working at the rear of the panel, disconnect the wiring plugs from the rheostat, then depress the securing clips, and push the rheostat out through the front of the panel **(see illustrations)**.
15 Refitting is a reversal of removal.

Heater blower motor switch
16 The switch is integral with the heater control panel, and cannot be removed independently.

17 Removal and refitting of the control panel is described in Chapter 3.

Roof console-mounted switches
18 Fold the sunvisors away from the roof console.
19 Remove the two securing screws, then lower the roof console, and disconnect the wiring plugs.
20 Depress the securing clips, and push the switch out through the front of the panel **(see illustration)**.
21 Refitting is a reversal of removal.

Door-mounted switches
22 Working at the front of the armrest assembly, unscrew the bolt securing the switch/door handle housing.
23 Carefully prise the switch/door handle housing from the front of the armrest **(see illustration)**.

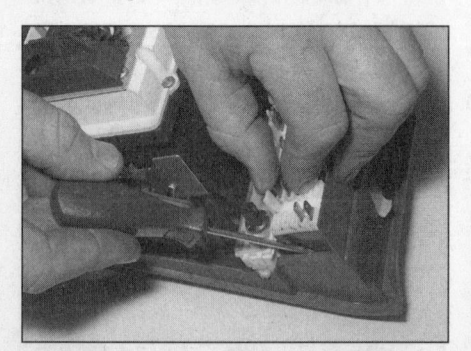

4.9 Depressing a facia-mounted push-button switch securing clip

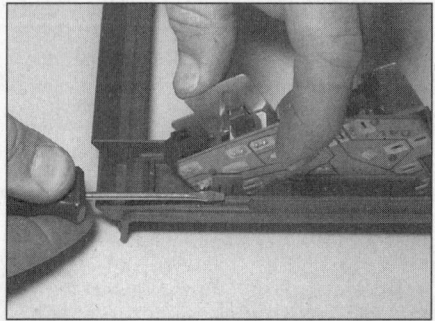

4.14a Depress the securing clip . . .

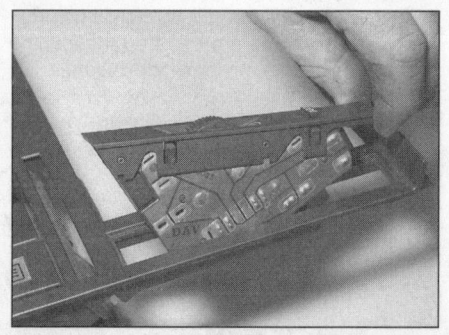

4.14b . . . and push the rheostat out through the front of the instrument panel

Body electrical systems 12•5

4.20 Depressing a roof console-mounted switch securing clip

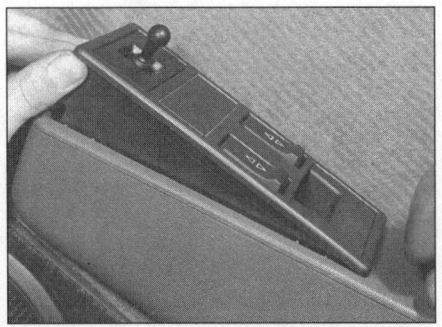

4.23 Prise the switch/door handle housing from the armrest

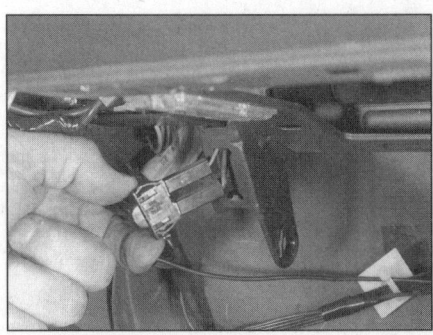

4.30 Removing the glovebox light switch

24 Disconnect the wiring plug from the switch, then release the securing clips, and push the switch out through the top of the housing.
25 Refitting is a reversal of removal.

Courtesy light switches

26 Open the door, then prise the rubber gaiter from the switch.
27 Remove the securing screw, then withdraw the switch from the door pillar. Disconnect the wiring connector as it becomes accessible.

> **HAYNES HINT** Tape the wiring to the door pillar, to prevent it falling back into the door pillar. Alternatively, tie a piece of string to the wiring, to retrieve it.

28 Refitting is a reversal of removal, but ensure that the rubber gaiter is correctly seated on the switch.

Glovebox light switch

29 To improve access, remove the glovebox, as described in Chapter 11.
30 Carefully lever the switch from the facia, and disconnect the wiring plug **(see illustration)**.
31 Refitting is a reversal of removal.

Luggage compartment light switch

32 Open the tailgate.
33 Remove the securing screws, and withdraw the rear luggage compartment trim panel to expose the lock striker.
34 Carefully lever the switch from the lock striker assembly on the rear body panel, and disconnect the wiring plug.
35 Refitting is a reversal of removal.

5 Bulbs (exterior lights) - renewal

1 Whenever a bulb is renewed, note the following points.
a) Disconnect the battery negative lead before starting work.
b) Remember that, if the light has just been in use, the bulb may be extremely hot.
c) Always check the bulb contacts and holder, ensuring that there is clean metal-to metal contact between the bulb and its live(s) and earth. Clean off any corrosion or dirt before fitting a new bulb.
d) Wherever bayonet-type bulbs are fitted (see Specifications), ensure that the live contact(s) bear firmly against the bulb contact.
e) Always ensure that the new bulb is of the correct rating, and that it is completely clean before fitting it; this applies particularly to headlight/foglight bulbs (see below).

Headlight

Models with single headlights

2 Working in the engine compartment, twist the headlight rear cover anti-clockwise, and remove it from the rear of the headlight **(see illustration)**.
3 Disconnect the wiring plug from the rear of the headlight bulb **(see illustration)**.
4 Release the spring clip by compressing its ends, then withdraw the bulb **(see illustration)**.
5 When handling the new bulb, use a tissue or clean cloth, to avoid touching the glass with the fingers; moisture and grease from the skin can cause blackening and rapid failure of this type of bulb. If the glass is accidentally touched, wipe it clean using methylated spirit.
6 Install the new bulb, ensuring that it locates correctly in the light unit. Secure the bulb in position with the spring clip, and reconnect the wiring plug.
7 Refit the cover to the rear of the light unit.

Models with twin headlights

8 Models with twin headlights have a combined main/dipped beam bulb in the main (larger) headlight housing, and an additional main beam bulb in the secondary (smaller) housing. Both bulbs have separate covers.
9 Working in the engine compartment, remove the relevant cover from the rear of the headlight. If a round cover is fitted, it can be removed by twisting anti-clockwise. If a square cover is fitted, release the spring clip, and withdraw the cover.
10 Proceed as described in paragraphs 3 to 7.

Front sidelight

11 Working in the engine compartment,

5.2 Remove the headlight rear cover ...

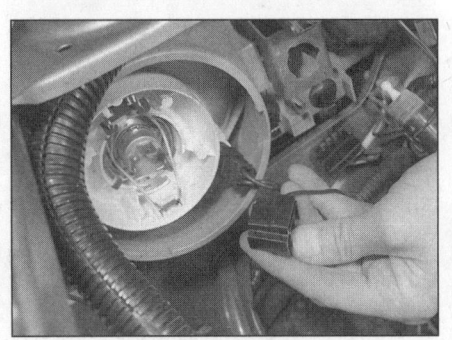

5.3 ... disconnect the wiring plug from the bulb ...

5.4 ... then remove the headlight bulb - model with single headlights

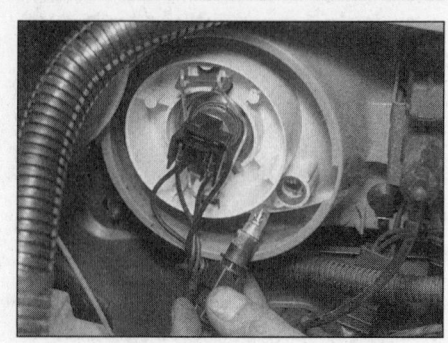

12•6 Body electrical systems

5.12 Removing a sidelight bulb

remove the cover from the rear of the headlight unit. On models with twin headlights, remove the larger cover. If a round cover is fitted, it can be removed by twisting anti-clockwise. If a square cover is fitted, release the spring clip, and withdraw the cover.
12 Pull the sidelight bulbholder from the rear of the headlight **(see illustration)**.
13 The bulb is a bayonet-fit in the bulbholder.
14 Refitting is a reversal of the removal procedure, ensuring that the bulbholder seal is in good condition.

Front direction indicator light

Front wing-mounted light

15 The indicator light unit is mounted next to the headlight unit.
16 Working in the engine compartment, at the rear of the direction indicator light unit, twist the bulbholder anti-clockwise to release it from the light unit.
17 The bulb is a bayonet-fit in the bulbholder.
18 Refitting is a reversal of removal, ensuring that the bulbholder seal is in good condition.

Front bumper-mounted light

19 The indicator light unit is mounted in the front bumper.
20 Push one side of the light unit into the bumper, then slide a thin screwdriver between the side of the light unit and the bumper, and push the screwdriver to release the light unit securing clip **(see illustration)**. If necessary, repeat the procedure to release the remaining securing clip.

21 Pull the light unit forwards from the bumper.
22 Release the securing clips and withdraw the lens from the light unit for access to the bulb.
23 The bulb is a bayonet-fit in the bulbholder.
24 Refitting is a reversal of removal, but ensure that the light unit securing clips are securely engaged.

Integral headlight/direction indicator assembly

25 Working in the engine compartment, at the rear of the headlight, twist the direction indicator light bulbholder, and pull it from the light unit **(see illustration)**.
26 The bulb is a bayonet-fit in the bulbholder.
27 Refitting is a reversal of removal, ensuring that the bulbholder seal is in good condition.

Front foglight

28 Working at the front of the light unit, unscrew the two securing screws, and withdraw the light unit from the bumper.
29 Disconnect the light unit wiring connector.
30 Twist the bulbholder anti-clockwise, and withdraw it from the light unit.
31 Pull the bulb from the bulbholder.
32 When handling the new bulb, use a tissue or clean cloth, to avoid touching the glass with the fingers; moisture and grease from the skin can cause blackening and rapid failure of this type of bulb. If the glass is accidentally touched, wipe it clean using methylated spirit.
33 Using a tissue or clean cloth, push the new bulb into position in the bulbholder, then refit the bulbholder and refit the light unit using a reversal of the removal procedure.

Front direction indicator side repeater light

34 Grip the light unit lens, and pull the light unit from the wing panel.
35 Pull the bulbholder from the rear of the light unit **(see illustration)**.
36 The bulb is a push-fit in the bulbholder.
37 Refitting is a reversal of removal, but ensure that the light unit is securely engaged with the wing panel.

Rear light cluster

38 Working in the luggage compartment,

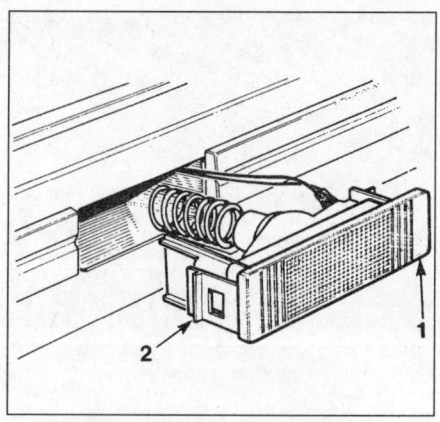

5.20 Front bumper-mounted front direction indicator light

1 Lens
2 Light unit securing clip

unscrew the wing nuts (two or three nuts, depending on model) securing the light unit to the rear body **(see illustration)**. Recover the washers.
39 Withdraw the light unit from the rear of the vehicle, and depress the retaining clips to release the bulbholder from the rear of the light unit **(see illustrations)**.
40 The bulbs are a bayonet-fit in the bulbholder.
41 Refitting is a reversal of the removal procedure.

Number plate light

42 Remove the securing screw, and withdraw the light unit from the rear bumper.
43 Disconnect the wiring plug from the light unit.
44 Where applicable, unclip the lens assembly from the light unit for access to the bulb.
45 Pull the festoon-type bulb from its contacts.
46 Fit the new bulb using a reversal of the removal procedure, but check the tension of the spring contacts, and if necessary bend them so that they firmly contact the bulb end caps.

5.25 Removing a direction indicator bulbholder - integral headlight/direction indicator assembly

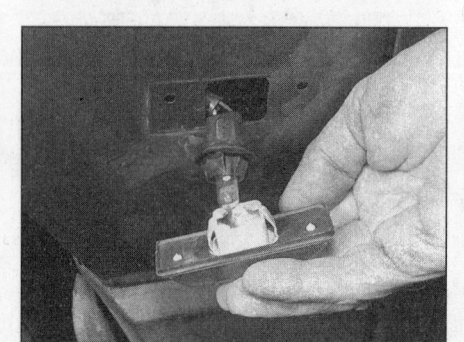

5.35 Removing the direction indicator side repeater light bulbholder

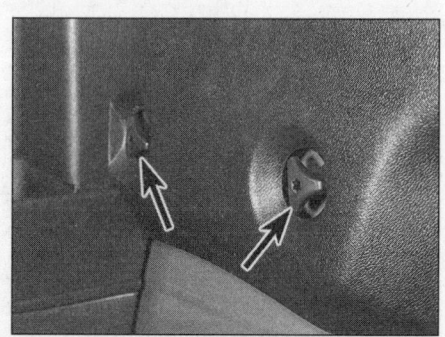

5.38 Rear light cluster securing nuts (arrowed)

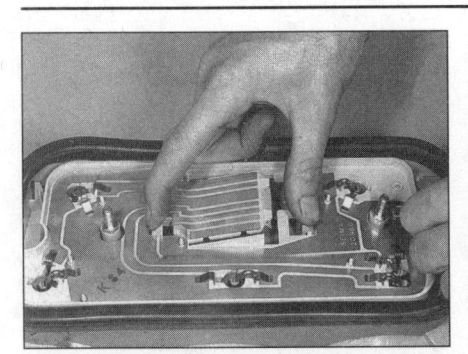

Body electrical systems 12•7

5.39a Release the bulbholder retaining clips ...

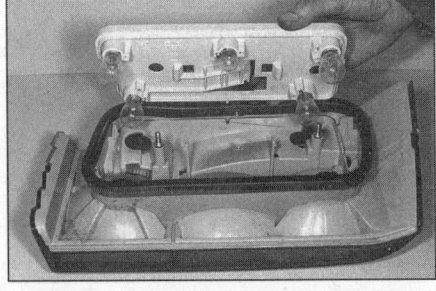

5.39b ... and withdraw the bulbholder

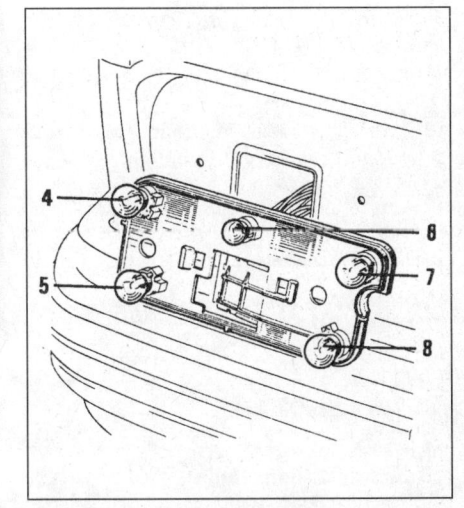

5.39c Rear light cluster bulb identification

4 Tail/stop light
5 Direction indicator light
6 Tail light
7 Rear foglight
8 Reversing light

6 Bulbs (interior lights) - renewal

General
1 Refer to Section 5, paragraph 1.

Courtesy light
2 Unclip the lens from the courtesy light, then unclip the bulb from its contacts **(see illustration)**.
3 Refitting is a reversal of removal.

Front map reading light
4 Fold back the sunvisors, then remove the securing screws and withdraw the roof console **(see illustration)**.

5 Working at the rear of the light, twist the bulbholder anti-clockwise and remove it **(see illustration)**.
6 The bulb is integral with the bulbholder.

Rear map reading light bulb
7 Pull the seat back side squab upwards to release the upper hook from the securing lug on the body, then unhook the lower end of the squab, and withdraw the squab.
8 Remove the two front screws, and the two side screws, and withdraw the parcel shelf side support panel. Disconnect the loudspeaker wiring.
9 Unclip the rear pillar trim panel for access to the rear of the light unit.
10 Working at the rear of the light, twist the bulbholder anti-clockwise and remove it.
11 The bulb is integral with the bulbholder.
12 Refitting is a reversal of removal.

Luggage compartment and glovebox light
13 Carefully prise the light assembly from the luggage compartment trim panel, or facia, as applicable, then unclip the bulb from its contacts **(see illustration)**.
14 Refitting is a reversal of removal.

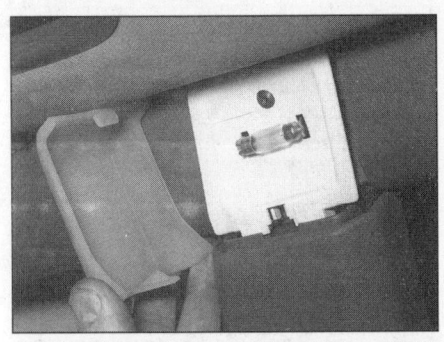

6.2 Unclip the lens from the courtesy light

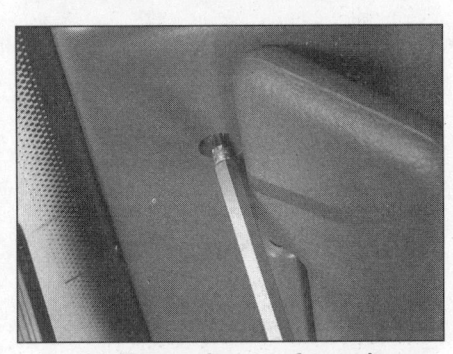

6.4 Unscrewing a roof console securing screw

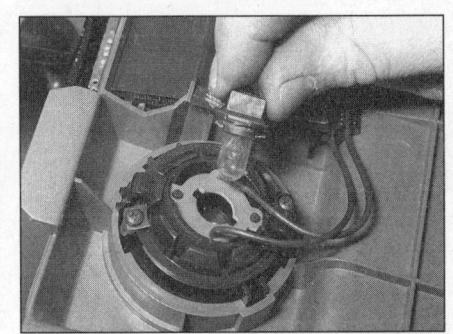

6.5 Removing the front map reading light bulb

6.13 Prise the luggage compartment light from the trim panel

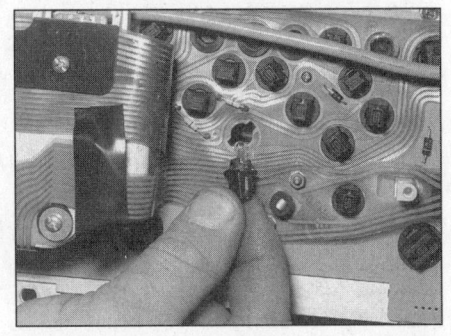

6.16 Removing an instrument panel bulb

6.19 Removing the clock/outside temperature display illumination bulb

12

12•8 Body electrical systems

Instrument panel illumination and warning light bulbs

15 Remove the instrument panel as described in Section 9.
16 Twist the relevant bulbholder anti-clockwise to remove it from the rear of the panel (see illustration).
17 The bulbs are integral with the bulbholders.
18 Refitting is a reversal of removal.

Clock/temperature display illumination bulb

19 Proceed as described previously in paragraphs 9 to 12 for the instrument panel bulbs (see illustration).

7 Exterior light units - removal and refitting

Note: *Disconnect the battery negative lead before removing any light unit, and reconnect the lead after refitting the unit.*

Headlight

Removal

1 On models with a front direction indicator light unit mounted next to the headlight, remove the direction indicator light unit as described later in this Section.
2 On models fitted with headlight wipers, remove the headlight wiper arms, with reference to Section 16.
3 Remove the radiator grille panel as described in Chapter 11.
4 Working in the engine compartment, remove the cover(s) from the rear of the headlight. If a round cover is fitted, it can be removed by twisting anti-clockwise. If a square cover is fitted, release the spring clip, and withdraw the cover.
5 Disconnect the wiring plugs from the bulbs mounted in the headlight unit.
6 Twist the headlight beam adjuster unit anti-clockwise, and carefully pull it from the headlight (see illustration). Move the adjuster to one side, leaving the control line connected.
7 Working at the top of the headlight, unscrew the upper bolts securing the light unit to the body front panel (see illustration).
8 Working inside the engine compartment, behind the headlight, unscrew the two headlight securing nuts, and withdraw the headlight forwards from the body panel (see illustrations). Note that, where applicable, it may be necessary to remove the headlight wiper motor for access to one of the securing nuts.

Refitting

9 Refitting is a reversal of removal, but on completion, have the headlight beam alignment checked with reference to Section 8.

7.6 Pull the headlight beam adjuster unit (arrowed) from the rear of the headlight

7.7 Headlight upper securing bolts (arrowed)

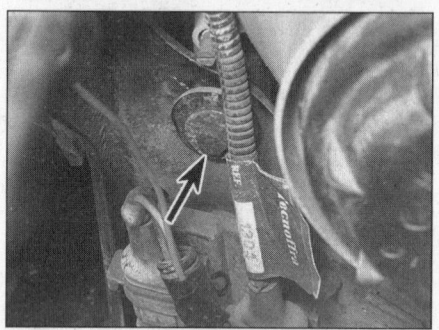

7.8a Headlight inner . . .

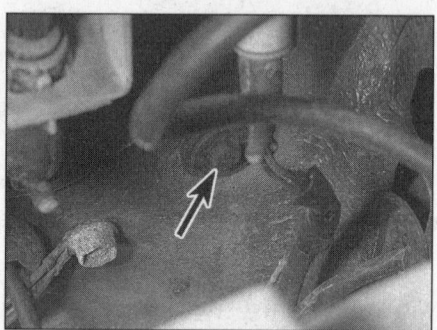

7.8b . . . and outer securing nuts (arrowed)

Front direction indicator light

Front wing-mounted light

10 Working in the engine compartment, at the rear of the direction indicator light unit, twist the bulbholder anti-clockwise to release it from the light unit.
11 Again working behind the headlight, unhook the indicator light unit retaining spring from the lug on the body, then withdraw the light unit forwards from the wing panel.
12 Refitting is a reversal of removal, but ensure that the retaining spring is securely engaged with the lug on the body.

Front bumper-mounted light

13 Removal and refitting is described as part of the bulb renewal procedure in Section 5.

Front foglight

14 Removal and refitting is described as part of the bulb renewal procedure in Section 5.

Front direction indicator side repeater light

15 Removal and refitting is described as part of the bulb renewal procedure in Section 5.

Rear light cluster

16 Removal and refitting is described as part of the bulb renewal procedure in Section 5.

Number plate light

17 Removal and refitting is described as part of the bulb renewal procedure in Section 5.

8 Headlight beam alignment - general information

1 Accurate adjustment of the headlight beam is only possible using optical beam-setting equipment, and this work should therefore be carried out by a Renault dealer or suitably-equipped workshop.
2 For reference, the headlights can be finely adjusted using a flat-bladed screwdriver to rotate the adjuster screws fitted to the rear of each light unit. The type and location of the adjuster screws varies depending on the type of light unit fitted (see illustrations). Prior to adjustment, ensure that the vehicle is unladen, and that the adjuster units (see below) are both set to position "0.
3 Each headlight unit is equipped with a four-position vertical beam adjuster unit - this can be used to adjust the headlight beam, to compensate for the load which the vehicle is carrying. An adjuster switch is provided on the facia. The adjuster knob should be positioned as follows according type, and the load being carried in the vehicle:

Position 0	Front seat(s) occupied, luggage compartment empty
Position 1	Front and rear seats occupied, luggage compartment empty
Position 2	Front and rear seats occupied and luggage compartment fully loaded
Position 3	Driver's seat only occupied and luggage compartment fully loaded

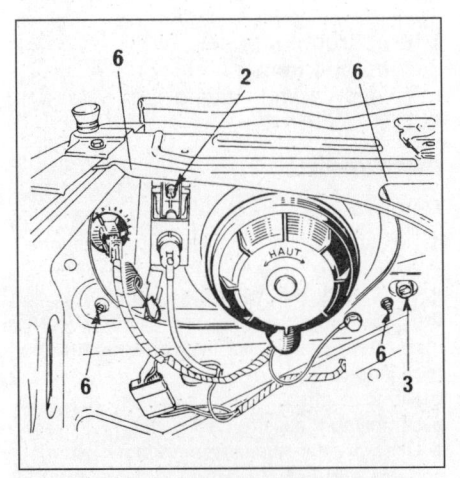

Body electrical systems 12•9

8.2a Early single headlight adjuster screws

2 Vertical beam adjuster screw
3 Horizontal beam adjuster screw
6 Headlight mounting nuts and bolts

Headlight adjustment system - components removal and refitting

Adjuster switch

4 Working in the passenger compartment, pull the knob from the switch (see illustration).
5 Depress the securing clip, and remove the switch surround (see illustration).

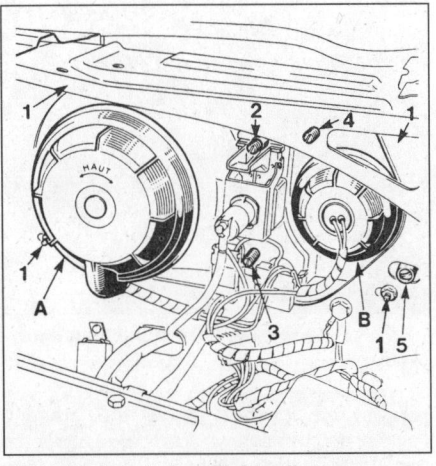

8.2b Early twin headlight adjuster screws

A Main headlight (main/dipped beam) unit
B Auxiliary (main beam) unit
1 Mounting nuts and bolts
2 Vertical beam adjuster screw (main unit)
3 Horizontal beam adjuster screw (main unit)
4 Vertical beam adjuster screw (auxiliary unit)
5 Horizontal beam adjuster screw (auxiliary unit)

6 Unscrew the switch securing screw(s), then depress the securing clip and push the switch from the facia (see illustrations).
7 Working in the engine compartment, at the rear of each headlight, twist the headlight beam adjuster unit anti-clockwise, and carefully pull it from the headlight (see illustrations).

8.2c On later models, the vertical beam adjuster screw can be reached through the hole in the body front panel

8 The control lines and adjusters must now be fed through the bulkhead into the passenger compartment. It should be possible to do this by removing the lower facia trim panel and reaching up behind the facia, but on some models, it may be necessary to remove the complete facia assembly (see Chapter 11).
9 Release the adjuster control lines from any clips and/or brackets in the engine compartment, noting the routing to ensure correct refitting.
10 Withdraw the complete adjuster assembly, complete with bulkhead grommet, from the engine compartment.

Refitting

11 Refitting is a reversal of removal, bearing in mind the following points..
 a) Make sure that the control lines are correctly routed and secured as noted before removal.
 b) Where applicable, refit the facia assembly with reference to Chapter 11.
 c) On completion, check the operation of the adjustment mechanism, and check the headlight beam alignment with reference to Section 8.

8.4 Pull the knob from the headlight adjuster switch

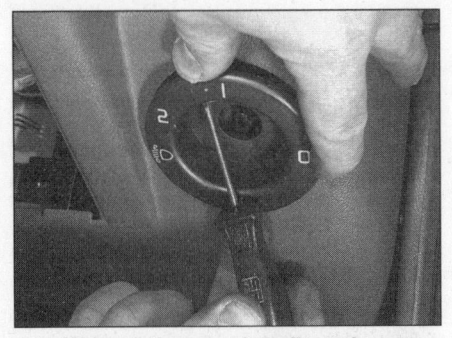

8.5 Depress the securing clip and remove the switch surround

8.6a Unscrew the securing screw . . .

8.6b . . . then depress the securing clip . . .

8.6c . . . and push the switch from the facia

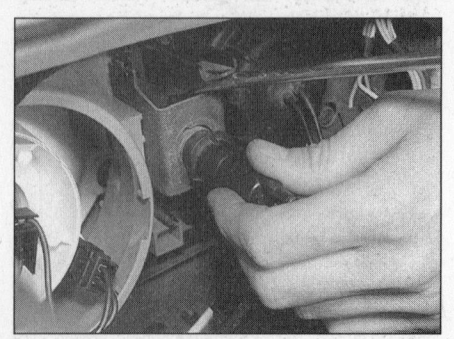

12•10 Body electrical systems

8.7a Twist the headlight beam adjuster . . .

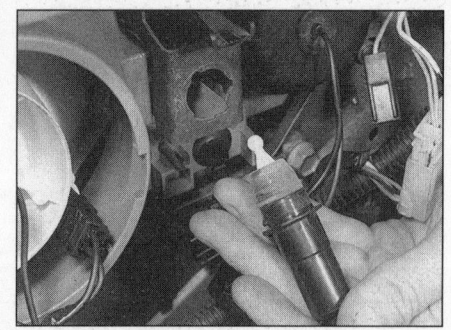

8.7b . . . to release it from the headlight

9 Instrument panel - removal and refitting

Removal

1 The instrument panel incorporates the heater control panel surround, and also the trip computer, clock/ temperature display unit, and radio display unit, where applicable.
2 Disconnect the battery negative lead.
3 Remove the steering wheel as described in Chapter 10.
4 Remove the steering column shrouds, with reference to Chapter 11 if necessary.
5 Carefully prise out the trim panel from the bottom of the heater control panel **(see illustration)**.
6 Working at the lower edge of the panel, remove the panel securing screws. Two screws are located above the steering column, and two screws are located under the trim panel, below the heater controls **(see illustrations)**.
7 Pull the instrument panel forwards from the facia, until the wiring plugs and, where applicable, the speedometer cable, and any hoses, can be disconnected **(see illustrations)**.

Refitting

8 Refitting is a reversal of removal, but refit the steering wheel with reference to Chapter 10.

10 Instrument panel components - removal and refitting

Instruments

Removal

1 Remove the instrument panel as described in Section 9.
2 Working at the rear of the panel, unscrew the securing screws, and withdraw the instrument panel pod from the main panel **(see illustration)**.
3 Remove the screws securing the lens assembly to the instrument panel.
4 Unscrew the relevant instrument securing screws and/or nuts, then disconnect the wiring circuits from the rear of the instrument. To release the circuit connectors, use a small plastic strip, and push the strip into the fold in the circuit, before hooking the circuit from the instrument **(see illustration)**. Take care, as the circuits are easily damaged. Also take note of which way round the circuits connect, to ensure correct refitting.
5 Withdraw the instrument from the front of the panel.

Refitting

6 Refitting is a reversal of removal, bearing in mind the following points.
  a) When reconnecting the circuit connectors, use the plastic strip, as during removal to push the connectors fully home. It is very important to ensure

9.5 Prise the trim panel from the bottom of the heater control panel

9.6a Instrument panel securing screws (arrowed) located above steering column

9.6b Instrument panel securing screws located below heater controls

9.7a Pull the instrument panel from the facia . . .

9.7b . . . and disconnect the wiring plugs

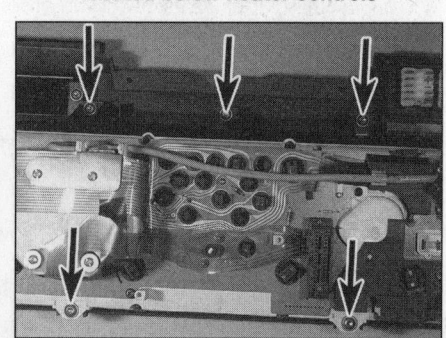

10.2 Instrument panel pod securing screws (arrowed)

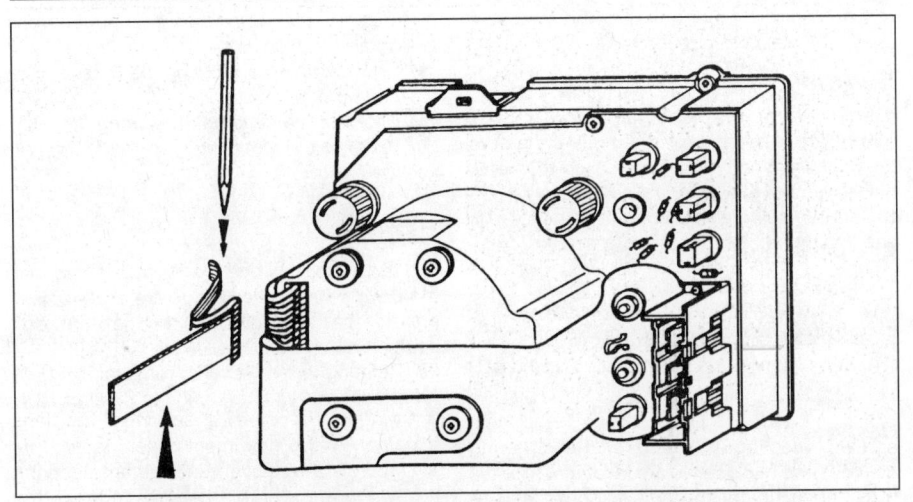

Body electrical systems 12•11

10.4 Method of releasing circuit connectors from rear of instruments

that the connectors are reconnected the correct way round, as noted before removal.
b) Where applicable, ensure that the metal connecting clip between the main printed circuit and the tachometer circuit is in place - this is vital to prevent a short circuit when the ignition is switched on.

Instrument panel bulbs

7 Remove the instrument panel as described in Section 9.
8 Twist the relevant bulbholder anti-clockwise to remove it from the rear of the panel.
9 The bulbs are integral with the bulbholders.
10 On completion, refit the instrument panel with reference to Section 9.

11 Cigarette lighter - removal and refitting

Removal

1 Two cigarette lighters may be fitted, one under the front ashtray flap, and one in the rear centre console.
2 Disconnect the battery negative lead, and pull out the cigarette lighter element.
3 Using a small flat-bladed screwdriver, prise out the cigarette lighter body, then disconnect the wiring plugs.

Refitting

4 Refitting is a reversal of removal.

12 Clock/temperature display unit - removal and refitting

Removal

1 The clock/temperature display unit panel is integral with the main instrument panel.
2 Remove the instrument panel as described in Section 9.
3 Working at the rear of the panel, remove the securing screws, and withdraw the clock/temperature display panel from the instrument panel **(see illustration)**.

Refitting

4 Refitting is a reversal of removal.

13 Horn(s) and compressor - removal and refitting

Horns
Removal

1 The horns are located on the front body panel, behind the front bumper.
2 Disconnect the battery negative lead.
3 Remove the front bumper as described in Chapter 11.
4 Unscrew the securing nut, then withdraw the horn and disconnect the wiring plug **(see illustration)**.

Refitting

5 Refitting is a reversal of removal.

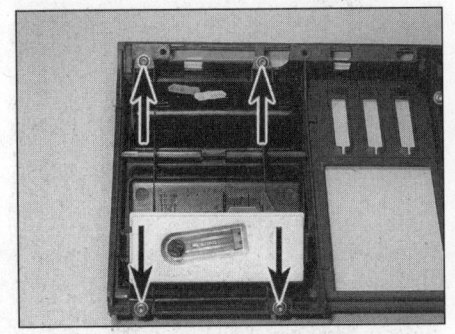

12.3 Clock temperature display unit securing screws (arrowed)

Compressor
Removal

6 Some models are fitted with air horns, powered by an electric compressor
7 Disconnect the battery negative lead.
8 To improve access, apply the handbrake, then jack up the front of the vehicle and support on axle stands (see *"Jacking and vehicle support"*).
9 The compressor is located under the left-hand wing panel.
10 Disconnect the wiring plug(s) and the air hoses, then unbolt the compressor from its mounting bracket.

Refitting

11 Refitting is a reversal of removal.

14 Speedometer cable/sensor - removal and refitting

Speedometer cable
Removal

Note: *On models with a cruise control system, the cable is in two sections, with the road speed sensor connected between the cable sections. In this case, the end of the relevant cable section must be disconnected from the speed sensor.*

1 Working in the engine compartment or under the vehicle, as applicable (the location varies according to gearbox type), unscrew the securing sleeve, and disconnect the end of the cable from the transmission.
2 Remove the instrument panel as described in Section 9.
3 Note the routing of the cable, and release the cable from any clips and/or brackets in the engine compartment.
4 Withdraw the cable through the bulkhead into the passenger compartment, and withdraw it from the vehicle.

Refitting

5 Refitting is a reversal of removal, but ensure that the cable is correctly routed as noted before removal.

13.4 Horn bracket securing bolt (arrowed)

12•12 Body electrical systems

Speedometer sensor

Removal

6 The location of the sensor in the transmission varies according to transmission type. Working in the engine compartment or under the vehicle, as applicable, disconnect the sensor wiring plug.
7 Unbolt or unclip the sensor, as applicable.

Refitting

8 Refitting is a reversal of removal.

15 Wiper arm - removal and refitting

Removal

1 Operate the wiper motor, then switch it off so that the wiper arm returns to the at-rest position.

> **HAYNES HiNT** Stick a piece of masking tape along the edge of the wiper blade, to use as an alignment aid on refitting.

2 Lift up the wiper arm spindle nut cover, then slacken and remove the spindle nut **(see illustration)**. Lift the blade off the glass, and pull the wiper arm off its spindle. If necessary, the arm can be levered off the spindle using a suitable flat-bladed screwdriver.

3 If a headlight wiper arm is being removed, unclip the washer nozzle from the arm.

Refitting

4 Ensure that the wiper arm and spindle splines are clean and dry, then refit the arm to the spindle. Where applicable, align the wiper blade with the tape fitted on removal.
5 Refit the spindle nut, tightening it securely, and clip the nut cover back into position.

16 Windscreen wiper motor and linkage - removal and refitting

Removal

1 Disconnect the battery negative lead.
2 Remove the wiper arms as described in Section 15.
3 Working at the base of the windscreen, remove the scuttle grille panel securing screws, and remove the grille panel **(see illustration)**.
4 Remove the securing screws, and withdraw the scuttle cover panel **(see illustrations)**.
5 Disconnect the wiper motor wiring plug **(see illustration)**.
6 Unscrew the two plastic nuts, and withdraw the heater control vacuum reservoir. Move the reservoir to one side, leaving the vacuum hoses connected.
7 Unscrew the bolts securing the wiper motor and the linkage, then manipulate the assembly out from the scuttle **(see illustrations)**.

8 If desired, the motor can be unbolted from the linkage as follows.
 a) Unscrew the nut securing the linkage to the motor spindle.
 b) Unscrew the three bolts securing the motor to the mounting plate, and remove the motor.

Refitting

9 If the motor has been removed, ensure that the drive links on the linkage are aligned, and ensure that the motor is in the "parked" position before refitting.
10 Refitting is a reversal of removal, but refit the wiper arms with reference to Section 15.

17 Tailgate wiper motor - removal and refitting

Removal

1 Disconnect the battery negative lead.
2 Remove the wiper arm, with reference to Section 15.
3 Open the tailgate, then remove the securing screws, and remove the trim panel from the rear of the tailgate.
4 Working outside the tailgate, unscrew the motor spindle nut, and recover the washer and trim plate.
5 Reach inside the tailgate and disconnect the wiper motor wiring connector **(see illustration)**.

15.2 Unscrewing a wiper arm spindle nut

16.3 Removing a scuttle grille panel securing screw

16.4a Removing a scuttle cover panel upper securing screw . . .

16.4b . . . and side securing screw

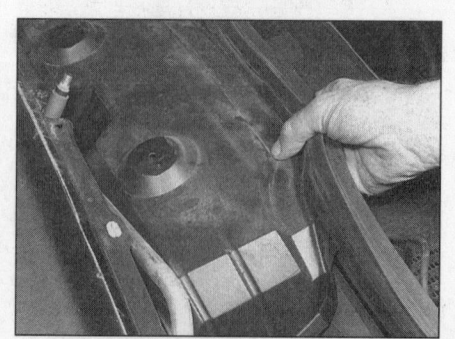

16.4c Removing the scuttle plastic cover panel

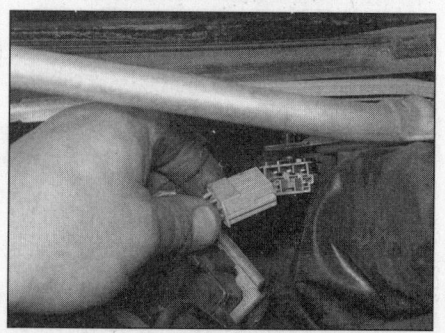

16.5 Disconnecting the wiper motor wiring plug

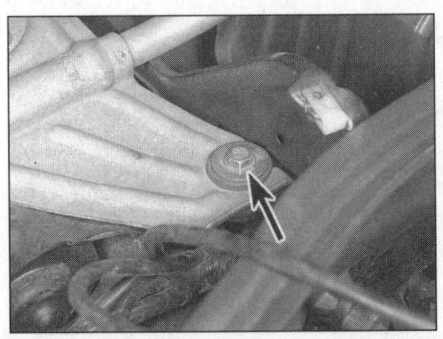

Body electrical systems 12•13

16.7a Unscrew the bolts securing the wiper motor . . .

16.7b . . . and the linkage . . .

16.7c . . . and withdraw the assembly

6 Unscrew the securing bolts, and manipulate the motor assembly out from the tailgate.

Refitting

7 Refitting is a reversal of removal, but ensure that the motor is in the "parked" position before refitting the wiper arm.

18 Washer system components - removal and refitting

Washer fluid reservoir

Removal

1 The reservoir is located behind the right-hand front wheel arch.
2 Empty the reservoir to minimise fluid spillage
3 Apply the handbrake, then jack up the front of the vehicle, and support securely on axle stands (see *"Jacking and vehicle support"*). Remove the right-hand roadwheel.
4 Remove the securing screws and clips, and withdraw the splash shield and the wheel arch liner to expose the reservoir.
5 Disconnect the wiring plug(s) from the washer pump(s) located in the reservoir.
6 Disconnect the fluid pipe(s) from the pump(s). Be prepared for fluid spillage if there is still fluid in the reservoir.
7 Unscrew the reservoir securing bolts, and manipulate the reservoir out from under the wing.

Refitting

8 Refitting is a reversal of removal.

Washer fluid pump

Note: *Prior to removing the pump, empty the contents of the reservoir, or be prepared for fluid spillage.*

Removal

9 Proceed as described in paragraphs 2 to 6.
10 Carefully ease the pump out of the sealing grommet.

Refitting

11 Refitting is a reversal of removal, but check the condition of the sealing grommet, and renew if necessary.

Windscreen washer jet

Removal

12 Remove the wiper arms, as described in Section 15.
13 Remove the securing screws, and withdraw the centre scuttle cover panel.
14 Working at the rear of the panel, disconnect the fluid hose from the washer nozzle, then squeeze the securing clips, and push the nozzle from the panel **(see illustration)**.

Refitting

15 Refitting is a reversal of removal.

Headlight washer jet

16 Lift the wiper arm, then unclip the nozzle from the arm, and disconnect the fluid hose.
17 Refitting is a reversal of removal.

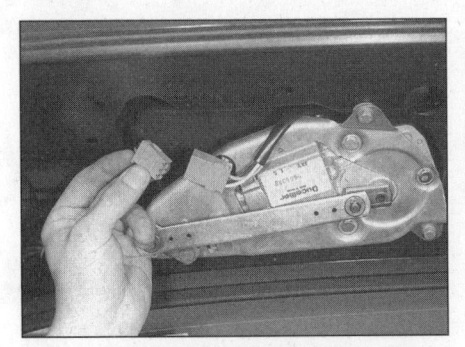

17.5 Disconnecting the tailgate wiper motor wiring plug

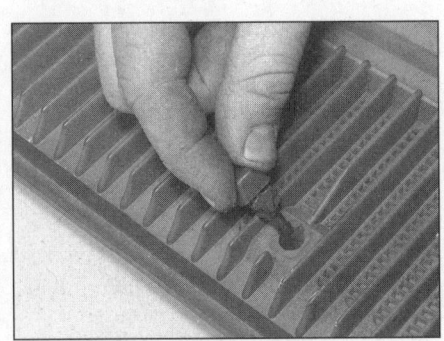

18.14 Removing a windscreen washer nozzle

19 Radio/cassette player - removal and refitting

Note: *On models with a security-coded radio/cassette player, once the battery has been disconnected, the unit cannot be re-activated until the appropriate security code has been entered. Do not remove the unit unless the appropriate code is known. The following information applies to radio/cassette players having standard DIN fixings.*

Removal

1 Disconnect the battery negative lead.
2 Where applicable, unclip the ashtray from the facia, beneath the radio/cassette player.
3 In order to release the retaining clips, two DIN removal tools will be required. These tools comprise two U-shaped rods, with cut-outs in the ends, which engage with the radio/cassette player securing clips (these tools are often supplied with the vehicle when new if a DIN standard audio unit is fitted). Suitable tools can easily be obtained from car accessory shops or audio specialists.
4 Slide the removal tools into the holes in the front of the radio/cassette player, until they are felt to engage with the securing clips.
5 Pull the unit from the facia using the tools, until the wiring connector(s) and aerial lead can be disconnected from the rear of the unit.

Refitting

6 Reconnect the wiring plug(s) and the aerial lead, then push the unit into its housing until the securing clips engage.
7 Where applicable, refit the ashtray, then reconnect the battery negative lead.

20 Loudspeakers - removal and refitting

Front door-mounted loudspeaker

Removal

1 Disconnect the battery negative lead.

12•14 Body electrical systems

20.3a Disconnecting the wiring from the front door-mounted loudspeaker

20.3b Removing a front door-mounted loudspeaker housing

2 Grip the loudspeaker cover panel, and turn it through a quarter-turn anti-clockwise to release it from the door trim panel.
3 Remove the securing screws, then lift the loudspeaker from the door and disconnect the wiring plugs. Withdraw the speaker housing from the door **(see illustrations)**.

Refitting

4 Refitting is a reversal of removal.

Front facia-mounted loudspeaker

5 At the time of writing, no information was available for the removal and refitting of the facia-mounted loudspeaker.

Rear loudspeaker

Removal

6 Unclip the loudspeaker cover panel from the top of the rear parcel shelf support.
7 Reach under the loudspeaker and disconnect the wiring plug(s).
8 Unscrew the four securing screws, and withdraw the loudspeaker.

Refitting

9 Refitting is a reversal of removal.

21 Radio aerial - removal and refitting

Removal

1 Disconnect the battery negative lead.
2 Remove the radio/cassette player, and disconnect the aerial lead from the rear of the unit, as described in Section 19.
3 Where applicable, remove the two securing screws, and withdraw the remote control central locking remote control receiver unit from the roof panel. Disconnect the wiring plug and remove the unit.
4 Alternatively, where applicable, remove the securing screw(s) and withdraw the front trim panel from the roof.
5 Working through the small aperture in the headlining, unscrew the aerial mounting nut and recover the washers.

6 Working in the radio/cassette player aperture, tie a length of string to the end of the aerial lead. The string will greatly ease refitting.
7 Lift the aerial from the roof panel, and pull the lead through behind the facia, up the door pillar, and out through the top of the roof. Note that it may be necessary to reach behind the interior trim panels to ease removal of the lead.
8 Untie the string from the end of the aerial lead.

Refitting

9 Refitting is a reversal of removal, but use the string to pull the aerial lead into position.

22 Cruise control system - general information

The cruise control system is vacuum-operated, and comprises the following components.
a) Vacuum pump.
b) Vacuum-operated throttle actuator.
c) Electronic control unit (ECU).
d) Road speed sensor.
e) Control switches mounted on the centre console, steering wheel, and brake and clutch pedals.

The system allows a constant road speed to be maintained without the need to operate the accelerator pedal. The desired "cruising" speed can be set manually at speeds above 30 mph (50 km/h).

The speed is controlled by moving the throttle lever, by means of the throttle actuator. The throttle actuator is supplied with vacuum from the vacuum pump. The vacuum pump is controlled by the electronic control unit according to information provided by the road speed sensor and control switches.

The system does not affect the operation of the throttle pedal (although the pedal will move in accordance with the movement of the throttle actuator), and the system can be overridden at any time by depressing the accelerator, brake or clutch pedals, or using the main control switch.

23 Cruise control system components - removal and refitting

Vacuum pump

Removal

1 The unit is located under the right-hand headlight unit, behind the front bumper.
2 Disconnect the battery negative lead.
3 Remove the front bumper as described in Chapter 11.
4 Unscrew the two vacuum pump securing nuts, and lift the assembly from the body panel **(see illustration)**.
5 Remove the four securing screws and withdraw the pump cover for access to the wiring and vacuum connections **(see illustration)**.
6 Disconnect the wiring plugs and the vacuum hoses, noting their locations to aid refitting, then withdraw the pump.

Refitting

7 Refitting is a reversal of removal, but ensure that the pump wiring plugs and vacuum hoses are correctly reconnected as noted before removal.

Throttle actuator

Removal

8 Disconnect the battery negative lead.
9 Disconnect the vacuum hose from the actuator **(see illustration)**.
10 Disconnect the throttle operating rod/cable from the actuator.
11 Unscrew the securing bolts, and remove the actuator complete with its mounting bracket.

Refitting

12 Refitting is a reversal of removal, but on completion, check the adjustment of the throttle operating rod/cable as follows.

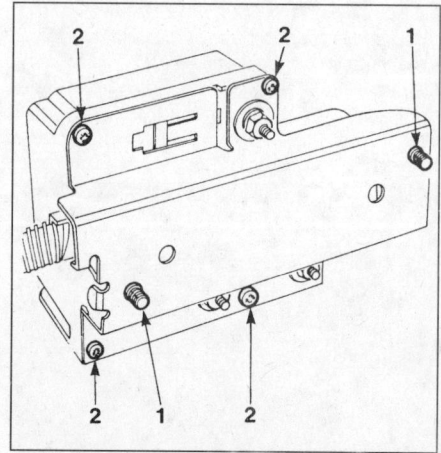

23.4 Cruise control system vacuum pump

1 Mounting studs
2 Cover securing screws

Body electrical systems 12•15

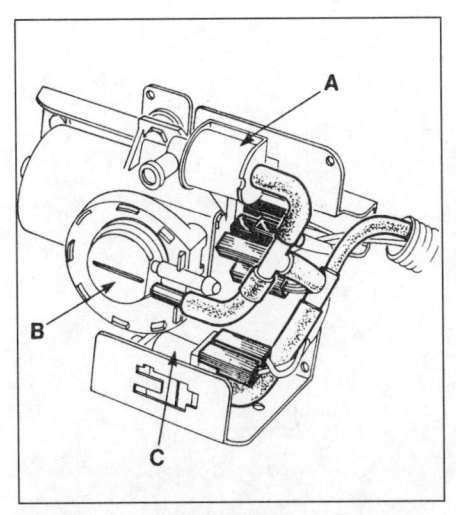

23.5 Cruise control system vacuum pump with cover removed to expose wiring and vacuum connections

A Solenoid valve
B Vacuum pump
C Solenoid valve

13 With the actuator in its rest (disengaged) position, and the throttle in the idle position, there should be 1.5 mm of free-play at the actuator end of the rod/cable.
14 On petrol engine models, the free-play can be adjusted by adjusting the length of the throttle operating rod.
15 On diesel engine models, the free-play can be adjusted by adjusting the position of the stop-clip in the slots on the throttle operating cable end fitting.

23.9 Cruise control throttle actuator (arrowed)

Electronic control unit
16 The unit is located behind the passenger's side of the facia, above the glovebox. For access to the control unit, remove the glovebox as described in Chapter 11.

Road speed sensor
17 On models fitted with an electronic speedometer, the road speed information is provided by the speedometer sensor on the transmission (see Section 14).
18 On models fitted with a conventional cable-driven speedometer, the sensor is incorporated in the speedometer cable (see Section 14).

Control switches
Steering wheel-mounted switches
19 No information was available for the removal and refitting of the steering wheel-mounted switches at the time of writing.

Centre console-mounted switch
20 Refer to Section 4.

Pedal switches
21 Refit to the procedure for stop-light switch removal and refitting in Chapter 9.

24 Trip computer - general information

Certain models are fitted with a trip computer, which provides data on fuel consumption and range. The unit analyses information supplied by a fuel flowmeter, the fuel gauge, and the speedometer/odometer.
The trip computer is mounted in the facia.

25 Voice synthesiser/warning systems - general information

Various voice synthesiser and warning systems may be fitted, depending on model. The systems operate from various sensors located around the vehicle. Any faults are most likely to be due to poor wiring connections or faulty switches.
In the event of a problem, consult a Renault dealer for advice.

12•16 Wiring diagrams

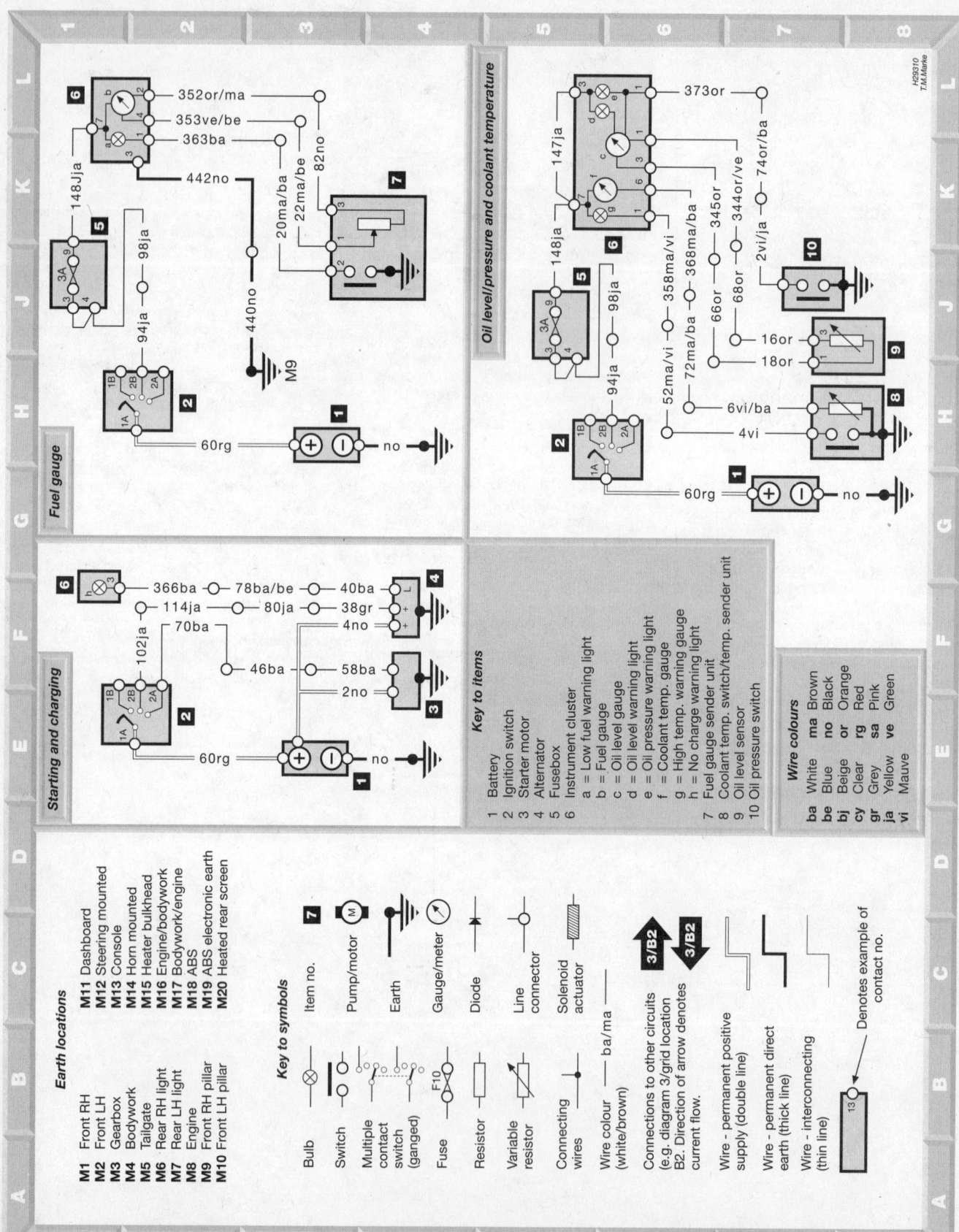

Diagram 1: Information for wiring diagrams, starting and charging, warning lights and gauges - typical

Earth locations

- M1 Front RH
- M2 Front LH
- M3 Gearbox
- M4 Bodywork
- M5 Tailgate
- M6 Rear RH light
- M7 Rear LH light
- M8 Engine
- M9 Front RH pillar
- M10 Front LH pillar
- M11 Dashboard
- M12 Steering mounted
- M13 Console
- M14 Horn mounted
- M15 Heater bulkhead
- M16 Engine/bodywork
- M17 Bodywork/engine
- M18 ABS
- M19 ABS electronic earth
- M20 Heated rear screen

Key to items

1. Battery
2. Ignition switch
3. Starter motor
4. Alternator
5. Fusebox
6. Instrument cluster
   - a = Low fuel warning light
   - b = Fuel gauge
   - c = Oil level gauge
   - d = Oil level warning light
   - e = Oil pressure warning light
   - f = Coolant temp. gauge
   - g = High temp. warning gauge
   - h = No charge warning light
7. Fuel gauge sender unit
8. Coolant temp. switch/temp. sender unit
9. Oil level sensor
10. Oil pressure switch

Wire colours

ba	White	ma	Brown
be	Blue	no	Black
bj	Beige	or	Orange
cy	Clear	rg	Red
gr	Grey	sa	Pink
ja	Yellow	ve	Green
vi	Mauve		

Key to symbols

- Bulb
- Switch
- Multiple contact switch (ganged)
- Fuse
- Resistor
- Variable resistor
- Connecting wires
- Wire colour — ba/ma (white/brown)
- Connections to other circuits (e.g. diagram 3/grid location B2. Direction of arrow denotes current flow.)
- Wire - permanent positive supply (double line)
- Wire - permanent direct earth (thick line)
- Wire - interconnecting (thin line)
- Item no.
- Pump/motor
- Earth
- Gauge/meter
- Diode
- Line connector
- Solenoid actuator
- Denotes example of contact no.

Wiring diagrams 12•17

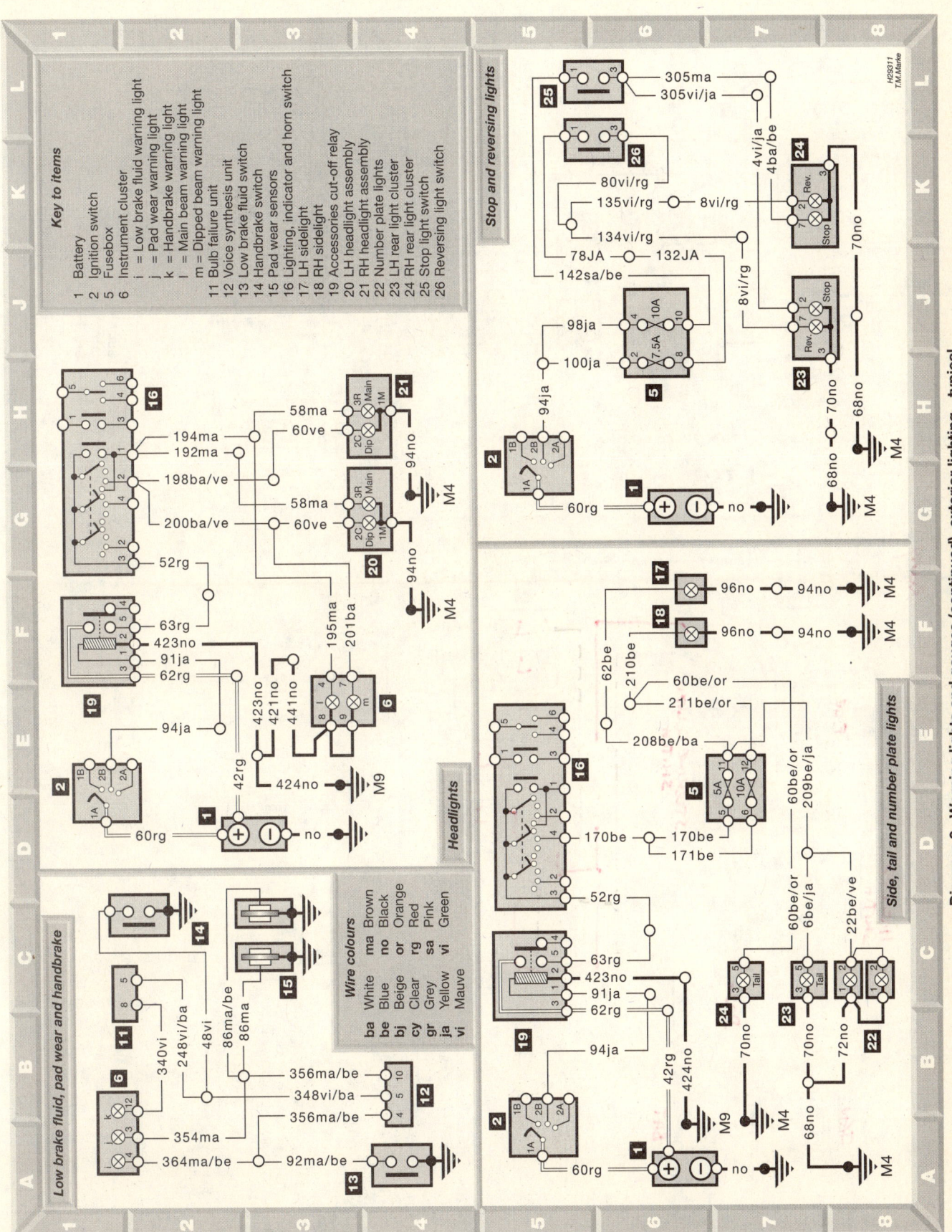

Key to items

1. Battery
2. Ignition switch
3. Fusebox
4. Instrument cluster
   - i = Low brake fluid warning light
   - j = Pad wear warning light
   - k = Handbrake warning light
   - l = Main beam warning light
   - m = Dipped beam warning light
5. Bulb failure unit
6. Voice synthesis unit
7. Low brake fluid switch
8. Handbrake switch
9. Pad wear sensors
10. Lighting, indicator and horn switch
11. LH sidelight
12. RH sidelight
13. Accessories cut-off relay
14. LH headlight assembly
15. RH headlight assembly
16. Number plate lights
17. LH rear light cluster
18. RH rear light cluster
19. Stop light switch
20. Reversing light switch

Wire colours

ba	White	ma	Brown
be	Blue	no	Black
bj	Beige	or	Orange
cy	Clear	rg	Red
gr	Grey	sa	Pink
ja	Yellow	vi	Mauve
		ve	Green

Diagram 2 : Warning lights and gauges (continued), exterior lighting - typical

- Stop and reversing lights
- Headlights
- Side, tail and number plate lights
- Low brake fluid, pad wear and handbrake

12•18 Wiring diagrams

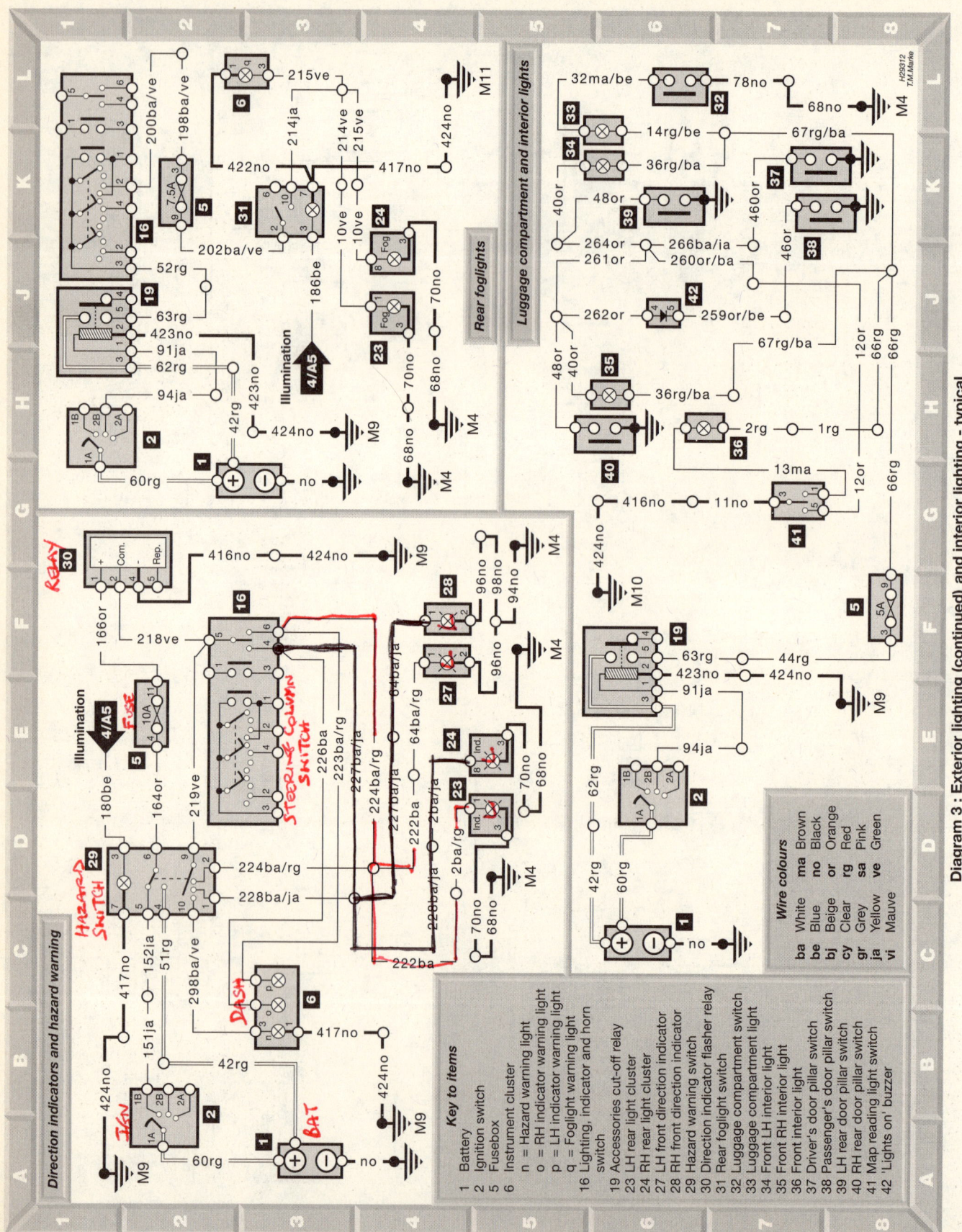

Diagram 3: Exterior lighting (continued) and interior lighting - typical

Wiring diagrams 12•19

Diagram 4: Interior lighting (continued), headlight and windscreen wash/wipe - typical

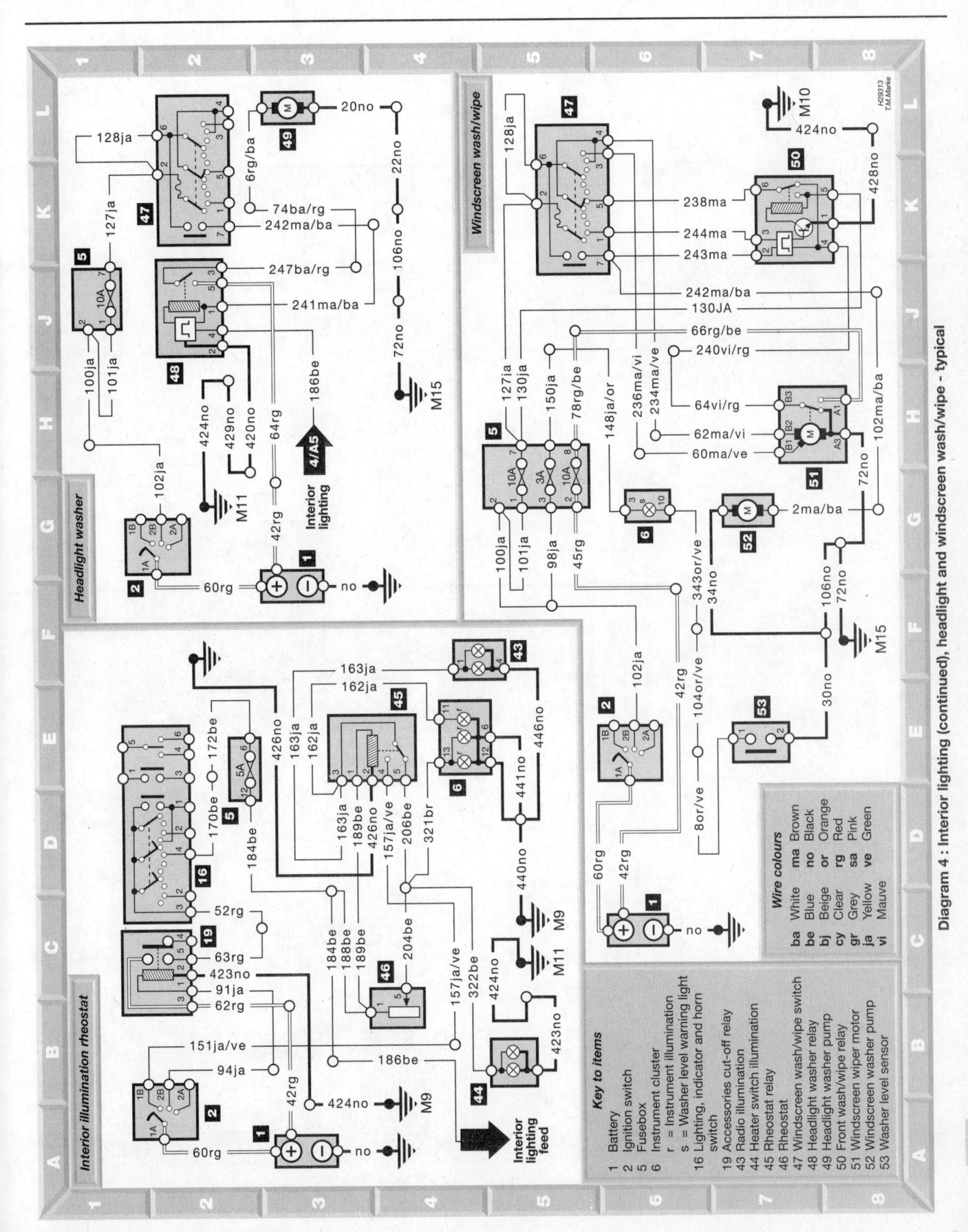

Wire colours

ba	White	ma	Brown
be	Blue	no	Black
bj	Beige	or	Orange
cy	Clear	rg	Red
gr	Grey	sa	Pink
ja	Yellow	ve	Green
vi	Mauve		

Key to items

1 Battery
2 Ignition switch
5 Fusebox
6 Instrument cluster
 r = Instrument illumination
 s = Washer level warning light
16 Lighting, indicator and horn switch
19 Accessories cut-off relay
43 Radio illumination
44 Heater switch illumination
45 Rheostat relay
46 Rheostat
47 Windscreen wash/wipe switch
48 Headlight washer relay
49 Headlight washer pump
50 Front wash/wipe relay
51 Windscreen wiper motor
52 Windscreen washer pump
53 Washer level sensor

12•20 Wiring diagrams

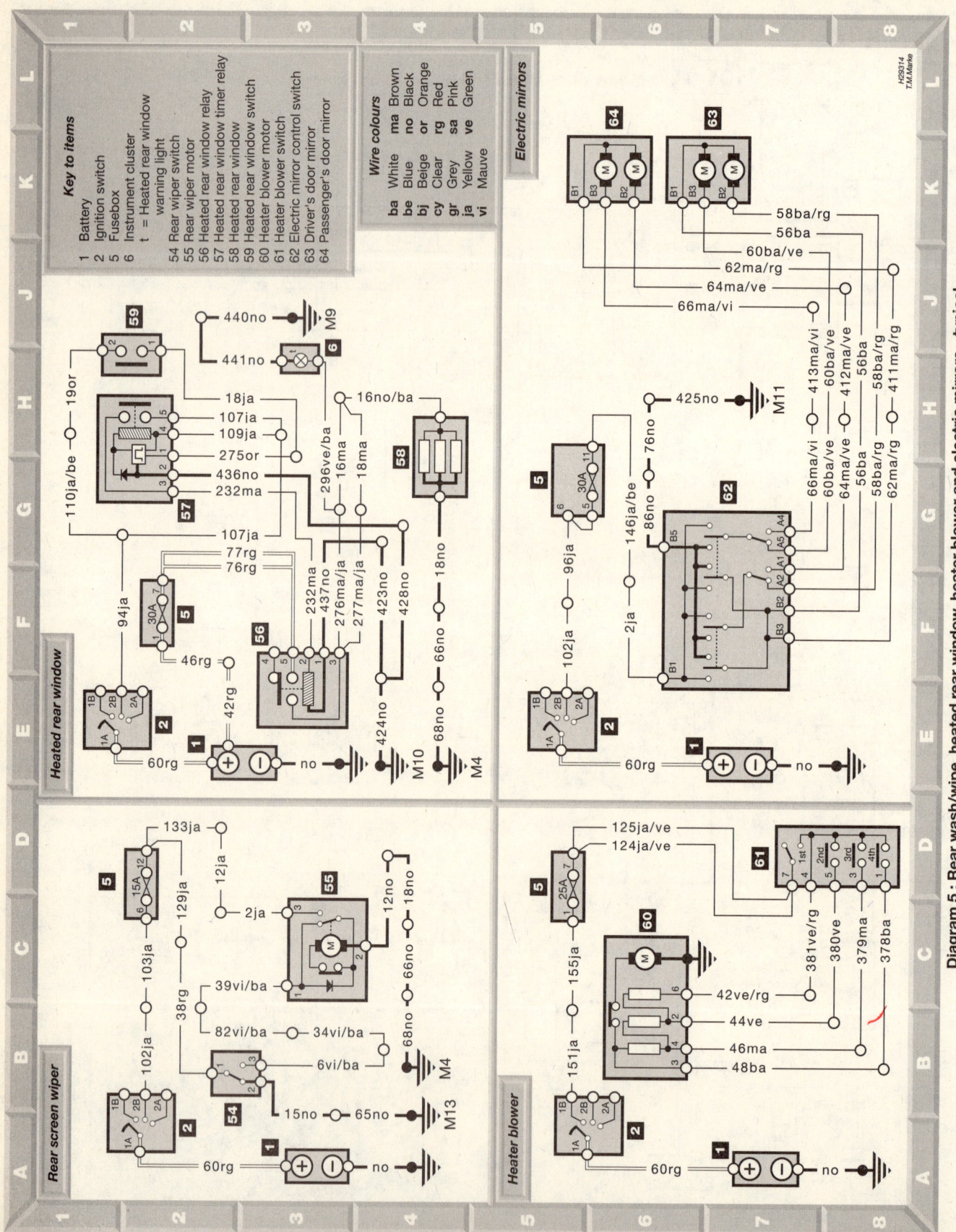

Diagram 5: Rear wash/wipe, heated rear window, heater blower and electric mirrors - typical

Wiring diagrams 12•21

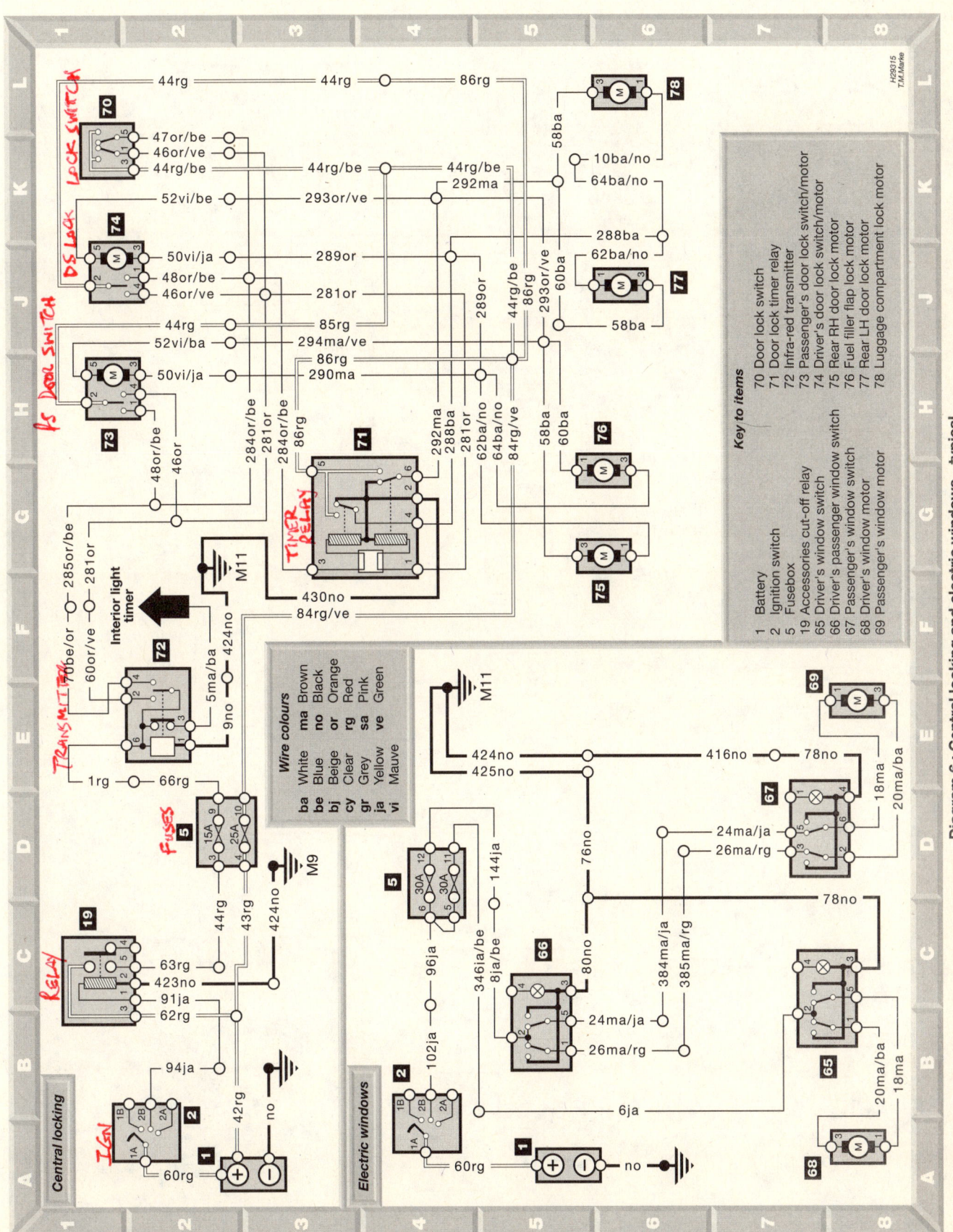

Diagram 6: Central locking and electric windows - typical

Key to items
1 Battery
5 Ignition switch
19 Fusebox
65 Accessories cut-off relay
66 Driver's window switch
67 Driver's passenger window switch
68 Driver's window motor
69 Passenger's window motor
70 Door lock switch
71 Door lock timer relay
72 Infra-red transmitter
73 Passenger's door lock switch/motor
74 Driver's door lock switch/motor
75 Rear RH door lock motor
76 Fuel filler flap lock motor
77 Rear LH door lock motor
78 Luggage compartment lock motor

Wire colours
ba White ma Brown
be Blue no Black
bj Beige or Orange
cy Clear rg Red
gr Grey sa Pink
ja Yellow ve Green
vi Mauve

12•22 Wiring diagrams

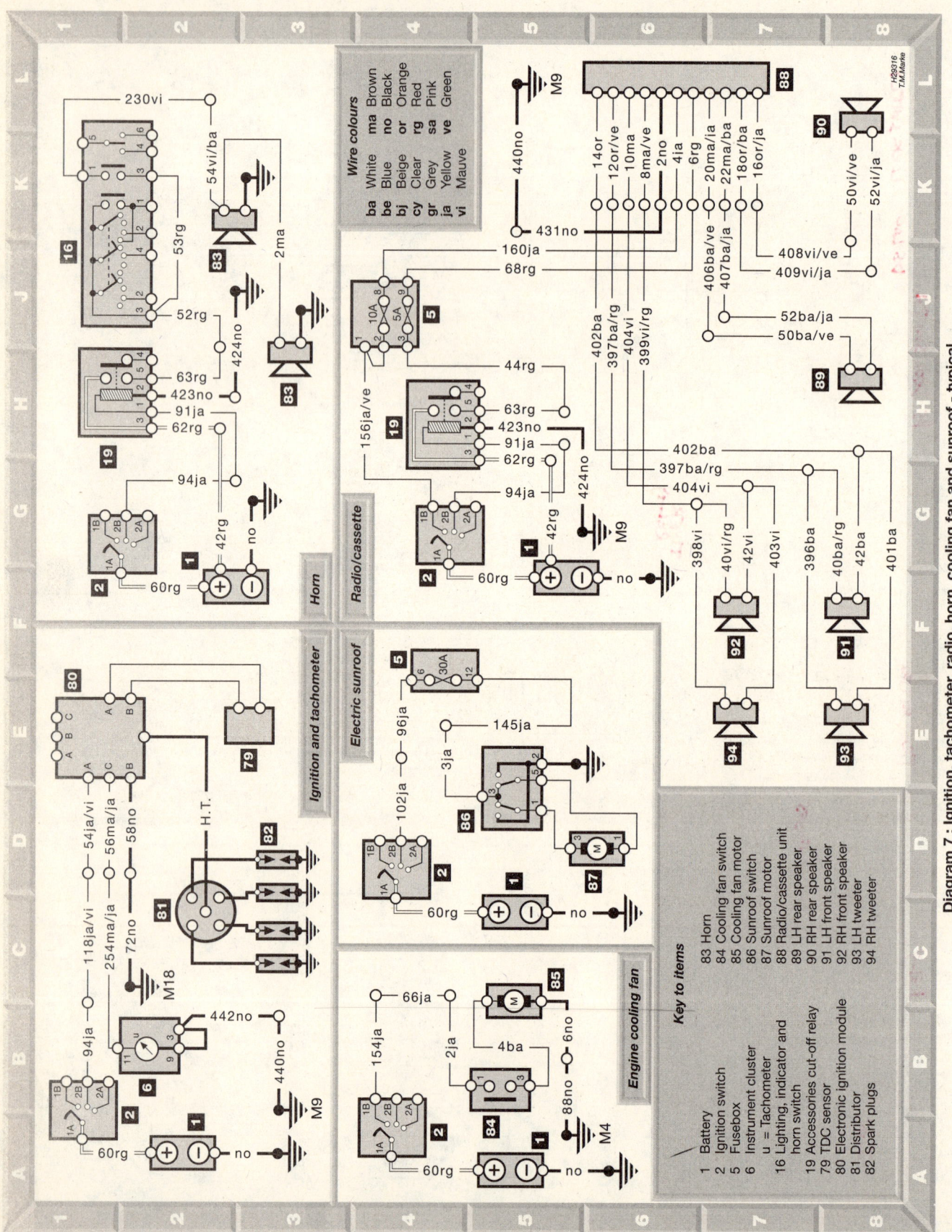

Diagram 7: Ignition, tachometer, radio, horn, cooling fan and sunroof - typical

Wire colours
ba White	ma Brown		
be Blue	no Black		
bj Beige	or Orange		
cy Clear	rg Red		
gr Grey	sa Pink		
ja Yellow	ve Green		
vi Mauve			

Key to items
1 Battery
2 Ignition switch
5 Fusebox
16 Lighting, indicator and horn switch
19 Accessories cut-off relay
79 TDC sensor
80 Electronic ignition module
81 Distributor
82 Spark plugs
83 Horn
84 Cooling fan switch
85 Cooling fan motor
86 Sunroof switch
87 Sunroof motor
88 Radio/cassette unit
89 LH rear speaker
90 RH rear speaker
91 LH front speaker
92 RH front speaker
93 LH tweeter
94 RH tweeter

u = Tachometer

Reference REF•1

Dimensions and weights .... **REF•1**
Conversion factors .... **REF•2**
Buying spare parts .... **REF•3**
Vehicle identification .... **REF•3**
General repair procedures .... **REF•4**
Jacking and vehicle support .... **REF•5**
Radio/cassette unit anti-theft system .... **REF•5**
Tools and working facilities .... **REF•6**
MOT test checks .... **REF•8**
Fault finding .... **REF•12**
Glossary of technical terms .... **REF•20**
Index .... **REF•26**

Dimensions and weights

Note: *All figures are approximate, and may vary according to model. Refer to manufacturer's data for exact figures.*

Dimensions
Overall length:
   "Phase I" models (up to 1989 model year) .... .4660 mm
   "Phase II" models (from 1989 model year) .... .4713 mm
Overall width (excluding wing mirrors)
   "Phase I" models (up to 1989 model year) .... .1780 mm
   "Phase II" models (from 1989 model year) .... .1806 mm
Overall height (unladen) .... .1415 mm
Wheelbase .... .2720 mm

Weights
Kerb weight:*
   Four-cylinder petrol engine models .... 1120 to 1235 kg
   Six-cylinder petrol engine models .... 1280 to 1350 kg
   Diesel engine models .... 1230 to 1270 kg

Weights (continued)
Maximum towing weight:**
   Unbraked trailer:
      Four-cylinder petrol engine models .... 570 to 615 kg
      Six-cylinder petrol engine models .... 650 to 675 kg
      Diesel engine models .... 595 to 625 kg
   Braked trailer:
      Four-cylinder petrol engine models .... 1300 to 1350 kg
      Six-cylinder petrol engine models .... 1500 kg
      Diesel engine models .... 1170 to 1200 kg
Maximum roof rack load .... 60 kg

*Depending on models and specification.
**Refer to a Renault dealer for exact recommendations

REF•2 Conversion factors

Length (distance)
Inches (in)	x 25.4	= Millimetres (mm)	x 0.0394	=	Inches (in)
Feet (ft)	x 0.305	= Metres (m)	x 3.281	=	Feet (ft)
Miles	x 1.609	= Kilometres (km)	x 0.621	=	Miles

Volume (capacity)
Cubic inches (cu in; in³)	x 16.387	= Cubic centimetres (cc; cm³)	x 0.061	=	Cubic inches (cu in; in³)
Imperial pints (Imp pt)	x 0.568	= Litres (l)	x 1.76	=	Imperial pints (Imp pt)
Imperial quarts (Imp qt)	x 1.137	= Litres (l)	x 0.88	=	Imperial quarts (Imp qt)
Imperial quarts (Imp qt)	x 1.201	= US quarts (US qt)	x 0.833	=	Imperial quarts (Imp qt)
US quarts (US qt)	x 0.946	= Litres (l)	x 1.057	=	US quarts (US qt)
Imperial gallons (Imp gal)	x 4.546	= Litres (l)	x 0.22	=	Imperial gallons (Imp gal)
Imperial gallons (Imp gal)	x 1.201	= US gallons (US gal)	x 0.833	=	Imperial gallons (Imp gal)
US gallons (US gal)	x 3.785	= Litres (l)	x 0.264	=	US gallons (US gal)

Mass (weight)
Ounces (oz)	x 28.35	= Grams (g)	x 0.035	=	Ounces (oz)
Pounds (lb)	x 0.454	= Kilograms (kg)	x 2.205	=	Pounds (lb)

Force
Ounces-force (ozf; oz)	x 0.278	= Newtons (N)	x 3.6	=	Ounces-force (ozf; oz)
Pounds-force (lbf; lb)	x 4.448	= Newtons (N)	x 0.225	=	Pounds-force (lbf; lb)
Newtons (N)	x 0.1	= Kilograms-force (kgf; kg)	x 9.81	=	Newtons (N)

Pressure
Pounds-force per square inch (psi; lbf/in²; lb/in²)	x 0.070	= Kilograms-force per square centimetre (kgf/cm²; kg/cm²)	x 14.223	=	Pounds-force per square inch (psi; lbf/in²; lb/in²)
Pounds-force per square inch (psi; lbf/in²; lb/in²)	x 0.068	= Atmospheres (atm)	x 14.696	=	Pounds-force per square inch (psi; lbf/in²; lb/in²)
Pounds-force per square inch (psi; lbf/in²; lb/in²)	x 0.069	= Bars	x 14.5	=	Pounds-force per square inch (psi; lbf/in²; lb/in²)
Pounds-force per square inch (psi; lbf/in²; lb/in²)	x 6.895	= Kilopascals (kPa)	x 0.145	=	Pounds-force per square inch (psi; lbf/in²; lb/in²)
Kilopascals (kPa)	x 0.01	= Kilograms-force per square centimetre (kgf/cm²; kg/cm²)	x 98.1	=	Kilopascals (kPa)
Millibar (mbar)	x 100	= Pascals (Pa)	x 0.01	=	Millibar (mbar)
Millibar (mbar)	x 0.0145	= Pounds-force per square inch (psi; lbf/in²; lb/in²)	x 68.947	=	Millibar (mbar)
Millibar (mbar)	x 0.75	= Millimetres of mercury (mmHg)	x 1.333	=	Millibar (mbar)
Millibar (mbar)	x 0.401	= Inches of water (inH$_2$O)	x 2.491	=	Millibar (mbar)
Millimetres of mercury (mmHg)	x 0.535	= Inches of water (inH$_2$O)	x 1.868	=	Millimetres of mercury (mmHg)
Inches of water (inH$_2$O)	x 0.036	= Pounds-force per square inch (psi; lbf/in²; lb/in²)	x 27.68	=	Inches of water (inH$_2$O)

Torque (moment of force)
Pounds-force inches (lbf in; lb in)	x 1.152	= Kilograms-force centimetre (kgf cm; kg cm)	x 0.868	=	Pounds-force inches (lbf in; lb in)
Pounds-force inches (lbf in; lb in)	x 0.113	= Newton metres (Nm)	x 8.85	=	Pounds-force inches (lbf in; lb in)
Pounds-force inches (lbf in; lb in)	x 0.083	= Pounds-force feet (lbf ft; lb ft)	x 12	=	Pounds-force inches (lbf in; lb in)
Pounds-force feet (lbf ft; lb ft)	x 0.138	= Kilograms-force metres (kgf m; kg m)	x 7.233	=	Pounds-force feet (lbf ft; lb ft)
Pounds-force feet (lbf ft; lb ft)	x 1.356	= Newton metres (Nm)	x 0.738	=	Pounds-force feet (lbf ft; lb ft)
Newton metres (Nm)	x 0.102	= Kilograms-force metres (kgf m; kg m)	x 9.804	=	Newton metres (Nm)

Power
Horsepower (hp)	x 745.7	= Watts (W)	x 0.0013	=	Horsepower (hp)

Velocity (speed)
Miles per hour (miles/hr; mph)	x 1.609	= Kilometres per hour (km/hr; kph)	x 0.621	=	Miles per hour (miles/hr; mph)

Fuel consumption*
Miles per gallon (mpg)	x 0.354	= Kilometres per litre (km/l)	x 2.825	=	Miles per gallon (mpg)

Temperature

Degrees Fahrenheit = (°C x 1.8) + 32   Degrees Celsius (Degrees Centigrade; °C) = (°F - 32) x 0.56

It is common practice to convert from miles per gallon (mpg) to litres/100 kilometres (l/100km), where mpg x l/100 km = 282

Buying spare parts REF•3

Spare parts are available from many sources, including maker's appointed garages, accessory shops, and motor factors. To be sure of obtaining the correct parts, it will sometimes be necessary to quote the vehicle identification number. If possible, it can also be useful to take the old parts along for positive identification. Items such as starter motors and alternators may be available under a service exchange scheme - any parts returned should be clean.

Our advice regarding spare parts is as follows.

Officially appointed garages

This is the best source of parts which are peculiar to your car, and which are not otherwise generally available (eg, badges, interior trim, certain body panels, etc). It is also the only place at which you should buy parts if the vehicle is still under warranty.

Accessory shops

These are very good places to buy materials and components needed for the maintenance of your car (oil, air and fuel filters, light bulbs, drivebelts, greases, brake pads, tough-up paint, etc). Components of this nature sold by a reputable shop are of the same standard as those used by the car manufacturer.

Besides components, these shops also sell tools and general accessories, usually have convenient opening hours, charge lower prices, and can often be found close to home. Some accessory shops have parts counters where components needed for almost any repair job can be purchased or ordered.

Motor factors

Good factors will stock all the more important components which wear out comparatively quickly, and can sometimes supply individual components needed for the overhaul of a larger assembly (eg, brake seals and hydraulic parts, bearing shells, pistons, valves). They may also handle work such as cylinder block reboring, crankshaft regrinding, etc.

Tyre and exhaust specialists

These outlets may be independent, or members of a local or national chain. They frequently offer competitive prices when compared with a main dealer or local garage, but it will pay to obtain several quotes before making a decision. When researching prices, also ask what "extras" may be added - for instance fitting a new valve and balancing the wheel are both commonly charged on top of the price of a new tyre.

Other sources

Beware of parts or materials obtained from market stalls, car boot sales or similar outlets. Such items are not invariably sub-standard, but there is little chance of compensation if they do prove unsatisfactory. In the case of safety-critical components such as brake pads, there is the risk not only of financial loss, but also of an accident causing injury or death.

Second-hand components or assemblies obtained from a car breaker can be a good buy in some circumstances, but this sort of purchase is best made by the experienced DIY mechanic.

Vehicle identification

Modifications are a continuing and unpublicised process in vehicle manufacture, quite apart from major model changes. Spare parts manuals and lists are compiled upon a numerical basis, the individual vehicle identification numbers being essential to correct identification of the component concerned.

When ordering spare parts, always give as much information as possible. Quote the car model, year of manufacture, body and engine numbers as appropriate.

The *Vehicle Identification Number (VIN)* plate is riveted to the body front panel under the bonnet, behind the right-hand headlight **(see illustrations)**. A single identification plate may be fitted, or a small oval plate (containing vehicle equipment information) and a larger rectangular plate (containing the VIN number) may be fitted. The VIN number may also be stamped into the top of the right-hand suspension turret in the engine compartment.

The *engine number* is situated on the exhaust manifold side of the cylinder block, and is stamped on a plate which is riveted to the front of the block. The plate is located at the front of the engine, below the exhaust manifold on four-cylinder engines, and on the cylinder block adjacent to the dipstick on six-cylinder engines **(see illustration)**.

Note: *The first part of the engine number gives the engine code - eg, "J7T".*

The *chassis number* is stamped into the top of the right-hand suspension turret, directly after the VIN number **(see illustration)**.

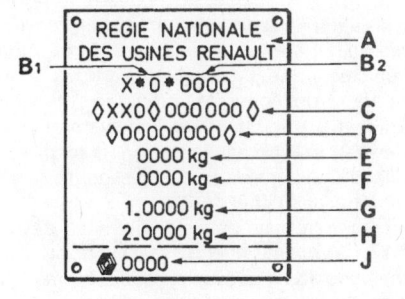

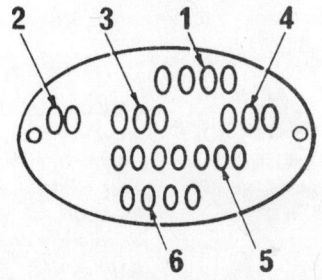

Engine number plate - four-cylinder engine

A Maker's name
B1 EEC number
B2 Vehicle type no.
C Maker's identification code
D Chassis number
E Gross vehicle weight
F Gross train weight
G Max. front axle weight
H Max. rear axle weight
J Model year (some countries)
1 Factory symbol
2 First figure - transmission type, second figure - special equipment
3 Basic spec. level
4 Extra factory-fitted equipment
5 Fabrication number
6 Model year (some countries)

Vehicle identification plate numbers

Vehicle identification plates

Chassis number on suspension turret

General repair procedures

Whenever servicing, repair or overhaul work is carried out on the car or its components, it is necessary to observe the following procedures and instructions. This will assist in carrying out the operation efficiently and to a professional standard of workmanship.

Joint mating faces and gaskets

When separating components at their mating faces, never insert screwdrivers or similar implements into the joint between the faces in order to prise them apart. This can cause severe damage which results in oil leaks, coolant leaks, etc upon reassembly. Separation is usually achieved by tapping along the joint with a soft-faced hammer in order to break the seal. However, note that this method may not be suitable where dowels are used for component location.

Where a gasket is used between the mating faces of two components, ensure that it is renewed on reassembly, and fit it dry unless otherwise stated in the repair procedure. Make sure that the mating faces are clean and dry, with all traces of old gasket removed. When cleaning a joint face, use a tool which is not likely to score or damage the face, and remove any burrs or nicks with an oilstone or fine file.

Make sure that tapped holes are cleaned with a pipe cleaner, and keep them free of jointing compound, if this is being used, unless specifically instructed otherwise.

Ensure that all orifices, channels or pipes are clear, and blow through them, preferably using compressed air.

Oil seals

Oil seals can be removed by levering them out with a wide flat-bladed screwdriver or similar tool. Alternatively, a number of self-tapping screws may be screwed into the seal, and these used as a purchase for pliers or similar in order to pull the seal free.

Whenever an oil seal is removed from its working location, either individually or as part of an assembly, it should be renewed.

The very fine sealing lip of the seal is easily damaged, and will not seal if the surface it contacts is not completely clean and free from scratches, nicks or grooves. If the original sealing surface of the component cannot be restored, and the manufacturer has not made provision for slight relocation of the seal relative to the sealing surface, the component should be renewed.

Protect the lips of the seal from any surface which may damage them in the course of fitting. Use tape or a conical sleeve where possible. Lubricate the seal lips with oil before fitting and, on dual-lipped seals, fill the space between the lips with grease.

Unless otherwise stated, oil seals must be fitted with their sealing lips toward the lubricant to be sealed.

Use a tubular drift or block of wood of the appropriate size to install the seal and, if the seal housing is shouldered, drive the seal down to the shoulder. If the seal housing is unshouldered, the seal should be fitted with its face flush with the housing top face (unless otherwise instructed).

Screw threads and fastenings

Seized nuts, bolts and screws are quite a common occurrence where corrosion has set in, and the use of penetrating oil or releasing fluid will often overcome this problem if the offending item is soaked for a while before attempting to release it. The use of an impact driver may also provide a means of releasing such stubborn fastening devices, when used in conjunction with the appropriate screwdriver bit or socket. If none of these methods works, it may be necessary to resort to the careful application of heat, or the use of a hacksaw or nut splitter device.

Studs are usually removed by locking two nuts together on the threaded part, and then using a spanner on the lower nut to unscrew the stud. Studs or bolts which have broken off below the surface of the component in which they are mounted can sometimes be removed using a stud extractor. Always ensure that a blind tapped hole is completely free from oil, grease, water or other fluid before installing the bolt or stud. Failure to do this could cause the housing to crack due to the hydraulic action of the bolt or stud as it is screwed in.

When tightening a castellated nut to accept a split pin, tighten the nut to the specified torque, where applicable, and then tighten further to the next split pin hole. Never slacken the nut to align the split pin hole, unless stated in the repair procedure.

When checking or retightening a nut or bolt to a specified torque setting, slacken the nut or bolt by a quarter of a turn, and then retighten to the specified setting. However, this should not be attempted where angular tightening has been used.

For some screw fastenings, notably cylinder head bolts or nuts, torque wrench settings are no longer specified for the latter stages of tightening, "angle-tightening" being called up instead. Typically, a fairly low torque wrench setting will be applied to the bolts/nuts in the correct sequence, followed by one or more stages of tightening through specified angles.

Locknuts, locktabs and washers

Any fastening which will rotate against a component or housing during tightening should always have a washer between it and the relevant component or housing.

Spring or split washers should always be renewed when they are used to lock a critical component such as a big-end bearing retaining bolt or nut. Locktabs which are folded over to retain a nut or bolt should always be renewed.

Self-locking nuts can be re-used in non-critical areas, providing resistance can be felt when the locking portion passes over the bolt or stud thread. However, it should be noted that self-locking stiffnuts tend to lose their effectiveness after long periods of use, and should be renewed as a matter of course.

Split pins must always be replaced with new ones of the correct size for the hole.

When thread-locking compound is found on the threads of a fastener which is to be re-used, it should be cleaned off with a wire brush and solvent, and fresh compound applied on reassembly.

Special tools

Some repair procedures in this manual entail the use of special tools such as a press, two or three-legged pullers, spring compressors, etc. Wherever possible, suitable readily-available alternatives to the manufacturer's special tools are described, and are shown in use. In some instances, where no alternative is possible, it has been necessary to resort to the use of a manufacturer's tool, and this has been done for reasons of safety as well as the efficient completion of the repair operation. Unless you are highly-skilled and have a thorough understanding of the procedures described, never attempt to bypass the use of any special tool when the procedure described specifies its use. Not only is there a very great risk of personal injury, but expensive damage could be caused to the components involved.

Environmental considerations

When disposing of used engine oil, brake fluid, antifreeze, etc, give due consideration to any detrimental environmental effects. Do not, for instance, pour any of the above liquids down drains into the general sewage system, or onto the ground to soak away. Many local council refuse tips provide a facility for waste oil disposal, as do some garages. If none of these facilities are available, consult your local Environmental Health Department, or the National Rivers Authority, for further advice.

With the universal tightening-up of legislation regarding the emission of environmentally-harmful substances from motor vehicles, most current vehicles have tamperproof devices fitted to the main adjustment points of the fuel system. These devices are primarily designed to prevent unqualified persons from adjusting the fuel/air mixture, with the chance of a consequent increase in toxic emissions. If such devices are encountered during servicing or overhaul, they should, wherever possible, be renewed or refitted in accordance with the vehicle manufacturer's requirements or current legislation.

Note: It is antisocial and illegal to dump oil down the drain. To find the location of your local oil recycling bank, call this number free.

OIL CARE – FOLLOW THE CODE
OIL BANK LINE
0800 66 33 66

Jacking and vehicle support REF•5

The jack supplied with the vehicle tool kit should only be used for changing the roadwheels - see *"Wheel changing"* at the front of this manual. When carrying out any other kind of work, raise the vehicle using a hydraulic (or "trolley") jack, and always supplement the jack with axle stands positioned under the vehicle jacking points.

When using a hydraulic jack or axle stands, always position the jack head or axle stand head under, or adjacent to one of the relevant wheel changing jacking points under the sills. Use a block of wood between the jack or axle stand and the sill - the block of wood should have a groove cut into it, in which the weld flange of the sill will locate **(see illustrations)**.

The front of the vehicle can be raised using a substantial wooden block located across both body side-members, just behind the front crossmember. Place the jack under the wooden block.

Do not attempt to jack the vehicle under the front crossmember, the sump, or any of the suspension components.

The jack supplied with the vehicle locates in the jacking points on the underside of the sills - see *"Wheel changing"* at the front of this manual. Ensure that the jack head is correctly engaged before attempting to raise the vehicle.

Never work under, around, or near a raised vehicle, unless it is adequately supported in at least two places.

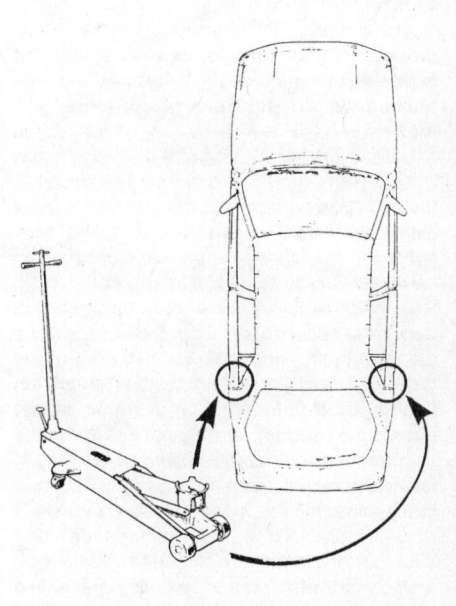

When jacking the rear of the vehicle using a trolley jack, position the jack head under the wheel changing jacking points

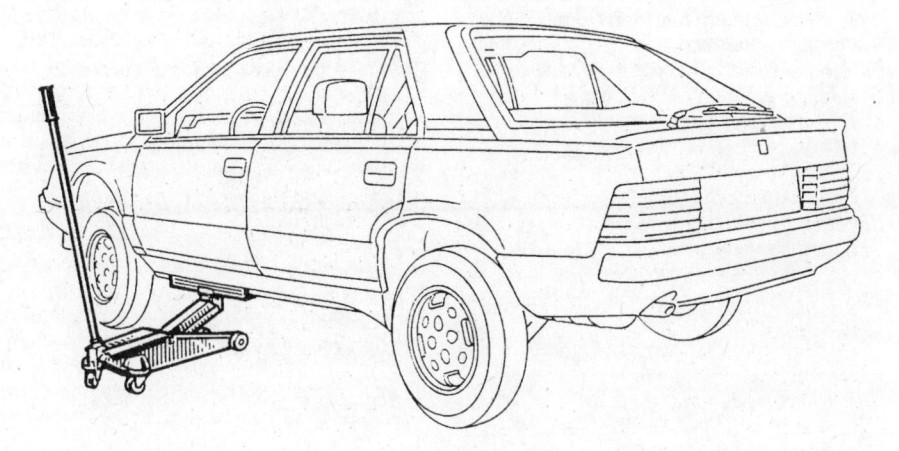

Use a block of wood with a groove cut in it, in which the weld flange of the sill will locate

Radio/cassette unit anti-theft system - precaution

The radio/cassette unit fitted to later models as standard equipment by Renault may be equipped with a built-in security code, to deter thieves. If the power source to the unit is cut, the anti-theft system will activate. Even if the power source is immediately reconnected, the radio/cassette unit will not function until the correct security code has been entered. Therefore if you do not know the correct security code for the unit, **do not** disconnect the battery negative lead, or remove the radio/cassette unit from the vehicle.

A number of different types of radio/cassette player may be fitted, with different methods of entering the security code.

If the incorrect code is entered a number of times, the unit will lock.

If the security code is lost or forgotten, seek the advice of your Renault dealer. On presentation of proof of ownership, a Renault dealer will be able to provide you with a new security code.

REF•6 Tools and working facilities

Introduction

A selection of good tools is a fundamental requirement for anyone contemplating the maintenance and repair of a motor vehicle. For the owner who does not possess any, their purchase will prove a considerable expense, offsetting some of the savings made by doing-it-yourself. However, provided that the tools purchased meet the relevant national safety standards and are of good quality, they will last for many years and prove an extremely worthwhile investment.

To help the average owner to decide which tools are needed to carry out the various tasks detailed in this manual, we have compiled three lists of tools under the following headings: *Maintenance and minor repair*, *Repair and overhaul*, and *Special*. Newcomers to practical mechanics should start off with the *Maintenance and minor repair* tool kit, and confine themselves to the simpler jobs around the vehicle. Then, as confidence and experience grow, more difficult tasks can be undertaken, with extra tools being purchased as, and when, they are needed. In this way, a *Maintenance and minor repair* tool kit can be built up into a *Repair and overhaul* tool kit over a considerable period of time, without any major cash outlays. The experienced do-it-yourselfer will have a tool kit good enough for most repair and overhaul procedures, and will add tools from the *Special* category when it is felt that the expense is justified by the amount of use to which these tools will be put.

Maintenance and minor repair tool kit

The tools given in this list should be considered as a minimum requirement if routine maintenance, servicing and minor repair operations are to be undertaken. We recommend the purchase of combination spanners (ring one end, open-ended the other); although more expensive than open-ended ones, they do give the advantages of both types of spanner.

☐ *Combination spanners:*
  Metric - 8 to 19 mm inclusive
☐ *Adjustable spanner - 35 mm jaw (approx.)*
☐ *Spark plug spanner (with rubber insert) - petrol models*
☐ *Spark plug gap adjustment tool - petrol models*
☐ *Set of feeler blades*
☐ *Brake bleed nipple spanner*
☐ *Screwdrivers:*
  Flat blade - 100 mm long x 6 mm dia
  Cross blade - 100 mm long x 6 mm dia
☐ *Combination pliers*
☐ *Hacksaw (junior)*
☐ *Tyre pump*
☐ *Tyre pressure gauge*
☐ *Oil can*
☐ *Oil filter removal tool*
☐ *Fine emery cloth*
☐ *Wire brush (small)*
☐ *Funnel (medium size)*

Repair and overhaul tool kit

These tools are virtually essential for anyone undertaking any major repairs to a motor vehicle, and are additional to those given in the *Maintenance and minor repair* list. Included in this list is a comprehensive set of sockets. Although these are expensive, they will be found invaluable as they are so versatile - particularly if various drives are included in the set. We recommend the half-inch square-drive type, as this can be used with most proprietary torque wrenches.

The tools in this list will sometimes need to be supplemented by tools from the *Special* list:

☐ *Sockets (or box spanners) to cover range in previous list (including Torx sockets)*
☐ *Reversible ratchet drive (for use with sockets)*
☐ *Extension piece, 250 mm (for use with sockets)*
☐ *Universal joint (for use with sockets)*
☐ *Torque wrench (for use with sockets)*
☐ *Self-locking grips*
☐ *Ball pein hammer*
☐ *Soft-faced mallet (plastic/aluminium or rubber)*
☐ *Screwdrivers:*
  Flat blade - long & sturdy, short (chubby), and narrow (electrician's) types
  Cross blade – Long & sturdy, and short (chubby) types
☐ *Pliers:*
  Long-nosed
  Side cutters (electrician's)
  Circlip (internal and external)
☐ *Cold chisel - 25 mm*
☐ *Scriber*
☐ *Scraper*
☐ *Centre-punch*
☐ *Pin punch*
☐ *Hacksaw*
☐ *Brake hose clamp*
☐ *Brake/clutch bleeding kit*
☐ *Selection of twist drills*
☐ *Steel rule/straight-edge*
☐ *Allen keys (inc. splined/Torx type)*
☐ *Selection of files*
☐ *Wire brush*
☐ *Axle stands*
☐ *Jack (strong trolley or hydraulic type)*
☐ *Light with extension lead*

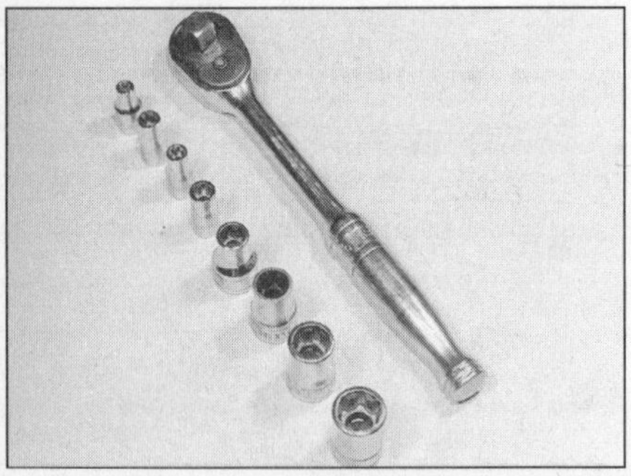

Sockets and reversible ratchet drive

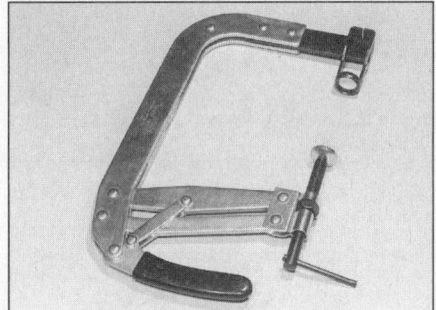

Valve spring compressor

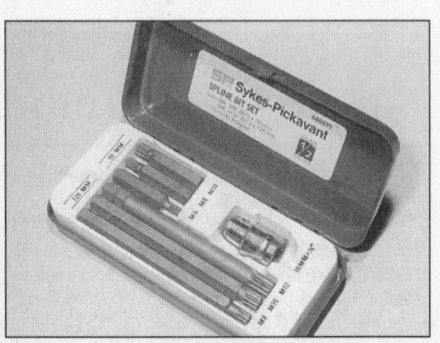

Spline bit set

Piston ring compressor

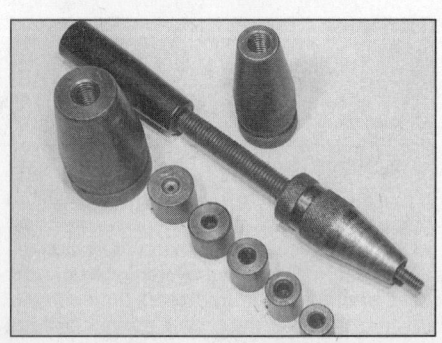

Clutch plate alignment set

Tools and working facilities REF•7

Special tools

The tools in this list are those which are not used regularly, are expensive to buy, or which need to be used in accordance with their manufacturers' instructions. Unless relatively difficult mechanical jobs are undertaken frequently, it will not be economic to buy many of these tools. Where this is the case, you could consider clubbing together with friends (or joining a motorists' club) to make a joint purchase, or borrowing the tools against a deposit from a local garage or tool hire specialist. It is worth noting that many of the larger DIY superstores now carry a large range of special tools for hire at modest rates.

The following list contains only those tools and instruments freely available to the public, and not those special tools produced by the vehicle manufacturer specifically for its dealer network. You will find occasional references to these manufacturers' special tools in the text of this manual. Generally, an alternative method of doing the job without the vehicle manufacturers' special tool is given. However, sometimes there is no alternative to using them. Where this is the case and the relevant tool cannot be bought or borrowed, you will have to entrust the work to a dealer.

☐ Valve spring compressor
☐ Valve grinding tool
☐ Piston ring compressor
☐ Piston ring removal/installation tool
☐ Cylinder bore hone
☐ Balljoint separator
☐ Coil spring compressors (where applicable)
☐ Two/three-legged hub and bearing puller
☐ Impact screwdriver
☐ Micrometer and/or vernier calipers
☐ Dial gauge
☐ Stroboscopic timing light
☐ Dwell angle meter/tachometer
☐ Universal electrical multi-meter
☐ Cylinder compression gauge
☐ Hand-operated vacuum pump and gauge
☐ Clutch plate alignment set
☐ Brake shoe steady spring cup removal tool
☐ Bush and bearing removal/installation set
☐ Stud extractors
☐ Tap and die set
☐ Lifting tackle
☐ Trolley jack

Buying tools

Reputable motor accessory shops and superstores often offer excellent quality tools at discount prices, so it pays to shop around.

Remember, you don't have to buy the most expensive items on the shelf, but it is always advisable to steer clear of the very cheap tools. Beware of 'bargains' offered on market stalls or at car boot sales. There are plenty of good tools around at reasonable prices, but always aim to purchase items which meet the relevant national safety standards. If in doubt, ask the proprietor or manager of the shop for advice before making a purchase.

Care and maintenance of tools

Having purchased a reasonable tool kit, it is necessary to keep the tools in a clean and serviceable condition. After use, always wipe off any dirt, grease and metal particles using a clean, dry cloth, before putting the tools away. Never leave them lying around after they have been used. A simple tool rack on the garage or workshop wall for items such as screwdrivers and pliers is a good idea. Store all normal spanners and sockets in a metal box. Any measuring instruments, gauges, meters, etc, must be carefully stored where they cannot be damaged or become rusty.

Take a little care when tools are used. Hammer heads inevitably become marked, and screwdrivers lose the keen edge on their blades from time to time. A little timely attention with emery cloth or a file will soon restore items like this to a good finish.

Working facilities

Not to be forgotten when discussing tools is the workshop itself. If anything more than routine maintenance is to be carried out, a suitable working area becomes essential.

It is appreciated that many an owner-mechanic is forced by circumstances to remove an engine or similar item without the benefit of a garage or workshop. Having done this, any repairs should always be done under the cover of a roof.

Wherever possible, any dismantling should be done on a clean, flat workbench or table at a suitable working height.

Any workbench needs a vice; one with a jaw opening of 100 mm is suitable for most jobs. As mentioned previously, some clean dry storage space is also required for tools, as well as for any lubricants, cleaning fluids, touch-up paints etc, which become necessary.

Another item which may be required, and which has a much more general usage, is an electric drill with a chuck capacity of at least 8 mm. This, together with a good range of twist drills, is virtually essential for fitting accessories.

Last, but not least, always keep a supply of old newspapers and clean, lint-free rags available, and try to keep any working area as clean as possible.

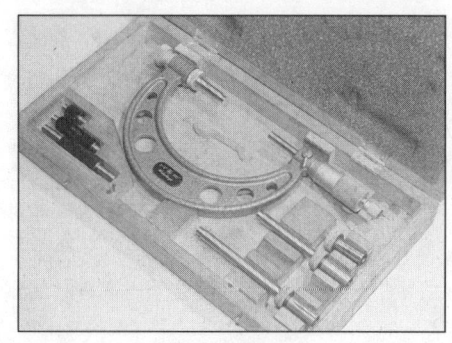

Micrometer set

Dial test indicator ("dial gauge")

Stroboscopic timing light

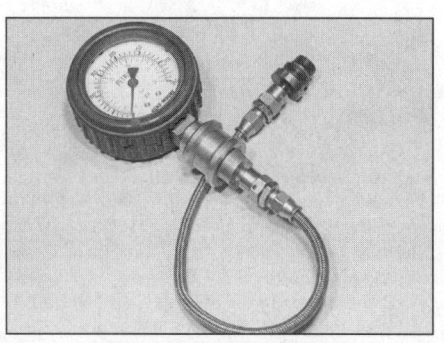

Compression tester

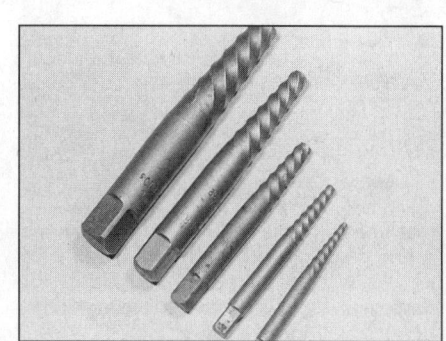

Stud extractor set

REF•8 MOT test checks

This is a guide to getting your vehicle through the MOT test. Obviously it will not be possible to examine the vehicle to the same standard as the professional MOT tester. However, working through the following checks will enable you to identify any problem areas before submitting the vehicle for the test.

Where a testable component is in borderline condition, the tester has discretion in deciding whether to pass or fail it. The basis of such discretion is whether the tester would be happy for a close relative or friend to use the vehicle with the component in that condition. If the vehicle presented is clean and evidently well cared for, the tester may be more inclined to pass a borderline component than if the vehicle is scruffy and apparently neglected.

It has only been possible to summarise the test requirements here, based on the regulations in force at the time of printing. Test standards are becoming increasingly stringent, although there are some exemptions for older vehicles. For full details obtain a copy of the Haynes publication Pass the MOT! (available from stockists of Haynes manuals).

An assistant will be needed to help carry out some of these checks.

The checks have been sub-divided into four categories, as follows:

1 Checks carried out **FROM THE DRIVER'S SEAT**

2 Checks carried out **WITH THE VEHICLE ON THE GROUND**

3 Checks carried out **WITH THE VEHICLE RAISED AND THE WHEELS FREE TO TURN**

4 Checks carried out on **YOUR VEHICLE'S EXHAUST EMISSION SYSTEM**

---

1 Checks carried out **FROM THE DRIVER'S SEAT**

Handbrake

☐ Test the operation of the handbrake. Excessive travel (too many clicks) indicates incorrect brake or cable adjustment.

☐ Check that the handbrake cannot be released by tapping the lever sideways. Check the security of the lever mountings.

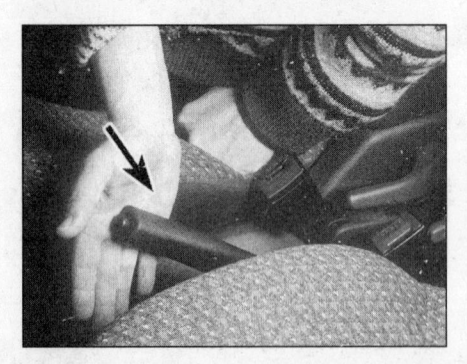

Footbrake

☐ Depress the brake pedal and check that it does not creep down to the floor, indicating a master cylinder fault. Release the pedal, wait a few seconds, then depress it again. If the pedal travels nearly to the floor before firm resistance is felt, brake adjustment or repair is necessary. If the pedal feels spongy, there is air in the hydraulic system which must be removed by bleeding.

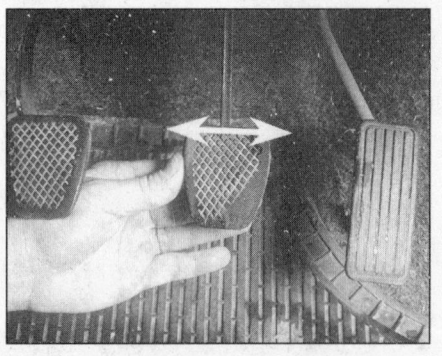

☐ Check that the brake pedal is secure and in good condition. Check also for signs of fluid leaks on the pedal, floor or carpets, which would indicate failed seals in the brake master cylinder.

☐ Check the servo unit (when applicable) by operating the brake pedal several times, then keeping the pedal depressed and starting the engine. As the engine starts, the pedal will move down slightly. If not, the vacuum hose or the servo itself may be faulty.

Steering wheel and column

☐ Examine the steering wheel for fractures or looseness of the hub, spokes or rim.

☐ Move the steering wheel from side to side and then up and down. Check that the steering wheel is not loose on the column, indicating wear or a loose retaining nut. Continue moving the steering wheel as before, but also turn it slightly from left to right.

☐ Check that the steering wheel is not loose on the column, and that there is no abnormal movement of the steering wheel, indicating wear in the column support bearings or couplings.

Windscreen and mirrors

☐ The windscreen must be free of cracks or other significant damage within the driver's field of view. (Small stone chips are acceptable.) Rear view mirrors must be secure, intact, and capable of being adjusted.

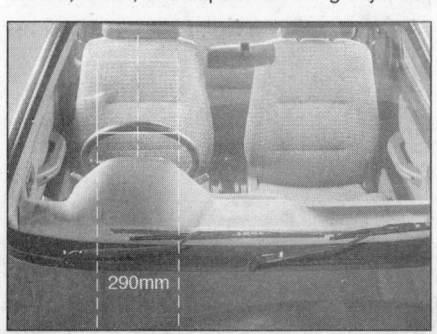

MOT test checks REF•9

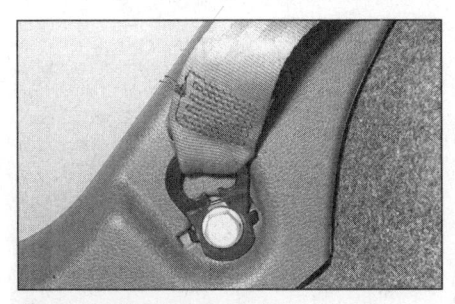

Seat belts and seats

Note: *The following checks are applicable to all seat belts, front and rear.*

☐ Examine the webbing of all the belts (including rear belts if fitted) for cuts, serious fraying or deterioration. Fasten and unfasten each belt to check the buckles. If applicable, check the retracting mechanism. Check the security of all seat belt mountings accessible from inside the vehicle.

☐ The front seats themselves must be securely attached and the backrests must lock in the upright position.

Doors

☐ Both front doors must be able to be opened and closed from outside and inside, and must latch securely when closed.

2 Checks carried out WITH THE VEHICLE ON THE GROUND

Vehicle identification

☐ Number plates must be in good condition, secure and legible, with letters and numbers correctly spaced – spacing at (A) should be twice that at (B).

☐ The VIN plate and/or homologation plate must be legible.

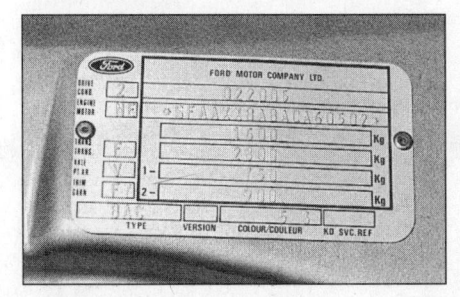

Electrical equipment

☐ Switch on the ignition and check the operation of the horn.

☐ Check the windscreen washers and wipers, examining the wiper blades; renew damaged or perished blades. Also check the operation of the stop-lights.

☐ Check the operation of the sidelights and number plate lights. The lenses and reflectors must be secure, clean and undamaged.

☐ Check the operation and alignment of the headlights. The headlight reflectors must not be tarnished and the lenses must be undamaged.

☐ Switch on the ignition and check the operation of the direction indicators (including the instrument panel tell-tale) and the hazard warning lights. Operation of the sidelights and stop-lights must not affect the indicators - if it does, the cause is usually a bad earth at the rear light cluster.

☐ Check the operation of the rear foglight(s), including the warning light on the instrument panel or in the switch.

Footbrake

☐ Examine the master cylinder, brake pipes and servo unit for leaks, loose mountings, corrosion or other damage.

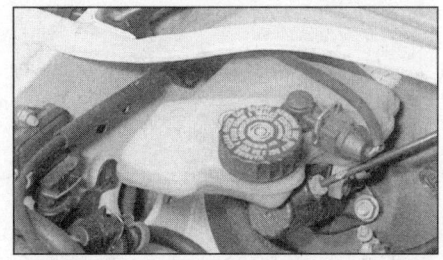

☐ The fluid reservoir must be secure and the fluid level must be between the upper (A) and lower (B) markings.

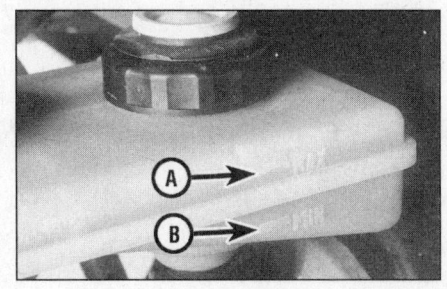

☐ Inspect both front brake flexible hoses for cracks or deterioration of the rubber. Turn the steering from lock to lock, and ensure that the hoses do not contact the wheel, tyre, or any part of the steering or suspension mechanism. With the brake pedal firmly depressed, check the hoses for bulges or leaks under pressure.

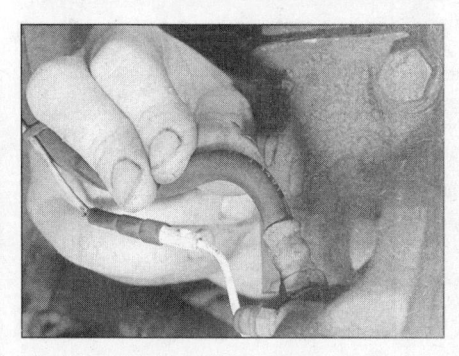

Steering and suspension

☐ Have your assistant turn the steering wheel from side to side slightly, up to the point where the steering gear just begins to transmit this movement to the roadwheels. Check for excessive free play between the steering wheel and the steering gear, indicating wear or insecurity of the steering column joints, the column-to-steering gear coupling, or the steering gear itself.

☐ Have your assistant turn the steering wheel more vigorously in each direction, so that the roadwheels just begin to turn. As this is done, examine all the steering joints, linkages, fittings and attachments. Renew any component that shows signs of wear or damage. On vehicles with power steering, check the security and condition of the steering pump, drivebelt and hoses.

☐ Check that the vehicle is standing level, and at approximately the correct ride height.

Shock absorbers

☐ Depress each corner of the vehicle in turn, then release it. The vehicle should rise and then settle in its normal position. If the vehicle continues to rise and fall, the shock absorber is defective. A shock absorber which has seized will also cause the vehicle to fail.

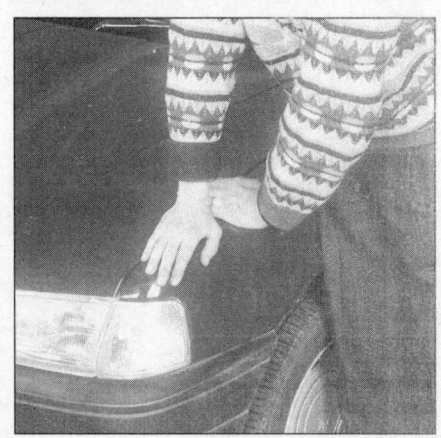

REF•10 MOT test checks

Exhaust system

☐ Start the engine. With your assistant holding a rag over the tailpipe, check the entire system for leaks. Repair or renew leaking sections.

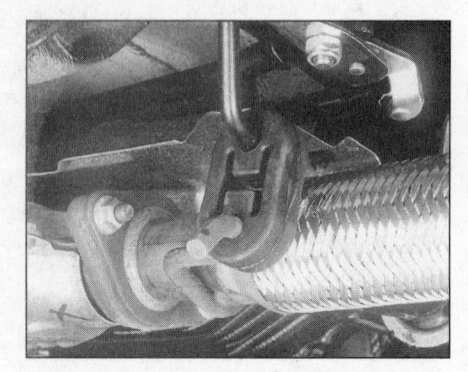

3 Checks carried out WITH THE VEHICLE RAISED AND THE WHEELS FREE TO TURN

Jack up the front and rear of the vehicle, and securely support it on axle stands. Position the stands clear of the suspension assemblies. Ensure that the wheels are clear of the ground and that the steering can be turned from lock to lock.

Steering mechanism

☐ Have your assistant turn the steering from lock to lock. Check that the steering turns smoothly, and that no part of the steering mechanism, including a wheel or tyre, fouls any brake hose or pipe or any part of the body structure.
☐ Examine the steering rack rubber gaiters for damage or insecurity of the retaining clips. If power steering is fitted, check for signs of damage or leakage of the fluid hoses, pipes or connections. Also check for excessive stiffness or binding of the steering, a missing split pin or locking device, or severe corrosion of the body structure within 30 cm of any steering component attachment point.

Front and rear suspension and wheel bearings

☐ Starting at the front right-hand side, grasp the roadwheel at the 3 o'clock and 9 o'clock positions and shake it vigorously. Check for free play or insecurity at the wheel bearings, suspension balljoints, or suspension mountings, pivots and attachments.
☐ Now grasp the wheel at the 12 o'clock and 6 o'clock positions and repeat the previous inspection. Spin the wheel, and check for roughness or tightness of the front wheel bearing.

☐ If excess free play is suspected at a component pivot point, this can be confirmed by using a large screwdriver or similar tool and levering between the mounting and the component attachment. This will confirm whether the wear is in the pivot bush, its retaining bolt, or in the mounting itself (the bolt holes can often become elongated).

☐ Carry out all the above checks at the other front wheel, and then at both rear wheels.

Springs and shock absorbers

☐ Examine the suspension struts (when applicable) for serious fluid leakage, corrosion, or damage to the casing. Also check the security of the mounting points.
☐ If coil springs are fitted, check that the spring ends locate in their seats, and that the spring is not corroded, cracked or broken.
☐ If leaf springs are fitted, check that all leaves are intact, that the axle is securely attached to each spring, and that there is no deterioration of the spring eye mountings, bushes, and shackles.

☐ The same general checks apply to vehicles fitted with other suspension types, such as torsion bars, hydraulic displacer units, etc. Ensure that all mountings and attachments are secure, that there are no signs of excessive wear, corrosion or damage, and (on hydraulic types) that there are no fluid leaks or damaged pipes.
☐ Inspect the shock absorbers for signs of serious fluid leakage. Check for wear of the mounting bushes or attachments, or damage to the body of the unit.

Driveshafts (fwd vehicles only)

☐ Rotate each front wheel in turn and inspect the constant velocity joint gaiters for splits or damage. Also check that each driveshaft is straight and undamaged.

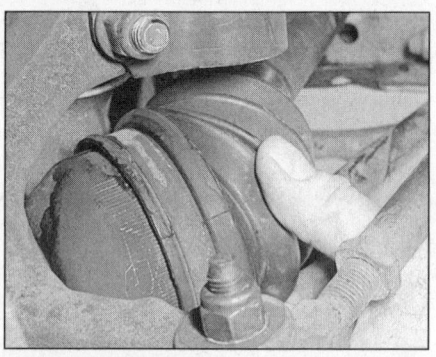

Braking system

☐ If possible without dismantling, check brake pad wear and disc condition. Ensure that the friction lining material has not worn excessively, (A) and that the discs are not fractured, pitted, scored or badly worn (B).

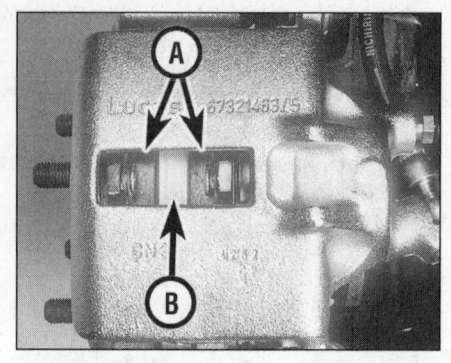

☐ Examine all the rigid brake pipes underneath the vehicle, and the flexible hose(s) at the rear. Look for corrosion, chafing or insecurity of the pipes, and for signs of bulging under pressure, chafing, splits or deterioration of the flexible hoses.
☐ Look for signs of fluid leaks at the brake calipers or on the brake backplates. Repair or renew leaking components.
☐ Slowly spin each wheel, while your assistant depresses and releases the footbrake. Ensure that each brake is operating and does not bind when the pedal is released.

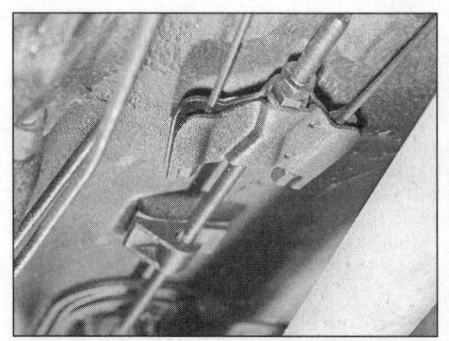

MOT test checks REF•11

□ Examine the handbrake mechanism, checking for frayed or broken cables, excessive corrosion, or wear or insecurity of the linkage. Check that the mechanism works on each relevant wheel, and releases fully, without binding.

□ It is not possible to test brake efficiency without special equipment, but a road test can be carried out later to check that the vehicle pulls up in a straight line.

Fuel and exhaust systems

□ Inspect the fuel tank (including the filler cap), fuel pipes, hoses and unions. All components must be secure and free from leaks.

□ Examine the exhaust system over its entire length, checking for any damaged, broken or missing mountings, security of the retaining clamps and rust or corrosion.

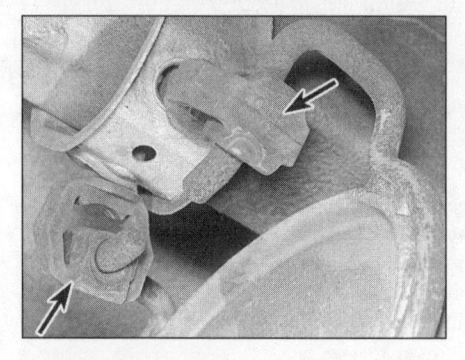

Wheels and tyres

□ Examine the sidewalls and tread area of each tyre in turn. Check for cuts, tears, lumps, bulges, separation of the tread, and exposure of the ply or cord due to wear or damage. Check that the tyre bead is correctly seated on the wheel rim, that the valve is sound and properly seated, and that the wheel is not distorted or damaged.

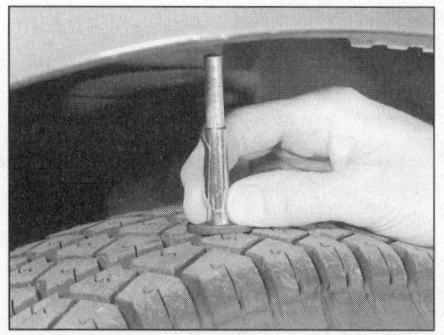

□ Check that the tyres are of the correct size for the vehicle, that they are of the same size and type on each axle, and that the pressures are correct.

□ Check the tyre tread depth. The legal minimum at the time of writing is 1.6 mm over at least three-quarters of the tread width. Abnormal tread wear may indicate incorrect front wheel alignment.

Body corrosion

□ Check the condition of the entire vehicle structure for signs of corrosion in load-bearing areas. (These include chassis box sections, side sills, cross-members, pillars, and all suspension, steering, braking system and seat belt mountings and anchorages.) Any corrosion which has seriously reduced the thickness of a load-bearing area is likely to cause the vehicle to fail. In this case professional repairs are likely to be needed.

□ Damage or corrosion which causes sharp or otherwise dangerous edges to be exposed will also cause the vehicle to fail.

4 Checks carried out on YOUR VEHICLE'S EXHAUST EMISSION SYSTEM

Petrol models

□ Have the engine at normal operating temperature, and make sure that it is in good tune (ignition system in good order, air filter element clean, etc).

□ Before any measurements are carried out, raise the engine speed to around 2500 rpm, and hold it at this speed for 20 seconds. Allow the engine speed to return to idle, and watch for smoke emissions from the exhaust tailpipe. If the idle speed is obviously much too high, or if dense blue or clearly-visible black smoke comes from the tailpipe for more than 5 seconds, the vehicle will fail. As a rule of thumb, blue smoke signifies oil being burnt (engine wear) while black smoke signifies unburnt fuel (dirty air cleaner element, or other carburettor or fuel system fault).

□ An exhaust gas analyser capable of measuring carbon monoxide (CO) and hydrocarbons (HC) is now needed. If such an instrument cannot be hired or borrowed, a local garage may agree to perform the check for a small fee.

CO emissions (mixture)

□ At the time of writing, the maximum CO level at idle is 3.5% for vehicles first used after August 1986 and 4.5% for older vehicles. From January 1996 a much tighter limit (around 0.5%) applies to catalyst-equipped vehicles first used from August 1992. If the CO level cannot be reduced far enough to pass the test (and the fuel and ignition systems are otherwise in good condition) then the carburettor is badly worn, or there is some problem in the fuel injection system or catalytic converter (as applicable).

HC emissions

□ With the CO emissions within limits, HC emissions must be no more than 1200 ppm (parts per million). If the vehicle fails this test at idle, it can be re-tested at around 2000 rpm; if the HC level is then 1200 ppm or less, this counts as a pass.

□ Excessive HC emissions can be caused by oil being burnt, but they are more likely to be due to unburnt fuel.

Diesel models

□ The only emission test applicable to Diesel engines is the measuring of exhaust smoke density. The test involves accelerating the engine several times to its maximum unloaded speed.

Note: *It is of the utmost importance that the engine timing belt is in good condition before the test is carried out.*

□ Excessive smoke can be caused by a dirty air cleaner element. Otherwise, professional advice may be needed to find the cause.

REF•12 Fault finding

Engine 1
- ☐ Engine fails to rotate when attempting to start
- ☐ Engine rotates, but will not start
- ☐ Engine difficult to start when cold
- ☐ Engine difficult to start when hot
- ☐ Starter motor noisy or excessively-rough in engagement
- ☐ Engine starts, but stops immediately
- ☐ Engine idles erratically
- ☐ Engine misfires at idle speed
- ☐ Engine misfires throughout the driving speed range
- ☐ Engine hesitates on acceleration
- ☐ Engine stalls
- ☐ Engine lacks power
- ☐ Engine backfires
- ☐ Oil pressure warning light illuminated with engine running
- ☐ Engine runs-on after switching off
- ☐ Engine noises

Cooling system 2
- ☐ Overheating
- ☐ Overcooling
- ☐ External coolant leakage
- ☐ Internal coolant leakage
- ☐ Corrosion

Fuel and exhaust systems 3
- ☐ Excessive fuel consumption
- ☐ Fuel leakage and/or fuel odour
- ☐ Excessive noise or fumes from exhaust system

Clutch 4
- ☐ Pedal travels to floor - no pressure or very little resistance
- ☐ Clutch fails to disengage (unable to select gears)
- ☐ Clutch slips (engine speed increases, with no increase in vehicle speed)
- ☐ Judder as clutch is engaged
- ☐ Noise when depressing or releasing clutch pedal

Manual transmission 5
- ☐ Noisy in neutral with engine running
- ☐ Noisy in one particular gear
- ☐ Difficulty engaging gears
- ☐ Jumps out of gear
- ☐ Vibration
- ☐ Lubricant leaks

Automatic transmission 6
- ☐ Fluid leakage
- ☐ Transmission fluid brown, or has burned smell
- ☐ General gear selection problems
- ☐ Transmission will not downshift (kickdown) with accelerator fully depressed
- ☐ Engine will not start in any gear, or starts in gears other than Park or Neutral
- ☐ Transmission slips, shifts roughly, is noisy, or has no drive in forward or reverse gears

Driveshafts 7
- ☐ Clicking or knocking noise on turns (at slow speed on full-lock)
- ☐ Vibration when accelerating or decelerating

Braking system 8
- ☐ Vehicle pulls to one side under braking
- ☐ Noise (grinding or high-pitched squeal) when brakes applied
- ☐ Excessive brake pedal travel
- ☐ Brake pedal feels spongy when depressed
- ☐ Excessive brake pedal effort required to stop vehicle
- ☐ Judder felt through brake pedal or steering wheel when braking
- ☐ Brakes binding
- ☐ Rear wheels locking under normal braking

Suspension and steering systems 9
- ☐ Vehicle pulls to one side
- ☐ Wheel wobble and vibration
- ☐ Excessive pitching and/or rolling around corners, or during braking
- ☐ Wandering or general instability
- ☐ Excessively-stiff steering
- ☐ Excessive play in steering
- ☐ Lack of power assistance
- ☐ Tyre wear excessive

Electrical system 10
- ☐ Battery will not hold a charge for more than a few days
- ☐ Ignition/no-charge warning light remains illuminated with engine running
- ☐ Ignition/no-charge warning light fails to come on
- ☐ Lights inoperative
- ☐ Instrument readings inaccurate or erratic
- ☐ Horn inoperative, or unsatisfactory in operation
- ☐ Windscreen/tailgate wipers inoperative, or unsatisfactory in operation
- ☐ Windscreen/tailgate washers inoperative, or unsatisfactory in operation
- ☐ Electric windows inoperative, or unsatisfactory in operation
- ☐ Central lockng system inoperative, or unsatisfactory in operation

Introduction

The vehicle owner who does his or her own maintenance according to the recommended service schedules should not have to use this section of the manual very often. Modern component reliability is such that, provided those items subject to wear or deterioration are inspected or renewed at the specified intervals, sudden failure is comparatively rare. Faults do not usually just happen as a result of sudden failure, but develop over a period of time. Major mechanical failures in particular are usually preceded by characteristic symptoms over hundreds or even thousands of miles. Those components which do occasionally fail without warning are often small and easily carried in the vehicle.

With any fault-finding, the first step is to decide where to begin investigations. Sometimes this is obvious, but on other occasions, a little detective work will be necessary. The owner who makes half a

Fault finding

dozen haphazard adjustments or replacements may be successful in curing a fault (or its symptoms), but will be none the wiser if the fault recurs, and ultimately may have spent more time and money than was necessary. A calm and logical approach will be found to be more satisfactory in the long run. Always take into account any warning signs or abnormalities that may have been noticed in the period preceding the fault - power loss, high or low gauge readings, unusual smells, etc - and remember that failure of components such as fuses or spark plugs may only be pointers to some underlying fault.

The pages which follow provide an easy-reference guide to the more common problems which may occur during the operation of the vehicle. These problems and their possible causes are grouped under headings denoting various components or systems, such as Engine, Cooling system, etc. The Chapter and/or Section which deals with the problem is also shown in brackets. Whatever the fault, certain basic principles apply. These are as follows:

Verify the fault. This is simply a matter of being sure that you know what the symptoms are before starting work. This is particularly important if you are investigating a fault for someone else, who may not have described it very accurately.

Don't overlook the obvious. For example, if the vehicle won't start, is there fuel in the tank? (Don't take anyone else's word on this particular point, and don't trust the fuel gauge either!) If an electrical fault is indicated, look for loose or broken wires before digging out the test gear.

Cure the disease, not the symptom. Substituting a flat battery with a fully-charged one will get you off the hard shoulder, but if the underlying cause is not attended to, the new battery will go the same way. Similarly, changing oil-fouled spark plugs for a new set will get you moving again, but remember that the reason for the fouling (if it wasn't simply an incorrect grade of plug) will have to be established and corrected.

Don't take anything for granted. Particularly, don't forget that a "new" component may itself be defective (especially if it's been rattling around in the boot for months), and don't leave components out of a fault diagnosis sequence just because they are new or recently-fitted. When you do finally diagnose a difficult fault, you'll probably realise that all the evidence was there from the start.

1 Engine

Engine fails to rotate when attempting to start

- [ ] Battery terminal connections loose or corroded (*"Weekly checks"*).
- [ ] Battery discharged or faulty (Chapter 5A).
- [ ] Broken, loose or disconnected wiring in the starting circuit (Chapter 5A).
- [ ] Defective starter solenoid or switch (Chapter 5A).
- [ ] Defective starter motor (Chapter 5A).
- [ ] Starter pinion or flywheel ring gear teeth loose or broken (Chapters 2A, 2B, 2C and 5A).
- [ ] Engine earth strap broken or disconnected (Chapter 5A).

Engine rotates, but will not start

- [ ] Fuel tank empty.
- [ ] Battery discharged (engine rotates slowly) (Chapter 5A).
- [ ] Battery terminal connections loose or corroded (*"Weekly checks"*).
- [ ] Ignition components damp or damaged - petrol models (Chapters 1 and 5B).
- [ ] Broken, loose or disconnected wiring in the ignition circuit - petrol models (Chapters 1 and 5B).
- [ ] Worn, faulty or incorrectly-gapped spark plugs - petrol models (Chapter 1).
- [ ] Preheating system faulty - diesel models (Chapter 5C).
- [ ] Choke mechanism incorrectly adjusted, worn or sticking - carburettor petrol models (Chapter 4A).
- [ ] Faulty fuel cut-off solenoid - carburettor petrol models (Chapter 4A).
- [ ] Fuel injection system fault - fuel-injected petrol models (Chapter 4B).
- [ ] Stop solenoid faulty - diesel models (Chapter 4C).
- [ ] Air in fuel system - diesel models (Chapter 4C).
- [ ] Major mechanical failure (eg camshaft drive) (Chapter 2A, 2B or 2C).

Engine difficult to start when cold

- [ ] Battery discharged (Chapter 5A).
- [ ] Battery terminal connections loose or corroded (*"Weekly checks"*).
- [ ] Worn, faulty or incorrectly-gapped spark plugs - petrol models (Chapter 1).
- [ ] Preheating system faulty - diesel models (Chapter 5C).
- [ ] Choke mechanism incorrectly adjusted, worn or sticking - carburettor petrol models (Chapter 4A).
- [ ] Fuel injection system fault - fuel-injected petrol models (Chapter 4B).
- [ ] Other ignition system fault - petrol models (Chapters 1 and 5B).
- [ ] Fast idle valve incorrectly adjusted - diesel models (Chapter 4C).
- [ ] Low cylinder compressions (Chapter 2A, 2B or 2C).

Engine difficult to start when hot

- [ ] Air filter element dirty or clogged (Chapter 1).
- [ ] Choke mechanism incorrectly adjusted, worn or sticking - carburettor petrol models (Chapter 4A).
- [ ] Fuel injection system fault - fuel-injected petrol models (Chapter 4B).
- [ ] Low cylinder compressions (Chapter 2A, 2B or 2C).

Starter motor noisy or excessively-rough in engagement

- [ ] Starter pinion or flywheel ring gear teeth loose or broken (Chapters 2A, 2B, 2C and 5A).
- [ ] Starter motor mounting bolts loose or missing (Chapter 5A).
- [ ] Starter motor internal components worn or damaged (Chapter 5A).

Engine starts, but stops immediately

- [ ] Loose or faulty electrical connections in the ignition circuit - petrol models (Chapters 1 and 5B).
- [ ] Vacuum leak at the carburettor/throttle body or inlet manifold - petrol models (Chapter 4A or 4B).
- [ ] Blocked carburettor jet(s) or internal passages - carburettor petrol models (Chapter 4A).
- [ ] Blocked injector/fuel injection system fault - fuel-injected petrol models (Chapter 4B).

Engine idles erratically

- [ ] Air filter element clogged (Chapter 1).
- [ ] Vacuum leak at the carburettor/throttle body, inlet manifold or associated hoses - petrol models (Chapter 4A or, 4B).
- [ ] Worn, faulty or incorrectly-gapped spark plugs - petrol models (Chapter 1).
- [ ] Uneven or low cylinder compressions (Chapter 2A, 2B or 2C).
- [ ] Camshaft lobes worn (Chapter 2A, 2B or 2C).
- [ ] Timing belt incorrectly tensioned - four-cylinder engines (Chapter 2A or 2C).
- [ ] Timing chain worn - six-cylinder engines (Chapter 2B).
- [ ] Blocked carburettor jet(s) or internal passages - carburettor petrol models (Chapter 4A).
- [ ] Blocked injector/fuel injection system fault - fuel-injected petrol models (Chapter 4B).
- [ ] Faulty injector(s) - diesel models (Chapter 4C).

REF•14 Fault finding

1 Engine (continued)

Engine misfires at idle speed
- [ ] Worn, faulty or incorrectly-gapped spark plugs - petrol models (Chapter 1).
- [ ] Faulty spark plug HT leads - petrol models (Chapter 1).
- [ ] Vacuum leak at the carburettor/throttle body, inlet manifold or associated hoses - petrol models (Chapter 4A or 4B).
- [ ] Blocked carburettor jet(s) or internal passages - carburettor petrol models (Chapter 4A).
- [ ] Blocked injector/fuel injection system fault - fuel-injected petrol models (Chapter 4B).
- [ ] Faulty injector(s) - diesel models (Chapter 4C).
- [ ] Distributor cap cracked or tracking internally - petrol models (where applicable) (Chapter 5B).
- [ ] Uneven or low cylinder compressions (Chapter 2A, 2B or 2C).
- [ ] Disconnected, leaking, or perished crankcase ventilation hoses (Chapter 4A, 4B or 4C).

Engine misfires throughout the driving speed range
- [ ] Fuel filter choked (Chapter 1).
- [ ] Fuel pump faulty, or delivery pressure low - petrol models (Chapter 4A or 4B).
- [ ] Fuel tank vent blocked, or fuel pipes restricted (Chapter 4A, 4B or 4C).
- [ ] Vacuum leak at the carburettor/throttle body, inlet manifold or associated hoses - petrol models (Chapter 4A or 4B).
- [ ] Worn, faulty or incorrectly-gapped spark plugs - petrol models (Chapter 1).
- [ ] Faulty spark plug HT leads - petrol models (Chapter 1).
- [ ] Faulty injector(s) - diesel models (Chapter 4C).
- [ ] Distributor cap cracked or tracking internally - petrol models (where applicable) (Chapter 5B).
- [ ] Faulty ignition coil - petrol models (Chapter 5B).
- [ ] Uneven or low cylinder compressions (Chapter 2A, 2B or 2C).
- [ ] Blocked carburettor jet(s) or internal passages - carburettor petrol models (Chapter 4A).
- [ ] Blocked injector/fuel injection system fault - fuel-injected petrol models (Chapter 4B).

Engine hesitates on acceleration
- [ ] Worn, faulty or incorrectly-gapped spark plugs - petrol models (Chapter 1).
- [ ] Vacuum leak at the carburettor/throttle body, inlet manifold or associated hoses - petrol models (Chapter 4A or 4B).
- [ ] Blocked carburettor jet(s) or internal passages - carburettor petrol models (Chapter 4A).
- [ ] Blocked injector/fuel injection system fault - fuel-injected petrol models (Chapter 4B).
- [ ] Faulty injector(s) - diesel models (Chapter 4C).

Engine stalls
- [ ] Vacuum leak at the carburettor/throttle body, inlet manifold or associated hoses - petrol models (Chapter 4A or 4B).
- [ ] Fuel filter choked (Chapter 1).
- [ ] Fuel pump faulty, or delivery pressure low - petrol models (Chapter 4A or 4B).
- [ ] Fuel tank vent blocked, or fuel pipes restricted (Chapter 4A, 4B or 4C).
- [ ] Blocked carburettor jet(s) or internal passages - carburettor petrol models (Chapter 4A).
- [ ] Blocked injector/fuel injection system fault - fuel-injected petrol models (Chapter 4B).
- [ ] Faulty injector(s) - diesel models (Chapter 4C).

Engine lacks power
- [ ] Timing belt incorrectly fitted or tensioned (Chapter 2A, 2B or 2C).
- [ ] Fuel filter choked (Chapter 1).
- [ ] Fuel pump faulty, or delivery pressure low - petrol models (Chapter 4A or 4B).
- [ ] Uneven or low cylinder compressions (Chapter 2A, 2B or 2C).
- [ ] Worn, faulty or incorrectly-gapped spark plugs - petrol models (Chapter 1).
- [ ] Vacuum leak at the carburettor/throttle body, inlet manifold or associated hoses - petrol models (Chapter 4A or 4B).
- [ ] Blocked carburettor jet(s) or internal passages - carburettor petrol models (Chapter 4A).
- [ ] Blocked injector/fuel injection system fault - fuel-injected petrol models (Chapter 4B).
- [ ] Faulty injector(s) - diesel models (Chapter 4C).
- [ ] Injection pump timing incorrect - diesel models (Chapter 4C).
- [ ] Brakes binding (Chapters 1 and 9).
- [ ] Clutch slipping (Chapter 6).

Engine backfires
- [ ] Timing belt/chain incorrectly fitted or tensioned (Chapter 2A, 2B or 2C).
- [ ] Vacuum leak at the carburettor/throttle body, inlet manifold or associated hoses - petrol models (Chapter 4A or 4B).
- [ ] Blocked carburettor jet(s) or internal passages - carburettor petrol models (Chapter 4A).
- [ ] Blocked injector/fuel injection system fault - fuel-injected petrol models (Chapter 4B).

Oil pressure warning light illuminated with engine running
- [ ] Low oil level, or incorrect oil grade ("Weekly checks").
- [ ] Faulty oil pressure sensor (Chapter 5A).
- [ ] Worn engine bearings and/or oil pump (Chapter 2A, 2B or 2C).
- [ ] High engine operating temperature (Chapter 3).
- [ ] Oil pressure relief valve defective (Chapter 2A, 2B or 2C).
- [ ] Oil pick-up strainer clogged (Chapter 2A, 2B or 2C).

Engine runs-on after switching off
- [ ] Excessive carbon build-up in engine (Chapter 2A, 2B or 2C).
- [ ] High engine operating temperature (Chapter 3).
- [ ] Faulty fuel cut-off solenoid - carburettor petrol models (Chapter 4A).
- [ ] Fuel injection system fault - fuel-injected petrol models (Chapter 4B).
- [ ] Faulty stop solenoid - diesel models (Chapter 4C).

Engine noises

Pre-ignition (pinking) or knocking during acceleration or under load
- [ ] Ignition timing incorrect/ignition system fault - petrol models (Chapters 1 and 5B).
- [ ] Incorrect grade of spark plug - petrol models (Chapter 1).
- [ ] Incorrect grade of fuel (Chapter 1).
- [ ] Vacuum leak at the carburettor/throttle body, inlet manifold or associated hoses - petrol models (Chapter 4A or 4B).
- [ ] Excessive carbon build-up in engine (Chapter 2A, 2B or 2C).
- [ ] Blocked carburettor jet(s) or internal passages - carburettor petrol models (Chapter 4A).
- [ ] Blocked injector/fuel injection system fault - fuel-injected petrol models (Chapter 4B).

Fault finding REF•15

1 Engine (continued)

Whistling or wheezing noises
- ☐ Leaking inlet manifold or carburettor/throttle body gasket - petrol models (Chapter 4A or 4B).
- ☐ Leaking exhaust manifold gasket or pipe-to-manifold joint (Chapter 4A, 4B or 4C).
- ☐ Leaking vacuum hose (Chapters 4A, 4B or 4C, 5B and 9).
- ☐ Blowing cylinder head gasket (Chapter 2A, 2B and 2C).

Tapping or rattling noises
- ☐ Worn valve gear or camshaft (Chapter 2A, 2B or 2C).
- ☐ Ancillary component fault (water pump, alternator, etc) (Chapters 3, 5A, etc).

Knocking or thumping noises
- ☐ Worn big-end bearings (regular heavy knocking, perhaps less under load) (Chapter 2A, 2B and 2C).
- ☐ Worn main bearings (rumbling and knocking, perhaps worsening under load) (Chapter 2A, 2B and 2C).
- ☐ Piston slap (most noticeable when cold) (Chapter 2A, 2B and 2C).
- ☐ Ancillary component fault (water pump, alternator, etc) (Chapters 3, 5A, etc).

2 Cooling system

Overheating
- ☐ Insufficient coolant in system (*"Weekly Checks"*).
- ☐ Thermostat faulty (Chapter 3).
- ☐ Radiator core blocked, or grille restricted (Chapter 3).
- ☐ Electric cooling fan or thermostatic switch faulty (Chapter 3).
- ☐ Inaccurate temperature gauge sender unit (Chapter 3).
- ☐ Airlock in cooling system (Chapter 3).
- ☐ Expansion tank pressure cap faulty (Chapter 3).

Overcooling
- ☐ Thermostat faulty (Chapter 3).
- ☐ Inaccurate temperature gauge sender unit (Chapter 3).

External coolant leakage
- ☐ Deteriorated or damaged hoses or hose clips (Chapter 1).
- ☐ Radiator core or heater matrix leaking (Chapter 3).
- ☐ Pressure cap faulty (Chapter 3).
- ☐ Coolant pump internal seal leaking (Chapter 3).
- ☐ Coolant pump-to-block seal leaking (Chapter 3).
- ☐ Boiling due to overheating (Chapter 3).
- ☐ Core plug leaking (Chapter 2A, 2B or 2C).

Internal coolant leakage
- ☐ Leaking cylinder head gasket (Chapter 2A, 2B or 2C).
- ☐ Cracked cylinder head or cylinder block (Chapter 2A, 2B or 2C).

Corrosion
- ☐ Infrequent draining and flushing (Chapter 1).
- ☐ Incorrect coolant mixture or inappropriate coolant (*"Weekly checks"*).

3 Fuel and exhaust systems

Excessive fuel consumption
- ☐ Air filter element dirty or clogged (Chapter 1).
- ☐ Faulty choke mechanism - carburettor petrol models (Chapter 4A).
- ☐ Fuel injection system fault - fuel-injected petrol models (Chapter 4B).
- ☐ Faulty injector(s) - diesel models (Chapter 4C).
- ☐ Ignition timing incorrect/ignition system fault - petrol models (Chapters 1 and 5B).
- ☐ Tyres under-inflated (*"Weekly checks"*).

Fuel leakage and/or fuel odour
- ☐ Damaged or corroded fuel tank, pipes or connections (Chapter 4A, 4B or 4C).
- ☐ Carburettor float chamber flooding (float height incorrect) - carburettor petrol models (Chapter 4A).

Excessive noise or fumes from exhaust system
- ☐ Leaking exhaust system or manifold joints (Chapters 1 & 4A, 4B or 4C).
- ☐ Leaking, corroded or damaged silencers or pipe (Chapters 1 and 4A, 4B or 4C).
- ☐ Broken mountings causing body or suspension contact (Chapter 1).

4 Clutch

**Pedal travels to floor -
no pressure or very little resistance**
- ☐ Badly stretched or broken cable (Chapter 6).
- ☐ Leak or other fault in clutch hydraulic system - where applicable (Chapter 6).
- ☐ Incorrect clutch adjustment (Chapter 6).
- ☐ Broken clutch release bearing or arm (Chapter 6).
- ☐ Broken diaphragm spring in clutch pressure plate (Chapter 6).

Clutch fails to disengage (unable to select gears)
- ☐ Incorrect clutch adjustment (Chapter 6).
- ☐ Clutch friction plate sticking on gearbox input shaft splines (Chapter 6).
- ☐ Clutch friction plate sticking to flywheel or pressure plate (Chapter 6).
- ☐ Faulty pressure plate assembly (Chapter 6).
- ☐ Clutch release mechanism worn or incorrectly assembled (Chapter 6).

Clutch slips (engine speed increases, with no increase in vehicle speed)
- ☐ Clutch friction plate linings excessively worn (Chapter 6).
- ☐ Clutch friction plate linings contaminated with oil or grease (Chapter 6).
- ☐ Faulty pressure plate or weak diaphragm spring (Chapter 6).

Judder as clutch is engaged
- ☐ Clutch friction plate linings contaminated with oil or grease (Chapter 6).
- ☐ Clutch friction plate linings excessively worn (Chapter 6).
- ☐ Faulty or distorted pressure plate or diaphragm spring (Chapter 6).
- ☐ Worn or loose engine or gearbox mountings (Chapter 2A).
- ☐ Clutch friction plate hub or gearbox input shaft splines worn (Chapter 6).

Fault finding

4 Clutch (continued)

Noise when depressing or releasing clutch pedal
- ☐ Worn clutch release bearing (Chapter 6).
- ☐ Worn or dry clutch pedal pivot (Chapter 6).
- ☐ Faulty pressure plate assembly (Chapter 6).
- ☐ Pressure plate diaphragm spring broken (Chapter 6).
- ☐ Broken clutch friction plate cushioning springs (Chapter 6).

5 Manual transmission

Noisy in neutral with engine running
- ☐ Input shaft bearings worn (noise apparent with clutch pedal released, but not when depressed) (Chapter 7).*
- ☐ Clutch release bearing worn (noise apparent with clutch pedal depressed, possibly less when released) (Chapter 6).

Noisy in one particular gear
- ☐ Worn, damaged or chipped gear teeth (Chapter 7).*

Difficulty engaging gears
- ☐ Clutch fault (Chapter 6).
- ☐ Worn or damaged gear linkage (Chapter 7).
- ☐ Worn synchroniser units (Chapter 7).*

Jumps out of gear
- ☐ Worn or damaged gear linkage (Chapter 7).
- ☐ Worn synchroniser units (Chapter 7).*
- ☐ Worn selector forks (Chapter 7).*

Vibration
- ☐ Lack of oil (Chapter 1).
- ☐ Worn bearings (Chapter 7).*

Lubricant leaks
- ☐ Leaking oil seal (Chapter 7).
- ☐ Leaking housing joint (Chapter 7).*

Although the corrective action necessary to remedy the symptoms described is beyond the scope of the home mechanic, the above information should be helpful in isolating the cause of the condition, so that the owner can communicate clearly with a professional mechanic.

6 Automatic transmission

Note: *Due to the complexity of the automatic transmission, it is difficult for the home mechanic to properly diagnose and service this unit. For problems other than the following, the vehicle should be taken to a dealer service department or automatic transmission specialist. Do not be too hasty in removing the transmission if a fault is suspected, as most of the testing is carried out with the unit still fitted.*

Fluid leakage
- ☐ Automatic transmission fluid is usually dark in colour. Fluid leaks should not be confused with engine oil, which can easily be blown on to the transmission by airflow.
- ☐ To determine the source of a leak, first remove all built-up dirt and grime from the transmission housing and surrounding areas using a degreasing agent, or by steam-cleaning. Drive the vehicle at low speed, so airflow will not blow the leak far from its source. Raise and support the vehicle, and determine where the leak is coming from. The following are common areas of leakage:
  a) Oil pan (Chapter 1 and 7B).
  b) Dipstick tube (Chapter 1 and 7B).
  c) Transmission-to-fluid cooler pipes/unions (Chapter 7B).

Transmission fluid brown, or has burned smell
- ☐ Transmission fluid level low, or fluid in need of renewal (Chapter 1 and 7B).

General gear selection problems
- ☐ Chapter 7B deals with checking and adjusting the selector cable on automatic transmissions. The following are common problems which may be caused by a poorly-adjusted cable:
  a) Engine starting in gears other than Park or Neutral.
  b) Indicator panel indicating a gear other than the one actually being used.
  c) Vehicle moves when in Park or Neutral.
  d) Poor gear shift quality or erratic gear changes.
- ☐ Refer to Chapter 7B for the selector cable adjustment procedure.

Transmission will not downshift (kickdown) with accelerator pedal fully depressed
- ☐ Low transmission fluid level (Chapter 1 and 7B).
- ☐ Incorrect selector cable adjustment (Chapter 7B).

Engine will not start in any gear, or starts in gears other than Park or Neutral
- ☐ Incorrect starter/inhibitor switch adjustment (Chapter 7B).
- ☐ Incorrect selector cable adjustment (Chapter 7B).

Transmission slips, shifts roughly, is noisy, or has no drive in forward or reverse gears
- ☐ There are many probable causes for the above problems, but the home mechanic should be concerned with only one possibility - fluid level. Before taking the vehicle to a dealer or transmission specialist, check the fluid level and condition of the fluid as described in Chapter 1, or 7B, as applicable. Correct the fluid level as necessary, or change the fluid and filter if needed. If the problem persists, professional help will be necessary.

7 Driveshafts

Clicking or knocking noise on turns (at slow speed on full-lock)
- ☐ Lack of constant velocity joint lubricant, possibly due to damaged gaiter (Chapter 8).
- ☐ Worn outer constant velocity joint (Chapter 8).

Vibration when accelerating or decelerating
- ☐ Worn inner constant velocity joint (Chapter 8).
- ☐ Bent or distorted driveshaft (Chapter 8).
- ☐ Worn right-hand driveshaft intermediate bearing (Chapter 8).

Fault finding

8 Braking system

Note: *Before assuming that a brake problem exists, make sure that the tyres are in good condition and correctly inflated, that the front wheel alignment is correct, and that the vehicle is not loaded with weight in an unequal manner. Apart from checking the condition of all pipe and hose connections, any faults occurring on the anti-lock braking system should be referred to a Renault dealer for diagnosis.*

Vehicle pulls to one side under braking
- Worn, defective, damaged or contaminated front or rear brake pads/shoes on one side (Chapters 1 and 9).
- Seized or partially-seized front or rear brake caliper/wheel cylinder piston (Chapter 9).
- A mixture of brake pad/shoe lining materials fitted between sides (Chapter 9).
- Brake caliper or rear brake backplate mounting bolts loose (Chapter9).
- Worn or damaged steering or suspension components (Chapters 1 and 10).

Noise (grinding or high-pitched squeal) when brakes applied
- Brake pad or shoe friction lining material worn down to metal backing (Chapters 1 and 9).
- Excessive corrosion of brake disc or drum - may be apparent after the vehicle has been standing for some time (Chapters 1 and 9).

Excessive brake pedal travel
- Faulty rear drum brake self-adjust mechanism (Chapter 9).
- Faulty master cylinder (Chapter 9).
- Air in hydraulic system (Chapter 9).
- Faulty vacuum servo unit (Chapter 9).
- Faulty vacuum pump - diesel models (Chapter 9).

Brake pedal feels spongy when depressed
- Air in hydraulic system (Chapter 9).
- Deteriorated flexible rubber brake hoses (Chapters 1 and 9).
- Master cylinder mountings loose (Chapter 9).
- Faulty master cylinder (Chapter 9).

Excessive brake pedal effort required to stop vehicle
- Faulty vacuum servo unit (Chapter 9).
- Disconnected, damaged or insecure brake servo vacuum hose (Chapters 1 and 9).
- Faulty vacuum pump - diesel models (Chapter 9).
- Primary or secondary hydraulic circuit failure (Chapter 9).
- Seized brake caliper or wheel cylinder piston(s) (Chapter 9).
- Brake pads or brake shoes incorrectly fitted (Chapter 9).
- Incorrect grade of brake pads or brake shoes fitted (Chapter 9).
- Brake pads or brake shoe linings contaminated (Chapter 9).

Judder felt through brake pedal or steering wheel when braking
- Excessive run-out or distortion of brake disc(s) or drum(s) (Chapter 9).
- Brake pad or brake shoe linings worn (Chapters 1 and 9).
- Brake caliper or rear brake backplate mounting bolts loose (Chapter 9).
- Wear in suspension or steering components or mountings (Chapters 1 and 10).

Pedal pulsates when braking hard
- Normal feature of ABS - no fault

Brakes binding
- Seized brake caliper piston(s) or wheel cylinder piston(s) (Chapter 9).
- Incorrectly-adjusted handbrake mechanism or linkage (Chapter 9).
- Faulty master cylinder (Chapter 9).

Rear wheels locking under normal braking
- Seized brake caliper piston(s) or wheel cylinder piston(s) (Chapter 9).
- Faulty brake pressure regulator (Chapter 9).

9 Steering and suspension

Note: *Before diagnosing suspension or steering faults, be sure that the trouble is not due to incorrect tyre pressures, mixtures of tyre types, or binding brakes.*

Vehicle pulls to one side
- Defective tyre ("Weekly checks").
- Excessive wear in suspension or steering components (Chapters 1 and 10).
- Incorrect front wheel alignment (Chapter 10).
- Accident damage to steering or suspension components (Chapters 1 and 10).

Wheel wobble and vibration
- Front roadwheels out of balance (vibration felt mainly through the steering wheel) (Chapter 10).
- Rear roadwheels out of balance (vibration felt throughout the vehicle) (Chapter 10).
- Roadwheels damaged or distorted (Chapter 10).
- Faulty or damaged tyre ("Weekly Checks").
- Worn steering or suspension joints, bushes or components (Chapters 1 and 10).
- Wheel bolts loose (Chapter 10).

Excessive pitching and/or rolling around corners, or during braking
- Defective shock absorbers (Chapters 1 and 10).
- Broken or weak coil spring and/or suspension component (Chapters 1 and 10).
- Worn or damaged anti-roll bar or mountings (Chapter 10).

Wandering or general instability
- Incorrect front wheel alignment (Chapter 10).
- Worn steering or suspension joints, bushes or components (Chapters 1 and 10).
- Roadwheels out of balance (Chapter 10).
- Faulty or damaged tyre ("Weekly Checks").
- Wheel bolts loose (Chapter 10).
- Defective shock absorbers (Chapters 1 and 10).

Excessively-stiff steering
- Lack of steering gear lubricant (Chapter 10).
- Seized track rod end balljoint or suspension balljoint (Chapters 1 and 10).
- Broken or incorrectly adjusted auxiliary drivebelt (Chapter 1).
- Incorrect front wheel alignment (Chapter 10).
- Steering rack or column bent or damaged (Chapter 10).

REF•18 Fault finding

9 Steering and suspension (continued)

Excessive play in steering
- [ ] Worn steering column universal joint(s) (Chapter 10).
- [ ] Worn steering track rod end balljoints (Chapters 1 and 10).
- [ ] Worn rack-and-pinion steering gear (Chapter 10).
- [ ] Worn steering or suspension joints, bushes or components (Chapters 1 and 10).

Lack of power assistance
- [ ] Broken or incorrectly-adjusted auxiliary drivebelt (Chapter 1).
- [ ] Incorrect power steering fluid level (*"Weekly Checks"*).
- [ ] Restriction in power steering fluid hoses (Chapter 10).
- [ ] Faulty power steering pump (Chapter 10).
- [ ] Faulty rack-and-pinion steering gear (Chapter 10).

Tyre wear excessive

Tyres worn on inside or outside edges
- [ ] Tyres under-inflated (wear on both edges) (*"Weekly Checks"*).
- [ ] Incorrect camber or castor angles (wear on one edge only) (Chapter 10).
- [ ] Worn steering or suspension joints, bushes or components (Chapters 1 and 10).
- [ ] Excessively-hard cornering.
- [ ] Accident damage.

Tyre treads exhibit feathered edges
- [ ] Incorrect toe setting (Chapter 10).

Tyres worn in centre of tread
- [ ] Tyres over-inflated (*"Weekly Checks"*).

Tyres worn on inside and outside edges
- [ ] Tyres under-inflated (*"Weekly Checks"*).
- [ ] Worn shock absorbers (Chapters 1 and 10).

Tyres worn unevenly
- [ ] Tyres out of balance (*"Weekly Checks"*).
- [ ] Excessive wheel or tyre run-out (Chapter 1).
- [ ] Worn shock absorbers (Chapters 1 and 10).
- [ ] Faulty tyre (*"Weekly Checks"*).

10 Electrical system

Note: *For problems associated with the starting system, refer to the faults listed under "Engine" earlier in this Section.*

Battery will not hold a charge for more than a few days
- [ ] Battery defective internally (Chapter 5A).
- [ ] Battery electrolyte level low - where applicable (*"Weekly Checks"*).
- [ ] Battery terminal connections loose or corroded (*"Weekly Checks"*).
- [ ] Auxiliary drivebelt worn - or incorrectly adjusted, where applicable (Chapter 1).
- [ ] Alternator not charging at correct output (Chapter 5A).
- [ ] Alternator or voltage regulator faulty (Chapter 5A).
- [ ] Short-circuit causing continual battery drain (Chapters 5A and 12).

Ignition/no-charge warning light remains illuminated with engine running
- [ ] Auxiliary drivebelt broken, worn, or incorrectly adjusted (Chapter 1).
- [ ] Alternator brushes worn, sticking, or dirty (Chapter 5A).
- [ ] Alternator brush springs weak or broken (Chapter 5A).
- [ ] Internal fault in alternator or voltage regulator (Chapter 5A).
- [ ] Broken, disconnected, or loose wiring in charging circuit (Chapter 5A).

Ignition/no-charge warning light fails to come on
- [ ] Warning light bulb blown (Chapter 12).
- [ ] Broken, disconnected, or loose wiring in warning light circuit (Chapter 12).
- [ ] Alternator faulty (Chapter 5A).

Lights inoperative
- [ ] Bulb blown (Chapter 12).
- [ ] Corrosion of bulb or bulbholder contacts (Chapter 12).
- [ ] Blown fuse (Chapter 12).
- [ ] Faulty relay (Chapter 12).
- [ ] Broken, loose, or disconnected wiring (Chapter 12).
- [ ] Faulty switch (Chapter 12).

Instrument readings inaccurate or erratic

Instrument readings increase with engine speed
- [ ] Faulty voltage regulator (Chapter 12).

Fuel or temperature gauges give no reading
- [ ] Faulty gauge sender unit (Chapters 3 and 4).
- [ ] Wiring open-circuit (Chapter 12).
- [ ] Faulty gauge (Chapter 12).

Fuel or temperature gauges give continuous maximum reading
- [ ] Faulty gauge sender unit (Chapters 3 and 4).
- [ ] Wiring short-circuit (Chapter 12).
- [ ] Faulty gauge (Chapter 12).

Horn inoperative, or unsatisfactory in operation

Horn operates all the time
- [ ] Horn contacts permanently bridged or horn push stuck down (Chapter 12).

Horn fails to operate
- [ ] Blown fuse (Chapter 12).
- [ ] Cable or cable connections loose, broken or disconnected (Chapter 12).
- [ ] Faulty horn (Chapter 12).

Horn emits intermittent or unsatisfactory sound
- [ ] Cable connections loose (Chapter 12).
- [ ] Horn mountings loose (Chapter 12).
- [ ] Faulty horn (Chapter 12).

Fault finding REF•19

Windscreen/tailgate wipers inoperative, or unsatisfactory in operation

Wipers fail to operate, or operate very slowly
- ☐ Wiper blades stuck to screen, or linkage seized or binding ("*Weekly Checks*" and Chapter 12).
- ☐ Blown fuse (Chapter 12).
- ☐ Cable or cable connections loose, broken or disconnected (Chapter 12).
- ☐ Faulty relay (Chapter 12).
- ☐ Faulty wiper motor (Chapter 12).

Wiper blades sweep over too large or too small an area of the glass
- ☐ Wiper arms incorrectly positioned on spindles (Chapter 12).
- ☐ Excessive wear of wiper linkage (Chapter 12).
- ☐ Wiper motor or linkage mountings loose or insecure (Chapter 12).

Wiper blades fail to clean the glass effectively
- ☐ Wiper blade rubbers worn or perished ("*Weekly Checks*").
- ☐ Wiper arm tension springs broken, or arm pivots seized (Chapter 12).
- ☐ Insufficient windscreen washer additive to adequately remove road film ("*Weekly Checks*").

Windscreen/tailgate washers inoperative, or unsatisfactory in operation

One or more washer jets inoperative
- ☐ Blocked washer jet (Chapter 12).
- ☐ Disconnected, kinked or restricted fluid hose (Chapter 12).
- ☐ Insufficient fluid in washer reservoir ("*Weekly Checks*").

Washer pump fails to operate
- ☐ Broken or disconnected wiring or connections (Chapter 12).
- ☐ Blown fuse (Chapter 12).
- ☐ Faulty washer switch (Chapter 12).
- ☐ Faulty washer pump (Chapter 12).

Washer pump runs for some time before fluid is emitted from jets
- ☐ Faulty one-way valve in fluid supply hose (Chapter 12).

Electric windows inoperative, or unsatisfactory in operation

Window glass will only move in one direction
- ☐ Faulty switch (Chapter 12).

Window glass slow to move
- ☐ Regulator seized or damaged, or in need of lubrication (Chapter 11).
- ☐ Door internal components or trim fouling regulator (Chapter 11).
- ☐ Faulty motor (Chapter 11).

Window glass fails to move
- ☐ Blown fuse (Chapter 12).
- ☐ Faulty relay (Chapter 12).
- ☐ Broken or disconnected wiring or connections (Chapter 12).
- ☐ Faulty motor (Chapter 12).

Central locking system inoperative, or unsatisfactory in operation

Complete system failure
- ☐ Blown fuse (Chapter 12).
- ☐ Faulty relay (Chapter 12).
- ☐ Broken or disconnected wiring or connections (Chapter 12).

Latch locks but will not unlock, or unlocks but will not lock
- ☐ Faulty switch (Chapter 12).
- ☐ Broken or disconnected latch operating rods or levers (Chapter 11).
- ☐ Faulty relay (Chapter 12).

One motor fails to operate
- ☐ Broken or disconnected wiring or connections (Chapter 12).
- ☐ Faulty motor (Chapter 11).
- ☐ Broken, binding or disconnected latch operating rods or levers (Chapter 11).
- ☐ Fault in door latch (Chapter 11).

REF•20 Glossary of technical terms

A

ABS (Anti-lock brake system) A system, usually electronically controlled, that senses incipient wheel lockup during braking and relieves hydraulic pressure at wheels that are about to skid.

Air bag An inflatable bag hidden in the steering wheel (driver's side) or the dash or glovebox (passenger side). In a head-on collision, the bags inflate, preventing the driver and front passenger from being thrown forward into the steering wheel or windscreen.

Air cleaner A metal or plastic housing, containing a filter element, which removes dust and dirt from the air being drawn into the engine.

Air filter element The actual filter in an air cleaner system, usually manufactured from pleated paper and requiring renewal at regular intervals.

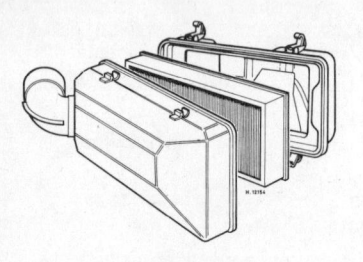

Air filter

Allen key A hexagonal wrench which fits into a recessed hexagonal hole.

Alligator clip A long-nosed spring-loaded metal clip with meshing teeth. Used to make temporary electrical connections.

Alternator A component in the electrical system which converts mechanical energy from a drivebelt into electrical energy to charge the battery and to operate the starting system, ignition system and electrical accessories.

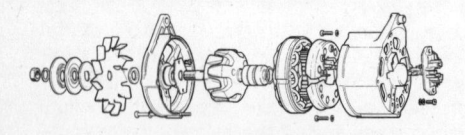

Alternator (exploded view)

Ampere (amp) A unit of measurement for the flow of electric current. One amp is the amount of current produced by one volt acting through a resistance of one ohm.

Anaerobic sealer A substance used to prevent bolts and screws from loosening. Anaerobic means that it does not require oxygen for activation. The Loctite brand is widely used.

Antifreeze A substance (usually ethylene glycol) mixed with water, and added to a vehicle's cooling system, to prevent freezing of the coolant in winter. Antifreeze also contains chemicals to inhibit corrosion and the formation of rust and other deposits that would tend to clog the radiator and coolant passages and reduce cooling efficiency.

Anti-seize compound A coating that reduces the risk of seizing on fasteners that are subjected to high temperatures, such as exhaust manifold bolts and nuts.

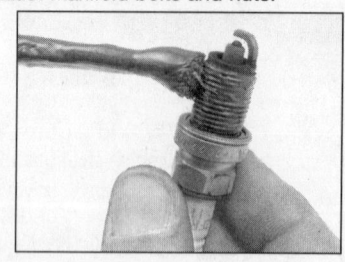

Anti-seize compound

Asbestos A natural fibrous mineral with great heat resistance, commonly used in the composition of brake friction materials. Asbestos is a health hazard and the dust created by brake systems should never be inhaled or ingested.

Axle A shaft on which a wheel revolves, or which revolves with a wheel. Also, a solid beam that connects the two wheels at one end of the vehicle. An axle which also transmits power to the wheels is known as a live axle.

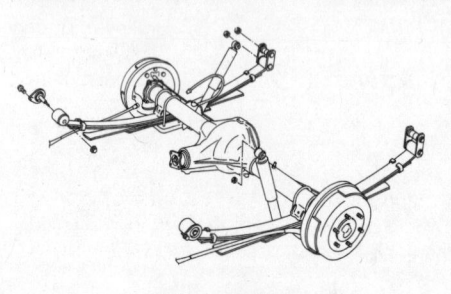

Axle assembly

Axleshaft A single rotating shaft, on either side of the differential, which delivers power from the final drive assembly to the drive wheels. Also called a driveshaft or a halfshaft.

B

Ball bearing An anti-friction bearing consisting of a hardened inner and outer race with hardened steel balls between two races.

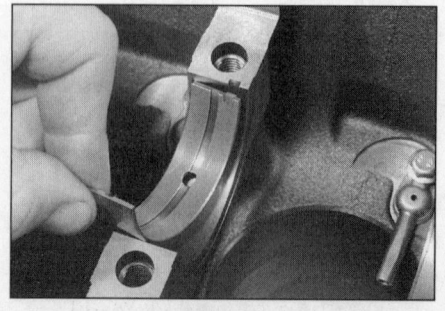

Bearing

Bearing The curved surface on a shaft or in a bore, or the part assembled into either, that permits relative motion between them with minimum wear and friction.

Big-end bearing The bearing in the end of the connecting rod that's attached to the crankshaft.

Bleed nipple A valve on a brake wheel cylinder, caliper or other hydraulic component that is opened to purge the hydraulic system of air. Also called a bleed screw.

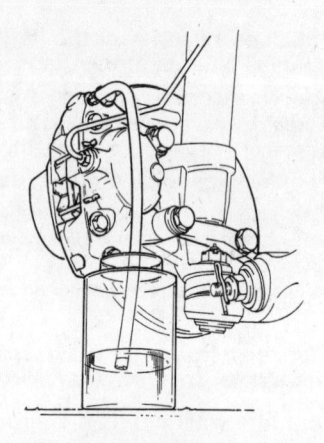

Brake bleeding

Brake bleeding Procedure for removing air from lines of a hydraulic brake system.

Brake disc The component of a disc brake that rotates with the wheels.

Brake drum The component of a drum brake that rotates with the wheels.

Brake linings The friction material which contacts the brake disc or drum to retard the vehicle's speed. The linings are bonded or riveted to the brake pads or shoes.

Brake pads The replaceable friction pads that pinch the brake disc when the brakes are applied. Brake pads consist of a friction material bonded or riveted to a rigid backing plate.

Brake shoe The crescent-shaped carrier to which the brake linings are mounted and which forces the lining against the rotating drum during braking.

Braking systems For more information on braking systems, consult the *Haynes Automotive Brake Manual*.

Breaker bar A long socket wrench handle providing greater leverage.

Bulkhead The insulated partition between the engine and the passenger compartment.

C

Caliper The non-rotating part of a disc-brake assembly that straddles the disc and carries the brake pads. The caliper also contains the hydraulic components that cause the pads to pinch the disc when the brakes are applied. A caliper is also a measuring tool that can be set to measure inside or outside dimensions of an object.

Glossary of technical terms REF•21

Camshaft A rotating shaft on which a series of cam lobes operate the valve mechanisms. The camshaft may be driven by gears, by sprockets and chain or by sprockets and a belt.

Canister A container in an evaporative emission control system; contains activated charcoal granules to trap vapours from the fuel system.

Canister

Carburettor A device which mixes fuel with air in the proper proportions to provide a desired power output from a spark ignition internal combustion engine.

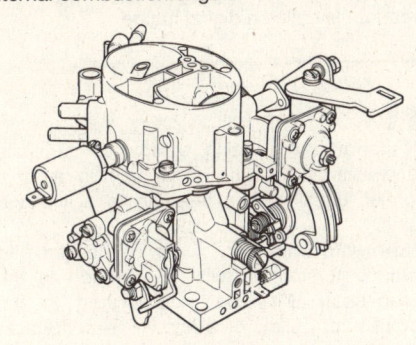

Carburettor

Castellated Resembling the parapets along the top of a castle wall. For example, a castellated balljoint stud nut.

Castellated nut

Castor In wheel alignment, the backward or forward tilt of the steering axis. Castor is positive when the steering axis is inclined rearward at the top.

Catalytic converter A silencer-like device in the exhaust system which converts certain pollutants in the exhaust gases into less harmful substances.

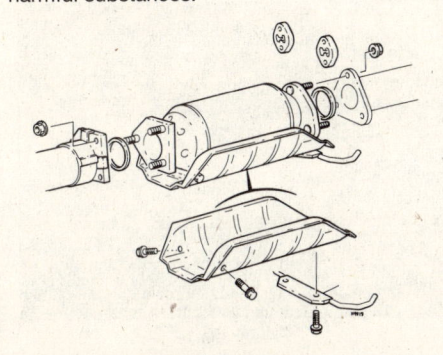

Catalytic converter

Circlip A ring-shaped clip used to prevent endwise movement of cylindrical parts and shafts. An internal circlip is installed in a groove in a housing; an external circlip fits into a groove on the outside of a cylindrical piece such as a shaft.

Clearance The amount of space between two parts. For example, between a piston and a cylinder, between a bearing and a journal, etc.

Coil spring A spiral of elastic steel found in various sizes throughout a vehicle, for example as a springing medium in the suspension and in the valve train.

Compression Reduction in volume, and increase in pressure and temperature, of a gas, caused by squeezing it into a smaller space.

Compression ratio The relationship between cylinder volume when the piston is at top dead centre and cylinder volume when the piston is at bottom dead centre.

Constant velocity (CV) joint A type of universal joint that cancels out vibrations caused by driving power being transmitted through an angle.

Core plug A disc or cup-shaped metal device inserted in a hole in a casting through which core was removed when the casting was formed. Also known as a freeze plug or expansion plug.

Crankcase The lower part of the engine block in which the crankshaft rotates.

Crankshaft The main rotating member, or shaft, running the length of the crankcase, with offset "throws" to which the connecting rods are attached.

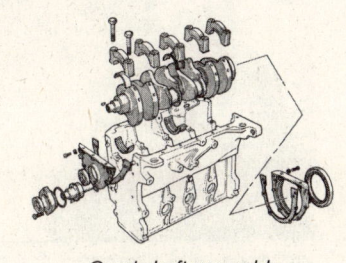

Crankshaft assembly

Crocodile clip See Alligator clip

D

Diagnostic code Code numbers obtained by accessing the diagnostic mode of an engine management computer. This code can be used to determine the area in the system where a malfunction may be located.

Disc brake A brake design incorporating a rotating disc onto which brake pads are squeezed. The resulting friction converts the energy of a moving vehicle into heat.

Double-overhead cam (DOHC) An engine that uses two overhead camshafts, usually one for the intake valves and one for the exhaust valves.

Drivebelt(s) The belt(s) used to drive accessories such as the alternator, water pump, power steering pump, air conditioning compressor, etc. off the crankshaft pulley.

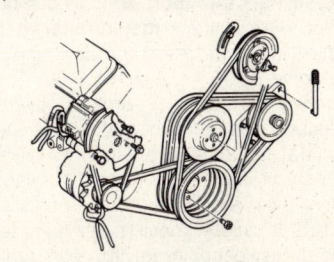

Accessory drivebelts

Driveshaft Any shaft used to transmit motion. Commonly used when referring to the axleshafts on a front wheel drive vehicle.

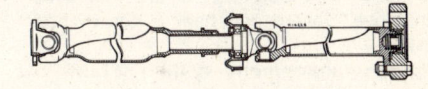

Driveshaft

Drum brake A type of brake using a drum-shaped metal cylinder attached to the inner surface of the wheel. When the brake pedal is pressed, curved brake shoes with friction linings press against the inside of the drum to slow or stop the vehicle.

Drum brake assembly

Glossary of technical terms

E

EGR valve A valve used to introduce exhaust gases into the intake air stream.

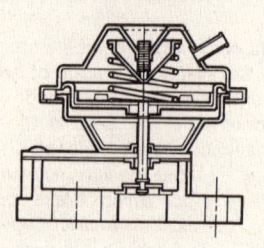

EGR valve

Electronic control unit (ECU) A computer which controls (for instance) ignition and fuel injection systems, or an anti-lock braking system. For more information refer to the Haynes Automotive Electrical and Electronic Systems Manual.

Electronic Fuel Injection (EFI) A computer controlled fuel system that distributes fuel through an injector located in each intake port of the engine.

Emergency brake A braking system, independent of the main hydraulic system, that can be used to slow or stop the vehicle if the primary brakes fail, or to hold the vehicle stationary even though the brake pedal isn't depressed. It usually consists of a hand lever that actuates either front or rear brakes mechanically through a series of cables and linkages. Also known as a handbrake or parking brake.

Endfloat The amount of lengthwise movement between two parts. As applied to a crankshaft, the distance that the crankshaft can move forward and back in the cylinder block.

Engine management system (EMS) A computer controlled system which manages the fuel injection and the ignition systems in an integrated fashion.

Exhaust manifold A part with several passages through which exhaust gases leave the engine combustion chambers and enter the exhaust pipe.

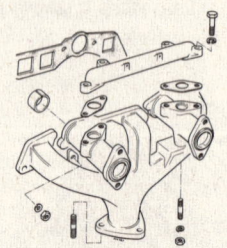

Exhaust manifold

F

Fan clutch A viscous (fluid) drive coupling device which permits variable engine fan speeds in relation to engine speeds.

Feeler blade A thin strip or blade of hardened steel, ground to an exact thickness, used to check or measure clearances between parts.

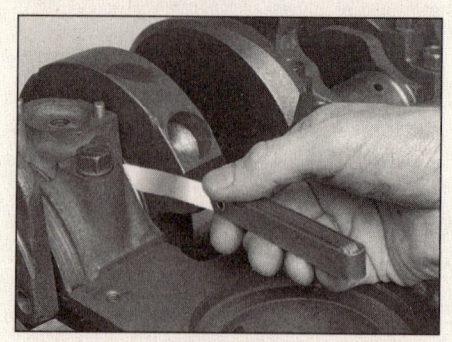

Feeler blade

Firing order The order in which the engine cylinders fire, or deliver their power strokes, beginning with the number one cylinder.

Flywheel A heavy spinning wheel in which energy is absorbed and stored by means of momentum. On cars, the flywheel is attached to the crankshaft to smooth out firing impulses.

Free play The amount of travel before any action takes place. The "looseness" in a linkage, or an assembly of parts, between the initial application of force and actual movement. For example, the distance the brake pedal moves before the pistons in the master cylinder are actuated.

Fuse An electrical device which protects a circuit against accidental overload. The typical fuse contains a soft piece of metal which is calibrated to melt at a predetermined current flow (expressed as amps) and break the circuit.

Fusible link A circuit protection device consisting of a conductor surrounded by heat-resistant insulation. The conductor is smaller than the wire it protects, so it acts as the weakest link in the circuit. Unlike a blown fuse, a failed fusible link must frequently be cut from the wire for replacement.

G

Gap The distance the spark must travel in jumping from the centre electrode to the side

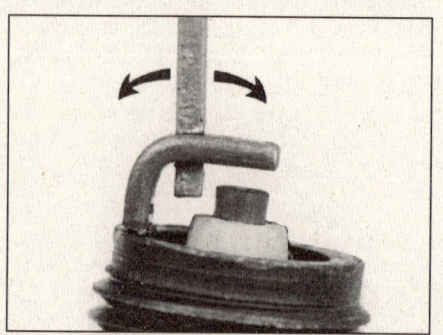

Adjusting spark plug gap

electrode in a spark plug. Also refers to the spacing between the points in a contact breaker assembly in a conventional points-type ignition, or to the distance between the reluctor or rotor and the pickup coil in an electronic ignition.

Gasket Any thin, soft material - usually cork, cardboard, asbestos or soft metal - installed between two metal surfaces to ensure a good seal. For instance, the cylinder head gasket seals the joint between the block and the cylinder head.

Gasket

Gauge An instrument panel display used to monitor engine conditions. A gauge with a movable pointer on a dial or a fixed scale is an analogue gauge. A gauge with a numerical readout is called a digital gauge.

H

Halfshaft A rotating shaft that transmits power from the final drive unit to a drive wheel, usually when referring to a live rear axle.

Harmonic balancer A device designed to reduce torsion or twisting vibration in the crankshaft. May be incorporated in the crankshaft pulley. Also known as a vibration damper.

Hone An abrasive tool for correcting small irregularities or differences in diameter in an engine cylinder, brake cylinder, etc.

Hydraulic tappet A tappet that utilises hydraulic pressure from the engine's lubrication system to maintain zero clearance (constant contact with both camshaft and valve stem). Automatically adjusts to variation in valve stem length. Hydraulic tappets also reduce valve noise.

I

Ignition timing The moment at which the spark plug fires, usually expressed in the number of crankshaft degrees before the piston reaches the top of its stroke.

Inlet manifold A tube or housing with passages through which flows the air-fuel mixture (carburettor vehicles and vehicles with throttle body injection) or air only (port fuel-injected vehicles) to the port openings in the cylinder head.

Glossary of technical terms REF•23

J

Jump start Starting the engine of a vehicle with a discharged or weak battery by attaching jump leads from the weak battery to a charged or helper battery.

L

Load Sensing Proportioning Valve (LSPV) A brake hydraulic system control valve that works like a proportioning valve, but also takes into consideration the amount of weight carried by the rear axle.

Locknut A nut used to lock an adjustment nut, or other threaded component, in place. For example, a locknut is employed to keep the adjusting nut on the rocker arm in position.

Lockwasher A form of washer designed to prevent an attaching nut from working loose.

M

MacPherson strut A type of front suspension system devised by Earle MacPherson at Ford of England. In its original form, a simple lateral link with the anti-roll bar creates the lower control arm. A long strut - an integral coil spring and shock absorber - is mounted between the body and the steering knuckle. Many modern so-called MacPherson strut systems use a conventional lower A-arm and don't rely on the anti-roll bar for location.

Multimeter An electrical test instrument with the capability to measure voltage, current and resistance.

N

NOx Oxides of Nitrogen. A common toxic pollutant emitted by petrol and diesel engines at higher temperatures.

O

Ohm The unit of electrical resistance. One volt applied to a resistance of one ohm will produce a current of one amp.

Ohmmeter An instrument for measuring electrical resistance.

O-ring A type of sealing ring made of a special rubber-like material; in use, the O-ring is compressed into a groove to provide the sealing action.

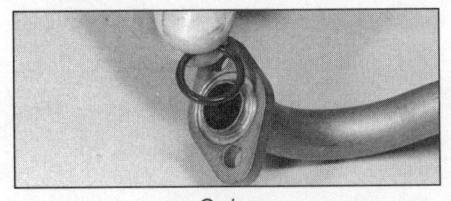

O-ring

Overhead cam (ohc) engine An engine with the camshaft(s) located on top of the cylinder head(s).

Overhead valve (ohv) engine An engine with the valves located in the cylinder head, but with the camshaft located in the engine block.

Oxygen sensor A device installed in the engine exhaust manifold, which senses the oxygen content in the exhaust and converts this information into an electric current. Also called a Lambda sensor.

P

Phillips screw A type of screw head having a cross instead of a slot for a corresponding type of screwdriver.

Plastigage A thin strip of plastic thread, available in different sizes, used for measuring clearances. For example, a strip of Plastigage is laid across a bearing journal. The parts are assembled and dismantled; the width of the crushed strip indicates the clearance between journal and bearing.

Plastigage

Propeller shaft The long hollow tube with universal joints at both ends that carries power from the transmission to the differential on front-engined rear wheel drive vehicles.

Proportioning valve A hydraulic control valve which limits the amount of pressure to the rear brakes during panic stops to prevent wheel lock-up.

R

Rack-and-pinion steering A steering system with a pinion gear on the end of the steering shaft that mates with a rack (think of a geared wheel opened up and laid flat). When the steering wheel is turned, the pinion turns, moving the rack to the left or right. This movement is transmitted through the track rods to the steering arms at the wheels.

Radiator A liquid-to-air heat transfer device designed to reduce the temperature of the coolant in an internal combustion engine cooling system.

Refrigerant Any substance used as a heat transfer agent in an air-conditioning system. R-12 has been the principle refrigerant for many years; recently, however, manufacturers have begun using R-134a, a non-CFC substance that is considered less harmful to the ozone in the upper atmosphere.

Rocker arm A lever arm that rocks on a shaft or pivots on a stud. In an overhead valve engine, the rocker arm converts the upward movement of the pushrod into a downward movement to open a valve.

Rotor In a distributor, the rotating device inside the cap that connects the centre electrode and the outer terminals as it turns, distributing the high voltage from the coil secondary winding to the proper spark plug. Also, that part of an alternator which rotates inside the stator. Also, the rotating assembly of a turbocharger, including the compressor wheel, shaft and turbine wheel.

Runout The amount of wobble (in-and-out movement) of a gear or wheel as it's rotated. The amount a shaft rotates "out-of-true." The out-of-round condition of a rotating part.

S

Sealant A liquid or paste used to prevent leakage at a joint. Sometimes used in conjunction with a gasket.

Sealed beam lamp An older headlight design which integrates the reflector, lens and filaments into a hermetically-sealed one-piece unit. When a filament burns out or the lens cracks, the entire unit is simply replaced.

Serpentine drivebelt A single, long, wide accessory drivebelt that's used on some newer vehicles to drive all the accessories, instead of a series of smaller, shorter belts. Serpentine drivebelts are usually tensioned by an automatic tensioner.

Serpentine drivebelt

Shim Thin spacer, commonly used to adjust the clearance or relative positions between two parts. For example, shims inserted into or under bucket tappets control valve clearances. Clearance is adjusted by changing the thickness of the shim.

Slide hammer A special puller that screws into or hooks onto a component such as a shaft or bearing; a heavy sliding handle on the shaft bottoms against the end of the shaft to knock the component free.

Sprocket A tooth or projection on the periphery of a wheel, shaped to engage with a chain or drivebelt. Commonly used to refer to the sprocket wheel itself.

Starter inhibitor switch On vehicles with an

REF•24 Glossary of technical terms

automatic transmission, a switch that prevents starting if the vehicle is not in Neutral or Park.
Strut See MacPherson strut.

T

Tappet A cylindrical component which transmits motion from the cam to the valve stem, either directly or via a pushrod and rocker arm. Also called a cam follower.
Thermostat A heat-controlled valve that regulates the flow of coolant between the cylinder block and the radiator, so maintaining optimum engine operating temperature. A thermostat is also used in some air cleaners in which the temperature is regulated.
Thrust bearing The bearing in the clutch assembly that is moved in to the release levers by clutch pedal action to disengage the clutch. Also referred to as a release bearing.
Timing belt A toothed belt which drives the camshaft. Serious engine damage may result if it breaks in service.
Timing chain A chain which drives the camshaft.
Toe-in The amount the front wheels are closer together at the front than at the rear. On rear wheel drive vehicles, a slight amount of toe-in is usually specified to keep the front wheels running parallel on the road by offsetting other forces that tend to spread the wheels apart.
Toe-out The amount the front wheels are closer together at the rear than at the front. On front wheel drive vehicles, a slight amount of toe-out is usually specified.
Tools For full information on choosing and using tools, refer to the Haynes Automotive Tools Manual.
Tracer A stripe of a second colour applied to a wire insulator to distinguish that wire from another one with the same colour insulator.
Tune-up A process of accurate and careful adjustments and parts replacement to obtain the best possible engine performance.
Turbocharger A centrifugal device, driven by exhaust gases, that pressurises the intake air. Normally used to increase the power output from a given engine displacement, but can also be used primarily to reduce exhaust emissions (as on VW's "Umwelt" Diesel engine).

U

Universal joint or U-joint A double-pivoted connection for transmitting power from a driving to a driven shaft through an angle. A U-joint consists of two Y-shaped yokes and a cross-shaped member called the spider.

V

Valve A device through which the flow of liquid, gas, vacuum, or loose material in bulk may be started, stopped, or regulated by a movable part that opens, shuts, or partially obstructs one or more ports or passageways. A valve is also the movable part of such a device.
Valve clearance The clearance between the valve tip (the end of the valve stem) and the rocker arm or tappet. The valve clearance is measured when the valve is closed.
Vernier caliper A precision measuring instrument that measures inside and outside dimensions. Not quite as accurate as a micrometer, but more convenient.
Viscosity The thickness of a liquid or its resistance to flow.
Volt A unit for expressing electrical "pressure" in a circuit. One volt that will produce a current of one ampere through a resistance of one ohm.

W

Welding Various processes used to join metal items by heating the areas to be joined to a molten state and fusing them together. For more information refer to the Haynes Automotive Welding Manual.
Wiring diagram A drawing portraying the components and wires in a vehicle's electrical system, using standardised symbols. For more information refer to the Haynes Automotive Electrical and Electronic Systems Manual.

REF•26 Index

Note: *References throughout this index are in the form - "Chapter number" • "page number"*

A

ABS - 9•15
Accelerator
   cable - 4A•5, 4B•2, 4C•10
   pedal - 4A•6, 4B•2, 4C•11
Aerial - 12•14
Air cleaner assembly - 4A•4, 4B•2, 4C•12
Air conditioning system - 3•1 et seq
   components - 3•9
   refrigerant - 1A•14, 1B•9
Air filter element - 1A•14, 1B•11
Alternator - 5A•3
   drivebelt - 5A•3
Anti-lock braking system - 9•15
Anti-roll bar - 10•5, 10•9
Anti-stall speed - 1B•7
Antifreeze - 0•12, 0•16, 1A•2, 1A•19, 1A•20, 1B•2, 1B•12, 1B•13
ATF - 0•16, 1A•2, 1A•11, 1A•17, 1B•2
Automatic transmission - 7B•1 et seq
   and engine - 2A•31, 2B•25, 2C•22
   differential shaft oil seal - 7B•9
   draining and refilling - 7B•3
   fault finding - REF•16
   fluid - 0•16, 1A•2, 1A•11, 1A•17, 1B•2
   fluid cooler - 7B•9
   fluid filter - 7B•5
   governor cable - 7B•6
   kickdown switch - 7B•6
   overhaul - 7B•12
   removal and refitting - 7B•10
   selector cable - 7B•6
   selector lever - 7B•9
   strainer - 1A•17
Auxiliary drivebelt - 1A•10, 1B•6

B

Battery - 0•15, 5A•2
Body electrical system - 12•1 et seq
Bodywork and fittings - 11•1 et seq
Bonnet - 11•5
   lock - 11•6
   release cable - 11•6
Boost pressure fuel delivery corrector - 4C•12
Bosch K-Jetronic injection system - 4B•6
Brake
   caliper - 9•9, 9•10
   disc - 9•7, 9•8
   drum - 9•9
   fluid - 0•12, 0•16, 1A•17, 1B•12
   pad - 1A•8, 1B•5, 9•4, 9•5
   pedal - 9•13
   pressure regulating valve - 9•14
   shoes - 1A•17, 1B•11, 9•6
Braking system - 9•1 et seq
   bleeding - 9•2
   fault finding - REF•17
   master cylinder - 9•12
   vacuum pump - 9•16
   vacuum servo unit - 9•12
   vacuum servo unit air filter - 9•13
   vacuum servo unit non-return valve - 9•13
   wheel cylinder - 9•12
Bulbs - 12•5, 12•7
Bumpers - 11•4

C

Cable
   accelerator - 4A•5, 4B•2, 4C•10
   bonnet release - 11•6
   clutch - 6•2
   handbrake - 9•14
Caliper (brake) - 9•9, 9•10
Camshaft - 2A•15, 2B•10, 2C•13
   cover - 2A•8
   oil seals - 2A•15, 2B•15, 2C•12
Capacities - 1A•2, 1B•2
Carburettor - 4A•6, 4A•7
Carpets - 11•2
Catalytic converter - 4B•18
Central locking system - 11•13
Centre console - 11•17
Charging system - 5A•1 et seq
Cigarette lighter - 12•11
Clock temperature display unit - 12•11
Clutch - 6•1 et seq
   check - 1A•9, 1B•6
   cable - 6•2
   fault finding - REF•15
   fluid - 0•12
   hydraulic system - 6•4
   master cylinder - 6•3
   pedal - 6•5
   release mechanism - 6•7
   removal and refitting - 6•6
   slave cylinder - 6•3
Coil springs - 10•5, 10•9
Compression test - 2A•6, 2B•4, 2C•4
Contents - 0•2
Control unit (ignition system) - 5B•4
Conversion factors - REF•2
Coolant - 0•12, 0•16, 1A•2, 1A•19, 1A•20, 1B•2, 1B•12, 1B•13
   pump - 3•5
   temperature switch (post-heating cut-off) - 5C•3
Cooling fan - 3•4
Cooling system - 3•1 et seq
   draining - 1A•19, 1B•12
   electric cooling fan - 3•4
   electrical switches - 3•4
   fault finding - REF•15
   filling - 1A•19, 1B•12
   flushing - 1A•19, 1B•12
   hoses - 3•2
   radiator - 3•2
   thermostat - 3•3
   water pump - 3•5
Crankcase stiffener casting - 2A•21, 2C•19
Crankshaft - 2A•23, 2B•16, 2C•20
   oil seals - 2A•22, 2B•16, 2C•19
   pulley - 2C•6
Cruise control system - 12•14
Cylinder block/crankcase - 2A•28, 2B•22, 2C•21
Cylinder head - 2A•17, 2A•19, 2B•11, 2B•15, 2C•15, 2C•19
   cover - 2C•5

Index REF•27

D
Dents - 11•2
Diesel engine - 2C•1 et seq
Differential shaft oil seal - 7A•2, 7B•9
Dimensions - REF•1
Disc (brake) - 9•7, 9•8
Distributor - 5B•3
Door - 11•6
  handles - 11•9
  lock - 11•9
Driveshafts - 8•1 et seq
  fault finding - REF•16
  overhaul - 8•7
  removal - 8•1
  rubber gaiters - 8•3
Drum (brake) - 9•9

E
Electric cooling fan - 3•4
Electric windows - 11•14
Electrical fault-finding - 12•2
Electrical system - 0•16
  check - 1A•13, 1B•8
  fault finding - REF•18
Electronic variable-rate suspension - 10•11
Emission control systems - 4B•17, 4C•12, 4C•13
Engine
  and automatic transmission, removal & refitting - 2A•31, 2B•25, 2C•22
  and manual transmission, removal & refitting - 2A•30, 2B•24, 2C•22
  camshaft - 2A•15, 2B•10, 2C•13
  compression test - 2A•6, 2B•4, 2C•4
  crankshaft - 2A•23, 2B•16, 2C•20
  crankshaft pulley - 2C•6
  cylinder block/crankcase - 2A•28, 2B•22, 2C•21
  cylinder head - 2A•17, 2A•19, 2B•11, 2B•15, 2C•15, 2C•19
  diesel - 2C•1 et seq
  engine/transmission mountings - 2A•22, 2B•16, 2C•19
  fault finding - REF•13
  flywheel/driveplate - 2A•21, 2B•16, 2C•19
  four-cylinder petrol - 2A•1 et seq
  intermediate shaft - 2A•22, 2C•19
  leakdown test - 2C•4
  oil - 0•11, 0•16, 1A•2, 1A•8, 1B•2, 1B•4
  oil cooler - 2A•32, 2B•25, 2C•19
  oil pump - 2A•20, 2B•6, 2C•19
  oil seals - 2A•15, 2A•22, 2B•15, 2B•16, 2C•12, 2C•19
  piston/connecting rod assembly - 2A•26, 2B•20, 2C•20
  piston rings - 2A•28, 2B•22, 2C•21
  removal and refitting - 2A•29, 2A•31, 2B•23, 2B•24, 2C•23
  rocker arms - 2A•15, 2B•10, 2C•13
  rocker covers - 2B•5
  six-cylinder petrol - 2B•1 et seq
  start-up after overhaul - 2A•32, 2B•25, 2C•24
  sump - 2A•19, 2B•15, 2C•19
  timing belt - 2A•11, 2C•7
  timing chain and sprockets - 2B•7
  valve clearances - 2A•9, 2B•5, 2C•5
Environmental considerations - REF•4
Exhaust manifold - 4A•9, 4B•13, 4C•11
Exhaust system, carburettor engines - 4A•1 et seq
Exhaust system - 4A•10, 4B•14, 4C•12
  check - 1A•13, 1B•8
  diesel models - 4C•1 et seq
  fault finding - REF•15
  fuel injection models - 4B•1 et seq

Exterior fittings - 11•14
Exterior lights - 12•8
Exterior mirror - 11•14

F
Facia - 11•17
Fast idle speed system - 5C•3
Fast idle thermostatic valve - 4C•3
Fault finding - REF•12
Fluid cooler - 7B•9
Fluid leak check - 1A•16, 1B•10
Fluids - 0•16, 1A•2, 1B•2
Flywheel/driveplate - 2A•21, 2B•16, 2C•19
Four-cylinder petrol engine - 2A•1 et seq
Front wheel alignment check - 1A•17, 1B•11
Fuel filter - 1A•18, 1B•10
  water draining - 1B•5
Fuel gauge sender unit - 4A•4, 4B•6, 4C•11
Fuel injection
  pump - 4C•5
  systems - 4B•3, 4B•6
Fuel injectors - 4C•9
Fuel pump - 4A•4, 4B•5
Fuel system
  accelerator cable - 4A•5, 4B•2, 4C•10
  accelerator pedal - 4A•6, 4B•2, 4C•11
  carburettor engines - 4A•1 et seq
  depressurisation - 4B•5
  diesel models - 4C•1 et seq
  fault finding - REF•15
  fuel injection models - 4B•1 et seq
  fuel gauge sender unit - 4A•4, 4B•6, 4C•11
  fuel tank - 4A•5, 4B•6, 4C•11
  priming and bleeding - 4C•3
Fuel tank - 4A•5, 4B•6, 4C•11
Fuses - 12•3

G
Gaiters
  driveshafts - 8•3
  steering gear - 10•13
Gearbox - see automatic or manual transmission
Gearchange linkage - 7A•2
Glossary of technical terms - REF•20
Glovebox - 11•17
Glow plugs - 5C•2
Governor cable - 7B•6

H
Handbrake - 1A•9, 1B•5
  cables - 9•14
  lever - 9•14
Handles - 11•9
Headlight - 12•8
  beam alignment - 12•8
Heating system - 3•1 et seq
  components - 3•7
Horn - 12•11
Hose check - 1A•16, 1B•10
HT coil - 5B•3
Hub bearing - 10•4, 10•8
Hub carrier - 10•3, 10•8
Hydraulic fluid - 0•12, 0•16, 1A•17, 1B•12

Index

Hydraulic pipes and hoses - 9•3
Hydraulic system, bleeding - 9•2

I

Idle speed - 1A•12, 1B•7
　mixture - 1A•12
Ignition HT coil - 5B•3
Ignition switch - 5A•5
　steering column lock - 10•12
Ignition system - petrol models - 5B•1 et seq
　control unit - 5B•4
　testing - 5B•2
Ignition timing - 5B•4
Injection timing - 4C•6, 4C•8
Injectors - 4C•9
Inlet manifold - 4A•9, 4B•12, 4C•11
Instrument panel - 12•10
Intercooler - 4B•17, 4C•12
Intermediate shaft - 2A•22, 2C•19, 10•13
　oil seals - 2A•15, 2C•12
Introduction - 0•4

J

Jacking - REF•5
Jump starting - 0•7

K

Kickdown switch - 7B•6

L

Leakdown test - 2C•4
Leaks - 0•9
Lights - 12•8
Lock
　bonnet - 11•6
　door - 11•9
　tailgate - 11•12
Loudspeakers - 12•13
Lower arm - 10•6, 10•10
　balljoint - 10•6
Lubricants - 0•16

M

Maintenance - see Routine maintenance and servicing
Maintenance procedures
　diesel models - 1B•4
　petrol models - 1A•7
Maintenance schedule
　diesel models - 1B•3
　petrol models - 1A•4
Major damage - 11•3
Manifolds - 4A•9, 4B•12, 4B•13, 4C•11
Manual transmission - 7A•1 et seq
　and engine, removal and refitting - 2A•30, 2B•24, 2C•22
　differential shaft oil seal - 7A•2
　fault finding - REF•16
　gearchange linkage - 7A•2
　oil - 0•16, 1A•2, 1A•16, 1A•17, 1B•2, 1B•10, 1B•11
　overhaul - 7A•4
　removal and refitting - 7A•3

Master cylinder - 9•12
Maximum speed - 4C•3
Minor damage - 11•2
MOT test checks - REF•8

N

No-load switch - 5C•2

O

Oil, engine - 0•11, 0•16, 1A•2, 1A•8, 1B•2, 1B•4
Oil cooler - 2A•32, 2B•25, 2C•19
Oil filter - 1A•8, 1B•4
Oil level sensor - 5A•5
Oil pressure warning light switch/pressure sensor - 5A•5
Oil pump - 2A•20, 2B•6, 2C•19
Oil seals (crankshaft) - 2A•22, 2B•16, 2C•19

P

Pads (brake) - 1A•8, 1B•5, 9•4, 9•5
Piston/connecting rod assembly - 2A•26, 2B•20, 2C•20
Piston rings - 2A•28, 2B•22, 2C•21
Power steering
　fluid - 0•13, 0•16
　pump - 10•15
　system - 10•15
Preheating system - diesel models - 5C•1 et seq
　control unit - 5C•2
Punctures - 0•8

R

Radiator - 3•2
　grille panel - 11•5
Radio aerial - 12•14
Radio/cassette player - 12•13
　anti-theft system - REF•5
Radius rod - 10•7, 10•10
Reference - REF•1 et seq
Refrigerant (air conditioning) - 1A•14, 1B•9
Relays - 12•3
Renix fuel injection system - 4B•9
Repair procedures - REF•4
Reverse gear interlock cable - 7A•2
Reversing light switch - 7A•3
Road test - 1A•18, 1B•12
Roadside repairs - 0•6
Roadwheel bolt check - 1A•14, 1B•9
Rocker arms - 2A•15, 2B•10, 2C•13
Rocker covers - 2B•5
Routine maintenance and servicing
　petrol models - 1A•1 et seq
　diesel models - 1B•1 et seq
Rust - 11•2

S

Safety first! - 0•5
Scratches - 11•2
Screen washer fluid - 0•13
Seat belts - 1A•13, 1B•8, 11•15
Seats - 11•14

Index

Selector cable - 7B•7
Selector lever - 7B•9
Servicing - see Routine maintenance and servicing
Shock absorber - 10•5, 10•8
Shoes (brake) - 1A•17, 1B•11, 9•6
Short circuit - 12•3
Six-cylinder petrol engine - 2B•1 et seq
Spare fuse check - 1A•17, 1B•11
Spare parts - REF•3
Spark plug renewal - 1A•15
Speedometer cable sensor - 12•11
Starter motor - 5A•4, 5A•5
Starting problems - 0•6
Starting system - 5A•1 et seq
Steering - 10•1 et seq
   angles - 10•17
   check - 1A•14, 1B•9
   column - 10•13
   column lock/ignition switch - 10•12
   fault finding - REF•17
   gear - 10•14
   gear gaiters - 10•13
   wheel - 10•11
Stop-light switch - 9•15
Stop solenoid - 4C•5
Sump - 2A•19, 2B•15, 2C•19
Sunroof - 11•14
Suspension - 10•1 et seq
   check - 1A•14, 1B•9
   electronic variable-rate - 10•11
   fault finding - REF•17
   lower arm - 10•6, 10•10
   lower arm balljoint - 10•6
   radius rod - 10•7, 10•10
   upper arm - 10•5
   upper arm balljoint - 10•5
Switches - 12•4
   stop light - 9•15

T

Tailgate - 11•12
   glass - 11•14
   lock - 11•12
   support struts - 11•12
   wiper motor - 12•12
TDC - 2A•7, 2B•4, 2C•4
Temperature display unit clock - 12•11
Thermostat - 3•3
Timing belt - 1A•19, 1B•12, 2A•11, 2C•7
   cover - 2B•6, 2C•6
   sprockets and tensioner - 2A•13, 2C•8
Timing chain and sprockets - 2B•7
Tools - REF•6

Towing - 0•9
Track-rod - 10•16
   balljoints - 10•16
Trim panels - 11•7, 11•16
Trip computer - 12•15
Turbocharger - 4B•15, 4C•11
   boost pressure - 1A•14, 1B•9
Tyres - 0•14, 0•16

U

Underbody views - 1A•6, 1A•7
Underbonnet check points - 0•10
Underbonnet views - 1A•5, 1A•6
Unleaded petrol - 4A•6, 4B•3
Upholstery - 11•2
Upper arm - 10•5
   balljoint - 10•5

V

Vacuum pump - 9•16
Vacuum servo unit - 9•12
   air filter - 9•13
   non-return valve - 9•13
Valve clearances - 2A•9, 2B•5, 2C•5
Vehicle identification - REF•3
Vehicle support - REF•5
Ventilation system - 3•6
   components - 3•7
Voice synthesiser - 12•15

W

Washer system components - 12•13
Water pump - 3•5
Weekly checks - 0•10 et seq
Weights - REF•1
Wheel
   alignment - 10•17
   changing - 0•8
   cylinder - 9•12
   steering - 10•11
Window
   glass - 11•10
   regulator - 11•10
Windscreen
   glass - 11•14
   wiper motor and linkage - 12•12
Wiper arm - 12•12
Wiper blades - 0•15
Wiring diagrams - 12•16
Working facilities - REF•6

Haynes Manuals – The Complete List

Title	Book No.
ALFA ROMEO	
Alfa Romeo Alfasud/Sprint (74 - 88) up to F	0292
Alfa Romeo Alfetta (73 - 87) up to E	0531
AUDI	
Audi 80 (72 - Feb 79) up to T	0207
Audi 80, 90 (79 - Oct 86) up to D & Coupe (81 - Nov 88) up to F	0605
Audi 80, 90 (Oct 86 - 90) D to H & Coupe (Nov 88 - 90) F to H	1491
Audi 100 (Oct 82 - 90) up to H & 200 (Feb 84 - Oct 89) A to G	0907
Audi 100 & A6 Petrol & Diesel (May 91 - May 97) H to P	3504
Audi A4 (95 - Feb 00) M to V	3575
AUSTIN	
Austin/MG/Rover Maestro 1.3 & 1.6 (83 - 95) up to M	0922
Austin/MG Metro (80 - May 90) up to G	0718
Austin/Rover Montego 1.3 & 1.6 (84 - 94) A to L	1066
Austin/MG/Rover Montego 2.0 (84 - 95) A to M	1067
Mini (59 - 69) up to H	0527
Mini (69 - Oct 96) up to P	0646
Austin/Rover 2.0 litre Diesel Engine (86 - 93) C to L	1857
BEDFORD	
Bedford CF (69 - 87) up to E	0163
Bedford/Vauxhall Rascal (86 - Oct 94) C to M	3015
BMW	
BMW 316, 320 & 320i (4-cyl) (75 - Feb 83) up to Y	0276
BMW 320, 320i, 323i & 325i (6-cyl) (Oct 77 - Sept 87) up to E	0815
BMW 3-Series (Apr 91 - 96) H to N	3210
BMW 3- & 5-Series (sohc) (81 - 91) up to J	1948
BMW 520i & 525e (Oct 81 - June 88) up to E	1560
BMW 525, 528 & 528i (73 - Sept 81) up to X	0632
CITROEN	
Citroën 2CV, Ami & Dyane (67 - 90) up to H	0196
Citroën AX Petrol & Diesel (87 - 97) D to P	3014
Citroën BX (83 - 94) A to L	0908
Citroën C15 Van Petrol & Diesel (89 - Oct 98) F to S	3509
Citroën CX (75 - 88) up to F	0528
Citroën Saxo Petrol & Diesel (96 - 01) N to X	3506
Citroën Visa (79 - 88) up to F	0620
Citroën Xantia Petrol & Diesel (93 - 98) K to S	3082
Citroën XM Petrol & Diesel (89 - 98) G to R	3451
Citroën Xsara (97 - 00) R to W	3751
Citroën ZX Diesel (91 - 98) J to S	1922
Citroën ZX Petrol (91 - 98) H to S	1881
Citroën 1.7 & 1.9 litre Diesel Engine (84 - 96) A to N	1379
FIAT	
Fiat 500 (57 - 73) up to M	0090
Fiat Bravo & Brava (95 - 00) N to W	3572
Fiat Cinquecento (93 - 98) K to R	3501
Fiat Panda (81 - 95) up to M	0793
Fiat Punto Petrol & Diesel (94 - Oct 99) L to V	3251
Fiat Regata (84 - 88) A to F	1167
Fiat Tipo (88 - 91) E to J	1625
Fiat Uno (83 - 95) up to M	0923
Fiat X1/9 (74 - 89) up to G	0273

Title	Book No.
FORD	
Ford Capri II (& III) 1.6 & 2.0 (74 - 87) up to E	0283
Ford Capri II (& III) 2.8 & 3.0 (74 - 87) up to E	1309
Ford Escort (Sept 80 - Sept 90) up to H	0686
Ford Escort & Orion (Sept 90 - 97) H to P	1737
Ford Escort Mk II Mexico, RS 1600 & RS 2000 (75 - 80) up to W	0735
Ford Fiesta (76 - Aug 83) up to Y	0334
Ford Fiesta (Aug 83 - Feb 89) A to F	1030
Ford Fiesta (Feb 89 - Oct 95) F to N	1595
Ford Fiesta Petrol & Diesel (Oct 95 - 97) N to R	3397
Ford Granada (Sept 77 - Feb 85) up to B	0481
Ford Granada & Scorpio (Mar 85 - 94) B to M	1245
Ford Ka (96 - 99) P to T	3570
Ford Mondeo Petrol (93 - 99) K to T	1923
Ford Mondeo Diesel (93 - 96) L to N	3465
Ford Orion (83 - Sept 90) up to H	1009
Ford Sierra 4 cyl. (82 - 93) up to K	0903
Ford Sierra V6 (82 - 91) up to J	0904
Ford Transit Petrol (Mk 2) (78 - Jan 86) up to C	0719
Ford Transit Petrol (Mk 3) (Feb 86 - 89) C to G	1468
Ford Transit Diesel (Feb 86 - 99) C to T	3019
Ford 1.6 & 1.8 litre Diesel Engine (84 - 96) A to N	1172
Ford 2.1, 2.3 & 2.5 litre Diesel Engine (77 - 90) up to H	1606
FREIGHT ROVER	
Freight Rover Sherpa (74 - 87) up to E	0463
HILLMAN	
Hillman Avenger (70 - 82) up to Y	0037
HONDA	
Honda Accord (76 - Feb 84) up to A	0351
Honda Civic (Feb 84 - Oct 87) A to E	1226
Honda Civic (Nov 91 - 96) J to N	3199
HYUNDAI	
Hyundai Pony (85 - 94) C to M	3398
JAGUAR	
Jaguar E Type (61 - 72) up to L	0140
Jaguar MkI & II, 240 & 340 (55 - 69) up to H	0098
Jaguar XJ6, XJ & Sovereign; Daimler Sovereign (68 - Oct 86) up to D	0242
Jaguar XJ6 & Sovereign (Oct 86 - Sept 94) D to M	3261
Jaguar XJ12, XJS & Sovereign; Daimler Double Six (72 - 88) up to F	0478
JEEP	
Jeep Cherokee Petrol (93 - 96) K to N	1943
LADA	
Lada 1200, 1300, 1500 & 1600 (74 - 91) up to J	0413
Lada Samara (87 - 91) D to J	1610
LAND ROVER	
Land Rover 90, 110 & Defender Diesel (83 - 95) up to N	3017
Land Rover Discovery Petrol & Diesel (89 - 98) G to S	3016
Land Rover Series IIA & III Diesel (58 - 85) up to C	0529
Land Rover Series II, IIA & III Petrol (58 - 85) up to C	0314
MAZDA	
Mazda 323 (Mar 81 - Oct 89) up to G	1608

Title	Book No.
Mazda 323 (Oct 89 - 98) G to R	3455
Mazda 626 (May 83 - Sept 87) up to E	0929
Mazda B-1600, B-1800 & B-2000 Pick-up (72 - 88) up to F	0267
MERCEDES BENZ	
Mercedes-Benz 190, 190E & 190D Petrol & Diesel (83 - 93) A to L	3450
Mercedes-Benz 200, 240, 300 Diesel (Oct 76 - 85) up to C	1114
Mercedes-Benz 250 & 280 (68 - 72) up to L	0346
Mercedes-Benz 250 & 280 (123 Series) (Oct 76 - 84) up to B	0677
Mercedes-Benz 124 Series (85 - Aug 93) C to K	3253
Mercedes-Benz C-Class Petrol & Diesel (93 - Aug 00) L to W	3511
MG	
MGA (55 - 62)*	0475
MGB (62 - 80) up to W	0111
MG Midget & AH Sprite (58 - 80) up to W	0265
MITSUBISHI	
Mitsubishi Shogun & L200 Pick-Ups (83 - 94) up to M	1944
MORRIS	
Morris Ital 1.3 (80 - 84) up to B	0705
Morris Minor 1000 (56 - 71) up to K	0024
NISSAN	
Nissan Bluebird (May 84 - Mar 86) A to C	1223
Nissan Bluebird (Mar 86 - 90) C to H	1473
Nissan Cherry (Sept 82 - 86) up to D	1031
Nissan Micra (83 - Jan 93) up to K	0931
Nissan Micra (93 - 99) K to T	3254
Nissan Primera (90 - Aug 99) H to T	1851
Nissan Stanza (82 - 86) up to D	0824
Nissan Sunny (May 82 - Oct 86) up to D	0895
Nissan Sunny (Oct 86 - Mar 91) D to H	1378
Nissan Sunny (Apr 91 - 95) H to N	3219
OPEL	
Opel Ascona & Manta (B Series) (Sept 75 - 88) up to F	0316
Opel Ascona (81 - 88) *(Not available in UK see Vauxhall Cavalier 0812)*	3215
Opel Astra (Oct 91 - Feb 98) *(Not available in UK see Vauxhall Astra 1832)*	3156
Opel Calibra (90 - 98) *(See Vauxhall/Opel Calibra Book No. 3502)*	
Opel Corsa (83 - Mar 93) *(Not available in UK see Vauxhall Nova 0909)*	3160
Opel Corsa (Mar 93 - 97) *(Not available in UK see Vauxhall Corsa 1985)*	3159
Opel Frontera Petrol & Diesel (91 - 98) *(See Vauxhall/Opel Frontera Book No. 3454)*	
Opel Kadett (Nov 79 - Oct 84) up to B	0634
Opel Kadett (Oct 84 - Oct 91) *(Not available in UK see Vauxhall Astra & Belmont 1136)*	3196
Opel Omega & Senator (86 - 94) *(Not available in UK see Vauxhall Carlton & Senator 1469)*	3157
Opel Omega (94 - 99) *(See Vauxhall/Opel Omega Book No. 3510)*	
Opel Rekord (Feb 78 - Oct 86) up to D	0543
Opel Vectra (Oct 88 - Oct 95) *(Not available in UK see Vauxhall Cavalier 1570)*	3158
Opel Vectra Petrol & Diesel (95 - 98) *(Not available in UK see Vauxhall Vectra 3396)*	3523

* Classic reprint

Title	Book No.
PEUGEOT	
Peugeot 106 Petrol & Diesel (91 - 01) J to X	1882
Peugeot 205 Petrol (83 - 97) A to P	0932
Peugeot 206 Petrol and Diesel (98 - 01) S to X	3757
Peugeot 305 (78 - 89) up to G	0538
Peugeot 306 Petrol & Diesel (93 - 99) K to T	3073
Peugeot 309 (86 - 93) C to K	1266
Peugeot 405 Petrol (88 - 97) E to P	1559
Peugeot 405 Diesel (88 - 96) E to N	3198
Peugeot 406 Petrol & Diesel (96 - 97) N to R	3394
Peugeot 505 (79 - 89) up to G	0762
Peugeot 1.7/1.8 & 1.9 litre Diesel Engine (82 - 96) up to N	0950
Peugeot 2.0, 2.1, 2.3 & 2.5 litre Diesel Engines (74 - 90) up to H	1607
PORSCHE	
Porsche 911 (65 - 85) up to C	0264
Porsche 924 & 924 Turbo (76 - 85) up to C	0397
PROTON	
Proton (89 - 97) F to P	3255
RANGE ROVER	
Range Rover V8 (70 - Oct 92) up to K	0606
RELIANT	
Reliant Robin & Kitten (73 - 83) up to A	0436
RENAULT	
Renault 5 (Feb 85 - 96) B to N	1219
Renault 9 & 11 (82 - 89) up to F	0822
Renault 18 (79 - 86) up to D	0598
Renault 19 Petrol (89 - 94) F to M	1646
Renault 19 Diesel (89 - 95) F to N	1946
Renault 21 (86 - 94) C to M	1397
Renault 25 (84 - 92) B to K	1228
Renault Clio Petrol (91 - May 98) H to R	1853
Renault Clio Diesel (91 - June 96) H to N	3031
Renault Espace Petrol & Diesel (85 - 96) C to N	3197
Renault Fuego (80 - 86) up to C	0764
Renault Laguna Petrol & Diesel (94 - 00) L to W	3252
Renault Mégane & Scénic Petrol & Diesel (96 - 98) N to R	3395
ROVER	
Rover 213 & 216 (84 - 89) A to G	1116
Rover 214 & 414 (89 - 96) G to N	1689
Rover 216 & 416 (89 - 96) G to N	1830
Rover 211, 214, 216, 218 & 220 Petrol & Diesel (Dec 95 - 98) N to R	3399
Rover 414, 416 & 420 Petrol & Diesel (May 95 - 98) M to R	3453
Rover 618, 620 & 623 (93 - 97) K to P	3257
Rover 820, 825 & 827 (86 - 95) D to N	1380
Rover 3500 (76 - 87) up to E	0365
Rover Metro, 111 & 114 (May 90 - 98) G to S	1711
SAAB	
Saab 90, 99 & 900 (79 - Oct 93) up to L	0765
Saab 900 (Oct 93 - 98) L to R	3512
Saab 9000 (4-cyl) (85 - 95) C to N	1686
SEAT	
Seat Ibiza & Cordoba Petrol & Diesel (Oct 93 - Oct 99) L to V	3571
Seat Ibiza & Malaga (85 - 92) B to K	1609

Title	Book No.
SKODA	
Skoda Estelle (77 - 89) up to G	0604
Skoda Favorit (89 - 96) F to N	1801
Skoda Felicia Petrol & Diesel (95 - 99) M to T	3505
SUBARU	
Subaru 1600 & 1800 (Nov 79 - 90) up to H	0995
SUZUKI	
Suzuki SJ Series, Samurai & Vitara (4-cyl) (82 - 97) up to P	1942
Suzuki Supercarry (86 - Oct 94) C to M	3015
TALBOT	
Talbot Alpine, Solara, Minx & Rapier (75 - 86) up to D	0337
Talbot Horizon (78 - 86) up to D	0473
Talbot Samba (82 - 86) up to D	0823
TOYOTA	
Toyota Carina E (May 92 - 97) J to P	3256
Toyota Corolla (Sept 83 - Sept 87) A to E	1024
Toyota Corolla (80 - 85) up to C	0683
Toyota Corolla (Sept 87 - Aug 92) E to K	1683
Toyota Corolla (Aug 92 - 97) K to P	3259
Toyota Hi-Ace & Hi-Lux (69 - Oct 83) up to A	0304
TRIUMPH	
Triumph Herald (59 - 71) up to K*	0010
Triumph TR2, TR3, TR3A, TR4 & TR4A (52 - 67)*	0028
Triumph TR5 & 6 (67 - 75)*	0031
Triumph Spitfire (62 - 81) up to X	0113
Triumph Stag (70 - 78) up to T	0441
VAUXHALL	
Vauxhall Astra (80 - Oct 84) up to B	0635
Vauxhall Astra & Belmont (Oct 84 - Oct 91) B to J	1136
Vauxhall Astra (Oct 91 - Feb 98) J to R	1832
Vauxhall/Opel Calibra (90 - 98) G to S	3502
Vauxhall Carlton (Oct 78 - Oct 86) up to D	0480
Vauxhall Carlton & Senator (Nov 86 - 94) D to L	1469
Vauxhall Cavalier 1600, 1900 & 2000 (75 - July 81) up to W	0315
Vauxhall Cavalier (81 - Oct 88) up to F	0812
Vauxhall Cavalier (Oct 88 - 95) F to N	1570
Vauxhall Chevette (75 - 84) up to B	0285
Vauxhall Corsa (Mar 93 - 97) K to R	1985
Vauxhall/Opel Frontera Petrol & Diesel (91 - Sept 98) J to S	3454
Vauxhall Nova (83 - 93) up to K	0909
Vauxhall/Opel Omega (94 - 99) L to T	3510
Vauxhall Vectra Petrol & Diesel (95 - 98) N to R	3396
Vauxhall/Opel 1.5, 1.6 & 1.7 litre Diesel Engine (82 - 96) up to N	1222
VOLKSWAGEN	
Volkswagen Beetle 1200 (54 - 77) up to S	0036
Volkswagen Beetle 1300 & 1500 (65 - 75) up to P	0039
Volkswagen Beetle 1302 & 1302S (70 - 72) up to L	0110
Volkswagen Beetle 1303, 1303S & GT (72 - 75) up to P	0159

Title	Book No.
Volkswagen Golf & Bora Petrol & Diesel (April 98 - 00) R to X	3727
Volkswagen Golf & Jetta Mk 1 1.1 & 1.3 (74 - 84) up to A	0716
Volkswagen Golf, Jetta & Scirocco Mk 1 1.5, 1.6 & 1.8 (74 - 84) up to A	0726
Volkswagen Golf & Jetta Mk 1 Diesel (78 - 84) up to A	0451
Volkswagen Golf & Jetta Mk 2 (Mar 84 - Feb 92) A to J	1081
Volkswagen Golf & Vento Petrol & Diesel (Feb 92 - 96) J to N	3097
Volkswagen LT vans & light trucks (76 - 87) up to E	0637
Volkswagen Passat & Santana (Sept 81 - May 88) up to E	0814
Volkswagen Passat Petrol & Diesel (May 88 - 96) E to P	3498
Volkswagen Polo & Derby (76 - Jan 82) up to X	0335
Volkswagen Polo (82 - Oct 90) up to H	0813
Volkswagen Polo (Nov 90 - Aug 94) H to L	3245
Volkswagen Polo Hatchback Petrol & Diesel (94 - 99) M to S	3500
Volkswagen Scirocco (82 - 90) up to H	1224
Volkswagen Transporter 1600 (68 - 79) up to V	0082
Volkswagen Transporter 1700, 1800 & 2000 (72 - 79) up to V	0226
Volkswagen Transporter (air-cooled) (79 - 82) up to Y	0638
Volkswagen Transporter (water-cooled) (82 - 90) up to H	3452
VOLVO	
Volvo 142, 144 & 145 (66 - 74) up to N	0129
Volvo 240 Series (74 - 93) up to K	0270
Volvo 340, 343, 345 & 360 (76 - 91) up to J	0715
Volvo 440, 460 & 480 (87 - 97) D to P	1691
Volvo 740 & 760 (82 - 91) up to J	1258
Volvo 850 (92 - 96) J to P	3260
Volvo 940 (90 - 96) H to N	3249
Volvo S40 & V40 (96 - 99) N to V	3569
Volvo S70, V70 & C70 (96 - 99) P to V	3573
AUTOMOTIVE TECHBOOKS	
Automotive Air Conditioning Systems	3740
Automotive Brake Manual	3050
Automotive Carburettor Manual	3288
Automotive Diagnostic Fault Codes Manual	3472
Automotive Diesel Engine Service Guide	3286
Automotive Electrical and Electronic Systems Manual	3049
Automotive Engine Management and Fuel Injection Systems Manual	3344
Automotive Gearbox Overhaul Manual	3473
Automotive Service Summaries Manual	3475
Automotive Timing Belts Manual – Austin/Rover	3549
Automotive Timing Belts Manual – Ford	3474
Automotive Timing Belts Manual – Peugeot/Citroën	3568
Automotive Timing Belts Manual – Vauxhall/Opel	3577
Automotive Welding Manual	3053
In-Car Entertainment Manual (3rd Edition)	3363

* Classic reprint

All the products featured on this page are available through most motor accessory shops, cycle shops and book stores. Our policy of continuous updating and development means that titles are being constantly added to the range. For up-to-date information on our complete list of titles, please telephone: (UK) +44 1963 442030 • (USA) +1 805 498 6703 • (France) +33 1 47 78 50 50 • (Sweden) +46 18 124016 • (Australia) +61 3 9763 8100

Notes

Notes

Preserving Our Motoring Heritage

The Model J Duesenberg Derham Tourster. Only eight of these magnificent cars were ever built – this is the only example to be found outside the United States of America

Almost every car you've ever loved, loathed or desired is gathered under one roof at the Haynes Motor Museum. Over 300 immaculately presented cars and motorbikes represent every aspect of our motoring heritage, from elegant reminders of bygone days, such as the superb Model J Duesenberg to curiosities like the bug-eyed BMW Isetta. There are also many old friends and flames. Perhaps you remember the 1959 Ford Popular that you did your courting in? The magnificent 'Red Collection' is a spectacle of classic sports cars including AC, Alfa Romeo, Austin Healey, Ferrari, Lamborghini, Maserati, MG, Riley, Porsche and Triumph.

A Perfect Day Out

Each and every vehicle at the Haynes Motor Museum has played its part in the history and culture of Motoring. Today, they make a wonderful spectacle and a great day out for all the family. Bring the kids, bring Mum and Dad, but above all bring your camera to capture those golden memories for ever. You will also find an impressive array of motoring memorabilia, a comfortable 70 seat video cinema and one of the most extensive transport book shops in Britain. The Pit Stop Cafe serves everything from a cup of tea to wholesome, home-made meals or, if you prefer, you can enjoy the large picnic area nestled in the beautiful rural surroundings of Somerset.

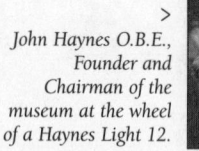

John Haynes O.B.E., Founder and Chairman of the museum at the wheel of a Haynes Light 12.

Graham Hill's Lola Cosworth Formula 1 car next to a 1934 Riley Sports.

The Museum is situated on the A359 Yeovil to Frome road at Sparkford, just off the A303 in Somerset. It is about 40 miles south of Bristol, and 25 minutes drive from the M5 intersection at Taunton.
Open 9.30am - 5.30pm (10.00am - 4.00pm Winter) 7 days a week, *except Christmas Day, Boxing Day and New Years Day*
Special rates available for schools, coach parties and outings Charitable Trust No. 292048